SPIE Volume 673

International Conference on Holography Applications

Dahang Wang
Chairman

Jingtang Ke, Ryszard J. Pryputniewicz
Editors

Sponsored by
COS—Chinese Optical Society
CSTAM—Chinese Society of Theoretical and Applied Mechanics
SPIE—The International Society for Optical Engineering
OSA—Optical Society of America
EPnA—European Photonics Association

In cooperation with
China Association for Science and Technology

2-4 July 1986
Beijing, China

Published and Distributed Worldwide by
SPIE—The International Society for Optical Engineering
P.O. Box 10, Bellingham, Washington 98227-0010 USA
Telephone 206/676-3290 (Pacific Time) • Telex 46-7053

The papers appearing in this book comprise the proceedings of the meeting mentioned on the cover and title page. They reflect the authors' opinions and are published as presented and without change, in the interests of timely dissemination. Their inclusion in this publication does not necessarily constitute endorsement by the editors or by SPIE.

Please use the following format to cite material from this book:
Author(s), "Title of Paper," *International Conference on Holography Applications,* Jingtang Ke, Ryszard J. Pryputniewicz, Editors, Proc. SPIE 673, page numbers (1987).

Library of Congress Catalog Card No. 87-60730
ISBN 0-89252-708-0

Printed in the United States of America.

INTERNATIONAL CONFERENCE ON HOLOGRAPHY APPLICATIONS

SPIE Volume 673

Contents

International Program Committee viii
Organizing Committee/Invited Speakers ix
Session Chairmen x
Introduction xii

SESSION 1. OPENING SPECIAL LECTURES. 1

673-35 **Applications of holography 1947-86,** P. Hariharan, CSIRO (Australia). 2

673-119 **Holography in China,** D.-H. Wang, Academia Sinica (China); Z.-J. Wang, Shanghai Institute of Optics and Fine Mechanics (China); D.-H. Hsu, Beijing Institute of Posts and Telecommunications (China); B.-X. Lu, Academia Sinica (China). 8

SESSION 2. FUNDAMENTALS AND DISPLAY IN HOLOGRAPHY I. 13

673-74 **Spatial multichannel holography: longitudinal multiplexed Fourier storage,** D. Zhao, W. Mei, Beijing Institute of Technology (China). 14

673-02 **Use of fringe scanning method in electron holographic interferometry,** T. Yatagai, K. Ohmura, Univ. of Tsukuba (Japan); S. Iwasaki, National Research Lab. of Metrology (Japan); A. Tonomura, J. Endo, S. Hasegawa, Hitachi, Ltd. (Japan). 19

673-07 **Holographic cinematography and its applications,** P. Smigielski, H. Fagot, F. Albe, Franco-German Research Institute of Saint-Louis (France). 22

673-130 **New methods of making flat holographic stereograms,** Z. Qu, J. Liu, Shanghai Institute of Laser Technology (China). 28

SESSION 3. FUNDAMENTALS AND DISPLAY IN HOLOGRAPHY II. 35

673-46 **Proposals of two reference beam double exposure holographic interferometer with double pulsed laser,** G. Lai, T. Yatagai, Univ. of Tsukuba (Japan). 36

673-47 **Self-enhancement effect of dynamic amplitude-phase holograms and its applications,** A. O. Ozols, Latvian SSR Academy of Sciences (USSR). 41

673-89 **New proposal for phase-only CGH and its application,** W.-F. Chen, X.-Y. Huang, X.-T. Ying, Y.-R. Jia, Fudan Univ. (China). 44

673-95 **Infrared holography using a CO_2 laser and its application to Si-wafer inspection,** G. Sun, Z. Yuan, Hefei Polytechnical Univ. (China). 49

673-49 **Picosecond time- and space-domain holography by photochemical hole burning,** P. Saari, R. Kaarli, A. Rebane, Academy of Sciences of the Estonian SSR (USSR). 53

673-85 **Experimental study of the holographic technique applied to supersonic cascade wind-tunnel,** J. Li, J. Yang, W. Wang, Y. Song, Institute of Engineering Thermophysics (China). 62

673-97 **Spatial moire hologram,** W.-S. Huang, J.-R. Guo, L. Zhou, Q.-Y. Sun, Shanghai Jiao Tong Univ. (China). 66

673-147 **New method of color holography,** K. Bazargan, Imperial College of Science and Technology (UK). 68

673-107 **Investigation of strains of objects by means of sandwich holographic interferometry,** L. Wang, C. Zhao, J. Ke, Zhengzhou Institute of Technology (China). 71

673-50 **Influence of medium nonlinearity on the properties of the steady-state hologram,** A. I. Khyzniak, I. L. Rubtsova, Ukranian Academy of Science (USSR). 77

SESSION 4. FUNDAMENTALS AND DISPLAY IN HOLOGRAPHY III. 83

673-22 **Holographic 3D display of x-ray tomogram,** K. Okada, T. Ose, Chiba Univ. (Japan). 84

673-96 **Application of holographic lens in head-up display,** H. Chen, B. Lu, Beijing Insitute of Technology (China). 89

673-129 **Method for making large rainbow holograms,** Z. Qu, Y. Xu, M. Li, X. Cai, Shanghai Institute of Laser Technology (China). 93

673-23 **Ohya large holography studio in a huge rock cavern,** K. Watanabe, Holotec, Inc. (Japan); A. Hirosawa, Institute of Scientifically Synchronized with New Life (Japan); T. Morie, Shimizu Construction Co., Ltd. (Japan); M. Suzuki, Fuji Photo Optical Co., Ltd. (Japan); H. Yamashita, Mazda Motor Corp. (Japan). 95

(continued)

673-03 **Holographic registration and spectroscopy of a modulated laser radiation,** V. N. Borkova, V. A. Zubov, A. V. Krajskij, P. N. Lebedev Physical Institute of the Academy of Sciences of the USSR. 100
673-109 **CGH made by dot matrix printer,** M. Zhao, D. Hsu, Z. Fei, Beijing Institute of Posts and Telecommunications (China). 105
673-118 **Making CGH by using electron beam,** Y. Yan, D. Yu, K.-F. Chin, Tsinghua Univ. (China). 107

SESSION 5. HOLOGRAPHIC INTERFEROMETRY I. 111
673-16 **Holography and speckle techniques for deformation measurements,** I. Yamaguchi, Institute of Physical and Chemical Research (Japan). 112
673-76 **New experimental technique to separate principal stresses—proportional exposure piled-plate holography,** H. Cao, S. Chen, W. Dong, Zhejiang Univ. (China). 120
673-17 **Analysis of an in-line object beam hologram interferometer,** M. Celaya, D. Tentori, R. Villagomez, CICESE (Mexico). 124
673-69 **Holographic measurement of an arc plasma field,** H. Ding, Beijing Institute of Aeronautics and Astronautics (China). 130
673-18 **Profile measurement on IC wafers by holographic interference,** J. A. Sun, A. Cai, G. Wade, Univ. of California/Santa Barbara (USA). 135
673-78 **New method of 3D quantitative analysis of holographic interferometry in the applications of solid biomechanics and vibration,** Z. Ding, N. Bao, Tongji Univ. (China). 144
673-145 **Vibration analysis of large-sized structure using cw laser holographic interferometry,** H. Yamashita, Mazda Motor Corp. (Japan). 150
673-19 **Numerical method for holographic optical fiber diagnostics,** S.-I. Himeno, Tokai College of Industry (Japan); M. Seki, Hokkaido Automotive Junior College (Japan); H. Mochizuki, Univ. of Electro-Communications (Japan). 154
673-98 **Application of holography to research reason of spurting water of the antique "Water Spurting Basin",** H. Zhu, G. Zhou, Lab. of Shanghai Museum (China). 160
673-20 **Hologram interferometer for film deposits evaluation,** D. Tentori, M. Celaya, CICESE (Mexico). 163
673-139 **Heat transfer studies by microholographic interferometry,** F. Bloisi, Univ. di Napoli (Italy); P. Cavaliere, Univ. di Palermo (Italy); S. Martellucci, Univ. Tor Vergata (Italy); R. Meucci, P. Mormile, G. Pierattini, J. Quartieri, Istituto di Cibernetica CNR (Italy); L. Vicari, Univ. di Napoli (Italy). 167

SESSION 6. HOLOGRAPHIC INTERFEROMETRY II. 175
673-101 **Laser holographic inspection of solder joints on printed circuit board (PCB),** J. Hong, W. Geng, L. Jiang, W. Xue, J. Ma, S. Zhao, J. Yang, Harbin Institute of Technology (China). 176
673-24 **Automatic measurements of the small angle variation using a holographic moire interferometry and a computer processing,** Y. Nakano, Hokkaido Institute of Pharmaceutical Sciences (Japan). 180
673-91 **Shearing holographic moire for strain patterns,** J. Fang, F. Dai, Tsinghua Univ. (China). 184
673-25 **Phase-only Fourier hologram as an optical matched spatial filter,** K. Chałasińska-Macukow, T. Nitka, Warsaw Univ. (Poland). 188
673-75 **Holographic interferometry using digital image processing technique for the measurement of 3D axisymmetric temperature field,** L. Han, S.-P. He, X.-P. Wu, Univ. of Science and Technology of China. 192
673-148 **Moire evaluation holography,** O. D. D. Soares, Univ. do Porto (Portugal); J. F. Fernandez, Univ. de Santiago (Spain). 198
673-26 **Some aspects of application of a double frequency interferometer for distance measurements,** J. J. Galiński, H. Z. Kowalski, J. Sanecki, Institute of Geodesy and Cartography (Poland). 207
673-116 **Application of holography in the distribution measurement of fuel spraying field in diesel engines,** W. X. He, Z. X. Li, Tianjin Univ. (China). 212
673-138 **White-light color holography and its applications,** F. T. S. Yu, Pennsylvania State Univ. (USA). 222

SESSION 7. HOLOGRAPHIC INTERFEROMETRY III. 235
673-28 **Automatic processing of holographic interference fringes to analyze the deflection of a thin plate,** S. Toyooka, Y. Iwaasa, H. Nishida, Saitama Univ. (Japan). 236
673-71 **Dynamic process study by preset light pulse method in holographic system with Faraday Rotator,** X. Wu, G. Ren, Z. Zhang, Z. Ye, S. Wang, Tianjin Univ. (China). 239
673-104 **Application of monowavelength pulsed laser holometry in the measurement of arc plasma,** J. Li, J. Liu, J. Lian, Y. Li, Tianjin Univ. (China). 243

673-140 **Computer-aided fringe analysis,** R. J. Pryputniewicz, Worcester Polytechnic Institute (USA). 250

673-102 **Fatigue detection based on the change of laser-produced diffraction patterns,** B. Xu, L. Li, X. Wu, Univ. of Science and Technology of China. 258

673-134 **Polychromatic speckle interference for edge enhancement of image,** S. Q. Shen, Beijing Institute of Posts and Telecommunications (China); X. P. Wu, Univ. of Science and Technology of China; F. P. Chiang, SUNY/Stony Brook (USA). 263

SESSION 8. HOLOGRAPHIC INTERFEROMETRY IV. 267

673-117 **Interferometric method for measuring optical spherical surfaces using holographic phase conjugate compensation,** Z. Zou, J. Liao, Y. Gu, Q. Gu, Changchun Institute of Optics and Fine Mechanics (China). 268

673-32 **Vibration analysis of automotive bodies using continuous wave laser holographic interferometry,** H. Yamashita, K. Sasanishi, I. Masamori, Mazda Motor Corp. (Japan). 272

673-86 **Real time grating shearing interferometry applied to investigating of evaporative convection in liquid drops,** Y. Xu, East China Technical Univ. of Water Resources (China); N. Zhang, B. X. Wang, Tsinghua Univ. (China). 276

673-33 **Holographic testing of gear tooth surface,** D. Y. Yu, T. Kondo, N. Ohyama, T. Honda, J. Tsujiuchi, Tokyo Institute of Technology (Japan). 283

673-72 **Optical testing using a point diffraction holographic interferometer,** W. Zhou, Z. Lu, Changchun Institute of Optics and Fine Mechanics (China). 289

673-111 **Method for measuring strain directly by laser objective speckles,** X. Zhou, L. Niu, Tsinghua Univ. (China). 293

673-131 **Advanced holographic scannings and applications,** C. S. Ih, L. Q. Xiang, Univ. of Delaware (USA). 296

673-34 **Industrial holography combined with image processing,** J. Schörner, H. Rottenkolber, W. Roid, Rottenkolber Holo-System GmbH (FRG); K. Hinsch, Univ. of Oldenburg (FRG). 303

SESSION 9. HOLOGRAPHIC APPLICATIONS IN MEDICINE. 311

673-08 **Multiplex holograms and their applications in medicine,** J. Tsujiuchi, Tokyo Institute of Technology (Japan). . . . 312

673-81 **Dynamic laser speckles and refractive measurements of the eye,** G. Chen, J. Lui, H. Tang, Hefei Polytechnical Univ. (China). 317

673-10 **Mechanical reaction of human skull bones to external load examined by holographic interferometry,** H. Podbielska, Technical Univ. of Wroclaw and Medical Academy (Poland); G. von Bally, Univ. of Münster (FRG); H. Kasprzak, Technical Univ. of Wroclaw (Poland). 321

673-149 **Holography in biomedical sciences,** G. von Bally, Univ. of Münster (FRG). 327

673-11 **Holography and the freedom of science—a welcome address to ICHA '86,** G. von Bally, Univ. of Münster (FRG). 337

673-93 **Study of optometry apparatus of laser speckles,** B.-C. Wang, K. Yao, X.-Q. Wu, Univ. of Science and Technology of China; C.-Y. Long, J.-Q. Shi, S.-Z. Shi, Anhui Univ. (China). 338

673-12 **Deformation measurement of lumbar vertebra by holographic interferometry,** T. Matsumoto, Osaka Prefectural Technical College (Japan); A. Kojima, R. Ogawa, Kansai Medical Univ. (Japan); K. Iwata, R. Nagata, Univ. of Osaka Prefecture (Japan). 340

SESSION 10. SPECKLE TECHNIQUES. 345

673-52 **Electronic speckle pattern interferometry,** O. J. Løkberg, Norwegian Institute of Technology (Norway). 346

673-87 **Measurement for dynamic deformation by mismatch white light speckle method,** Z. Cao, F. Chen, R. Fang, P. Chen, Tongji Univ. (China). 354

673-54 **Speckle velocimetry applied to wake flows,** I. Grant, G. H. Smith, Heriot-Watt Univ. (UK). 358

673-108 **Oblique-optical-axis speckle photography used for measuring 3D displacements of practical engineering structures,** Y. He, Y. Tan, C. Ku, Xi'an Jiaotong Univ. (China). 366

673-55 **Measuring rotating component strains using ESPI,** R. W. T. Preater, City Univ. (UK). 373

673-70 **New method of surface roughness measurement using far field speckle,** R.-P. Liu, P.-S. Liu, Beijing Institute of Technology (China). 377

673-56 **Forming phase diffusers using speckle,** M. Kowalczyk, Univ. of Warsaw (Poland). 382

673-94 **Study of speckle pattern and surface roughness measurement with image processing technique,** Q.-C. Zhang, B.-Q. Xu, X.-P. Wu, Univ. of Science and Technology of China. 388

(continued)

SESSION 11. HOLOGRAPHIC APPLICATION IN MEDICINE AND SPECKLE TECHNIQUES. 395
673-79 **Method of determining minute deformation and displacement of skull by speckle photography,** F. Liu, West China Univ. of Medical Science (China); S. Wang, Chengdu Institute of Radio Engineering (China). 396
673-04 **Manifestation of Gabor's holographic principle in various evolutionary stages of the living material,** P. Greguss, Technical Univ. of Budapest (Hungary). 402
673-14 **Holographic testing of human vision—shapes differentiation,** B. Smoliński, Warsaw Technical Univ. (Poland) **Holographic testing of human vision—optical correlator matched with human eye,** M. Komarnicki, B. Smolińska, Warsaw Technical Univ. (Poland). 412
673-142 **Study of speckle multiple-shearing interferometry,** Y. Ma, J. Ke, Zhengzhou Institute of Technology (China). 416

SESSION 12. HOLOGRAPHIC RECORDING MATERIALS AND DEVICES I. 425
673-38 **Optimum holographic disk scanners with bow-locus corrections,** Y. Ishii, K. Murata, Hokkaido Univ. (Japan). 426
673-133 **Hologram filters for optical image processing,** K. Murata, Hokkaido Univ. (Japan). 434
673-59 **Dichromated gelatin holographic scanner,** C. Liegeois, R. Piel, X-IAL (France); P. Meyrueis, Ecole Nationale Supérieure de Physique de Strasbourg (France). 439
673-40 **Oblique incidence interferometer using holographic optical elements,** M. Suzuki, M. Kanaya, T. Saito, K. Yasuda, Fuji Photo Optical Co., Ltd. (Japan). 446
673-122 **Formation of amplitude grating in real-time holographic recording medium BSO crystal,** Y. Yan, W.-S. Wu, Y.-L. Chen, Shanghai Jiao Tong Univ. (China). 451
673-42 **Noteworthy qualities of holographic grating in the design of analytical spectrometer,** J. P. Laude, Jobin Yvon (France). 455
673-114 **Development of high resolution zone plate by holography and ion-etching techniques,** Y.-G. Su, S.-J. Fu, X.-M. Tao, Y.-L. Hong, Y.-W. Zhang, Univ. of Science and Technology of China. 459
673-115 **Study of the narrow band holographic reflection filters,** D. Liu, W. Tang, J. Chou, W. Huang, Beijing Normal Univ. (China). 463

SESSION 13. HOLOGRAPHIC RECORDING MATERIALS AND DEVICES II. 471
673-39 **Thermoplastic holographic camera for industrial nondestructive testing: data acquisition and image analysis,** C. Liegeois, P. Meyrueis, Ecole Nationale Supérieure de Physique de Strasbourg (France); J. Fontaine, Ecole Nationale Supérieure des Arts et Industries de Strasbourg (France). 472
673-60 **Holographic aquaculture,** R. Ian, E. King, Advanced Environmental Research Group (USA). 479
673-61 **Holographic storage properties of electro-optic crystals,** E. Krätzig, R. A. Rupp, Univ. Osnabrück (FRG). 483
673-125 **Fe-doped $LiNbO_3$ crystal for applications in white-light information processing,** S. Wang, J. Zhang, Y. Peng, Y. Zhao, C. Bao, Beijing Normal Univ. (China). 489
673-63 **Microscopic and macroscopic measurement of holographic emulsions,** R. Aliaga, H. Chuaqui, Univ. Católica de Chile. 491
673-80 **Study on the performance of the liquid crystal light valve,** Z.-Y. Chen, R.-C. Ji, H.-Q. Zhao, Fudan Univ. (China). 495

SESSION 14. OPTICAL INFORMATION PROCESSING I. 501
673-43 **Coherent optics in environmental monitoring,** K. Hinsch, Univ. Oldenburg (FRG). 502
673-88 **Application of computer-generated oriented speckle screen to image processing,** G. G. Mu, X. M. Wang, Z. Q. Wang, Nankai Univ. (China). 508
673-44 **Hybrid real-time correlation system of aerial stereo photographs,** R. T. Hong, J. Tsujiuchi, N. Ohyama, T. Honda, Tokyo Institute of Technology (Japan). 512
673-123 **Real-time optical logical operations using grating filtering,** Z.-P. Fu, D.-H. Hsu, B. Wang, Beijing Institute of Posts and Telecommunications (China). 517
673-83 **Determining the phase function in a 2D Fourier spectrum,** K.-S. Xu, X.-D. Xu, Fudan Univ. (China). 519
673-45 **Principle of a photonical computer with biological and medical applications,** A. M. Landraud, Univ. Pierre et Marie Curie (France). 526
673-124 **Data processing on particle field hologram,** X. He, J. Yao, Z. Zhu, Dalian Institute of Technology (China). 532

SESSION 15. OPTICAL INFORMATION PROCESSING II. 537
673-65 **Use of synthetic holograms in coherent image processing for high resolution micrographs of a conventional transmission electron microscope (CTEM),** E. Reuber, Fritz-Haber-Institut der Max-Planck-Gesellschaft (FRG); S. Boseck, B. Schmidt, H. Block, Univ. of Bremen (FRG). 538
673-103 **New type of holographic encoding filter for correlation—a lensless intensity correlator,** G. G. Mu, X. M. Wang, Z. Q. Wang, Nankai Univ. (China). 546
673-100 **Optical tomography—a new holographic inverse filter technique for 3D x-ray imaging,** Y. Jiang, L. Chen, Suzhou Univ. (China); E. N. Leith, Univ. of Michigan (USA). 550

673-112 **Holographic match filtering method with high diffraction efficiency,** G. Yang, S. Fu, Y. Su, Univ. of Science and Technology of China. 554

673-99 **Synthetic phase-only holographic filter for multiobject recognition,** X.-Y. Su, G.-S. Zhang, L.-R. Guo, Sichuan Univ. (China). 557

SESSION 16. OPTICAL INFORMATION PROCESSING III. 561

673-113 **Optical sine transformation,** G. Yang, J. Gong, J. Zhang, J. Chen, Y. Ho, Univ. of Science and Technology of China. 562

673-37 **Generalized Laplace's operators realized by computer generated spatial filters,** A. Kalestyński, Warsaw Technical Univ. (Poland). 567

673-73 **Improved holographic technique for imaging through fog,** Y. Zhang, Z. C. Li, Tianjin Univ. (China). 570

673-77 **Density pseudocolor encoder by Bragg effect,** J. Xie, Y. Zhao, D. Zhao, M. Yu, Beijing Institute of Technology (China). 575

673-90 **Image quality improvement on variable spatial-rotation high resolution electron micrograph by optical means,** X. Shen, Shanghai Institute of Mechanical Engineering (China); S. H. Zheng, F. H. Li, Institute of Physics, Academia Sinica (China). 578

673-121 **Holographic fingerprint lock system,** Y. Wang, Y. Chang, First Institute of Ministery of Public Security (China). 584

Author Index 588

INTERNATIONAL PROGRAM COMMITTEE

Conference Chairman
Dahang Wang
Academia Sinica, China

Program Committee Chairman
Zhijiang Wang
Shanghai Institute of Optics and Fine Mechanics, China

Program Committee Cochairmen
G. von Bally, University of Münster, FRG
H. J. Caulfield, University of Alabama in Huntsville, USA
J. Tsujiuchi, Tokyo Institute of Technology, Japan

Secretary General
Da-Hsiung Hsu, Beijing Institute of Posts and Telecommunications, China

Members

N. Abramson, Sweden
R. Dändliker, Switzerland
P. Greguss, Hungary
Youquan Jia, China
Boxiang Lu, China
T. Nakajima, Japan
O. D. D. Soares, Portugal
D. Vukicévic, Yugoslavia
Yi-Mo Zhang, China
Songling Zhuang, China
J. Bulabois, France
A. E. Ennos, UK
Da-Hsiung Hsu, China
Jingtang Ke, China
G. Molesini, Italy
H. Nishihara, Japan
M. Soskin, USSR
Runwen Wang, China
Zhi-Ming Zhang, China
Kuofan Chin, China
H. J. Frankena, Netherlands
Y. Ishii, Japan
T. Kubota, Japan
Guo-Guang Mu, China
Zhi-Min Qu, China
Yu-Shan Tan, China
Mei-Wen Yu, China
Xin-Gen Zhou, China

SPIE Volume 673

ORGANIZING COMMITTEE

Chairman
Bao-Cheng Nie, Shanghai Institute of Laser Technology, China

Cochairman
Jingtang Ke, Zhengzhou Institute of Technology, China

INVITED SPEAKERS

Nils Abramson, Sweden*
H. J. Caulfield, USA*
Pal Greguss, Hungary
P. Hariharan, Australia
Yuming He, China
K. Hinsch, FRG
Y. Y. Hung, USA*
Charles S. Ih, USA
Y. Ishii, Japan
Tung H. Jeong, USA
Y. G. Jiang, China
J. Li, China
Fuxiang Liu, China
Ole S. Løkberg, Norway
Ryszard J. Pryputniewicz, USA
Zhiming Qu, China
E. Reuber, FRG
Jumpei Tsujiuchi, Japan
Gert von Bally, FRG
Dahang Wang, China
Ke Xu, China
Ichirou Yamaguchi, Japan
Guoguang Yang, China
Francis T. S. Yu, USA
Zhenshu Zou, China

*Paper not available for publication.

INTERNATIONAL CONFERENCE ON HOLOGRAPHY APPLICATIONS

SPIE Volume 673

SESSION CHAIRMEN

Session 1—Opening Special Lectures
Da-Hsiung Hsu, Beijing Institute of Posts and Telecommunications, China

Session 2—Fundamentals and Display in Holography I
Kuofan Chin, Tsinghua University, China

Session 3—Fundamentals and Display in Holography II
Nils Abramson, Royal Institute of Technology, Sweden
Guo-Guang Mu, Nankai University, China

Session 4—Fundamentals and Display in Holography III
Guo-Guang Yang, University of Science and Technology of China

Session 5—Holographic Interferometry I
D. Vukicévic, The University of Zagreb, Yugoslavia
Jingtang Ke, Zhengzhou Institute of Technology, China

Session 6—Holographic Interferometry II
Xin-Gen Zhou, Tsinghua University, China

Session 7—Holographic Interferometry III
O. D. D. Soares, Universidade do Porto, Portugal
Erzhen Sheng, Beijing Institute of Optoelectronics, China

Session 8—Holographic Interferometry IV
Y. Y. Hung, Oakland University, USA
F. T. S. Yu, The Pennsylvania State University, USA

Session 9—Holographic Applications in Medicine
Pal Greguss, Technical University of Budapest, Hungary
Zhi-Min Qu, Shanghai Institute of Laser Technology, China

Session 10—Speckle Techniques
O. J. Løkberg, Norwegian Institute of Technology, Norway
Tongshu Lian, Beijing Institute of Technology, China

Session 11—Holographic Applications in Medicine and Speckle Techniques
R. J. Pryputniewicz, Worcester Polytechnic Institute, USA
Yu-Shan Tan, Xian Jiaotong University, China

Session 12—Holographic Recording Materials and Devices I
Tung H. Jeong, Lake Forest College, USA
Mei-Wen Yu, Beijing Institute of Technology, China

Session 13—Holographic Recording Materials and Devices II
Charles S. Ih, University of Delaware, USA
Y. Ishii, Hokkaido University, Japan

Session 14—Optical Information Processing I
G. von Bally, University of Münster, FRG
Zhi-Ming Zhang, Fudan University, China

Session 15—Optical Information Processing II
Yi-Mo Zhang, Tianjin University, China

Session 16—Optical Information Processing III
H. Podbielska, Technical University of Wroclaw, Poland
Quwu Gu, Changchun Institute of Optics and Fine Mechanics, China

INTRODUCTION

This volume contains the proceedings of the International Conference on Holography Applications, held in Beijing, the modern capital of China, July 2 to July 4, 1986. The idea for this conference was conceived during the 1982 ICH held in Dubrovnik, Yugoslavia. The proposal to hold the ICHA '86 in China was warmly received by the Chinese Optical Society and the Chinese Society of Theoretical and Applied Mechanics and gained the support of SPIE, OSA, EPnA, and the Science Panel of the Parliamentary Assembly of the Council of Europe. The result was most gratifying.

The conference was attended by over 250 researchers from 20 countries presenting 114 papers. Twenty-four of these papers were invited. The presentations at the ICHA '86 embraced the fields of holography, holographic display, recording materials, status and trends in development of holography, and laser speckle methods. Countless discussions between the members of this international group of researchers vividly demonstrated the interest and timeliness of the material covered and stimulated further growth in the field of holography.

These discussions were held not only in the meeting rooms of the famous Beijing Friendship Hotel, but also during numerous site trips generously arranged by the Chinese hosts. As intriguing now as 700 years ago when Marco Polo visited, China continued to astonish the international visitors with her spectacular beauty, enormous population, and technological achievement. The site trips were thoughtfully arranged and allowed those who participated to encounter friendly people, a bit of roughing, and a real insight into China's history, tradition, and modern culture. Each of these trips offered a stimulating, fascinating, informative and unique experience—modern cities, ancient capitals, scenic wonders, and the vast diversity of this great land. Particularly, it was astonishing to see the wealth of ancient remembrances and the bustling, modern style of Beijing.

Beijing is absolutely fascinating with its 3000 year old history. There, the "Sacred Way," lined with full-size stone figures of animals and mystical figures, led the conference participants to the fabled Ming Tombs, the last resting place of 13 emperors from the Ming Dynasty with its spectacular 4 story underground marble tomb of Ding Ling, buried in 1620. Then, a visit to one of man's most remarkable accomplishments—the Great Wall, the only man-made object visible from space, begun in 260 BC and stretching 3000 miles across some of the most rugged mountains and beautiful valleys in the world. The visit to the Summer Palace was also unforgettable with its more than 1000 rooms, 2000 foot long painted and lacquered lakeside promenade, and marble boathouse. But all this was pale compared to the Forbidden City, the imposing architectural masterpiece consisting of six main palaces with over 9000 rooms from which 24 Chinese emperors ruled the immense empire. Prior to 1900, commoners entering the Forbidden City were on a one-way passage to certain death, but now the experience was quite safe. The sights and examples of history are innumerable in Beijing alone and in China as a whole. No matter where the conference participants were or what they had seen, there was always a delightful relaxation at the dining table. Foods were varied and unique, but none of the dishes was as exciting as the Beijing Duck ... in Beijing.

All in all, the topical content of the ICHA '86, together with the people and places encountered, made this conference a very rewarding, informative, and unforgettable experience. Therefore, it is with deep gratitude that the undersigned thank the sponsoring organizations for support of this conference, and the speakers and participants for their efforts to make the conference one of a kind.

Thank you very much.

Jingtang Ke
Zhengzhou Institute of Technology, China

Ryszard J. Pryputniewicz
Worcester Polytechnic Institute, USA

Session 1

Opening Special Lectures

Chairman
Da-Hsiung Hsu
Beijing Institute of Posts and Telecommunications, China

Applications of holography 1947-86

P. Hariharan

CSIRO Division of Applied Physics, Sydney, Australia 2070

Abstract

Holography has now become an established technique. While some of its early applications failed to fulfil their initial promise, others have opened up major commercial possibilities. This paper will briefly review the growth of holography and attempt an objective assessment of some of its current applications.

Introduction

It is now almost forty years since the concept of holographic imaging was put forward by Gabor in 1947.[1] This period has seen holography grow into an established technique with a remarkably wide range of applications.

During these forty years holography has also had its ups and downs. When Gabor originally proposed holographic imaging, it was with the limited objective of obtaining increased resolution in electron microscopy. This particular application has still to be achieved, and it was no surprise therefore that, despite several seminal papers by workers in the 1950s, interest in holography gradually declined.

This period of dormancy was followed in the mid 1960s by a sudden growth of activity, triggered off by a number of breakthroughs. These included the development of the off-axis reference beam technique by Leith and Upatnieks and the availability of lasers which made it possible to record holograms of diffusely reflecting objects with appreciable depth.[2,3] In addition, around the same time, Denisyuk described the first volume reflection holograms, which reconstructed an image with a white light source.[4]

Applications which followed in rapid succession included high-resolution imaging of aerosols,[5] imaging through diffusing and aberrating media,[6] multiple imaging,[7,8] multicolour imaging,[9] holographic optical elements[10] and computer-generated holograms.[11] Other fields which were investigated included holographic information storage,[12] image deblurring and character recognition,[13,14] and acoustic and microwave holography. Perhaps the most significant application to come out of this period was holographic interferometry, which was discovered almost simultaneously by several independent groups.[15-19]

During the next decade some of these applications established themselves, but others did not fulfil their initial promise because of practical problems. These failures led to a certain disillusionment. However, this period was, on the whole, one of steady growth. It also saw several new developments which have opened up major commercial possibilities.

After almost forty years, holography has now reached a stage at which it appears possible to make an objective assessment of some of its applications. This paper will attempt such a review and try to suggest some possible directions for future growth.

Recent trends

Recent trends suggest that the areas in which holography is likely to find an increasing number of applications in the next few years are art and advertising, high resolution imaging, information storage and security coding, holographic optical elements and holographic interferometry. This paper will therefore concentrate on developments in these areas.

Art and advertising

Some very spectacular applications of holography as a medium for art and advertising were demonstrated at quite an early stage. These included the production of a life-size hologram of the Venus de Milo and portraits with pulsed lasers. However, these applications represented a scientific tour de force rather than viable technologies, and further progress was held up by three main problems. These were the need for a laser or a bright point source of monochromatic light to illuminate the hologram, the fact that the reconstructed image was relatively dim and could be viewed only in subdued light and the fact that the image was reconstructed in a single colour, that of the source used to illuminate the hologram.

Some progress was made towards the solution of these problems in the sixties. This period saw the development of multicolour imaging using reflection holograms,[20,21] the development of new recording materials such as photopolymers[22] and dichromated gelatin[23,24] and the development of low-noise reversal bleach techniques for commercial photographic emulsions.[25-27]

A major advance was the invention by Benton in 1969 of the rainbow hologram.[28] While it took some time for the practical advantages of the rainbow hologram to be appreciated, it ultimately led to many new applications. One was the white-light holographic stereogram of Cross.[29] This was a composite hologram built up from a number of views of the subject from different angles in the horizontal plane. It has the great advantage that these views can be recorded with a motion picture camera in white light, and subject movement presents no problems. Another was the production of rainbow holograms which, when illuminated with a white light source, produced bright multicolour images.[30] The combination of these techniques has made possible the production of multicolour holographic stereograms up to 2 metres by 1 metre in size which exploit the full potential of holography as a display medium.

Another valuable technique which was developed during the 1970s was the production of copies of holograms by embossing.[31] Embossed copies can be produced extremely cheaply in very large quantities on a thin sheet of plastic backed with an evaporated metal coating and give a very bright image when viewed by reflection. Embossed holograms have opened up a rapidly expanding field of applications ranging from novelties and greeting cards to book and album covers. Perhaps the best known example is the National Geographic cover which sold 11 million copies.

An application which has attracted considerable attention is holographic movies. The major problem is that the holograms recorded have to be small for economy, while the image has to be large enough for comfortable viewing by a number of people.

Various systems have been proposed for holographic movies. The most advanced is probably that developed by Komar in the USSR.[32] In this system, a series of image holograms are recorded with a pulsed laser on 70 mm film using a lens with an aperture of about 200 mm. The same lens is then used to project the reconstructed images on a holographic screen which forms a number of images of the pupil of the projection lens in the viewing space. Each of these imaged pupils constitutes a viewing zone at which an observer can see a three-dimensional image.

High-resolution imaging

One of the interesting features of a hologram is that it can reconstruct an image whose resolution is limited only by diffraction effects over a considerable depth. Because of this, one of the first successful applications of holography was in the study of aerosols.[5] Measurements on moving microscopic particles distributed over an appreciable volume are not possible with a conventional microscope because of its very limited depth of field. Pulsed laser holography makes it possible to store a high-resolution image of the whole particle field at a given instant. The stationary image reconstructed by the hologram can then be studied in detail with a microscope.

Holography can also be applied to obtain an image which is unaffected by lens aberrations and even by the presence of aberrating media in the optical path.[6] For this, the object transparency is illuminated with coherent light and a hologram is recorded of the aberrated image wave using a collimated reference beam. When the hologram is illuminated by the conjugate to the original reference beam which, in this case, is merely the same collimated reference beam reflected back along its original path, it reconstructs the conjugate to the object wave. The conjugate wave has the same phase errors as the original object wave but with the opposite sign. Accordingly, when this conjugate wave propagates back through the optical system, these phase errors cancel out exactly, resulting in an aberration-free image. Generation of such a phase-conjugate wave in real time is now possible with a photorefractive crystal such as bismuth silicon oxide (BSO) as the recording medium.[33]

Production of a phase-conjugate wave in this fashion can also be looked upon as an example of degenerate four-wave mixing. However, this takes us into applications which are in the realm of nonlinear optics rather than holography.

One possible application of holographic high-resolution imaging is in the production of photolithographic masks for semiconductor devices.[34] Holographic imaging could overcome the problem of designing and producing lenses which can give very high resolution over a large field.

Information storage

The development of holography was greatly stimulated by the application of the concepts of communication theory to optics. Early work in this area led to the development of spatial filtering techniques using coherent light. In turn, holographic techniques were applied to the synthesis of complex filters which could be used for image enhancement and restoration as well as matched spatial filters for optical pattern recognition.[13,14]

The fact that information stored in a hologram is not easily corrupted by local surface damage also led at an early stage to research on holographic memories.[12] The results were largely limited by the lack of a suitable recording material, and this application now appears to have been overtaken by improvements in semiconductor memories and other techniques such as the optical disk.

However, holographic information storage still appears to have advantages for some specialized applications. One is in the reduction of space requirements for archival storage of documents, far below what is possible with microfilms.[35] Another is for storage of multicolour material, such as maps and motion pictures.[36,37]

An interesting application which has been described recently is for storing 3-dimensional information.[38] Normally when patients start orthodontic treatment, a plaster mould is made of their teeth. This mould is used for comparison with moulds made at later stages. As many as 400 000 moulds are made every year in Britain alone and their storage is a serious problem. Holograms of these moulds take up very little space and allow accurate measurements to be made in three dimensions. They also have the advantage that two holograms can be readily superposed for comparing the images.

An application of holographic information storage with considerable commercial importance is security coding for credit and identity cards. A simple method which has already been adopted by some of the major credit cards, involves an embossed hologram incorporated in the card to provide an additional safeguard against forgery.

Cash is replaced in the cardphone, now widely used in Britain, by a credit card with a number of holograms imprinted on its surface.[39] These are illuminated by a light source in the cardphone and detected and counted to work out how many units the user has left. This number is shown on a liquid crystal display. As the call progresses, the holographic patterns are erased, one by one, by a focused heat source.

Embossed holograms are obviously only the first step in security coding since they do not provide complete protection against forgery, and it is likely that they will be replaced by volume holograms. More sophisticated systems are being developed which provide a much higher degree of security and these will, no doubt, find many applications.

Holographic optical elements

Diffraction gratings formed by recording an interference pattern in a suitable light-sensitive medium are commonly known as holographic gratings. Because holographic gratings are free from ghosts and exhibit very low levels of scattered light they have replaced conventional ruled gratings for many spectroscopic applications. Such high-quality holographic gratings are recorded on photoresist layers coated on optically worked blanks.[40] The triangular groove profile required to blaze the grating for maximum diffraction efficiency is obtained by aligning the photoresist layer obliquely to the fringe pattern.[41] After development, the surface of the photoresist is coated with an evaporated metal layer.

Holographic scanners are another recent development.[42,43] Their most promising application is in point-of-sale terminals where they appear to have solved many of the problems associated with mirror scanners. Typically, the scanner consists of a disc with a number of holograms recorded on it. Rotating the disc about an axis perpendicular to its surface causes the reconstructed image spot to scan the image plane in the desired pattern. A straight-line scan is obtained by using an auxiliary reflector, so that the output beam is incident normal to the scanned surface.[44]

More generalized holographic optical elements (HOEs) are finding several specialized applications because they make possible unique configurations and functions. One of their main advantages is that, unlike conventional optical elements, their function is largely independent of substrate geometry. In addition they can be much lighter, since they can be produced on quite thin substrates. Another advantage is the possibility of recording several holograms in the same layer to produce multifunction elements. In addition, holograms can correct system aberrations, so that separate corrector elements are not required.

One of the most successful applications of holographic optical elements has been in head-up displays for high-performance aircraft.[45] The optical system in such a head-up display projects an image of the instruments at infinity along the pilot's normal line of vision, so that he can monitor critical functions while looking straight ahead through the windscreen. Head-up displays using holographic optical elements have the advantage that they are much lighter and more compact than those using conventional optics and can be fitted more easily into the limited space available.

An attractive possibility is the use of computer-generated holograms as holographic optical elements.[46] With these it should be possible to realize quite unusual optical components and achieve geometrical transformations as well as phase transformations. However, the main application of computer-generated holograms so far has been in interferometric tests of aspheric optical surfaces.[47,48] The hologram in this case is a representation of the interferogram that would be obtained if the wavefront from the desired aspheric surface were to interfere with a tilted plane wavefront. This hologram is placed in the plane in which the surface under test is imaged. The superposition of the actual interference pattern and the computer-generated hologram then produces a moire pattern which gives the deviation of the actual wavefront from the ideal computed wavefront. Techniques such as electron-beam recording on photoresist layers coated on optically worked substrates are now being used for the production of computer-generated holograms of very high quality for such tests.

A simple method of building up any desired image light distribution which is adequate for many applications is by means of a multifacet hologram.[49] The surface of the hologram is divided into a number of small elements, each of which contains a blazed grating of a specified spatial frequency and orientation. When this hologram is illuminated by a plane or spherical wave, the light incident on a particular facet is diffracted to a specific location in the image plane. The image is thus built up of small patches of light with near 100 percent efficiency.

Holographic interferometry

Holographic interferometry has many unique applications because holography permits storing a wavefront for reconstruction at a later time. Wavefronts which were originally separated in time or space or which originally were of different wavelengths can be compared by holographic interferometry. In addition, holographic interferometry makes possible interferometric measurements on rough surfaces.

Double exposure holographic interferometry has found wide application in non-destructive testing, in particular for detecting cracks and areas of poor bonding in composite structures. An allied area of applications has been in medical and dental research to study the deformations of anatomical structures and implants under stress. Other applications have been in aerodynamics, heat transfer and plasma diagnostics.[50]

Techniques such as sandwich holography have been used to eliminate unwanted rigid body displacements and simplify the interpretation of the fringes.[51] These techniques have made it possible to apply holographic interferometry to studies on even such large and massive objects as a milling machine.[52]

Holographic interferometry has now become a standard and vitally important part of the manufacturing and testing process in the aerospace industry. Turbine blades, wings and even tyres are routinely subjected to holographic inspection. Automobile manufacturers are also applying this technique to improve engine performance and reduce noise from transmissions and door panels.

Real-time holographic interferometry is also finding wide application because of the development of techniques for rapid in-situ processing[53] as well as the development of recording systems using photothermoplastic materials.[54] The latter completely eliminate the need for wet processing and make it possible to record and view a holographic interferogram in less than a minute.

Double-pulsed ruby lasers have opened up many applications for holographic interferometry on the shop floor. These include the study of transient deformations due to impact loading.[55] Objects rotating at extremely high speeds can also be studied with the help of an optical derotator consisting of an inverting prism which rotates at half the speed of the object.[56]

Holographic interferometry can also be used to produce an image modulated by a fringe pattern corresponding to contours of constant elevation with respect to a reference plane. Three basic methods of holographic contouring are available; these involve changing either the wavelength used or the angle of illumination or, alternatively, the refractive index of the medium in which the object is immersed.[57-59]

Yet another application of holographic interferometry has been in studies of vibrating surfaces. Real-time holographic interferometry can be used to identify the resonant frequencies,[19] but the most widely used technique is time-average holographic interferometry which yields fringes which can be used to map the vibration amplitude.[60] Observations of the phase of the vibration as well as measurements of much smaller vibration amplitudes are possible with stroboscopic holographic interferometry,[61-63] as well as with other techniques based on modulating either the phase or the amplitude of the reference beam.[64]

An application for which holographic interferometry has always appeared to be an ideal tool is quantitative strain analysis. However, its use for this purpose has been held up so far by the problems of fringe interpretation and data reduction.[65] Heterodyne holographic interferometry[66] and, more recently, digital phase-stepping holographic interferometry [67]now make possible direct measurements of the optical phase difference at a uniformly spaced network of points covering the interference pattern with very high accuracy. Because of the speed with which data can be acquired and manipulated, the digital phase-stepping technique opens up many new applications for holographic interferometry.[68,69]

Conclusions

It is clear from this brief survey that holography has already established itself in several rapidly expanding areas. We can look forward confidently to many completely new applications in the next few years.

References

1. Gabor, D., "A New Microscopic Principle," Nature, Vol.161, pp. 777-778. 1948.
2. Leith, E.N., and Upatnieks, J., "Reconstructed Wavefronts and Communication Theory," J. Opt. Soc. Amer., Vol. 52, pp. 1123-1130. 1962.
3. Leith, E.N., and Upatnieks, J., "Wavefront Reconstruction With Diffused Illumination and Three-Dimensional Objects," J. Opt. Soc. Amer., Vol. 54, pp. 1295-1301. 1964.
4. Denisyuk, Yu.N., "Photographic Reconstruction of the Optical Properties of an Object in its Own Scattered Radiation Field," Sov. Phys. Dokl., Vol. 7, pp. 543-545. 1962.

5. Thompson, B.J., Ward, J.H., and Zinky, W.R., "Applications of Hologram Techniques for Particle Size Analysis," Appl. Opt., Vol. 6, pp. 519-526. 1967.
6. Kogelnik, H., "Holographic Image Projection Through Inhomogeneous Media," Bell Syst. Tech. J., Vol. 44, pp. 2451-2455. 1965.
7. Lu, S., "Generating Multiple Images for Integrated Circuits by Fourier-Transform Holograms," Proc. IEEE, Vol. 56, pp. 116-117. 1968.
8. Groh, G., "Multiple Imaging by Means of Point Holograms," Appl. Opt., Vol. 7, pp. 1643-1644. 1968.
9. Pennington, K.S., and Lin, L.H., "Multicolor Wavefront Reconstruction," Appl. Phys. Lett., Vol. 7, pp. 56-57. 1965.
10. Schwar, M.R.J., Pandya, T.P., and Weinberg, F.J., "Point Holograms as Optical Elements," Nature, Vol. 215, pp. 239-241. 1967.
11. Lohmann, A.W., and Paris, D.P., "Binary Fraunhofer Holograms Generated by Computer," Appl. Opt., Vol. 6, pp. 1739-1748. 1967.
12. Anderson, L.K., "Holographic Optical Memory for Bulk Data Storage," Bell Lab. Record, Vol. 46, pp. 318-325. 1968.
13. Stroke, G., and Zech, R.G., "A Posteriori Image-Correcting 'Deconvolution' by Holographic Fourier-Transform Division," Phys. Lett., Vol. 25A, pp. 89-90. 1967.
14. Vander Lugt, A., Rotz, F.B., and Klooster, Jr., A., "Character Reading by Optical Spatial Filtering." In Optical and Electro-Optical Information Processing, ed. J.T. Tippett, D.A. Berkowitz, L.C. Clapp, C.J. Koester and A. Vanderburgh Jr., pp.125-141. 1965.
15. Brooks, R.E., Heflinger, L.O., and Wuerker, R.F., "Interferometry With a Holographically Reconstructed Comparison Beam," Appl. Phys. Lett., Vol. 7, pp. 248-249. 1965.
16. Burch, J.M., "The Application of Lasers in Production Engineering," The Production Engineer, Vol. 44, pp. 431-442. 1965.
17. Collier, R.J., Doherty, E.T., and Pennington, K.S., "Application of Moire Techniques to Holography," Appl. Phys. Lett., Vol. 7, pp. 223-225. 1965.
18. Haines, K.A., and Hildebrand, B.P., "Contour Generation by Wavefront Reconstruction," Phys. Lett., Vol. 19, pp. 10-11. 1965.
19. Stetson, K.A., and Powell, R.L., "Interferometric Hologram Evaluation and Real-Time Vibration Analysis of Diffuse Objects," J. Opt. Soc. Amer., Vol. 55, pp. 1694-1695. 1965.
20. Lin, L.H., Pennington, K.S., Stroke, G.W., and Labeyrie, A.E., "Multicolor Holographic Image Reconstruction with White Light Illumination," Bell. Syst. Tech. J., Vol. 45, pp.659-660. 1966.
21. Upatnieks, J., Marks, J., and Federowicz, R., "Color Holograms for White Light Reconstruction," Appl. Phys. Lett., Vol. 8, pp. 286-287. 1966.
22. Jenney, J.A., "Holographic Recording with Photopolymers," J. Opt. Soc. Amer., Vol. 60, pp. 1155-1161. 1970.
23. Shankoff, T.A., "Phase Holograms in Dichromated Gelatin," Appl. Opt., Vol. 7, pp. 2101-2105. 1968.
24. Lin, L.H., "Hologram Formation in Hardened Dichromated Gelatin Films," Appl. Opt., Vol. 8, pp. 963-966. 1969.
25. Lamberts, R.L., and Kurtz, C.N., "Reversal Bleaching for Low Flare Light in Holograms, Appl. Opt., Vol. 19, pp. 1342-1347. 1971.
26. Buschmann, H.T., "The Production of Low Noise, Bright, Phase Holograms by Bleaching," Optik, Vol. 34, pp. 240-253. 1971.
27. Hariharan, P., "Reversal Processing Technique for Phase Holograms," Opt. Commun., Vol. 3, pp. 119-121. 1971.
28. Benton, S.A., "Hologram Reconstructions with Extended Incoherent Sources," J. Opt. Soc. Amer., Vol. 59, pp. 1545-1546. 1969.
29. Benton, S.A., "Holographic Displays - a Review," Opt. Eng., Vol. 14, pp.402-407. 1975.
30. Hariharan, P., Steel, W.H., and Hegedus, Z.S., "Multicolor Holographic Imaging with a White Light Source," Opt. Lett., Vol. 1, pp. 8-9. 1977.
31. Bartolini, R., Hannan, W., Karlsons, D., and Lurie, M., "Embossed Hologram Motion Pictures for Television Playback, Appl. Opt., Vol. 9, pp. 2283-2290. 1970.
32. Komar, V.G., "Progress on the Holographic Movie Process in the USSR." In Three-Dimensional Imaging, Proc. SPIE, Vol. 120, ed. S.A. Benton, pp. 127-144. 1977.
33. Huignard, J.P., Herriau, J.P., Aubourg, P., and Spitz, E., "Phase-Conjugate Wavefront Generation via Real-Time Holography in Bi_{12} SiO_{20} Crystals," Opt. Lett., Vol. 4, pp. 21-23. 1979.
34. Levenson, M., "Applications of Phase Conjugation to Photolithography," Holosphere, Vol. 10, Nos. 7,8, p. 7. July/August 1981.
35. Vagin, L.N., Nazarova, L.G., Arseneva, T.M., and Vanin, V.A., "Holographic Miniaturization of Scientific and Technical Documents," Opt. and Spectrosc., Vol. 38, pp. 571-573. 1975.
36. Gale, M.T., Knop, K., and Russell, J.P., "A Colour Micro-Storage and Display System Using Focused Image Holograms," Opt. and Laser Technol., Vol. 7, pp. 234-236. 1975.
37. Yu, F.T.S., Tai, A., and Chen, H., "Archival Storage of Color Films by Rainbow Holographic Techniques," Opt. Commun., Vol. 27, pp. 307-310. 1978.
38. Pepper, A., "Storage Solution for Orthodontics," Holosphere, Vol. 12, No. 6, p.45. Winter 1984.
39. Pepper, A., "Holographic Telephones Tested in Britain," Holosphere, Vol. 12, No. 2, pp. 3,6. February 1983.
40. Rudolph, D., and Schmahl, G., "Verfahren zur Herstellung von Rontgenlinsen und Beugungsgittern," Umschau in Wissenschaft und Technik, Vol. 7, p. 225. 1967.
41. Sheridon, N.K., "Production of Blazed Holograms," Appl. Phys. Lett., Vol. 12, pp. 316-318. 1968.
42. Cindrich, I., "Image Scanning by Rotation of a Hologram," Appl. Opt., Vol. 6, pp. 1531- 1534. 1967.

43. McMahon, D. H., Franklin, A.R., and Thaxter, J.B., "Light Beam Deflection Using Holographic Scanning Techniques, Appl. Opt., Vol. 8, pp. 399-402. 1969.
44. Kramer, C.J., "Holographic Laser Scanners for Nonimpact Printing," Laser Focus, Vol. 17, No. 6, pp. 70-82. June 1981.
45. McCauley, D.G., Simpson, C.E., and Murbach, W.J., "Holographic Optical Element for Visual Display Applications," Appl. Opt., Vol. 12, pp. 232-242. 1973.
46. Bryngdahl, O., "Computer Generated Holograms as Generalized Optical Components," Opt. Eng., Vol. 14, pp. 426-435. 1975.
47. MacGovern, A.J., and Wyant, J.C., "Computer-Generated Holograms for Testing Optical Elements," Appl. Opt., Vol. 10, pp. 619-624. 1971.
48. Wyant, J.C., and Bennett, V.P., "Using Computer-Generated Holograms to Test Aspheric Wavefronts," Appl. Opt., Vol. 11, pp. 2833-2839. 1972.
49. Case, S.K., Haugen, P.R., and Løkberg, O.J., "Multifacet Holographic Optical Elements for Wavefront Transformations," Appl. Opt., Vol. 20, pp. 2670-2675. 1981.
50. Vest, C.M., Holographic Interferometry, Wiley, New York, 1979.
51. Abramson, N., "Sandwich Hologram Interferometry: a New Dimension in Holographic Comparison," Appl. Opt., Vol. 13, pp. 2019-2025. 1974.
52. Abramson, N., "Sandwich Hologram Interferometry. 4: Holographic Studies of Two Milling Machines," Appl. Opt., Vol. 16, pp. 2521-2531. 1977.
53. Hariharan, P., and Ramprasad, B.S., "Rapid In-Situ Processing for Real-Time Holographic Interferometry," J. Phys. E: Sci. Instrum., Vol. 6, pp. 699-701. 1973.
54. Urbach, J.C., and Meier, R.W., "Thermoplastic Xerographic Holography," Appl. Opt., Vol. 5, pp. 666-667. 1966.
55. Gates, J.W.C., Hall, R.G.N., and Ross, I.N., "Holographic Interferometry of Impact-Loaded Objects Using a Double-Pulse Laser," Opt. and Laser Technol., Vol. 4, pp. 72-75. 1972.
56. Stetson, K.A., "The Use of an Image Derotator in Hologram Interferometry and Speckle Photography of Rotating Objects," Exp. Mech., Vol. 18, pp.67-73. 1978.
57. Hildebrand, B.P., and Haines, K.A., "The Generation of 3-Dimensional Contour Maps by Wavefront Reconstruction," Phys. Lett., Vol.21, pp. 422-423. 1966.
58. Menzel, E., "Comment to the Methods of Contour Holography," Optik, Vol. 40, pp. 557-559. 1974.
59. Tsuruta, T., Shiotake, N., Tsujiuchi, J., and Matsuda, K., "Holographic Generation of Contour Maps of Diffusely Reflecting Surfaces by Using Immersion Method," Japan J. Appl. Phys., Vol. 6, pp.661-662. 1967.
60. Powell, R.L., and Stetson, K.A., "Interferometric Vibration Analysis by Wavefront Reconstruction," J. Opt. Soc. Amer., Vol. 55, pp. 1593-1598. 1965.
61. Archbold, E., and Ennos., A.E., "Observation of Surface Vibration Modes by Stroboscopic Hologram Interferometry," Nature, Vol. 217, pp. 942-943. 1968.
62. Shajenko, P., and Johnson, C.D., "Stroboscopic Holographic Interferometry," Appl. Phys. Lett., Vol. 13, pp. 44-46. 1968.
63. Watrasiewicz, B.M., and Spicer, P., "Vibration Analysis by Stroboscopic Holography," Nature, Vol. 217, pp. 1142-1143. 1968.
64. Aleksoff, C.C., "Temporally Modulated Holography," Appl. Opt., Vol. 10, pp. 1329-1341. 1971.
65. Briers, J.D., "The Interpretation of Holographic Interferograms," Opt. and Quant. Electron., Vol. 8, pp. 469-501. 1976.
66. Dändliker, R., "Heterodyne Holographic Interferometry." In Progress in Optics, Vol. 17, ed. E. Wolf, pp. 1-84, North-Holland, Amsterdam, 1980.
67. Hariharan, P., Oreb, B.F., and Brown, N., "Real-Time Holographic Interferometry: a Microcomputer System for the Measurement of Vector Displacements," Appl. Opt., Vol. 22, pp. 876-880. 1983.
68. Hariharan, P., and Oreb, B.F., "Two-Index Holographic Contouring: Application of Digital Techniques," Opt. Commun., Vol. 51, pp. 142-144. 1984.
69. Hariharan, P., and Oreb, B.F., "Stroboscopic Holographic Interferometry: Application of Digital Techniques," Opt. Commun., (to appear shortly).

HOLOGRAPHY IN CHINA

Wang Da-heng, Wang Zhi-jiang,
Hsu Da-hsiung, Lu Bo-xiang

(Chinese Program Committee of ICHA'86, China)

Abstract

This paper briefly reviews the recent developments on holography in China,such as the application of holography to interferometry, optical information processing, information storage, 3-D display, optical elements and nondestructive testing etc.

Introduction

In recent years,considerable progress has been made in the application of holography to a variety of problems. Study on holography in China started ever since the introduction of laser into holography by E. N. Leith and J. Upatnieks. But only in late 1960s and early 1970s holography raised much attention of scientists and engineers in China in the fields of optics, applied physics, mechanics and in various fields, such as interferometry, optical information storage, information processing, 3-D display, optical elements and nondestructive testing etc. Many papers were presented. In this paper, we will give a brief introduction of holography in China. However,the literatures cited are fairly representative but by no means complete.

3D Display

In China, much interest has been paid in the 3D images reconstructed in Lippmann holograms, Benton holograms, and printing embossed holograms.

Some dichromated gelatin reflection holograms of high resolution & high diffraction efficiency have been made and studied by Qu Zhi-ming et al (Shanghai Inst. of Laser Technology) and Fu Zi-ping et al (Beijing Inst. of Posts & Telecomm.). The Shanghai Inst. of Laser Tech. have also studied multiplex holograms, rainbow holograms, and tried to create 3D arts products such as "The Big Lantern" at the cooperation of Shanghai Lantern Instrument Factory. Silver halide holograms of high diffraction efficiencies have been studied and fabricated by Gu Qu-wu's group(Changchun Inst. of Optics and Fine Mech.), Zhao Lin et al (Computer Science Inst.,Academia Sinica), Yuan Wei-ben et al(Tianjin Computer Inst.), Zhang Guang-yong(Beijing Normal Inst.). They have succeeded in making silver halide holograms of 40% diffraction efficiency without resorting to bleaching process. Chen Ming-yi et al(Shanghai Univ. of Scien. & Tech.) have made effective studies in the principles and techniques of 360° panoramic astigmatic rainbow holograms (1). The method to make off-axis Fresnel holograms of large depth objects has been developed by Ge Wan-fu, Xiong Bing-hen (Changsha Railway Inst.). Based on the principle of the periodicity of the coherent length in a multiple-mode laser, they succeeded in making Fresnel holograms of very large object (4.5 m long) and objects of very large depth (8.2m)(2). Large depth holograms have also been studied by Lin De-hong and Jiang Ya-guang's group(Suzhou Univ.).Techniques of printing embossed hologram have been studied by Ge Wan-fu et al (Changsha Railway Inst.) , Yu Mei-wen's group (Beijing Inst. of Tech.), Lin Xiang-zhi, Liu Hong-du et al (Beijing Univ.)

Meanwhile the holographic 3D display has been spreading among the public. International exhibitions have been held in Shanghai, Beijing, etc. The first national holography exhibition sponsored by the Chinese Optical Society, was held in Beijing in 1984 with 30,000 visitors and continued in Xingjiang、Suzhou in 1985, and in Shanxi (city & country-side), Hebei in 1986.

Holographic Optical Elements

Applications of holography has been found in many other fields which bear no evident relationships on the 3D display property.

The researchers and engineers in Shanghai Inst. of Optical Instruments and Suzhou Univ. have made high efficiency holographic gratings used for various spectral instruments. They also designed multiplex gratings illuminated by white light projector to obtain 2D rotating rainbow shows. The multiplex gratings may also be used in photographic,cine-, or television cameras to obtain special rainbow effects. Zhang Guang-yong(Beijing Normal Inst.),Guo Lu-rong and her group(Sichuan Univ.), Zhang Jing-jiang et al(Beijing Normal Univ.), Mu Guo-guang and his group (Nankai Univ.) have studied and succeeded in using gratings for pseudocolor encodings(3)(4)(5)(6)(7). Yu Mei-wen,Chen Huang-ming and their group have made extensive studies in the principles and applications of holographic gratings and lenses,including the automatic design of lens systems including holographic elements (8). Hsu Da-hsiung and his group (Beijing Inst.of Posts & Telecommunications)studied the holographic lens wavelength division multi-plexers used in fiber communications and optical sensors.Since the holographic dichromated gelatin lens coupler made with Argon laser (4800Å) must be operated at 8500Å or 13000 Å,the aberrations caused by the wavelength shifts were carefully corrected by the aberration balancing method(9).Chin Guo-fan and his group(Tsinghua Univ.)have made extensive studies on the principles and applications of the computer generated holographic elements.They made CGH to generate standard aspherical wavefronts for aspherical lens testing.They also made CGH to generate nonspherical objectives to compensate aberrations in making holographic gratings. They have developed computer originated holo-lens for SAR signal processing system which combined the effects of cylindrical lens and conical lens through the same hologram (10). Liang Tie-cheng et al(ASI 613 Inst.) have studied and fabricated head-up displays with size 220X240 mm and efficiency 70%.Jiang Ya-guang et al(Suzhou Univ.) have studied and fabricated holographic lens for processing color photographs(11). Holographic composite lens developed by Yu Mei-wen and her group (BIT) , involving a speckle random phase shifter can be used in rainbow hologram recording systems and real-time image processing systems for pseudocolor encoding with two colors and encoding for the difference between photographs. Zhang Yi-mo, Zhao Xue-shan(Tianjin Univ.) have made extensive studies on the Fourier-transforming properties of holographic lenses.They succeeded in using holographic lenses to realize information storage,pattern recognition,etc.(12).Chin Guo-fan,Du Lin and their group(Tsinghua Univ.),engineers in Beijing Posts Inst., Guo Wen-qi, Ye Quan-shu(Nanking Univ.)made extensive studies on the principles and applications of holographic and CGH scanners(13). Hu Jia-shen et al (Changchun Inst. of Optics and Fine Mechanics) have developed a type of computer generated hologram called computer generated data film used in the optical processor for reflective tomography based on coherent processing of Doppler frequency shifted echoes from rotating objects (14).

Holographic Information Storage

A set of equipments for supermicrofiche holography storage has been developed by Yuan Wei-ben and his group (Tianjin Computer Inst.), Zhao Lin et al(Computer Science Inst. of academic Sinica). It consists of the high diffraction efficiency silver halide holographic medium, the supermicrofiche holography recorder, duplicator, reader, copier,and searcher. The size of the holographic medium(plate or film)is 6"X4",so that a magazine(sextodecimo)of 3010 pages can be recorded on the same single plate with 3010 small Fourier transform holograms. The diffraction efficiency reaches 40% without bleaching process. As a special feature a vibration isolation stand without any shock-absorber has been designed and used in the recorder.

Extensive studies on the principles and applications of the holographic information storage have been carried out. Xu Ming,Zhen Yong et al(Beijing Polytechnic Univ.), Xu Shu-xiang et al suggested and fabricated the"speckle phase plate"used in the holographic recording optical systems so as to improve the quality of reconstructed images. Fu Zi-ping et al(Beijing Inst. of Posts & Telecommunications)have made speckle holographic lenses with large relative aperture working as a Fourier transforming lens to record the Fourier transform holograms (15). Lu Bo-xiang et al (Physics Inst. of Academia Sinica) studied the principles and techniques of the thermoplastic recording medium to develop real-time recording materials.

Holographic Interferometry

Holographic interferometry has well been developed as an important field of holography in China. Chen Ji-shu et al (Tianjin Univ.) have studied the method of time-averaged hologram in examining the steady vibration modes (caused by "earthquake") of typical hydraulic structure models (16). The main advantages of holographic interferometry technique lies in the fact that the surfaces of the vibrating objects neednot be optically flat and complex vibrating objects can also be studied. Yuan Ge et al(Beijing Aeromechanics Inst.) used the double exposure techniques to measure the density and temperature distribution of the air current in the supersonic wind tunnel. Huang Gong-yu (Dalian Inst. of Tech.) used in-line holograms for measuring the sizes and the distribution of the silica particles in lubricants. The pulsed ruby laser (1.2 ms pulse length)was used as a source to get sufficient resolution. Chen Ming-yi et al (Shanghai Univ. of Scien. & Tech.) proposed an effective holographic technique Fourier transform holographic double exposure for the measurement of velocity of a uniformly moving object.

HNDT techniques have also been developed in China. Ke Jing-tang & his colleagues (Zhengzhou Inst. of Tech.) have developed a panoramic (360°)holographic interferometer for holographic nondestructive tire testing. By means of double exposure,6 holograms are obtained in each test, with fringe patterns clearly recorded, from which information on hidden defects in a tire may be collected at one stroke (17). Tan Yu-shan(Xi'an Jiaotong Univ.) et al have studied and proposed a rapid quantitative analysis of 3D displacements of large complex objects such as machine tools by means of holography interference(18)(19). Wang Yu-zhu and his colleagues(Harbin Univ. of Scien. & Tech.) and engineers in Shanghai Museum have made very nice studies on the vibration properties of Ancient Chinese Relics. They made time-averaged holograms of ancient bells and the "Water Spurting Basin" (a relic of Ming Dynasty, 1368-1644) and examined their vibration modes. Wang Shi-fan (Chendu Inst. of Radio Engineering) and Liu Fu-xiang(Sichuan Med. Inst.) have developed a method to determine the tooth rotation and translation directly from interferometric hologram providing sufficient data and informations for dentists (20). Fu Zi-ping et al (Beijing Inst. of Posts & Telecomm.) have studied and proposed a new real-time holographic recording material —— dichromated polyvinyl alcohol and given good experimental results of holographic interference, information storage, and character recognition (21). Lu Tuan-sun, Liu Wen-hu (Fujian Normal Univ.),Zhang Hong-jun et al (Inst. of Physics, Academia Sinica) Xu Liang-ying(Shanghai Inst. of Ceramics, Academia Sinica) have studied real-time optical correlation processing with Fe:$LiNbO_3$ crystal (22).

Further Researches and Applications

Researches and applications such as principles of holographic materials , applications in optical information processing, metrology, technical improvements,etc., have been made in China.

Wang Zhi-jiang, Wang Neng-he et al (Shanghai Inst. of Optics and Fine Mechanics) have studied one step rainbow hologram using aspherical mirror with large aperture and large relative aperture(23). Liu Shu-hui, Li Zhi-ming (Zhejiang Univ.) have studied and used a lens-slit-lens system (imaging system) in astigmatic one step rainbow holographic process and obtained large depth,clear,distortionless,deepscene 3D images(24). Chen Yan-song et al(Physics Inst., Academica Sinica) have studied the application of lensless Fourier transform to 3D measurements(25). Zhang Zhi-ming, Yin Xuan-tong et al(Fudan Univ.)have studied and proposed laser scanner for computer generated holograms (26). Wang Neng-he(Shanghai Inst. of Optics & Fine Mech.)Zhuang Song-lin(Shanghai Optical Instrument Research Inst.) have proposed a white light achromatic interferometer system for recording and reconstruction of 2D color complex object (27). Zhu Yi-qing, Lin De-hong, Wang Ce (Suzhou Univ.) have proposed a method of speckle shearing component (28). Yang Guo-guang, Zuo Jian, Wu Ruo (Graduate School,Univ. of Scien. & Tech. of China, Beijing)have proposed a method for image differentiation with spatial incoherent light to supress coherent noises, using Fourier transform holographic filter from diffraction grating type differentiation masks(29). Fu Xin-ding et al (Shanghai Inst.of Metallurgy), Cai Xue-giang, Qu Zhi-ming et al(Shanghai Inst. of Laser Technology)have studied microfabrication of holographic blazed grating by reactive ion beam etching(30). Zhang Chun-pin, Zhang Guan-yin et al(Nankai Univ.) have made experimental and theoretical investigations on optical and spectral properties of dilute development of holographic silver halide plates(31). Mu Guo-guang, Gong Qian, Li Zheng-ming (Nankai Univ.) have made theoretical analysis and experimental demonstrations on the image processing with bleached silver halide plates(32). An automatic development control system for thermoplastic recording materials has been designed at the Institute of Physics, Academia Sinica,Beijing. By using this system the development process can be controlled and stopped automatically at the optimum moment at high efficiency of diffraction. Computer-generated hologram is the combination of holography and computer science.Similar to CGH is the computer-generated Moiré. It was mainly developed for profile testing by A. W. Lohmann etc. A new identification system has been developed at the Institute of Physics, Academia Sinica, Beijing by using the technology of computer-generated Moiré. The informations needed to be identified, such as signatures or photographs, are coded in the grid pattern. Nothing can be seen by merely inspecting the grid itself.However the information will reappear in the form of Moire fringes when a reference grid is superimposed on such grid with the information coded. This identification system might be used with credit cards, I. D. card etc.

Conclusions

This paper reviews recent developments on holography in the People's Republic of China. The important fields of application are the 3D display, the HOE,the information storage, the holographic spectral diffraction gratings and the holographic interferometry. Researches are mainly carried out in universities and institutes, but more and more people have got interested in holography.

References

1. Hsuan Chen and Ming-yi Chen,"Fabrication of a 360 astigmatic rainbow hologram", Applied Optics,vol.22, No.16, 1983, 2474
2. Ge Wan-fu, Xiong Bing-hen,"The method for making a hologram of great depth", Acta Opt. Sinica. vol.5, No.7, July 1985, 600
3. Guo Lu-rong,Su Xian-yu,Wang Zhi-heng,Chen Zhen-pei,Chen Ze-xian,"White light pseudocolor encoding by phase modulating an image",Conference Digest,ICO-13, Sapporo'84, PI-19
4. Guo Lu-rong et al. A.O.S., vol.4, No.2, 1984, 145
5. Zhang Jing-jiang,Wang Shu-ying,Liu Da-he, "Pseudocoloring of an encoded phase picture", SPIE,1985Los Angels,Proceedings, vol.523, 341

6. Zhang Jing-jiang et al, A.O.S., vol.5, No.10, 1985, 944
7. Mu Guo-guang et al,(Image processing with bleaching silver halide plate", A.O.S.,vol.4, No.8, 1984, 693
8. Yun Gong, Yu Mei-wen, A.O.S., vol.5, No.6, 1985, 488
9. Xu Zou,Wang Ben,Fu Zi-ping,Hsu Da-hsiung "Holographic wavelength demultiplexer for optical fiber communications",Proceedings of Sino-Japanese Joint Meeting on Optical Fiber Science & Electromagnetic Field Theory, Beijing 1985, China
10. Lu Da, Chin Guo-fan. "Composition and optimum design of computer originated HOE(COHOE)" A.O.S., vol.5, No.7, 1985, 599
11. Jiang Ya-guang et al,"Simplified color image process. Syst. using a dichro. gelat.holo. elem." Appl. Opt. vol.21, No.17, 1982, 3138
12. Zhao Xue-shan, Zhang Yi-mo,"Fourier transform system using holographic lenses", A.O.S., vol.3, No.4, 1983, 410
13. Ye Quan-shu, Gao Wen-qi," A laser scanner applicable to scanning any graphics", A.O.S., 1983,4,314
14. Jia-shen Hu,G.D.Zhang, "CGH for simulating experiments of coherent Doppler tomography", Proc. SPIE, vol.437, 1983, 191
15. Fu Zi-ping, Hsu Da-hsiung, " Speckle phase-shift holographic lens for information storage ", A.O.S., vol.5, No.3, 1985, 281
16. Chen Ji-shu,He Bo-sen et al,"Laser holo.interferometry applied to the invest. of earthquake-resis." Proceedings of the inter. con. on exper. mechanics, Beijing, 1985, 347
17. Ke Jing-tang et al,"New type of holographic nondes. tire tester",Proc.SPIE vol.370,Oct. 4-8, 1982, 249
18. Yu-shan Tan,E.W.Smith,"Quantitative analytical tech. using holo. in study. structural charac. of machine tools", Annals of the CIRP, vol.32/1/1983, 269
19. Yu-shan Tan, E.W.Smith, Digest, 1983 Inter. Con. on laser, China 1983, 400
20. Wang Shi-fan,Liu Fu-xiang, "Determination of the tooth rot. & trans. produced by orthodontic forces using a sing. interf. hologram",Journal of Chengdu Inst. of Radio Eng." No.1, Mar, 1985, 92
21. Fu Zi-ping, Zhang Ju-qin, Hsu Da-hsiung, A.O.S.,vol.4, No.12, 1984, 1101
22. Lu Tuan-sun, Liu Wen-hu, Zhang Hong-jun, Dai Jian-hua et al,Chinese Journal of Lasers, vol.12, No.1, 1985
23. Wang Neng-he, Wo Ming-zhen, Lu Deng-wu, Wang Zhi-jing, AOS,2,1, 1982, 52
24. Liu Shu-hui, Li Zhi-ming, AOS, vol.5, No.10, 1985, 890
25. Chen Yan-song et al, AOS, vol.3, No.1, 1981, 227
26. Yin Xuan-tong,Xia Shao-feng,Jia Yu-shun,Zhang Zhi-ming,A.O.S., vol.2, No.4, 1982, 322
27. Wang Neng-he, Zhuang Song-lin, A.O.S., vol.4, No.1, 1984, 32
28. Zhu Yi-qing, Lin De-hong, Wang Ce, AOS, vol.5, No.11, 1985, 1025
29. Yang Guo-guang et al, AOS, vol.5, No.1, 1985, 42
30. Fu Xin-ding et al, Cai Xue-qiang et al, AOS, vol.5, No.1, 1985, 49
31. Zhang Chun-pin et al, AOS, vol.4, No.5, 1984, 444
32. Gong Qian,Li Zheng-ming,Mu Guo-guang, "Image processing with bleaching silver halide plate", Acta Optica Sinica, vol.4, No.8, 1984, 693

Session 2

Fundamentals and Display in Holography I

Chairman
Kuofan Chin
Tsinghua University, China

Spatial multichannel holography: longitudinal multiplexed Fourier storage

Zhao, Dazun and Mei, Wenhui

Optical Engineering Department, Beijing Institute of Technology
P.O.Box 327, Beijing, China

Abstract

A holographic method for storing two or more logitudinally arranged journal page transparencies at a quasi-Fourier spot is developed. Adequate redundancy and high storage density can be expected. Theoretical analysis and experimental demonstration are given.

Introduction

Holographic techniques for recording two or more pages of data or pictures at one hologram have been known for a long time.[1] In those holograms, different pages are recorded with different carrier frequencies, thus dynamic range of the recording material should be shared by all pages, accordingly, the diffraction efficiency of each page is reduced. That shortcoming has been excluded in so called "longitudinal multiplexed hologram" which we proposed earlier.[2] The recording scheme of that kind of hologram is illustrated in Figure 1. A large diffusing plate ("multichannel generator") G illuminated by laser beam serves as a light source for the object, a set of transparencies. Those teanspareucies are longitudinally arranged behind G with sufficient distances among themselves. In the Figure, three such transparencies P1, P2, P3 are shown. After passing through P1, P2, P3, the light wave carries the information of all three transparencies and forms an object beam which is recorded on a holographic plate H in cooperation with a reference beam R. The hologram thus made is called a "longitudinal multiplexed Fresnel hologram". It has been shown[2], when reconstructing the hologram with the conjugate reference beam R*, images of P3, P2, P1 will appear at planes P3, P2, P1 successively. Each image suffers certain aberrations but essentially no cross-talk background from other images. To evaluate the aberrations, we have assumed Fresnel approximation is acceptable for wave propagation between any two planes of P1, P2, P3 and H. The ensemble average of intensity distributions of each image is then elaborated. The term "ensemble average" here means the diffusing plate G is considered as a sample of a random diffusing plates ensemble with some definite statistical properties. The intensity ensemble average of image P3 is:

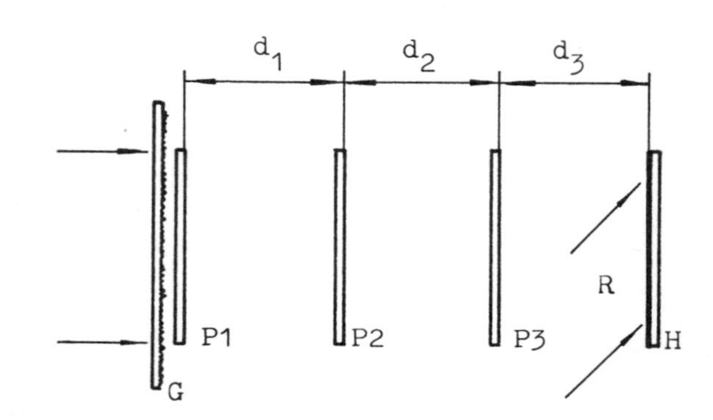

Figure 1. Recording scheme for longitudinal multiplexed Fresnel hologram.

$$\langle | u_3' (\vec{r}_3) |^2 \rangle = t_3^2 (\vec{r}_3) \otimes | \widetilde{D}_H(\vec{r}_3) |^2 \tag{1}$$

where $\vec{r}_3$ is the position vector at P3, u_3' is the complex amplitude of the reconstructed wavefront at that plane, "$\langle ... \rangle$" is a sign of ensemble average, t_3 is the amptitude transmittance of transparency P3 which is assumed to be real, "$\otimes$" stands for an operation of convolution and

$$\widetilde{D}_H (\vec{r}_3) = \int D_H (\vec{r}) \exp (-jk\vec{r}_3 \cdot \vec{r} / d_3) \, d\vec{r} \tag{2}$$

is the Fourier transform of $D_H(\vec{r})$ with $\vec{r}$ being the position vector at the hologram, D_H being the aperture function of the hologram and k being the wave number. Equation (1) shows the "average image" of P3 suffers an "incoherent" point spread function (PSF) of $|\widetilde{D}_H|^2$. In other words, according to Equation (2), $\langle |u_3'|^2 \rangle$ is equivalent to an image formed by an diffraction limited incoherent imaging system with a pupil function of D_H . When the area of the hologram is big enough, the PSF is essentially a δ - (Dirac delta) function. It should be noticed, however, since the hologram is actually recorded with only one sample of the diffusing plates ensemble, the reconstructed image is not $\langle |u_3'|^2 \rangle$ but $|u_3'|^2$, thus it suffers speckle noise in addition to the PSF. A diffusing plate with sufficient size usually makes the size of speckle granules be smaller than the required resolution length. The ensemble average intensity at P2 is:

$$\langle | u'_2(\vec{r}_2)|^2\rangle = t_2^2(\vec{r}_2) \otimes |\tilde{D}_2(\vec{r}_2)|^2 \tag{3}$$

where suffix "2" denotes the corresponding quantities at plane P2, $\tilde{D}_2$ is the Fourier transform of t_3 :

$$\tilde{D}_2(\vec{r}_2) = \int t_3(\vec{r}_3)\exp(-jk\vec{r}_2\cdot\vec{r}_3/d_2)d\vec{r}_3 \tag{4}$$

when $\tilde{D}_H$ can be taken as a δ-function. Equations (3) and (4) show another equivalent incoherent imaging process. This time the equivalent system has a pupil function of $t_3(\vec{r}_3)$. This fact implys that in order to have acceptable image quality, the recorded transparencies should be transparent at their most parts and have low spatial frequency compnents only. Fortunately, journal pages often meet these requirements. More complicated but similar expressions can be deduced for ensemble average at P1.

Figure 2 shows three photos of enlarged fractions of reconstructed images from a quintuplicate hologram. Those images appeared at planes P1 (left photo), P3 (middle photo) and P5 (right photo) respectively.

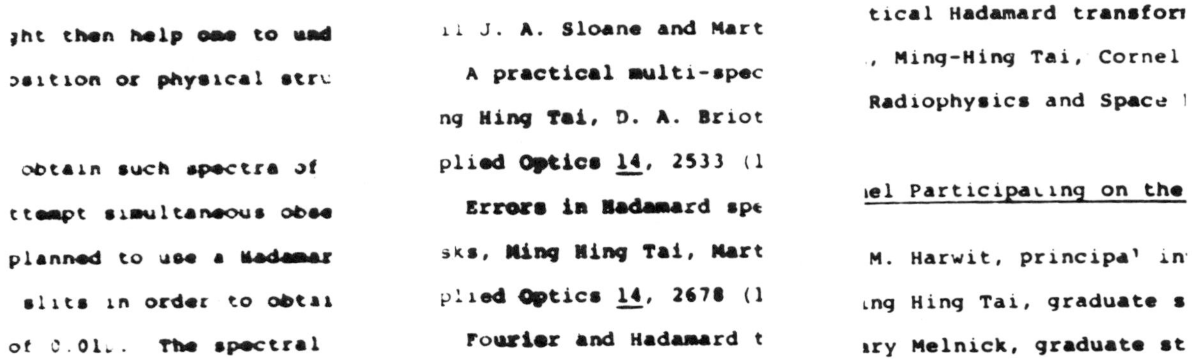

Figure 2. Images reconstructed from a quintuplicate hologram.

Merits of longitudinal multiplexed holograms as comparing with multiple-exposured multiplexed holograms are:
(a) No dynamic range competition among recorded pages;
(b) No diffraction efficiency competition among reconstnucted images;
(c) Compatible with multiple-exposure approach, i.e., a multiple-direction longitudinal multiplexed hologram could be made if necessary.
Those merits are resulted from the fact that for all recorded pages, only "one" wave front is recorded on the hologram and only "one" wave is reconstructed from the hologram.

Longitudinal multiplexed holograms have their own shortcomings, too. For examples, the recorded objects are confined within some kinds as mentioned above and the storage density is poor. The latter point is caused by the broad bandwidth of the object wave due to diffuse illumination. Ordinary optical Fourier transform to such an object wave could not reduce the size of the light spot much even if the transform lens has a big relative aperture. A novel kind of multichannel hologram is thus proposed in this paper to deal with this problem.

The principle of longitudinal multiplexed quasi-Fourier storage

The basic idea to get a smaller storage spot is to compensate the diffuse effect of the diffusing plate, after the multichannel object beam has been formed.

A random phase plate G is first prechosen as the compensation plate, then a Fresnel hologram, F, of G is made with the setup shown in Figure 3, where S is a monochromatic point source. It is obvious that when F is reconstructed with the conjugate reference beam R'*, the wave field between F and G will reappear but in reverse propagation direction. After passing through G, that wave will focus at S due to the compensation. Hologram F can thus be used as a multichannel generator for recording the storage hologram. Figure 4 shows the recording system. The transparencies to be recorded (two of them, P1 and P2 are shown in the Figure) are located between F and G, another holographic plate H is placed behind G, the distance between H and G is same as that between S and G in Figure 3. The storage hologram is recorded on H.

Mathmetical description of those processes are given briefly as follows. In order to

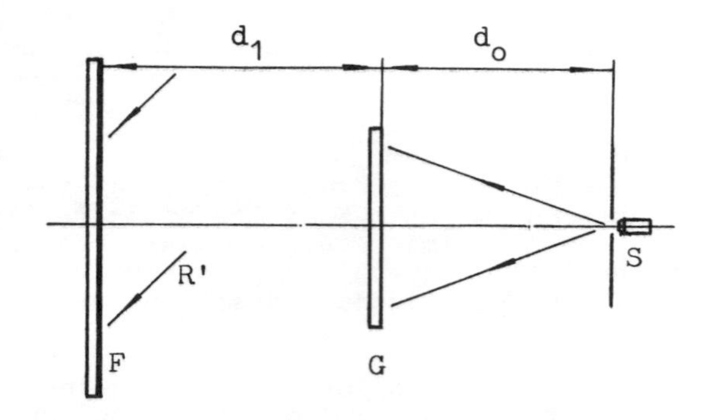

Figure 3. Preparation of multichannel generator from compensation plate

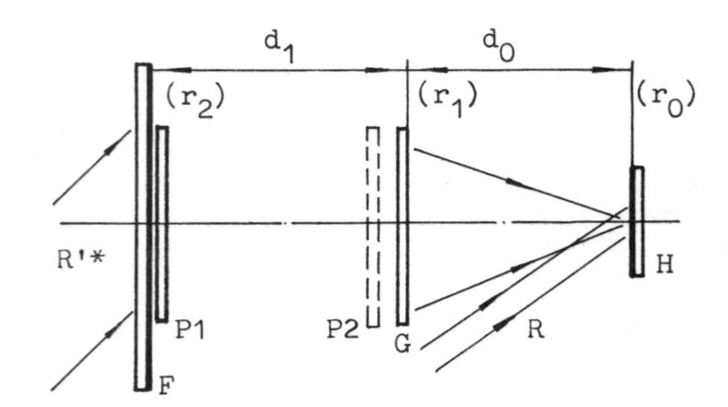

Figure 4. Recording of quasi-Fourier hologram

concentrate our attention on the function of the compensation, we assume only one transparency, P1, which is in contact with F, is recorded. In addition, all unimportant constant coefficients are dropped whenever possible. Let the position vectors at planes G, F, H be expressed by $\vec{r}_1$, $\vec{r}_2$, $\vec{r}_o$ respectively and the transmittance of the random phase plate be:

$$G(\vec{r}_1) = \exp(j\varphi(\vec{r}_1)) \tag{5}$$

The wavefront recorded on F is a speckle field:

$$u_2(\vec{r}_2) = V^*(\vec{r}_2)\exp(jkr_2^2/2d_1) \tag{6}$$

where

$$V^*(\vec{r}_2) = \int \exp(j\varphi(\vec{r}_1))\cdot\exp(jkr_1^2(1/2d_o+1/2d_1))\cdot\exp(-jk\vec{r}_1\cdot\vec{r}_2/d_1)\,d\vec{r}_1 \tag{7}$$

During recording the storage hologram, the illumination field for P1 is

$$u^*_2(\vec{r}_2) = V(\vec{r}_2)\exp(-jkr_2^2/2d_1) \tag{8}$$

After passing through P1, the wavefront back to G is:

$$u_1'(\vec{r}_1) = \int t(\vec{r}_2)\,V(\vec{r}_2)\cdot\exp(jkr_1^2/2d_1)\cdot\exp(-jk\vec{r}_2\cdot\vec{r}_1/d_1)\,d\vec{r}_2 \tag{9}$$

where t is the amplitude transmittance of P1. Finally, the wavefront passes through G and arrives at H:

$$u_o'(\vec{r}_o) = \exp(jkr_o^2/2d_o)\int t(\vec{r}_2)V(\vec{r}_2)\cdot V^*(\vec{r}_2+\vec{r}_o d_1/d_o)\,d\vec{r}_2 \tag{10}$$

Now let us discuss the spatial distribution of u_o' . Corresponding to a Fourier compnnent of frequency $\vec{f}$ of t, $T(\vec{f})$, u'_o has its own component as:

$$u'_{of}(\vec{r}_o) = CT(\vec{f})\int G^*(r_1'-\lambda d_1 f)\,G(\vec{r}_1')\exp(-jk\vec{r}_1'\cdot(\vec{r}_o-(d_1+d_o)\lambda\vec{f})/d_o)\,d\vec{r}_1' \tag{11}$$

where C is a quadric phase coefficient. As expected, for $\vec{f}=0$, component u'_{of} is a δ-function centered at $\vec{r}_o=0$. The compensation is complete. Alongside the growth of $|\vec{f}|$ starting from zero, the compensation goes worse and the width of u'_{of} expanses. Meanwhile, the "center" of u'_{of} shifts to $\vec{r}_o=(d_1+d_o)\lambda\vec{f}$, where u'_{of} has a peak until $|\lambda d_1 f|$ is greater than the corelation length of $G(\vec{r}_1)$. Based on the similarities and differences between u'_o and the Fourier spectrum of t, we call u'_o a "quasi-Fourier spot". While the expansion of Fourier components makes the spot size bigger, it also makes the recorded hologram become redundant and have better dust/scratch proof performance. In addition, the spectrum expansion also makes a fraction of information of high frequencies go to the central region of the spot, thus an aperture can be placed in front of H during recording in order to reduce the storage area. Contrary to the case of conventional Fourier hologram, it does not reduce the resolution power apparently.

If the quasi-Fourier hologram is recorded within an aperture of $D(\vec{r}_o)$ and reconstructed with the conjugate reference beam R^*, after passing through G, the reconstructed wave produces a field at plane P1 as:

$$u_2''(\vec{r}_2) = \exp\,(jkr_2^2/2d_1) \int t(\vec{r}_2')\; V^*(\vec{r}_2')\; h\,(\vec{r}_2', \vec{r}_2)\; d\vec{r}_2' \tag{12}$$

where

$$h(\vec{r}_2', \vec{r}_2) = \int D(\vec{r}_o) V(\vec{r}_2' + \vec{r}_o d_1/d_o)\; V^*(\vec{r}_2 + \vec{r}_o\, d_1/d_o)\; d\vec{r}_o \tag{13}$$

is the amplitude PSF of whole imaging process for P1 if the modulation by speckle field V* and the quadric phase coefficient are ignored. This PSF may also be written as:

$$h(\vec{r}_2', \Delta\vec{r}) = \int D(\;(\vec{r}-\vec{r}_2')d_o/d_1)\; V\,(\vec{r}\,)\; V^*(\vec{r} + \Delta\vec{r})\; d\vec{r} \tag{14}$$

with

$$\Delta\vec{r} = \vec{r}_2 - \vec{r}_2' \tag{15}$$

It can be seen that this PSF is spatial variant but always has a peak at $\Delta\vec{r}=0$. If the diameter of D is much greater than the corelation length of V, in other words, if the area of D is big enough to contain sufficient number of speckle granules (note that the smaller the ratio d_o/d_1 is, the more granules can be contained), h is essentially a δ-function. Furthermore, as we have pointed out,[2] when an object is modulated by a random field (it is V* in the present case as shown in Equation (12)), the filtering by an imperfect coherent imaging process is much less harmful than when the object is not modulated, if the speckle noise is negligible. Thus a good image quality can be expected. This conclusion can also be confirmed by statistical estimation. In fact, if the G/F combination is considered as a sample of a G/F ensemble , and if

$$\langle V(\vec{r})\; V^*\;(\vec{r}\,')\rangle = \delta(\vec{r}-\vec{r}\,') \tag{16}$$

holds, the ensemble average of $|u_2''|^2$ is:

$$\langle |u_2''(\vec{r}_2)|^2 \rangle \;=\; t^2(\vec{r}_2) \tag{17}$$

Experiments and results

The experiments are implemented under following conditions: original light source, He-Ne laser; size of phase plate, 70mm x 70mm; size of the first hologram F, 90 mm x 90 mm; d_1=100mm do = 40mm. Figure 5 shows two reconstructed images of a resolution test chart. (a) is from

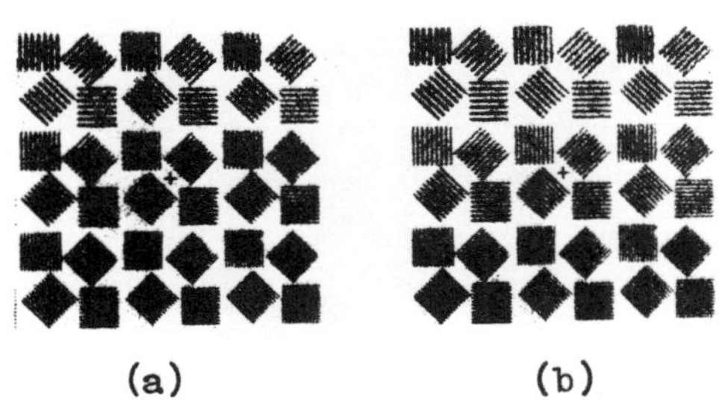

(a) (b)

Figure 5. Images from holograms with different aperture diameters.
(a) 3mm, (b) 10mm.

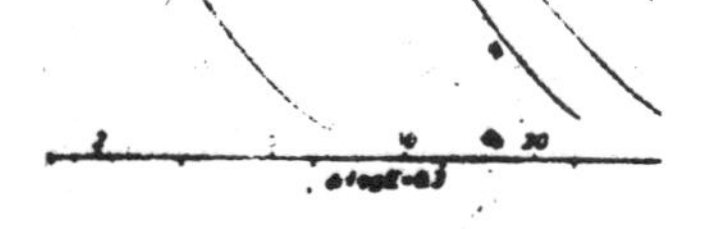

Figure 6. Reconstructed images from a longitudinal duplicate quasi-Fourier hologram.

a hologram with an aperture diameter of 3mm and (b) is from one with aperture diameter 10mm. No much difference in resolution powers can be seen on the images. It means that most part of the object information is concentrated at a circle of 3mm diameter on the hologram. Figure 6 shows two images reconstructed from a longitudinal duplicate quasi-Fourier hologram of 3 mm diameter. Distance between P1 and P2 is about 70 mm.

Conclusions

Both theoretical analysis and experimental demonstrations have confirmed the feasibility of longitudinal multiplexed quasi-Fourier storage. This technique not only improves the storage density, but also keeps the merits of longitudinal multiplexed Fresnel hologram.

References

1. Collier, R.J., Burckhardt, C.B., Lin, L.H., Optical Holography, Academic Press, 1971.
2. Zhao, D., Mei, W., "Spatial multichannel holography: longitudinal multiplexed hologram", submitted to Acta Optica Sinica.

Use of fringe scanning method in electron holographic interferometry

Toyohiko Yatagai and Katsuyuki Ohmura
Institute of Applied Physics, University of Tsukuba
Tsukuba Science City, Ibaraki 305, Japan

Shigeo Iwasaki
National Research Laboratory of Metrology
Tsukuba Science City, Ibaraki 305, Japan

Akira Tonomura, Junji Endo and Shuji Hasegawa
Advanced Research Laboratory, Hitachi, Ltd.
Kokubunji, Tokyo 185, Japan

Abstract

Phase analysis technique for reconstructed images of an electron holographic microscope is described. The fringe scanning method is applied to gain high sensitivity in phase detection. An example of measuring a magnetic field of a fine particle is presented. The measurement accuracy in the case of median filtering is about 1/70 fringe.

Introduction

Holographic techniques have been applied to the field emission electron microscope.[1-3] Thickness variations and magnetic field distributions in an electron microscopic region have been detected by the electron holographic interferometry. Recently Takeda et al applied the FFT method of the subfringe analysis for electron holographic fringes.[4] We discuss here the use of the fringe scanning technique [5] to fringe analysis of an electron interference hologram.[6] This gives us high sensitivity and high spatial resolution in phase measurement. Quantitative phase measurement technique in electron holographic interferometry is very useful for magnetic field measurement in microscopic region as well as small thickness variation evaluation.

Electron holography and optical reconstruction

In an optically reconstructed image of an electron hologram, the phase of the transmitted electron beam is reconstructed as well as its amplitude. This means that interference contour fringes of the transmitted wavefront on an in-focus electron micrograph can be obtained if a plane wave is superimposed on the reconstructed image.

Figure 1 shows an optical reconstruction system for the interference microscope based on the electron holography. Two collimated laser beams coherent each other illuminate a hologram, so that the plus first order and the minus first order reconstructed images make a phase-difference amplified interference pattern by a factor of 2. Because of noise in reconstructed images, combination of the first order and the zero order diffracted waves is used to obtain an interferogram without phase amplification in the present experiment.

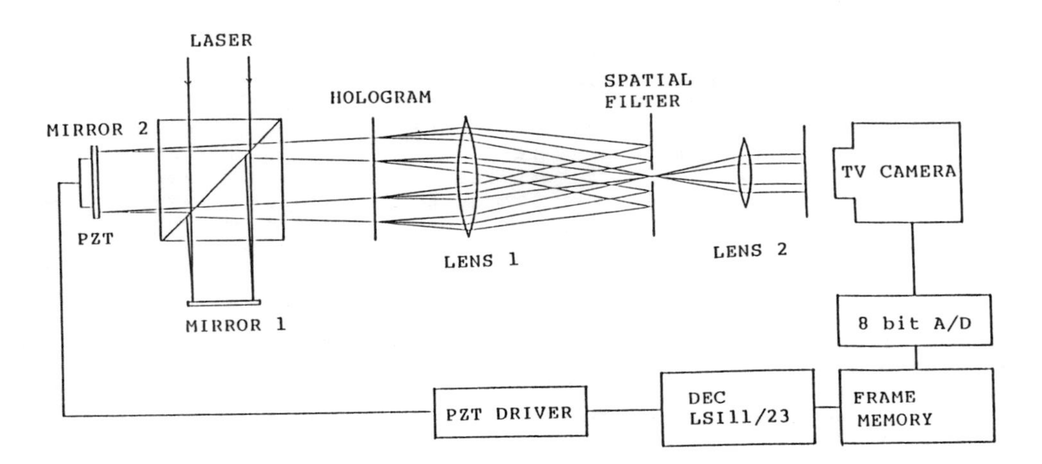

Fig. 1 Schematic diagram of hologram reconstruction and fringe analysis.

Fig. 2 Magnified version of an electron hologram.

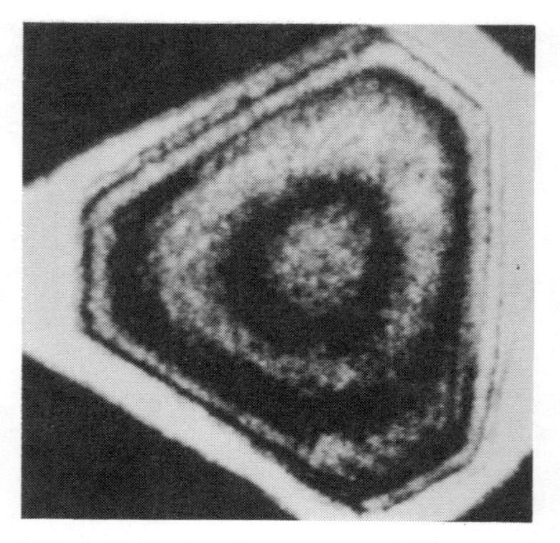

Fig. 3 Interference pattern of a magnetic particle.

Figure 2 shows an electron hologram, which is recorded by a 125 kV field-emission electron microscope. The specimen is a magnetic cobalt particle mounted on a carbon thin film. Its interferogram by using the zero and the first diffraction order is shown in Fig. 3.

The phase difference between an object beam and a reference beam is caused by two sources; thickness variation and magnetic flux. The effective refractive index, n, of a nonmagnetic specimen can be derived as

$$n = 1 + V_o/2\,\phi_o \qquad (1)$$

where V_o is the mean inner potential of the specimen and ϕ_o is the initial electron potential. Thus the phase change due to thickness variation of d is given by

$$\Delta\Psi = n\,d \qquad (2)$$

An amount of phase difference due to the magnetic flux is described by

$$\Delta\Psi = -\int e/h\, B_n\, dS \qquad (3)$$

where e is the electron charge, h is Planck's constant. According to Eq. (3), 2π phase difference or one fringe in an interferogram corresponds to an a closed magnetic flux of h/e = 4×10^{-15} Wb.

Phase analysis

Phase modulation of the phase of this reference plane wave provides to the fringe scanning phase detection technique. The phase of the reference beam is modulated by a PZT driven mirror. The amount of phase modulation is controlled by a computer. An interference fringe pattern is detected with a high resolution and a low distortion TV camera and is stored in a frame memory with a 512x512x16 bit resolution. This frame memory is connected to a computer memory with a DMA bus. Thus, 8 or 16 interference fringe patterns with different phases are transfered into a computer memory. According to the fringe scanning phase detection technique, summations with arbitrary sine and cosine weights are calculated and the inverse tangent of the ratio of the calculated summations gives the phase of the interference fringe.

Figure 4 shows interferograms with $\pi/2$ reference phase difference. According to the fringe scanning phase detection algorithm, the principal values of the phase are calculated. The calculated phase distribution is shown in Fig. 5. Phase irregularity due to speckle noise is observed in the phase profile. By using a median filtering window of 3x3 pixels, this type of phase irregurality is reduced. The noise level or the accuracy limit of the present method is estimated to be from 1/50 to 1/70 fringe spacing. In the case of phase amplification, higher diffraction orders are used to make interference fringes.

Conclusion

We described the importance of the phase sensitive detection method in the electron holographic microscope. The fringe scanning phase detection technique is applied to a phase-amplified holographic fringe pattern recorded a magnetic field distribution. To verify the proposed method, we analyze a cobalt fine particle. The spatial resolution is less than 100 A and the accuracy of phase analysis estimated is about 1/50 fringe.

Reference

1. A. Tonomura, J. Endo, and T. Matsuda, "An Application of Electron Holography to Interference Microscope," Optik, 53, 143 (1979).
2. A. Tonomura, T. Matsuda, H. Tanabe, N. Osakabe, J. Endo, A. Furukawa, K. Shinagawa, and H. Fujiwara, "Electron Holography Technique for Investigating Thin Ferromagnetic Films," Phys. Rev., B 25, 6799 (1981).
3. A. Tonomura, "Application of Electron Holography Using a Field-Emission Electron Microscope," J. Electron Microsc., 33, 101 (1984).
4. M. Takeda and Q-S. Ru, "Computer-Based Highly Sensitive Electron-Wave Interferometry," Appl. Opt., 24, 3068 (1985).
5. J. H. Bruning, D. R. Herriot, D. P. Rosenfeld, A.D. White and D.J. Brangaccio, "Digital Wavefront Measuring Interferometer for Testing Surfaces and Lenses," Appl. Opt.,13, 2693 (1974).
6. T. Yatagai, K. Ohmura, S. Iwasaki, S. Hasegawa, J. Endo and A. Tonimura, "Quantitative Phase Analysis in Electron Holographic Interferometry," Appl. Opt. to be published.

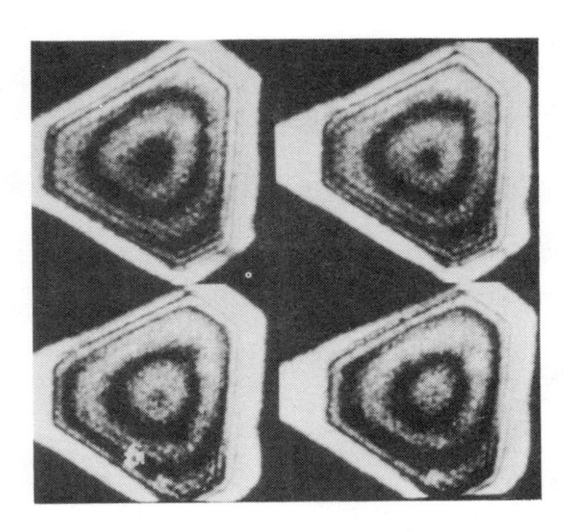

Fig. 4 Interferometric micrograms with different reference phases.

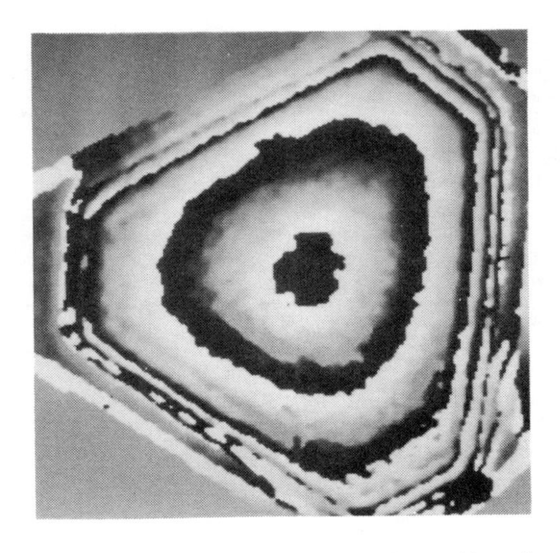

Fig. 5 Calculated phase distribution.

Holographic cinematography and its applications

P. Smigielski *), H. Fagot, F. Albe

Franco-German Research Institute of Saint-Louis (ISL)
68301 Saint-Louis (France)

Abstract

Important progresses were achieved for the first time: 1) recording of single-exposure cineholograms of living bodies on 126-mm films, at a repetition rate of 25 holograms per second with the help of a frequency-doubled pulse YAG-laser; 2) recording of double-exposure cineholograms of reflecting moving objects for medical and industrial applications. Limitations of 3-D movies are described.

Introduction

Holography is expected to become a potential tool for achieving in near future scientific and industrial three-dimensional cinematography. First, we shall describe the use of 126-mm films for recording cineholograms of living bodies at a frequency of 25 Hz. Limitations of 3-D holographic movies for entertainment purposes will be described. Thereafter, we shall report on double-exposure cineholography (interferometric cineholography) of reflecting moving objects for non-destructive testing and medical applications.

Two types of holographic cinematography

For many years efforts have been made to apply holography to cinematography, but it has been only recently that significant experiments were performed: 1976 in the Soviet Union[1], 1981 in the United States[2] and 1983 in France[3]. The question arises as to the reason why cineholography has come so late into practical use. To answer this question, we must distinguish between two methods of holographic recording.

1. Cineholography in transmitted light of transparent objects

In the holographic method of recording transparent objects only the shadows of the objects are visualized and not the objects themselves. Consequently, it is practically the whole energy of the illuminating wave which strikes the film when transparent objects are investigated (air flows, shock waves, plasmas, ...). Therefore, the recording can be done with a laser having a relatively low energy per pulse. Moreover, no high coherence length is required. These are the reasons why cineholography based on the recording of transparent objects made its way rapidly in the scientific field (particle analysis, for example). In effect, lasers with features such as a very high repetition rate, low energy per pulse and low coherence length have always been commercially available.

2. Cineholography of objects scattering light by reflection

This case is characterized in that the light waves scattered as a result of the reflection of the illuminating wave by the object are recorded. This means that the surface of the object is visualized. Only an extremely small part, a few thousandths of the quantity of light emanating from the laser, arrives on the hologram. Therefore, 3-D motion pictures based on the recording of large moving objects (living bodies, for example) scattering light by reflection call for the use of lasers delivering 25 or more high-energy pulses per second and having a high coherence length. Those lasers were not commercially available up to 1985. These are the reasons why we have built in our laboratory a frequency-doubled pulse YAG-laser well adapted for 3-D movies of living bodies and for holographic non-destructive testing on aircraft and medical applications.

Cineholography of living bodies[4]

A continuous film-transport camera which can be operated with a 126-mm film was especially adapted to our experiment. The continuous run of the film is a very classical technique used in high-speed photography. Each hologram is exposed to one laser pulse. The synchronization of the laser repetition rate (25 Hz) with the feed rate of the film allows a series of jointed holograms to be obtained on the moving film. The film can run continuously because the exposure time involved is only of the order of 15 ns. In effect, in order to obtain holograms of good quality, the film must not cover a distance, during the exposure

*) Also, associate professor at ENSPS, Louis Pasteur University, Strasbourg, France.

time, which exceeds one tenth of the wavelength of the laser beam. This leads to a tolerable maximum feed rate of approximately 3 m/s whereas the feed rate of the 126-mm film equals 0.25 m/s (hologram size: width 100 mm, height 10 mm). Therefore, in contrast to conventional cinematography, the intermittent film motion is not mandatory.

The holographic recording arrangement used is a conventional one. A small part of the illuminating wave is used to form the reference beam which illuminates directly the film. By superimposing approximately equally the light beam scattered from the scene with the reference beam, microinterferences are generated on the 126-mm film. The latter has a high resolution and is sensitive to the green light (wavelength λ = 0.532 µm). The green light is obtained after the infrared beam of light (λ = 1.06 µm) emitted from the YAG-laser has gone through a frequency-doubling crystal. We get 30 to 50 mJ per pulse in the green light. After photographic processing the film is solely illuminated with the reference beam. This occurs in the same way as in the recording set-up, but the laser used is not necessarily the same. Normally a cw-laser is used (argon laser λ = 0.514 µm) operating at a wavelength which approaches that of the recording laser (λ = 0.532 µm) to avoid geometrical aberrations. By using a helium-neon red laser the image quality is still good enough. The 126-mm film allows us in the reconstruction process to have a stereo picture of the scene by direct viewing with both eyes without wearing special glasses. It must be noticed that the film runs continuously during this reconstruction process.

Several movies were recorded. The good stability of the coherence of the YAG-laser allows the recording of very long film lengths without problems. But for exhibition purposes we recorded only 10 to 30 m film lengths.

1. "Christiane et les holobulles" is a 40-s duration movie which exhibits a young lady who produces soap bubbles and projects them in the direction of the observer (Fig. 1 and 2).

2. "La Belle et la Bête" is a 80-s duration fiction movie with several sequences with music and speech. It was presented for the first time in Paris, in the Museum of Holography on June 19, 1986. One image taken from the film representing a painting of Belle is shown in Figure 3.

This movie was directed by P.Smigielski (ISL) and A.M.Christakis (Museum of Holography, Paris) with the artistic assistance of Alexander and Daniele (Australia) and recorded in our laboratory at Saint-Louis on March 1986 by H.Fagot (ISL) with the assistance of F.Albe, A.Stimpfling, R.Meyer, J.Schwab, H.Royer, S.Reilhan, B.Tritsch, from the optical department of ISL.

Remark. All the movies were recorded in day-light (without darkening the laboratory).

Interferometric cineholography[5]

The double pulse of the coherent beam of light required for the holographic interferometry is generated by two successive pumpings of the same YAG-laser for time intervals greater than 3 ms or is produced by triggering twice the intra-cavity Pockels cell during one pumping for time intervals from 0 to 100 µs. For the first time we recorded holographic films by double-exposure of non-transparent phenomena using a conventional 35-mm camera specially adapted for holographic recording. Here, stops on the film for each hologram are necessary.

The laser and the camera were operated at a frequency of 10 Hz. Several scientific movies were recorded.

- "Holoparleur" (holo-loudspeaker) shows the deformation and displacements of a loudspeaker diaphragm of 25 cm in diameter vibrating at frequencies of 20 Hz and 70 Hz (Δt = 4 ms)(Fig. 4, 5).

- "Holocrâne" (holo-skull) visualizes "in vivo" the vertex cranii of a bald-headed man to study the correlation between the deformation of the skull and the mechanism of breathing (Δt = 10 ms)(Fig. 6, 7).

- "Holofrein" (holo-brake). Δt = 50 µs. This movie was recorded recently to study the deformation in time of a car block-brake rubbing against a rotating disc. It is an industrial application conducted in cooperation with a french company. Figure 8 shows a double-exposure hologram recorded with a double-pulse ruby laser prior to the recording with the YAG-laser.

The combined use of interferometry and cineholography is of great interest for the 4-D investigations of physical events varying with time such as the deformations of materials and structures. In particular the aeronautical industry is interested in this technique which is expected to allow non-destructive testing (maintenance checking) in quasi-real time. With one YAG-laser working at a low repetition rate (10 Hz), the method is recommended for the investigation of deformations and/or vibrations occuring at low or very low frequencies in the fields of mechanics, physiology or heat events. But, by using two pulse YAG-lasers with their beams well superimposed, the time interval separating the two pulses can be chosen at will and the study of quasi-steady objects vibrating at all frequencies is possible. Moreover, it is possible to use our cineholographic camera in connection with holographic endoscopy[6] or with our new compensation method when rigid-body motion of any amplitude occurs[7].

Limitations of cineholography

The technique of cineholography as described in this paper will meet the demand made by the scientists and the industrialists. This is not the case, however, for entertainment cinematography (3-D movies). The use of pulse lasers for recording pictures leads to serious drawbacks:

- the depth of the scenes is limited to a few meters because of the limited coherence length of the pulse laser,

- the velocity of the objects under recording must be very low: 1 m/s in the direction of the camera with a pulse duration of 20 ns.

- full-color cineholography cannot be envisaged within a foreseeable period even if it is theoretically possible,

- the film actors must protect their eyes from the intense laser light.

Problems are also encountered in the reconstruction process performed either with lasers (presence of a "speckle" in the reconstructed images, but there are some solutions) or in white light (reduction of the depth of the scene). The main problem is the visualization of the film by many people:

- the use of a reflecting holographic screen is theoretically possible (see Komar experiment[1]), but there are some serious drawbacks,

- one can use a film larger than 126 mm. Why not a film of 1 meter length ? But how many people can watch such a film at the same time ?

It seems that cineholography for entertainment purposes will stay a long time either a funny curiosity to see in museums or fairgrounds or a prestige advertising.

Outlook on the future

A solution to the problem inherent in 3-D cinematography would be to use holography not for the recording step, but only in the reconstruction process, as this is the case for multiplex holograms, for instance. Then, holography would be the strongest candidate tomorrow in the field of 3-D movies, because it might enable the spectators to observe the 3-D nature of the scene without wearing glasses. Moreover, spectators might change their point of sight by simple motion.

But the future of cineholography will be in industrial and medical applications.

References

1. Komar, V.G., "Principle of the holographic cinematography", 1st European Congress on Optics Applied to Metrology, Strasbourg, France, 1977.
2. Decker, A.J., "Holographic cinematography of time-reflecting and time-varying phase objects using a Nd:YAG-laser", Optics Letters, Vol. 7, No. 3, March 1982.
3. Smigielski, P., Fagot, H., Albe, F., "Cineholography", OPTO 84 Congress, Paris, May 1984, ICO-13 Congress, Sapporo, Japan, August 1984, 16th Int. Congress on High-Speed Phot., Strasbourg, France, August 1984, Optical Metrology, NATO Workshop, Viana do Castelo, Portugal, July 1984, La Recherche, Paris, No. 103, February 1985.
4. Smigielski, P., Fagot, H., Albe, F., "Progress in holographic cinematography", 2nd Intern. Technical Symposium on Optical and Electro-Optical Applied Science and Engineering. Progress in Holographic Applications, Cannes, France, December 1985. SPIE Vol. 600.
5. Fagot, H., Smigielski, P., "Cineholographie et interférométrie", C.R. Acad. Sc., Paris, t. 302, Vol. II, No. 4, 1986.

6. Albe, F., Fagot, H., Smigielski, P., "Use of an endoscope for optical fiber holography", 2nd Int. Technical Symposium on Optical and Electro-Optical Applied Science and Engineering. Progress in Holographic Applications, Cannes, France, December 1985, SPIE, Vol. 600.
7. Stimpfling, A., Smigielski, P., "A new method for compensating and measuring any motion of 3-D objects in holographic interferometry: recent developments", 2nd Int. Technical Symposium on Optical and Electro-Optical Applied Science and Engineering, Progress in Holographic Applications, Cannes, France, December 1985, SPIE, Vol. 600.

Figure 1. "Christiane et les holobulles": before the recording.

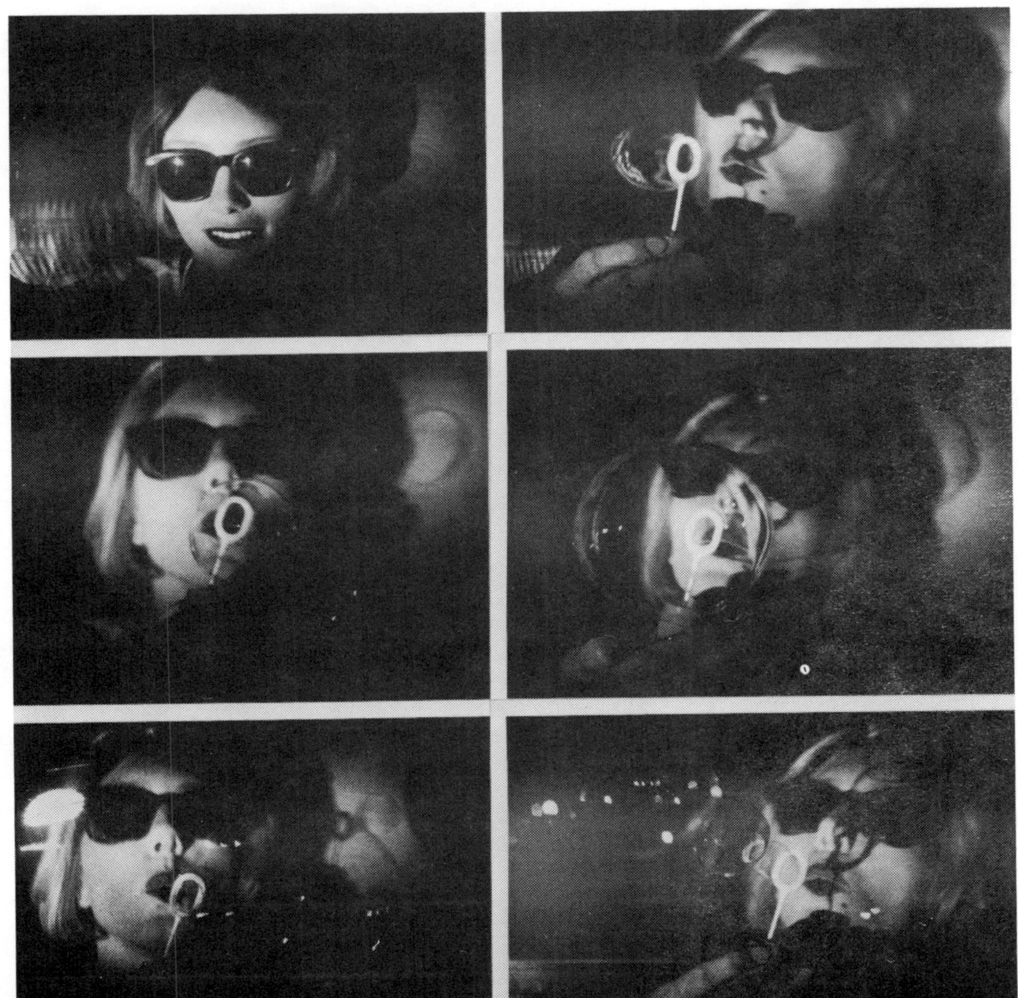

Figure 2. Pictures taken from the 126-mm holofilm "Christiane et les holobulles".

Figure 3. Picture taken from the 126-mm holofilm "La Belle et la Bête".

Figure 4. "Holoparleur": recording arrangement with the 35-mm camera.

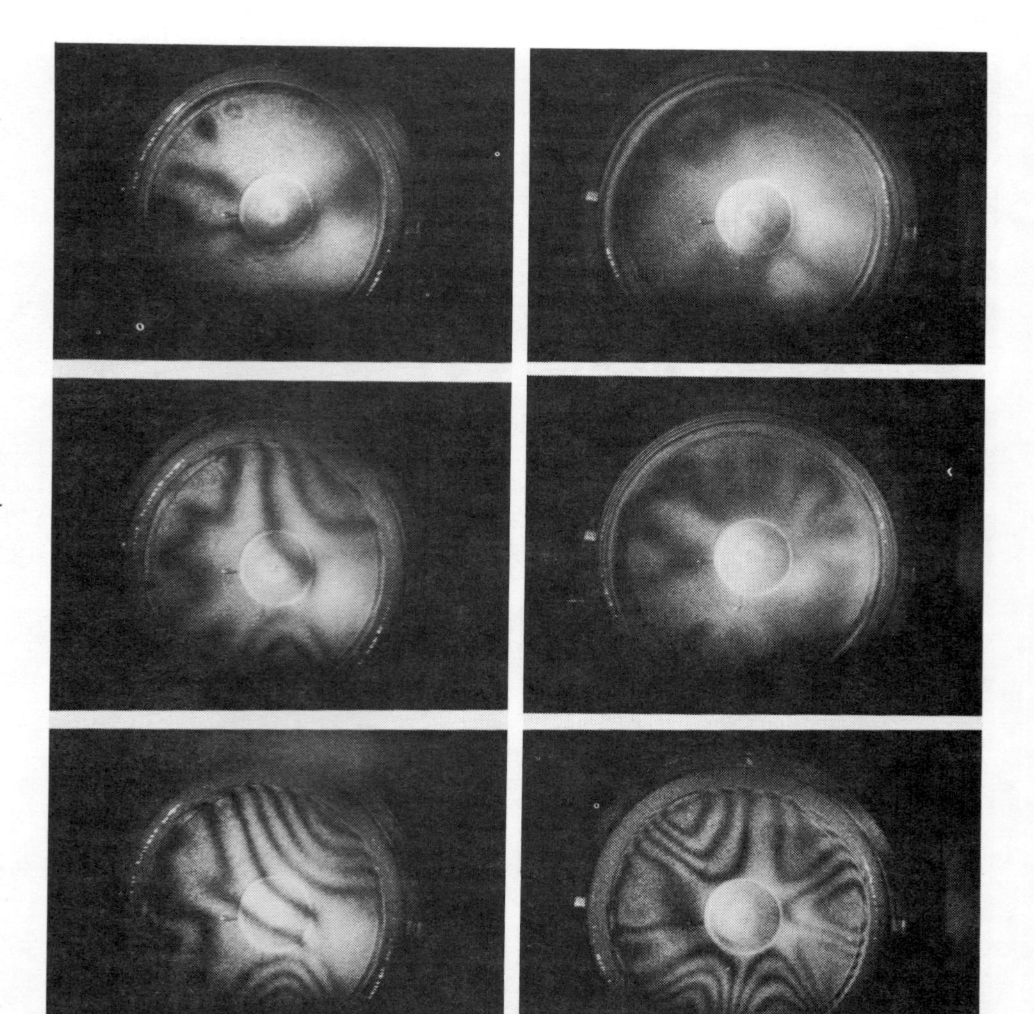

Figure 5. Reconstructed images taken from the 35-mm holofilm "Holoparleur". Weakly excited loudspeaker. Double-exposure 2 × 10 Hz, Δt = 4 ms.

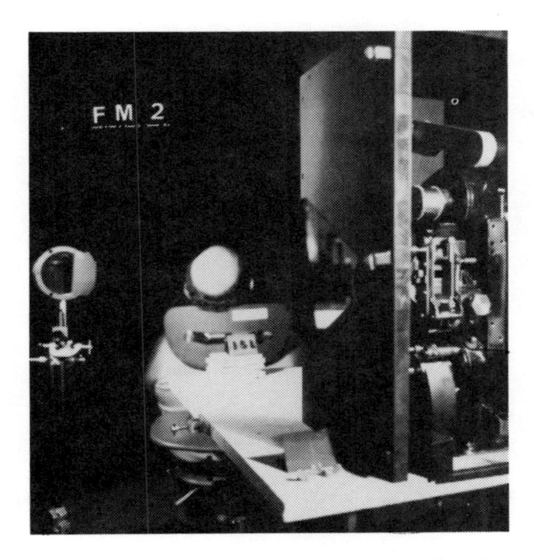

Figure 6. "Holocrâne": view of the bald head and 35-mm camera before holographic recording.

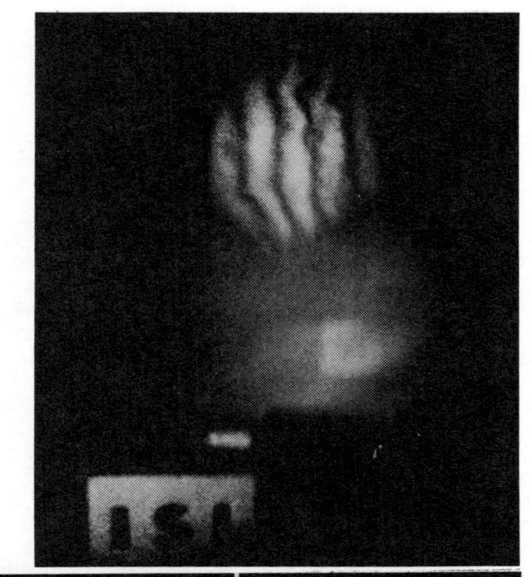

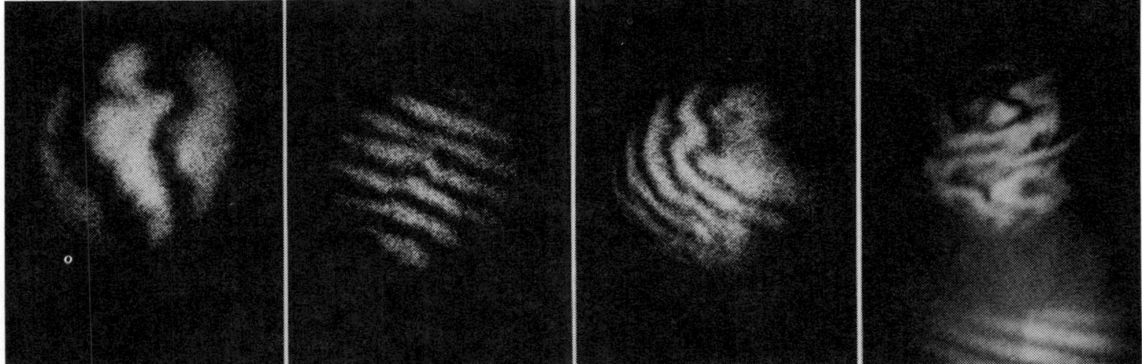

Figure 7. Pictures taken from the 35-mm film "Holocrâne". Steady living head. Double-exposure 2 × 10 Hz, Δt = 10 ms.

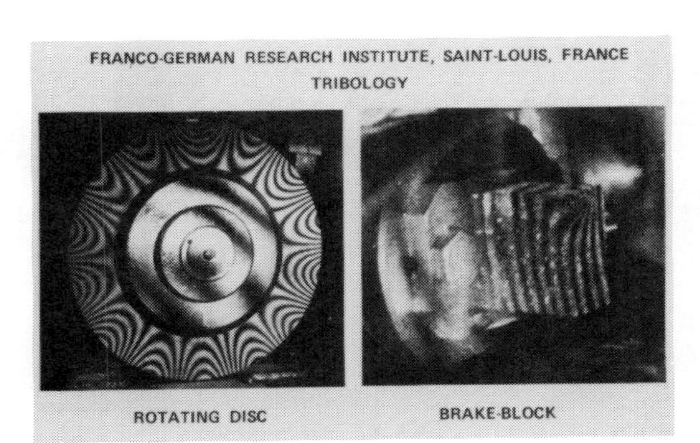

Figure 8. Tribology. Double-exposure holography (ruby laser).

New Methods of Making Flat Holographic Stereograms

Qu Zhimin, Liu Jingsheng

Shanghai Institute of Laser Technology, Shanghai, China

Abstract

In this paper several new methods for making holographic stereograms are presented. Wide angle, white light viewable flat stereograms with higher quality images, i.e, higher resolution, free of distortions and aberrations within a considerable viewing range, could be created.

Introduction

The holographic stereogram provides a versatile and effective means of displaying 3D imagery. Since the invention of the cylindrical holographic stereogram by L.Cross in 1973 , applications such as adverticement, education, entertainment, art, etc , have been developed.. Numourous scenes including movie stars and politicians have ever been photographed by this method. There was ever a portrait of a girl shown in very good effect. Apart from being applied to visual enjoyment, several researchers have examined seriously the possibility of applying the technique in 3D displaying of medical X-ray imagery and 3D dynamic motion analysis[1,2,3,4].

For the time being, people are very interested in the holographic stereogram sythesized from a sequence of 2D photographic transparencies.

The conventional holography is very limited. With the help of a contineous wave laser,only stationary objects can be recorded due to its insufficient power. Although some moving objects could be recorded by a pulsed laser, the dimentions and kinds of objects are restricted.Furthermore, the pulsed laser is very expensive and has strict requirements for conditions. In contrast, holographic stereogram does not require laser illumination and direct holographic recording of the object, the generation of the photographic transparencies and the reconstruction of the final hologram could be accomplished in white light. It is only the intermediate sythesizing procedures that are carried out with the help of laser holographic technology. For this reason, the restriction on object motion is eased and a much broader range of object imagery can be covered, for example, large outdoor scenes, portraits, animals etc.

Although the cylindrical format of the Cross holographic stereogram is able to show scene motion effectively, many consider it limiting. First, the perspective and time-smear distortions are unavoidable, and the unavoidable vertical lines make one uncorfortable when viewing. Secondly, it is not easy to display and requires a special display setup. In contrast, the flat format is much easier and more convenient to display, and distortion-free viewing without vertical lines is apparently possible over quite a large range of viewing distances[1]. Some researchers have worked out several methods of producing flat holographic stereograms[5,6,7,8]. The development of this new type makes this form of display more versatile and effective and can increase its commercial appeal.

However, by present methods of making the flat format it is not easy in fact to obtain high-quality stereograms. Pricryl suggested a method of making flat stereograms,[6] but it has shortcomings. A rotating(or shifting and tilting) part must be designed to shift and tilt the reference source together with the plate SSH(subsidiary sythetic hologram) as a rigid body. This requirement makes the setup complicated, though it can be accomplished. Haines[7] worked out another method. In order to reconstruct real images from the slit elements made by the common Cross method, a large cylindrical, or spherical or paraboloical reflector is compelled to be employed so as to obtain the naccessary reconstructing wave of large dimensions. This is quite difficult to realize. S.A.Benton developed a method of producing an achromatic flat stereogram.[5] However, it is difficult to obtain a flat stereogram with a large viewing angle. Furthermore, he employed diffuse screen as a midium of distributing the imaging rays of the transparencies(used as object light field when recording) over the intermediate hologram surface. Thus laser speckle noise is unavoidably produced, the image resolution is lowered. Phase modulation plate was ever suggested to replace the diffuse screen. However it is in practice difficult to obtain such a element with high quality.

In the following section, we describe in detail several new methods which we developed to overcome these defects and shortcomings. The diffuse screen is not used in our approaches, so image resolution can be expected to improve. Secondly, the image is free of distortions and aberrations within a proper range of viewing distances. Last, the setups we employ are simple, no strict conditions are required.

The Principles of the Approaches

I Approach I

First, we suggest a method of making flat stereograms with the help of a holographic element which can produce multiple images.

In order to obtain the necessary transparencies, let a camera take pictures from a series of equally spaced positions along a horizontal line(rail), as shown in Fig.1. A sufficient number of perspectives are taken so that the desired information about the changes of the object profile can be obtained. The suitable distance of the camera from the object and the number of pictures being taken depend on the object size and profile.

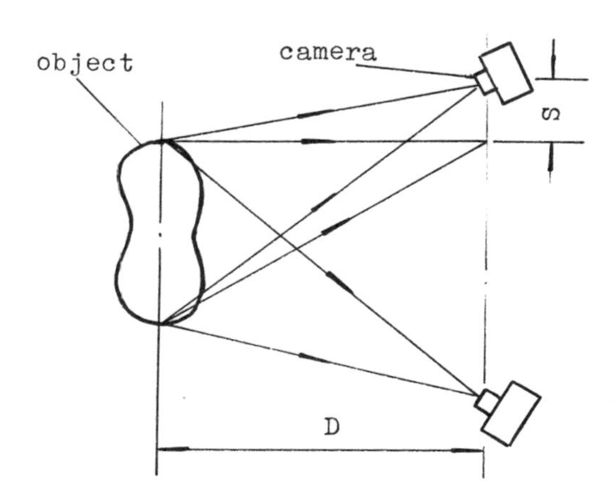

Fig.1. Taking pictures of object along a line.

Once the photos have been taken, they are used to construct an intermediate hologram H_1 as shown in Fig.2(a). All the transparencies are successively recorded on H_1 which consists of individual basic holograms where each one records one corresponding image of the transparency.

Fig.2(a) illustrates how the image of the transparencies are recorded on H_1, the intermediate hologram. The transparency is illuminated with laser light and projected by lens L onto the HE. The HE (a holographic element) is purposely prepared, it can produce multiple images. In this way, the holographic element focuses the object light that the HE receives to every part of the intermediate hologram. The mask with a short slit is used to protect the recording medium except the area being recorded. During the recording process, the mask is shifted and the transparency is successively projected so that all the transparencies are recorded on adjacent basic holograms of the intermediate hologram H_1.

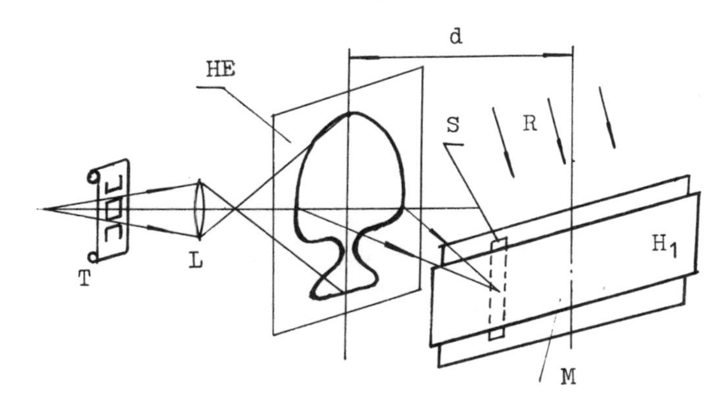

Fig.2(a) Recording of intermediate hologram, where T--transparencies, L--projecting lens, HE--holographic element producing multiple images, R--reference beam, M--mask, S--slit, H_1--intermediate hologram.

The recorded intermediate hologram can be reconstructed by the conjugate reference beam. Another recording plate is put across the reconstructed images, and the final rainbow hologram is thus created with the help of another reference beam, as shown in Fig.2(b).

The specially prepared HE is made by means of a glass rod array whose dimension is a little larger than the intermediate hologram. As illustrated in Fig.3, the laser beam is diverged by a glass rod lying along X axis direction onto another glass rod lying along Y axis direction. After going through a glass rod array, the beam is diverged by many rods and overlapping field is formed at a distance d where a recording plate is placed. The converging reference beam is directed from the above. Thus, the HE is created. It should be noted here that the function of HE is similar to a diffuse screen. Their difference is that the speckle noise produced by the HE is much smaller than that by the diffuse screen.

II Approach II

Now we suggest another method of producing flat stereograms. The first step is the preparation of a series of photographs taken at equally spaced discrete points along a circle, as shown in Fig.4.

Fig.5(a) diagramatically illustrates the recording process of the intermediate hologram H_1. We can see that the lens L projects the transparency T illuminated by a laser beam onto a plane at distance d from the recording medium H_1. A mask with a hole of the required diameter is used to protect the H_1 except the area being recorded. The recording film H_1 and the mask M are kept in touch with each other and placed near the focus of the illuminating beam modulated by the transparency T. The mask M is in cylindrical shape and its curvature axis is on the imaging plane T' of T. The reference light comes from left of H_1 and is focused at a point O' on the axis of the mask. Let O' be separated from the image T' of the transparency T spatially. The distance d is sele-

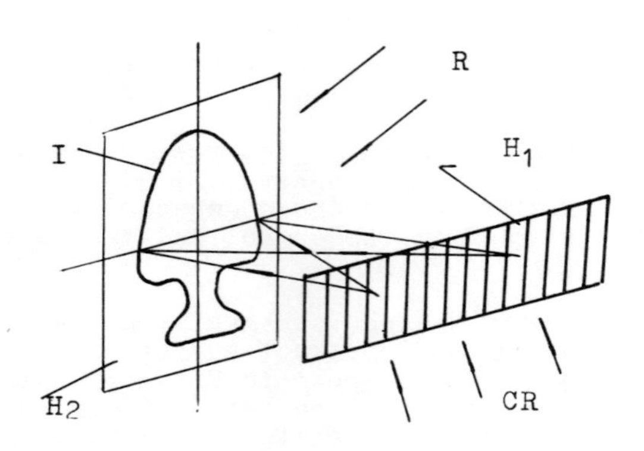

Fig.2(b) Creation of final hologram. I--image, R--reference beam, H_1--intermediate hologram, CR--conjugrat reconstructing beam, H_2--final hologram.

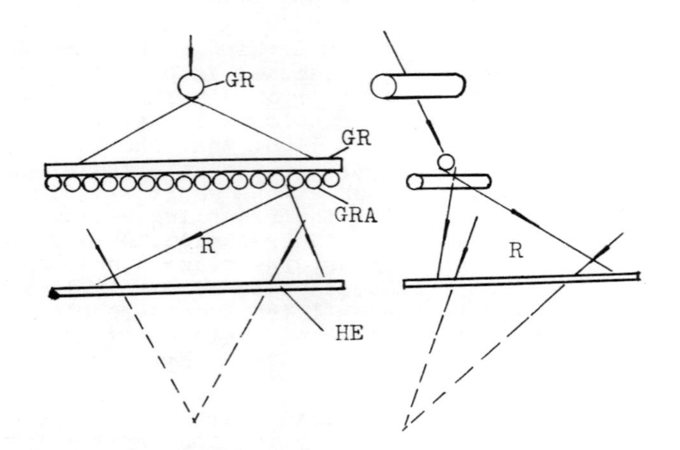

Fig.3. Construction of holographic element, where GR--glass rod, GRA--glass rod array, HE--holographic element required, R--reference beam.

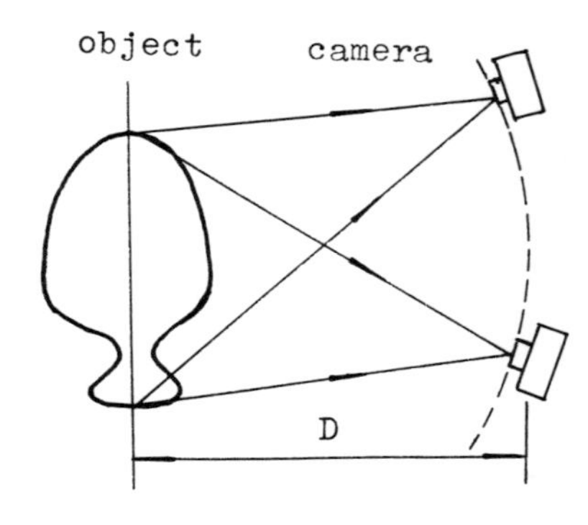

Fig.4. Taking pictures of object along a circle.

cted as that of viewing the final hologram. During recording, the recording medium H_1 is shifted and kept in the same shape as that of the mask M, and the transparencies are projected one by one. Thus we have recorded all the transparencies on an array of small holograms. When reconstructing as shown in Fig.5(b), all the basic holograms on the intermediate hologram will

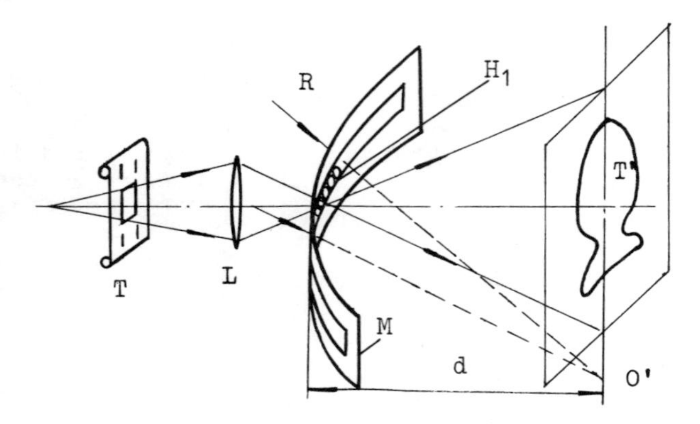

Fig.5(a). Recording of intermediate to hologram, where T--transparency, L--lens, M--mask, H_1--intermediate hologram, R--reference beam, T'--image of T, d--distance from T' to H_1.

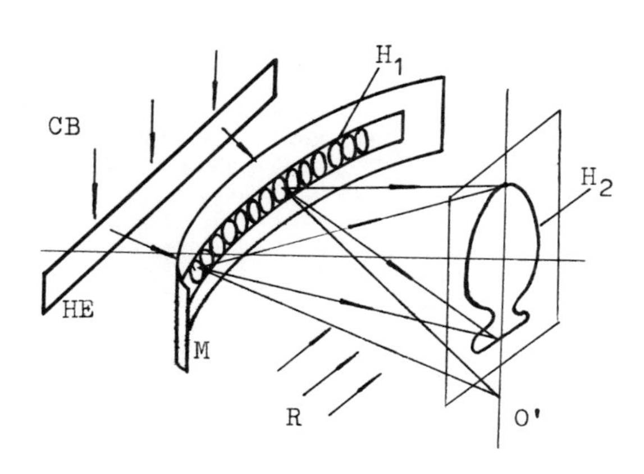

Fig.5(b).Creation of final hologram, where HE--strip of zone plate, M--mask R--reference beam, C--constructing CB--collimated beam, H_2--final hologram, I--image reconstructed.

be reconstructed simultaneously and produce a sythetic field of real images overlapping one another at the oringinal location of T'. It should be noted here that we employed a strip of zone plate, which is placed near the intermediate hologram. Its function is that it helps to creat a converging wave of large dimention due to its focusing effect. Thus a collimated light is transformed into converging light focused at the same point as the reference beam does on the right side.

The construction of the zone plate of large dimention can be done with help of a point source and a collimated wave.

III Approach III

Next, we have developed a more satisfactory method for making a flat stereogram. Up to this time, all the previous methods mentioned above employ transmission intermediate holograms as a medium to reconstruct at the same time all the basic holograms to make the final flat or cylindrical format of stereograms. Although they have advantages, complicated optics or mechnical rotating systems are compelled to be employed by each of the previous methods. Now we suggest a method of making flat stereograms by means of a reflection intermediate hologram. This method has many advantages. First the optical system is very simple and the requirements for experiment conditions are low. Secondly, the final flat sythetic hologram is in principle free of distortions and almost free of aberration. It is a complete image plane hologram. Thirdly, it is easy to obtain the reference beam and reconstructing beam employed. Large reflectors are not needed. Forthly, a natural color rainbow stereogram can be constructed by this method.

The preparation of the transparencies is the same as that of Approach II.

Fig.6 illustrates diagramatically the optical system employed in this method. We can see it

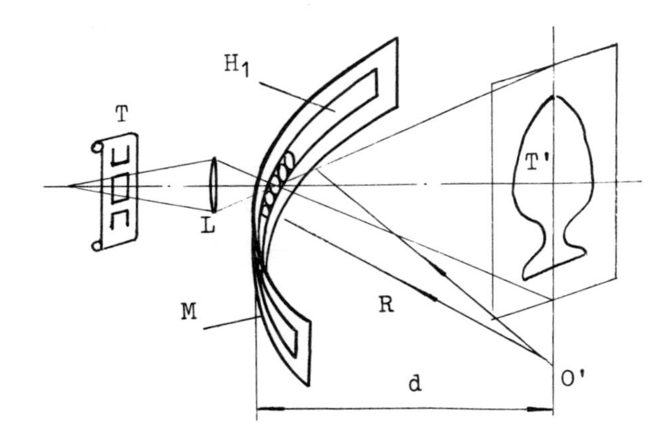

Fig.6(a).Recording of intermediate hologram, where T--transparency, L--projecting lens, H_1--intermediate hologram, M--mask, d--distance from image T' to recording medium H_1, R--reference beam, T'--image of T.

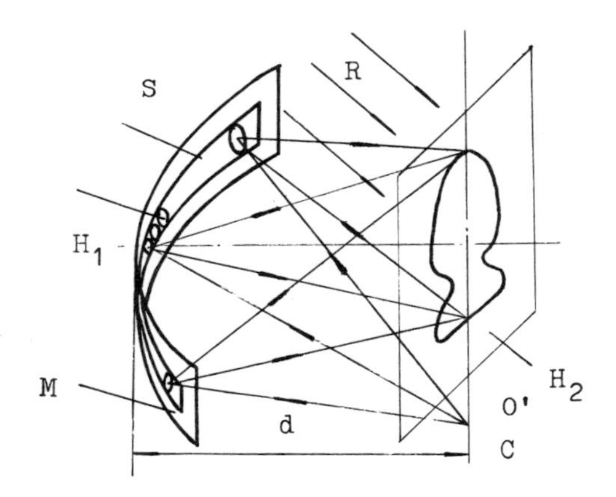

Fig.6(b).Creation of final hologram H_1--intermediate hologram, S--slit, M--mask, C--reconstructing, beam, H_2--final hologram, R--reference beam.

is similar to that of Approach II. The great difference between them is that the reference beam and the reconstructing beam here start from the focusing point O' of that in Approach II.The beams can be either a point source diverging onto the intermediate hologram or a line source. You can select according to your convenience. So in this way, a reflection intermediate hologram is made, the recording process is quite similar to that in Approach II. Fig.6(a),(b) illustrates the recording apparatus and the process for reconstructing and making of a final hologram.

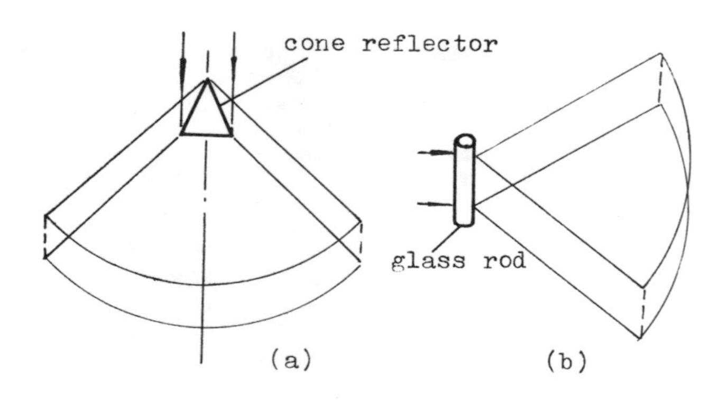

Fig.7. Creation of reconstructing beam (a) by using a cone reflector, (b) by using a glass rod, where GR--glass rod, CR--cone reflector.

Two alternatives are suggested to construct the reconstructing beam of large diverging angle for a wide viewing angle stereogram. The first one is accomplished by an introduction of a cone reflector or others.[7] It is equvalent to a line source illuminating a cylindrical zone, as shown in Fig.7(a). The other is more simple, only one glass rod is used for reconstructing, as shown in Fig.7(b). But the wave-front quality may be poorer. If it is adopted, a small modification of the optical system has to be done. This means that the reference beam and the image of the transparency T should interchange their positions; that is, let the reference beam illuminate perpendicularly the intermediate hologram, and the transparency image is projected off the center O' on the optical axis, So does the reconstructing beam. The reason is as follows. If the reconstructing beam is not incident at 90° angle, it will be diverged into a tilted plane, therefore all the basic holograms can not be illuminated simutaneously.

IV Approach IV

Last, we describe an alternative of the approach III, which possesses advantages over the former one. This arrangement is quite similar to that of the approach III. However, there are differences. The film in approach III is replaced by crystal $LiNbO_3$ recording medium,[3,4] as shown in Fig.8. The crystal is cut into small thin plates, for instence, 12 X 12 X 0.5 mm^3. These

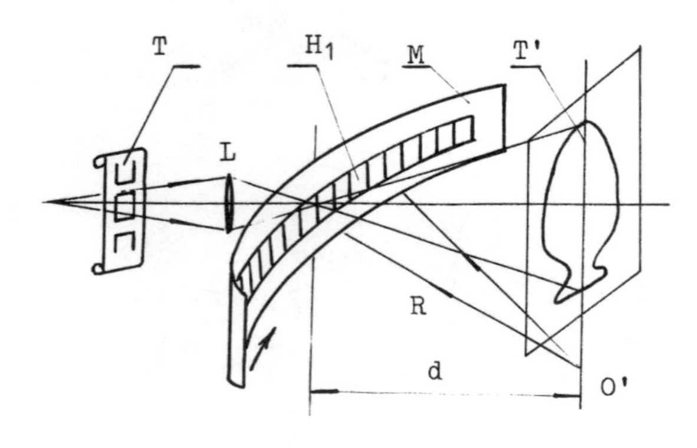

Fig.8. Recording of intermediate hologram; T--transparency, L--projecting lens, M--mask, H_1--intermediate hologram, d--distance from image T' to recording medium, R--reference beam, T'--image of transparency.

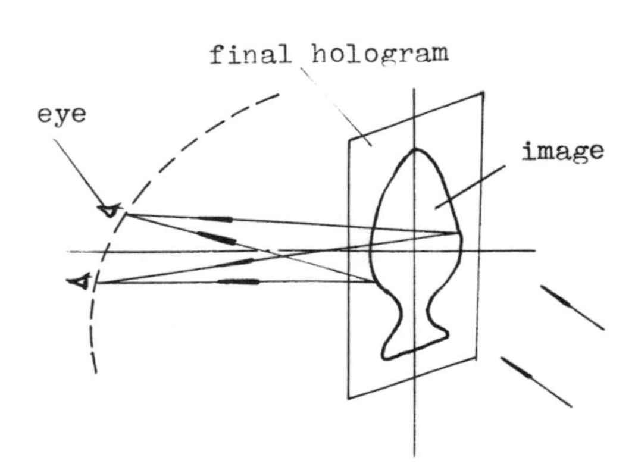

Fig.9. Viewing final hologram where d--distance from image to viewing circle.

plates are sticked together side by side onto a cylindrical surface whose axis is the same as that in Approach III. A rotating arm is connected rigidly to them. When recording the intermediate hologram(here the small plates), for a new basic hologram(plate), just rotate the arm a small angle with the axis at the same position as that of the mask M in approach III. And the left procedures are the same as that in Approach III.

The advantages of this alternative is as follows:
1. High overlapping precision of real images reconstructed can be easily achieved;
2. Chemical processing is not naccessary;
3. Natural color flat stereograms may be obtained;
4. No consumption of recording medium except the final hologram, LiN_bO_3 can be utilized repetitively.

We believe that this method has great potential for practical applications due to the simple constructing process, low cost, high image quality,and it is the best one of all.

In the next section, we shall concentrate ourselves to qualitative analysis of distortions and aberrations caused by practical unperfectness and approximations.

Analysis of Distortions and Aberrations

In principle, there exist no distortions covering a wide viewing range. And aberrations are quite small for complete image plane holograms. Here we mainly analyze the distortions and aberrations of Approach III. In approach III, appropriate control of the film(recording medium) is the key to avoiding the degredation of the sythetic image.

First let's analyze various distortions and aberrations and find out their possible origins and its magnitudes, so that measures are taken to reduce their unfavorable influences. The imaging defects can be divided into two classes[1]: geometric distortions and holographic aberrations. The former includes perspective distortion, time-smear distortion and nonlinear magnification. The latter consists mainly of image blur due to chromatic aberration and extended source illumination. Our final holograms are completeimage plane holograms due to enough imaging depth. They don't suffer color blur. There two types of defects have been studied by several researchers. Geometric image distortion results when the observer's eye is not at the same location as the camera entrance pupil when the perspective pictures are taken, or when the image seen by the observer does not correspond to the appropriate perspective views. By our methods, holographic stereograms free of perspective and time-smear distortions may be produced.

Now suppose the observer's eye is located in the focal plane, i.e, the intermediate hologram surface, the image seen will exhibit neither perspective nor time-smear distortions appearing in the Cross cylindrical hologram. The reason is that in this case all the image rays corresponding to a single basic hologram are intercepted by the eye pupil as shown in Fig.9, thus both distortions are eliminated. Furthermore if the diameter of the basic hologram is a little smaller than that of the eye pupil, there will be no exit pupil gaps across the image horizontally, and the vertical lines in the cross hologram are at the same time eliminated. However, as the observer's eye moves away from these special positions in either direction, the above-mentioned distortions of perspective, time-smear, and the vertical line all appear. Of course, their appearance is not abrupt, and approximately distortion-free viewing is possible over quite a large range of viewing distances. The nonlinear magnification can only come from the projecting lens here. If the lens is suitably selected, this defect can be reduced to minimum.

Now, we analyze the influences of film deformation, especially local ones. Deformations may influence the image quality[3,4] and the overlapping precission. Therefore the control of the film deformation is the key to the construction of the sythetic stereograms with high quality. Measures, such as using thick substrate film[3,4] , sticking it to a rigid transparent base with the same curvature as the mask, selecting high-quality smooth mask, and keeping the film and the mask in good touch, etc, may be taken so as to reduce its influences.

General Conclusion

In this paper, we have suggested several methods of making flat holographic stereograms of large image width and viewing angle. We show that these methods have many advantages over the previous ones. The approach III and IV are analyzed in detail. The results indicate that with the help of simple optics, a flat holographic stereogram with a sharp image, wide viewing range especially viewing angle, free of distortions can be produced. It is also possible to make a natural color flat stereogram. 3D imagery with superior image quality can be displayed. If the methods are combined with embossed holography, flat holographic stereograms can be mass made.

References

1. "3-D Imaging With Holographic Stereograms" L.Huff etal, SPIE 402, P.38.
2. "A Survey of Holographic Stereograms" S.A.Benton SPIE 367, P.15.
3. "Handbook of Optical Holography" H.J.Caulfield, 1979.
4. "Holographic Recording Materials" H.M.Smith. Spring Verlag, 1977.
5. S.A.Benton "Holographic Products and Processes", U.S.Pat.
6. Prikryl, "Holographic Sythesis", U.S.Pat. 4, 509, 818, Apr., 9,1985.
7. Haines, "Holograms Created from Cylindrical Hologram Masters",U.S.Pat.4,339,168.Jul13,1982.
8. S.A.Benton "Achromatic Holographic Stereograms",OSA Meeting Paper, Oct. 1981

Session 3

Fundamentals and Display in Holography II

Chairmen
N. Abramson
Royal Institute of Technology, Sweden
Guo-Guang Mu
Nankai University , China

Proposals of two-reference-beam double exposure holographic interferometer with double pulsed laser

Lai Guanming and T.Yatagai

Institute of Applied Physics, University of Tsukuba
Tsukuba Science City, Ibaraki 305, Japan

Abstract

A system that can be applied to high speed phenomena, especially with solid diffusely scattering objects is proposed, with which an automatic measurement of high precision can be realized.

Introduction

As it is well known, Double Exposure Holographic Interferometry is widely used in measuring the displacement of the object between the two exposures. By the use of opto-electronic techniques such as fringe scanning and heterodyne techniques,the automated interpretation of interference patterns is achieved. So, it is reasonable to find a way to use the opto-electronic techniques together with the double exposure holographic interferometry. In addition, double pulses emitted by a double pulsed laser, such as a ruby laser, can be used as two reference beams, if we use an appropriate interferometer.

In the following, the principles of holographic interferometry with diffusely reflecting objects are reviewed, which is useful to solve the problems followed, but it will be done as simple as possible. The resetting problem and the effects of wave length difference are examined, because we use a pulsed laser to record the hologram and use a continuous laser to reconstruct it. Then experimental results will be reported and an estimation of accuracy limitation will be considered. At last,proposals for switching the two pulses the interval of the double pulses are made.

Basic principle of holographic interferometry

We consider breifly about coherent imaging of a rough surface(see Fig.1). The surface roughness can be described by a complex reflection coefficient $\eta(x)$ with statistical properties. By assuming that the coherent imaging system can be described by a space invariant amplitude impulse function $h(x)$, we can write the analytic signal in the image plane like this:

$$V(X_i) = \int \bar{O}(X_o)\eta(X_o)h(X_i - MX_o)dX_o \tag{1}$$

where $\bar{O}(x)$ represent the macro property of the reflecting object,and $\eta(x)$ can be considered as its micro characteristic. The intensity in the image plane appears as the so called speckle pattern due to the random variations of the phases of reflectivity. By considering that the correlation peak of the surface roughness is very narrow compared with the resolution of the imaging system, its auto-correlation function may be approximated by a Delta function.

$$\langle \eta(X)\eta^*(X')\rangle = \delta(X-X') \tag{2}$$

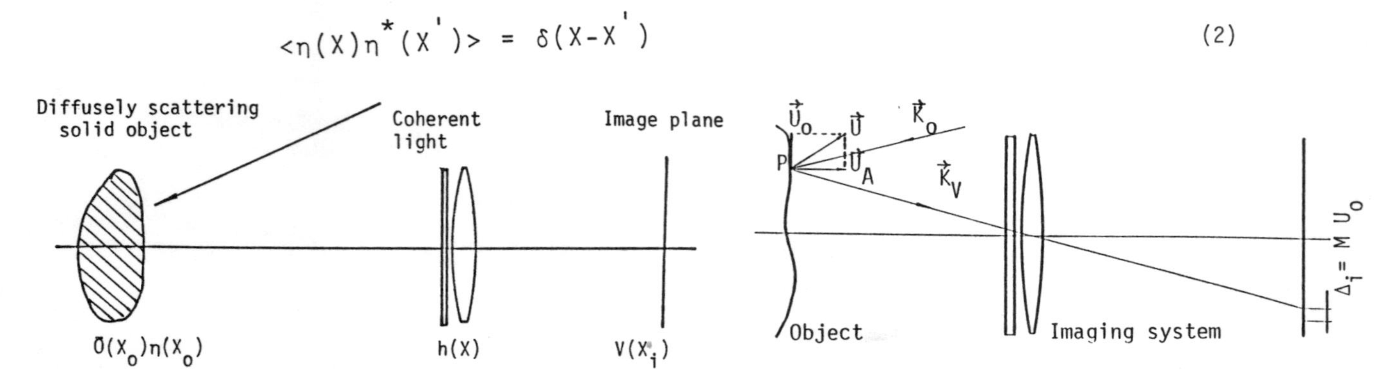

Fig.1 A diagram of coherent imaging system

Fig.2 Schematic setup of holographic interferometer

So, taking ensemble averaging over a large number of diffusers (generally,200 speckles when we use a TV camera), the autocorrelation function of the image amplitude is found to be

$$<V(X_i)V^*(X_i')>=\int |\bar{O}|(X_o)h(X_i-MX_o)h^*(X_i'-MX_o)dX_o \quad (3)$$

If the object intensity is assumed to change slowly with respect to the resolution of the imaging system, the shape of the correlation is essentially determined by the autocorrelation of the impulse response function

$$C_h=\int h(X_i-MX_o)h^*(X_i-MX_o)dX_o \quad (4)$$

The basic setup for holographic interferometer of objects with diffusely scattering surface is shown schematically in Fig.2. Suppose that two states of the objects' surface differ by a slight displacement $\vec{U}$. it can be considered like this. First, the displacement changes the phase of light reflected from a point p on object by

$$\Phi(X_i) = \vec{U}\cdot(\vec{K}_v - \vec{K}_o) \quad (5)$$

Second, the speckle field in the image plane is shifted by a Δ_i because of the transverse displacement $\vec{U}_o$, which will cause a reduction of the fringe contrast. So, V2 can be simply written as

$$V_2(X_i)= V_1(X_i+\Delta_i)\exp(-i\Phi(X_i)) \quad (6)$$

Then the ensemble average of V1 and V2 can be written as:

$$<V_1(X_i)V_2^*(X_i)>=|\bar{O}|^2\exp(-i\Phi(X_i))C_h(\Delta_i) \quad (7)$$

As long as the autocorrelation Ch of the impulse response function do not go to too small, the interferogram representing the out of plane displacement will be found.

Resetting problem and compensation of different wavelengths

It has been shown by some researchers[1,2] that in order to reduce the sensitivity to hologram resetting error, reference sources must be close together. In this circumstance, the aberrations introduced by hologram and imaging system can be neglected because the two waves to be compared will nearly get the similar distortion, and the statistical properties of the two states are completely the same. The main problem here is the lateral shift of speckles, which may be caused by lateral displacement of the object and the hologram resetting error and the different wavelengths between recording and reconstruction. The interference fringes are visible only as long as the mutual shift of the speckle pattern is smaller than the speckle size. For a circular aperture, the speckle size is given by

$$(\Delta X)_s = 4\pi \frac{\lambda}{F} \quad (8)$$

where F is the numerical aperture of the imaging system. On the other hand, when the reference sources are close together, two cross-reconstructions will overlap to the main images. In order to make the desired and undesired images uncorrelated, such a condition as to insure the lateral separation of the cross-reconstructions much larger than the speckle size.

The optical setup used in our preparing experiment is demonstrated in Fig.3. A concave lens and a convex lens is used to collimate the reference beams, because objective lens will be damaged or damage other optical surfaces by pulse laser's high power density. A beam-splitter and two reflecting mirrors is used to produce two reference beams. The angle between them can be adjusted by one of the mirrors.

It can be reconstructed with the same optical setup used in recording (see Fig. 4). A He-Ne continuous laser is used in the expriment. The relative optical length can be introduced by a Piezo electric element. The TV camera is focused on the reconstructed images, or to say the interference fringe is located on the objects. The image taken from TV camera through A/D converter is stored in image memory, then the image data is read and processed by a micro computer.

Fringes appeared on the holographic plate can be detected by the image sensor. By calculating the spatial frequency of the intensity or to say fringe distance, we can control the angle between the two reference beams. Besides, we can use it to calibrate the PZT precisely. If we vary the code transffered to PZT driver,the fringes will move. If we calculate the phase difference, then the line showing the relative phase versus voltage code can be obtained. In our experiment linearity is RMS=0.8% (see Fig.5).

Fringe analysis and limitation for the determination of the interference phase

It is stated before, that if cross-reconstructions are made uncorrelated with desired images, they will only exist as a background. Fringe scanning and heterodyne methods can be used. In our special condition the total intensity in the image becomes

$$\langle I(X_i)\rangle = 2|O_1|^2+2|O_2|^2+2|O_1O_2^*|\cos(\Phi(X_i)+\Psi_k) \tag{9}$$

The first term is contributed by cross-reconstructions. The best fringe visibility can be 0.5, but it will not arrive 0.5 owing to mutual speckle shifts. Then the phase can be calculated conventionally.[8]

The presence of speckles in the inteferogram of objects with diffusely scattering surfaces gives rise to a statistical error for the phase measurement. The statistical error caused by speckle noise under the condition of reduced visibility has been considered by Ruedi Thalmann and Rene Dandliker. The result is that the mean square value of statistical error is inversely proportional to the number of detected speckles, and it is also related to the mutual transverse displacement of speckles.

$$\langle \delta\Phi^2 \rangle = \frac{6 - C_h(2\Delta_i)}{2(N+1)\, C_h^2(\Delta_i)} \tag{10}$$

Note that for the interference of two desired images without cross-reconstructions, the first term of the numerator is not 6 but 1.

If N=200, C_h =0.7, the statistical error is found to be about 10 degrees, relative to one 36th fringe. Because the resolution of a TV camera is very high,the speckle number within a pixel is limitted, the phase measurement precision is limitted.

Fig.6 shows the picture of a raw fringe pattern, the cross section of the phase obtained by the fringe scanning method is plotted in Fig.7. A large discontinuity appeared, owing to the large difference of intensity within the aperture. The phase from the bottom to the top is about 1.5 fringes, the average fluctuation is about 1/35 λ . If more precise measurement is necessary, heterodyne method with large enough detector will be appropriate.

Proposals to switch the two reference beams

By the use of electrooptic components,such as ADP and KDP with a polarized interferometer, it can be realized to let pulse one go this way and let pulse two go the other way, as demonstrated in figure 8. It can be done like this: the E-O is prepared to rotate the polarization direction by 90 degrees,then the first pulse will be splitted as the reference one by a polarized beam splitter, when the first beam is detected by a photodetector set near the hologram, then the E-O driver will be trigger to turn off before the second beam is emitted. when the E-O is not loaded, the second pulse will go as the reference two. The interferometer can be similar to that used in preparing experiments, except that a E-O cell synchronous to pulsed laser by the signal of a photo-detector is added, and the beam splitter is a polarized one.

Two acousto-optic modulator can also be used to achieve spatial modulation (see Fig.9). The light which propagates in the original direction of incidence has the original frequency, while the first order diffracted beam is shifted by the sound frequency fs. The diffracted light will have a frequency lower or higher than that of the incident light according to the relative direction of the sound wave and the incident beam. The good features of this modulator are : the ratio of the two beams can be adjusted by the sound intensity, and the optical path can be altered by turning the sound signal on and off.

Conclusions

The resetting problem has been studied. The resetting sensitivity in our experiment is less than 5 degrees. There is no problem to use a pulsed laser to record the waves and use continuous laser to reconstruct them. It is found theoretically and experimentally that with a TV camera 1/30 λ can be obtained. At the end, two setups are proposed, which are appropriate to use double pulses of a pulsed laser as two reference beams required in the two-reference-beam double exposure holographic interferometer.

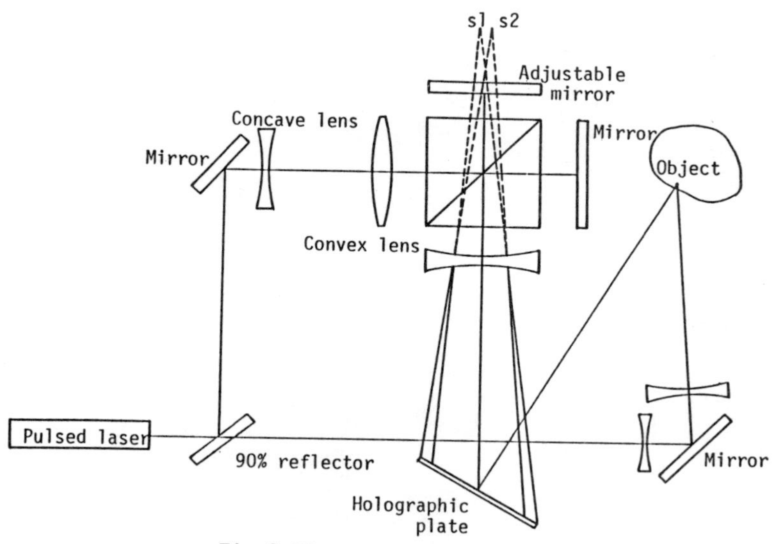

Fig.3 Diagram of two reference beams holographic interferometer with pulsed laser

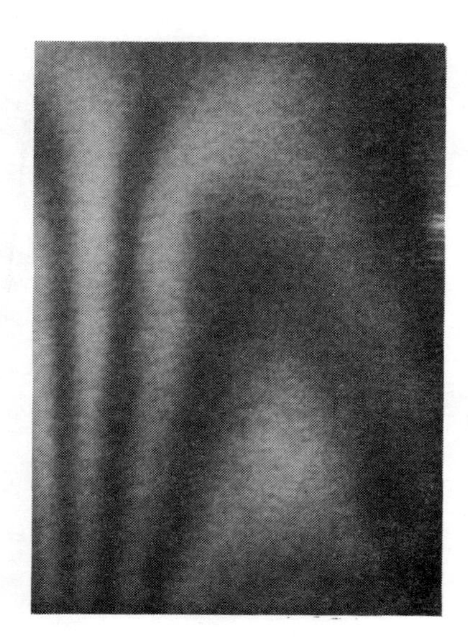

Fig.6 A picture of raw interference pattern

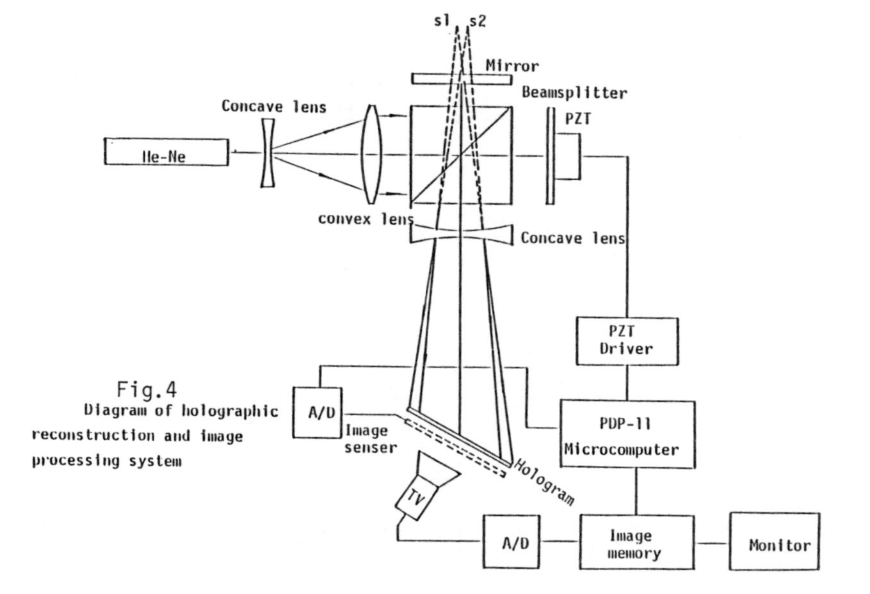

Fig.4
Diagram of holographic reconstruction and image processing system

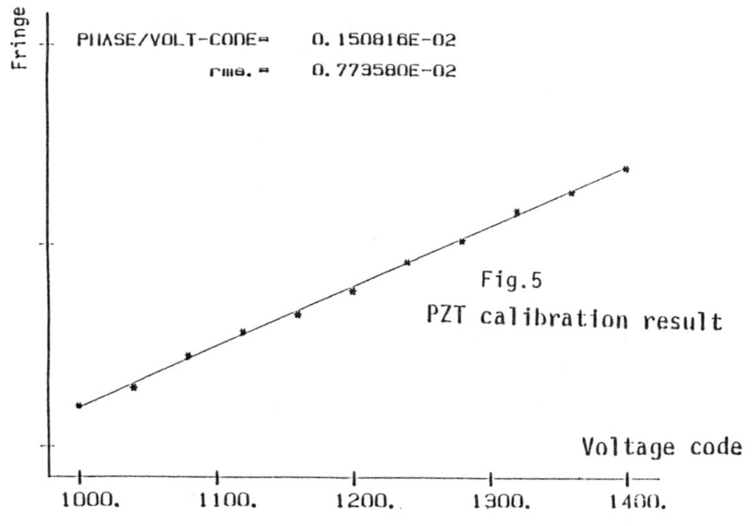

Fig.5
PZT calibration result

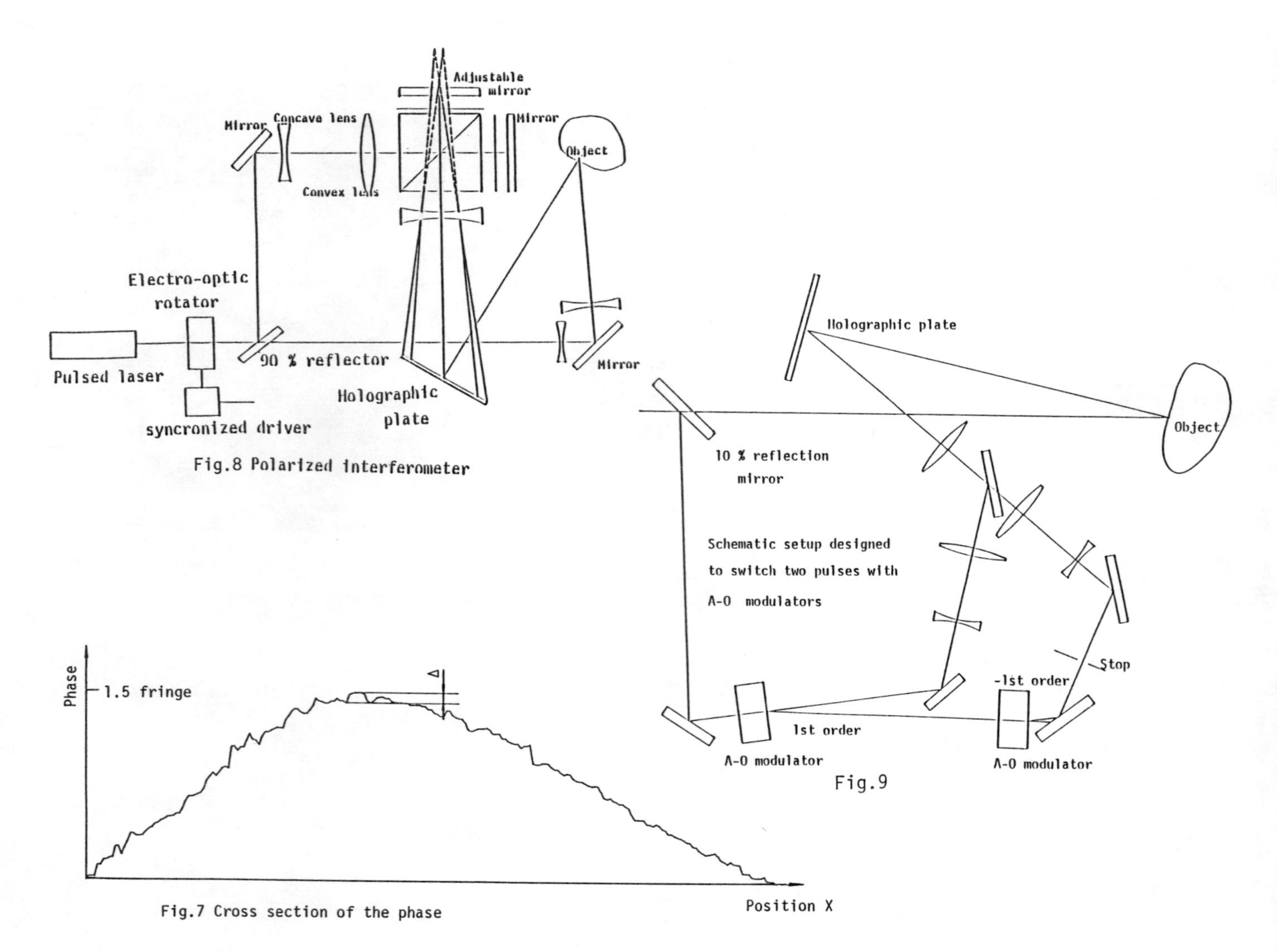

Fig.8 Polarized interferometer

Fig.9

Fig.7 Cross section of the phase

<u>References</u>

1. R. Dandliker, R. Thalmann, J.-F. Willemin, "Fringe interpolation by two-reference-beam holographic interferometry: reducing sensitivity to hologram misalignement", Opt. Commun. 42, 301-306 (1982).
2. R. Thalmann, R. Dandliker, "High resolution viedo-processing for holographic interferometry applied to contouring and measuring deformations", European Conference on Optics, Optical Systems and Applications, SPIE 492, 299-306 (1984).
3. R. Dandliker, "Heterodyne holographic interferometry", Progress in Optics, Vol. 17, pp. 1-84.
4. R. Dandliker, R. Thalmann, "Heterodyne and quasi-heterodyne holographic interferometry", Opt. Eng. 24, 824-831 (1985).
5. S. Nakadate, " Shearing heterodyne interferometry using acoustooptic light modulators", Applied Optics 24, No. 18 (1985).
6. H. Bjelkhagen, " Pulsed sandwich holography", Applied Optics 16, pp1727, (1977).
7. N. Abramson, H. Bjelkhagen," Pulsed sandwich holography. 2: Practical application", Applied Optics 17, pp.187 (1978).
8. P.Hariharan, B.F.Oreb, and N.brown, "Real time holographic interferometry : a microcomputer system for measurement of vector displacements", Appl. Opt., Vol.22, pp876-880, 1983.

The self-enhancement effect of dynamic amplitude-phase holograms and its applications

A.O.Ozols

Institute of Physics, Latvian SSR Academy of Sciences,229021, Riga,Salaspils-1,USSR

Abstract

The mechanism of the self-enhancement effect of dynamic amplitude-phase,amplitude and phase holograms in different materials is considered as well as its applications to the photorefraction investigation in electrooptic crystals and to the modification of holographic recording and readout.

Introduction

Self-enhancement (SE) of a dynamic hologram is the effect of its diffraction efficiency (η) increase during the readout by one of the recording beams. SE is analogous to the recollection process in the brain of a human being 1 . The SE effect is of interest as a new method of recording 2 and nondestructive readout 2,3 in dynamic volume holography as well as investigation method in photophysics 4,5 .

The SE effect was for the first time reported in 1973 by Gaylord et al 6 . They observed the SE of phase holograms in $LiNbO_3$ crystals. After that, the observation of SE in electrooptic crystals (mainly $LiNbO_3$) was reported in numerous papers,but,until 1984, to our knowledge, this was the only experimental paper fully devoted to the study of SE in electrooptic materials. In 1977 the SE of amplitude-phase holograms in irradiation-coloured NaCl:Ca crystals was reported by Kravets et al 7 . In 1979 I described the SE of amplitude holograms in additively coloured KBr crystals 8 . A detailed description of the SE effect properties in $LiNbO_3$:Fe crystals was given by us in 1984 4. In 1985 we observed for the first time and studied the SE effect of amplitude-phase holograms in As_2S_3 amorphous semiconductor films 9 . The SE effect is quite general,however,it is little investigated both experimentally and theorethically. This paper is devoted to the description of the main features of SE according to our results.

Coherent and incoherent self-enhancement of holograms

SE takes place for the readout beam incidence angles equal or close to the Bragg angle ($LiNbO_3$:Fe 4-6 , KBr 8 , As_2S_3 9) or for arbitrary readout beam incidence angle (NaCl:Ca 7 , KBr 8). The first kind of SE, which we call the "coherent SE" is based on the formation of complementary holographic gratings (CHG) by incident and diffracted waves as proposed by Staebler and Amodei 10 . The second kind of SE is due to the peculiarities of the photochemical reaction used for the recording and can be performed with spatially incoherent light ("incoherent SE"). As far as the origin and the properties of SE are different in these cases we consider them seperately. To describe the SE effect we use SE factor $\xi=\eta/\eta_0$ (η_0 -initial η).Generally, the $\xi(t)$ dependence (t-time) has a maximum (ξ_{max}). The ξ_{max} value is a measure of the SE efficiency. Mainly, the elementary holograms- holographic gratings were investigated.

We observed the coherent SE effect in the case of amplitude-phase holograms in KBr at $\lambda_1=\lambda_2=514.5$nm. (λ_1 and λ_2 - the wavelengths of the recording and readout light, respectively, recording due to photobleaching of F-centres, $\xi_{max}=2.8$) and As_2S_3 amorphous films (photodarkening effect, $\xi_{max}=10$ at $\lambda_1=\lambda_2=514.5$ nm and $\xi_{max}=350$ at $\lambda_1=514.5$ nm, $\lambda_2=632.8$ nm) as well as in the case of phase holograms in $LiNbO_3$:Fe crystals (photorefraction effect, $\xi_{max}=50$ at $\lambda_1=\lambda_2=632.8$nm). ξ increases when η_0 decreases. ξ depends on the recording geometry, period of a hologram,Λ,and on the intensities of the recording beams. The SE effect is observed in volume holograms as well as in intermediate holograms and never - in thin holograms 2,4,9 .

The calculation of light intensity spatial distribution inside the amplitude-phase hologram,created by the zero order and minus first order diffracted waves based on the Kogelnik's coupled wave approach 11 showed the formation of three CHG - one amplitude CHG and two phase CHG's. The amplitude CHG (at local recording) is phase-shifted by 0 in the case of photodarkening and by π - in the case of photobleaching with respect to the light intensity pattern. This means that for amplitude holograms the coherent SE is possible only in the case of photobleaching. The two phase CHG's have phase shifts $\Phi\pm\pi/2$, Φ describing the nonlocality of the recording (i.e, the hologram is shifted by Φ with respect to the recording light intensity pattern). As the amplitude of the CHG is decreasing with increasing of the propagation depth,the formation of the CHG means not only the change of the initial hologram but also its inclination 2 .

The vectorial picture of CHG model enabled us to explain the main features of the coherent SE and, after simple calculations, to estimate from SE measurements for the $LiNbO_3$: Fe crystal the free mean path of the photovoltaic electrons (≈ 80 nm), the quantum efficiency ($\approx 10^{-4}$) and the product of the electron mobility and their lifetime in the conduction band ($\approx 10^{-9} cm^2/V$). The SE effect allows to determine the direction of optical axis, because ξ depends on the readout beam orientation: ξ is larger for the beam with a positive wavevector projection on the optical axis.

We have investigated the incoherent SE of amplitude holograms in the case of $F \rightarrow x$ reaction based KBr crystal photobleaching ($\lambda_1 = \lambda_2 = 632.8$ nm, $\xi_{max} = 3.6$). There are 4 reasons for SE in this case. 1) The modulation increase in photobleaching process which is enhanced in KBr by nonmonotonic photobleaching intensity dependence. 2) The modulation increase due to the $X \rightarrow F$ reaction in the minima of a hologram with simultaneous stability of X - centres in the maxima (this effect is important in the saturation stage). 3) The effective modulation increase as a result of photobleaching nonlinearity due to the change of hologram profile from sinusoidal to rectangular. 4) The decrease of a hologram optical density D_0 when $D_0 > 0.87$. The effect of coherent SE for $\lambda_1 = \lambda_2 = 632.8$ nm is much weaker.

The incoherent SE can be regarded as a light-induced development of a hologram. The well-known fixation procedure of holograms in $LiNbO_3$ crystals 12 can also be regarded as incoherent SE.

Self-enhancement of image holograms

The SE of image holograms is also possible. The coherent SE allows to reach much higher ξ_{max} than incoherent one but it has two drawbacks: 1) ξ_{max} decreases 5-10 times in the case of coherent SE of image holograms when compared with elementary holograms., 2) The coherent SE depends on spatial frequency (the maximum SE is reached at $\Lambda = 15\ \mu m$ for 3 mm thick $LiNbO_3$:Fe crystals and at $\Lambda = 1\ \mu m$ for 11 μm thick As_2S_3 films), therefore, the restored image is, generally, distorted.

However, if the spatial frequency range of an image is not too large (as, for instance, in the case of transparencies), its quality after the coherent SE is acceptable. The quality of the self-enhanced images is better for the reflection holograms 3 and in the case of incoherent SE.

The SE effect of image holograms can be used for nondestructive readout of dynamic holograms although this application is limited by ξ decrease after reading the ξ_{max}. SE can be also regarded as a new, two-step holographic recording method (the first step - a conventional two beam recording until small initial diffraction efficiency η_0 is reached, the second step - recording by a reference beam only) which can be more advantageous when the recording time or energy is at the first step limited.

Conclusion

The investigation of the SE effect is far from to be completed. There rigorous theory of coherent SE is absent except for SE of phase transmission holograms developed by Kuchtarev 1 . As the recording material properties greatly influence the features of SE the further experimental work in this field is necessary for good understanding of the SE effect and for its practical applications.

References

1. Kukhtarev N.V., "The Effect of Nonstationary Self-Enhancement During the Readout of Volume Holograms", Ukr.Fiz. Zhurn., vol.23, No12, pp. 1947-1953, 1978.
2. A.O.Ozols., "Self-Enhancement of Holograms in the Materials with a Complex Photorefraction", Proc.of the 5 th USSR Conference on Holography, Riga, USSR, November 12-14, 1985. Edited by the Institute of Physics, Latvian SSR Academy of Sciences,Riga, part I, pp. 116- 117. 1985.
3. Vartanyan E.S., Gulanyan E.Kh., Ovsepyan R.K., "On Some Features of Data Recording and Readout in Lithium Niobate", Kvantovaya elektronika, vol.7, No 2,pp. 435-437. 1980.
4. Reinfelde M.J., Ozols A.O., Shvarts K.K., "Volume Holographic Gratings and Self-Enhancement Process in $LiNbO_3$:Fe Crystals", Izv.Akad.Nauk Latv.SSR.Ser.Fiz.i Tehn.Nauk,No 6 pp.88 -96. 1984.
5. Shvarts K.K., Ozols A.O., Reinfelde M.J., "Dynamic Self-Enhancement of Phase Holograms in $LiNbO_3$:Fe Crystals", Ferroelectrics, vol.63, No 1-4, pp. 309-314. 1985.
6. Gaylord T.K., Rabson T.A., Tittel F.K., Quick C.R., "Self-Enhancement of $LiNbO_3$ Holograms", J.Appl.Phys., vol.44, No2, pp. 896-897. 1973.
7. Kravets A.N., Kasimov M.K., Tschumakov A.V., "Self-Enhancement of Holographic Recording in NaCl:Ca Crystals", Optika i Spektroskopia,vol.43,No6,pp.1180-1182. 1977.

8. Ozols A.O., "Self-Enhancement of Amplitude Holograms in Additively Colored KBr Crystals", Izv.Akad.Nauk Latv.SSR.Ser.Fiz.i Tehn. Nauk, No 3, pp. 45-52. 1979.
9. Reinfelde M.J., Ozols A.O., Shvarts K.K, "Self-Enhancement of Holographic Recording in Amorphous As_2S_3", Izv.Akad.Nauk Latv.SSR, No 3, pp. I28-I3I. 1986.
IO. Staebler D.I., Amodei J.J., "Coupled Wave Analysis of Holographic Storage in $LiNbO_3$", J.Appl.Phys., vol.43, No 3, pp. I042- I049. 1972.
II. Kogelnik H., "Coupled Wave Theory for Thick Hologram Gratings", Bell System Techn. Journ., vol.48, No 9, pp. 2909-2947. 1969.
I2. Staebler D.I., "Ferroelectric Crystals". In: Holographic Recording Materials, Springer Verlag, pp.IOI-I32. 1977.

A NEW PROPOSAL FOR PHASE — ONLY CGH AND ITS APPLICATION

Wei-feng Chen, Xiang-yang Huang,
Xuan-tong Ying, Yu-run Jia

Lab of Laser Physics and Optics, Department
of Physics, Fudan University

Abstract

A new coding method for a CGH with pure phase based on the decomposition of any complex vector into two symmetrically situated conponent vectors with equal amplitude is proposed. In this method only two cells are needed for each sampling point as compared with the conventional types with three or four cells of Burckhardt's and Lee's configurations. Theoretical analysis of this proposal is given and some preliminary experimental results are presented.

Introduction

Over the past two decades, several different coding methods for CGHs were proposed. In 1966, Brown and Lohmann[1] described a detour phase method of making binary CGH for complex spatial filtering . In 1970 W.H. Lee[2] and C.B.Burckhardt[3] successively proposed four and three resolution cells for each sampling point, which could modulate both the amplitude and the phase for a complex vector. In Lee's method, between each two cells there is a quadrature phase delay based on Cartesian coordinates; in Burckhardt's there is a $2\pi/3$ phase delay based on the unit vectors with seperation of 120°. As a result to each sampling point two cells in Lee's and one cell in Burckhardt's are always possessing zero transmission, so there is some space in CGH which has no contribution in amplitude, but must be remained only for the phase relations. Because a carrier frequency must be used in preserving these methods, the wavefronts reconstructed from CGH are off the optical axis and the diffraction efficiency is low. An another type of CGH is Kinoform, which uses the form of reliefs recorded on film to record the phase variations of complex wavefronts. Without a carrier frequency, the wavefronts reconstructed are centered on the optical axis and have a high diffraction efficiency. However, the Kinoforms does not record the amplitude variations of the complex function.

By analysing these methods above, we suggest a new pure phase coding for CGHs based on the decomposition of any complex vector into two symmetrically situated component vectors with equal amplitude. It combines all the advantages of the above methods. In this method only two cells are needed for each sampling point. They can modulate both of the phase and the amplitude of a complex vector. In this method there has no cell with zero transmission as in Lee's or Burckhardt's method and hence the SBW is increased. Also a carrier frequency is not necessary, the wavefronts reconstructed from such CGH are centered on the optical axis. The diffraction efficiency of the proposed method is high.

Principle

The theory of complex vectors demonstractes that any complex vector $A\exp(j\theta)$ can be composed of two components $A_1\exp(j\theta_1)$ and $A_2\exp(j\theta_2)$, as shown in fig. 1, we obtain

$$A\exp(j\theta) = A_1\exp(j\theta_1) + A_2\exp(j\theta_2) \quad (1)$$

where the four variables A_1, A_2, θ_1, θ_2 are related and only two of them can be chosen independently. If we choose

$$\theta_2 - \theta = \theta - \theta_1 \quad ,$$
$$A_1 = A_2 = 1/2 \quad , \quad (2)$$

then

$$A\exp(j\theta) = 1/2\ [\exp(j\theta_1) + \exp(j\theta_2)] \quad . \quad (3)$$

It means that the two components with the same amplitude are symmetrically located at the two sides of the complex vector. The phase difference of the two components can modulate the amplitude of complex vector from 0 to 1, and the phase of the complex vector can be varied from 0 to 2π. So that the two components with the same amplitude can compose any complex vector $A\exp(j\theta)$. Then we can get

$$\theta = (\theta_1 + \theta_2)/2 \quad ,$$
$$A = \cos\ (\theta_2 - \theta_1)/2 \quad . \quad (4)$$

Alternatively, if the $A\exp(j\theta)$ is known, we can get

$$\theta_1 = \theta - \arccos A \quad ,$$
$$\theta_2 = \theta + \arccos A \quad . \quad (5)$$

So in our coding method, only two cells with pure phase (i.e., having the same amplitude) are used. They are recorded on a film in the form of reliefs by conventional bleaching techniques.

According to the sampling theory, the width dx of two cells must be satisfied the following equation

$$1/2dx = f_{max} \quad (6)$$

where f_{max} is the maximam frequency of a bandlimited function for the object field.

In the special case when the amplitude of a complex vector is a constant, only one cell is needed and it is then reduced to the Kinoforms.

Experimental Procedure

The phase variation of the sampling cells is recorded on the holographic film by exposing, developing and bleaching. The bleaching solution in our experiment makes the phase variation relies mainly on the thickness of the film. The relationship between the phase variation and the thickness of the gelatin can be simply given by the following equation:

$$\theta = 2\pi(n-1)\Delta d/\lambda \tag{7}$$

where θ is the phase, n is the index of refraction of the emulsion, Δd is the difference in thickness between the exposed and unexposed areas through bleaching. Measuring the phase values with different exposure times, we obtain an experimental curve of θ and E and fitted with a microprocessor which is written as

$$E = E_0 \left(A_0 + A_1\theta + A_2\theta^2 + A_3\theta^3 + A_4\theta^4\right) \tag{8}$$

where E_0 is minimum exposure time, A_0, A_1, A_2, A_3 and A_4 are constants. In our fitting, we get $A_0 = 0.0199$, $A_1 = 0.295$, $A_2 = 4.32\times10^{-3}$, $A_3 = -6.84\times10^{-6}$, $A_4 = -2.69\times10^{-9}$

The CGH is fabricated by a laser scanning system[4-5] as shown in Fig. 2. The block diagram of the experimental procedures is shown in Fig. 3. The complex function is sampled by a computer. Each sampled value $A_{nm}(x,y)$ is coverted into θ_{nm1} and θ_{nm2} following the equation (2) and then determines the exposure times t_{nm1} and t_{nm2} for the laser scanning system according to the curve of θ — E. By carefully controlling the experimental procedures including bleaching and keeping stable of the laser beam, reproducible results of the phase for our CGH could be obtained.

Preliminary Experimental Results

We fabricated two kinds of CGH: (1) Imaging CGH, (2) Differential filter.

(1) Imaging CGH

We made two imaging CGHs. One is a intensity distribution of cosine: cos $2\pi fx$. Where $f = 0.5\ mm^{-1}$ is the space frequency, x is the space position. Putting the imaging CGH at the input plane of a convenient optical information processing system, we get the intensity distribution at the output plane as shown in Fig. 4. Experimental result was

$$I/I = 10\%,\ f' = (0.500 \pm 0.005)\ mm^{-1}.$$

The other is a grey-scale with four levels. The ratio of their intensities is designed as follows

$$I_1:I_2:I_3:I_4 = 3.0:1.0:2.0:4.0.$$

Fig. 5 (a) shows the intensity distribution measured at the output plane. Fig. 5 (b) shows the photo of this CGH taken at the output plane. The rate is

$$I_1:I_2:I_3:I_4 = 3.2:1.0:2.0:5.1.$$

(2) Differential filter (one-dimension)

A one-dimension differential filter was made by putting the CGH at the Fourier Plane in a conventional optical information processing system. We can get the processing result at the output plane. Fig. 6 (a) shows the processing result of a rectangular aperture with width L = 1.9 mm where w = 96μ is the differential line width. S/N = 100:3 is the ratio of signal to noise. Fig 6 (b) is a photo of processing result for the Chinese word "川" and the English capital letter "F".

Conclusion

(1) The proposed phase-only CGH can modulate both the amplitudes and the phases with the advantage of on-axis reconstruction, high SBW and high diffraction efficiency.

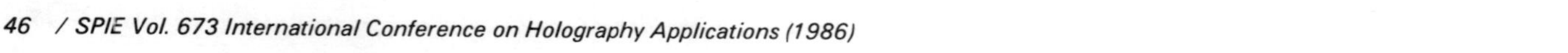

(2) we have fabricated the phase only CGHs which work quite well.

(3) Further investigation will be carried on by improving the errors in the phase measurement and some other interesting applications.

Acknowledgments

The authors wish to thank Prof. Zhi-ming Zhang for his encouragement and helpful discussions.

References

1. Brown, B.R. and Lohmann, A.W., Appl. Opt., vol. 5, P. 967. 1966.
2. W.H.Lee, Appl. Opt., vol. 9, P. 639 1970.
3. Burckhardt, C.B., Appl. Opt., vol. 9, P. 1949. 1970.
4. S.H. Lee and Leger, J.R., Opt, Eng., vol. 18, P. 511. 1979.
5. X.T.Ying et al., Acta optica Sinica, vol.2, P. 315. 1982.

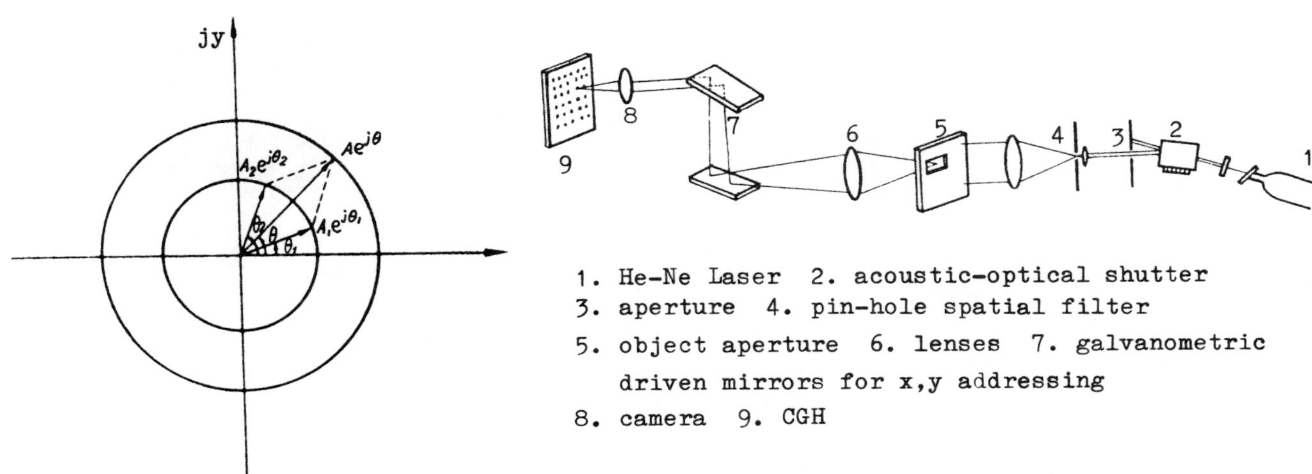

1. He-Ne Laser 2. acoustic-optical shutter
3. aperture 4. pin-hole spatial filter
5. object aperture 6. lenses 7. galvanometric driven mirrors for x,y addressing
8. camera 9. CGH

Fig. 1 Graphical composition of the two component vectors of the same amplitude

Fig. 2 Schematic diagram of the laser scanner

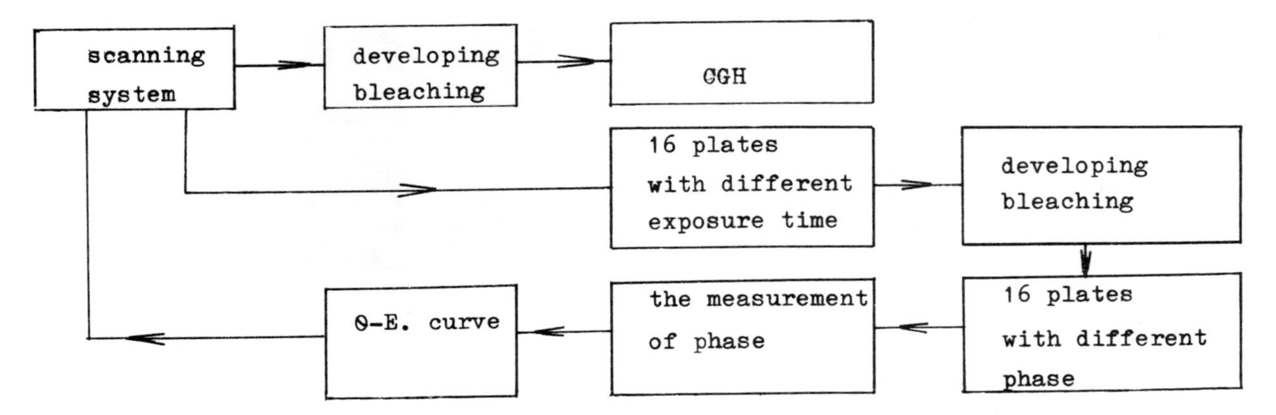

Fig. 3 The block diagram of the experiment at procedures

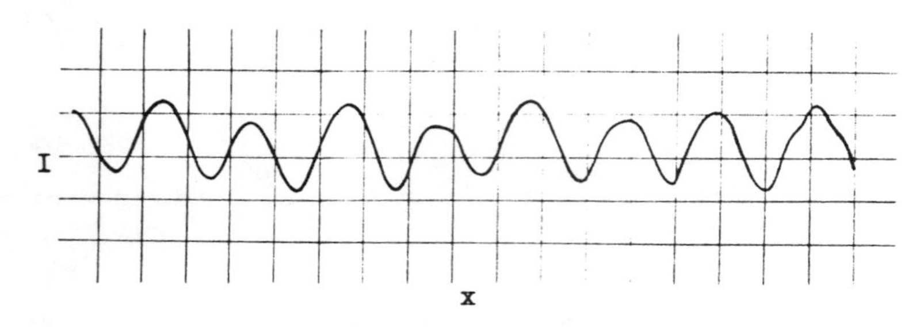

Fig. 4 The intensity destribution of imaging CGH of cos2 fx at the output plane

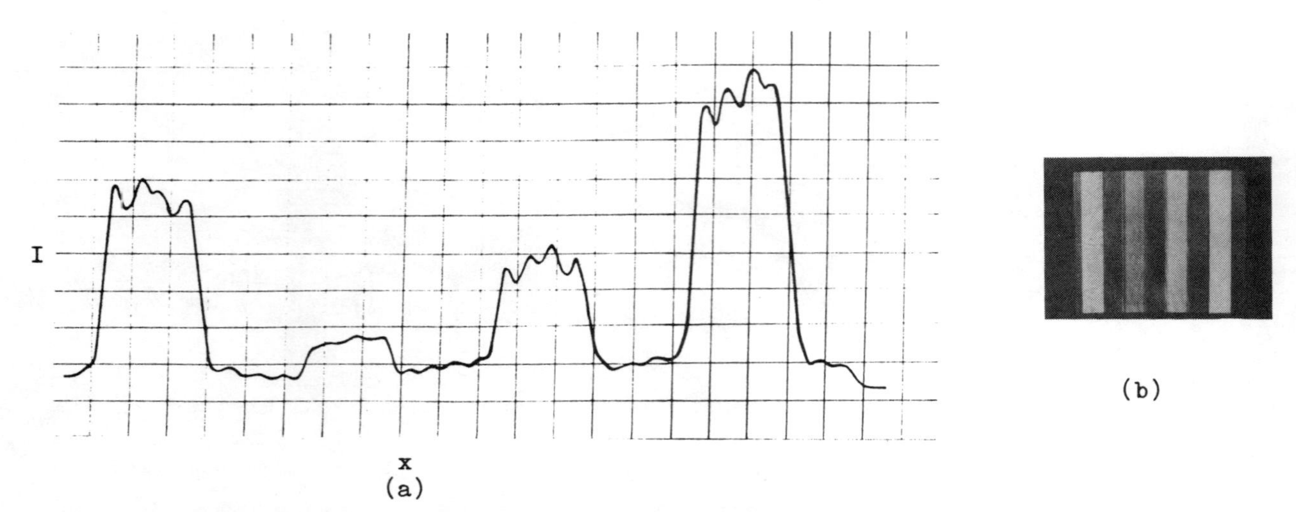

Fig. 5 Photo of the reconstructed grey-scale with four levels and the intensity distribution at the output plane

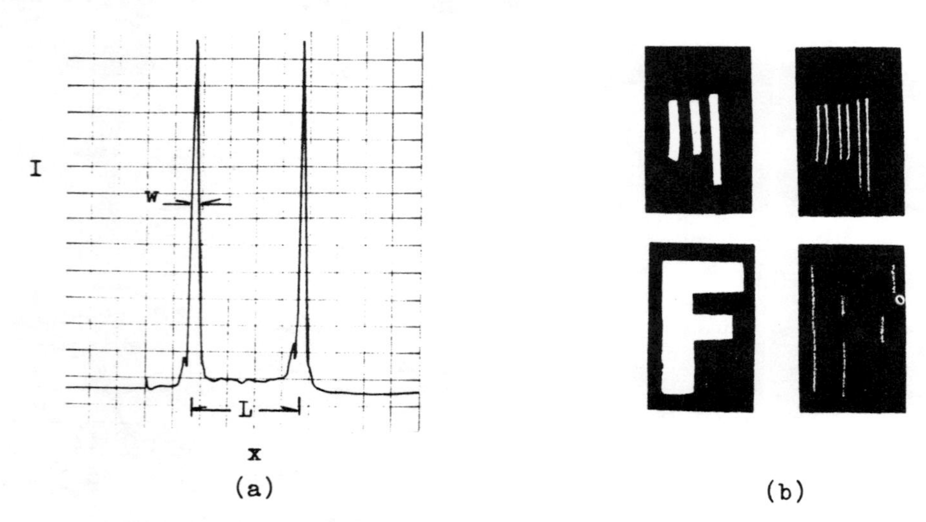

Fig. 6 The differentiation filtering processing result

Infrared holography using a CO_2 Laser and its application to Si-wafer inspection

Guanchao Sun, Zijuen Yuan

Institute of Laser Technology, Hefei Polytechnical University, Anhui, China.

Abstract

The results of the development of IR holography at 10.6μm wavelength and its application to inspection of nontransparent metal object and transparent Si-wafer are presented. Polyvinyl alcohol film and oil film have been used as recording media.

Introduction

CO_2 Laser with the wavelength of 10.6μm made it possible to spread holographic methods to infrared(IR) spectral range. The holographic recording of objects opaque to visible light , such as Si-wafer can be realized, and non-distructive inspection of them can be made. Because the wavelength is over 10 times longer than that of visible light, the sensitivity of the setup to the environmental vibrations is decreased.The use of such a longer wavelength in holographic interferometry also makes the detection of large displacements or deformations easy.

One of the important problems of realizing holography at 10.6μm is the search of new recording media with improved characteristics. The first infrared holography was reported by Chivian et al.[1] in 1969. They used a thermochromic material Cu_2HgI_4 as a recording medium. Izawa et al.[2] used a polymeric film containning a photo-chromic spiropirans painted on a glass substrate.Liquid crystal was used by Simpson et al.[3],Sakusaba et al.[4].

Recently(1982), T.Ya.Chelidze et al.[5]made use of IR holography to plasma diagnostics. In 1984, J.Lewandowski et al[6]made use of CO_2 Laser to produce dynamic holograms on an oil film on glass substrate.

This paper reports an experimental study on the IR holography at 10.6μm using thin polyvinyl alcohol(PVA) film and thin oil film as recording media. With this system we have successfully recorded the holograms of a metal ring and Si-wafers of semiconductor devices.

Experiment

Holographic system

A stablized single mode CW CO_2 Laser was used to obtain a Laser beam at a wavelength of 10.6μm.Its maximum output power was 8W. Two types of setups are shown in figure 1,(a)&(b):

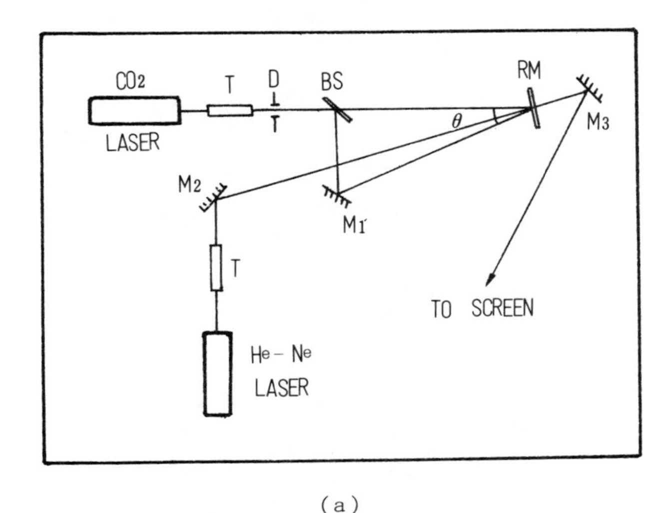

(a)

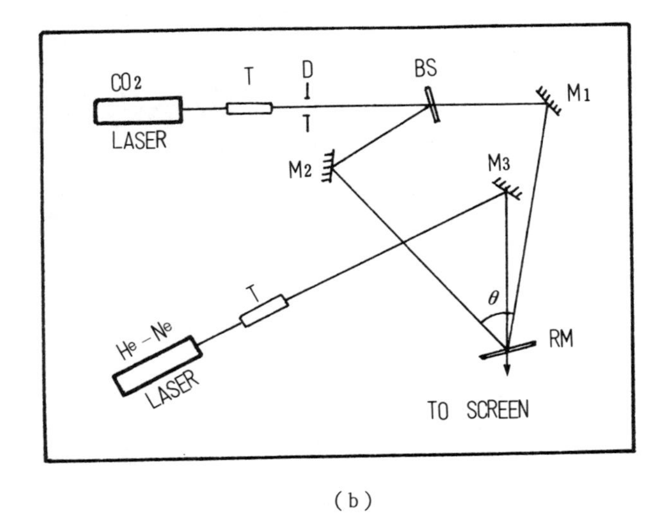

(b)

Figure 1. Two types of experimental setups for studying characteristics of recording media.

the light was expanded to 18mm diameter by a telescopic system made from two gold plated reflecting mirrors. The expanded beam was limited by a diaphragm of proper diameter. Then the beam was splitted into two parts of approximately equal intensity by a plane-parallel Ge plate. Beams were directed at the recording medium by gold plated plane mirrors. The reconstructing beam of the He-Ne Laser was expanded by conventional telescopic system. The recording medium was illuminated by this He-Ne beam continuously, and the variation of the recorded hologram could be monitored through that of the visible reconstruction on the screen in real-time.

Recording media

Two kinds of recording media, thin film of polyvinyl alcohol(PVA) and thin film of oil on a glass plate of 2mm thickness, have been used. The preparation was as follows: at first the glass substrate was uniformly smeared with several drops of water solution of PVA(10% by weight). Thickness could be controlled by the density of the solution. Then the wet substrate was set on its place and began to record. At the same time, it was dried by the recording beams.During the formation of the hologram, the reconstruction appeared on the screen simultaneously. Its intensity increased with time and eventually reached a steady state, then the process was accomplished. In our experiment the thickness of the film was about 1μm.

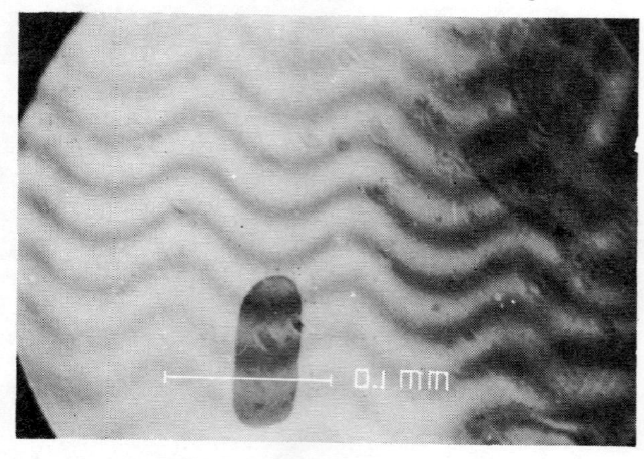

Figure 2. Interferomicrograph of IR holographic grating recorded on PVA film. The wavelike curves are interference fringes which display the profiles of the grating.

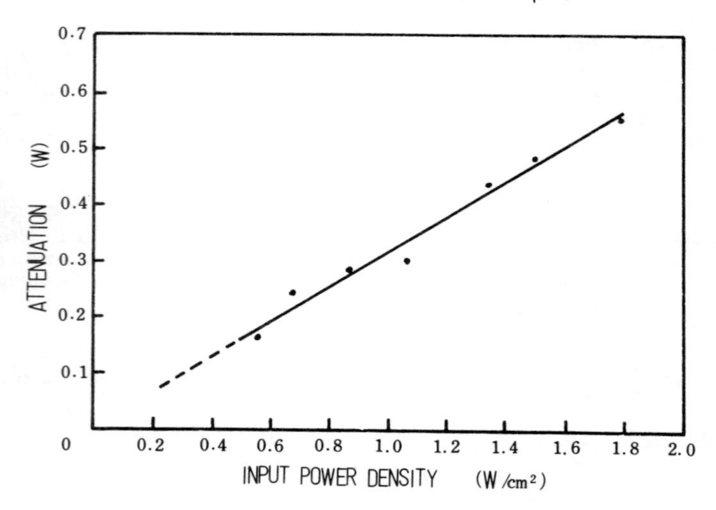

Figure 3. Characteristics of film attenuator. composition: polyvinyl + 5% polyacryl. film thickness: about 15μm. Laser beam diameter: about 18mm.

Figure 2. is the interferomicrograph of IR holographic grating. As shown, the grating just parallel to y-axis with a period of 0.1mm. The wavelike curves along x-axis are interference fringes which display the profiles of the grating. They appear to be distorted sinusoid.

The preparation of oil film was the same as PVA. Instaed of water, gasoline was used as diluant. Oil film is a real time recording medium[8].

Attenuator

In the experiment, we discovered that the commercially available polyvinyl film, which was used for wrappage primarily, could be used as an attenuator when inserted into the IR beam. Analyzing with an IR spectroscopic analyzer, we found that it contained 5% polyacryl and there were no high absorption peaks at 10.6μm wavelength, but only moderate absorption. Changing the number of films inserted, the CO_2 Laser beam power could be adjusted in a wide range, without apparent influence on its stability.

For a given film thickness, there was a maximum beam power density, greater than that the film would be destroyed. Thinner film could withstand higher power density. Increasing the power density, the attenuation increased linearly within the range of our experiment. Figure 3. shows the experimental results of a given film attenuator.

Results

Diffraction efficiency vs. spatial frequency

The diffraction efficiency at reconstruction is defined as the ratio of the intensity of reconstructed first order image to that of the incident reconstructing beam. Measurements of diffraction efficiency η with He-Ne Laser beam were made for both oil film and PVA film. The curve shown in Figure 4 is $\sqrt{\eta}$ vs. spatial frequency f[8]. In the figure the measureable η up to the spatial frequency of 60 cycles/mm is given. For PVA the curve η is similar but the η values higher than 30% at a spatial frequency of 10 cycle/mm have been measured, ie. higher than that of oil film, but there are also samples with lower values. Different heating conditions, thickness of films, etc. make the profile of the holographic grating different, and thus the η different as well. These problems call for further study.

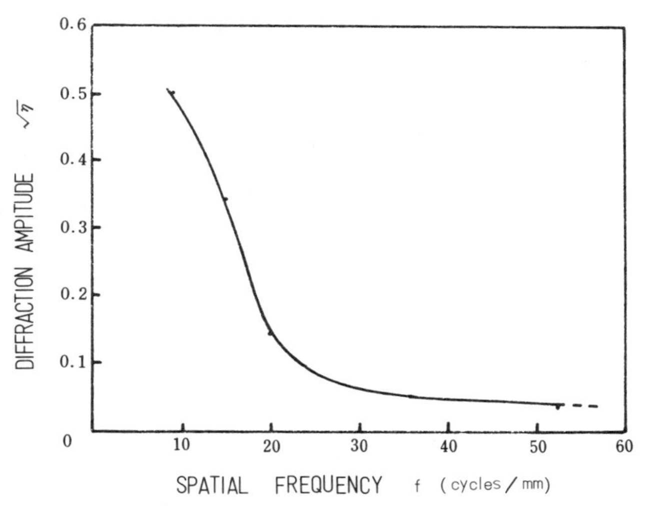

Figure 4. Diffracting amplitudes of oil film. (thickness: about 1μm)

Linearity

One of the causes degrading image quality is the inherent nonlinearity of the phase holograms. According to W. Rioux et al[7], we have chosen to charaterize a good linearity of the recording by a high ratio of the intensity of the reconstructed first order image to that of the higher orders. There are several restrictions on this problem. Under these conditions the intermodulation noise can be ignored. In our experiment, when the diffraction efficiency is 1% or less, the second order diffracting point almost cannot be seen(Fig.5). So that, according to Cathy's limitation, we can think all of the exposure parameters linear, and thus the inherent nonlinearity noise will not be considered.

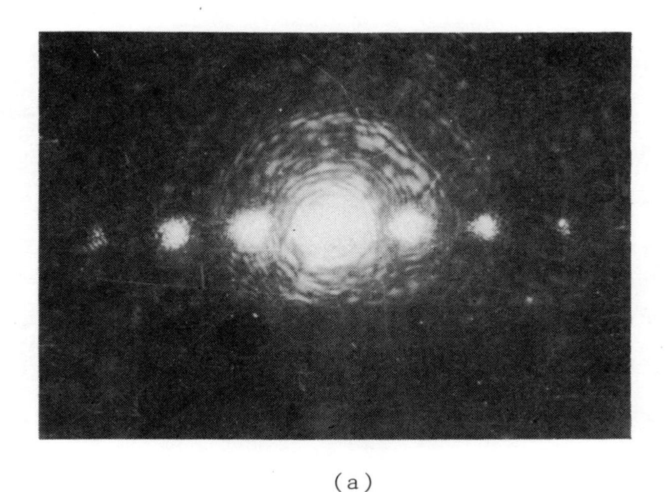

(a)

For higher diffraction efficiency;

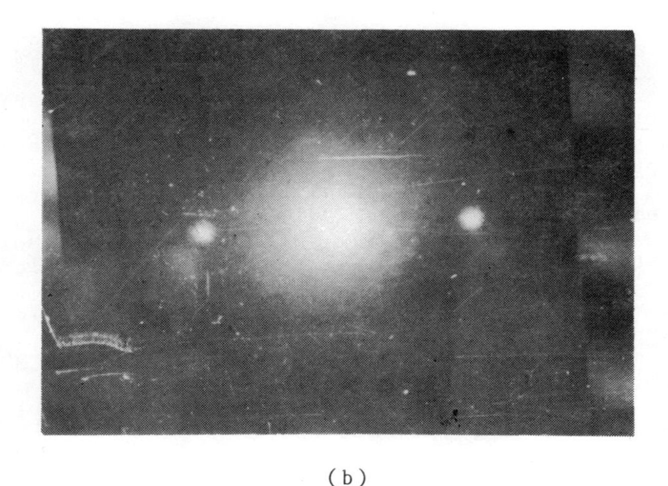

(b)

Efficiency less than 1%.

Figure 5. Reconstructions by He-Ne Laser beam.

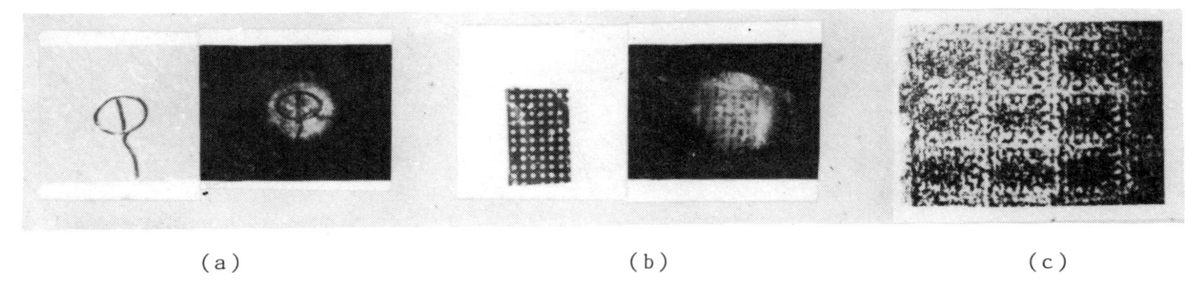

(a) (b) (c)

Fignre 6. Reconstructed images of the IR holograms.

(a) The metal ring and its reconstructed image; (b) A Si-wafer and its image; (c) Photo-magnified image of another Si-wafer. The size of each square is 1x1.5mm².

Holograms

Using this system and recording media, we have successfully recorded the holograms of two kinds of objects, a metal ring which is opaque to IR radiation and Si-wafer of semiconductor devices, which is transparent to IR beam. As shown in Figure 6: (a) is a metal ring; (b) is a Si-wafer. The left is the object and the right is its reconstructed image. A collimated CO_2 Laser beam was used for recording, θ corresponding to the spatial frequency of 17 cyc./mm . The reconstructing beam was also collimated. Under such conditions the lateral magnification of the reconstructed image was unity. Good image resolution and higher brightness can be observed in these photographs. Figure 6 (c) is the reconstructed image of another Si-wafer of a semiconductor device. Each square is an element of the wafer ($1x1.5mm^2$). It was photographically magnified to show the limiting resolution.

Conclusions

We have used the PVA film and oil film as recording media, and recorded the IR holograms. The spatial resolution of PVA film as well as oil film was limited to approximate 60 cyc./mm. The low diffraction efficiency as spatial frequency increased over 20 cyc./mm are caused by thermo-conduction of the film. We believed that it could be improved by using pulsed laser.

Generally we can ignor the inherent nonlinearity. But in actual applications we should make a compromise between linearity, i.e. signal to noise ratio, and diffracting efficiency.

The speckle noise degrade the image resolution heavily. In our system, there are two kinds of speckle, i.e. IR speckle in recording process and the reconstruction speckle. These call for improvement.

The preparation of PVA film and oil film with dilution method makes it easy to control and the film with large area.

References

1. Chivian, J. S.,et al., "Infrared Holography at 10.6μm", Appl. Phys. Let., Vol.15, pp. 123-125. 1969.
2. Izawa, T., et al., "Infrared Holography with Organic Photochromic films", Appl. Phys. Let., Vol.15, pp. 201-203. 1969.
3. Simpson, W. A.,et al., "Real time Visual Reconstruction of Infrared Holograms", Appl. Opt., Vol. 9, pp. 499-501. 1970.
4. Sakusabe, T., etal., "Infrared Holography with Liquid Crystals", Jap. J. Appl. Phys. Vol. 10, pp. 758-761. 1971.
5. Chelidze, T. Ya., et al., "Infrared Holography Using a Pulsed CO_2 Laser and Its Application to Electron Density Measurement in Plasma", Proceedings of the International Conference on Lasers'82., pp. 651-658.
6. Lewandowski, J., et al., "Real-time Interferometry Using IR Holography on Oil Films", Appl. Opt., Vol. 23, pp. 242-246. 1984.
7. Rioux, M. et al., "Plastic Recording Media for Holography at 10.6μm", Appl. Opt., Vol. 16, pp. 1876-1879. 1977.
8. Cormier, M., et al., "Holographie en Infrarouge sur de Minces Couches d'huile", Appl. Opt., Vol. 17, pp. 3622-3626.

Picosecond time- and space-domain holography by photochemical hole burning

P. Saari, R. Kaarli, and A. Rebane

Institute of Physics, Academy of Sciences of the Estonian SSR,
142 Riia Street, 202400 Tartu, USSR

Abstract

In this paper, the concept of holographic storage and reproduction of optical signals is generalized for the case of a spectrally highly selective recording medium, which in addition to fixing the spatial intensity distribution of the incident light is also able to memorize its intensity spectrum. We demonstrate persistent storage, recall and conjugation of picosecond light signals from various model objects by making use of coherent optical responses in photochemically active media. A simple linear theory of holographic storage and playback of both the spatial and the temporal behaviour of the signal is shown to well describe the experimental results obtained by utilizing octaethylporphin-doped polystyrene at 1.8 K as a spectrally selective recording material. The proposed method of time- and space-domain holographic recording provides the storage and reproduction of time-dependent optical signals with a duration of 10^{-8} to 10^{-13} sec. This presents unique possibilities for ultrahigh-speed data-storage and optical signal-processing.

Introduction

In ordinary holography, in common photographic materials a stationary spatial interference pattern between the overlapping reference and object beams is recorded and all time variances of the signal field are lost. Real-time holography in non-linear media can handle time-dependent fields, whereas the time dependence of the signal is not stored. In Ref.[1-5] it was shown that spectrally highly selective photosensitive materials, which in addition to the spatial intensity distribution of the incident light can also fix its intensity spectrum, provide the possibility of holographic storage and playback of both spatial and temporal dependences of the optical signal. Experiments in frozen organic impurity-doped matrices undergoing photochemical burning of spectral holes reveal diffraction efficiencies (up to 50% from the transmitted probe beam), high enough for practical applications.

The purpose of the present paper is to survey the experimental results on recording and playing back spatial and temporal characteristics of ultrashort light signals by the proposed method of time- and space-domain holography and to demonstrate the accordance between the results obtained and the theoretical predictions. In Section 2 we first introduce photochemical holeburning (PHB) which is responsible for the storage of spatial and temporal dependences of incident light. Secondly, we introduce a new phenomenon - the photochemically accumulated stimulated photon echo (PASPE) that is responsible for the playback of the time-dependent holographic image of the object in the form of the PASPE-signal. In Section 3 we give a brief account of the linear theory of time- and space-domain holography in spectrally highly selective media. Then, after listing main experimental details, we discuss the features of direct and phase-conjugated holographic images of light signals formed from picosecond pulses by model transparencies. Particularly, we demonstrate the possibility of complete elimination of wave-front distortions by using the conjugated image (earlier accomplished in nonlinear media; see Ref.[6] and references therein). Further we consider a PASPE hologram of a coin illuminated by a picosecond pulse that demonstrates the far-reaching prospects of this type of holography, finally suggesting some practical applications in data storage and processing.

Burning and probing of spectral holes by picosecond pulses

Photochemical hole burning (PHB),[7,8] in addition to its well-known applications in high-resolution laser spectroscopy of impurities in solid matrices,[9,10] has also been considered a promising method for optical data storage,[11-14] since PHB allows the data bits to be attached not only to different spatial locations of an optical memory but also to different optical frequencies within an inhomogeneously broadened impurity absorption band. To compare with conventional photochromics, the materials undergoing PHB are also self-developing, but monochromatic illumination bleaches the sample within only a very narrow region of the inhomogeneously broadened impurity absorption band located in the vicinity of the excitation wavelength. The width of the obtained hole in transparency spectrum is determined by the homogeneous linewidth of the photoactive impurities, which for the materials known from PHB experiments until now is of the order of 10^{-2} to 10^{-4} cm^{-1}, while the inhomogeneous broadening may exceed the homogeneous one up to 10^5 times. Traditionally, in both spectroscopic and data-storage applications, PHB is performed by burning spectral holes through tunable monochromatic laser excitation in inhomogeneously broadened absorption bands of impurity molecules. Successively, these persistent holes are detected by fine scanning of the excitation over the modified inhomogeneous spectra either to perform the readout of spectral data bits or to obtain homogeneous spectral characteristics of impurity molecules.

In Ref.[15], a radically different approach, designated time-domain frequency-selective data storage, is proposed in which all the data bits assigned to a given spatial location are written in parallel by a temporally modulated laser pulse. We have demonstrated experimentally that time-domain PHB can be carried out by sequences of coherent picosecond pulses, with the spectral width 2 orders of magnitude larger than the homogeneous hole width (0.05 cm^{-1}).[1] As the duration of the applied sequences of burning pulses is less than the impurity-molecule phase relaxation time T_2, the complex shape of the resulting hole corresponds to the Fourier spectrum of the pulse sequence, i.e., the width of the hole envelope is reciprocal to the pulse duration, whereas the fine spectral structure (containing up to 10^2 peaks) of the hole originating from the interference of the burning pulses is determined by the number and the intervals of the pulses in the sequence (Fig.1).[2]

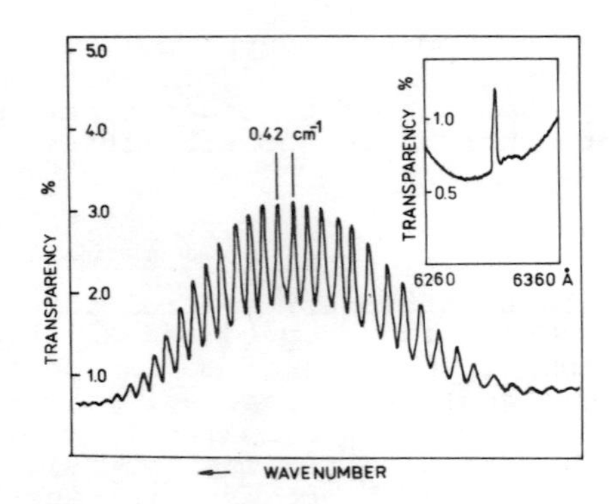

Figure 1. Transparency spectrum measured after PHB exposure to the sequence of picosecond pulses. Insert - burned in photochemical hole and its pseudophonon sideband (1 Å resolution), main frame - fine structure of the hole (0.075 cm^{-1} resolution).

Figure 2. Temporal response of the sample and the shape of the applied PHB sequence (below).

Consequently, we can conclude that materials undergoing PHB form a new class of photosensitive materials - spectrally highly selective photocromics with spectral selectivity 10^{-2} to 10^{-4} cm^{-1}.

In Ref.[2,3] we have also shown that passing a single arbitrarily weak probe picosecond pulse through a sample with such a hole in its inhomogeneous absorption spectrum will result in delayed echo pulses that completely reproduce the burning-pulse sequence and have very high intensity compared with the transmitted probe pulse (Fig.2). These delayed pulses were interpreted as the temporal response of the sample acting as a linear filter with a burned-in persistent spectral transparency grating. On the other hand, in terms of time-domain coherent phenomena, the physical nature of these coherent optical responses lies in free-induction decay under weak excitation. For that reason, we have termed the observed phenomenon, *photochemically accumulated stimulated photon echo* (PASPE), in that it is most closely related to the accumulated three-pulse stimulated photon echo.[16,17] However, PASPE displays several special features. *First*, although the stimulated photon echo can occur only within the relaxation time limits of light-induced transient frequency-domain population gratings, the lifetime of PASPE is determined by the very long lifetime (at least many hours, maybe even years) of PHB photoproducts, and so the time available to recall the frozen-in echo is practically infinite. This allows the spectral gratings in PASPE to be accumulated to a very high contrast, which in turn results in very high relative-intensity echo signals. *Second*, because of the accumulation effect, PASPE experiments can be performed well under modest linear excitation conditions, and, in the first approximation, a simple theory based on a linear dielectric susceptibility approach can be established.[18] *Third*, since only weak readout pulses are required, the replica of the signal, once it is stored, can be recalled from the sample many times before distortions may appear. These special features make PASPE useful in determining impurity excited-state phase relaxation times.[4,5]

But, from our viewpoint, one of the most interesting applications of PASPE lies in the persistent holographic recording of both temporal and spatial characteristics of ultrashort light pulses.[18,19] Let us suppose that the PHB sequence comparises an object pulse with arbitrary temporal and spatial amplitude and phase distribution and a plane-wave reference pulse, which is short enough to be considered a δ pulse as compared with the signal. In Ref.[18], where the theory of time- and space-domain holography is developed by extending the formalism of ordinary holography into the fourth dimension, we have demonstrated that, if the phase relaxation time T_2 of the PHB media is long enough compared with the duration of the signal, exhaustive information about the signal can be stored by means of spatial and spectral distribution of the PHB bleaching effect and later reproduced by PASPE.

Again, bearing in mind the special features of PASPE, it should be stressed that PASPE holograms, unlike the echo holograms in ordinary resonant media,[15,20,21] are able to play back object scenes with their temporal dependence at arbitrary moments after storage.

The purpose of the present paper is to report the experimental results on recording temporal and spatial characteristics of ultrashort light signals by using PHB and on playing back their holographic images by using PASPE as well as to demonstrate accordance of the results with theoretical predictions.

Theory

Following Ref.[18] we outline here the derivation of formulas governing the holographic process. Consider a plate with dimensions ($2x_{max}$, $2y_{max}$, d) containing photochemically active dye molecules having zero-phonon absorption lines (ZPL's) of homogeneous width T_2^{-1} around frequency ω_0. Let the plate be illuminated from the $z < 0$ side, as shown in Fig.3, by an object pulse

$$S(r, t) = s(x, y, t - z/c) \exp[i\omega_0(t - z/c)], \qquad (1)$$

with a spectral width $\Delta\omega_s$ and a duration $t_s \ll T_2$. If the front edge of the object pulse reaches the plate at moment $t = 0$, the trailing edge of the pulse will leave the plate at $t = t_s$. Let us further suppose that, with the delay t_R relative to the front edge of the object pulse, the plate is also illuminated by a short plane-wave reference pulse, which is passed through the plate at a small angle θ with respect to the Z axis (all the approximations made and inequality relations used have been discussed in Refs.[18] and [19]; see also Fig.4). If the reference pulse has a duration small enough compared with that of the object pulse, it can be considered a δ pulse and can be written as

$$R(r, t) = R_0\delta(t - \hat{n}_R\cdot r/c - t_R) \exp[i\omega_0(t - \hat{n}_R\cdot r/c - t_R)], \qquad (2)$$

where $\hat{n}_R = (-\sin\theta, 0, \cos\theta) \sim (-\theta, 0, 1)$ is a unit vector in the direction of the reference pulse.

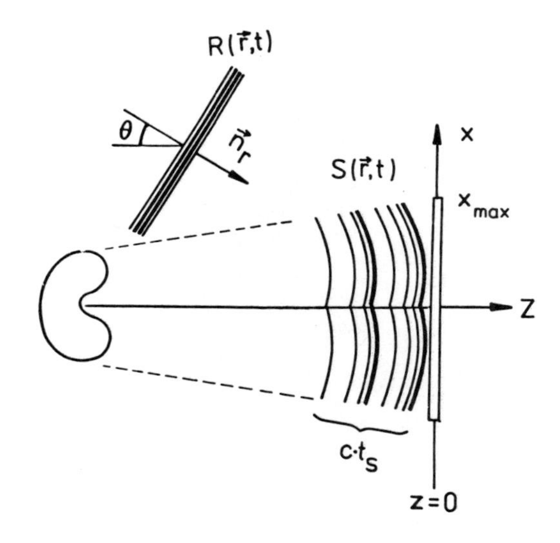

Figure 3. Diagram of the proposed holographic-recording procedure. Object pulse $S(r,t)$ reaches the hologram plate at the moment $t = 0$; reference pulse $R(r, t)$ is shown to be delayed, i.e., $t_R > t_s$.

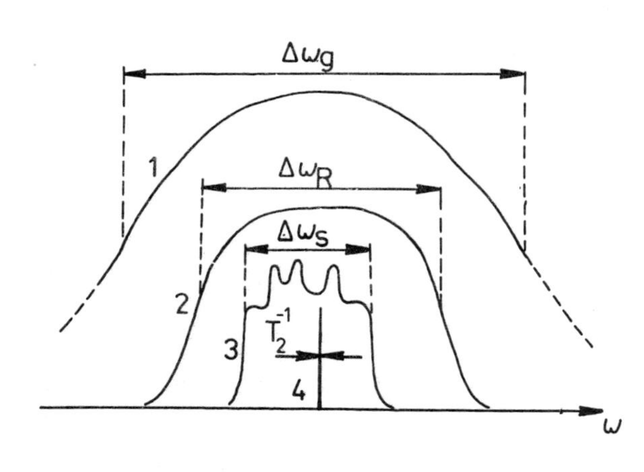

Figure 4. Relations among spectral widths of 1, inhomogeneous absorption band; 2, reference pulse; 3, object pulse; and 4, homogeneous ZPL linewidth.

The intensity spectrum of the sequence of object and reference pulses acting jointly on the medium is given by

$$I(x, y, \omega) = R_0^2 + |\bar{s}(x, y, \omega - \omega_0)|^2 + R_0\bar{s}(x, y, \omega - \omega_0) \exp[-i\omega(x\Theta/c - t_R)] + $$
$$+ R_0\bar{s}^*(x, y, \omega - \omega_0) \exp[i\omega(x\Theta/c - t_R)], \quad (3)$$

where $\bar{s}$ stands for the Fourier image.

Optical properties of the medium of the plate are described by means of its dielectric permittivity

$$\varepsilon(r, \omega) = \varepsilon_0 + (\sigma c/2\pi\omega) \int d\omega' g(r, \omega')/(\omega' - \omega + iT_2^{-1}), \quad (4)$$

where σ is the integral absorption cross section of the ZPL, T_2^{-1} is the homogeneous width of the ZPL, and $g(r, \omega)$ is the inhomogeneous distribution function, altered in the course of being recorded, that depends on spatial coordinates within the recording medium and is defined so that $g(r, \omega)d\omega$ is the number of impurity centers in the unit volume of the medium that have their ZPL frequencies in the interval $(\omega; \omega + d\omega)$. The constant nonresonant part of the permittivity will be further taken as $\varepsilon_0 = 1$. Assuming that T_2 is much longer than the overall duration of the pair of object and reference pulses, the permittivity (Eq.(4)) can be written in the form

$$\varepsilon(r, \omega) = 1 - (\sigma c/2\omega)\{ig(r, \omega) - \hat{H}[g(r, \omega)]\}, \quad (5)$$

where the symbol $\hat{H}$ means the Hilbert transformation

$$\hat{H}[f(\omega)] \equiv \pi^{-1} \oint d\omega' f(\omega')/(\omega' - \omega). \quad (6)$$

If the intensity of object and the reference pulses is moderate so that nonlinear-saturation and power-broadening effects can be overlooked, the PHB bleaching effect resulting from the resonant excitation can be considered proportional to intensity (Eq.(3)). If we further assume that the plate has a considerable resonant optical density and that the spectrally selective bleaching does not change it significantly, the inhomogeneous distribution function may be expressed as

$$g(r, \omega) = g_0\{1 - m\sigma\eta I(x, y, \omega) \exp[-g_0\sigma z]\}, \quad (7)$$

where g_0 is the density of centers before the PHB, η is a coefficient of PHB efficiency, and m is an integer counting the number of applied sequences of reference and object pulses, i.e. taking into account the accumulation process during the storage.

Using Eqs.(5) and (7), we can write the complex linear transmittance function of the plate as follows:

$$K(x, y, \omega) = \exp[-d(i\omega/c + g_0\sigma/2)]\{1 + (\kappa/2)(1 + i\hat{H})[I(x, y, \omega)]\}, \quad (8)$$

where $\kappa = m\sigma\eta$ and the exponential factor describes the attenuation and phase shift of the output signal after crossing the plate of thickness d. One can see that the second term in Eq. (8) contains, as a hologram, exhaustive information about the object pulse. Considering the plate now as a spectral and spatial filter, one can calculate the response of the plate for an arbitrary read-out field. In this case we use a read-out pulse with properties identical to those if the reference pulse; the resulting optical signal in the output of the hologram is

$$F^{out}(r, t) = f_0(r, t) \exp[i\omega_0(t + x\Theta/c)] + f_s(r, t) \exp[i\omega_0(t + t_R)] + $$
$$+ f_s(r, t) \exp[i\omega_0(t - t_R + 2\Theta x/c)], \quad (9)$$

where

$$f_0(r, t) = [1/(R_0\kappa) + R_0/2]\delta(t + \Theta x/c) + R_0^{-1}Y(t + \Theta x/c) \int s(x, y, \tau)s^*(x, y, \tau - t - \Theta x/c)d\tau,$$
$$f_s(r, t) = Y(t + \Theta x/c)s(x, y, t + t_R),$$
$$f_s^*(r, t) = Y(t + \Theta x/c)s^*(x, y, t_R - t - 2\Theta x/c). \quad (10)$$

In writing Eq.(9) we have dropped the constant factor $R_0^2 \cdot \exp[-\sigma g_0 d/2]$ common for all three terms and assume that $z = +0$; in Eqs.(10), $Y(\tau)$ is the Heaviside unit step function. Equations (9) and (10) show that, as in the case of ordinary two-dimensional holograms, in the output of the spectral hologram there will appear three different pulsed-light waves (see Fig.5). The first term, with amplitude f_0, propagates in the direction $\hat{n}_R$ and describes a transmitted readout pulse with a distorted train determined by the autocorrelation function of the signal. The second term, with amplitude f_s, propagates along the Z axis and represents the playback of the stored object signal as a virtual image of the event recorded. The last term shows the possibility of recalling the conjugated image of the object with reversed time behaviour.

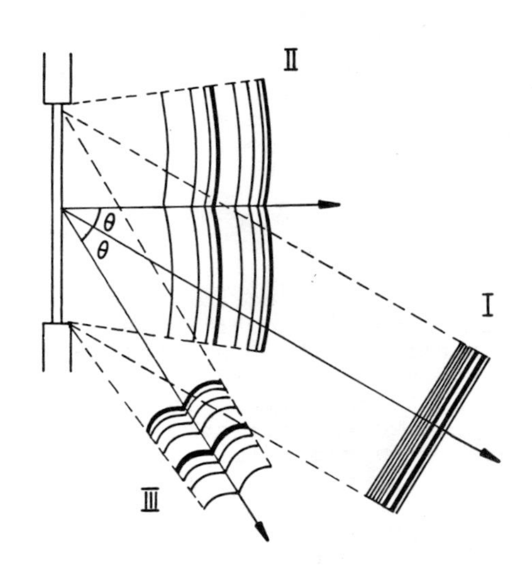

Figure 5. Diagram showing three different kinds of readout pulses: I, passed-through pbobe pulse; II, reproduced object pulse; and III, conjugated object pulse. In the case of a three-dimensional hologram, the conjugated object occurs only if the probe pulse is applied in a direction opposite the direction of the reference pulse.

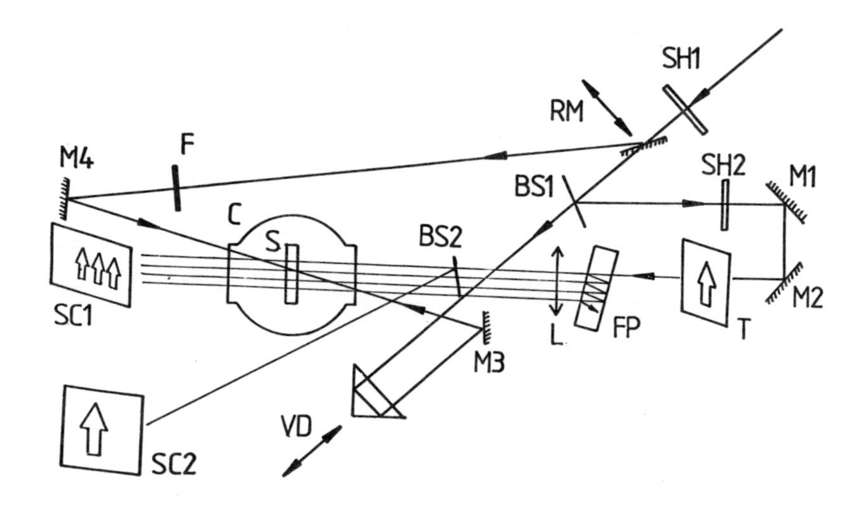

Figure 6. Schematic diagram of experimental setup for recording spatially modulated picosecond signals. Beam splitter BS1 divides the expanded input picosecond laser beam between reference and object channels; mirrors M1 and M2 direct the object beam through the transparency T, the Fabry-Perot étalon FP, and the lens L with focal length 2 m; optical delay OD is used to place the reference pulse in the 50-psec intervals between object pulses; movable mirrors M5 and M6 have been inserted to obtain conjugated images by passing the probe pulse in the opposite direction; echo signals are visualized on screens SC1 and SC2. C, cryostat; S, sample; F, neutral-density filter; SH1 and SH2, shutters.

In Eqs.(10), the causality function $Y(\tau)$ guarantees that no output signal appears at any point behind the hologram before the readout pulse has arrived. It also cuts off a part of the reproduced object pulse in dependence on the delay of the reference pulse during the process of writing the hologram: If $t_R \leq -|\Theta x_{max}|$ is valid, i.e., the reference pulse came before the object pulse, the latter is completely reproduced but the conjugated object pulse is entirely cut off; if, on the contrary, $t_R \geq |\Theta x_{max}| + t_s$, the nonconjugated object pulse is absent; finally, in an intermediate case of overlapping object and reference pulses $-|\Theta x_{max}| \leq t_R \leq |\Theta x_{max}| + t_s$, both conjugated and nonconjugated object pulses are partially reproduced. It can also be shown that, if the condition of two-dimensional hologram $\Theta^2 d \leq 2\pi c/\omega_0$ is violated so that the synchronism of writing and readout pulses must be considered, the conjugated object pulse can be reproduced only if a counter-propagating probe pulse with the direction $-\hat{n}_R$ is applied.

Experiment

In our experiments we utilized a Rhodamine 6G laser synchronously pumped by an actively mode-locked Ar-ion laser (Spectra - Physics Model 171). The picosecond dye laser provided 82-MHz repetition-rate pulses with 2 - 3 - psec duration (5 - 6 cm^{-1} spectral FWHM) at 100-mW average output power. Picosecond time analysis was performed by means of a synchroscan streak camera coupled to an optical multichannel analyzer. This system provided 20-psec temporal resolution.[22]

To study the spatial properties of time- and space-domain holography, a plane wave-front reference beam was applied at an angle of 6° with respect to the beam of the object pulses (see Fig.6). After PHB exposures, probe pulses were passed through the hologram either in the direction coinciding with the reference beam or in the opposite direction. The spatial and temporal structure of the picosecond signal was formed by passing the expanded laser beam through a transparency and an étalon. The étalon was tilted by 45° with respect to the beam axis so that the passed-through beam consisted of spatially separated reflections with the contour of the transparency, whereas the 50-psec delay between the reflections was determined by the double pass-through time of the étalon. The wave fronts of the object arrows to be recorded on the hologram were curved by a 2-m focusing lens. The spatial structure of picosecond signals was photographed as the pulses were scattered from sheets of scale paper used as screens to block the output object and conjugated object beams.

As a spectrally selective medium for recording time and space holograms by PHB, we used polystyrene doped with octaethylporphin at concentrations of $10^{-4} - 10^{-3}$ M. The inhomogeneously broadened 0-0 impurity absorption band was 200 cm^{-1} FWHM and had a maximum at 617 nm. Homogeneous ZPL widths were less than 0.05 cm^{-1} at 1.8 K. Samples were prepared in blocks with a thickness of 0.3 - 1.0 cm and an optical density of 1 - 3 and were contained in a liquid-He cryostat with pass-through optical windows. To resord high-contrast spectral holograms, PHB exposures over 0.1 J/cm^2 were needed. Depending on the average intensity of the incident light, the PHB effects of $10^9 - 10^{11}$ identical sequences of writing pulses were accumulated.

Results and discussion

In order to gain some insight into the temporal responses of the PASPE holograms obtained, let us first consider streak-camera experiments in which all beams were collinear and spatially unmodulated.[3,4] In this geometry ($\theta = 0$; see theory-section) all output pulses propagate collinearly and can be collected onto the entrance slit of the camera in order to measure the entire response of the hologram directly. Figure 7 represents the result in the case when PHB was performed with the reference pulse delayed with respect to the object pulse by 150 psec. Since $t_R = 150$ psec $> t_S = 50$ psec here, in full accordance with the theory, the time-reversed replica of the object appears in the response after the 150-psec delay (Fig.7b). To demonstrate the high efficiency of PASPE Fig.8 represents the result in the case when PHB was performed with an exponentially decaying pulse sequence under near-optimum conditions. One can see that the overall intensity of the three echo pulses following the first reading pulse is nearly the same as that of the latter. As far as the applied pulse sequence can arbitrarily be divided into signal and reference parts in this experiment, in terms of holography the result is nothing else but the restoration of the object by its time-domain fragment.

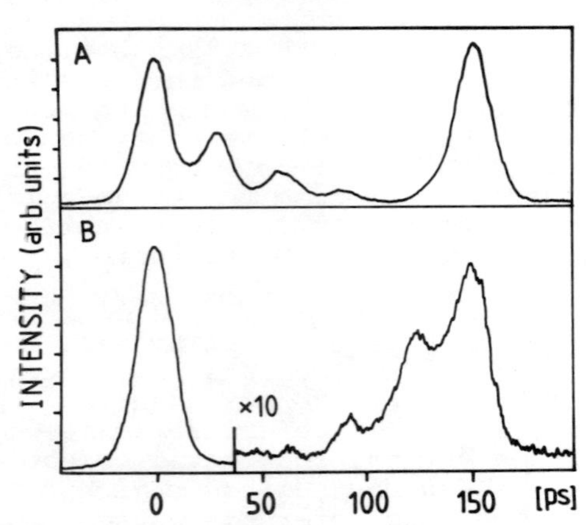

Figure 7. a, Streak-camera images of the applied PHB sequence comprising an unsymmetrical object pulse and a reference pulse; b, the passed-through probe pulse followed by the time-reversed PASPE signal. The object pulse was formed by a étalon from the laser pulse. The apparent 20-psec FWHM of both reference and probe pulses is determined by streak-camera time resolution.

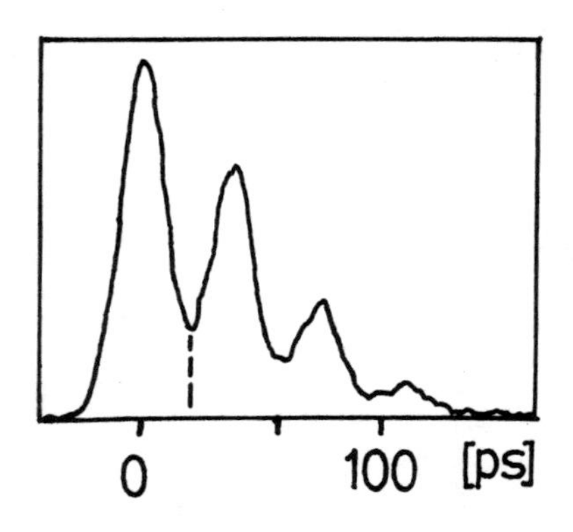

Figure 8. Streak-camera image of PASPE-signal under near optimum conditions.

Let us consider now experiments carried out in accordance with the full scheme of time- and space-domain holography, i.e., experiments with both spatially and temporally modulated picosecond signals and variable delayed reference pulses. Figure 9(a) gives an idea of the model signal used. It should be recalled that the arrows were 1-mm-thick wave packets flying one after another. To discuss the responses of holograms obtained at various settings of the reference-pulse delay (Fig.9(b) and 9(c)), let us also keep in mind that, according to Section 3 the nonconjugated image of the retarded part and the conjugated image of the advanced part of the signal pulse should be reproduced in the case of the forward-propagating probe-beam. Figure 9(b) indicates that all temporally separated spatial components of the object were completely restored in the case when the reference pulse had been applied first. Further, in Figs.9(c)-9(d) one can see that the signal components preceding the reference pulse in the recording procedure are absent in the restored image, showing the temporal selectivity of the hologram. All observed images correspond only to the nonconjugated signal wave. The reason lies in the fact that the holograms were recorded in a thick sample ($\theta^2 d \sim 2\pi c/\omega_0$).

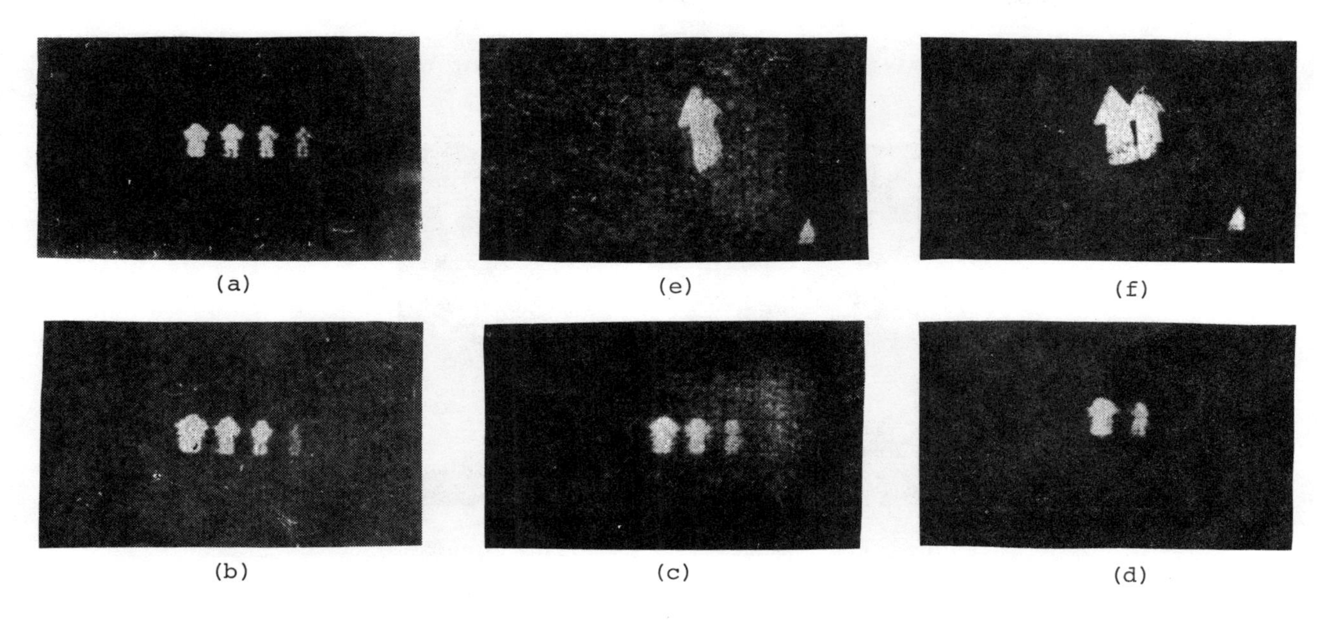

Figure 9. Photograph images of (a) the pulses used for the PHB and (b) those of echo signals reproduced from the hologram after PHB exposures with the reference pulse applied before the first object pulse, (c) and (e) between the first and the second object pulses, (d) and (f) between the second and the third object pulses, and (e) between the third and the fourth object pulses. Image of conjugated echo signals (e) and (f) photographed from the screen SC2 with the probe pulse passed through in the opposite direction.

In accordance with the theory, conjugated images appeared when a counterpropagating picosecond probe beam was applied (Fig.9(e) and 9(f)). In that case, on the contrary, an increase of the reference-pulse delay resulted in the successive appearance of the preceding components of the object that were played back in a reversed temporal sequence. Note that the conjugated images are magnified and that the diffraction patterns present in Fig.9(a)-9(d) are absent here. These effects were also expected from the theory, as the conjugated images are reversed not only temporally but also spatially.

To demonstrate the conjugation effect more explicitly, the experimental setup was modified by replacing the lens and the étalon by a distorter, which scattered the input object-beam contour over an angle of 4° (see Fig.10). The distorted object image was recorded on the hologram simultaneously with the delayed reference pulse. Afterward, with the help of a counterpropagating probe beam, a conjugated picosecond PASPE signal was produced. PASPE pulses traveled through the distorter in the opposite direction and reconstructed the initial object image of transparency (Fig.10(b)). It should be noted that the temporal distortions of picosecond object pulses were also compensated for.

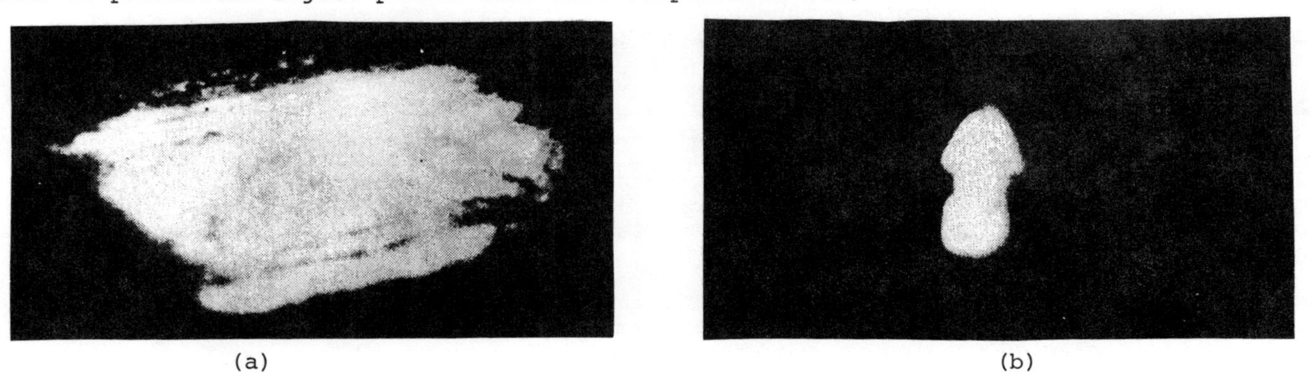

Figure 10. (a) Distorter object image photographed at the input of the cryostat, (b) reproduced object image obtained after conjugated echo pulses are passed through the distorter in the opposite direction.

As the last step of our experiments, we recorded a PASPE hologram of a coin illuminated by a 2-psec-duration pulse. The pulse scattered from the coin modeled an arbitrary spatially and temporally modulated object pulse that passed through the recording medium at an angle of 10° and some 20 psec later than the plane wave-front reference pulse. The recorded hologram was later illuminated, as in previous experiments, by a beam of 2-psec-duration forward-propagation probe pulses. The reproduced image of the coin was photographed from behind the cryostat (Fig.11). Although no temporal analysis of the image was performed, on

Figure 11. Image of a coin photographed from the hologram.

the basis of the theoretical and experimental data presented the reconstruction of the time structure of the object wave can also be stated. It could be partially proved by illuminating the hologram simultaneously by the probe and the object waves. A constructive interference between the object waves scattered from the coin and those reproduced from the hologram was observable, i.e., the reproduced object wave had the same spatial- and temporal-phase structure as the wave scattered from the object.

Conclusions

In conclusion we suggest some applications of time- and space-domain holography in optical data-processing:

1. Mathematical operations on picosecond signals:

a) algebraic summation making use of successive recording of summands. The signals to be subtracted must be recorded through a half-wave retarder;

b) time- and space-domain convolution of events by making use of one signal for reading the hologram of the other;

c) temporal and spatial frequency filtration of events via passing the signal through the filter-hologram;

d) 4-dimensional wave conjugation (wave-front spatial conjugation as well as time reversal of the signal).

2. Shaping of pico- and femtosecond pulses into signals of given temporal and spatial structure by passing them through an appropriately synthesized hologram.

3. Identification of picosecond events - generalization of holographic method of image recognition. If the signal to be identified coincides with one of the reference signals recorded, the focused response gives a δ-like pulse in the respective point of the focal plane.

4. Parallel writing and reading of data bits into the frequency-selective memories on PHB. To estimate the actual extension of optical memories at the expense of the frequency coordinate in [23] over 1600 holes were burnt and detected in parallel.

Acknowledgements

The authors are indebted to K. Rebane for constant interest in this research and for useful discussions as well as to J. Kikas for discussing the PHB problems. We also wish to thank I. Renge for preparing samples.

References

1. Rebane, A. K., Kaarli, R. K., and Saari, P. M., "Hole-burning by coherent sequencies of picosecond pulses," Opt. Spectrosk., Vol. 55, pp. 405-407. 1983.

2. Rebane, A. K., Kaarli, R. K., and Saari, P. M., "Dynamic picosecond holography produced by means of photochemical hole burning," JETP Lett., Vol. 38, pp. 383-386. 1983.

3. Rebane, A., Kaarli, R., Saari, P., Anijalg, A., and Timpmann, K., "Photochemical time-domain holography of weak picosecond pulses," Opt. Commun., Vol. 47, pp. 173-176. 1983.

4. Rebane, A. and Kaarli, R., "Picosecond pulse shaping by photochemical time-domain holography," Chem. Phys. Lett., Vol. 101, pp. 317-319. 1983.

5. Rebane, A. K., Kaarli, R. K., and Saari, P. M., "Burning and probing photochemical holes with picosecond pulses," J.Mol. Struc., Vol. 114, pp. 343-345. 1984.

6. Carlson, N. W., Babbit, W., and Mossberg, T. W., "Storage and phase conjugation of light pulses using stimulated photon echoes," Opt. Lett., Vol. 8, pp. 623-625. 1983.

7. Gorokhovskii, A. A., Kaarli, R. K., and Rebane, L. A., "Hole burning in the contour of a pure electronic line in a Shpolskii system," JETP Lett., Vol. 20, pp. 216-218. 1974.

8. Kharlamov, B. M., Personov, R. I., and Bykovskaya, L. A., "Stable 'gap' in absorption spectra of solid solutions of organic molecules by laser irradiation," Opt. Commun., Vol. 12, pp. 191-193. 1974.

9. Rebane, L. A., Gorokhovskii, A. A., and Kikas, J. V., "Low-temperature spectroscopy of organic molecules in solids by photochemical hole burning," Appl. Phys., Vol. B29, pp. 235-250. 1982.

10. Friedrich, J. and Haarer, D., "Photochemical hole-burning: spectroscopic study of relaxation processes in polymers and glasses," Angew. Chem. Int. Ed. Engl., Vol. 23, pp. 113-140. 1984.

11. Szabo, A., "Frequency selective optical memory," U.S. Patent No. 3896420. 1975.

12. Castro, G., Haarer, D., Macfarlane, R. M., and Trommsdorff, H. D., "Frequency selective optical data storage system," U.S. Patent No. 4101976. 1978.

13. Rebane, K. K., "Laser study of inhomogeneous spectra of molecules in solids," in Proceedings of the International Conference on Lasers '82 (STS, McLean, Va., 1982), pp. 340-345.

14. Szabo, A., in Proceedings of the International Conference on Lasers '82 (STS, McLean, Va., 1980), pp. 374-379.

15. Mossberg, T. W., "Time-domain frequency-selective optical data storage," Opt. Lett., Vol. 7, pp. 77-79. 1982.

16. Hesselink, W. H. and Wiersma, D. A., "Picosecond photon echoes stimulated from an accumulated grating," Phys. Rev. Lett., Vol. 43, pp. 1991-1994. 1979.

17. Hesselink, W. H. and Wiersma, D. A., "Photon echoes, stimulated from an accumulated grating: theory of generation and detection," J. Chem. Phys., Vol. 75, pp. 4192-4197. 1981; de Vries, H. and Wiersma, D. A., "Numerical simulations of accumulated stimulated photon echoes," J. Chem. Phys., Vol. 80, pp. 657-666. 1984.

18. Saari, P. and Rebane, A., "Time- and space-domain holography of pulsed light fields in spectrally selective photoactive medium," Proc. Acad. Sci. Estonian SSR, Vol. 33, pp. 322-332. 1984.

19. Saari, P. M., Kaarli, R. K., and Rebane, A. K., "Holography of spatial-temporal events," Kvantovaya Elektron., Vol. 12, pp. 672-682. 1985.

20. Shtyrkov, E. I. and Samartsev, V. V., "Dynamic holograms on the superposition states of atoms," Phys. Status Solidi, Vol. A45, pp. 647-655. 1978.

21. Denisyuk, Yu. N., "Holography and its prospects," in Problems in Optical Holography, Yu. N. Denisyuk, ed. (Nauka, Leningrad, 1981), pp. 7-27.

22. Freiberg, A. and Saari, P., "Picosecond spectrochronography," IEEE J. Quantum Electron., Vol. QE-19, pp. 622-629. 1983.

23. Kikas, J., Kaarli, R., Rebane, A., "Multifrequency photochemical holeburning in impurity spectra studied by time-domain detection," Proc. Acad. Sci. Estonian SSR, Vol. 33, pp. 124-127. 1984.

An experimental study of the holographic technique applied to supersonic cascade wind-tunnel

Li Jianyi, Yang Jun, Wang Wuyi, Song Yaozu

Institute of Engineering Thermophysics, Academia Sinica
P.O. Box 2706, Beijing 100080, P.R. CHINA

Abstract

This is a recent technical report of the optical holographic technique applied to the supersonic cascade wind-tunnel. The experimental methods, techniques and results are presented.

Introduction

The supersonic cascade wind-tunnel is an important facility to study the aero-thermodynamics of turbomachinery. The flow parameters are usually measured by means of probes from point to point. The accuracy is reduced by the interference of probes with the flow. At some places the measurement data can not be obtained because these places are not accessible for the probes. Normally, optical schlieeren and interference methods can measure the flow field without any interfere but in our tunnel we can use plexiglass window only we must have a lot of windows for various cascades, the area of the optical window of the test section is quite large, and the complekity and high cost of the processing of the window, the application of normal optical interference methods on the experimental study of wind tunnel to be restricted by the poor conditions listed above. However, the holograms of the supersonic cascade wind-tunnel flow field can be obtained clearly by means of the optical holographic technique.[1]

The contour of density, gradient of density and second derivative of density inside every cascade channel are obtained from the holograms of the flow field and then by means of the information the related parameters of the flow field, e.g. density, temperature, pressure etc., are deduced from physical states of the flow field. Therefore, the application of optical holographic technique on the experimental study of aero-thermodynamics of turbomachinery has a bright future.

Principle and device of the experiment

Fig1. shows the principle of the experiment of optical holographic technique applied to the supersonic cascade wind-tunnel and the installation. The supersonic cascade wind-tunnel to be tested is designed by our institute, the air supply is intermittent, test section dimension: width 220 mm, height 290 mm, nozzle Mach number 1.3, 1.5 etc., Re number $2x10^6$-$3x10^6$, every cascade has 7-9 blades[2]. There is no density change in beam direction (test section width direction). Fig.2.shows that the indices of refraction of the test flow field are two-dimensional distribution n (x.y).

A 22HD double-pulse ruby laser from Apollo Laser Company U.S.A. was used as a light source for the holography. It has an output energy of 0.68 J, the wavelength is 6943Å, the pulse separation is 1-500 μs, the pulse width is 20 ns, the coherence length is more than 2 meters. A He-Ne laser with an output power of 10 mw at 6328Å was used as an adjustment light source. This holographic system based on the Schlieren system with the testing field of 500 mm in diameter. After the laser passes through beam splitter S, the reflection beam is used as the reference beam, and the transmited beam is used as the object beam. After the object beam passes through test section, reflected by the mirror M_4. The testing section is focused on the recording plane H by lens L_2. In order to make the optical path length of reference beam and object beam approxiamtely equal, the reference beam must be reflected several times. After the lens L_4 the collimate reference beam is formed and overlapped with object beam at 30^0 angles on H-plane so the image hologram is obtained. The holographic optical system is fixed on a shock proofed table in order to enhance the quality of hologram. By applying different holographic techniques, different holograms of the flow field can be obtained by this holographic optical system. Different information of the flow field can also be obtained. At present, we adopt two methods:

(1) Single exposure holographic method

Exposure is taken after wind-tunnel started and flow stablized to obtain the single exposure hologram of flow field.

(2) Double exposure holographic method

The first is taken when the wind-tunnel is running at the stabilized flow, the second exposure is taken immediately after the wind-tunnel shut down. The infinite holographic interferogram could be obtained from this hologram.

To turn the mirror M_9 in a very small angles δ between two exposures. The finite fringe holographic interferogram of flow field could be obtained.

Experimental technique

In order to obtain a satisfactory hologram. We must not only design a reasonable optical path, but also solve technical problems which may influence the quality of hologram. In experiments three factors which influence the quality of the holograms are studied:

(1) The problem to obtain equal-path of object beam and reference beam

In order to guarantee the qualily of hologram. When construct the holographic system, the most important thing is to make the optical path of object beam and reference beam to be equal. To achieve that, first set some reflect mirrors, let the reference beam reflects several times let object beam and reference beam have the same geometric path. Second, because of π phase jump of light waves and change in optically thinner-optically denser medium boundary surface, so in the optical system, the number of reflection of the object beam and reference beam have to be the same, or the difference between them have to be an even number. The clarity of hologram can be raised in this way.

(2) The choice of the energy ratio of the object and reference beam

The visibility V of the fringes defined by Michelson when $I_1=I_2$ [3], a maximum value V=1 indicates perfect coherence. But in reality, because of film works in the non-linear region, produced non-linear distortion results in high-order diffracted image. So the required first order diffracted efficiency become lower. Experimental study shows that under our experimental condition, when use Agfa-10E75 film, clear hologram can be obtained by choosing the strength ratio of these two beams in the range of 1:5 to 1:8. So we draw the conclusion that suitable choice of energy proportion of object and reference beam according to the special experimental environment is a key factor to improve image quality and reconstruct efficiency.

(3) Adjustment of polarization effect

Polarization effect is one of the key factors to get the best visibility of the hologram. In our experimental installation, because of irregular polarization function of the plexiglass window, after object beam passes the window. The space distributions of polarization dizection are not consistent. Examine with polarization plate, unequal bright and dark variation appear at beam cross section, so on hologram some parts have high diffracted efficiency, some parts have low diffracted efficiency. To overcome polarization effect on hologram, besides choosing the high quality material for the window, fix a quarter wavelength plate in the exit laser beam can make object and reference beams to be round polarization beams.

Analysis of experimental results

Using different methods to obtain flow field holograms including different field information, adopt different reconstruction method the distribution of density, density gradient, and second decivative of density in the flow field can be obtained separately.

(1) Holographic schlieren and shadowgraph reconstruction

By reconstruction system shown in Fig3. shadowgraph can be obtained from single exposure hologram. It shows second derivative of density in the flow field. Fix a knife edge in the focus plane of reconstrution system, schlieren image can be obtained. It shows the gradient of density in the flow field. By which we can get the information about shock structure and location, this result are the same as the result of common schlieren method. But the merit of holography is to obtain shadowgraph and schlieren pictures in the same optical system at the same time.

(2) Holographic interferogram reconstruction

By reconstruction system shown in Fig4. infinite fringe interferogram from double exposure hologram can be obtained as shown in Fig5. Equal density distribution in the flow field can be obtained as shown in Fig6. Using the reconstruct system shown in Fig4 to reconstruct finite fringe interferogram, the finite fringe interferogram shown in Fig7 can be obtained, it provides experimental basis for the quantitative analysis of the flow field.

Certainly, for quantitative analyse of hologram, in order to obtain other parameters of flow field, still rely on further study.

In summary, experimental study shows optical holography is an important method to study flow phenomenon inside supersonic cascade wind-tunnel. It will provide meaningful experimental data for theoretical study of the aero-thermodynamics of turbomachinery.

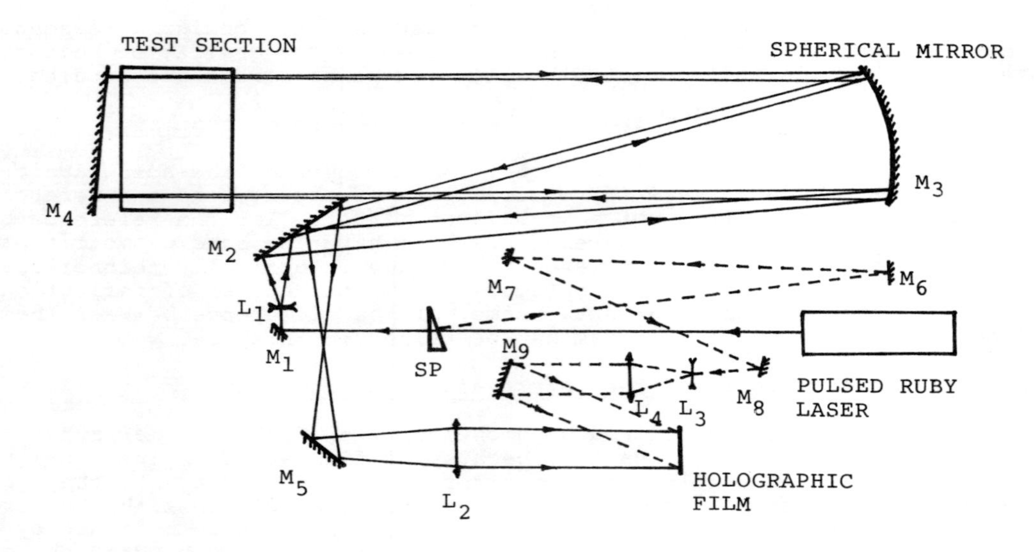

Fig.1. Holographic System.

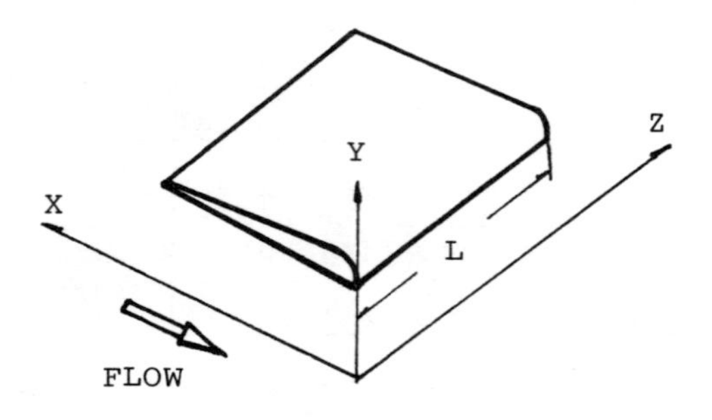

Fig.2. Principle of the experiment.

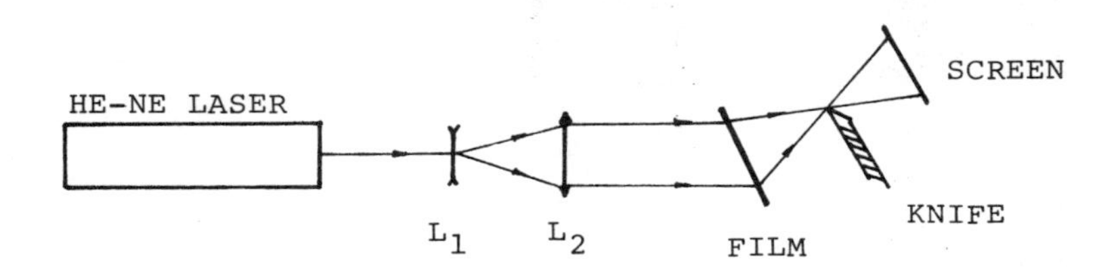

Fig 3. Reconstruction system(a).

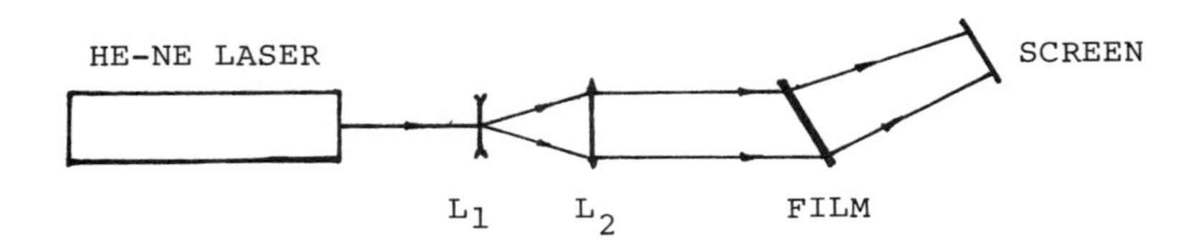

Fig.4. Reconstruction system (b).

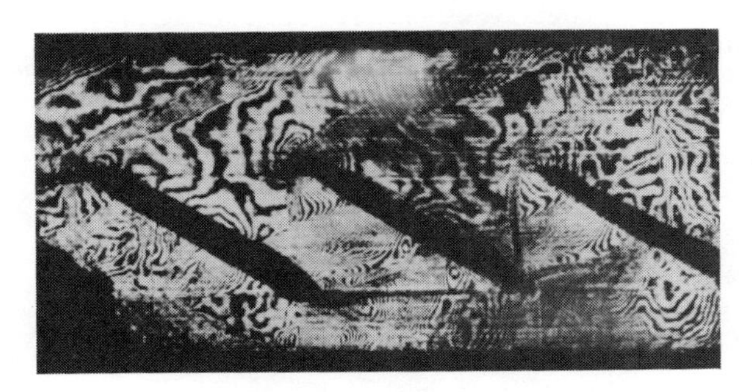

Fig.5. Infinite fringe interferogrom.

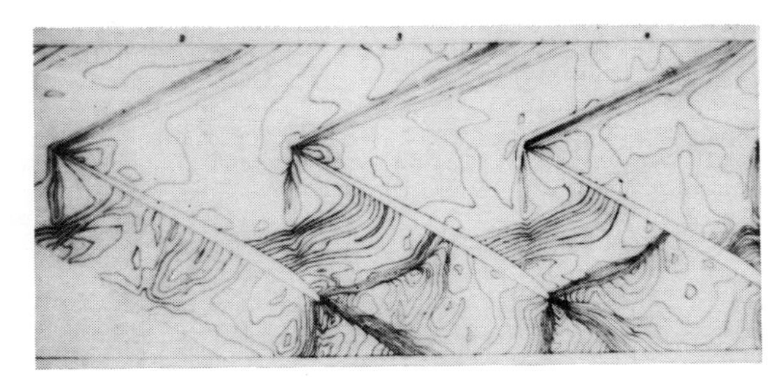

Fig.6. Equal density distribution.

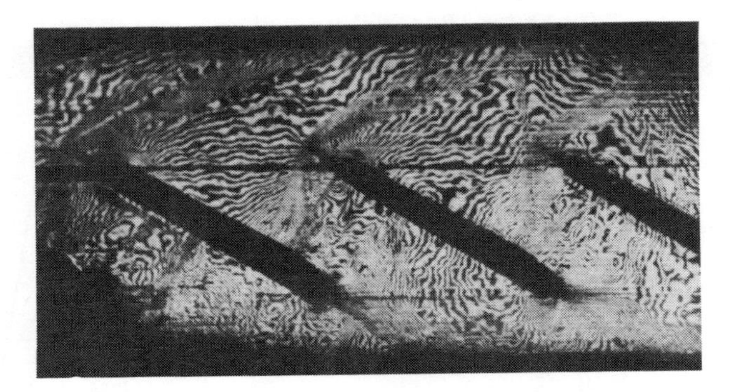

Fig.7. Finite fringe interferogram.

References

1. C.M. Vest, Holographic Interferometry, John wiley and Sons, New York 1979.
2. Yu Shen, The Design of a Supersonic Cascade Wind-Tunnel, 1984, 5.
3. R.J. Collier, C.B. Burckhardt, L.H. Lin, Optical Holography, New York and London 1971.
4. W. Merzkirch, Flow Visualization, Academic Press New York 1974.

Spacial Moiré hologram

Huang Wei-shi, Guo Jia-rong, Zhou Liang, Sun Qi-yue

Department of Applied Physics, Shanghai Jiao Tong University, PRC

Abstract

An interesting spacial moiré hologram and its reconstruction were recommended. Some properties of reconstruction moiré and its prospect of application were also discussed.

A spacial moire hologram is given in this paper. The object of the hologram is two groups of transparant and opaque periodical or quasi-periodical grating, grid, Fresnal zone plate and so on. In order to form different kinds of spacial moire patterns, a certain spacing between two gratings is needed. Besides, they would be parallel, tilted or twist to form different structures of spacial moiré. When reconstruction of the hologram, in the modulated optical field, there are not only the original periodical or quasi-periodical structure, but also the spacial bit frequency pattern ---- the spacial moiré fringe can be obtained.

It is well known that holographic images usually have parallax effect enhanced by the increasing of the spacing in a scene; and moire pattern obtained by superposition of periodical structures has translation amplifying effect too. Now, if these two well-known properties are synthesized by the spacial moiré hologram, then an interesting parallax amplification effect will be available. While reconstructing the hologram, the position and configurations of the moire fringe will change with the observer's position. When constructing, if reasonablely arrange the gratings each other to make a variety of shapes and orientations of moiré pattern, then we will find a very interesting spacial moire hologram image during observing. This kind of hologram can be used for ornament.

The principles of spacial moire holograms are as follows:

1. The formation of moiré pattern

If a grating I is placed in close contact with a grating II (if there is a separation between them, it can be considered that the projection of the front grating I is in contact with the grating II), a moire pattern is formed. The mathematical equations of it can be obtained by solving the n order No. equation of the family of curves of the grating I and m order No. equation of the family of curves of the grating II simulaneously, i.e. the trace equation N=m-n of the intersecting parts of these two families of curves represents the moiré pattern. The simplest situation of moire pattern is the superposition between two line gratings, which are configurated in such a way as shown in the figures.

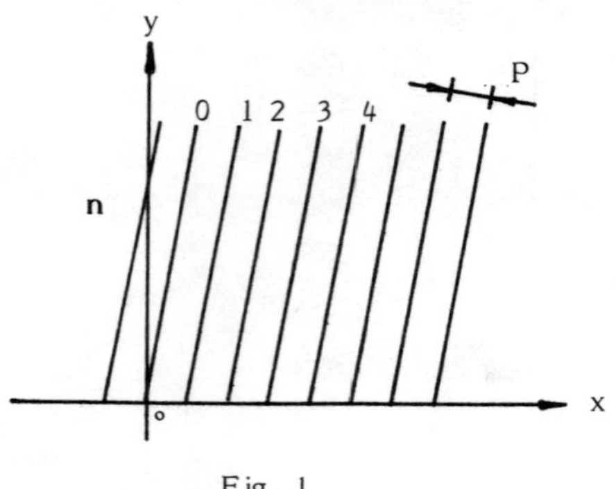

Fig. 1

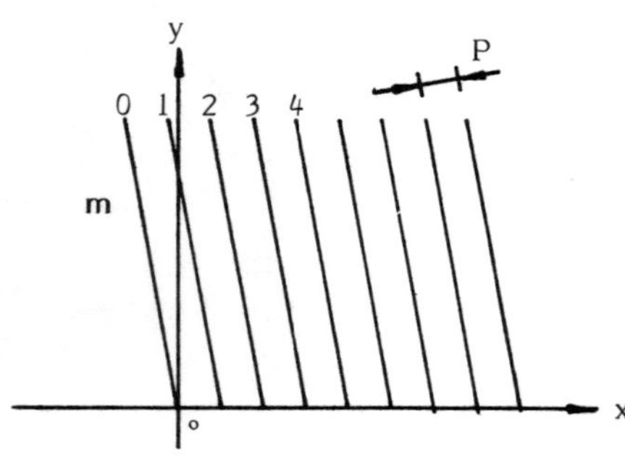

Fig. 2

The order No. equations of the two gratings and the moiré order No. equation respectively are as follows:

$$n = \frac{x\cos\theta - y\sin\theta}{p_1} \qquad (1)$$

$$m = \frac{x\cos\theta + y\sin\theta}{p_2} \qquad (2)$$

$$N = n - m = \left(\frac{1}{p_1} - \frac{1}{p_2}\right) x \cos\theta - \left(\frac{1}{p_1} + \frac{1}{p_2}\right) y \sin\theta \tag{3}$$

$$y = \frac{p_2 - p_1}{p_1 + p_2} x \operatorname{ctg}\theta - N \frac{p_1 p_2}{(p_1 + p_2)\sin\theta} \tag{4}$$

From the equation (4), we can see that the moire pattern is a line family, whose slope is $\frac{p_2 - p_1}{p_2 + p_1} \operatorname{ctg}\theta$ and whose intercepts are $\frac{N p_1 p_2}{(p_1 + p_2)\sin\theta}$ where N=0, ±1, ±2, ······.

2. The movement of moire patterns

The moiré hologram is constructed from two gratings with certain space configuration and the moiré pattern is the trace of intersecting points of them, so due to the parallax effect of the hologram, we can see that the positions of the grating images will change with the position of observation, consequently, the moiré patterns will move. The rate of change is inversely proportional to the grating spacial period and directly proportional to the spacing between the two gratings.

After all, spacial moire holograms combine the characteristics of moiré effect and holography. In applying the embossing techniques, one can construct a surface distribution of the moire effect, instead of a space one, and it can become one of the ornamental articles for commercial or family use. In the other hand, it is also possible to use it in measuring.

References

1. Huang Wei-shi, Grating and Moiré Fringe Techniques, (to be published).
2. Yu Mei-wen, Optical Holography and Information Processing, The Industries of National Defence Press 1984.
3. Huang Wei-shi et al., Journal of Zhejiang University, No.3, pp. 128-137. 1980.

A New Method of Colour Holography

Kaveh Bazargan

Optics Section, Blackett Laboratory, Imperial College of Science & Technology, London SW7 2BZ

Abstract

A new method of producing multicolour holograms is proposed. The method is based on the technique of dispersion compensation, and exploits the wavelength selectivity of volume holograms. The colour saturation of the resultant image increases with increasing wavelength selectivity of the hologram.

Introduction

Multicolour holograms have been produced using different recording techniques[1,2]. In our review of colour holography[1] we briefly pointed to our work in a new technique. In this report we present the details, starting with a brief review of dispersion compensation.

Dispersion Compensation

The elegant method of dispersion compensation was first proposed in 1966[3,4,5]. Figs. 1 & 2 illustrate the method: In Figure 1a, a simple transmission hologram of a point object is recorded using a wavelength near the centre of the visible spectrum, eg. green. When the resultant hologram is illuminated with a reconstruction beam identical to the original reference beam (Figure 1b), an undistorted and unaberrated image is observed. When white light illumination is used with the same reconstruction geometry (Figure 2a), a laterally and longitudinally dispersed image is observed. The lateral dispersion can be corrected by 'predispersing' the reconstruction beam using a diffraction grating (Figure 2b). The fringe spatial frequency of the compensating grating must be equal to the average spatial frequency of the hologram. The final image is sharp and achromatic, even though longitudinal dispersion is present. It can be shown that the distortions and the aberrations in the component images are also significantly reduced by dispersion compensation[6].

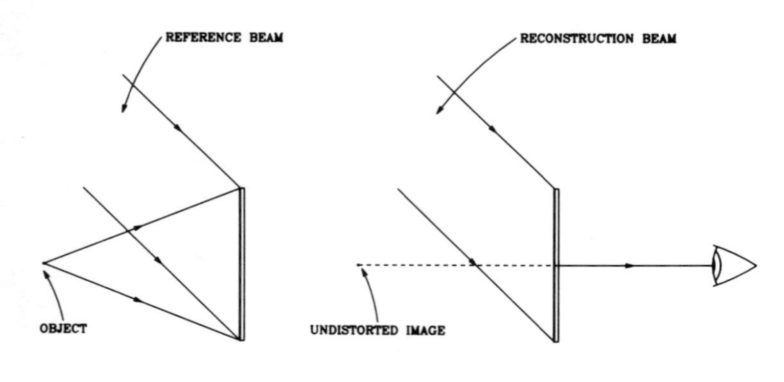

Figure 1. Reconstruction of a sharp, undistorted image by a coherent reconstruction beam.

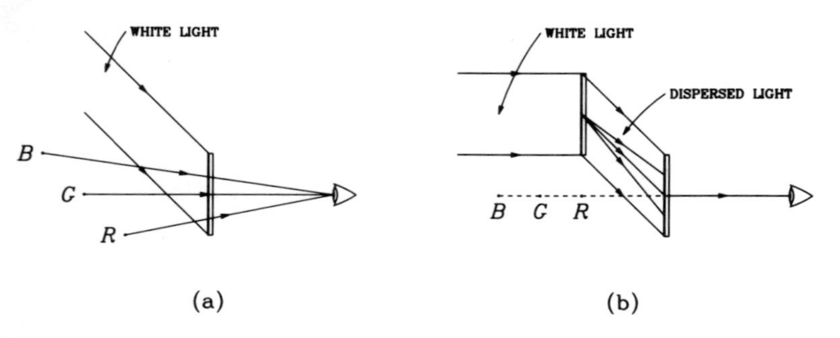

Figure 2. Image reconstruction by a polychromatic reconstruction beam (a) without and (b) with a compensating grating.

Burckhardt[4] proposed the use of a miniature Venetian blind structure between the two diffracting elements (Figure 3) so that the two elements can be placed close to one another without the undiffracted portion of the reconstructing light reaching the observer. (A suitable miniature Venetian blind structure for the method is "Light Cotrol Film," available from: Industrial Optics Division, 3M Center, St. Paul, Minnesota 55144, USA.) We have proposed a compact hologram viewing device[7,6] based on this method (Figure 4).

Multicolour Holograms

In the above description, we have assumed that the diffraction grating and the hologram have negligible wavelenth selectivity, and therefore the image observed is achromatic. Suppose now that the recording medium has a significant thickness (Figure 5), and that a green recording wavelength is again used. When the dispersion-compensated image is reconstructed using a thin compensating grating, the hologram selectively diffracts a portion of the incident light. Assuming that the thickness and the average refractive index of the hologram are unchanged by processing, then the image observed has a green tint, the saturation depending on the wavelength selectivity of the hologram, in this case green (Figure 6). A multicolour image is obtained by superimposing several such holograms each recorded with a different wavelength. To ensure that the resultant component images are superposed, it is necessary to adjust the reference angle in each case such that the average spatial frequencies of the holograms are equal, and equal to that of the compensating grating (Figure 7). In other words, the reference angle in each case must be equal to the angle at which that wavelength emerges from the compensating grating. The component holograms may either be recorded on the same emulsion or on several adjacent emulsions.

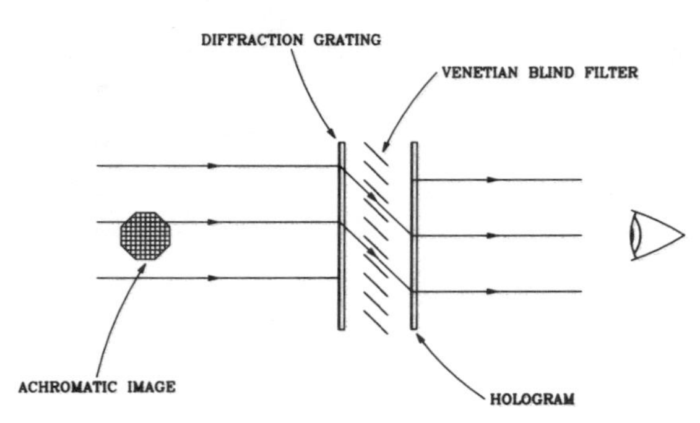

Figure 3. Prevention of the undiffracted reconstruction beam reaching the observer by using a 'Venetian blind' structure.

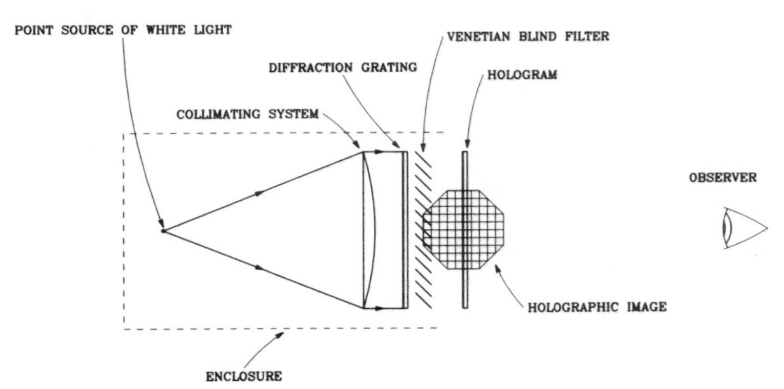

Figure 4. A compact hologram viewing device based on dispersion compensation.

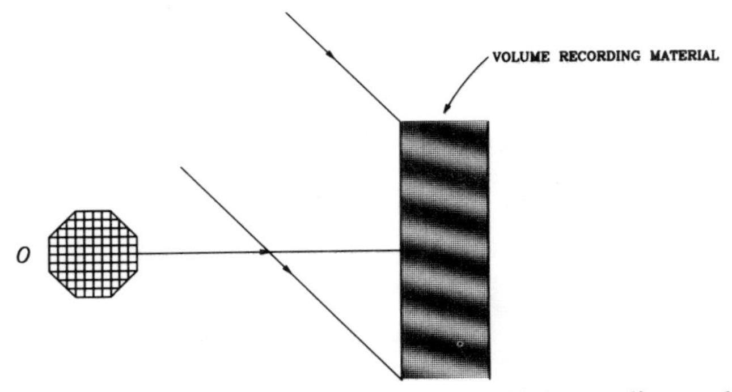

Figure 5. Recording a holographic image in a thick recording medium. The *average* orientation of the fringes is shown.

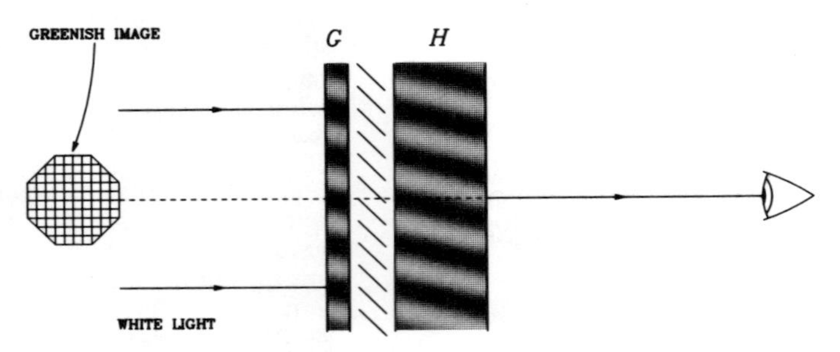

Figure 6. Viewing the image produced in a dispersion-compensated system comprising a thin grating and a thick hologram. A non-achromatic image is observed.

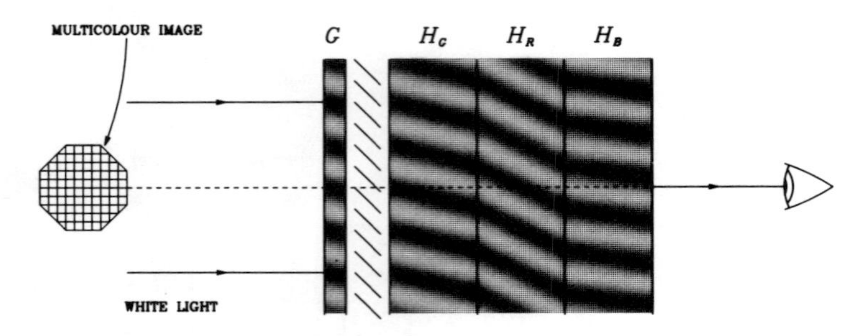

Figure 7. A dispersion-compensated multicolour image produced by superimposing several monochromatic images.

We have produced multicolour images using this technique[6]. The colour saturation of the images obtained is low, but can be increased by using a thicker recording material, or a higher refractive index modulation in the recording material. Holograms made with this method are compatible with the compact viewing device proposed.

An interesting property of this type of hologram is that, unlike other multicolour techniques, spurious images are not produced when the wavelength selectivity of the component holograms decreases - light of any wavelength diffracting from any component hologram produces an image superimposed on other images so that the effect of, say, red light diffracting from the 'green' hologram is simply to reduce the colour saturation of the composite image.

Conclusions

We have proposed a new method of producing multicolour display holograms. In conjunction with the compact hologram viewing device previously proposed by the author[7,6], the method may facilitate comfortable viewing of high quality multicolour images.

References

1. Bazargan, K., "Review of Colour Holography," Proc. SPIE (Optics in Entertainment), Vol. 391, pp. 11-18. 1983.
2. Hariharan, P., "Colour Holography," Progress in Optics, Vol. XX, E. Wolf, ed., North Holland, 1983, pp. 265-324.
3. Paques, H., "Achromatization of Holograms," Proc. IEEE, Vol. 54, pp. 1195-1196. 1966.
4. Burckhardt, C. B., "Display of Holograms in White Light," Bell Sys. Tech. J., Vol. 45, pp. 1841-1844. 1966.
5. De Bitetto, D. J., "white-Light Viewing of Surface Holograms by Simple Dispersion Compensation," Appl. Phys. Lett., Vol. 9, pp. 417-418. 1966.
6. Bazargan, K., "Techniques in Display Holography," PhD Dissertation, University of London, 1986.
7. Bazargan, K., "A Hologram Viewing Device," British Patent 8303465, US Patent 465620. Filed 1982.

Investigation of strains of objects by means of sandwich holographic interferometry

Linli Wang , Caifu Zhao , Jingtang Ke

Holographic Interferometry Lab., Zhengzhou Institute of Technology
10 Wenhua Road, Zhengzhou, Henan, People's Republic of China

Abstract

In this paper, the authors derived general formulas for computation of in-plane displacements and stresses resulted from which, measurements being carried out by sandwich holographic interferometry. Computation of stresses due to in-plane deformations by means of derivatives of second order of contours of equal displacements, or second-order derivatives of the fringes, are also presented.

Introduction

Holographic interferometry is an important means of measuring small displacements and deformations, methods most commonly used and more or less sophisticated are double-exposure and real-time holographic interferometry, among others. The method of double exposure is to record on a single holographic plate the changes which take place at an object between two exposures. The reconstruction of the hologram is instrumental to attainment of interference fringes brought about by deformation of the object. The fringes are known as "frozen", as against the "moving" fringes obtained at the surface of an object through real-time holographic interferometry, the image of the object on the hologram prior to deformation of the object is compared with that after deformation. The latter facilitates observation of the condition of the object undergoing the change. Sandwich holographic interferometry was first presented by Nils Abramson in 1974 [1, 4]. He also made studies on in-plane displacements of a rigid body [5].

Besides having the advantages of both double-exposure and real-time method, sandwich holographic interferometry is functional in eliminating rigid body displacement of an object, so that the interference fringes on the surface of an object thus obtained are due to its deformation between the two exposures. Furthermore, the method is easy to use and is instrumental to direct determination of the direction of deformation and displacement of an object. For holographic interferometry in general, in order to obtain deformation of the object in question, and the stresses produced in it due to deformation, it is necessary to attain fringes resulting from mutual interference of wave front of light from the object before and after its being deformed, or isopleths of displacements, from which to determine the derivatives. More or less accurate results are therefore obtainable only in case of densely spaced interference fringes. To get results closely agreeing with actual deformation of an object, sandwich holographic interferometry can be applied without requiring very dense fringes, thanks to the basic principles of computing displacements and strains of object inherent to sandwich holographic interferometry.

Principles

The optical scheme for sandwich holographic interferometry is shown in Fig.1. The light source is He - Ne laser. Process of testing is shown in Fig.2. Information related to the object prior to deformation is recorded on two pieces of plates on the sandwich holographic plate holder, as seen in Fig.2(a). Information of the object subsequent to deformation is recorded on two other plates, as shown in Fig.2(b). The second plate subject to first exposure and first plate subject to second exposure are processed and superimposed on one another before being placed on a particular holder [1], to be reconstructed by the original reference beam to get sandwich holographic interfergram. For the case of simple deformation, as seen in Fig.2(c), interference fringes covering the surface of the object may be partly or altogether removed through turning the holder, so as to obtain displacements and deformations of the object. In general, turning of the plate holder will bring about changes in fringe pattern covering the object, by which the fringes due to rigid body displacement of the object may be eliminated and the direction of deformation directly determined, as shown in Fig.2(d).

If the in-plane displacement of an object is very small, the magnitude of planar displacement may be directly measured by means of sandwich holographic interferometry. In case of slight displacement of the object Δx along the x-axis between the two exposures, as shown in Fig.2, reconstruction of sandwich holographic interfergram by means of the original reference beam is instrumental to bringing forth fringes parallel to the y-axis on the object observed, as shown in Fig.3.

Turn the sandwich holographic plate holder in clockwise direction around the y-axis through an angle of ϕ_2, the parallel fringes will vanish when viewed in the positive direction of the y-axis. From geometry and the law of refraction, we have

$$\mathrm{tg}\,\eta = 2\cdot\Delta x/d$$
$$n.\sin\eta = \sin\phi_2 \qquad (1)$$

hence

$$\sin\eta = (\sin\phi_2)/n$$

but

$$\mathrm{tg}\,\eta = \sin\eta/(1 - \sin^2\eta)^{\frac{1}{2}}$$

so that

$$\Delta x = 0.5\, d \sin\phi_2/(n^2 - \sin^2\phi_2)^{\frac{1}{2}} \qquad (2)$$

in which n and d --- respectively index of refraction and thickness of plate glass;
η --- angle included between straight lines PP_3 and P_3P_4, as shown in Fig.2.

Similarly, in case of slight in-plane displacement Δy of the object along the y-axis, the sandwich holographic interfergram being turned around the x-axis and fringes parallel to x-axis will vanish, thus we have:

$$y = 0.5\, d \sin\psi_2/(n^2 - \sin^2\psi_2)^{\frac{1}{2}} \qquad (3)$$

in which ψ_2 --- angle through which the sandwich holographic interfergram is rotated around the x-axis.

If the object observed has in-plane displacement Δx along the x-axis, and the holder is rotated around the x-axis, interference fringes on the object cannot be eliminated, but oblique fringes will appear and there will also be fringes parallel to the x-axis on the reference object, as seen in Fig.4. Here we have

$$q = (\lambda/2).(2Ln)/(\psi_2 d) \qquad (4)$$

$$\Delta x = q.\mathrm{tg}\,\gamma \qquad (5)$$

in which λ --- wave length of laser;
L --- distance from the object to the holder;
ψ_2 --- angle through which the holder is rotated around the x-axis;
γ --- angle included between interference fringes and the x-axis after turning the holder;
q --- spacing between fringes.

From reference [1] we know

$$\phi_1 = \frac{1}{2}\,\mathrm{arc\,tg}\,\frac{d}{L\left[(n/\sin\phi_2)^2 - 1\right]^{\frac{1}{2}}}$$

or

$$L.\mathrm{tg}\,2\phi_1 = \frac{d}{\left[(n/\sin\phi_2)^2 - 1\right]^{\frac{1}{2}}}$$

Inserting Eq.(2) into the above expression to get:

$$\Delta x = \frac{1}{2}\, L\, \mathrm{tg}\, 2\phi_1 \qquad (6)$$

From reference [2] we know

$$\phi_1 = (\lambda/2q).\mathrm{tg}\,\gamma$$

Inserting Eq.(4) and the above expression into Eq.(6), we have:

$$\Delta x = \frac{1}{2}\, L\, \mathrm{tg}\,(d/Ln)\psi_2.\mathrm{tg}\,\gamma \qquad (7)$$

Knowing the thickness and index of refraction of the plate glass, distance of the object from the plate holder, the angle included between the interference fringes and the x-axis, and the angle through which the holder is turned, the in-plane displacement Δx can be determined.

Similarly, with fringes parallel to the y-axis on the object observed, as the holder is turned around the y-axis, we have:

$$\Delta y = \tfrac{1}{2} L \, tg(d/Ln) \, \phi_2 \cdot tg \, \gamma' \tag{8}$$

in which Δy --- planar displacement of object along y-axis;
γ' --- angle included between oblique fringes and y-axis;
ϕ_2 --- angle through which the holder is turned around y-axis.

Let the displacement of the object along the x-axis be u, and that along the y-axis be v, then

$$u = \tfrac{1}{2} L \cdot tg \, (d/Ln) \psi_2 \cdot tg \, \gamma \tag{9}$$

$$v = \tfrac{1}{2} L \cdot tg \, (d/Ln) \, \phi_2 \cdot tg \, \gamma' \tag{10}$$

then

$$\frac{\partial u}{\partial x} = \tfrac{1}{2} L \cdot tg \, (d/Ln) \psi_2 \cdot \frac{\partial(tg\gamma)}{\partial x} \tag{11}$$

By numerical method of gradual approach, we then get the approximate functions of interference fringes represented by $f_1(x)$, $f_2(x)$ ... $f_n(x)$, so that

$$\partial(tg \, \gamma_1)/\partial x = f_1''(x), \; \ldots , \; \partial(tg \, \gamma_n)/\partial x = f_n''(x) \tag{12}$$

When the holder is turned around the y-axis, the approximate functions of interference fringes obtained by using numerical method of gradual approach are $g_1(y)$, $g_2(y)$, ... $g_n(y)$, and we have:

$$\frac{\partial v}{\partial y} = \tfrac{1}{2} L \cdot tg \, (d/Ln) \, \phi_2 \cdot \frac{\partial(tg\gamma')}{\partial y} \tag{13}$$

$$\partial(tg \, \gamma'_1)/\partial y = g_1''(y), \; \ldots , \; \partial(tg \, \gamma'_n)/\partial y = g_n''(y) \tag{14}$$

and the values of normal stresses at corresponding points on the object are, from the theory of elasticity, as follows:

$$\sigma_{x_i} = \frac{E}{1-\mu^2} L \cdot tg \left[(d/Ln)\right] \left[\psi_2 f_i''(x) + \mu \, \phi_2 \, g_i''(y)\right] \tag{15}$$

$$\sigma_{y_i} = \frac{E}{1-\mu^2} L \cdot tg \left[(d/Ln)\right] \left[\phi_2 g_i''(y) + \mu \psi_2 \, f_i''(x)\right] \tag{16}$$

Let the inverse functions of $f_i(x)$ and $g_i(y)$ be respectively $\tilde{f}_i(y)$ and $\tilde{g}_i(x)$, then

$$\frac{\partial u}{\partial y} = - L \; tg \left[(d/Ln)\right] \phi_2 \cdot \frac{\tilde{f}_i''(y)}{\left[\tilde{f}_i'(y)\right]^2} \tag{17}$$

$$\frac{\partial v}{\partial x} = - L \; tg \left[(d/Ln)\right] \phi_2 \cdot \frac{\tilde{g}_i''(x)}{\left[\tilde{g}_i'(x)\right]^2} \tag{18}$$

From the theory of elasticity, shearing stresses at corresponding points on the object is:

$$\tau_{xy} = \frac{-E}{2(1+\mu)} L \cdot tg \left[(d/Ln)\right] \left\{ \phi_2 \frac{\tilde{f}_i''(y)}{\left[\tilde{f}_i'(y)\right]^2} + \psi_2 \frac{\tilde{g}_i''(x)}{\left[\tilde{g}_i'(x)\right]^2} \right\} \tag{19}$$

Hence, so long as approximate functions of fringe curves are acquired by means of numerical method of gradual approach, and the values of angle through which the plate holder turns is known, stresses in the object examined may be determined by using Eq.(15),(16) and

(19).

For the case of flexure of thin plate, we have from the theory of elasticity:

$$\sigma_x = (12\ M_x/t^3)z \tag{20}$$

$$\sigma_y = (12\ M_y/t^3)z \tag{21}$$

in which σ_x and σ_y are respectively normal stresses along the x-axis and y-axis, t is the thickness of the plate, as shown in Fig.5, and M_x and M_y are respectively bending moment along the x-axis and y-axis.

$$M_x = -D\ (\partial^2 w/\partial x^2 + \mu\ \partial^2 w/\partial y^2)$$

$$= D\ (\frac{1}{r_1} + \mu \frac{1}{r_1'}) \tag{22}$$

$$M_y = -D\ (\partial^2 w/\partial y^2 + \mu\ \partial^2 w/\partial x^2)$$

$$= D\ (\frac{1}{r_1'} + \mu \frac{1}{r_1'}) \tag{23}$$

in which w --- deflection;
r_1, r_1' ---respectively radius of curvature of the thin plate under flexure, in the plane XOZ and YOZ;
μ --- Poisson ratio;
D --- stiffness of thin plate under flexure.

Insert values of r_1 and r_1' obtained by referring to [2] into Eq.(22),(23) and (20), (21) for direct determination of stresses in the thin plate. In the case of short beam restrained at the shorter edges only, stresses and strains may be determined by using the following formulas [2], as deformation in the direction of short edges is very small.

$$M = (d/2Ln)(\psi_2 EI/r_2) \tag{24}$$

$$\varepsilon = (Mt)/(2EI)$$

$$= (d/2Ln)(\psi_2 \cdot t/2)(1/r_2) \tag{25}$$

$$\sigma = (d/2Ln)(\psi_2 E \cdot t/2)(1/r_2) \tag{26}$$

in which M, ε, σ --- respectively the bending moment, strain and stress;
E --- modulus of elasticity;
I --- moment of inertia of the cross section;
ψ_2 --- angle through which the plate holder is turned;
r_2 --- radius of curvature of interference fringe when the holder is turned through an angle of ψ_2.

Sandwich holographic interfergram for circular plate fixed along the periphery and subject to concentrated load at the center is shown in Fig.6(a). By turning the plate holder around the y-axis and x-axis, the interfergrams become those shown in Fig.6(b) and (c).

Experiments and results

The specimen used in the test is a circular plate of diameter d = 35 mm and thickness t= 8 mm. The interference fringes when the plate is subject to diametrically opposite compressive forces are shown in Fig.7.

Theoretical values for the case of circular plate subject to diametrically opposite compressive forces:

As shown in Fig.8, stresses along x-axis and y-axis at any point N on the chord ML are:

$$\sigma_x = -\frac{2P}{\pi}\left(\frac{\sin^2\theta\cos\theta}{r} + \frac{\sin^2\theta_1\cos\theta_1}{r_1}\right) + \frac{2P}{\pi d}$$

$$\sigma_y = -\frac{2P}{\pi}\left(\frac{\cos^3\theta}{r} + \frac{\cos^3\theta_1}{r_1}\right) + \frac{2P}{\pi d}$$

$$\tau_{xy} = -\frac{2P}{\pi}\left(\frac{\sin\theta\cos^2\theta}{r} + \frac{\sin\theta_1\cos^2\theta_1}{r_1}\right)$$

and the strains are:

$$\varepsilon_x = -\frac{2P}{\pi E}\left[\left(\frac{\sin^2\theta\cos\theta}{r} + \frac{\sin^2\theta_1\cos\theta_1}{r_1}\right) - \mu\left(\frac{\cos^3\theta}{r} - \frac{\cos^3\theta_1}{r_1}\right) - \frac{1}{d}(1-\mu)\right]$$

$$\varepsilon_y = -\frac{2P}{\pi E}\left[\left(\frac{\cos^3\theta}{r} + \frac{\cos^3\theta_1}{r_1}\right) - \mu\left(\frac{\sin^2\theta\cos\theta}{r} + \frac{\sin^2\theta_1\cos\theta_1}{r_1}\right) - \frac{1}{d}(1-\mu)\right]$$

The theoretical values of strains along y-axis at all points on the chord ML and the approximate functions of interference fringes obtained by using numerical method of gradual approach are inserted into the computational formula (13) for sandwich holographic interferometry, to get the corresponding experimental values of strains, as shown in Fig.9.

A comparison of experimental results and theoretical values will reveal that variation of strains does agree closely with each other, errors being very small.

Conclusions

Experimental results show that precision of approximate functions of fringes directly affects precision of stresses and strains determined. In actual measurements, therefore, the most important problem is to choose an appropriate approximate function, so that experimental results may reflect the actual condition as accurately as possible.

References

1. Abramson,N., Appl.Opt., 13, 1974, 2019.
2. ditto, 14, 1975, 981.
3. ditto, 15, 1976, 200.
4. ditto, 16, 1977, 2521.
5. ditto, 18, 1979, 2870.

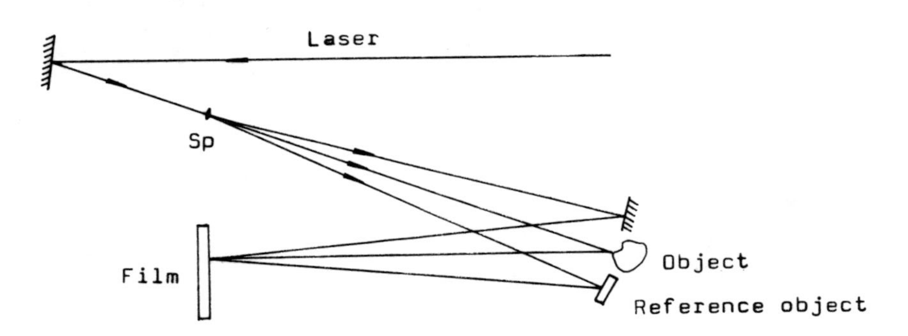

Fig.1 Optical scheme for sandwich holographic interferometry

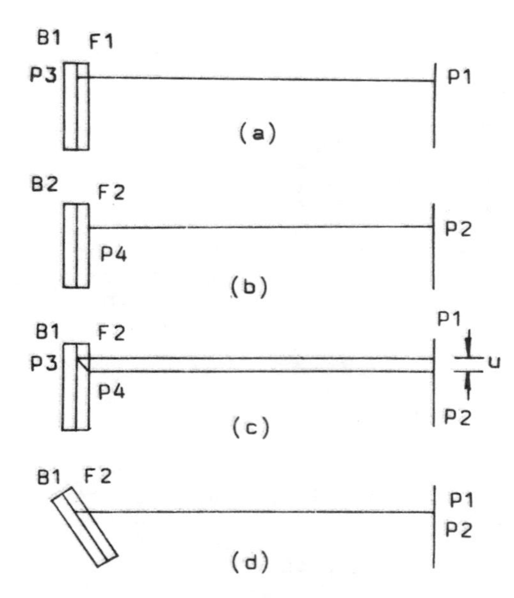

Fig.2 Process of testing in sandwich holographic interferometry

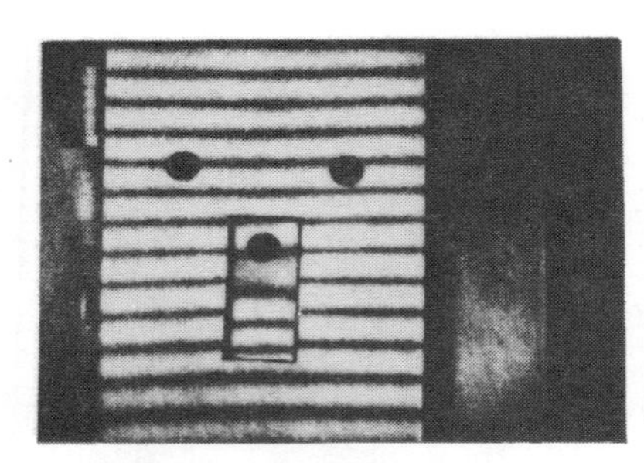

Fig 3 Interference fringe pattern due to displacement x along the x-axis

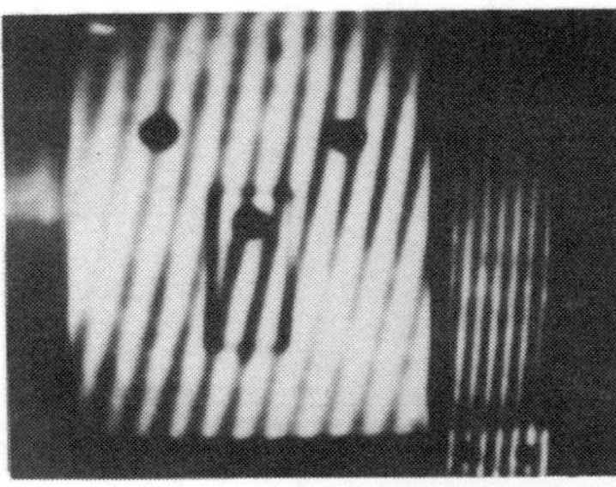

Fig. 4 Fring pattern due to rotation of plate holder around x-axis

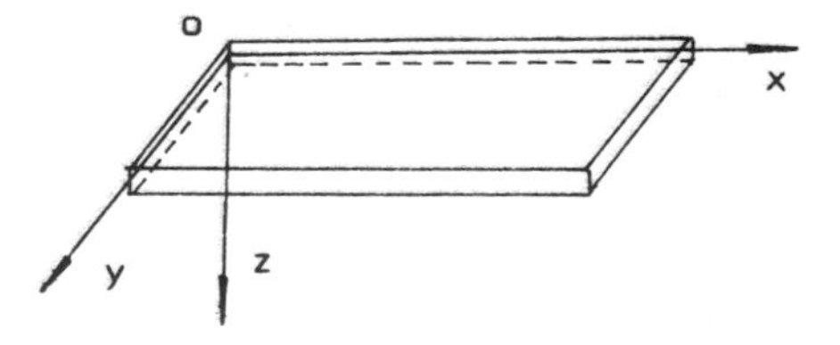

Fig.5 Bending of thin plate

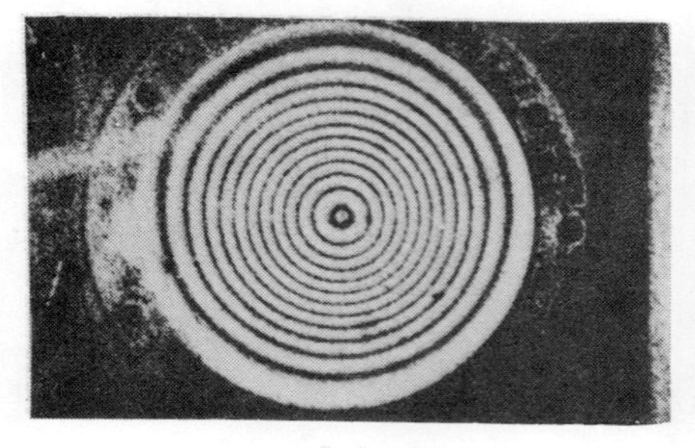

(a)

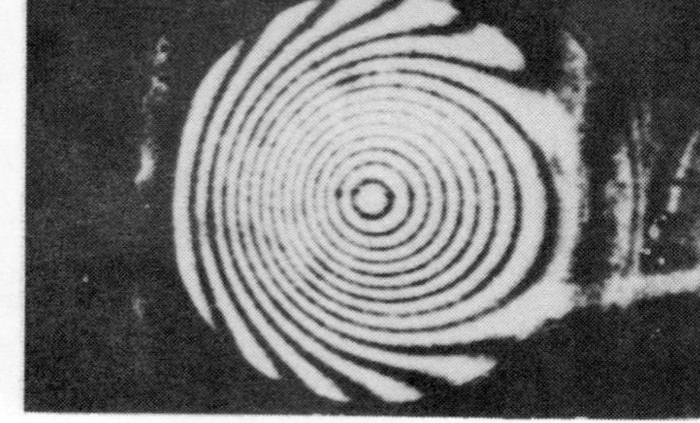

(b)

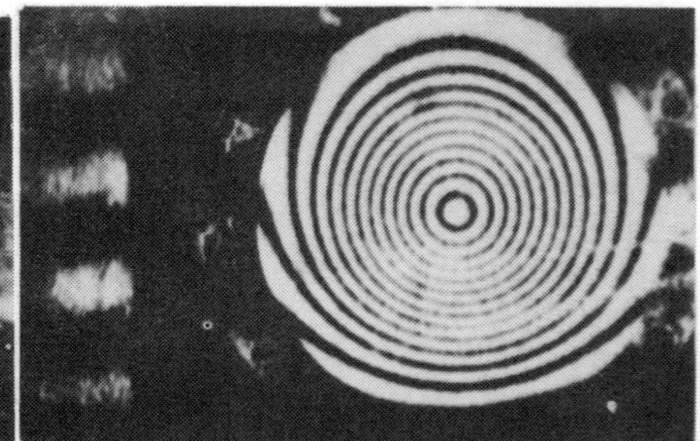

(c)

Fig.6 Sandwich holographic interference fringe pattern for circular plate fixed along the periphery and subject to concentrated load at the center

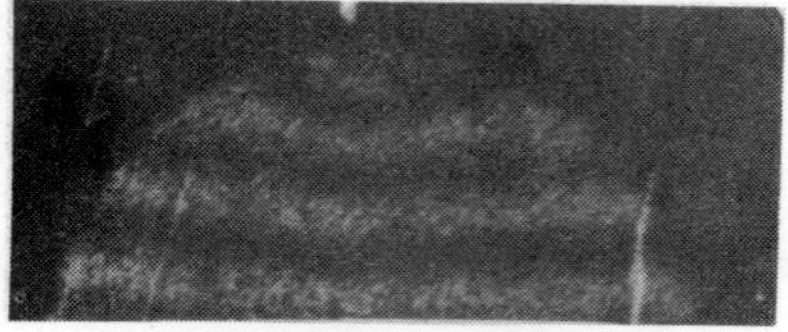

Fig.7 Interference fringe pattern for circular plate subject to diametrically opposite compressive load

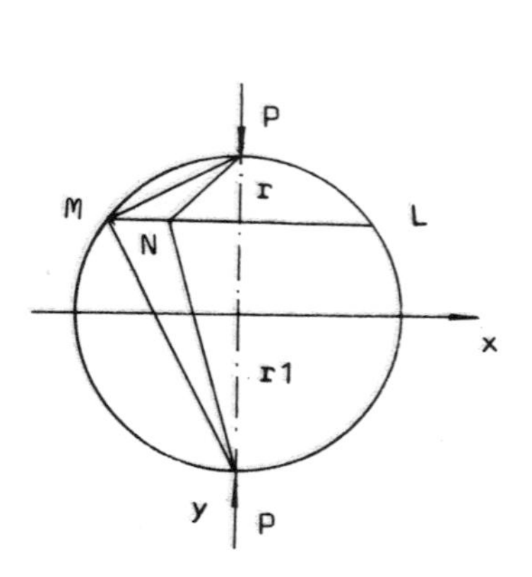

Fig.8 Geometry

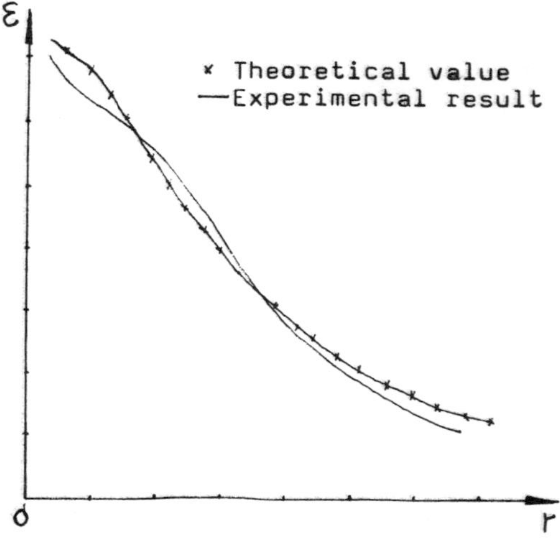

Fig.9 Strain curves

The influence of medium nonlinearity on the properties of the steady-state hologram

A. I. Khyzniak, I. L. Rubtsova

Department of Quantum Electronics, Institute of Physics Ac. Sci. Ukr. SSR,
Prospect Nauki 144, Kiev, USSR

Abstract

The influence of the third-order nonlinearity of a medium containing a steady- state holographic grating on the grating properties is investigated theoretically. It is shown that the angular position and the value of diffraction efficiency maximum can be radically changed. When utilizing the grating in optical scheme, its characterisrics appeare to behave in a jump-like way.

1. Nowadays holographic volume phase gratings (HVPG) are widely used as selective elements in laser systems, where they are irradiated by intense laser radiation.[1] It is of particular interest to have a clear understanding of HVPG characteristics changes caused by the nonlinear optical properties of the HVPG storage medium. Thus the purpose of our analysis is to investigate the influence of the third-order nonlinearity in a HVPG storage medium with a local response on the HVPG diffraction characteristics.

2. The dielectric permittivity of the HVPG storage medium is given by:

$$\mathcal{E} = \mathcal{E}_o + 2\gamma \cos(\bar{\varkappa}\bar{r}) + \alpha|\bar{E}|^2 , \tag{1}$$

where γ is a modulation amplitude of the grating, α is the third-order nonlinearity coefficient and E is an electric component of the light field.

Let us suppose that a plane wave propagates within a slab of such a medium (see Figure 1a). Transmitted (3) and diffracted (1) waves form a new dynamic grating due to nonlinear medium response. The process of these waves propagation may be described by the system of differential equations, which can be obtained as in[2]:

$$\beta_{1,3}\frac{da_{1,3}}{d\mathfrak{z}} = - i \left[\left(|a_{1,3}|^2 + 2|a_{3,1}|^2 \right) a_{1,3} + B\, a_{3,1}\, e^{\mp i\delta\mathfrak{z}} \right] , \tag{2}$$

where $a_i = \mathcal{E}_i(z)/\sqrt{J}$; $i = 1,3$; $J = |\mathcal{E}_3(o)|^2$ is the intensity of wave 3 at the entrance of the medium ; $\mathcal{E}_i$ is the amplitude of the i-th wave ; $B = \gamma/\alpha J$; $\mathfrak{z} = k_o \alpha Jz/2\mathcal{E}_o$ $k_o = 2\pi/\lambda$; $\beta_i = k_{iz}/k_o$; $\delta = \left[\sqrt{k_o^2 - (\bar{k}_3 \mp \bar{\varkappa})^2_{xy}} - k_{3z}\right]/ k_o \alpha J$. The same signs of β_1 and β_3 correspond to transmission HVPG, and the opposite - to the reflection HVPG.

The system of equations (2) has a conservation law

$$\beta_1 |a_1|^2 + \beta_3 |a_3|^2 = I . \tag{3}$$

Using (3) we reduce the system of equations (2) to one equation. For the diffracted wave ($Y = \beta_1 |a_1|^2$) we obtain:

$$\frac{dY}{d\mathfrak{z}} = \pm\sqrt{4B^2 Y(I - Y) - (AY^2 / 2 + CY)^2} , \tag{4}$$

and the relation between the phases and intensities of the interacting waves:

$$\cos(\phi)\sqrt{Y(I - Y)} = (AY^2/2 + CY)/2B , \tag{5}$$

where $\phi = \varphi_3 - \varphi_1 - \delta\mathfrak{z}$, φ_i - is the phase of the i-th wave, $A = \beta_1\beta_3(4/\beta_1\beta_3 - 1/\beta_1^2 - 1/\beta_3^2)$, $C = \beta_1\beta_3\,\delta + I\,(1/\beta_3^2 - 2/\beta_1\beta_3)$.
The equation (4) is solved in terms of elliptic Jacobi functions.

3. As intensity of incident wave tends to zero (or $B \to \infty$), solution of system (2)

coincides with the well-known Kogelnik's solution, which was received for the steady-state HVPG3. Figure 1b shows diffraction efficiency (DE) of HVPG vs detuning from the Bragg angle $\Delta\theta$ and the thickness z of the medium, calculated when $\alpha > 0$ and $\beta_1 = \beta_3$ for the B = 1, 0.5, 0.25. The value of the grating thickness d is optimal for weak incident signal (DE =1, $\Delta\theta = 0$ and $\Phi = \pi/2$).

As shown at the Figure 1b, dynamic grating presence is the medium results in the DE decrease, and DE maximum shifts from the Bragg angle. When $B > 0.5$ solution of the system (2)gives:

$$Y = \left[1 - \mathrm{cn}\left(\frac{k_o \gamma z}{\beta}, \frac{\gamma}{2B} \right)\right] /2 \tag{6}$$

if detuning $\Delta\theta = 0$.

As follows from (5) and (6), DE reaches unity when the condition $\Phi = K(1/B)$ is satisfied at the exit of the grating ($K(1/B)$ is an entire elliptic integral). Thus the decrease of B leads to the difference between $K(1/B)$ and $\pi/2$. When $B < 0.5$, DE is always less than unity and achieves its maximum value with detuning from the Bragg angle.

4. The phase shift between the interference pattern and dielectric permittivity grating in a nonlinear medium is known to control the direction and efficiency of energy transfer. The energy transfer efficiency (ETE) reaches its maximum value when the phase schift equals $\pi/2$. Expression (1) can be rewritten as:

$$\mathcal{E} = \mathcal{E}_o + \Delta\mathcal{E}' \cos(\varkappa \tau + \xi) , \tag{7}$$

where the function
$$\xi = \mathrm{arctg}\left[\sqrt{Y(I - Y)} \sin(\Phi)/(B + \sqrt{Y(I - Y)} \cos(\Phi)\right] \tag{8}$$

denotes the spatial shift between the dielectric permittivity and the steady-state grating fringes. The phase difference of interacting waves $\varphi_1 - \varphi_3$ thus indicates spatial shift between the interference pattern and steady-state grating fringes.

Figure 1c shows the value of ξ (solid line) and $\varphi_1 - \varphi_3$ (dashed line) vs the thickness of the grating for B = 1, 0.5, 0.25. Thus the relative position of the curves on Figure 1c characterizes ETE vs the thickness of the medium. As mentioned above, B represents the ratio between steady- state and dynamic components of HVPG. So, when the steady-state component of the HVPG is greater than the dynamic one, the energy transfer occures in an optimal way (Figure 1c, B = 1, $\Phi = \pi/2$) and 100% ETE is achieved with the thickness d (Figure 1b, B = 1). Evidently, the increase of HVPG dynamic component leads to the DE decrease at the optimal thickness d due to the energy transfer delay (Figure 1b, B = 0.5, 0.25). Until the steady-state grating component is greater that the dynamic one ($B \geq 0.5$), the DE diminishing can be compensated by the thickness increase. If $B < 0.5$ dynamic grating becames so strong, that it is possible to change the direction of energy transfer and DE will never reach unity. The change of energy transfer direction gives an upper limit for the maximum value of DE. In this case the DE maximum can be reached with detuning from the Bragg angle.

5. When utilizing the nonlinear storage medium in a scheme shown in Figure 2a (diffracted wave is reflected backward to the HVPG), the four wave mixing is realized. Such a problem is of interest for intracavity frequency selection of laser spectrum.

When the incident waves 1 and 3 are directed at the Bragg angle and the counter-propagating interaction is neglible, the four wave mixing is described by the following system of equations:

$$\frac{da_{1,4}}{d\eta} = a_{3,2} (a_1 a_3^* + B + a_2 a_4^*)$$

$$\frac{da_{2,3}}{d\eta} = a_{1,4} (a_1^* a_3 + B + a_2 a_4^*)$$

where $a_i = \mathcal{E}_i/\sqrt{J_o}$; J_o is the total intensity of the interacting waves ; $\eta = F/(2B\mathcal{E}_o \cos(\Phi))$; $F = k_o \gamma z$; $B = \alpha/\gamma J_o$.

Using four conservation laws

$$|a_{1,2}|^2 + |a_{3,4}|^2 = I_{1,2} \qquad (10,11)$$

$$|a_1 a_3^* + B + a_2^* a_4|^2 = |C_o|^2 \qquad (12)$$

$$IT + T^2 - 2B \cdot \mathrm{Re}(C_o) = C \qquad (13)$$

we reduce the system (9) to one equation:

$$\frac{dT}{d\gamma} = \pm\sqrt{4B^2|C_o|^2 - (T^2 + IT - C)^2} \qquad (14)$$

where $T = |a_4|^2 - |a_1|^2$; $I = I_1 - I_2$.

Worse noting that conservation law (12) signifies constancy of resulting HVPG amplitude within the nonlinear medium. This assumption is correct for local medium response, even when there is no steady-state HVPG ($\gamma = 0$). If the response of the medium is nonlocal and $\gamma = 0$, the dynamic grating phase is conserved. Equation (14) is solved with the help of elliptic integral:

$$\gamma = \mathcal{M} \cdot F(\varphi / k), \qquad (15)$$

where parameters $\mathcal{M}$, F and k depend on the boundary conditions.

6. Figures 2b,c,d exhibit the reflectivity $R = |a_4(o)|^2 / |a_3(o)|^2$ of the optical system illustrated by Figure 2a: Figure 2b represents computed reflectivity dependence vs optical length F, Figure 2c shows R as a function of mirror reflectivity r, and Figure 2d illustrates dependence of R on the incident light intensity J_o. From the Figures it follows that device reflectivity depends in a jump-like way on the parameters mentioned above and this scheme can operate in a bistable regime. For example, there are two branches in the dependence of reflectivity R on the incident light intensity. These branches are realized when $F \lessapprox \pi/2$ and $F \gtrapprox \pi/2$ correspondingly. Each of them can stabilize the laser power within a range of incident intensity variations when utilizing this scheme as laser reflecting element. The jump-like transition between the branches may be used for the switching of laser oscillation regimes.

7. Thus the nonlinearity of the medium containing steady-state HVPG is shown to affect essentially on optimal conditions for its efficient operation. When HVPG is utilized as illustrated in Figure 2a, the medium nonlinearity results in a jump-like dependence of device reflectivity R on the device parameters.

Acknolegments

The authors wish to thank Ph. M. Yatsyuk and M. V. Vasnetsov from Institute of Physics Ac. Sci. Ukr. SSR for encouragment and helpful discussion on the topic.

References

1. Soskin, M. S., Taranenko, V. B., "Holographic Selector-Telescope", Sov. Techn. Phys. Lett., Vol. 2, pp. 99-103. 1979.
2. Vinetsky, B. L., Kuhtarev, N. V., Odulov, S.G., Soskin, M. S. "Dynamic Self-Diffraction of the Coherent Light Beams", Sov. J. : Usp. Phys. Nauk, Vol. 129, N1, pp. 113-137. 1979.
3. Kogelnik, H., "Coupled Wave Theory for Thick Hologram Gratings", Bell. Syst. Techn J., Vol.48, pp. 2808-2947. 1969.

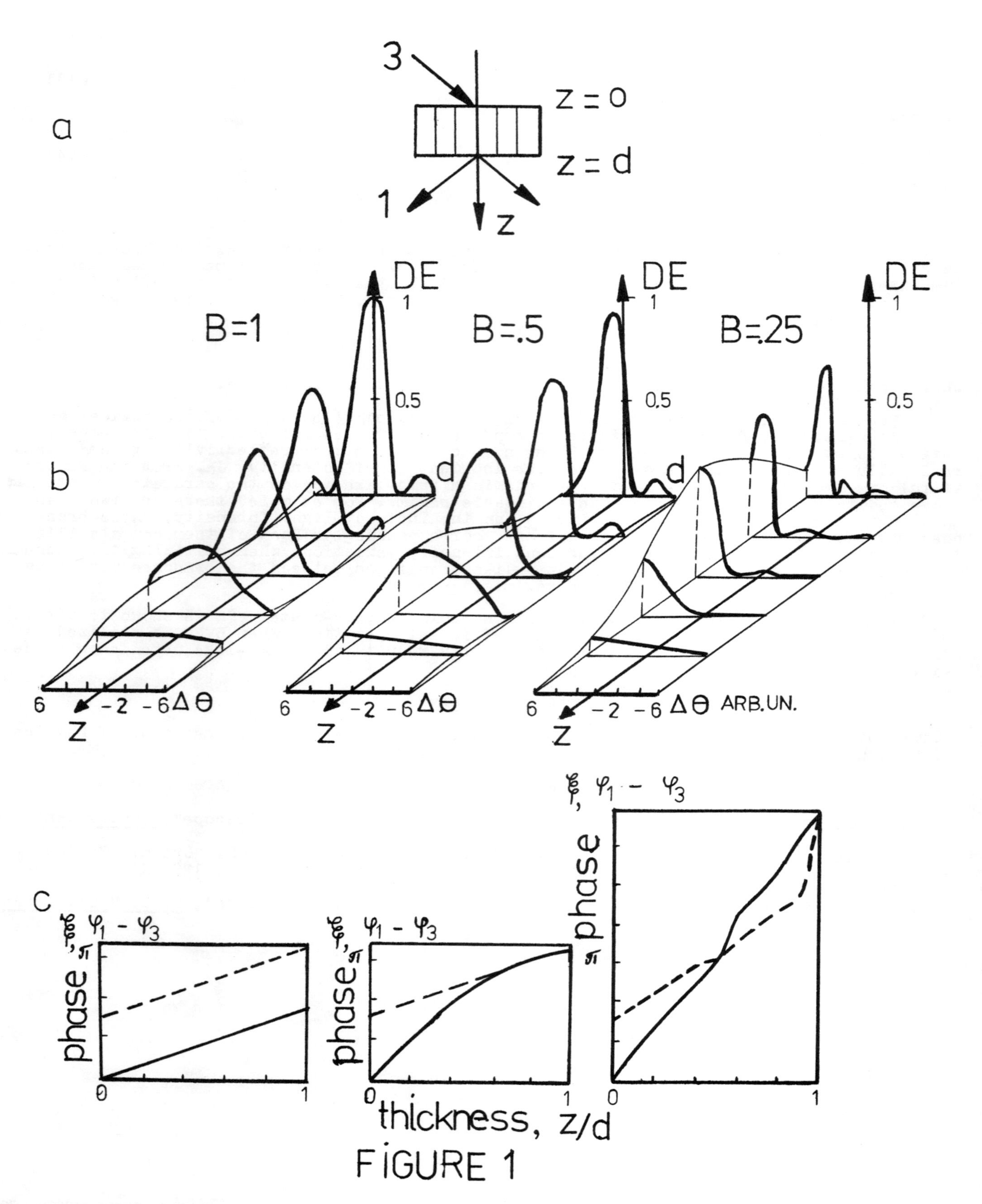

3
z = o
a
z = d
1
z
DE
1
0.5
B=1
B=.5
B=.25
b
d
6
-2
-6
ΔΘ
ARB.UN.
z
c
phase
φ₁ - φ₃
π
0
1
thickness, z/d

FIGURE 1

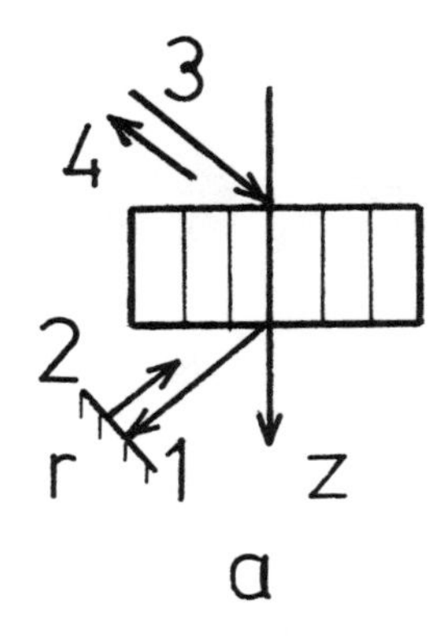

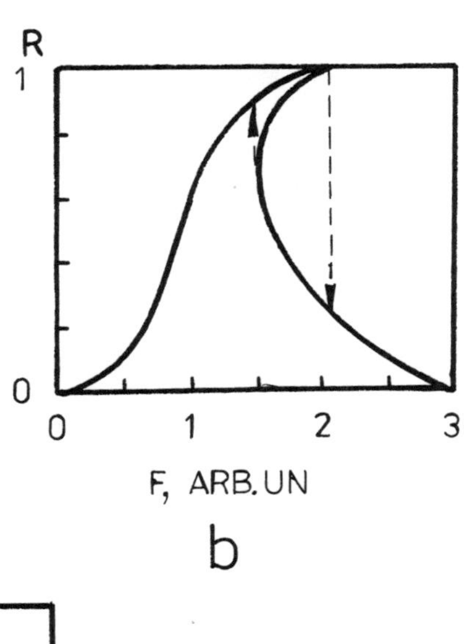

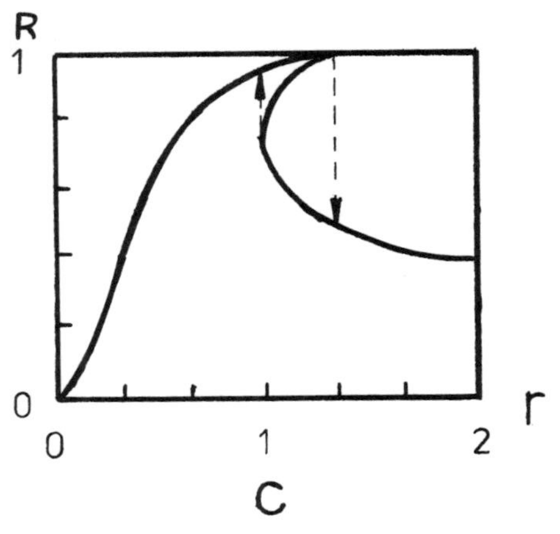

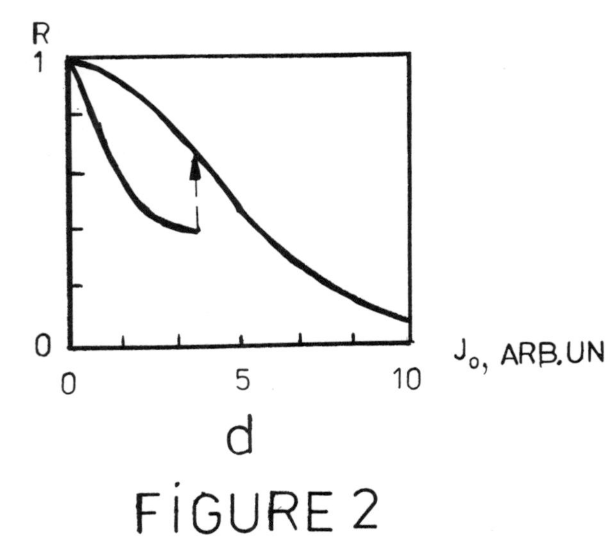

FIGURE 2

Session 4

Fundamentals and Display in Holography III

Chairman
Guo-Guang Yang
University of Science and Technology of China

Holographic three-dimensional display of X-ray tomogram

Katsuyuki Okada and Teruji Ose

Department of Image Science & Engineering
Faculty of Engineering, Chiba University
1-33, Yayoi-cho, Chiba, Japan

Abstract

In the field of medical diagnosis, many X-ray tomograms are taken at the different slice planes of affected parts and organs so as to grasp their three dimensional structures. Multiple recording hologram in which all slices are recorded on single holographic material is used for this purpose. To avoid the faults of usual incoherent superposition and to make possible white light reconstruction, we present a new recording method for making a multiple recording hologram. The hologram is made by two step process and can be reconstructed by white light illumination. Some example of hologram is also shown.

1. Introduction

Holography is one of the most attractive method to demonstrate an image in 3-D. If holography is conbined with x-ray imageing, it is possible to display the internal structure of a human body in 3-D space, which is useful for medical diagnosis and education.

Two types of holographies can be used for this pourpose. One is the multiplex hologram, or cylindrical holographic stereogram[1-3]. But to make multiplex hologram, it is necessary some special designed X-ray camera, or some complex computer processing. The other is the multiple recording hologram. Multiple recording hologram can be made from x-ray computer tomgraphic images and these images are available in more general hospital.

2. Conventional multiple recording hologram

Multiple recording hologram is made from about 20 or more CT images whose slice planes are different. Fig.1 shows a schematic diagram of the optical system for recording a conventional multiple recording hologram. In this figure, A is the first transparency of the CT image with a diffuser and these are illuminated behind the diffuser. Then expose the hologram.

Next, remove the first CT transparency and the second slice of CT image is placed at B in fig.1, and expose the second image. Repeating the similar process, all the CT images are recorded sequentially on single holographic plate.

In the reconstruction using monochromatic illumination, all the CT images are reconstructed simultaniously, and a 3-D image can be obtained. However this simple multiple recording method have the following fault.

1. Because of incoherent multiple superposition, the diffraction efficiency of hologram tends to be low, and back groud noise to be large.

2. Monochomatic light is need in its reconstruction.

To avoid the faults of the conventional multiple recording hologram, we present a new recording method for making a multiple recording hologram.

3. Hologram recording process

The procedure for making the holgoram is divided in two steps, similar to rainbow holograms. In the first step, a master hologram is recorded by the optical setup shown in Fig.2. The first transparency of X-ray tomogram is placed at A in fig.2 and it is recorded on the holographic plate through the slit. This slit limits the

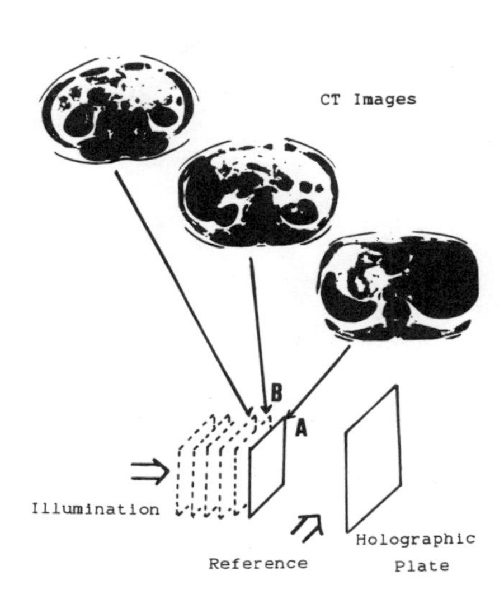

Fig.1 Schematic diagram of the optical system for recording a conventional multiple recording hologram

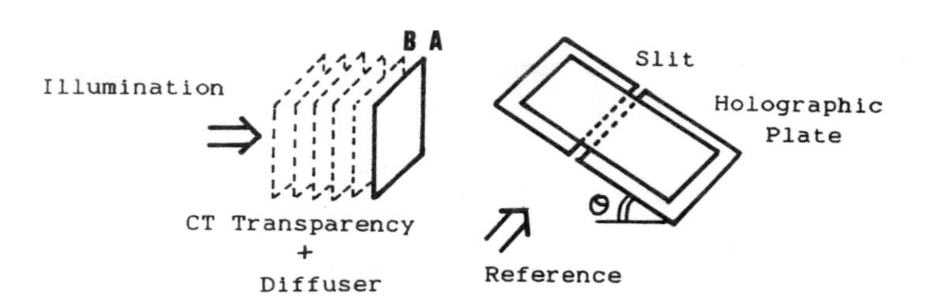

Fig.2 Schematic diagram of the optical system for recording a master hologram

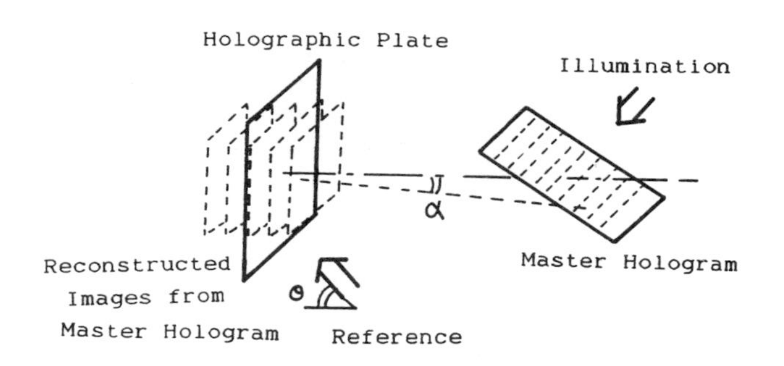

Fig.3 Schematic diagram of the optical system for recording a final hologram

recording area to avoid incoherent superposition. After the first exposure, the transparency is replaced by the second transparency and the slit is moved by the distance equal to its width, and the second transparency is recorded on the hologram plate. This process is repeated and the master hologram is made in which many hologrms of transparencies are reconded in order.

In the second step, as shown in Fig.3, the master hologram is illuminated by the conjugate illumination, the real images of the transparencies are reconsturcted. These images are recorded on another hologram plate.

When the final hologram is illuminated by the monochromatic conjugate light, the real image of the master hologram is reconstructed (fig.4(a)). If we places our eyes in this image of master hologram, we can not see all the CT images simultaniously.

White light is used in place of the above monochromatic light, many reconstructed images of the master hologram are overlapped each other with a amount of shift of their position according to wavelength of light because of color dispersion, as shown in fig.4(b). If we places our eyes in the overlapped area of this colored image, we can view all of the images of trancparency simultaniously. This hologram is made without incoherent superposition, so high diffraction efficency and low noise image can be obtainid and the number of transparencies is not limited.

4. Distortion of the image and its compensation

The usual hologram is on the promise that whole the image is viewed by one color. But the images of this hologram is constructed from many colored images. Because of difference of the reconstructing wavelength, some distortion is occured in the reconstructed images. However this distortion is determined by the arrengements of the hologram recording setup, and the distortion can be predicted. So that this distortion would be cancelled by optimizing the arrangement of recording optincal system.

At first we consider the relations of condition to overlap the images of the master hologram. For simplify, we assume the reference beam of the final hologram is a

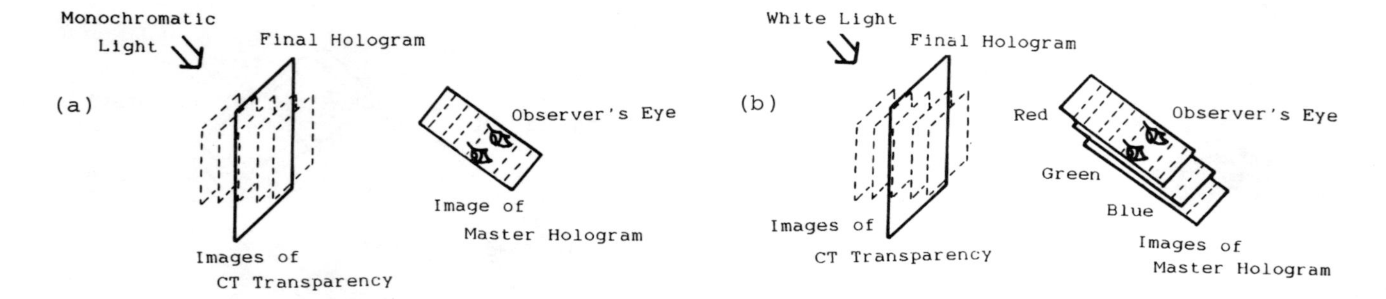

Fig.4 Reconstruction of a new multiple recording hologram
(a) Monochromatic light illumination (b) White light illumination

parallel one. Fig.5 is the side view of the optical setup in the recording of the final hologram. Let the element hologram on the master hologram positioned at (R_{o1}, y_{o1}) be reconstructed at the position (R'_{o1}, y'_{o1}). The relations between them are obtained from the Chanpagnue's imaging equation[4] as follows.

$$\lambda R_{o1} = \lambda' R'_{o1} \tag{1}$$

$$\frac{1}{\lambda}\left(\frac{x_o}{R_{o1}} - \sin\theta_r\right) = \frac{1}{\lambda'}\left(\frac{x'_o}{R'_{o1}} - \sin\theta_r\right) \tag{2}$$

$$\frac{1}{\lambda}\left(\frac{y_o}{R_{o1}} - \sin\theta_r\right) = \frac{1}{\lambda'}\left(\frac{y'_o}{R'_{o1}} - \sin\theta_r\right) \tag{3}$$

Where θ_r is the angle of incidence of the reference beam of the final hologram. Equations (1)-(3) give the relation between R_o' and y_o', as folows.

$$\frac{y_o}{R_{o1}} = \sin\theta_r \left(1 - \frac{R'_{o1}}{R_{o1}}\right) \tag{4}$$

Simplify this equation, then

$$\frac{y_o}{R_{o1} - R'_{o1}} = \sin\theta_r = \tan\alpha \tag{5}$$

This relation means that if the master hologram is inclined by the angle of α to the optical axis. Then the image of the master hologram will overlap. This relation is similar as the relation of achromatic angle in achromatic holographic stereogram reported by Dr. Benton[5].

Next we consider the reconstructed position of the CT transparency, and the magnification of the image. In the recording of the final hologram, the real image of CT image constructed by the master hologram is positioned at the distance R_{o2} from the final hologram (fig.6(a)). In the reconstruction, this image is reconstructed by the light of wavelength λ' and its position is R'_{o2} (fig.6(b)). From Chanpagne's equations, eqations (6) and (7) are obtained.

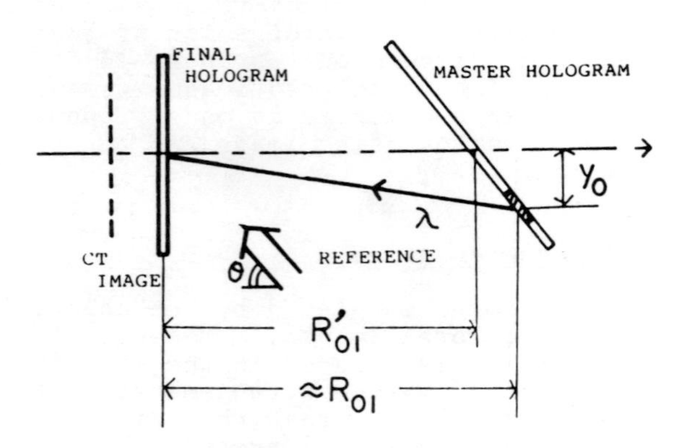

Fig.5 Side view of recording of a final hologram

$$\frac{x'_o}{x_o} = 1 \tag{6}$$

$$R_{o2} = \frac{\sin\theta_r}{\sin\theta_r - \dfrac{y_o}{R_{o1}}} R'_{o2} \tag{7}$$

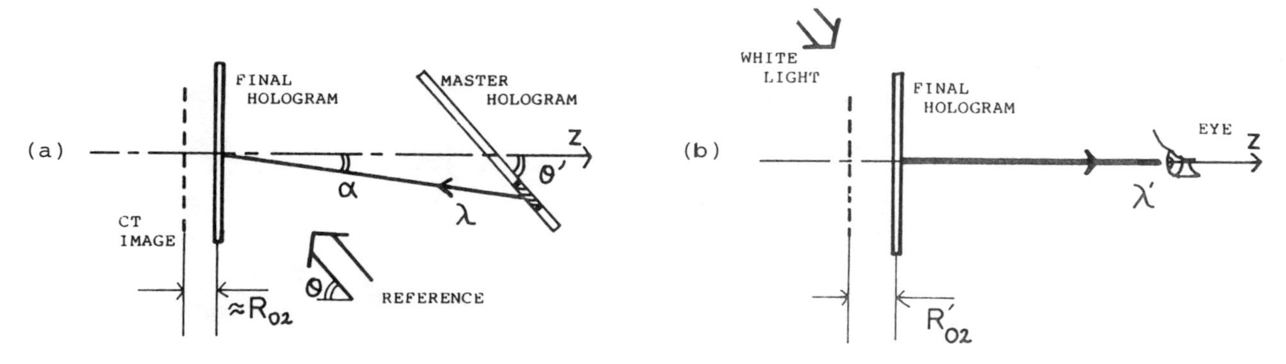

Fig.6 Magnification of image in the depth direction
(a) Recording of a final hologram (b) Reconstruction of a final hologram

The magnification for horizontal direction is always unity, that is the size of the image is not changed through hologram recording and reconstructing process. And the magnification for the depth direction is given by eq.(7).

These equations mean the distortion arises only in the depth direction of the image. So that the compensation for the distortion can easily be made by changing the position of transparency in the recording step of the master hologram.

5. Example of hologram

Fig.7 is a schematic diagram of the optical setup to make a master hologram. One of the problem to make this hologram is the precise setting of the CT transparency. So we use a personal computer to set the transparencies, to move the slit and to control the shutter.

We made some examples of hologram. Fig.8 shows 12 slices of the CT images of a thighbone and a hipbone. In this figure, upper left shows the CT image of the upper side in a human body and the bottom right shows the lower side. The spaceing between the slices is 5 mm. From these photographs, we can hardly understand the 3-D structure of the joint.

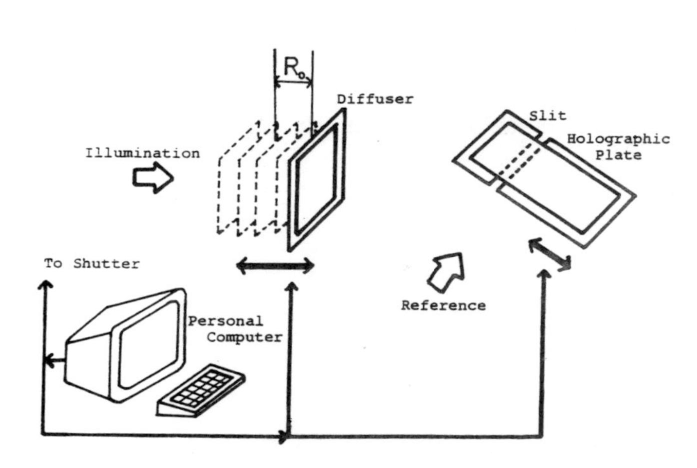

Fig.7 Schematic diagram of the computer controled optical system for recording a master hologram

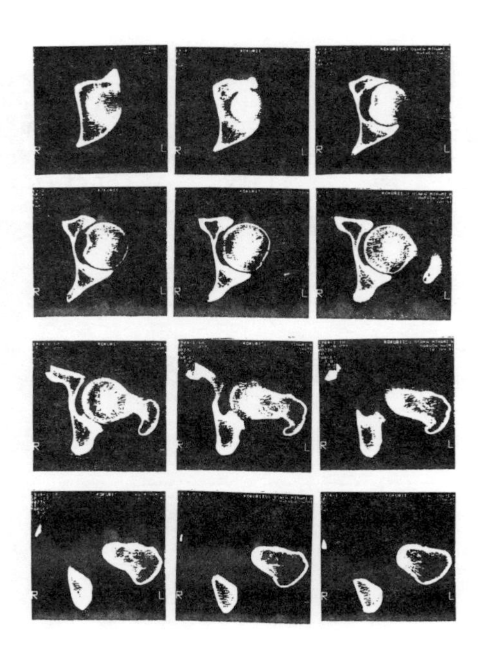

Fig.8 Images of X-ray tomogram

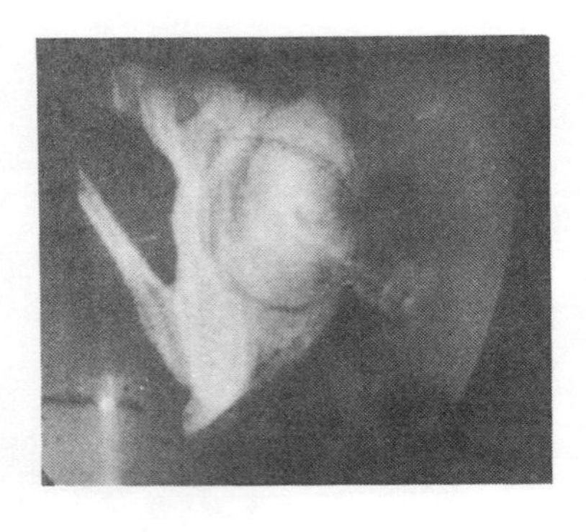

(a)

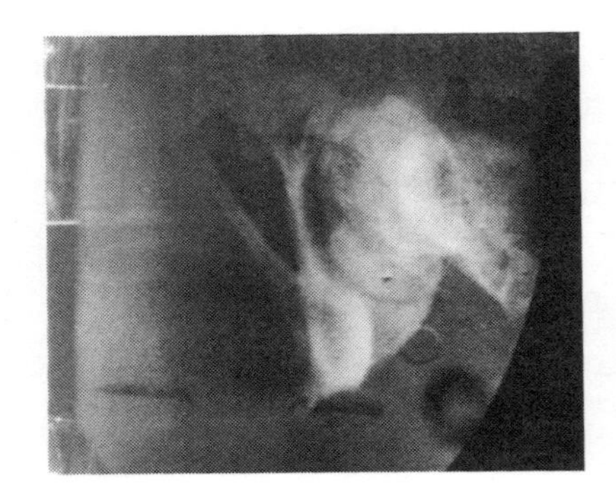

(b)

Fig.9 Photographs of the image reconstructed from a multiple recording hologram

Fig.9 is the photograph of the hologram, where (a) is photographed from left and (b) is photographed from right. This hoogram is mae from 20 images of X-ray tomogram. In the reconstructed image, difference of color represents the position in the depth direction, i.e. in this example, the red part represents the upper side of the bone and the blue parts represents the lower part.

6. Cnclusion

In this paper we present a new recording procedure for making a multiple recording hologram. This holgram displays inside structures of human body in 3-D space. Hologram making process propoposed here does not include incoherent superposition, therefore high diffrection efficiency and low noise images can be obtained and the limitation of the number of recording planes is that of the length of hologram plate. Because of color dispersion, the reconstructed image is a little distorted, so that we present a method of compensation for the distortion. This hologram can be used not only for medical purpose but also for the display of general computer generated 3-D images as a color hologram.

References

1) Okada, K., Honda T. and Tsujiuchi, J., Proc. SPIE, Vol.212 (1979) 28
2) Huff, L. and Fusek, R., Proc SPIE, Vol.215 (1980) 39
3) Grossmann, M., Meyrueis, P. and Fontaine, J., Holography in Medicine and Biology, Edited by von Bally, G., (Springer Verlag, 1979) 110
4) Champagne, E. B., J. Opt. Soc. Am., 57 (1967) 51
5) Benton, S. A., Proc SPIE, Vol.391 (1983) 2

Application of holographic lens in head-up display

Chen Huangming, Lu Bo

Department of Optical Engineering, Beijing Institute of Technology
Beijing, China

Abstract

Based on the holographic theory, the optical properties and characteristics of aberrations for reflective holographic lens have been described in detail in this paper. The aberration distribution of holographic lens is confirmed by experiments. According to the aberration characteristics of reflective holographic lens, calculation and analysis on asymmetrical spheric system have been performed to compensate aberrations of holographic lens. The formulae of ray tracing and calculated aberrations of asymmetrical spheric system are deduced. The imaging properties of asymmetrical optical system are studied.

Introduction

Head-up display (HUD) is an advanced device used in many fields as aircraft etc. It can provide sufficient information for pilots and reduce observing fatigue. Optical system plays an important role in HUD, which provides all information acceptable to the pilot. Although conventional optical system is easy to manufacture and assemble, the action of HUD is limited by its small field of view. The diffractive optical system of HUD using holographic lens as the combiner has recently received considerable attention due to its larger field of view.

Imaging principle of holographic lens

Holographic lens is a kind of imaging element based on diffraction theory. The imaging principle can be described by following equations: (Figure 1.)

$$\frac{1}{d_I} = \frac{1}{d_C} + u \left(\frac{1}{d_O} - \frac{1}{d_R} \right) \tag{1}$$

$$\frac{y_I}{d_I} = \frac{y_C}{d_C} + u \left(\frac{y_O}{d_O} - \frac{y_R}{d_R} \right) \tag{2}$$

$$\frac{z_I}{d_I} = \frac{z_C}{d_C} + u \left(\frac{z_O}{d_O} - \frac{z_R}{d_R} \right) \tag{3}$$

where d_O and d_R are the distances from two point sources O and R to the origin of coordinates, u is the ratio of reconstruction wavelength to construction wavelength, d_O and d_R are taken positive value if the direction from origin to two construction point sources keeps the same as that of light propagation, negative value is taken in opposite case. The focal length of holographic lens is usually defined as:

$$\frac{1}{f'} = u \left(\frac{1}{d_O} - \frac{1}{d_R} \right) \tag{4}$$

Line OO and OR are defined as the object space axis and the the image space axis respectively. Ween the distance of object is infinite, the relevant imaging equations can be expressed as:

$$\cos B_I = \cos B_C + u \left(\cos B_O - \cos B_R \right) \tag{5}$$

$$\cos r_I = \cos r_C + u \left(\cos r_O - \cos r_R \right) \tag{6}$$

where B_j and r_j are the angles between the light propagated direction and the positive of coordinate axis Y.

Holographic lens used in the optical system of HUD is in reflective model. It can be considered as a common lens, a grating and a mirror with diffrent medium layers. Diffraction efficiency, imaging quality and transparency are the main criteria applied to judge the optical performance of holographic lens.

Focal surface distribution and the determination of ideal image plane in holographic lens

The focal length formula of holographic lens is obtained by Fresnel condition, which means d_O, d_R are large enough, and y,z must be small enough. Actually, the size of holographic lens is only tiny smaller than d_O or d_R. Therefore, d_O and d_R are not only considered as the distances from constuction point sources to origin of coordinates as shown in Figure 1, but should be regarded as the distances between construction point sources and a certain point on holographic lens. The distances vary with the position of the point. Thus the focal length of the holographic lens is not a constant. Based on the analysis above, the focal surface distribution of holographic lens is calculated and analyzed in detail. The conclusions are:

1. In the case of coaxial construction, if $d_O \neq |d_R|$, objective and image focal surfaces are all curved, but the curved directions are not the same. When $d_O = d_R$, the focal surfaces are flat planes. (Figure 2.)
2. The focal length of holographic lens is symmetrically distributed to optical axis under the condition of coaxial construction.
3. In off-axis construction model, the focal surface of holographic lens is a curved one tilted towards the optical axis. (Figure 3.)

These conclusions show the vital differences between co-axis and off-axis holographic lens. These two kind of holographic lens differ greatly in imageing and aberration performances. After discussed the focal surface distribution, it shows the importance of the choice of a plane near focal surface as an image plane corresponding to objective plane at infinity. It should be considered that the point symmetrically about the center of image plane should correspond to the same field of view. Many image planes may be found on this view, but the difference among them is very limited. The calculation shows that the tilt of image plane increases with the off-axis angle (bend angle).

Aberration analysis and experiment of holographic lens

Two main differences of holographic lens compared with common lens in imaging characteristics are:

1. Tilted image surface.
2. Asymmetric distribution of aberrations.

In order to correctly evaluate the image quality of holographic lens, the image distribution of sagittal object must be calculated along with meridional object. The calculation shows that the aberration distribution of holographic lens has many outstanding features:

1. Holographic lens has larger curvature of field, axial astigmatism and axial coma.
2. The aberration values between symmetric fields of view are greatly different.
3. The asymmetry and aberrations increase rapidly with off-axis angle.
4. The aberrations of holographic lens are improved in symmetric recording than in asymmetric recording.
5. The aberrations reduce with the increase of the distance of construction point source in symmetric recording.

A holographic lens is made with 45 mm in diameter, 125 mm of focal length, and 45° off-axis angle. The experiment aims only at providing the aberration features of holographic lens. Figure 4. is the photographs of aberrations for axial light beam and shows the singnificance of the axial astigmatism and axial coma.

Ray tracing of asymmetric spherical optical system

The calculation of aberrations for asymmetric spherical optical system is very complex. The asymmetric spherical system includes two cases, the decentration and the tilt. However, they may be reduced to one. In general, they only serve as the factors affecting image quality to be controlled and are seldom used as the variables for correcting aberrations. But those factors are used for correcting aberrations in holographic optical system. According to the imaging characteristics of holographic lens, the spherical surfaces of the relay lens for correcting aberrations of holographic lens are only reqireed to be decentred and tilted in the meridional section. When the imaging characteristics of the asymmetric relay lens are calculated, the optical axis of the object and image space are defined as follows (Figure 5.), calculating from the center O of real stop for line OP along geometrical optical axis (optical axis in coaxis of spherical surfaces) by using Gauss formulae and obtained line AB, optical axis of object space, tracing a ray along line AB through relay lens by ray-tracing formula and obtained line CD, optical axis of image space. The significance of this definition is that the rays with the same field of view and aperture are assured to pass the real stop.

Design of holographic optical system

It is well known that perfect point-to-point imagery can be obtained with a holographic lens provided that the lens is constructed with two wavefronts centered at the desired object at image points. However, a phrase reflective holographic lens which is thus constructed has a limited field of view. The reason is that the other points on object plane depart from Bragg condition which relates angle to wavelength for optimum efficiency. The diffraction efficiency is weakened by the departure from Bragg angle. In order to achieve high efficiency over a large field of view, the holographic lens must be constructed with wavefronts centered at points located in the desired entrance and exit pupils of the lens. The chief-rays in reconstruction are all diffracted at Bragg angle and high efficiency is obtained. A desired field of view can be covered if the distance of entrance pupil is properly chosen.

The principal factors which must be considered in the design of holographic optical system of HUD are:

1. The dimension of holographic optical system must be compatible within the limitations of aircraft cockpit. In addition, it is desirable that the opto-mechanical configuration causes a minimum obscuration to both the overlooking and the instrument panel. These requirements govern the determination of the aperture, off-axis angle, focal length of holographic lens, the aperture and the focal length of relay lens.
2. The eye relief of holographic optical system is determined by the distance from the point to the holographic lens.
3. There are five methods used to compensate aberrations of plane holographic lens: reduction of off-axis angle, introduction of holographic lens with curve substrate, use of relay holographic lens, construction with wavefront-aberrations, and insert of asymmetric relay lens.

There are individual characteristics of correcting aberrations for these methods. As the determination of off-axis angle is affected by the geometry of aircraft cockpit, it is usually taken in the range of 30° to 50°. The use of curved holographic lens and the construction with wavefront-aberrations are of advantages to reduce image tilt and asymmetry of distributions, but the design and manufacturing technology are more complicated. Relay holographic lens has the advantages of simple structure, light weight, easy to design and manufacture but with the disadvantage of limited ability for the aberration compensation. In our design, the asymmetric relay lens is only used for compensating the aberrations of plane holographic lens.

As with the conventional optics, the design of holographic optical system is accomplished by computer-based ray-tracing. The procedure tends to be more complex because of the off-axis operation of holographic lens. According to the characteristics of holographic optical system, a program is specially written to calculate the performance of holographic optical system. The program can calculate the distributions of imaging light beam with arbitrary positions and apertures (limited to less than desired maximum entrance pupil) in both meridional and sagittal directions through the entrance pupil. It can calculate the performance of both the whole system and the holographic lens. The program processes wide suitability.

The research on conventional asymmetric optical system is not perfect at present. There are many problems to be considered. The effect of individual variable for for asymmetric relay lens on aberrations of the system is studied in detail in our work. The characteristics that aberrations of plane holographic lens are compensated with asymmetric relay lens are found as follows:

1. The aberration values of the system can be adjusted as the curvature and the thickness of lens changed, but those parameters could hardly affect the relative distributions of aberrations corresponding to different field of view. The changes also significantly govern the optical performances of relay lens, therefore, they can hardly act in correcting aberrations in the case of strict limited configuration.
2. The major corrected aberrations are curvature of field, astigmatism and coma in the system.
3. The use of cement lens is not imperative because of the unattention to correct chromatic aberration.
4. The off-axis characteristics of reflective holographic lens in the optical system lead to non-symmetrical distortion which can not be corrected for plane holographic lens only with asymmetric relay lens. The desired distortion pattern must be artificially generated on the CRT in order to provide an undistorted image to the eyes. The optical method must be combined with the electronical one in order to meet the required accuracy of the display. A holographic optical system of HUD has been designed. The values of all kind of aberrations meet the required accuracy.

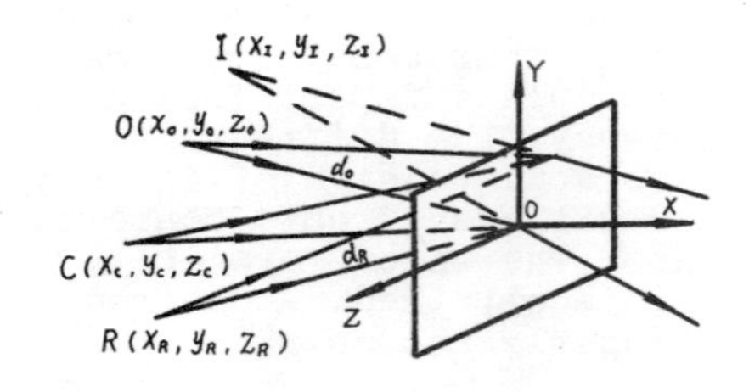

Figure 1. Imaging principle of holographic lens.

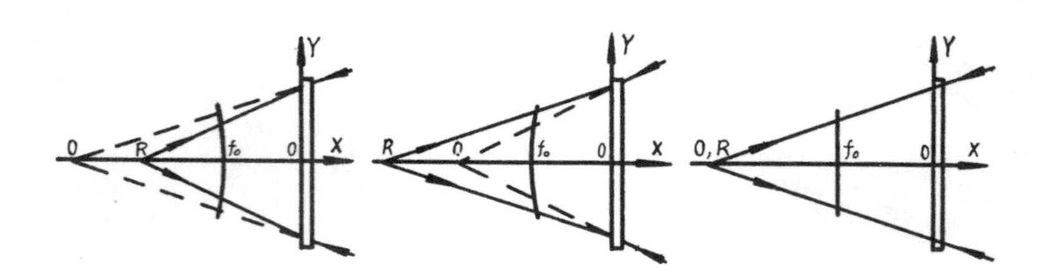

Figure 2. Focal surface distribution of co-axis holographic lens.

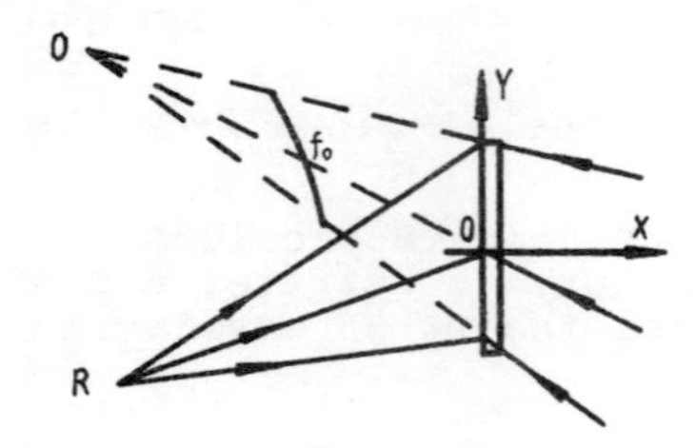

Figure 3. Focal surface distribution of off-axis holographic lens.

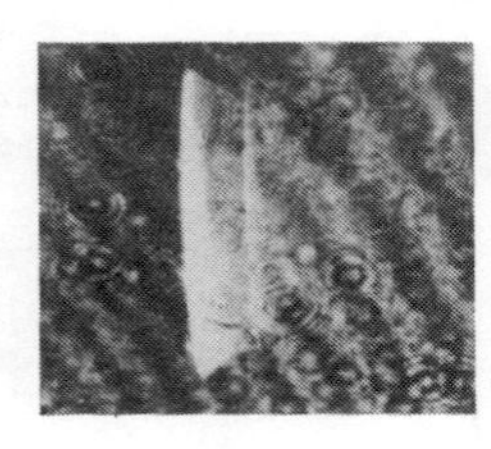
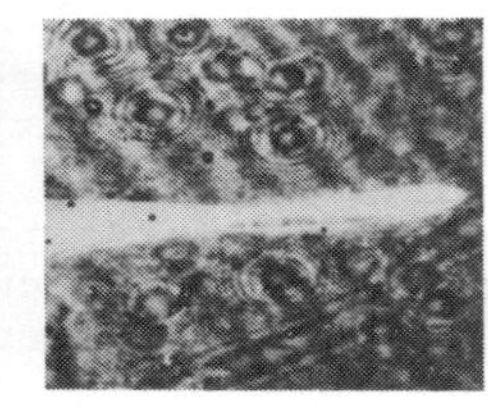
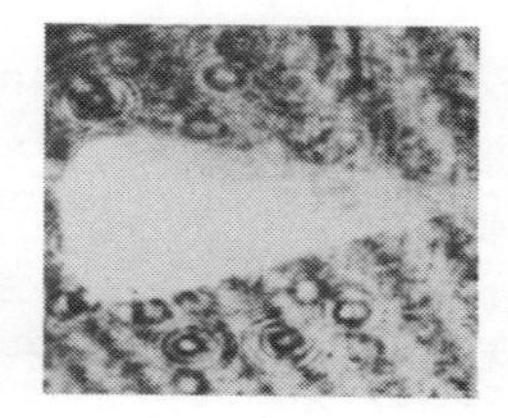

Figure 4. Aberrations of axial light beam in holographic lens.

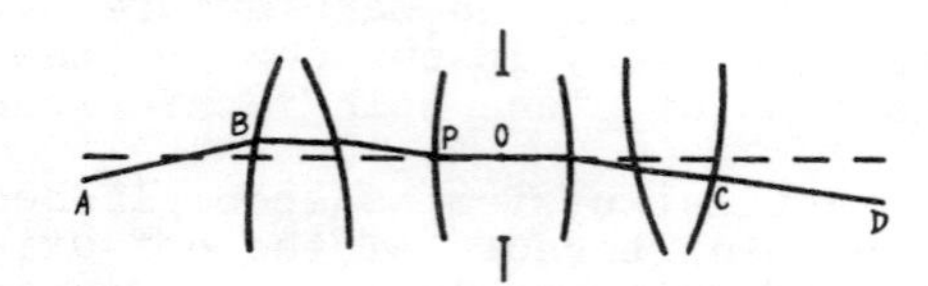

Figure 5. Determination of optical axis of object and image space.

A Method for Making Large Rainbow Holograms

Qu Zhimin, Xu Yingming, Li Meiyue, Cai Xueqang

Shanghai Institute of Laser Technology, Shanghai, China

Abstract

In this paper a practical method for making large rainbow holograms by using a common glass rod is introduced. It is able to make white-light transmission holograms as large as 1000 X 1000 mm^2.

Introduction

For many years people have been concentrating their mind in the principle and technique in order to seek for true 3D enjoyment.

Initial hologram can only be reconstructed by laser light, its applications are limited, even though it can offer obvious parallax. Combining the method presented by russian scientist N.Denisyuk [1] with DCG material, the living image can be reconstructed by white light. Because the sensitivity of DCG material is very low, it is difficult to obtain large holograms. In 1969, S.A.Benton presented a new method called rainbow holography[2]. Using this method a bright image can be obtained by white light. As the man's eyes are horizontal,an excellent horizontal parallax can be observed, even though the information of vertical parallax is lost. Unfortunately, according to Benton's schematic diagram, it is more difficult to obtain large size rainbow holograms. By using H.Chen's[3] one-step method, the procedure of making rainbow holograms is simplified, but, in practice, it is not easy to obtain the image with wide view angle and large size. However, in many cases, such as advertisement, propagenda, education, entertainment, art, ornamentation and so on, holograms with quite large size have to be made so as to get better visual effect. In this paper a practical method for making large rainbow holograms by using common glass rod is introduced. By means of this method it is rather easy to get a large rainbow hologram of 500 X 500 mm^2 or more than 1000 X 1000 mm^2.

The Principle

S.A.Benton's two-step method can be outlined as follows: First, a common hologram, called the master hologram, is obtained by using the method of off-axis holography. Second, a narrow slit is set in front of the master hologram, the real image of the master hologram is reconstructed through the slit. And then the second hologram is recorded by means of this real image beam. This second hologram is the so-called rainbow hologram. Obviously, in order to get rainbow hologram with wide view angle, big object, large size, this master hologram is bound to be rather large. For this reason the plane or spherical wave illuminating the hologram has to be correspondingly expanded. In addition, the slit in front of the master hologram obviously attenuates the intensity of reconstructed image. Although, in principle, it can be realized, in actual practice it is very difficult to make a large rainbow hologram with high quality.

In view of the fact that our purpose is 3D display, the problem is not complicated. In fact, we make the first transmission hologram, the master hologram, in a conventional manner, and then, the master hologram is reconstructed by a cylindrical beam for making the second rainbow hologram. If only the wavefront of the cylindrical beam is reasonably designed, ,it is still able to obtain a satisfactory reconstructed image.

Actually, if the scheme recording the master hologram is shown in Fig.(1), from the basic equations of hologram the relationships among the reconstruction point source $C(X_c, Y_c, Z_c)$, the object $O(X_o, Y_o, Z_o)$, reference source $R(X_R, Y_R, Z_R)$ and conjugate real image $I_{Re}(X_{Re}, Y_{Re}, Z_{Re})$ can be derived:

$$Z_{Re} = \frac{Z_o Z_R Z_c}{Z_o Z_R - Z_c Z_R + Z_c Z_o} \quad (1)$$

$$X_{Re} = \frac{X_c Z_o Z_R - X_o Z_c Z_R + X_R Z_c Z_o}{Z_o Z_R - Z_c Z_R + Z_c Z_o} \quad (2)$$

$$Y_{Re} = \frac{Y_c Z_o Z_R - Y_o Z_c Z_R + Y_R Z_c Z_o}{Z_o Z_R - Z_c Z_R + Z_c Z_o} \quad (3)$$

here, suppose that the holograms are recorded and reconstructed in the same wavelength without amplification. If the reference wave to record the master hologram is plane wave, that is

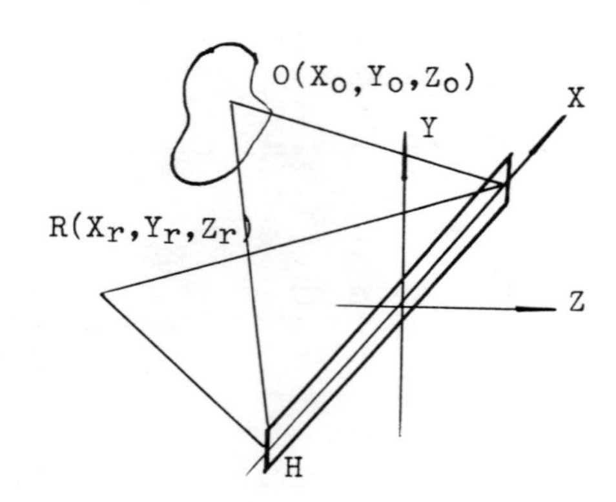

Fig.1. Schematic diagram of recording master hologram.

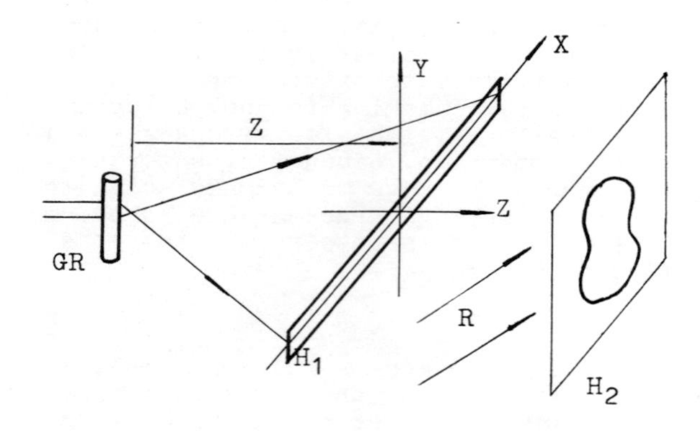

Fig.2. Schematic diagram of recording rainbow hologram.

$Z_R = \infty$, we have

$$Z_{Re} = \frac{1}{1/Z_c - 1/Z_o} \tag{4}$$

$$X_{Re} = \frac{X_c/Z_c - X_o/Z_o}{1/Z_c - 1/Z_o} \tag{5}$$

$$Y_{Re} = \frac{Y_c/Z_c - Y_o/Z_o}{1/Z_c - 1/Z_o} \tag{6}$$

If the master hologram is reconstructed by cylindrical wave, as shown in Fig.(2), for recording rainbow hologram H_2, it is plane wave in the virtical direction, so we have $Z_c = Z_{c\perp} = \infty$, By using three equations(4), (5),(6), we can obtain the coordinates of the conjugate image in virtical direction.

$$Z_{Re\perp} = -Z_o \tag{7}$$

$$X_{Re\perp} = X_o \tag{8}$$

$$Y_{Re\perp} = Y_o \tag{9}$$

In the horizontal direction one has $Z_c = Z_{c\parallel}$, in this case, the coordinates of conjagate image in the horizontal direction are

$$Z_{Re\parallel} = \frac{1}{1/Z_{c\parallel} - 1/Z_o} \tag{10}$$

$$X_{Re\parallel} = \frac{X_c/Z_{c\parallel} - X_o/Z_o}{1/Z_{c\parallel} - 1/Z_o} \tag{11}$$

$$Y_{Re\parallel} = \frac{Y_c/Z_{c\parallel} - Y_o/Z_o}{1/Z_{c\parallel} - 1/Z_o} \tag{12}$$

From Fig.(2), we can see that it is easy to satisfy the condition $X_c = Y_c = 0$. Moreover, the optical arrangement is so designed as to satisfy the following condition:

$$|Z_{c\parallel}| \gg |Z_o|$$

one has

$$Z_{Re\parallel} = -Z_o \tag{13}$$

$$X_{Re\parallel} = X_o \tag{14}$$

$$Y_{Re\parallel} = Y_o \tag{15}$$

It is evident that these results are the same as the equations(7),(8),(9), in the virtical direction.

Discussions

By using above two-step method with a common glass rod, there is a little loss of the object beam, which comes from the master hologram, and optical elements are very simple and available. If only the master hologram has high quality, even without bleaching, we are able to obtain a satisfactory large rainbow hologram of 500 X 500 mm^2 or more than 1000 X 1000mm^2 with a 30 mw He-Ne laser. The experiments have demonstrated that it is good and practical method for making large rainbow holograms with a broad viewing angle.

References

1. Y.N.Denisyuk, Soviet Phys. - Doklady, 7, 543(1962)
2. S.A.Benton, J.O.S.A. 59, 1545 A (1969)
3. H.Chen Optics Letters, 2, 85-87 (1978)

Ohya Large Holography Studio in a huge rock cavern

K. Watanabe (Holotec, Inc.)
A. Hirosawa (Institute of Scientifically Synchoronized with New Life)
T. Morie (Shimizu Construction Co., Ltd.)
M. Suzuki (Fuji Photo Optical Co., Ltd.)
H. Yamashita (Mazda Moter Corporation)

Introduction

There is a great demand recently for a large size hologram, 2meters width for instance, in which a brand-new car, a life-size statue and sometimes a group of the them, or interior spaces could be taken. Especially, in the world of advertisement or art, there is a big demand for a large hologram. The larger hologram needs the larger space for its suitable holography. This is the report of a large underground studio built in Japan.

Outline of the studio

The large holography studio was built in a district called Ohya, which is located about a hundred kilometers north of Tokyo near the famous resort Nikko as shown in Figure 1.

In Ohya, special stones for house building called Ohya-stone have been quarried since three hundred years ago. There remains a huge cavern under the ground where the stones have been quarried out for such a long time(Figure 2.). The space is now used for warehouses or underground experimental rooms. The holography studio was built as one of these utilizations of underground caverns.

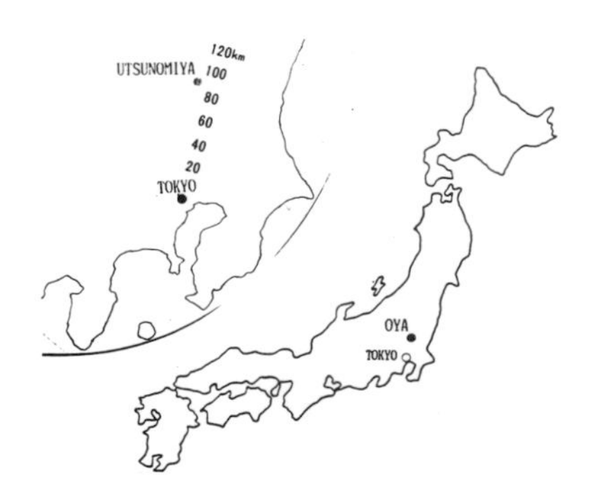

Figure 1. Map of Ohya

Figure 2. The underground cavern of Ohya

The conditions required for a large holography studio are:

1. No vibration should be transmitted from outside.
2. Vibrations should be absorbed.
3. There should be no air disturbed.
4. It could be completely dark.
5. Electric power should be available.
6. Water should be available.

The nature provides all these conditions for the Ohya underground cavern.

The size of the studio now used is 13m(w) X 7.6m(l) X 3.4m(h), where even two cars can be taken at the same time. The hologram of 2m X 1m can be developed here. The conceptual view of the hologram studio is shown in Figure 3

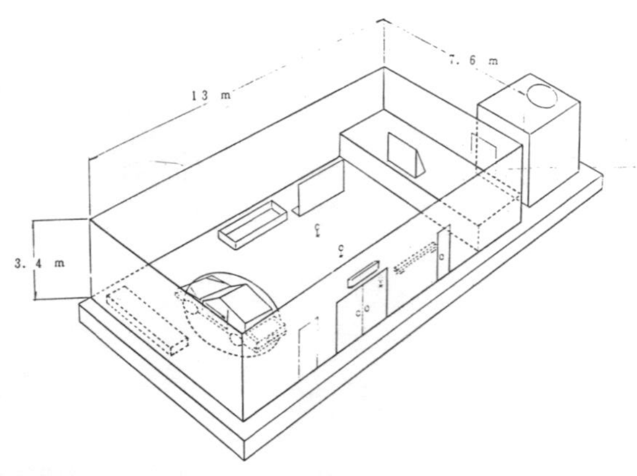

Figure 3. Conceptual view of the Ohya Holography studio. Perspective view of the studio.

Characteristics of the underground cavern subject to vibration

A part of the Ohya-stone cavern is shown here(Fig.4). The Ohya-stone mainly consists of zeolite as shown in Table 1.. It was made from volcanic ash being settled and coagulated forming a huge bed rock.

Figure4. A part of the cavern

Silicic Acid	66.96	SiO_2
Oxided Iron	1.85	Fe_2O_3
Aluminum	12.55	Al_2O
Manganesc	0.06	MnO_2
Magnesium	0.47	MgO_3
Potassium	2.35	K_2O
Sodium	2.87	Na_2O
Water	11.02	H_2O

Table 1 . Ingredients of Ohya-stone

The rock mass as illustrated in Fig5 is lying over the Ohya area, where the entire mass is continuous with the thickness of 300m, featuring a huge solid mass which is stable against vibration.
The measured results of desplacement, velocity and acceleration are 1/10-1/100 of those of the regular sedimentary foundation.

It is well known that a rock mass is more tolerant against earthquake compared to the regular sedimentary foundation. The state of seismic vibration was compared by numerical methods between a building on sedimentary foundation and a tunnel in a rock mass using the seismic wave of the earthquake occurred in Tokachi-offshore. The results shows that the tunnel at 75m depth sways only 4mm while the 20m high building sways 10cm(Fig. 6).

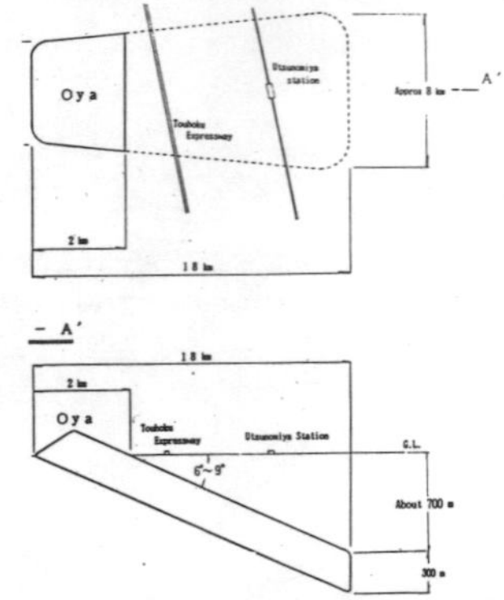

Figure 5. The Plane and side view of the Ohya-stone layers

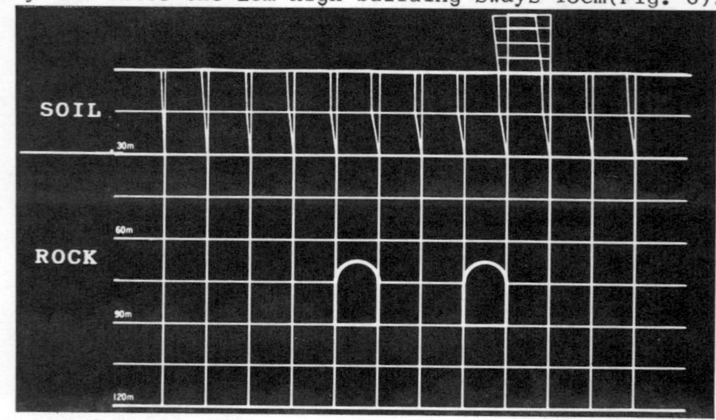

Figure 6. Displacements of surface and underground at earthquake

The porous structure of Ohya-stone, having the porosity of 40% while that of an ordinary rock is less than 20%, is considered to contribute effectively in reducing the vibration. The seismic energy transmiting the rock mass will be transformed, through the expansion and contraction of flexible pores with air, from potential energy to thermal energy and finally emitted absorbing the seismic vibration as illustrated in Fig. 7..

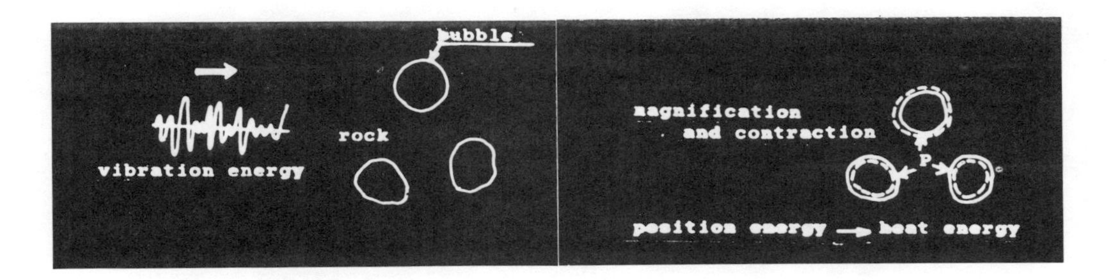

Figure 7. Dry Ohya-stone Wet Ohya-stone, containing wet Ohya-stone in the layer.

Experimental results and discussions appraisal of the conditions provided by the Ohya-stone cavern

The bed rock of Ohya is ideal as a vibration-proof material. Adding to that, the shape of the cavern could be easily decided by excavating the rock.
Whether this studio is available for interference measure, especially against vibrations, depends on the degree of the rock absorbing the exciting vibrations.The rock base here is hardly resonant with vibration because, as was mentioned , the layer is huge and the condition inside is similar to clay formations.
It is possible, therefore, to use CW Laser to measure vibrations. The Fig.8 shows an example that is the vibration mode of a car measured with Ar Laser. A big exciting force is produced by a big vibrator,but its repulsion is absorbed in the rock mass.

Figure 8. Vibration mode of a car

As the large Hologram needs long exposure time, the air disturbance is the big problem. But here, the rock itself is quite large, thermal capacity is big and the moisture on the rock surface doesn't change throughout the year. These conditions help the air to be stable in this studio.
To see the rock completely static, and to measure air disturbance, Michelson Interferometer with He-Ne Laser as the light source was used to measure the background vibration of our holography studio. The interference fringe did not move almost at all. I was found that this place was suitable for hologram studio.

This Fig.9. shows the example when seven big statues were taken in 2.0m X 1.0m Fresnel hologram.
No air disturbance occurred and a bright image was reconstructed.

Figure 9. the reconstructed image of Shichifukujin(seven lucky statues)

Darkness of the studio can be easily gained as the cavern is in the rock. Electric power is also easily supplied because this place has been used as a huge food storehouse. Water cooled Ar or Kr Laser of 5 to 10 watt needs 10 to 20 liters of water a minute. Water was the problem of the first stage, but solved soon by digging a well. It provides enough water to cool the device now(Fig. 10,11).

With Ar Laser and Kr Laser, colored Fresnel type large hologram of a car was taken.

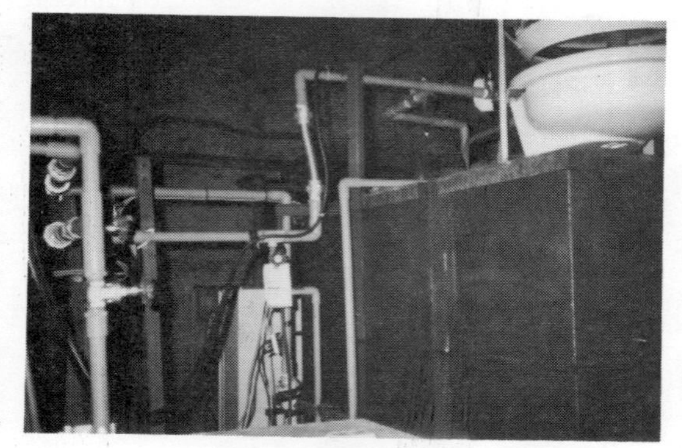

Figure 10. Photography of water equipment

Figure 11.

There is a big turn-table in this studio and it is used to turn the object such as a group of people (Fig. 12). It is also used for the original picture of the multiplex hologram of a large object. Turning speed adjusted to the multiplex hologram is one turn per minute. The diameter is 4 meters.

Fig. 13 shows an example of the reconstructed image of the multiplex hologram produced from the original picture taken in this studio.

Figure 12. Picture of the turntable

Figure 13. Reconstructed image of Sankaijuku in multiplex hologram

This studio could be used for making a large rainbow hologram because it works as if there were a big vibration-proof foundation. Lippman hologram with a big D C G plate was taken here.

Conclusions and Acknowledgments

The studio has been proved to satisfy all the conditions necessary to take the desired hologram pictures. Although the drawbacks of this studio is the relatively high humidity and the law temperature, they can be prevented by treating the wall surface with mortar paints, and setting equipments firmly.

We would like to express our gratitude to professor J. Tsujiuchi of Tokyo Institute of Technology.

Holographic registration and spectroscopy of a modulated laser radiation

V.N.Borkova, V.A.Zubov, A.V.Krajskij

P.N.Lebedev Physical Institute of the Academy of Sciences of the USSR,
Moscow, Leninsky prospect, 53, USSR

Abstract

The principle of recording the modulated optical signals in the holographic scheme used a nonstationary reference wave with linear change of frequency along the coordinate and in time in the detection plane is considered. The analysis of the signal spectrum reproduction is presented. The description of the experimentally realized scheme to form the nonstationary reference wave, of the holographic registration and reconstruction scheme is given. The expression of the instrumental function for recieving the signal spectrum pattern is presented. The results of measuring the spectrum components intensity distribution are presented. The examples of oscillating chemical reaction and of two liquids diffusion illustrate the scheme operation.

A nonstationary reference wave used in the process of holographic detection enables a complete recording and processing of a modulated optical signal. This detection method makes it possible to obtain information about time variation of the signal, about its spectrum, and about the variation of the signal phase at the reconstruction stage. An analysis of the spectral composition of a modulated signal is of great interest for it contains information about the object. Systems of this type may be applied to study processes in which optical properties of the medium, the refractive index in particular, change. Such processes can be initiated, for example, by changes in the inner structure of the medium. Another possible application is the study of biological objects either in the scattering of light of extremely low frequencies, or while measuring the parameters of the motion of biological objects (or their parts) by Doppler technique. This method can also be applied for comparing two independent lasers in the spectral range of interest.

The scheme is based on the following principle. There is laser radiation modulated due to the interaction with an object or a medium whose optical properties vary with time. This radiation is directed onto a hologram. The hologram is simultaneously irradiated with a nonstationry reference wave. In the simplest case, the wave is characterized by linear frequency variation along the section in the detection plane. In this case each section detects, against a certain background, an interference pattern of a specific spectral component of the signal [1]. In a more complicated case considering in present paper, the frequency of the reference wave varies in the detection plane linearly along the coordinate and in time. In this case each spectral component is recorded on a photoplate as a part of a cylindrical Fresnel zone plate [2]. The use of a nonstationary reference wave with linear frequency variation along the coordinate and in time appears to be more convenient for spectroscopy as compared to the simplest case. The distribution of information about each spectral component over the photoplate surface enhances the diffraction efficiency, widens the dynamical range of the detected intensities, and simplifies the spectrum recording procedure. The two schemes correlate with each other almost in the same way as the holography of focused image correlates with Fresnel holography.

We consider the scheme operation. The hologram is illuminated with a modulated light signal

$$E(x,t) = E(t).\exp\left[-i\omega t+i(\omega/c)x.\cos\theta\right] \quad . \tag{1}$$

This radiation is directed onto the hologram at an angle θ . The nonstationary wave described by the expression

$$E_r(x,t) = E_r.\exp\left[-i\omega t -i(\omega/c)axt-i(\omega/c)bt^2\right] \tag{2}$$

is used as a reference one. This wave has a linear frequency variation along the coordinate and in time in the detection plane. After an exposure and photochemical treatment, one obtains a hologram whose amplitude extinction is given by

$$t(x) \sim t_r + gE_r \int_{-T/2}^{T/2} E(t).\exp\left[i(\omega/c)axt+i(\omega/c)bt^2\right].\exp\left[i(\omega/c)x.\cos\theta\right]dt + c.c. \tag{3}$$

The field of the signal E(t) is recorded at every moment on the hologram as an interference pattern. The resulting interference pattern is, therefore, averaged over the exposure time T . It is convenient to rewrite the expression for the amplitude transparency of the hologram, taking into account the apectral representation of the signal

$$E(t) = \int E(\Omega).\exp\left[-i\Omega t\right]dt$$

where $E(\Omega)$ is the field spectrum. As a result we get

$$t(x) \sim t_r + gE_r \int \left\{ E(\Omega).\exp\left[-i\frac{\omega a^2}{4cb}\left(x - \frac{c\Omega}{\omega a}\right)^2\right] . \int_G \exp\left[i\frac{\omega a^2}{4cb}(\xi - x)^2\right] d\xi \right\} d\Omega + \text{c.c.} \quad (4)$$

where we substituted $\xi = (c\Omega/\omega a) - 2bt/a$. Integration is performed over the region G , which is an interval

$$\left[(c\Omega/\omega a) - 2bT/a\ ,\ (c\Omega/\omega a) + 2bT/a\right] \quad (5)$$

and T is the total detection time. To represent the result in more simple form, we shall assume that the radiation $E(\Omega) = \delta(\Omega - \Omega_0)$ is modulated monochromatically. In this case the amplitude transparency of the hologram, which is given by expression (4), contains two factors. One describes the field due to a cylindrical Fresnel zone plate, the other, due to the diffraction of a plane wave on a cutting diaphragm. Thus, every spectral component of the signal is recorded on the hologram as a part of a cylindrical Fresnel zone plate, and all the elements of these parts are displaced along the x-axis at a distance proportional to the frequency of the component.

The experimental set-up was realized as follows (Fig. 1). The chahhel, containing telescopic system L1, L2, formed a reference wave of the desired structure due to the displacement at a constant speed v of the first element L1 of the system. The diaphragm D3, which performs spatial filtration, increases the homogeneity of the reference wave. The radiation is directed onto the hologram H, spaced at the double focal length from element L2 of the system. The front of the reference wave is rotated about the axis passing through the focal plane of the second element L2 and the projection of this axis is linearly displaced along the hologram formed by the plane radiation wave. The boundary of the irradiated spot on the hologram remains unshifted. This leads to frequency variation along the coordinate and additional temporal frequency variation for a specific hologram area, which are linear in the first approximation. The second channel contains telescopic system L3, L4 and an object Ob, which modulates the radiation. The diaphragm D4 plays the role of a spatial filter and is, in fact, the input slit of the spectral apparatus. The width of the slit during the experiments was chosen a normal one.

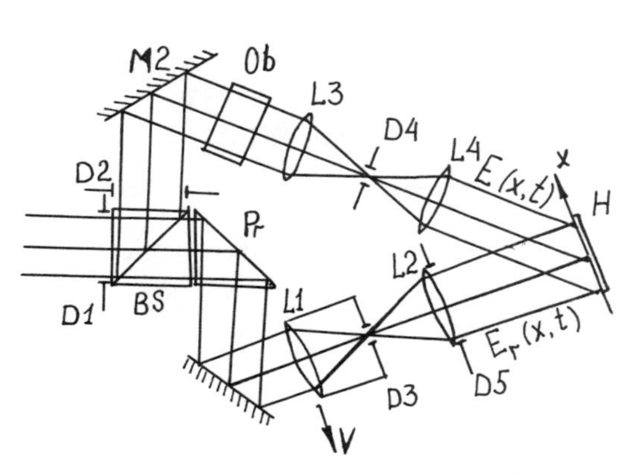

Fig. 1. Experimental set-up for holographic detection, using a nonstationary reference wave with linear frequency variation along the coordinate and in time in the detection plane.

Information about the signal spectrum is obtained at the reconstruction stage during the irradiation of the hologram H by a stationry plane wave conjugated with the signal wave at an average in the direction (Fig. 2)

$$E_p(x,t) = E_p.\exp\left[-i\omega t - i(\omega/c)x.\cos\theta\right] \quad (6)$$

The signal component of the reconstructed field is formed at a distance

$$\rho = (\omega/c).(2b/a^2)$$

from the hologram. It is described in the Fresnel approximation by the expression

$$E(u) \sim gE_pE_r(\pi/b).\exp\left[i(\omega/c)\rho\right] \int \left\{E(\Omega) \int_{-T/2}^{T/2} \exp\left\{i\left[(\omega/c)au - \Omega\right]t\right\}dt\right\}d\Omega \quad . \quad (7)$$

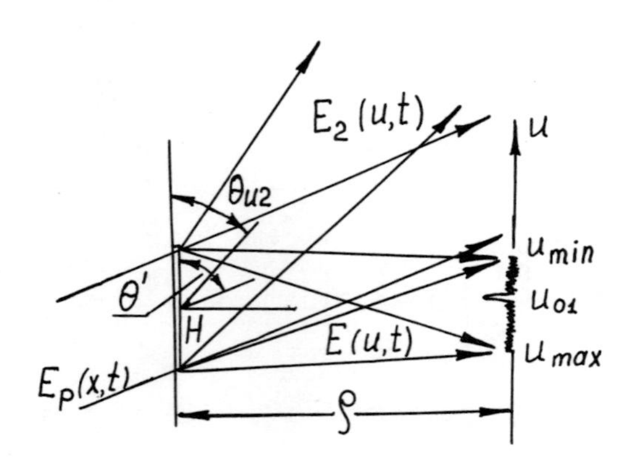

Fig. 2. The experimental set-up to obtain information about the spectrum of a modulated optical signal recorded on a hologram.

Integration with respect to t yields the function sinc , which determins the position of the reconstructed image and resolution. At the maximum of the function sinc equation (7) takes the form

$$E(u) \sim gE_pE_r(\pi/b)2x_o \cdot \exp\left[i(\omega/c)\rho\right] \cdot E(\Omega) \quad (8)$$

where $\Omega = (\omega/c)au$. Hence, each cylindrical Fresnel zone plate forms the real image of a certain spectral component, i.e. the picture of the light field spectrum $E(\Omega)$ is formed in the image plane. Further, the spectrum of the modulated optical signal can be detected by usual means.

The spectral characteristics of the scheme can be estimated by analysing the radiation modulated by a single frequency Ω_o . We have in this case

$$E(t) = \exp\left[-i\Omega_o t\right] \quad ,$$

$$E(\Omega) = 2\pi \cdot \delta(\Omega - \Omega_o) \quad , \quad (9)$$

$$t(x) \sim t_r + gE_r(T/2b) \cdot \exp\left\{-i\left[(\omega/c)ax - \Omega_o\right]^2 / \left[4(\omega/c)b\right]\right\} \cdot \exp\left[i(\omega/c)x \cdot \cos\theta\right] + c.c.$$

in the range determined by relation (5). The center of the zone plate for radiation modulated with frequency Ω_o lies at the point

$$x\Omega_o = \lambda\Omega_o/2\pi a \quad .$$

When the spot on the hologram is uniforly irradiated and the displacement vT of the element L1 does not exceed the hologram size $2x_o$, one can use the following estimates:

$$\nu_{+mo} = a(-x_{mo})/\lambda \quad , \quad \nu_{-mo} = a(x_{mo})/\lambda$$

where $\nu_{\pm mo}$ are the upper and lower frequencies. This gives the dispersion range

$$\Delta\nu = 2ax_o/\lambda \quad (10)$$

which is determined by the frequency interval in reference wave. The instrumental function broadens at a distance of avT/λ from the hologram boundary. In the remaining part of the dispersion range the spectral resolution due to a finite detection time within the Rayleigh criterion is described by (see (9)) $(\Omega - \Omega_o)T/2 = \pm\pi$ and is equal to

$$\delta\nu = 1/T \quad .$$

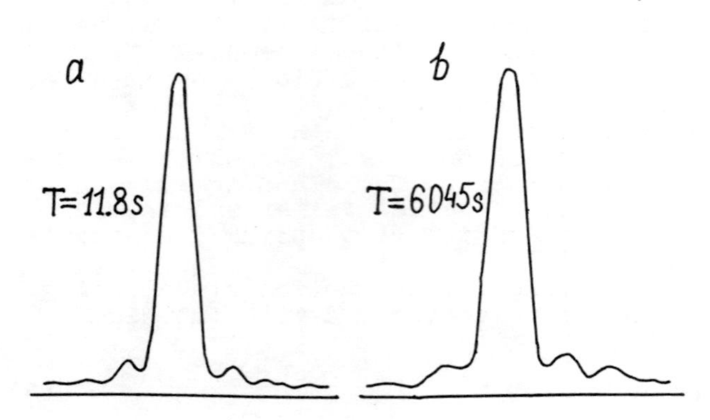

Fig. 3. An example of the instrumental function obtained for total registration time T = 11.8 s and T = 6045 s.

We experimentally realized the following characteristics: the resolution

$$\delta\nu \sim 0.2 \text{ Hz} - 0.2 \text{ mHz} \; ,$$

depending on the detection time; the number of resolved spectral elements is about 100 on both sides of the zeroth (nonmodulated) component.

Fig. 3 shows the instrumental function. This curves were obtained in the absence of signal modulation, the exposure time was equal to T = 11.8 s (Fig. 3a) and T = 6045 s (Fig. 3b). The widths of the instrumental functions at half height determined from the experimental spectra with due regard to the dispersion value were equal to

$$\delta\nu_{exp} \simeq 0.077 \text{ Hz and } \delta\nu_{exp} \simeq 0.168 \text{ mHz}$$

which is in good agreement with calculated values

$$\delta\nu_{calc} = 0.075 \text{ Hz and } \delta\nu_{calc} = 0.147 \text{ mHz} \quad .$$

The problems concerned with photometric measurements were experimentally studies, since accurate recording of the spectral component intensity can be affected by several mechanisms, the nonlinearity of the photomaterial S-shaped curve, in particular. In this experiment we used signals with rectangular modulation and variable relative pulse duration. The spectrum of this signal is easy to compute. The intensity of the N-th spectral component with respect to the intensity of the unshifted component is given by

$$I_{calc} = I_N/I_o = \left\{ \mathrm{sinc} \left[\pi N \tau / T \right] \right\}^2 \qquad (12)$$

where τ is the pulse duration, T is the modulation period. Typical experimental and calculated data for different relative pulse durations are listed in Fig. 4. All harmonic intensities are normalized to the intensity of the zeroth component. One may deduce that the spectrum behaviour is described enough correctly. An analysis of experimental errors characterizing the deviation from the calculated values reveals that there is some systematic error, since deviations of one sign prevail over deviations of the other sign. This error can be estimated as an arithmetic mean of the deviations from the calculated values. For the three spectra given in Fig. 4 these errors are equal to -0.005 , -0.003 and -0.035 , respectively. Along with the systematic error, the results obtained contain a random error (the scatter of experimental points around the mean value), which may be estimated from the variation of the general dispersion σ , calculated from Pirson's distribution with a probability of 0.98. The corresponding ranges are $0.008 \leq \sigma \leq 0.017$, $0.006 \leq \sigma \leq 0.011$ and $0.026 \leq \sigma \leq 0.059$ with a maximum intensity equals 1 .

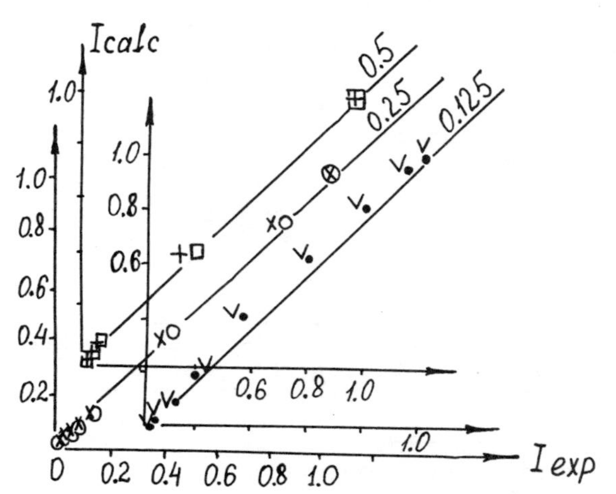

Fig. 4. A comparison of experimental and calculated data about intensity distribution in spectrum by rectangular modulation and variable relative pulse duration.

Here we give some examples which illustrate the potentialities of the proposed holographic spectroscopy technique.

One of the study objects was an oscillating chemical reaction - the Belousov-Zhabotinskij reaction [3] . Reactions of this type represent in biophysics the models of biological clocks. We used the well known solution, characterized by the transition of manganese ions from two-valent to three-valent state. The concentration of manganese sulphate was about 0.001 M , the thickness of the sample layer l = 50 mm. Fig. 5 shows the time behaviour of this chemical reaction in the absence of synchronization in the reaction volume. The result is obtained for the whole volume. We carried out a series of experiments, and all the results are plotted in the Figure. There are also shown the exposure times and the accuracy of measuring the frequency. The abscissa axis plots the time from the beginning of the reaction, and the ordinate axis plots the frequency shift caused by change of the optical properties of the medium in time. We note that the frequency shift amounts to 1.7 mHz, which corresponds to the refractive index variation rate

$$dn/dt = -(\lambda/l) . \Delta\nu \simeq -2.10^{-8} \ s^{-1} \quad .$$

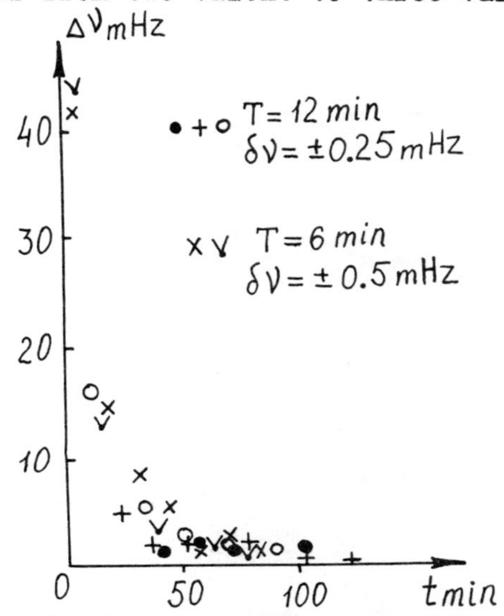

Fig. 5. The modulation of optical radiation in the course of a chemical reaction. The time is measured from the reaction beginning.

To synchronize the reaction through the whole volume we stirred the solution at a frequency of about 1 Hz. The amplitude modulation of the solution transparency was about 10 - 15 % (Fig. 6), which might cause the increase of the first spectral harmonic less that 4 % of the intensity of the zeroth harmonic. Fig. 6 also gives an example of the oscillating reaction spectrum. The spectrum consists of a number of components, whose intensities are high and different for positive and negative frequency shifts. This means that in the course of the reaction there is a significant unharmonic phase modulation. The effect of amplitude modulation can be neglected, for it is rather small. Estimates performed for two very different concentrations of the active substance are presented in Table.

Another possible application of the holographic spectroscopy technique deals with studying the diffusion of two liquids. If the direction of the instrument dispersion coincides with the interface, lying in the horizontal plane, there appears a two-dimensional

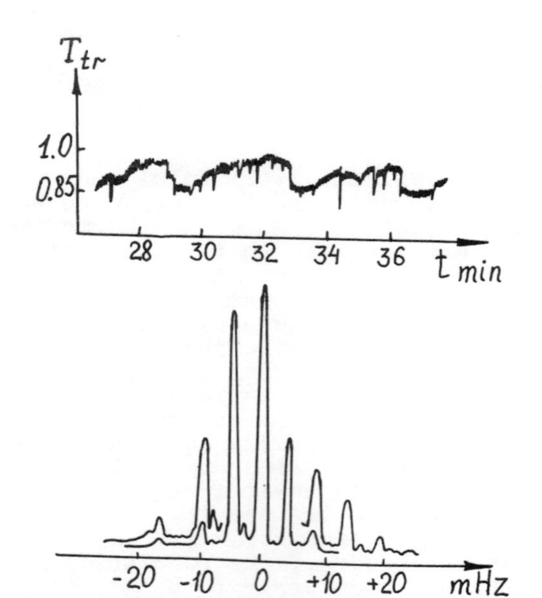

Fig. 6. The amplitude modulation of the solution transparency during the oscillating chemical reaction (a) and the spectrum of optical signal transmitted through a solution where an oscillating chemical reaction takes place (b).

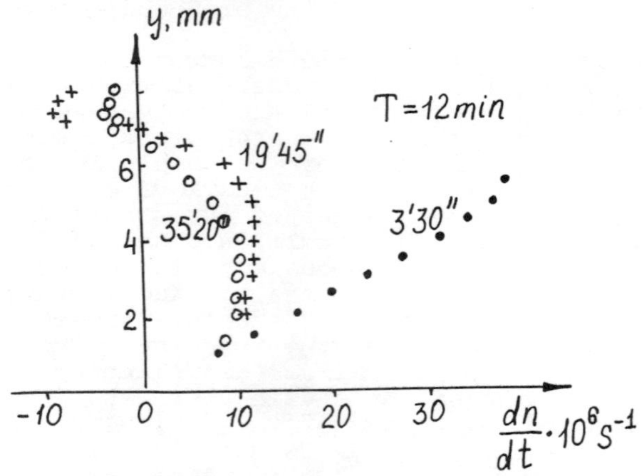

Fig. 7. The rate of the refractive index variation in the diffusion process of two liquids as a function of the coordinate plotted in the direction normal to the interface.

Table

T	$\Delta\nu = 1/T$	$\Delta\nu_{exp}$	$\Delta n \cdot 1/\lambda$	Δn
s	mHz	mHz		
222	4.5	4.6	1.2	19×10^{-6}
85	11.8	12.1	0.6	7.5×10^{-6}

pattern in the plane of spectrum detection. This pattern shows the distribution of the spectrum in the direction normal to the interface (vertical direction). This disrtibution refers to the plane, which is optically conjugated with the plane of spectrum formation. Because a cylindrical zone plate is formed on the hologram surface, which acts along the frequency axis, the reconstructed light beam becomes astigmatic. This means that optical conjugation takes place only in the vertical direction. The plane, which is optically conjugated with the plane of spectrum formation, can be shifted to any position of interest, to the output plane of the sample cell, in particular. Fig. 7 gives the experimental data for the diffusion of two liquids: 1.5 M solution of sulphuric acid and water. The exposure time equal 12 min, the cell thickness is 1 mm. The horizontal axis plots the rate of the refractive index variation in the diffusion process, the vertical axis plots the coordinate normal to the interface of liquids. The presented dependences correspond to different times measured from the beginning of the diffusion process.

The analysis of experimental data shows that omitting complicated calculations and processing of numerous interferogram, it is possible to get the function

$$dn/dt = f(t,y)$$

directly by processing a single hologram.

References

1. Borkova V.N., Zubov V.A., KrajskijA.V. Holographic recording of time dependent variable optical signals in the schemes with nonstationary reference wave, in book: Holographic processing of information based on the use of nonstationary fields (P.N.Lebedev Phys. Ins. of Ac. of Sci. of USSR, v. 131), Moscow, Nauka Publ., 1982, p. 68 (in Russion).

2. Borkova V.N., Zubov V.A., KrajskijA.V. Holographic detection of modulated optical signals with the use of nonstationary reference wave, having the coordinate and time frequency variation. Preprint of P.N.Lebedev Phys. Ins. of Ac. of Sci. of USSR № 165, Moscow, 1984 (in Russian).

3. Zhabotinskij A.M., Concentration autooscillations. Moscow, Nauka Publ., 1974 (in Russian).

THE CGH MADE BY DOT MATRIX PRINTER

ZHao Mingxi, Hsu Dahsiung, Fei Zhuzeng

Depart. of Appl. Physics, Beijing Institute of Posts & Telecommunications, Beijing, China.

Abstract

The conventional CGH is made by a graph plotter, in this paper a new way to make CGH by dot Matrix printer has been introducted. The CGHs of 16X16 samples of character 'F' 32X32 samples of Chinese word '光' and 64X64 samples of Chinese word '华' and their reconstructed images are given. The method is useful in industry science and instruction.

1 Introduction

Computer holography is based on the combination of digital computing and of modern optics. It began in 1965 at IBM, when Prof. Lohmann's laser broke down. He wanted to make holograms. In this emergency situation, he made holograms with the help of a computer instead of a laser. Hence the period of the computer-generated hologram (CGH)[2] began. The conventional CGH is made by a graph plotter. We tried to use a new way to make CGHs with the dot matrix printer. First, the computer we used is Apple-2, which controlls the dot matrix printer, and the printer produces clear CGHs which reconstructs original characters, and now we also use IBM personal computer and Brother M2024 printer to make CGHs.

2 C G H

One of the advantages of CGH is that its fabrication merely needs the data of mathematical description of the object, it can be realizable without the object. Consequently ,with the computer we can create optical elements, which cannot be fabricated by conventional methods[1] .The CGH can widely be applied in various fields such as optical information processing[3] ,interferometry[1] ,and laser beam scanning[4] ,etc.

CGHs can roughly be classified into two types. One is Detour phase hologram and the other is a computer-generated interferogram. Besides there are several varieties of CGH types for different coding. In order to test our method of making CGHs we choose the Lohmann-3 type of Detour phase and Burckhardt's method.

3 The main points of our method

Our computers are the personal computer APPLE-2 and IBM-XT, using BASICA . For points of the 1,2,3,4,7,8 the normal method is used but for points of the 5,6 we use the dot matrix printer instead of the graph plotter.

1. Inputting the data of an object to computer
2. Adding the random phase factor to the data
3. Making 2-D. F.F.T.
4. Coding according to Lohmann type or to Burckhardt
5. Processing the coding data for dot matrix printer
6. Typing out the CGH by dot matrix printer
7. Photoreduction
8. Reconstruction by the optical reconstruction system

4 The experiments and results

In order to show our method, we do some following experiments.

1. We choose N=16 and the object has 16X16 samples of character ' F ' . Fig. 1 is the CGH before being photoreducted. The image reconstructed from the CGH is shown in Fig. 2
2. N=32, the object has 32X32 samples of Chinese ' 光 '. The image is also reconstructed from the CGH . It is shown in Fig.3

For 1.2 the CGHs are made by Apple-2 and FAX-100 printer.

3. N=32, the object is Chinese '光' with 32X32 samples. The CGH is made by JB-3001 Panasonic Computer and JB-3022 Dot Matrix Printer.
4. N=64, the object is Chinese '华' with 64X64 samples.

For 1, 2, 3, 4 the CGHs are made according to Lohmann type.

5. N=64, the object is Chinese '华' with 64X64 samples. But the CGH is made according to Burckhardt's method. A part of the CGH is shown in Fig. 4.

For 4, 5, the CGHs are made by IBM-XT Personal Computer and Brother M2024 Printer.

5 Advantages

1. Now personal computers are widely used in institutes and universities in China. They are typically with a dot matrix printer , but without a graph plotter. Our method is useful in popularizing computer holography.

Prof. W. H. Lee said that the CGH is not only used in industry and science but also in instruction. Our method is useful in all fields mentioned.

2. CGHs can be produced more rapidly by a printer than by a plotter.

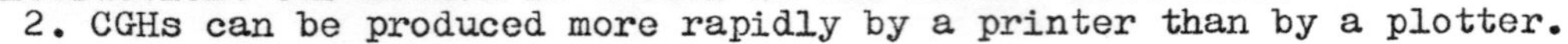

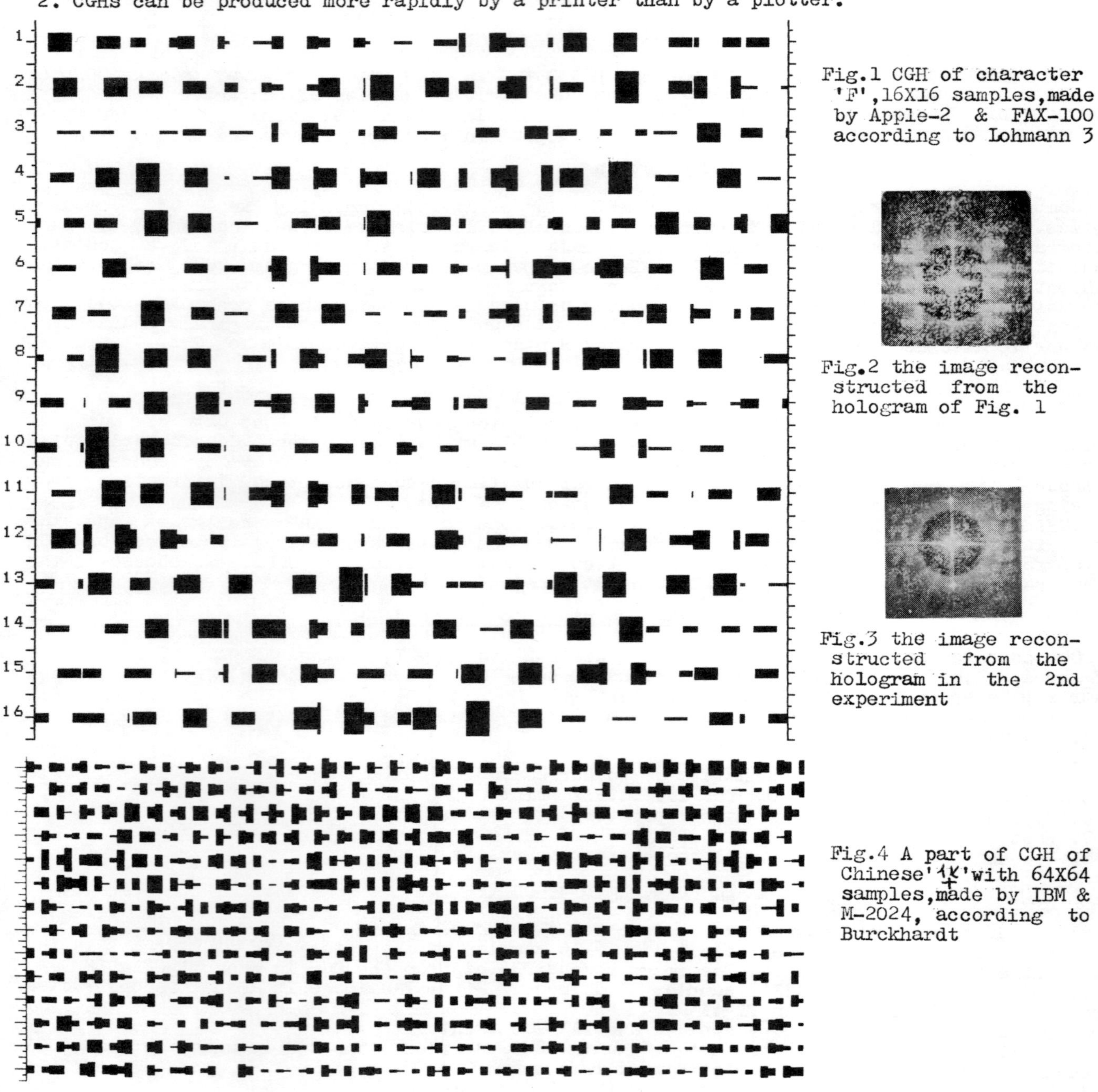

Fig.1 CGH of character 'F',16X16 samples,made by Apple-2 & FAX-100 according to Lohmann 3

Fig.2 the image reconstructed from the hologram of Fig. 1

Fig.3 the image reconstructed from the hologram in the 2nd experiment

Fig.4 A part of CGH of Chinese'化'with 64X64 samples,made by IBM & M-2024, according to Burckhardt

References

1. W. H. Lee Progress in Optics, ed by E. Wolf, Vol.XVI, 1978.
2. Yu Zuliang and Chin Kuofan, Computer Generated Holography (in Chinese), 1984.
3. B. R. Brown and A. W. Lohmann, Appl. opt. , 5, 96, 1966.
4. O. Bryngdahl and W. H. Lee, Appl. Opt.,15, 183, 1978.

Making CGH By Using Electron Beam

Yan Yingbai Yu Dongxiao Chin kuo-fan

Department of Precision Instruments, Tsinghua University, Beijing, China

Abstract

The Electron-beam Lithography is a newly developed microfabrication technique. The principle and technology of making CGH by E-beam were discussed. The influences of error on the E-beam Generated Holograms have been analysed. The experimental results of the differential filters and a CGH scatter-plate have also been presented.

Introduction

Since the invention of Computer-generated Hologram by Lohmann in 1966 most CGH were fabricated by the process of plotting, photoreducing or laser-scanning. Recently, a new approach that is to make a CGH by using an electron-beam on an E.B.scanning machine. Or say it is an E.B.CGH. This method has the superiority in obtaining the high spatial frequency, less errors and enhancing the production efficiency. So it is a method of high potentiality and good prospect.

We have tried to make such a hologram on a China made DB-3 E.B.scanning machine. Although it has only limited functions; fortunately, we have obtained some preliminary good results and have tried to reduce the errors by a program of correction.

Principle of fabrication

The principle of fabricating E.B.CGH is just like the conventional CGH. Take the Fourier-transform hologram of Lohmann type as an example to describe the principle and method of the fabrication of E.B.hologram.

Just as the procedures of conventional CGH, first, we have to take samples of the object function and then we have the discrete function of the object:

$$U_s(m, n) = \sum_{m=-M/2}^{M/2} \sum_{n=-N/2}^{N/2} u(mx_o, ny_o) \tag{1}$$

where M,N are the numbers of cells.

Calculate the function by FFT program on a digital computer, then we have the function of the frequency domain.

$$\bar{U}_{jk} = \sum_{m=-M/2}^{M/2} \sum_{n=-N/2}^{N/2} u_{mn} \exp(-2\pi i(mj/M + nk/N)) \tag{2}$$

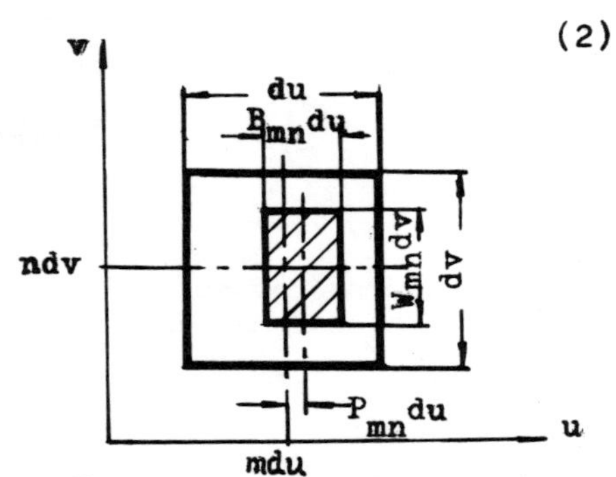

Fig.1. A typical sampling cell in a detour phase hologram.

Code the complex function by phase delay method initiated by Prof. Lohmann. Just as shown in Fig.1. Use a rectangular aperture to express the complex value of a cell. The size of the aperture represents the amplitude, the position of that is the phase. That is:

$$P_{mn} = \phi_{mn}/2\pi ; \quad B_{mn} = a_{mn} ; \quad W_{mn} = 1/2 \tag{3}$$

The data, after processing, is used as the input for the E.B.scanner. So the E.B.scanner is controlled by the computer of the scanner. An exposure on a chromine photoresist plate is obtained by the beam. After a series of processing, such as processing the plate, film hardening, gelatin removing and drying, an E.B.CGH is obtained.

Put it on an optical processor, a reconstruction of the object is displayed.

Experimental results

Several kinds of hologram have been made, such as:

1. Fourier-transform CGH of Lohmann type. Fig.2 (a) shows the local magnification of a Chinese character "巾",fabricated by the scanning of an E.B., (b) shows the reconstruction of this Chinese character "巾". The hologram is produced under the following conditions:

Accelerating voltage:	30 kV
Size of the scanning spot:	0.3 µm
The field of scanning:	2x2 mm^2
Scanning frequency:	100 kC
Photoresist:	PSS possitive photoresist

Scan the aperture shown in Fig.1. by an E.B. controlled by a computer and a program for correcting errors is prepared. Evidently, the displayed image is very clear, lines are quite sharp.

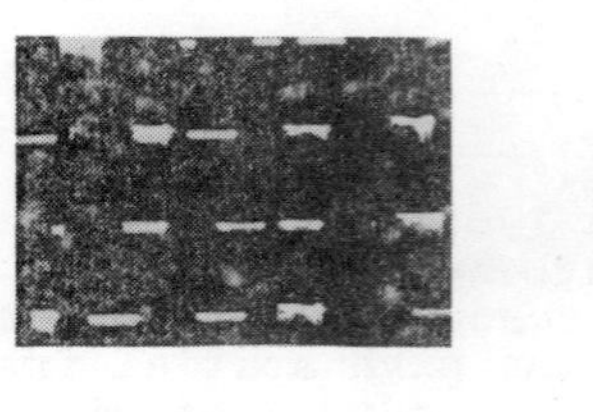

(a) (b)

Fig.2.(a) The E.B.CGH magnified by an electron-microscope. (b) the reconstruction of a Chinese character "巾".

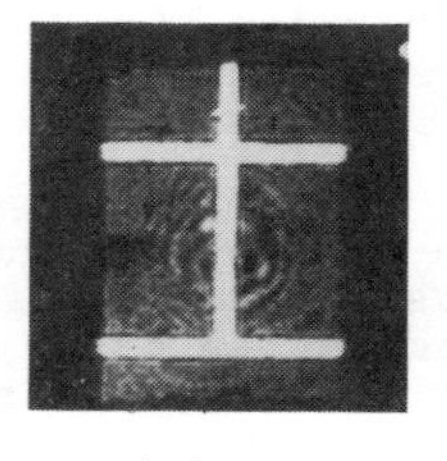

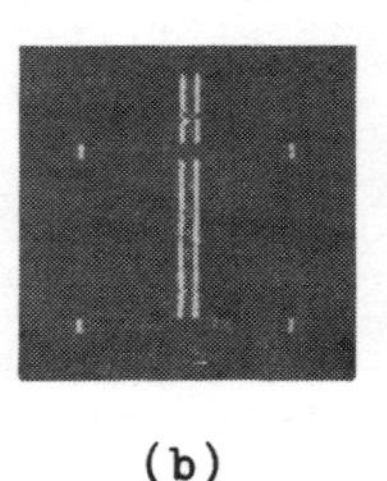

(a) (b)

Fig.3.(a) The original object "土".(b) The differentiated image after 1-D differentiation.

2. A spatial filter for differentiation.A spatial filter for differentiation is used for enhancing an image.A spatial filter of a one dimensional object can be written as:

$$H(u,v) = 2\pi i\, vx \tag{4}$$

For amplitude: That is $A(u,v) = 2\pi|v|\, x$

For phase: That is
$$\phi(u,v) = \begin{matrix}\pi/2 \\ -\pi/2\end{matrix} \tag{5}$$

It can have more higher carrier frequency for the E.B.CGH than that of a conventional one. It also has the advantages of 1. a larger image can be processed; 2. high signal/noise ratio; 3. aliasing of different orders is not easily produced. In Fig.3.(a) the original object "土"(a Chinese character) is shown. (b) shows the differentiated image after one dimensional differentiation. Fig.4. (a) and (b) show the letter " G " and its two dimensional differentiated image respectively.

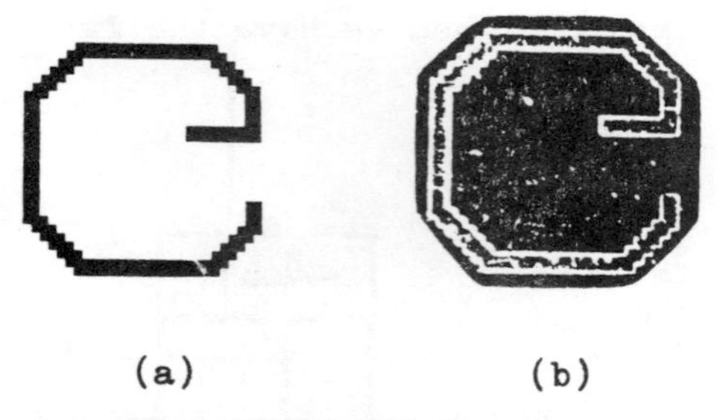

(a) (b)

Fig.4.(a) The letter "G". (b) It's two dimensional differentiated image.

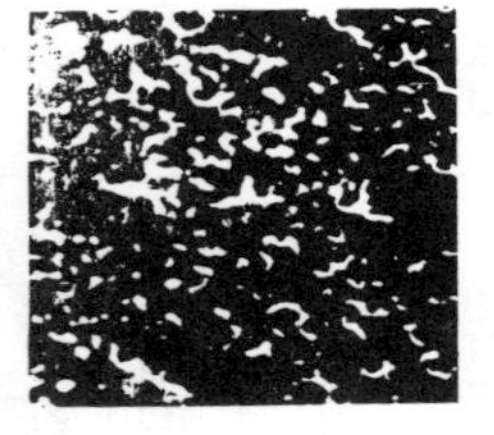

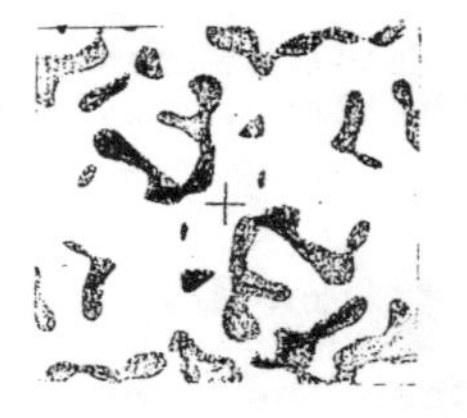

(a) (b)

Fig.5.(a) The local magnification of an electro-microscopic image of a CGH scattering plate. (b) The local magnified image of an optical holographic scattering plate.

3. CGH scattering plate.

The "scattering plate" is the core component of a scattering interferometer , it is suitable to be made by an E.B. scanner. Using a group of random number repeatedly many times by a scanning program , to construct the aperture of a cell of Lohmann type,then a 1024x1024 central symmetrical uniformly distributed hologram is obtained. This approach in comparison with the conventional two exposure optical hologram is more convinient and versatile. Fig.5.(a) shows the local magnification of an electro-microscopic image of a CGH scattering plate, 2x2 mm^2 , which is scanned by a E.B.spot of 0.5 µm in diameter. (b) is the local magnified image of an optical holographic scattering plate. Evidently, both are very similar.

Fig.6. shows the spectrun of the scattering plate. The non-uniformity of the scattered field is influenced by the quantity of the storage of the computer, for, in this case, only a group of 64x64 random number has been used repeatedly.

If the quantity of storage can be increased, surely the quality will be largely improved.

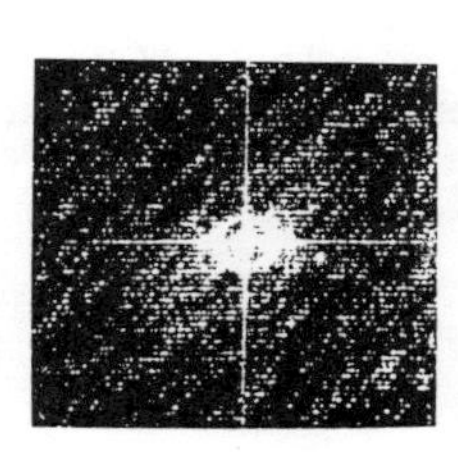

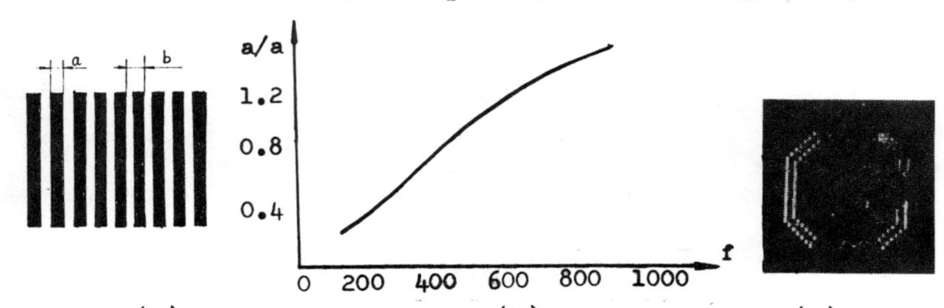

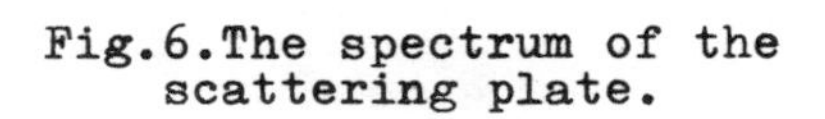

Fig.6.The spectrum of the scattering plate.

Fig.7.(a).The pattern of a "rectangular target" scanned by an E.B.(b).The relation between $\Delta a/a$ and f. (c).A second order differentiated error is existed in a first order differentiated image.

Error correction

As the fabrication of a CGH is very complicated, so there are many sources of error. The main errors from the E.B.scanning machine are the neighbouring effect of E.B., the shift of E.B., the positioning of E.B.,etc. While the technological errors are the quantization error, the variation of spot size and the variation of the lines due to etching. The comprehensive influence of those errors is the chief reason of the deformation of the image. The total effect can be easily measured by a computer analogued "optical target", and then be compensated and corrected automatically. Good results are obtained in our experiments.

Fig.7.(a) is the pattern of a "rectangular target" scanned by an E.B.. The minimum line width is a, while the spatial frequency is f=1/b. If the frequency f is fixed, a is changed by the reason of the variation of spot size, the variation of lines due to etching, etc. Then, the period of lines is changed. Let Δa represents the total error between two neighbour lines. If the Δa is added to the period then two lines will just get into touch.

When the step of the E.B. s is 0.25μm, the size of the E.B. spot r is 0.3 μm and the positioning error Δ_3 is 0.3 μm, then the precision of measuring is

$$\Delta = \sqrt{(s/2)^2 + (r/2)^2 + \Delta_3^2} = 0.3 \ \mu m \tag{6}$$

The measured error is closely related to the spatial frequency, as shown in Fig.7(b). Based on this relation, a correction program is designed for automatic compensation and correction.

The compensation effect can be shown by the following example. When error is existed for a differentiation filter, if a filter function is:

$$H(v) = H_1(v) + \Delta(v) \tag{7}$$

$\Delta(v)$ is the error functin.

Neglecting the effect of high order, expand $\Delta(v)$, then the distribution of the differentiated image is:

$$g(x,y) = g_1(x,y) + C_2 \, g_2(x,y) \tag{8}$$

From the above equation, it is clear that there is the second order derivative $g_2(x,y)$ in the first order differentiated image $g_1(x,y)$. This conclusion, of course, can be extened to the case of a two dimensional differentiation.

Fig.7(c) shows a second order differentiated error is existed in a first order differentiated image. Compared with a corrected image(Fig.4(b)), it is shown that double lines are existed. This means the error has not been completely corrected.

Conclusion

The above experiments showed that an E.B.CGH has a series of advantages and considerable potentiality.

For the wavelength used in E.B. is short, the size of spot is small , the scanning speed is fast, and the scanning is controlled automatically by a computer, so the approach of E.B.scanning has the advantages of high sptial bandwidth product, less error, and high efficiency in fabrication.

References

1. Honeywell Company:"Computers and E-Beams", pp.54, Photonics Spectra, Jan. 1984.
2. K.M.Leung and L.C.Lindquist," E-Beam Computer Holograms for Aspheric Testing", SPIE. Vol.215, pp.72, 1982.
3. Steven K. Case and S.M.Arnold, " E-Beam Generated Holographic Masks for Optical Vector-Matrix Multiplication " (to be published).
4. S.M.Arnold, " E-Beam Fabrication of CGH" (to be published).

Session 5

Holographic Interferometry I

Chairmen

D. Vukicévic
The University of Zagreb, Yugoslavia
Jingtang Ke
Zhengzhou Institute of Technology, China

Holography and Speckle Techniques for Deformation Measurements

Ichirou Yamaguchi

The Institute of Physical and Chemical Research
351-01 Wako, Saitama, Japan

Abstract

By using holographic interferometry and speckle techniques we can measure deformation of diffusely reflecting surfaces with sensitivities related to laser wavelengths. Although these techniques use interference of diffusely reflected laser light and are similar to each other in optical arrangements, they are mutually complementary in various points. In this paper these relationships are discussed in view of fringe formations and recent advances using optoelectronic tools for practical applications.

Introduction

Holographic interferometry and speckle techniques enable us to measure deformation of diffusely reflecting surfaces. All of these techniques use interference of diffusely reflected laser light and similar optical arrangements are employed in them. However, they deliver different information in the form of interference fringes. Both shape and contrast of these fringes play a very important role for sensitivity and range of measurement. In this paper we first investigate the fringe formations in holographic interferometry and speckle techniques and discuss the factors affecting shape and contrast of the fringes.

Parallel to the studies on fringe formations, there have recently been developed many techniques and apparatuses for extracting quantitative information that is required in practical applications. These techniques, which owe much to advances in optoelectronics and computers, comprises both automatic reading of the fringes to give distributions of deformation and direct derivation of deformation value at a fixed point. Typical examples of these techniques are surveyed in the second part of the paper.

Shape and Contrast of Fringes

Surveys on basic principles

Holographic interferometry directly detects the change in the phase of the light scattered from an object surface. The interference fringes represent the contour lines of the phase change caused by surface displacement.

Speckle techniques can be classified into two categories. The first, called speckle interferometry, generates the contour lines of surface displacement as a kind of moire fringes between a pair of speckle patterns obtained before and after object deformation. In this case interferometric set-ups are used so that the phase information may be preserved in each of the speckle patterns. The second category of speckle techniques is based on detection of speckle displacement caused by object deformation. In speckle photography the speckle displacement is detected by optical processing of the doubly exposed negatives of speckle patterns before and after object deformation, called specklegrams, while in the speckle correlation method it is evaluated from correlation analysis of the video signals resulting from electronic scanning of speckle patterns. An advantage of these methods is simplicity in their optical systems, which need no interferometer as used in holographic interferometry and speckle interferometry.

General relations[1]

Quantities representing the patterns observed in each of the techniques are summarized in Table 1. All the quantities are expressed in terms of the cross-correlation function of the complex amplitudes at the plane of observation before and after object deformation

$$C_{12}(\overline{q},q) \equiv \langle U_1(q)U_2^*(q+\overline{q})\rangle, \qquad (1)$$

where the averaging is performed over a statistical ensemble of microscopic structure of surfaces. This averaging means physically a spatial averaging over many speckles. The modulus of the correlation function are related to the intensity cross-correlation function through the equation

Table 1. Quantities representing the patterns

Methods	Observed Patterns
Holographic Interferometry	$C_{11}(0,q)+C_{22}(0,q)+2ReC_{12}(0,q)$
Speckle Interferometry	$Re\,[C_{12a}(0,q)C^{\dagger}_{12b}(0,q)]$
Speckle Photography	$Re\{\mathfrak{F}[C_{12}(\bar{q},q_o)]\}$
Speckle Correlation	$\lvert C_{12}(\bar{q},q_o)\rvert^2$

note) 1. Suffices a and b show arms of interferometers. 2. Re: real part.
3. $\mathfrak{F}$: Fourier transform.

$$\langle I_1(q)I_2(q+\bar{q})\rangle = \langle I_1(q)\rangle\langle I_2(q+\bar{q})\rangle + \lvert\langle U_1(q)U_2^*(q+\bar{q})\rangle\rvert^2 = C_{11}(0,q)C_{22}(0,q+\bar{q}) + \lvert C_{12}(\bar{q},q)\rvert^2, \quad (2)$$

which holds for the complex gaussian random variables. The conditions for equation (2) are:
(1) fine structure of surface is far beyond resolution of optical systems.
(2) the standard deviation of its microscopic height should exceed the light wavelength so as to make the phase of the scattered light to be distributed uniformly between $-\pi$ and π.

The intensity cross-correlation function generally shows such a behavior as shown in Figure 1. The position of the peak means speckle displacement caused by surface deformation, while the heights of the peak as compared with that of the function without deformation $\langle I_1(q)I_1(q+\bar{q})\rangle$ shows speckle decorrelation accompanying the displacement. The evolution of this behavior remarkably depends on the sort of surface deformation and optical configurations. In a special case speckles do not move and show only decorrelation. This state is called "boiling". In another case the decorrelation occurs very slowly, as called "pure translation."

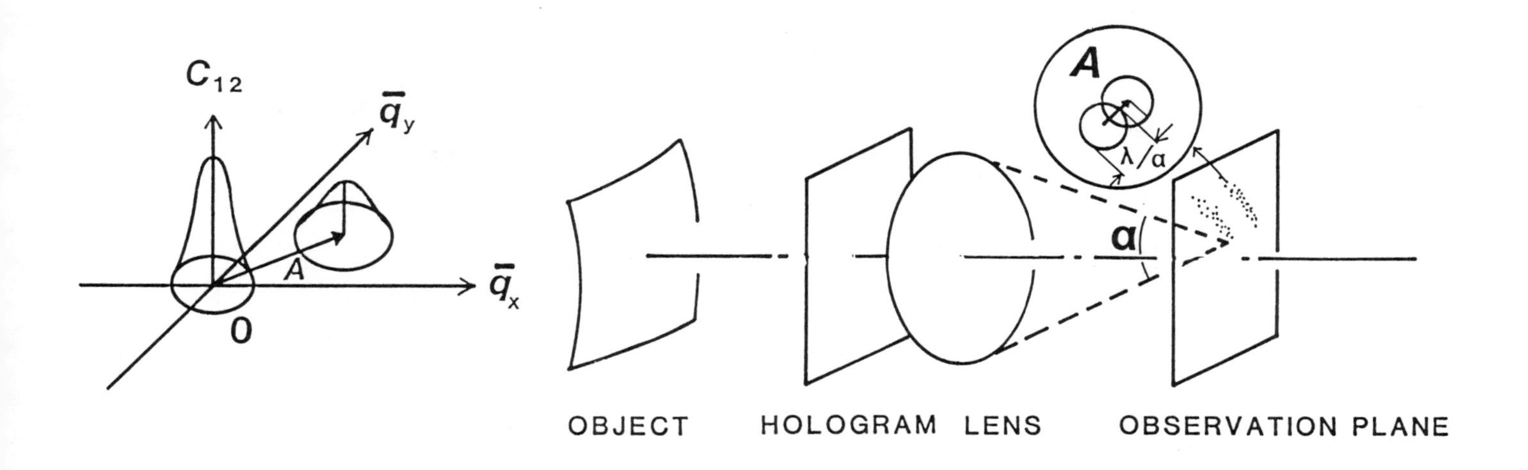

Figure 1. Behavior of intensity cross-correlation function

Figure 2. Fringe formation in holographic interferometry

Holographic interferometry

The fringes observed in holographic interferometry represent the contour lines of displacement component along the bisector between the unit vector directed from the object point to the center of wavefront curvature of the incident laser beam and that directed to the point of observation. Therefore, under normal incidence and normal observation the interference fringes deliver the distribution of out-of-plane displacement. These fringes are normally observed on object surfaces and called fringes of equal thickness or Fizeau fringes, according to an analogy with classical interferometry where a mirror surface and an extended light source are used. However, if the displacement is uniform, the fringe structure is observed not on the object surface but in its far field. These fringes of equal inclination, normally concentric rings called the Haidinger rings, represent the translation of surface.

Contrast of fringes observed in holographic interferometry depends on the overlap of identical speckles before and after object deformation. If there is no overlap of identical speckle, we observe no fringe structure, because the phase relation between different speckles is random. As consequences, fringes can be clearly observed only if displacement of speckle induced by object deformation is smaller than its mean size as shown in Figure 2. Degree of the overlap can be changed by adjusting aperture size to vary the mean speckle size, while the speckle displacement is independent of the aperture size. If speckles do not move even under object deformation, that is, at the boiling state, then the overlap of identical speckles is perfect except for decorrelation effect and we can observe fringes of high contrast. Since the speckle displacement is not uniform in space depending on the kind of object deformation and optical configuration, the degree of the overlap and thus the fringe contrast also depends remarkably on a position of observation. This is an origin of fringe localization. However, if we reduce the aperture size to broaden the region of the overlap too much, then speckle noise in the fringes becomes distinct and fringe clarity cannot be reasonably evaluated only from the fringe contrast.

In the presence of displacement lateral to an axis of observation system, fringe contrast on an object surface decreases, because speckle displacement at the plane conjugate to the surface is always equal to the lateral object displacement multiplied by imaging magnification. This speckle displacement can be compensated for at the defocused region by a contribution from object tilt, which always vanishes at the conjugate plane. The distance between the object and the localization plane is proportional to the product of the lateral displacement and the spatial frequency of fringes along the direction of the lateral displacement. If the distance exceeds the focal depth of the imaging system, it becomes difficult to identify the positional correspondence between fringes and object.

Speckle interferometry

Speckle interferometry yields fringes that also represent the contours of displacement component in the direction depending on optical systems, all of which are interferometers as shown in Figure 3. These fringes, which result from the nonlinear superposition of the speckle patterns before and after object deformation, correspond to moire fringes between the speckle patterns. Object deformation gives rise to a change in the optical path difference of the interferometers and thus a change in the brightness of each speckle of the resultant pattern. If the change is equal to an integer multiple of a wavelength, then the brightness remains unchanged. On the other hand, if the change is an odd multiple of half a wavelength, the brightness is inverted. Multiplication of these patterns results in the fringes which connect the region of equal change in the phase difference.

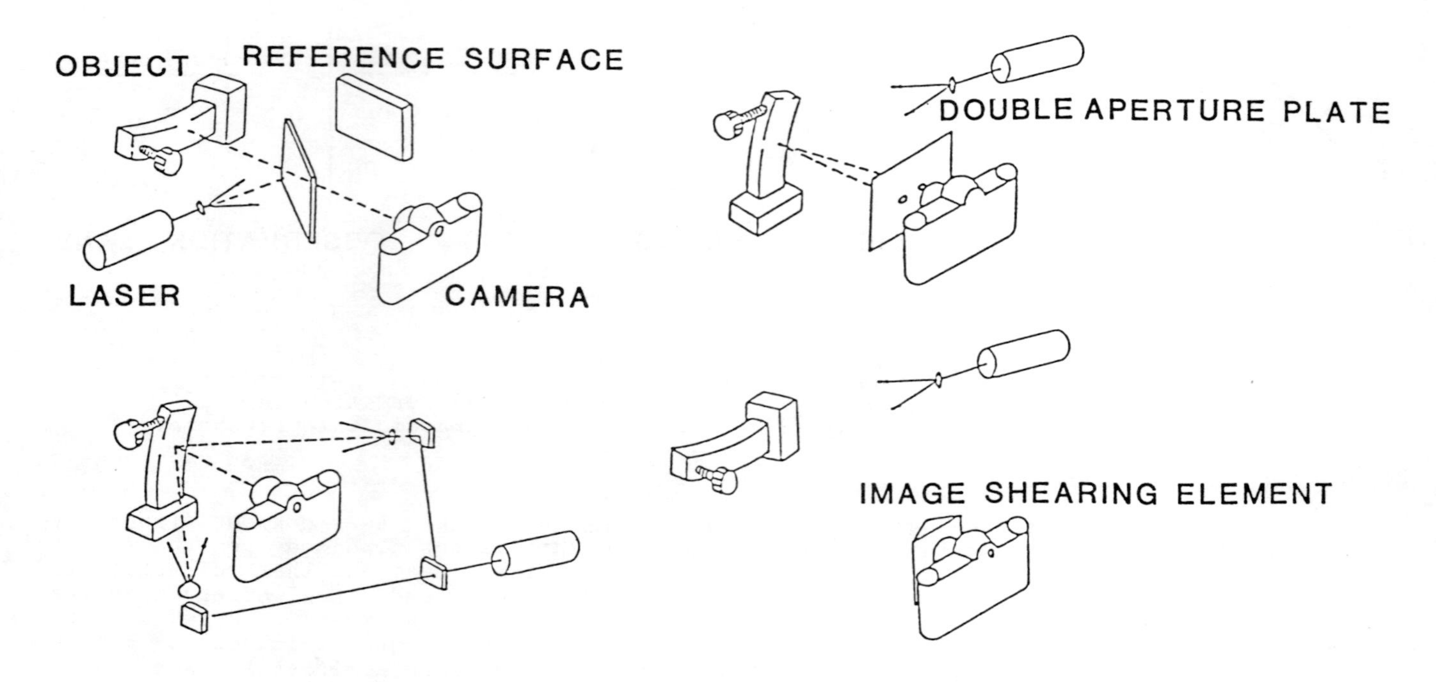

Figure 3. Arrangements of speckle interferometry

By speckle interferometry we can obtain contours of out-of-plane displacement by the reference light method, those of in-plane displacement by the dual beam or the double aperture methods, and those of displacement derivatives by the shearing method. Contrast of fringes observed by these techniques also depends on the ratio of speckle displacement to speckle size. The reason is as follows. In order to get the brightness change of speckle only caused by object deformation, the identical speckles in each arm of the interferometer have to be superposed before and after object deformation.

Since focused images of the object are superposed in speckle interferometry, only in-plane displacement as compared with the mean speckle size affects the contrast of fringes. Besides, these fringes have a bias level higher than that in holographic interferometry that corresponds to the first term of the right-hand side of equation (2), and thus fringe clarity is generally lower than in holographic interferometry. This bias can be removed, however, either by spatial filtering used in the double aperture and the shearing methods or by electronic subtraction of patterns undertaken in electronic speckle pattern interferometry.

Speckle photography

The fringes observed in speckle photography represent speckle displacement caused by object deformation. The Young fringes, that appear from pointwise Fourier transformation of a specklegram, give both the direction and magnitude of speckle displacement except for the sign of the displacement. As mentioned previously, speckle displacement is proportional to in-plane displacement if the specklegram is focused on the object, while a defocused specklegram additionally gives information on object tilt. The contrast of the Young fringes depends on the speckle decorrelation accompanying the displacement, provided that the speckle displacement is much smaller than the diameter of the read-out beam. The decorrelation arises from the speckle displacement at the lens aperture, which is especially amplified by object tilt.

The contrast of the contour fringes observed in a spatial filtering arrangement for specklegrams is dependent on the ratio of the point spread of the filter to the magnitude of speckle displacement. It is not easy, however, to get these fringes distinctly.

Quantitative and Automatic Measurements

For practical applications of these techniques quantitative and automatic evaluation of the fringe patterns have been undertaken.

Holographic interferometry

In holographic interferometry digital image processing has been applied for extracting and interpolating the fringes. In typical cases these procedures consist of the following steps; low pass filtering for removing speckle noise, shading correction for compensating for uneven illumination or reflectivity, binarization, thinning, and determination of fringe order.[2] Fringe order, however, cannot be determined in a fully automatic way. For this purpose, the interactive system is desirable, where an operater can suggest or correct the order timely.

In optical heterodyne method,[3,4] object waves before and after deformation are recorded on a plate with different reference beams. Reconstruction is performed by using the double reference beams whose frequencies are shifted from each other, for example, by using a Bragg cell. In the real time method, we can also shift the frequency of the reconstructing beam. Then fringes in the reconstrued image flow continuously and intensity at each point shows temporally sinusoidal variation whose phase is proportional to displacement at the point except for integer multiple of wavelength. The point of detection is moved successively from a reference point while the detected phase is integrated to deliver the displacement relative to that at the reference point. Since the amplitude of the sinusoidal intensity variation at each point is temporally constant, the heterodyne method is not affected by local irregularities in image brightness.

In fringe scanning method or phase shifting interferometry,[5] fringes in the reconstructed image are moved in steps by changing the phase difference between the double reference beams at least in three steps, for example, by driving a PZT mirror. Each of the images is stored in a digital computer and added with different weights depending on the phase. The results are transformed by inverse trigonometric functions to yield fringe phase at each image point.

Speckle interferometry

Digital image processing has been used for electronic speckle pattern interferometry.[6,7] The fringe scanning method has also been applied to speckle interferometry where a PZT mirror is introduced into one arm of the interferometer.[8,9]

Speckle photography

In speckle photography the Young fringes are especially suited for automatic analysis because of their simple structure. Two ways are available for this purpose. The first is to project the intensity on to one axis for calculating the period alone the axis and fringe orientation.[10,11] In Figure 4[11] the projection is accomplished by using a linear image sensor and the fringes are rotated by a Dove prism under microcomputer control. The second way of the automatic analysis is to perform the two-dimensional fast Fourier transform of the image data directly. This is simple in mechanical arrangement but needs a more powerful computer.

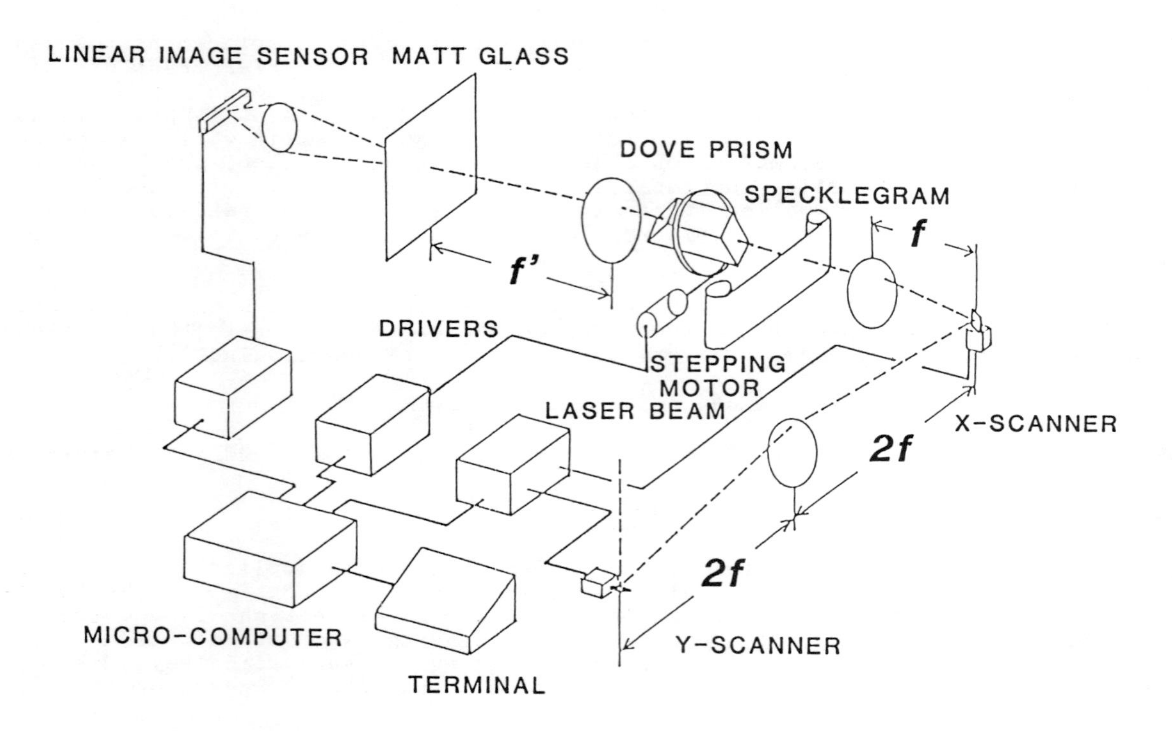

Figure 4. Arrangement of an automatic specklegram analyzer

Speckle correlation method

The speckle correlation method delivers the value of displacement at an object point quickly by computing speckle displacement from the cross-correlation of video signals.

The basic arrangement of speckle correlation using a linear image sensor is shown in Figure 5. A narrow laser beam is directly incident on an object and the sensor is oriented along the direction of speckle displacement. By the arrangement, in-plane translation a_x along the sensor or tilt Ω_y about the axis perpendicular to the sensor can be determined from the relation[12]

$$A_X = a_x\left(\frac{L_0 \cos^2\theta_s}{L_s \cos\theta_o} + \cos\theta_o\right) + \Omega_y L_0\left(\frac{\cos\theta_s}{\cos\theta_o} + 1\right) \text{ and } A_Y = 0, \tag{3}$$

where L_s is the radius of wavefront curvature of the incident beam, θ_s incident angle, L_o the distance of the sensor from the object, and θ_o the direction of the sensor. Figure 5 also shows output signals of the sensor before and after speckle displacement and the cross-correlation function between these signals. Considering high contrast of speckle, the signals are transformed into binary signals by clipping about the average levels prior to correlation computation. This clipping reduced computation time to one fourth. Figure 5 further contains an experimental result from translation measurement, which was obtained by using a linear sensor of 15 μm pitch and 1024 elements. Since we illuminated the object by a divergent beam, object translation is amplified in speckle displacement. The amplification could be attained up to 15 times.

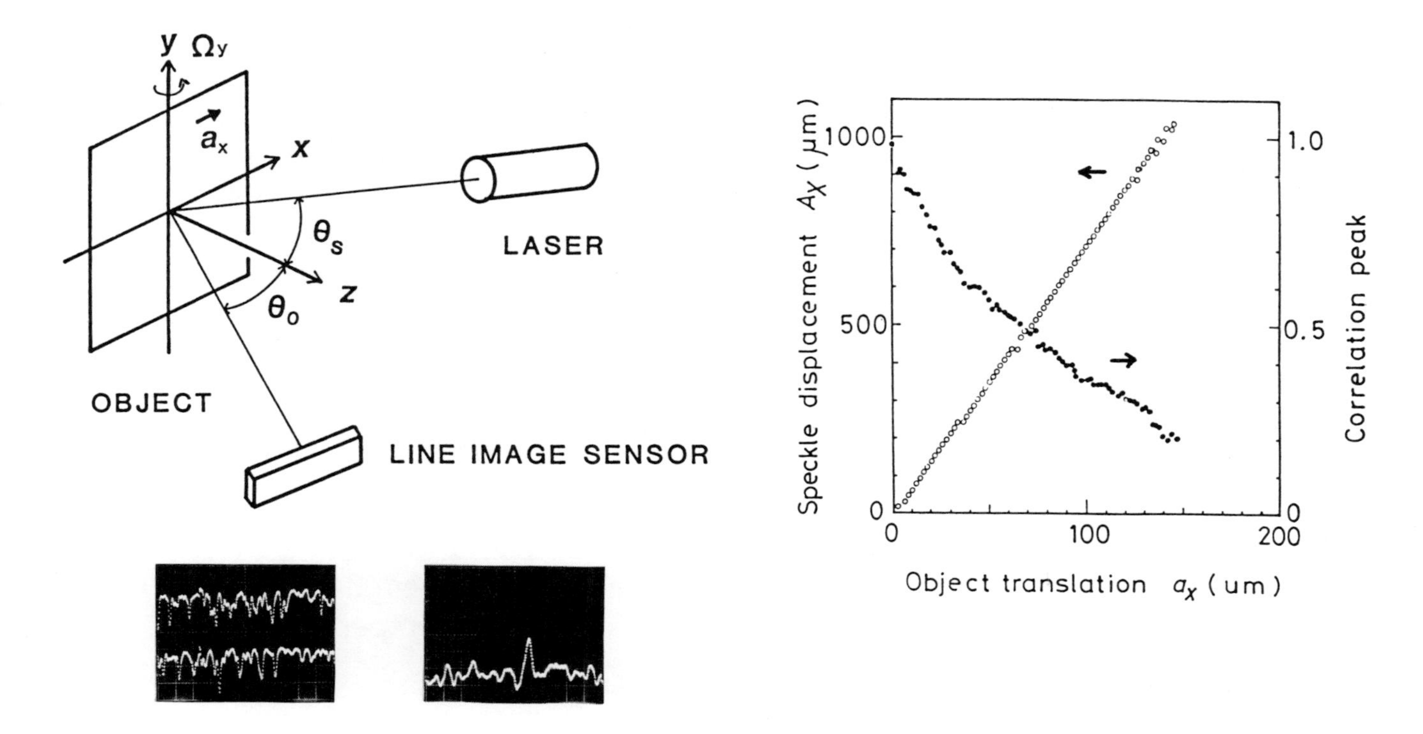

Figure 5. Arrangement, waveforms, and a measurement result of speckle correlation using a linear image sensor

The measurement of object tilt can be regarded as an extension of optical lever which uses a mirror illuminated by a narrow laser beam. The present method has more freedom for detector position and higher accuracy due to statistical averaging over many speckles.

Since speckle displacement receives different contributions from translation, rotation, and strain components of an object, one of the components can be separated by comparing speckle displacements at different positions or for different illuminating beams. For an example of this differential system, we developed a laser speckle strain gauge that is so arranged as to eliminate the contributions from rigid body motion as shown in Figure 6. Speckle displacements at two positions symmetrical about the surface normal are detected in the direction connecting these positions. The difference between these displacements is given by

$$\triangle A_X = A_X(\theta)-A_X(-\theta)=-2\varepsilon_x L\tan\theta-2a_z\sin\theta, \quad (4)$$

where ε_x is the strain in the direction connecting the sensors, and a_z the out-of-plane translation. If a_z is so small that $|a_z|\ll|\varepsilon_x|L/\cos\theta$, the strain is given by

$$\varepsilon_x=-\triangle A_x/L\tan\theta \quad (5)$$

independent of other rigid body motion. Experimental comparisons with a resistance strain gauge resulted in good agreement with the theory. This strain gauge has a sensitivity of ten microstrains with a gauge length of 1 mm, where we took the distance and direction of sensor such as L = 300 mm, θ = 45° and image sensors of 15 μm pitch and several hundred elements.

Since it took a few seconds to compute strain value with a microcomputer, we first applied the strain gauge to measuring slowly varying strain, for example, thermal expansion of materials such as metal and plastics. We could detect such an expansion of metal of order 10^{-5} caused by temperature change equal to 1°C as shown in Figure 6.

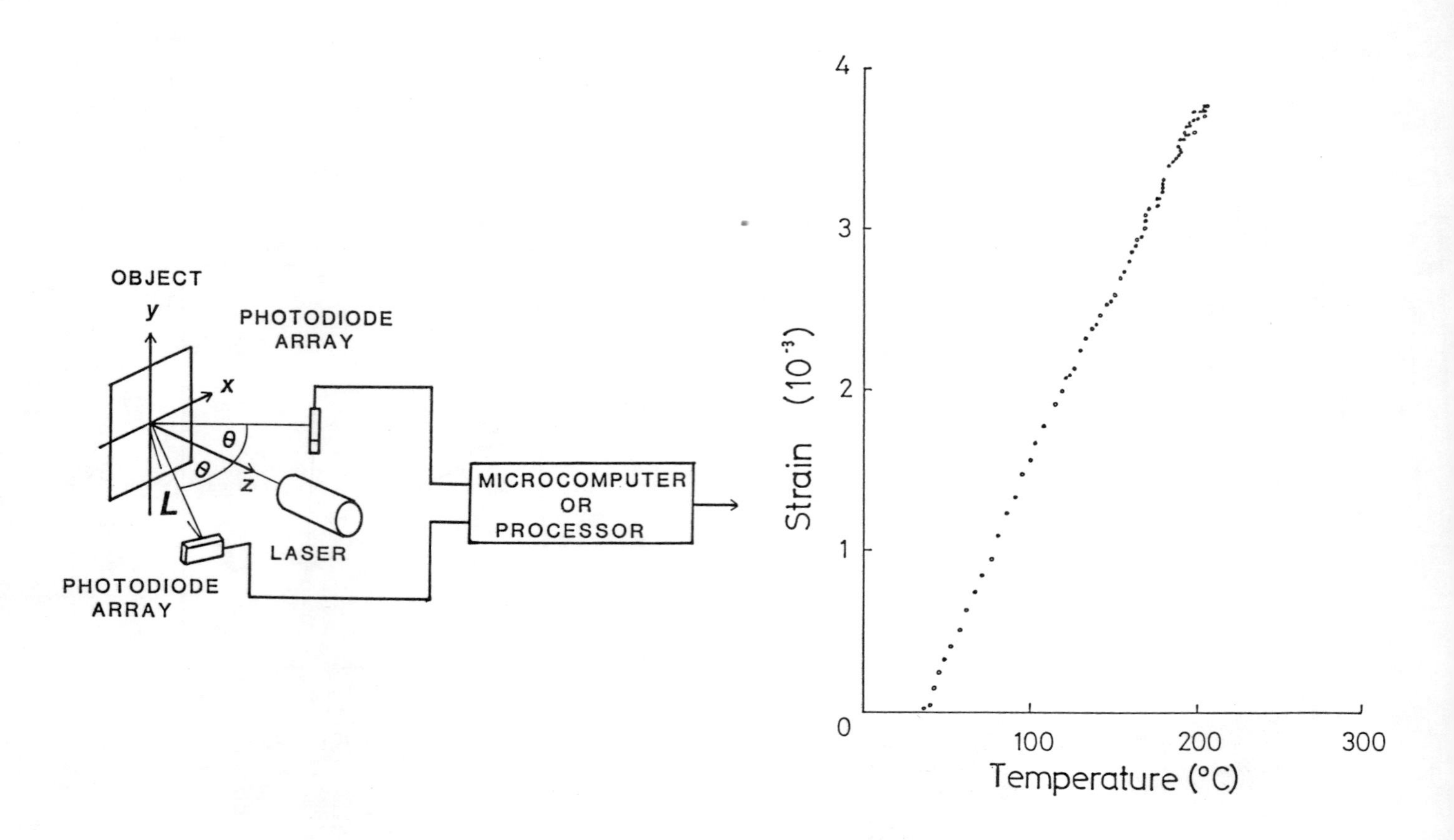

Figure 6. Laser-speckle strain gauge and a result of thermal expansion measurement

The sensitivity of the speckle correlaion method depends on geometries of the optical system, while the range of measurement is not critical since we can measure an arbitrarily large amount of object translation or strain by integrating the incremental speckle displacements caused by such a small amount of translation or strain for which the decorrelation is not yet so serious, although the illuminated region may be changed. The possible signal decorrelation that could be caused by speckle displacement lateral to the image sensor would be suppressed either by adopting a sensor having a width more than the pitch or by sampling the image sensor outputs more quickly.

Acceleration of the speckle correlation method

The speckle correlation method was limited by computation time of speckle displacement. This time has recently been reduced down to less than a millisecond by adopting a spatial filtering detector, which consists of both a photodiode array and its control circuit and delivers a voltage directly proportional to displacement of a speckle pattern normal to the array.[13] The basic principle of the detector is depicted in Figure 7. The diode array is subject to mutual connection, which is changed cyclically in a time sequence. This realizes virtual movement of a comb-shaped detector with constant velocity. A stationary speckle pattern gives rise to a sinusoidal ouput, whose phase is now shifted in proportion to displacement of the pattern. This phase shift is detected by an up-down counter into which both the output and a reference signal corresponding to the output for a stationary pattern are simultaneously led. The output of the counter is converted into an analog voltage. The voltages from the pair of detecters are differentially amplified to deliver the strain signal.

Although the spatial filtering detector has raised the response of the speckle strain gauge up to a few kilohertz, it has also manifested a disadvantage. It is the extinction of the sinusoidal signal that occurs frequently and makes it impossible to determine its phase. This extinction arises when the pattern on the detector happens to lose its component of the spatial frequency equal to that of the array. This phenomenon cannot be avoided because of randomness of speckle patterns. Therefore, it would be the best for accelerating the speckle strain gauge to construct a high-speed binary correlator for processing image sensor signals.

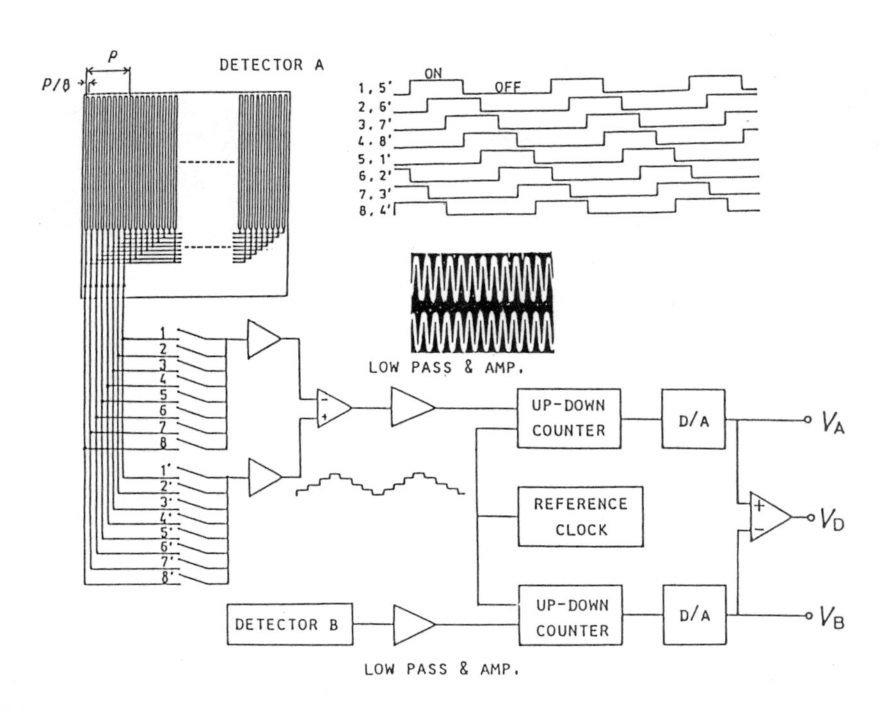

Figure 7. Schematics of spatial filtering detectors used in the speckle strain gauge

Conclusions

In summary holographic interferometry and speckle techniques are similar to each other in both optical set-ups and functions although the delivered information and the range of measurement are different. In practical uses we have to select an adequate method and the optimum manner of signal processing considering the required quantities and given conditions. At present, quantitative, real time, and automatic measurements are limited to a point by point mode. However, if digital image processing circuits or devises are developed, which enable to compute the local image correlations at the video rate, then these techniques could also be implemented to instruments that can be used outside laboratories.

References

1. Yamaguchi, I., "Fringe Formations in Deformation and Vibration Measurements Using Laser Light," in Progress in Optics, ed. Chap.5, North-Holland, 1985.

2. Nakadate, S., Magome, N., Honda, T., and Tsujiuchi, J., "Hybrid Holographic Interferometer for Measuring Three-dimensional Deformations," Opt. Eng., Vol.24, pp.246-252, 1981.

3. Sommargren, G. E., "Double-exposure Holographic Interferometry Using Commonpath Reference Wave," Appl. Opt., Vol.16, pp.1736-1741, 1977.

4. Dändliker, R. and Thalmann, R., "Heterodyne and Quasi-heterodyne Holographic Interferometry," Opt. Eng., Vol.24, pp.824-831, 1985.

5. Hariharan, P., Oreb, B. F., and Brown, N., "Real-time Holographic Interferometry: A Microcomputer System for the Measurement of Vector Displacements," Appl. Opt., Vol.22, pp.876-880, 1983.

6. Hurden, A. P. M., "Vibration Mode Analysis Using Electronic Speckle Pattern Interferometry," Opt. Laser Technol., Vol.14, pp.21-25, 1982.

7. Nakadate, S., Yatagai, T., and Saito, H., "Computer-aided Speckle Pattern Interferometry," Appl. Opt., Vol.22, pp.237-243, 1983.

8. Nakadate, S. and Saito, H., "Fringe Scanning Speckle-pattern Interferometry," Appl. Opt., Vol.24, pp.2172-2180, 1985.

9. Creath, K., "Phase-shifting Speckle Interferometry," Appl. Opt., Vol.24, pp.3053-3058, 1985.

10. Ineichen, B., Eglin, P., and Dändliker, R., "Hybrid Optical and Electronic Image Processing for Strain Measurements by Speckle Photography," Appl. Opt., Vol.19, pp.2191-2195, 1980.

11. Yamaguchi, I., Mizuno, S., and Saito, H., "Development of an Automatic Analyzer of Speckle Photographs," Kogaku, Vol.11, pp.583-588, 1982.

12. Yamaguchi, I., "Simplified Laser-speckle Strain Gauge," Opt. Eng., Vol.21, pp.436-440, 1982.

13. Yamaguchi, I., Furukawa, T., Ueda, T., and Ogita, E., "Accelerated Laser Speckle Strain Gauge," Opt. Eng., Vol.25, pp.671-676, 1986.

A New Experimental Technique To Separate Principal Stresses ------ Proportional Exposure Piled-plate Holography

Cao Hongsheng   Chen Shiming   Dong Wei

Mechanics Department, Zhejiang University
Hangzhou, China

Abstract

Propotional exposure piled-plate holography, a new experimental technique to separate principal stresses for both static and dynamic photoelastic holography is described, solves the problem of principal stress separation in multi-frame recording dynamic photoelastic holography. Applying the new technique to the multi-frame recording dynamic photoelastic holography, the authors recorded eight separated dynamic stress patterns for four different transients of a plan plate subjected to an impacting load. The dynamic stresses of the plate are analysed.

Introduction

Multi-frame recording dynamic photoelastic holography is a desirable way to study dynamic problems that are unrepeatable and destructive such as explosive load and crack propagation etc.. To separate isochromatic and isopachic fringes in single-frame recording dynamic photoelastic holography, Holloway[1] developed a reflecting photoelastic holography, Qin[2], Lallemand and Lagarde[3] applied the FRADY rotator in the photoelastic holography, getting the separated dynamic isochromatic and isopachic fringes individually. Though these methods mentioned above are effective in single-frame recording dynamic photoelastic holography, they are difficult to be used in the multi-frame recording dynamic photoelastic holography, and the problem has remained for years.

Based on Cao and Wang[4], the authors developed a new experimental technique, propotional exposure piled-plate holography, to separate isochromatic and isopachic fringes in multi-frame recording dynamic photoelastic holography.

Description of the method

The general expression for circularly polarized light-field intensity pattern in double-exposure holographic photoelasticity is given by

$$I \propto K + 2K\cos\frac{\pi h}{\lambda}(A' + B')(\sigma_1 + \sigma_2)\cos\frac{\pi h}{\lambda}C(\sigma_1 - \sigma_2) + \cos^2\frac{\pi h}{\lambda}C(\sigma_1 - \sigma_2) \quad (1)$$

where

$$K = \frac{t_1}{t_2} = \frac{\text{exposure time with model unstressed}}{\text{exposure time with model stressed}}$$

h ----- the thickness of the model in the unstressed state

λ ----- the wavelength of the illumination

A , B , C ----- the stress-optic constants of the model material

σ_1 , σ_2 ----- principal stresses in the model

There are two extreme situations which deserve special consideration.

1. Consider the situation depicted in Figure 1.

The pattern may be interpreted as consisting of the independent superposition of the isochromatic and isopachic fringe families with half-order fringe shifts in the isopachic fringes.

2. The othor extreme situation occurs when the isochromatic and isopachic fringes are parallel.

Assume the sum and difference of the principal stresses are linear functions of distance in the neighborhood of a half-order isochromatic fringe. Then the isochromatic function ψ and isopachic function Φ can be expressed as:

$$\psi = \cos\pi(\tfrac{1}{2} + \xi) \quad (2)$$

$$\Phi = \cos\{\pi(\tfrac{1}{2} + 2\Omega\xi) + \theta\} \tag{3}$$

where ξ ----- the dimensionless distance between adjacent isochromatic fringes,
Ω ----- the ratio of the spatial frequencies of the isopachic fringes to the isochromatic fringes,
θ ----- the phase angle of the isopachic function relative to the isochromatic functipn at $\xi = 0$.

Substitute eq.(2) and eq.(3) into eq.(1), get

$$I \propto f(K, \xi, \Omega, \theta) \; . \tag{4}$$

A system of curves of eq.(4) can be drawn by the computer. The mismatch(or shift) between the real and conditional isopachic fringes is discussed.

(a) The effects of the phase angle versus the shift of the fringe are shown in Figure(2).

(b) The effects of the ratio of exposure K versus the shift the fringe are shown in Figure 3.

(c) The effects of the ratio of the spatial frequencies of the fringes versus the fringe shift are shown in Figure 4.

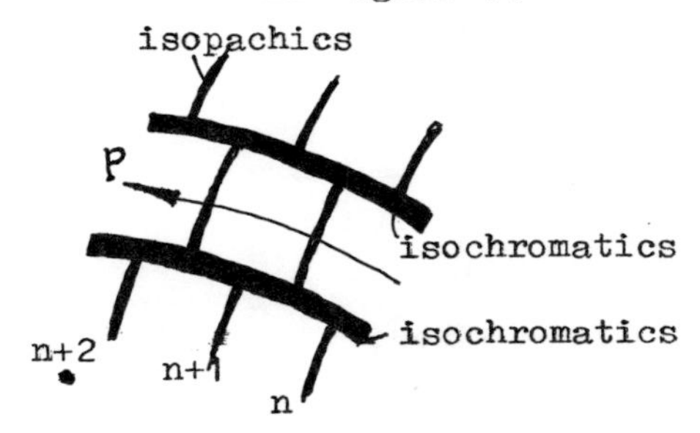

Figure 1. Two sets of fringes are nearly orthogonal.

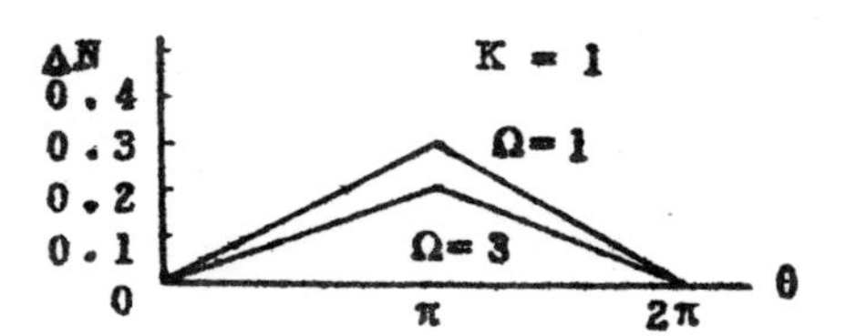

Figure 2. The effects of the phase angle versus the fringe shift ΔN.

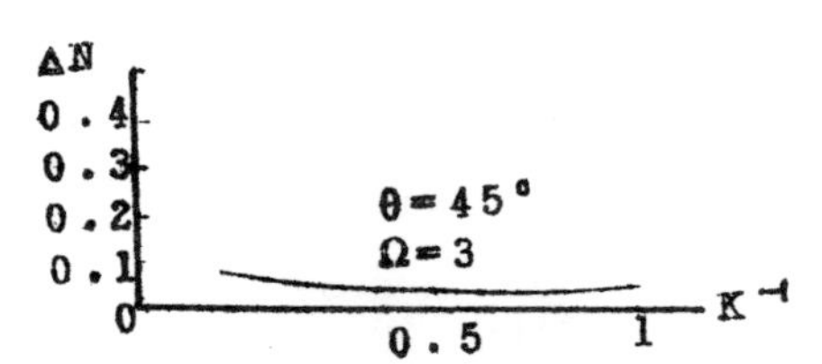

Figure 3. K versus ΔN.

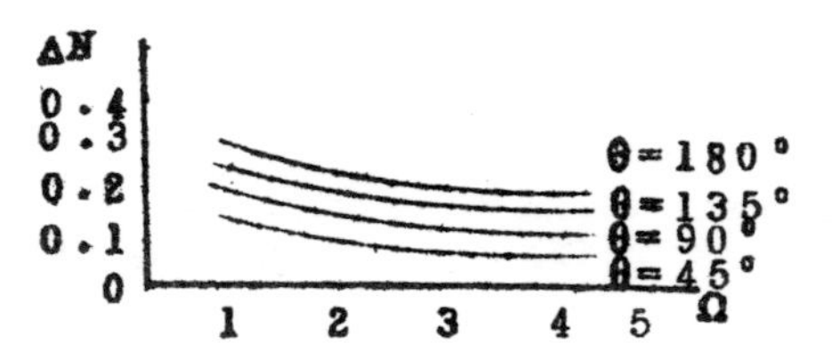

Figure 4. Fringe spatial frequency ratio versus the fringe shift ΔN.

The result of the analyses showes that ΔN is less than half order of the isopachic fringe. The relative error of the principal stresses , due to ΔN is:

$$|e|_{\sigma_{1,2}} = \left| \frac{\Delta N}{N_p \pm f_c N_c / f_p} \right| \tag{5}$$

where N_p ----- isopachic fringe order,

N_c ----- isochromatic fringe order,

f_p ----- value of isopachic fringe of the model material,

f_c ----- value of isochromatic fringe of the model material.

The experimental procedure is described as follows:

The first holoplate is exposed in time T before the applied load, then the second holoplate is piled on behind the first one. The second exposure time T is taken under the applied load. In order to decrease the effect of the last term $\cos^2 \frac{\pi h}{\lambda} C \cdot (\sigma_1 - \sigma_2)$ in eq.(1), the ratio of exposure K may be optimized. In practice, K= 1.5~2.5 is better. On the second holoplate, the isochromatic fringes are recorded, and on the first holoplate, the distinguishable conditional isopachic fringes are recorded. No additional special equipment is required, separated isochromatic fringes and conditional isopachic fringes are recorded synchronously. (See Figure 5)

Because half-order fringe shift occurs when an isopachic fringe crosses an isochromatic fringe, the identification of the isopachic fringe should be based on the isochromatic pattern getted in the piled-plate process.

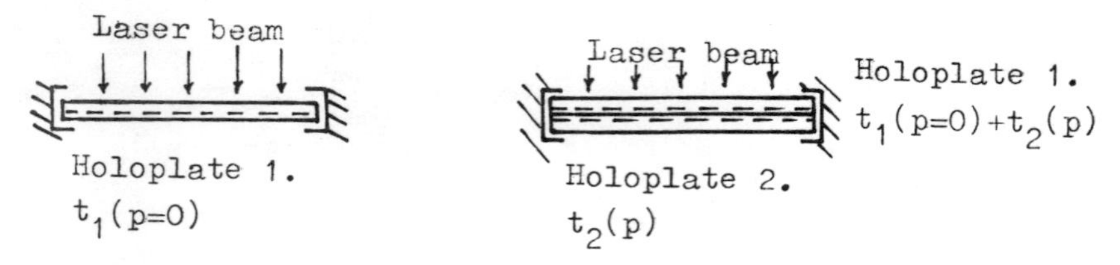

Figure 5. Piled-plate steps

Experimental result

1. Static experiment

Considering the extreme situation when two fringe families are parallel, the experiment of a pure bending beam is discussed and shown in Figure 6. and Figure 7. Good agreement with the theroy solution is achieved.

(a) Isochromatics (b) Isopachics

Figure 6. A pure bending beam isochromatic pattern and isopachic pattern.

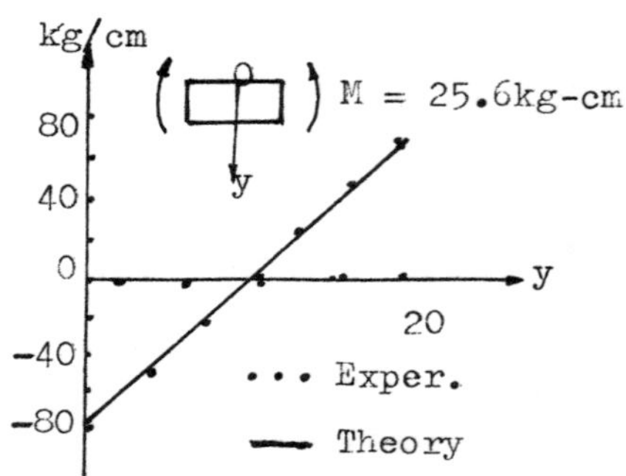

Figure 7. The distribution of principal stress.

2. Dynamic experiment

The optical system of the multi-frame recording holography is shown in Figure 8. A sequentilly pulsed laser is generated by the order. Each light pulse which may be diffracted after passing through the acousto-optic deflector, is divided into object beam and reference beam. At last four frames of isopachic pattern and four frames of isochromatic pattern at four different transients are obtained.(Figure 10.,Figure 11.)

The model material is epoxy resin(f_{pd}= 8.84kg/cm·ord, f_{cd}= 15.74kg/cm·ord). The sketch of the model and loading is shown in Figure 9. The impacting load is m v = 9.18× 10^{-2} kg·m/s.

Figure 12. and Figure 13. are dynamic stress order distribution and dynamic stress distribution on section A-A at different transients under the impacting load.

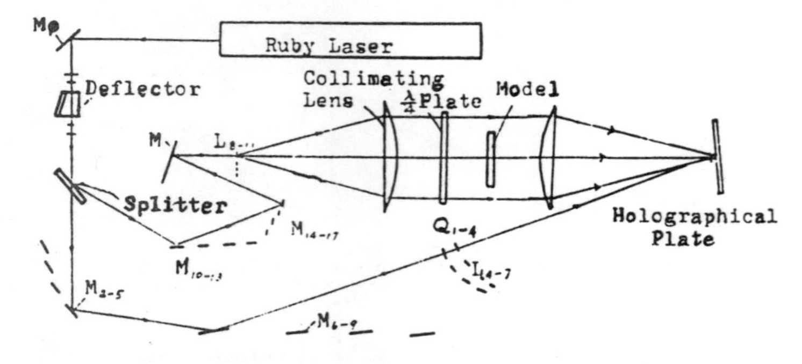

Figure 8. Optical arragement of multi-frame recording holography.

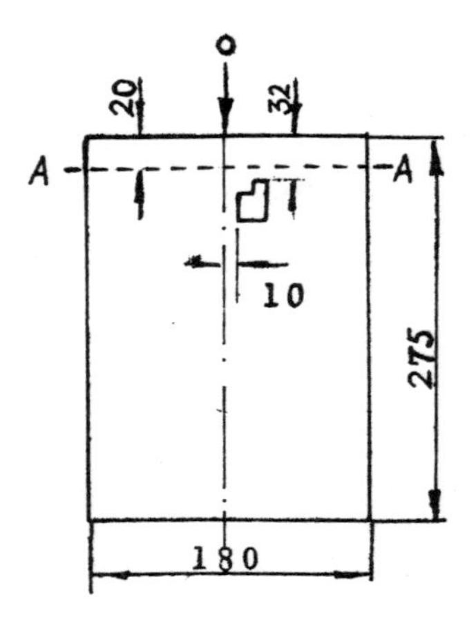

Figure 9. The sketch of the model and impacting load.

Figure 10. Dynamic isochromatic patterns under the impacting load.

Figure 11. Dynamic isopachic patterns under the impacting load.

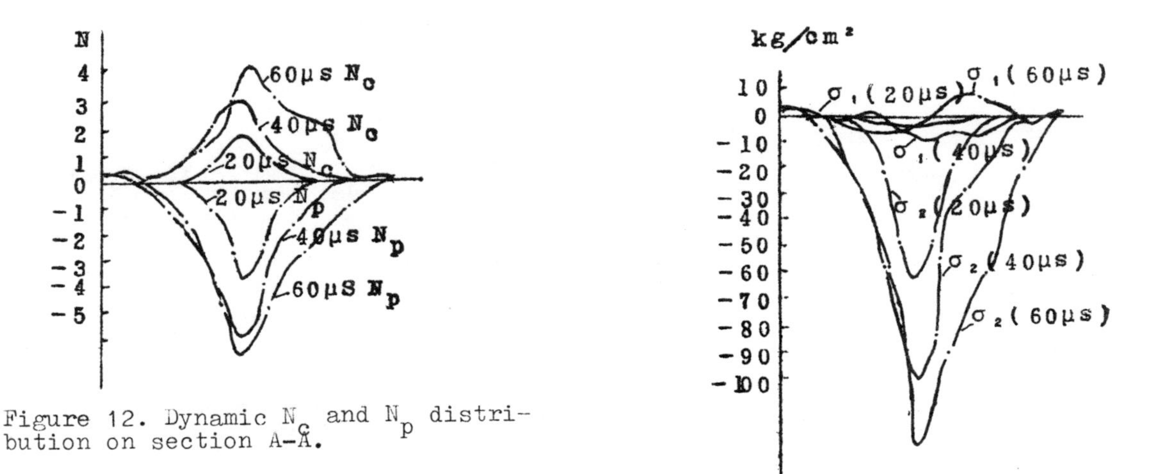

Figure 12. Dynamic N_c and N_p distribution on section A-A.

Figure 13. Dynamic stress distribution on section A-A.

Conclutions

1. It is convenient to obtain the isochromatic pattern and conditional isopachic pattern at the same stress situation by the propotional exposure piled-plate holography. The conditional isopachic pattern is nearly equal to the real isopachics in the region where two sets of isopachic and isochromatic fringes are nearly orthogonal. But there is a shift where the isopachics and isochromatics are parallel. The largest shift is less than a half-order.

2. The method of propotional exposure piled-plate holography may be used to solve the problem of the dynamic stress fringe separation in the multi-frame recording dynamic holographic photoelasticity. It is desirable to study dynamic problems that are unrepeatable.

References

1. Holloway,D.C. and T.A.M. Report 349 Univ. of Illinois Urban Il 1971.
2. Qin,Y.W., Exper. Mech., 21(8), 389-393, 1981.
3. Lallemand,J.P. and Lagard A., Exper. Mech., 22(5), 174-179, 1982.
4. Cao,H.S. and Wang,Z.L., Proceedings of International Conference on Experimental Mechanics (Beijing), 399-404, 1985.

Analysis of an In-Line Object Beam Hologram Interferometer

Martin Celaya, Diana Tentori and Ricardo Villagomez

Optics Department, Applied Physics Division, CICESE Research Center
P.O. Box 2732, 22830, Ensenada, B. C. Mexico

Abstract

The performance characteristics of an In-line Object Beam Hologram Interferometer working with a diffuse object are analyzed. This analysis considers two criteria optical path difference and the correlation of homologous points at the image plane.

Introduction

The In-Line Object Beam Hologram Interferometer (IOBHI), proposed by N. Abramson[1] can be very useful for several applications. It has been used in the inspection of phase objects, thickness[2], verification of rectilinear displacements[3], etc. Its wide applications are a result of the intrinsec properties of this optical arrangement. In this work the performance characteristics of this setup are analyzed. Its range of work depends on the optical path difference required to obtain interference fringes and the correlation present between homologous points for a particular numerical aperture of the viewing system[4,5]. In this type of hologram inteferometer the light source illuminating the object, the object itself, the hologram and the optical viewing system are aligned. As object a transmitting diffuser plate is used located in the middle between the point source used to illuminate the object and the vertex of the aperture of the optical viewing system.

Theory

The optical arrangement used for the In-Line Object Beam Hologram Interferometer (IOBHI) is shown in Fig. 1. The diffuser plate has coordinates (xo, yo), and the complex amplitude distribution over it represents the object wave front. It is initially at position 0 and later at position 0', after having experienced a rigid movement. The illumination source S, the object and the optical viewing system are aligned. At a distance So behind the object it is located the illumination source and at a distance Ho in front of the object we have the observation system. The radius of the aperture of the optical viewing system is denoted by ρ.

If P is a point on the diffuser plate in its original position and P' is the same point after a rigid displacement given by vector (Dx, Dy, Dz) has taken place, and considering that P, **D** << So, Ho holds. The optical path difference between both trajectories is given by[6] Eq. (1).

$$\mathrm{OPD} = \left(\frac{1}{So^2} - \frac{1}{Ho^2}\right)(xo^2 + yo^2)\frac{D_z}{2} + xoDx\left(\frac{1}{So} + \frac{1}{Ho}\right) + yoDy\left(\frac{1}{So} + \frac{1}{Ho}\right) \quad (1)$$

From the geometry shown in figure 1, and considering Eq. 1, it can be easily seen that for So = Ho, that is for a symmetric position of the light source and the observer with respect to the object, the optical path difference will be always equal to zero for an in-plane rotation, and in the approximation of Eq. 1 for a longitudinal translation or a tilt. This arrangement is only sensitive to lateral displacements of the diffuser plate. This conclusion is valid just with respect to the shape of the fringe pattern. Its fringe visibility will be affected by the presence of other rigid displacements.

Analysing OPD as well as fringe contrast allows us to determine the range of work of this optical arrangement for hologram interferometry. To do so we will consider the mathematical equations related with both parameters. It has been demonstrated that the intensity distribution at the image plane of the viewing system is given by[5]

$$I\ (xi,\ yi) \approx Io\ (1 + \gamma \cos \emptyset) \quad (2)$$

where

$$\emptyset = 2\Pi\ (\mathrm{OPD})/\lambda = 2\Pi\ \{(S_2 + H_2) - (S_1 + H_1)\}/\lambda \quad (3)$$

and

$$\gamma = J_1\ (\Pi A)\ /\ (\Pi A) \quad (4)$$

where $\emptyset$ is the phase difference between interfering wavefronts and γ gives us the fringe contrast. J_1 is a Bessel function of the first order whose argument is given by

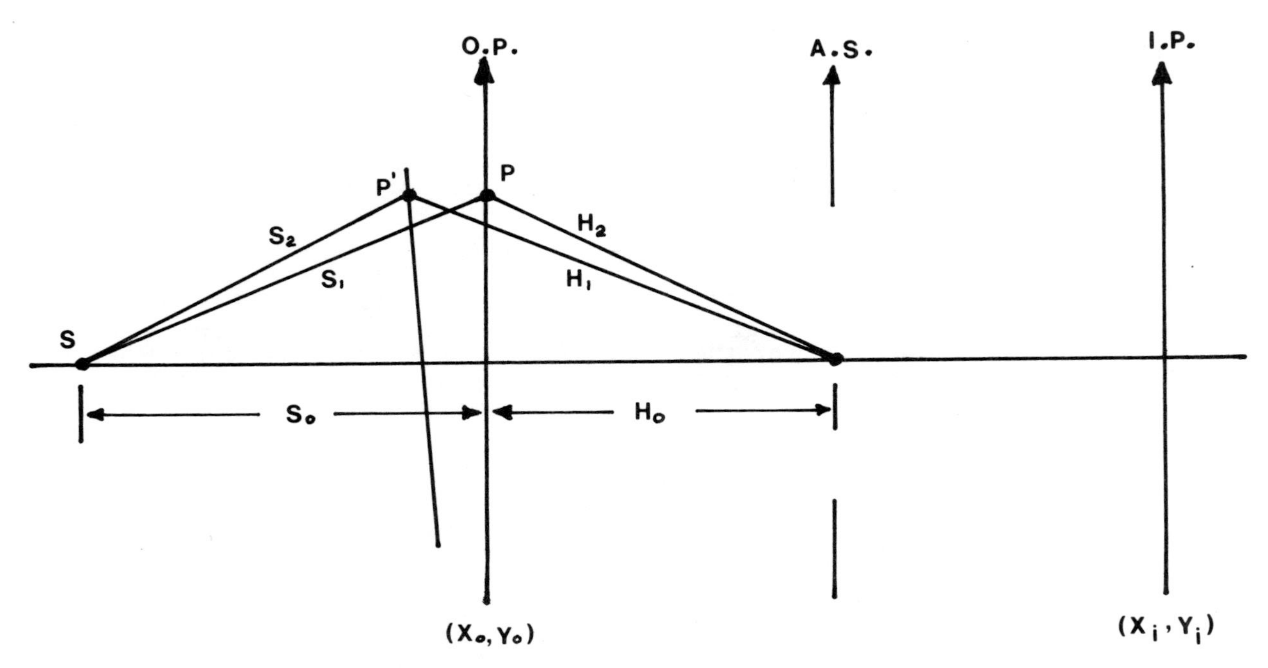

Fig. 1 Diagram presenting the definition of the variables and planes along the object beam.

$$A = \frac{2\rho}{\lambda Ho} (A_x{}^2 + A_y{}^2)^{1/2} \qquad (5)$$

where λ is the wavelenght of light and

$$Ax = x_i \, Dz / MHo - Dx; \quad Ay = y_i Dz/MHo - Dy \qquad (6)$$

here M = Hi/Ho is the lateral amplification.

In order to analyze the behaviour of the optical path difference and the fringe contrast for Ho equal to So, it results convenient to consider that the light source and the vertex of the aperture of the optical viewing system are located on the focus of an ellipse as it is shown in Fig. 2. If the object experiences a rigid body motion, the observation point P will now be located at a P'; given by the vector position **P' = P + D.** Since under the geometry shown in Figs. 1, 2 this interferometer is completely insensitive to in-plane rotation, we will analyze the range of work considering in-line translations, tilt around the x axis, and in plane translation along the y axis.

Using the properties of an ellipse, the optical path difference between points P and P' will be equal to the difference between points P' and P", where P" is a point located on the ellipse with coordinates (0, P + Δy, Dz). From Fig. 1.b we have that

$$OPD = (S_2 + H_2) - (S_1' + H_1') \qquad (7)$$

Since P" and P' are at the same distance on the Z axis and considering a first order approximation

$$S_2 = |S_1 + \Delta y \hat{j}| \approx S_1 + P \frac{\Delta y}{So} \qquad (8)$$

and

$$H_2 = |\bar{S}_1' + \Delta y \hat{j}| \approx H_1 + \frac{P \Delta y}{-Ho} \qquad (9)$$

The optical path difference for So = Ho can be written as

$$OPD = 2\ P\Delta y = \lambda\ (2n+1)/2; \qquad n = o, \mp 1, \mp 2, \dots \tag{10}$$

In order to get the first fringe at point P that is the fringe of order n=0, from Eq. 10 we have that

$$\Delta y \geqslant \frac{\lambda Ho}{4P} \tag{11}$$

for tilt, in-plane translation or a longitudinal displacement. In what follows we consider the specific value of Δy for each one of these movements. The relation between Δy and other parameters is obtained from Fig. 2, making the additional assumption that Ho = So >> P.

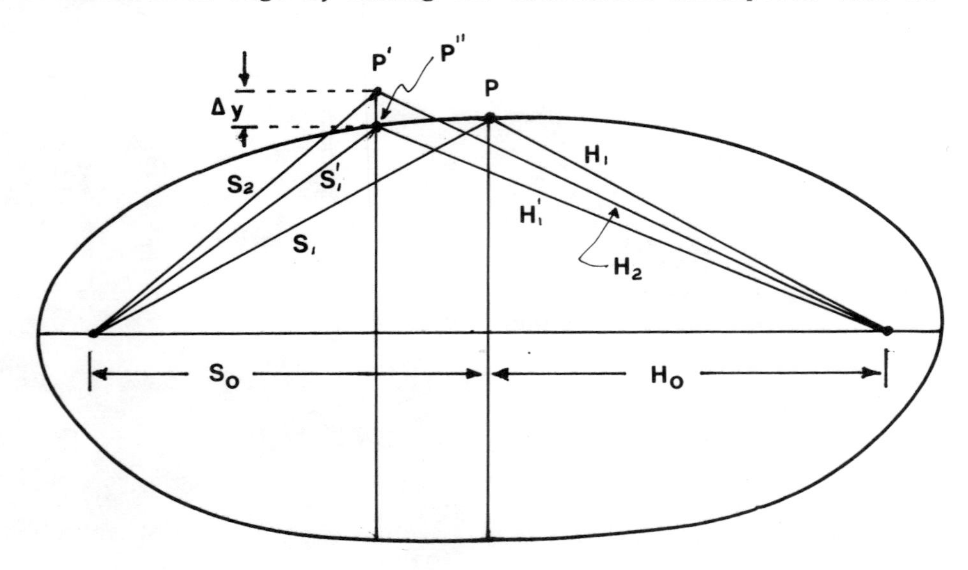

Fig. 2 Diagram of a displacement between homologous points using as reference an ellipse with the point source on one focus and the vertex of the aperture stop of the viewing system on the other focus.

In-plane Translation

For this movement,

$$Dz = 0$$

and,

$$\Delta y = Dy \tag{12}$$

Substituting in Eq. (11) we obtain that to get the first interference fringe

$$Dy \geqslant \frac{\lambda Ho}{4P} \tag{13}$$

In-Line Translation

In this case, Dz keeps constant for any point P, and

$$\Delta y = \frac{P}{2}\left(\frac{Dz}{Ho}\right)^2 \tag{14}$$

Using Eqs. 11 and 14, we have that in order to get the first interference fringe Dz must satisfy

$$Dz \geqslant \frac{Ho}{P}\left(\frac{\lambda Ho}{2}\right)^{1/2} \tag{15}$$

Comparing this relation with the limit value for Dy required to obtain the first fringe for in-plane translation (Eq. 14), we have that

$$Dz \gg Dy \left(\frac{8Ho}{\lambda}\right)^{1/2} \quad (16)$$

i.e. this arrangement is much more sensitive to in plane translations than to in line translations.

Tilt

We have that β and Dz are related by

$$Dz = P \text{ sen } \beta \quad (17)$$

and from Fig. 2

$$\Delta y = \frac{Dz^2}{2P} \quad (18)$$

Using Eqs. 11 and 18, the condition to obtain the first fringe is given by

$$Dx \gg \left(\frac{\lambda Ho}{2}\right)^{1/2} \quad (19)$$

Comparing this condition again with the limit condition for in plane translation we have that

$$Dz \gg (2PDy) \quad (20)$$

i.e. this interferometer is more sensitive to in-plane translation than to tilt, being also more sensitive to tilt than to in line displacements.

Fringe Visibility

At this point it is important to analyze the behaviour of fringe visibility for each type of movement. As it was shown in a previous work, fringe visibility under the geometry here proposed will remain constant for tilt, for different values of the numerical aperture. So if we want to use this interferometer in an insensitive mode for tilt it is necessary tocheck the condition given by Eq. 20. Fringe contrast for other types of movements will depend on the movement experienced by the object and the geometry of the experiment i.e. specifically on the radius ρ of the aperture of the optical viewing system. Hence it is necessary to compare the conditions on the fringes given by the OPD criteria with the conditions for fringe visibility for in plane translation. It is also necessary to consider the influence of the presence of an in line translation or an in plane rotation on fringe visibility although in the second case, OPD is equal to zero. To do so, at first we consider the condition to get the first fringe for a given value of the visibility in the case of an in line movement, after this we obtain the variation of the visibility for a combination of an in line and an in plane translation and later for an in-plane rotation combined with an in plane translation. In order to get an acceptable value for the visibility of the first fringe in an in line translation we must consider the value of Dz, substituting Eqs. 5 and 6 in the argument of the Bessel function associated to fringe contrast. This argument must be smaller than one half the radius of the first zero of this function. Under these circumstances

$$Dz \lesssim \frac{.61}{2} \frac{\lambda Ho^2}{2\rho P} \quad (21)$$

The critical value of Dz for observing the first fringe (Eq. 15) that at the same time satisfies the visibility condition (Eq. 21) gives us a relation for distance Ho = So as a function of the radius of the aperture stop of the viewing system,

$$Ho = \frac{2}{\lambda} \left(\frac{\rho}{.61} \right)^{1/2} \quad (22)$$

From this equation we have that since λ (the wavelenght of light) is small, the value of Ho will be very large for normal values of ρ. This means that for practical conditions, for a large range of Dz, circular fringes will not detected.

Lateral Plus In-Line Displacement

If a lateral displacement Dy is combined with an in line displacement Dz the argument of he Bessel function associated to the fringe contrast γ given by Eq. 4 will be given by

$$A = \frac{2\rho}{\lambda Ho}[(xiDz/MHo)^2 + (yiDz/MHo - Dy)^2]^{1/2} \quad (23)$$

Since OPD is only sensitive to in plane translations, an interference pattern formed by straight fringes will be obtained. This fringe pattern will be modulated by fringe contrast γ. From Eq. 23 it can be seen that for Dz = 0; i.e. a pure lateral displacement, fringe contrast keeps the same for the whole image plane, as has been shown in a previous work[5]. In this case fringe contrast is given by

$$\gamma = \frac{J_1(2\Pi\rho Dy/\lambda Ho)}{2\Pi\rho Dy/\lambda Ho}$$

If an in line displacement is introduced; i.e. $Dz \neq 0$, the contrast variation follows Eq. 4 with A given by Eq. 23. From these relations we can notice that fringe contrast will decrease radially until it goes to zero. The radius of the first zero is given by

$$r = [(\frac{xi}{M})^2 + (\frac{yi}{M} + \frac{Dy\ Ho}{Dz})^2]^{1/2} \quad (24)$$

where

$$r = 0.61\ \lambda Ho^2/\rho Dz$$

This equation defines a circle whose center is going to be shifted from the optical axis (0,0), to (0, HoDy/Dz). Furthermore, since the radius of the first zero is inversely proportional to the aperture stop radius of the optical viewing system, for some values of Ho and Dz we can reduce the aperture radius to get fringes over the whole object. Following this procedure a fringe pattern keeping the same shape as before will be observed. The increment in size of speckles only increases its characteristic noise.

A practical situation where this considerations can be helpfull is obtained when this interferometric arrangement is used to measure film thickness[2]. In this case a step is produced in the film deposit. This additional OPD produces a shift of the fringe pattern obtained when a lateral displacement of the hologram is made. The fringe pattern obtained in the region where the film deposit has been removed is shifted with respect to the fringe pattern from the region that has not changed. If an additional in line displacement is introduced, increasing the radius of the aperture stop of the optical viewing system, the radius of the first zero of the Bessel function will be reduced. This radial contrast variation is going to be observed shifted in the region where the film has been removed. This shift may allow us to distinguish which fringe in one pattern belongs to a given fringe in the second pattern. Otherwise a continuous step must be produced to observe the correspondence between both fringe patterns.

In Plane Rotation with In Plane Displacement

When an in plane rotation is combined with an in-plane translation Dy, the argument of the Bessel function related with fringe contrast is given by

$$A = \frac{2\rho}{\lambda Ho}\{2(1-\cos\alpha)[(\frac{xi}{M} - \varepsilon)^2 + (\frac{yi}{M} - \frac{Dy}{2})^2]\}^{1/2} \quad (25)$$

where

$$\varepsilon = Dy\frac{\text{sen}\ \alpha}{2(1-\cos\ \alpha)},$$

and α is the rotation angle.

Here again we observe a radial decrease of the fringe contrast. The radius of the circle corresponding to the first zero of the Bessel function is obtained when

$$r = [(\frac{xi}{M} - \varepsilon)^2 + (\frac{yi}{M} - \frac{Dy}{2})^2]^{1/2} \quad (26)$$

satisfies

$$r = \frac{0.61\lambda Ho}{\rho\sqrt{1-\cos\alpha}} \quad (27)$$

In this case the center of the circle will be shifted from the axis of rotation to a point (ε, Dy/2).. From Eqs. 26 and 27 we can notice that here again the radius of the first zero of the Bessel function is inversely proportional to the radius of the optical aperture the viewing system. Hence, a similar behaviour will be observed in this case.

Conclusions

In this work an IOBHI has been analized for a symmetrical position of the point source illuminating the object and the optical viewing system, with respect to the object position. It has been demonstrated that under the conditions here described, the shape of fringes is not going to be affected by in line, tilt and in plane rotation. This interferometer is only sensitive to lateral displacements (in plane displacements). Due to this behaviour this optical arrangement has been used for the verification of micrometric rectilinear movements[3]. In this work an analysis of fringe contrast variation at the image plane has also been presented. In this analysis two different cases has been considered: a lateral displacement combined with an in line displacement and a lateral displacement combined with an in plane rotation. Tilt contribution has not been analyzed since it has been previously demonstrated that it does not produce a variation of fringe contrast. From the theoretical analysis here presented a similar behaviour of fringe contrast variation is obtained for a lateral displacement combined with a longitudinal displacement or an in plane rotation. In both cases a radial decrease of fringe contrast is obtained. Finally a practical application of these results to find out the correspondence between the shifted patterns obtained when this interferometer is applied to thickness measurement has also been proposed.

References

1.- Nils Abramson, "Making and Evaluation of Holograms". Academic Press, Chap. 4, pp 151, (1981).

2.- D. Tentori and M. Celaya, "Film Deposit Assesment with Hologram Interferometry". Applied Optics, 25, No. 16 (1986) (in press).

3.- M. Celaya, D. Tentori, C. Lopez, C. G. Lopez, "Automatic evaluation of angular movements and linearity with holographical methods". SPIE Proceedings, Vol. 353, pp. 114 (1982).

4.- M. Celaya and D. Tentori, "Analysis of Speckle Correlation Variation in Hologram Interferometry to Distinguish Rigid Body Rotation". Amer. Inst. of Phys., Proc. No. 65, subseries on Optical Science and Engineering No. 1, pp. 219-226 (1980).

5.- M. Celaya and D. Tentori, "Fringe Contrast Variation in Hologram Interferometry with the Numerical Aperture of the Viewing System". Applied Optics 25, No. 16 (1986) (in press).

6.- Norman L. Hecht, John E. Minard, David Lewis and R. L. Fusek. Applied Optics 12, No. 11, 2665 (1973).
H. K. Liu, R. L. Kurtz.
Opt. Eng. 16, No. 12, 176 (1977).

Holographic Measurement of an Arc Plasma Field

Ding Hanquan

Department of Physics, Beijing Institute of Aeronautics and Astronautics, China

Abstract

A method using single-wavelength holographic interferometry for measurements of density and temperature distributions of an axial symmetric argon arc plasma flow field is presented. Calculating formulae and experimental results are given.

Introduction

The axial symmetric flow field with high temperature and high brightness often appears in such areas as thermophysics and welding, for which the study on the parameter distribution of this kind of flow field will provide valuable data. In our experiment, we choose argon welding arc as an object for study because it has a typical character of high temperature and high brightness. Furhtermore, since it is in the ionized state, the arc's physical properties differ from those in normal states, resulting in increased difficulties in analysis, measurement and calculation.

At present, a unified thermometric scale does not exist for the measurement of such high temperatures. In the diagonisis of plasma, methods of probe, of spectroscopy and of electromagnetic waves are widely used, but all of them have their own limitations. The method discussed in this article will complement the others.

Principle and Calculation

1. The calculation of refractive index field.

Let's take a cross section of an arc axial symmetric flow field shown in Figure 1 as the calculation plane, the radius of the flow field being R and Z being a constant. Assume that there is no refraction when the parallel light passes through the flow field along x

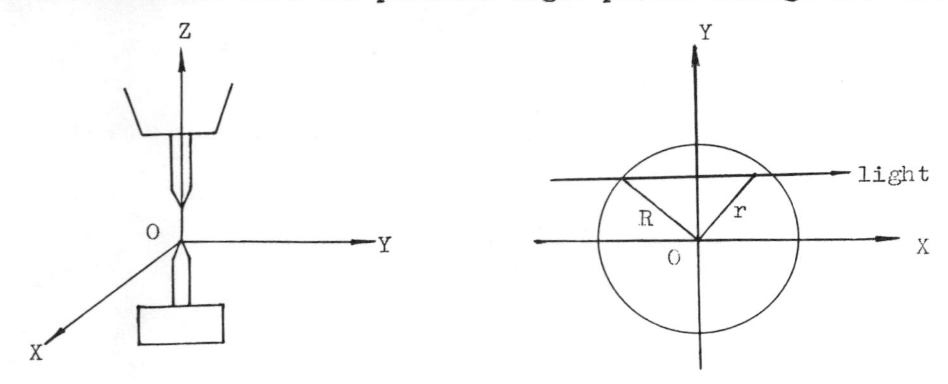

Figurel. The schematics of calculation for arc plasma flow field

axis. According to the principle of interferometry measurement, it is easy to obtain that on the path of the light which passes through a point y, the value of interference fringe shift caused by the difference between disturbed field (arc is ignited) and static field (environmental gas) is:

$$S(y) = \frac{1}{\lambda} \int_{r=y}^{R} \nu(r) \frac{d r^2}{r^2 - y^2} \tag{1}$$

where $\nu(r)=n(r)-n_o$; $n(r)$ and n_o are the refractive indices of the disturbed and static fields at point r respectively, while λ is the wavelenght of the illuminating light.

We divide the disturbed round section into N ring bands whose widths are all W. The radii of the ring bands are

$$0=r_0 < r_1 < r_2 < \dots r_{N-1} < r_N=NW=R$$

Assuming the $\nu(r)$ in each ring band depends on r^2 linearly, i.e.,

$$\nu(r) = \nu_\mu + \frac{\nu_{\mu+1} - \nu_\mu}{r^2_{\mu+1} - r^2_\mu}(r^2 - r^2_\mu) \quad , \quad r_\mu \leq r \leq r_{\mu+1}$$

and then substitute this relation into eq.(1) and calculate for $y=r_i$, we have:

$$\lambda S(r_i) = \sum_{\mu=i}^{N-1} \int_{r_\mu}^{r_{\mu+1}} [\nu_\mu + \frac{\nu_{\mu+1}-\nu_\mu}{r^2_{\mu+1}-r^2_\mu}(r^2 - r^2_\mu)] \frac{dr^2}{\sqrt{r^2 - r_i^2}}$$

$$= \sum_{\mu=i}^{N-1} [(\nu_\mu - \frac{\nu_{\mu+1}-\nu_\mu}{r^2_{\mu+1}-r^2_\mu} r^2_\mu) \int_{r_\mu}^{r_{\mu+1}} \frac{dr^2}{\sqrt{r^2-r_i^2}} + \frac{\nu_{\mu+1}-\nu_\mu}{r^2_{\mu+1}-r^2_\mu} \int_{r_\mu}^{r_{\mu+1}} r^2 \frac{dr^2}{\sqrt{r^2 - r_i^2}}]$$

Since

$$\int_{r_\mu}^{r_{\mu+1}} \frac{dr^2}{\sqrt{r^2 - r_i^2}} = 2(\sqrt{r^2_{\mu+1} - r_i^2} - \sqrt{r^2_\mu - r_i^2})$$

$$= 2W[\sqrt{(\mu+1)^2 - i^2} - \sqrt{\mu^2 - i^2}] = 2W C_0(i,\mu)$$

and

$$\int_{r_\mu}^{r_{\mu+1}} r^2 \frac{dr^2}{\sqrt{r^2 - r_i^2}} = \frac{2}{3}[(r^2_{\mu+1} + 2r_i^2)\sqrt{r^2_{\mu+1} - r_i^2} - (r^2_\mu + 2r_i^2)\sqrt{r^2_\mu - r_i^2}]$$

$$= \frac{2}{3} W^3 [(\mu+1)^2 \sqrt{(\mu+1)^2 - i^2} - \mu^2\sqrt{\mu^2 - i^2} + 2i^2(\sqrt{(\mu+1)^2 - i^2} - \sqrt{\mu^2 - i^2})]$$

$$= \frac{2}{3} W^3 C_1(i,\mu)$$

we have

$$\frac{\lambda S_i}{2W} = \sum_{\mu=i}^{N-1} \{ \nu_\mu [(1+\frac{\mu^2}{2\mu+1}) C_0(i,\mu) - \frac{C_1(i,\mu)}{3(2\mu+1)}] + \nu_{\mu+1} [\frac{C_1(i,\mu)}{3(2\mu+1)} - \frac{\mu^2 C_0(i,\mu)}{2\mu+1}] \}$$

$$= \nu_i \frac{3(i+1)^2 C_0(i,i) - C_1(i,i)}{3(2i+1)} + \sum_{\mu=i+1}^{N-1} \nu_\mu \frac{3(\mu+1)^2 C_0(i,\mu) - C_1(i,\mu)}{3(2\mu+1)} + \sum_{\mu=i}^{N-1} \nu_{\mu+1} \frac{C_1(i,\mu) - 3\mu^2 C_0(i,\mu)}{3(2\mu+1)}$$

After some manipulation, we have

$$\nu_i = \frac{\frac{3\lambda}{4W} S_i - \sum_{\mu=i+1}^{N-1} \nu_\mu B(i,\mu)}{A(i)} , \quad i = 0, 1, 2, \cdots, N-1 \qquad (2)$$

where $A(i)=\sqrt{2i+1}$, $B(i,u)= \frac{[(\mu+1)^2 - i^2]^{3/2}}{2\mu+1} + \frac{[(\mu-1)^2 - i^2]^{3/2}}{2\mu-1} - \frac{4\mu(\mu^2 - i^2)^{3/2}}{4\mu^2 - 1}$

Thus with the values of S_i of different points measured from the holographic interferogram, the corresponding ν_i's can be calculated from eq. (2).

2. The relation-ship between the refractive index and other parameters of the plasma gas.
Since the refractive index of plasma gas can be deduced from its holograp-hic interferogram, its other parameters, such as number density and temperature, may be obtained if we could find the connections between the refractive index and those parameters.

The refractive index of neutral atomic or molecular gaseous medium is given by the well-known Gladstone-Dale equation:

$$n_a - 1 = K_a \rho_a = K_a m_a N_a$$

where K_a is the G-D constant determined by the properties of the gas and the wavelength of light. ρ_a and N_a are the mass density and number density of the gas respectively.

In standard state we have $n_s - 1 = K_a \rho_s = K_a m_a N_0$, N_0 is the Loschmidt constant.

From Cauchy formula, [1]there is

$$n_s-1 = A(1+B/\lambda^2).$$

where A and B are constants connected with the properties of the gas, therefore

$$n_a-1 = \frac{A}{N_0}(1+\frac{B}{\lambda^2})N_a$$

For argon atom, $A=27.92\times10^{-5}$, $B=5.6\times10^{-3}$. the wavelength of ruby laser in $\lambda=0.6943$um, $N_0=2.6868\times10^{19}$ cm^{-3}. With these data, we then get

$$n_a-1 = 1.051\times10^{-23}N_a$$

The G-D constant of argon ion and that of argon atom have the following relation:[2]

$$K_+ = \frac{2}{3}K_a$$

so that the refractive index of argon ion is

$$n_+-1 = 0.7006\times10^{-23}N_+$$

where N_+ is the number density of argon ion.

The refrative index of electron is

$$n_e-1 = -\frac{e^2\lambda^2}{2\pi m_e c^2}N_e = -2.162\times10^{-22}N_e$$

where N_e is number density of electron; $e=4.8032\times10^{-10}$e.s.u.; $m_e=9.1035\times10^{-28}$g; $c=2.9979\times10^{10}$cm/s; $\lambda=0.6943$um.

Hence the refractive index of argon arc plasma is

$$n-1 = 1.051\times10^{-23}N_a+0.7006\times10^{-23}N_+-2.162\times10^{-22}N_e.$$

Before arc is ignited, the refractive index of environmental air is[1]

$$N_o-1 = \{(64.328+\frac{29498.1}{146-1/\lambda^2}+\frac{255.4}{41-1/\lambda^2})\frac{p[1+(1.049-0.0157t)\times10^{-6}p]}{720.883(1+0.003661t)}-\frac{0.0624-0.000680/\lambda^2}{1+0.003661t}f\}\times10^{-6}$$

In our experiment, $\lambda=0.6943$um; p=760mmHg; t=20°C, and steam pressure f=0. Therefore,

$$n_0-1 = 2.711\times10^{-4}$$

and

$$\nu_i = n-n_0 = 1.051\times10^{-23}N_a+0.7006\times10^{-23}N_+-2.162\times10^{-22}N_e-2.711\times10^{-4} \quad (3)$$

Only considering the first order ionization, we have

$$N_+ = N_e \quad (4)$$

Supposing that thermodynamical equations can be applied in a small region, by pressure formula

$$p = (N_a+N_++N_e)kT \quad (5)$$

and Saha ionization equation

$$\frac{N_+N_e}{N_a} = \frac{U_+}{U_a}\cdot 2\cdot\frac{(2\pi m_e kT)^{3/2}}{h^3}\exp(-\varepsilon_i/kT) \quad (6)$$

where $h=6.626\times10^{-27}$ erg.s; $k=1.380\times10^{-16}$erg/K; $U_+=4+2\exp(-2062/T)$; $U_a=1+60\exp(-162500/T)$; $\varepsilon_i=2.525\times10^{-11}$erg; T is absolute temperature. We also have

$$N_a=6.665\times10^{21}/T+8.688\times10^{21}\nu_i+2.356\times10^{18} \quad (7)$$

$$N_+ = N_e = 3.349\times10^{20}/T - 4.344\times10^{21}\nu_i - 1.178\times10^{18} \quad (8)$$

$$(3.349\times10^{4}/T - 4.344\times10^{5}\nu_i - 1.178\times10^{2})^2 \cdot [1 + 60\exp(-162500/T)]$$
$$= 4.824T^{3/2}(6.665\times10^{4}/T + 8.688\times10^{4}\nu_i + 23.56)[4+2\exp(-2062/T)]\exp(-1.828\times10^{5}/T) \quad (9)$$

With the help of a computer, we can calculate the temperature distribution of plasma and density distributions of argon atoms, ions, or electrons through eqs. (2),(7),(8) and (9). Actually, other parameters of the plasma, such as conductivity, current density and Lorentz force, can be easily obtained from the above calculated parameters.

Experimental Methods and Results

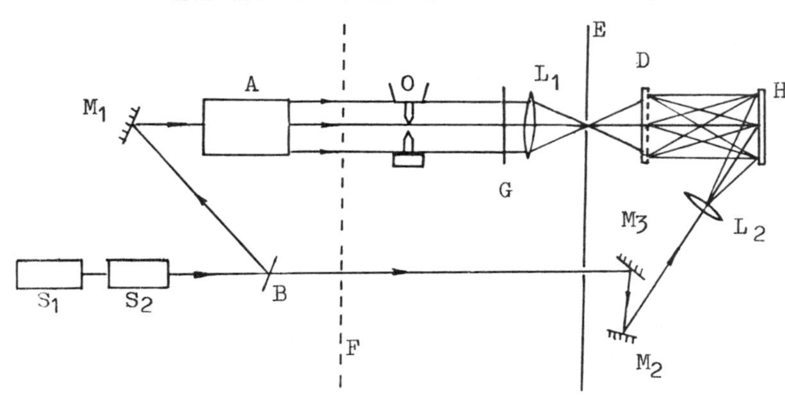

Figure 2. Schematic diagram of the optical paths arrangement.

Figure2. shows our experimental set-up. S_1 is He-Ne laser used for adjusting the resonator of the pulsed ruby laser S_2 and the whole optical setup. B and M are splitter and mirror respectively. A is a collimating lens. The collimated beam carries the information of the arc field andarrives at diffusion screen D, and hologram plate H without obstacle. The intense arc field is filtered by G(Filter) -L_1(lens) -pinhole(E is a plate with two pinholes) and only a very small amount of 'noise' is able to reach H. The transparent glass F is used to prevent arc's splashing from damaging optical components. G is both a filter and a protecter for L_1. O is arc source, L_2 is a expanding lens.

Table 1. The Experimental Datae and Results

(a) Current 15A, Arc length 1.95 mm.

Z = 0.89 mm						R = 4.56 mm				
r_i(mm)	0.00	0.46	0.91	1.37	1.82	2.28	2.73	3.19	3.65	4.10
S_i(-1)	2.743	2.696	2.611	2.492	2.341	2.147	1.819	1.340	0.681	0.166
ν_i(-10-4)	2.8804	2.6282	2.5414	2.5293	2.4962	2.4610	2.4017	2.3702	1.5110	0.4349
T_i(°K)	13047	8723	4726	4406	3977	3269	2582	2358	746	446
N_a(10^{18}cm-3)	0.3643	0.8368	1.5521	1.6648	1.8444	2.2431	2.8409	3.1108	9.8328	16.447
N_e	0.0989	0.0021	0	0	0	0	0	0	0	0

Z = 1.90 mm.						R = 4.46 mm				
r_i(mm)	0.00	0.45	0.89	1.34	1.78	2.23	2.68	3.12	3.57	4.02
S_i(-1)	2.552	2.490	2.392	2.292	2.152	1.927	1.570	1.089	0.563	0.188
ν_i(-10-4)	2.6429	2.6338	2.4832	2.4191	2.3813	2.3579	2.3496	1.9620	1.1589	0.5031
T_i(°K)	9384	8998	3572	2735	2434	2279	2229	1130	597	456
N_a(10^{18}cm-3)	0.7705	0.8085	2.0535	2.6820	3.0137	3.2186	3.2908	6.9914	12.287	16.086
N_e	0.0058	0.0033	0	0	0	0	0	0	0	0

(b) Current 30A, Arc length 3.40 mm

Z = 0.60 mm						R = 4.75 mm				
r_i (mm)	0.00	0.48	0.95	1.43	1.90	2.38	2.85	3.32	3.80	4.27
S_i (-1)	2.870	2.831	2.751	2.637	2.486	2.293	2.040	1.722	1.202	0.508
ν_i (-10-4)	2.6380	2.4949	2.3982	2.3382	2.2818	2.2712	2.2634	2.1826	2.0940	1.2776
T_i (°K)	9188	3757	2557	2165	1893	1850	1818	1558	1349	642
N_a(10^{18}cm-3)	0.7895	1.9524	2.8687	3.3881	3.8749	3.9650	4.0348	4.7081	5.4376	11.426
N_e	0.0044	0	0	0	0	0	0	0	0	0

After the setup is adjusted, the hologram is exposed in undisturbed field. Then the hologram is exposed again when argon gas is given and arc is ignited. If we hope to get a finite fringe interferogram, we slightly adjust the pitch angle of A before the second exposure to get reference background fringes.

S_i is read out on selected cross sections from the hologram and ν_i is calculated by eq.(2) and T_i, N_a, $N_e=N_+$ by eqs.(7) - (9). The datae are listed in Table 1.

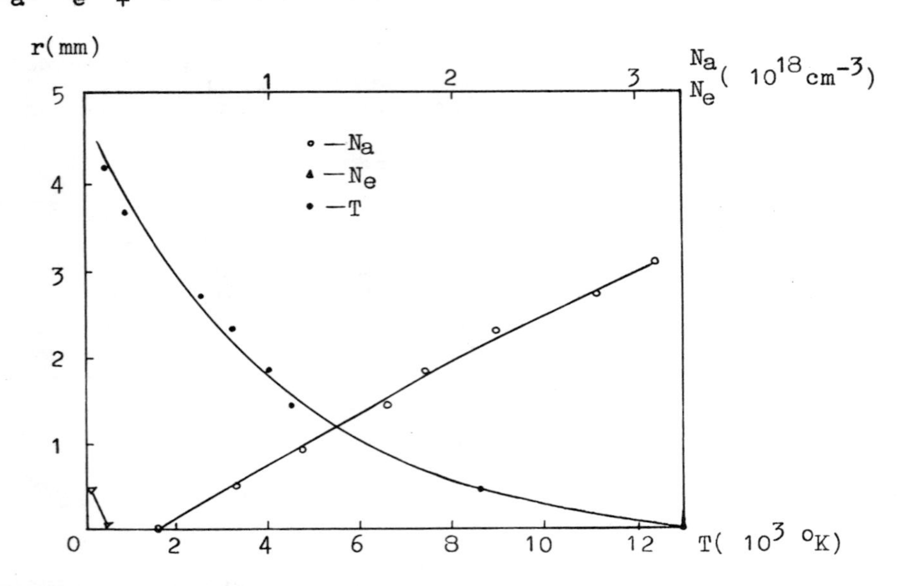

Figure 3. Radial distribution of temperature and density.

The radial distribution of temperature and density on the cross section Z=0.89mm and R=4.56mm is shown in Figure 3.

We studied the temperature field of alcohol burner carefully using eq.(2) and compare the theoretical results with experimental datae. The results show that eq.(2) is correct, and our datae in Table 1 are consistent with published results.

From the experimental datae we learn that, under the conditions of our experiment, the argon is in the plasma state when the temperature is over 8000°K. The equations of state of the ideal gas can still be applied below 8000°K. According to eq. (9), the temperature of argon gas under normal pressures is a singlevalued function of the refractive index when the temperature is lower than 17000°K.

References

1. Allen, C.W., Astrophysical Quantities, University of London, Athlone Press, 1973, 37.
2. Vest., C.M., Holographic Interferometry, John Wiley & Sons, New York, 1979, 353.

Profile Measurement on IC Wafers by Holographic Interference

Jing Ao Sun, Anni Cai and Glen Wade

Department of Electrical and Computer Engineering, University of California
Santa Barbara, CA 93106, Phone: (805) 961-2508

Abstract

We present here a new technique of holographic interferometry for profile inspection of integrated circuit (IC) wafers. With this technique the lateral resolution of the profilometer will not be affected by the limitations on optical aperture. In addition, there are no stringent requirements on the spot size of the laser irradiation. A principal advantage of this system is that it needs no magnifying elements to give reasonable resolution for lateral measurements. Also the system will work even if multiple reflections exist.

Introduction

In IC devices the dimensions of line geometries determine the resistances and capacitances of the circuit and, as a result, are crucial to the performance of the devices. For a variety of reasons (over or under exposure in fabrication, for instance), the dimensions of the lines may differ from what is designed. Consequently, monitoring the width and depth of the lines has become an important aspect in semiconductor processing as the lines in IC devices approach one micron or less.

Traditional methods to measure the line profiles are based on image formation with ordinary optical microscopes. However these methods have been found to be inadequate when the line dimensions approach 1.5 μm or less[1], because the optical image of such small structures is dominated by diffraction.[2–4] It is difficult to define physical edges in such images. Although scanning electron microscopes possess high resolvability, they suffer from the drawbacks of requiring vacuum operation and not being completely non-destructive.

In order to measure profiles for lines of one micron or less considerable efforts have been made on improving the resolution of coherent scanned microscopes. A resolution of 0.25 μm has been obtained by using confocal laser scanning.[5] This works only when the depth of the line is considerably larger than the depth of field of the confocal scanning microscope[6].

It is possible also to measure linewidth in the spatial frequency domain.[7–11] However, this method is too sensitive to multiple reflections to provide adequate accuracy for wafer measurements of submicron dimensions.

In this paper we present a new technique for wafer inspection by holographic interferometry. In this system there are no stringent requirements on the spot-size of the laser irradiation. A principal advantage is that it needs no magnifying elements to give reasonable resolution for lateral measurements. Also the system will work even if multiple reflections exist.

Basic Theory

We will first give a simple description of some basic problems involved in measuring the profiles of lines in order to provide a basis for discussing the new technique.

Suppose, for example, we need to measure the width of lines with periodic spacing. If only a few low-order spatial frequencies of the pattern are collected by the optical system, the image of the pattern will exhibit prominent fringes (see Fig. 1). It is difficult to locate the physical edges precisely in such cases.[12] In order to reduce fringe effects the aperture of the optical system must be large enough to include high spatial-frequency components. The Rayleigh resolution of linewidth measurement is limited by the aperture of the optical system.

It is well known that the diameter of a laser beam is diffraction-limited to no less than half a wavelength and can not be treated as infinitesimal. As illustrated in Fig. 2, when the laser beam simultaneously impinges on both surfaces at top and bottom levels, interference between the two components of light s_1 and s_2 occurs. As the linewidth approaches a dimension comparable with that of the beam, interference will produce detrimental effects on the measured data.

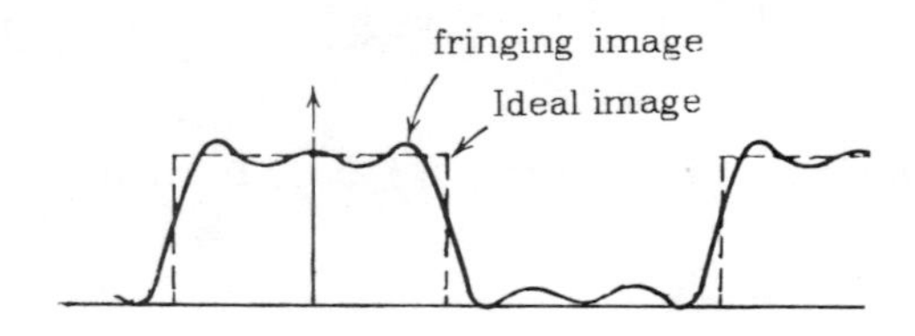

Figure 1. A fringing image formed from using only a few lower spatial frequency components.

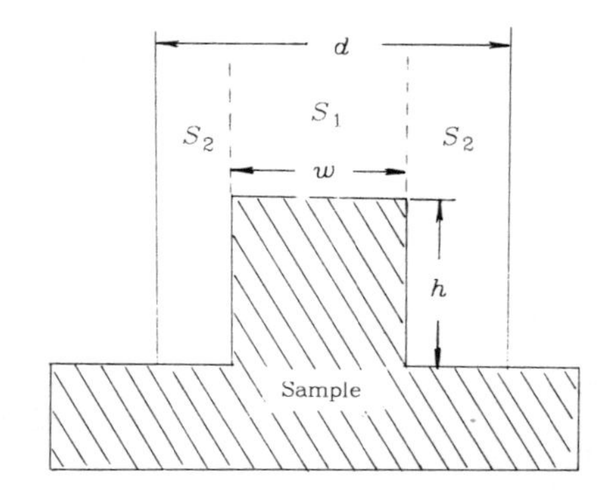

Figure 2. Schematic illustration of the relation between scanning beam and sample. d is the diameter of the laser beam, s_1 and s_2 are two components of the light impinging on the top and bottom levels of the surface.

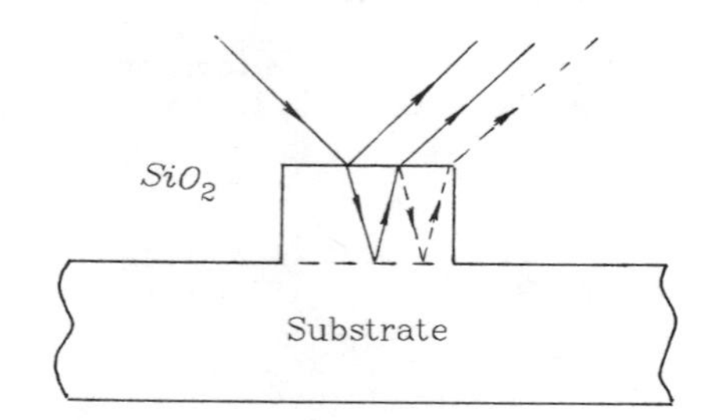

Figure 3. Multiple reflections within the layers of a test sample.

Figure 3 shows the existence of multiple reflections between two interfaces when the line is made of a material which is not opaque. In such a case additional diffraction will occur.

The above difficulties lead us to search for solutions. The zero spatial-frequency component in the reflected light carries information about the linewidth. This can be immediately seen by writing the expression for the intensity spectrum of the light reflected from a sample such as that shown in Fig. 2:

$$|\bar{U}(f_x)|^2 = r_b^2\delta^2(f_x) + (r_b-r_t)^2w^2 \left[\frac{\sin\pi f_x w}{\pi f_x w}\right]^2 - 2r_b(r_b-r_t)w\delta(f_x)\frac{\sin\pi f_x w}{\pi f_x w}\cos\frac{4\pi}{\lambda}h \qquad (1)$$

where w is the linewidth we are interested in, r_b and r_t are the reflectivities of the surfaces at the bottom and top levels, h is the depth of the line profile, f_x is the spatial frequency and λ is the wavelength of the laser.

If we can develop a technique canceling out one of the two components of reflected light, Eq. (1) then becomes

$$|\bar{U}(f_x)|^2 = \left(r_t w \frac{\sin\pi f_x w}{\pi f_x w}\right)^2 \qquad (2)$$

This is the equivalent of making r_b equal to zero. The interference between the light components reflected from the top and bottom surfaces respectively is eliminated and the zero spatial-frequency component $|\bar{U}(0)| = r_t w$ is proportional to the linewidth w. Since the diffraction angle corresponding to zero spatial-frequency is always zero, the linewidth measurement can be done without any limitation on the optical aperture of the system by only detecting the zero spatial-frequency component.

We have developed a system based on these principles. The basic optical setup is shown in Fig. 4(a).

In Figure 4(a), M' is the image position of the mirror surface. The light reflected from mirror surface M can be seen as if it is from M'. In order to show the dimensional relationship between the beam spot and test sample explicitly, two legs of the interferometer are redrawn in Fig. 4(b) in such a fashion that the signal beam and the reference beam are collimated. When the laser beam is scanned at the position that the light spot impinges on both the top and bottom surfaces simultaneously, a fringe pattern is formed on the surface of the beam splitter by interference between the reference beam

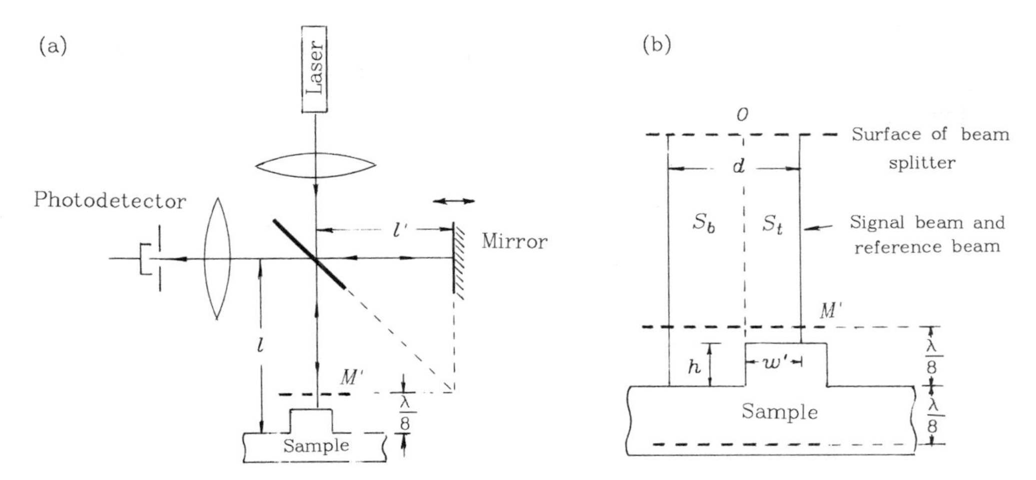

Figure 4. (a) Basic optical set up of the system
(b) Relative position between light spot and line surface.

s_r and two parts of signal beam s_b and s_t. The intensity of the pattern at a given point can be written as

$$I = (s_b+s_t+s_r)(s_b+s_t+s_r)^* \tag{3}$$

The intensity I_0 of the zero-order fringe appearing at the center position 0 is given by

$$I_0 = \frac{p_0}{2d^2} [d^2+r_t^2 w'^2+r_b^2(d-w')^2 + 2r_b r_t(d-w')w' \cos \frac{4\pi}{\lambda} h$$

$$+ 2r_t w'd \cos\frac{4\pi}{\lambda}(\ell-\ell'-h) + 2r_b(d-w')d \cos\frac{4\pi}{\lambda}(\ell-\ell')] \tag{4}$$

where w' is the lateral dimension of the top surface covered by the light spot, p_0 is the laser power, r_t and r_b are the reflective coefficients of the top and bottom surfaces, and d, h, ℓ and ℓ' are as shown in Fig. 4. The detailed derivation of this equation is given in Appendix A. In order to obtain the information about w', the mirror position is set for $(\ell-\ell') = \pm \lambda/8$ (see Fig. 4 (b)). Substituting $(\ell-\ell') = \pm \lambda/8$ in Eq. (4), we obtain the fringe intensities corresponding to the two mirror positions:

$$I_{01} = \frac{p_0}{2d^2} [d^2+r_t^2 w'^2+r_b^2(d-w')^2 + 2r_b r_t(d-w')w'\cos \frac{4\pi}{\lambda} h + 2r_t dw' \sin \frac{4\pi}{\lambda}h] \tag{5}$$

and

$$I_{02} = \frac{p_0}{2d^2} [d^2+r_t^2 w'^2+r_b^2(d-w')^2 +2r_b r_t(d-w')w'\cos \frac{4\pi}{\lambda}h - 2r_t dw' \sin \frac{4\pi}{\lambda}h] \tag{6}$$

respectively. Because of the quadratic phase relationship between s_r and s_b, the term for the interference between them (that is, the last term in Eq. (4)) no longer appears in both Eqs. (5) and (6).

Note that the only difference between Eqs. (5) and (6) is the opposite sign of the last term. Therefore after I_{01} and I_{02} are detected as currents i_1 and i_2 a simple subtraction will cancel these interference terms from the data and provide the information needed

$$i = i_1 - i_2 = \alpha(p_1 - p_2) = 2\alpha p_0 r_t A \frac{w'}{d} \sin \frac{4\pi}{\lambda}h \tag{7}$$

where α (Ampere/Watt) is the sensitivity of the photodetector, p_1 and p_2 are the light powers collected by the photodetector and A is the area of the pinhole in front of the photodetector.

Contributions from the light reflected by the bottom surface have been completely eliminated in i. Thus the reflectivity of the bottom surface has been made equivalent to zero. When the laser beam scans along the x direction, the data expressed by Eq. (7) is a function of x and is proportional to the convolution of the light spot and the top surface of the line profile as illustrated by Fig. 5.

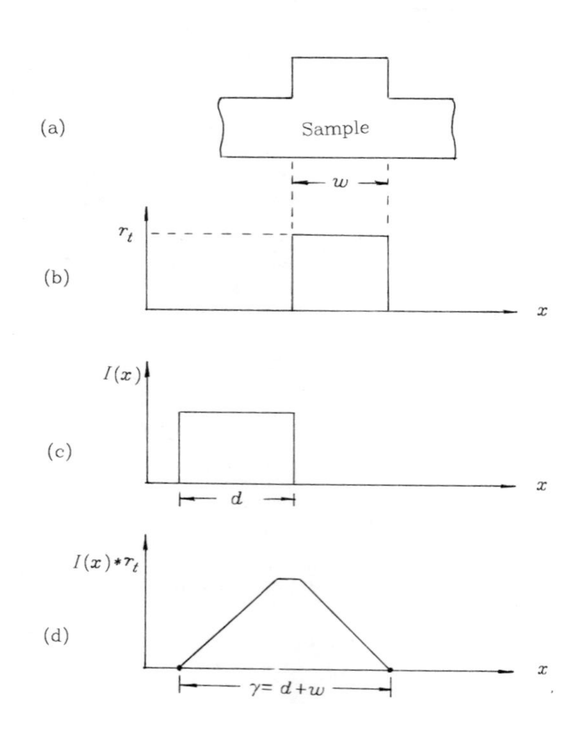

Figure 5. (a) Sample.
(b) Reflectivity of the top surface of sample. (c) Intensity distribution of the laser spot.
(d) Convolution of (b) and (c).

A level detector is used in the system to determine the distance γ between two zero-level points (see Fig. 5(d)). The top line-width is given by

$$w = \gamma - d$$

The factors r_t and sin $4\pi h/\lambda$ in Eq. (7) only affect the sensitivity of the system. In practice, however, it is possible for some devices to have h near $n\lambda/4$ (where n = 1,2,3,...), thus making sin $4\pi h/\lambda$ approach zero. To avoid this situation a frequency adjustable nitrogen laser λ_1 = 337 nm, λ_2 = 308 nm and λ_3 = 248 nm can be used in the system. As graphically demonstrated by Fig. 6 the value of $|\sin 4\pi h/\lambda| \geq 0.4$ can be achieved for $h \geq 8.1$ nm by choosing different wavelengths. This range of values of h is large enough to cover all values of concern in actual devices.

It is vital to be able to measure the width at both the top and the bottom of the structure. In IC devices, the bottom width generally represents an electrically significant dimension. In this technique, the bottom linewidth can be determined by measuring the distance between lines. This measurement is carried out in a manner similar to the above procedure except that the $\pm$ $\lambda/8$ mirror position are aligned with respect to the level of the top surface.

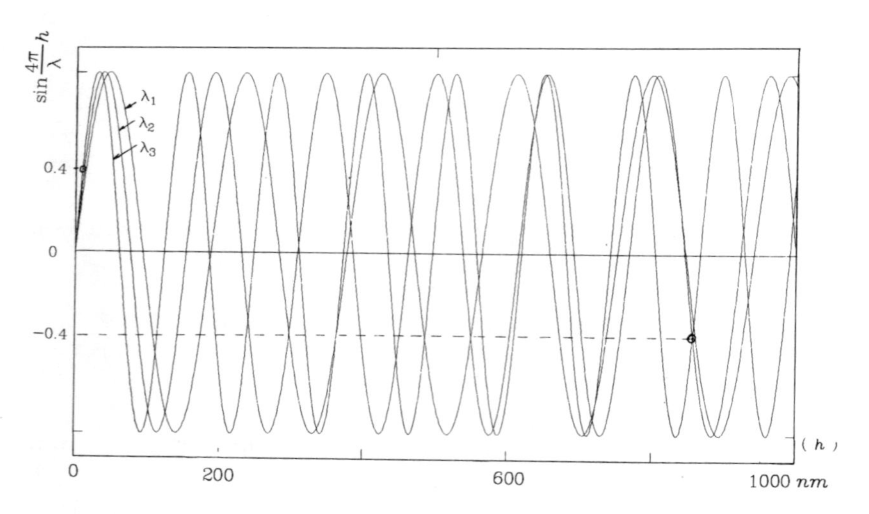

Figure 6. Plots of sin $4\pi h/\lambda$ with λ_1 = 377 nm, λ_2 = 308 nm and λ_3 = 248 nm.

Note that if h is greater than 8.1 nm, sin $4\pi h/\lambda \geq 0.4$ can be achieved by choosing different values of λ.

The $\pm\ \lambda/8$ mirror alignments in the system are performed after the system is adjusted to bring the mirror to the position where ℓ and ℓ' are equal. When the system is aligned with equal ℓ and ℓ', the distance $\Delta\ell$ (see Fig. 7) between the center 0 and the first maximum 0' of the interference pattern is measured. The diffraction angle ϑ becomes

$$\vartheta = \tan^{-1}\left(\frac{\Delta\ell}{\ell}\right) \tag{9}$$

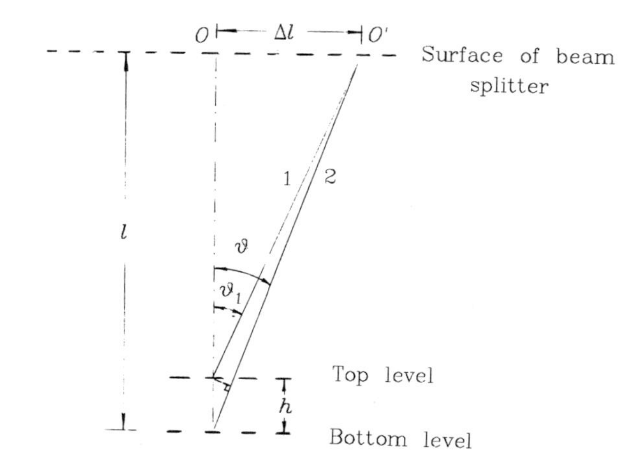

Figure 7. Geometry of the interference pattern. 0 is the center of the pattern. 0' is the location of the first maximum. The image of the mirror surface coincides with the bottom level.

In the far field, beam components 1 and 2 are approximately parallel, i.e., $\vartheta_1 \approx \vartheta$. The phase difference between the two components at point 0' is $\Delta\phi = 2\pi(h+h\cos\vartheta)/\lambda$. In order to get a maximum at 0', $\Delta\phi$ must be 2π. Therefore the depth h of the line profile can be calculated by

$$h = \frac{\lambda}{1+\cos\vartheta} \tag{10}$$

Significance of the Suggested Technique

There are several important features of this system. First, only information carried by the zero-order component of the reflected light is utilized in the linewidth measurement. Since this information is detected from the normal direction, the lateral resolution of the system will not be affected by the optical aperture.

Second, as derived in Appendix B, when multiple reflections are taken into account, Eq. (7) becomes

$$i = i_1 - i_2 = 2\alpha p_0 |r_{te}| A \frac{w'}{d} \sin\left(\frac{4\pi}{\lambda}h - \psi\right) \tag{11}$$

where $r_{te} = |r_{te}| e^{j\psi}$ is the reflectivity of the top surface when multiple reflections exist. Comparing Eq. (11) with Eq. (7), we can see that multiple reflections only affect the sensitivity of the system. i is still proportional to the width w' of the top surface covered by the light spot. Therefore, the positions of the two zeros in Fig. 5(d), as well as the measured line width w, will not be affected by multiple reflections.

Third, the interference between the light components reflected from the top and bottom surfaces has been eliminated during data acquisition. As a result, the ordinary requirements on the light spot size is relaxed in this system.

It has been experimentally demonstrated that the dominant noise factor in a surface probing system using coherent light is shot noise in the photodetector.[13 16] The effective shot noise current is expressed as

$$I_n^2 = 2e\alpha\Delta f\left(\frac{p}{2}\right) \tag{12}$$

where e is the electronic charge (1.6×10^{-19} coulombs), Δf is the operational bandwidth of the signal detecting circuit (30 Hz) and p is the light power actually collected by the photodetector. When signals i_1 and i_2 are subtracted, the noise in both measurements are added. From Eqs. (7) and (12), we obtain the signal to noise ratio as follows:

$$\frac{S}{N} = \frac{i^2}{I_{n_1}^2 + I_{n_2}^2} = \frac{\alpha(p_1 - p_2)^2}{e\Delta f(p_1 + p_2)} \tag{13}$$

Since $\alpha \cong e\eta/\hbar\omega$, the above equation corresponds to the quantum-noise limited detection, where η is the quantum efficiency of the photodetector, $\hbar$ is Planck's constant and ω is the angular frequency of the laser light.

If we take the laser light power to be p_0 = 0.1 μW over an area of 3μmx3μm, (which corresponds to a power of about $5\times10^3 W/m^2$ on the sample surface), a pinhole with a diameter of 1mm, and α = 0.3, r_t = 0.7, r_b = 0.3, and $|\sin 4\pi h/\lambda| = 0.4$, then w_{min} = 0.1 μm corresponds to S/N = 1.5. This gives an estimate of the detection limit.

It has been experimentally demonstrated in systems where depth-measurement operation is similar in principle to that of the system shown in Fig. 4(a) that a depth resolution of about a thousandth of the laser wavelength can be achieved[13].

Conclusion

We have theoretically shown that this system is capable of overcoming or partially overcoming certain problems frequently encountered in submicron profile measurements. In this system the lateral resolution will not be affected by limitation on the optical aperture. In addition, there are no stringent requirements on the spot size of the laser irradiation. A principal advantage of this system is that it needs no magnifying elements to give reasonable resolution for lateral measurements. Also the system will work even if multiple reflections exist.

References

1. J.T. Lindom et al., "Scanned Laser Imaging for Integrated Circuit Metrology," Submitted to Proceeding of Royal Society, Series A, July 1985.

2. D. Nyyssonen, "Linewidth Measurement with an Optical microscope: the Effect of Operating Conditions on the Image profiles," Applied Optics, Vol. 16, pp. 2223-2230, 1977.

3. D. Nyyssonen and J.M. Jerke, in Proceeding, International Electronic Devices Meeting, Washington, D.C., 4-6 Dec. p. 437, 1978.

4. D. Nyyssonen, in Development in Semiconductor Microlithography III, SPIE, Bellingham, Washington, 1978, Vol. 135, pp. 115-119.

5. The manual of specification of SiScan-1 Wafer Inspection System.

6. S.D. Bennett et al., "Integrated Circuit Metrology Using Confocal Optical Microscopy," Phil Trans. of Royal Society, Series A, March 1986.

7. H.L. Hasdon and N. George, "Developments in Semiconductor Microscopy," SPIE, Bellingham, Wash., 1976, Vol. 80, pp. 54-63.

8. H.P. Kleinknecht and Meier, "Linewidth Measurement on IC Masks and Wafers by Grating Test Pattern," J. of Applied Optics, Vol. 19, No. 4, pp. 525-533, Feb. 1980.

9. W.A. Bosenberg and H.P. Keinknencht, "Linewidth Measurement on IC Masks by Diffraction from Grating Test Pattern," J. of Solid State Technology, pp. 110-115, Oct., 1982.

10. W.A. Bosenberg and H.P. Kleinknecht, "Linewidth Measurement on IC Wafer by Diffraction from Grating Test Pattern," J. of Solid State Technology, pp. 79-85, July 1983.

11. H.S. Damar et al., "Diffraction Characterization for Processing, Monitoring, Linewidth Measurement and Alignment," SPIE Vol. 470, Optical Microlithography: Technology for Next Decade, pp. 157-167, 1984.

12. C.J.R. Sheppard and Choudhury, "Image formation in the scanning microscope," Optica Acta, Vol. 24, No. 10, pp. 1051-1073, 1977.

13. R.I. Whitman and A. Korpel, "Probing of Acoustic Surface Perturbation by Coherent Light," Applied Optics, Vol. 8, No. 8, pp. 1567-1576, Aug. 1969.

14. S. Sizgoric and A. Gundjian, "An Optical Homodyne Technique for Measurement of Amplitude and Phase of Subangstrom Ultrasonic Vibration," Proceeding of IEEE, Vol. 57, pp. 1313-1314, 1969.

15. T. Kwaataal, "Contribution to the Interferometric Measurement of Subangstrom Vibration," Rev. Sci. Instrum, Vol. 45, No. 1, pp. 39-41, Jan. 1974.

16. T. Kwaataal, B.J. Lygmes and G.A. Pijll, "Noise Limitation of Michelson Laser Interferometer," J. Phys. D: Appl. Phys., Vol. 13, pp. 1005-1015, 1980.

Appendix A Derivation of Eq. (4)

For simplicity, assume the light spot is a square of side d with a uniform intensity distribution. The amplitude U of the incident light in each leg of the interferometer is related to the total laser power p_0 by

$$\frac{U^2}{2} d^2 = \frac{1}{2} p_0 \qquad \text{(A-1)}$$

where the factor 1/2 in front of p_0 is introduced due to the beam splitter. The amplitudes of reflected light, U_1, U_2 and U_3 on the surfaces at the bottom and top levels of the sample as well as on the mirror surface, can be expressed as $r_b p_0^{1/2}/d$, $r_t p_0^{1/2}/d$, $p_0^{1/2}/d$ respectively, where r_b and r_t are the reflection coefficients of the bottom and top surfaces. Since the beam splitter is located in the far field of the reflective surfaces, the individual pattern formed by each of the beams at the surface of beam splitter can be expressed as a sinc function:

$$\bar{U}_i(f_{x_i}) = U_i w_i \frac{\sin \pi f_{x_i} w_i}{\pi f_{x_i} w_i} \qquad (i = 1,2 \text{ and } 3) \qquad \text{(A-2)}$$

where

$$w_i = \begin{cases} d-w' & i = 1 \\ w' & i = 2 \\ d & i = 3 \end{cases}$$

$$f_{x_i} = \begin{cases} \dfrac{x}{\lambda \ell} & i = 1 \\[2ex] \dfrac{x}{\lambda(\ell-h)} & i = 2 \\[2ex] \dfrac{x}{\ell'} & i = 3 \end{cases}$$

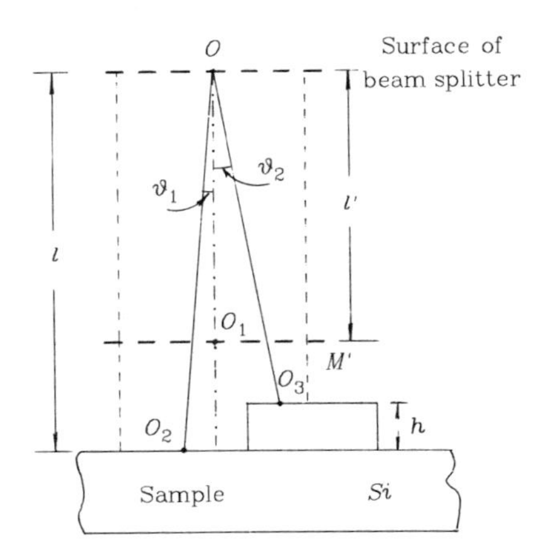

Figure A. The geometric relationship between beams. O_1 is the center of the image of reference beam. The center of the interference pattern is defined by O_1. O_2 and O_3 are centers of the illuminated surfaces at the bottom and top levels respectively.

and h, w', ℓ' are shown in Fig. 4. If the value of $\bar{U}_i(f_{x_i})$ at the center 0 of beam splitter is denoted as U_{io}, the intensity at point 0 is given by

$$I_0 = \frac{1}{2}(U_{10}e^{j\phi_1}+U_{20}e^{j\phi_2}+U_{30}e^{j\phi_3})(U_{10}e^{j\phi_1}+U_{20}e^{j\phi_2}+U_{30}e^{j\phi_3})^* \qquad \text{(A-3)}$$

where $\phi_1 = \frac{4\pi}{\lambda}\frac{\ell}{\cos\vartheta_1}$, $\phi_2 = \frac{4\pi}{\lambda}\frac{\ell-h}{\cos\vartheta_2}$ and $\phi_3 = \frac{4\pi}{\lambda}\ell'$, ϑ_1 and ϑ_2 are indicated in Fig. A.

In our system $h \leq 1$ μm, d is chosen as 3 μm, whereas ℓ is of the order of centimeters. Therefore ϑ_1 and ϑ_2 are very near to zero. Consequently, $\phi_1 \approx \frac{4\pi}{\lambda}\ell$, $\phi_2 \approx \frac{4\pi}{\lambda}(\ell-h)$ and $U_{io} \approx U_i w_i$. Substituting for ϕ_i and U_{io} in Eq. (A-3) we obtain for the intensity I_0

$$I_0 = \frac{1}{2}\left[\frac{r_b(d-w')}{d}p_0^{\frac{1}{2}}e^{j\frac{4\pi}{\lambda}\ell} + \frac{r_t w'}{d}p_0^{\frac{1}{2}}e^{j\frac{4\pi}{\lambda}(\ell-h)} + p_0^{\frac{1}{2}}e^{j\frac{4\pi}{\lambda}\ell'}\right]$$

$$\left[\frac{r_b(d-w')}{d}p_0^{\frac{1}{2}}e^{j\frac{4\pi}{\lambda}\ell} + \frac{r_t w'}{d}p_0^{\frac{1}{2}}e^{j\frac{4\pi}{\lambda}(\ell-h)} + p_0^{\frac{1}{2}}e^{j\frac{4\pi}{\lambda}\ell'}\right]^*$$

$$= \frac{p_0}{2d^2}[d^2+r_t^2w'^2+r_b^2(d-w')^2 + 2r_br_t(d-w')w'\cos\frac{4\pi}{\lambda}h$$

$$+ 2r_tw'd\cos\frac{4\pi}{\lambda}(\ell-\ell'-h) + 2r_b(d-w')d\cos\frac{4\pi}{\lambda}(\ell-\ell')] \qquad \text{(A-4)}$$

This is Eq. (4) in the paper.

Appendix B Derivation of Eq. (11)

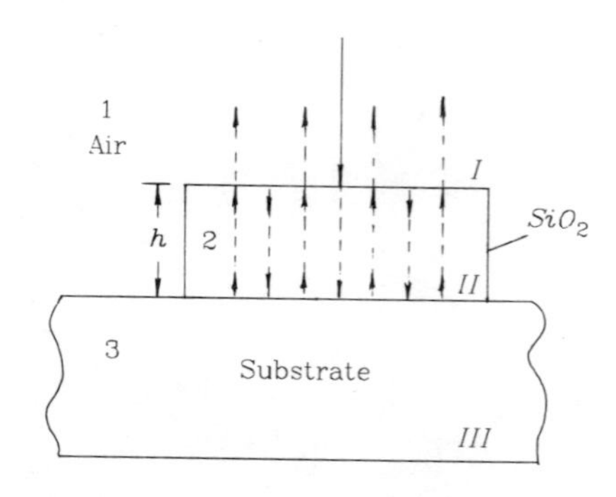

Figure B. Multiple reflections between interfaces I and II.

Figure B shows two layers in air. Assume that the thickness of the substrate is much larger than the wavelength in it and there is no reflection from interfaces III to II. This means that only multiple reflections between interfaces I and II are considered. In Fig. B, r_{ij} and t_{ij} are the reflectivity and transmittance from region i to j respectively.

Suppose the complex amplitude for the incident wave upon interface I is $U_o e^{j\phi}$. At this interface part of the light, $r_{12}U_0e^{j\phi}$, is reflected back and the remainder $t_{12}U_0e^{j\phi}$ is transmitted into region 2. The transmitted wave will be partially reflected back from interface II to I with phase delay $4\pi h/\lambda'$, and this part is equal to

$r_{23}t_{12}U_0e^{j(\phi+\frac{4\pi}{\lambda'}h)}$, where λ' is the wavelength in region 2. At interface I this back reflected wave is divided into a transmitted part $t_{21}r_{23}t_{12}U_0e^{j(\phi+\frac{4\pi}{\lambda'}h)}$ and a reflected part

$r_{21}r_{23}t_{12}U_0 e^{j(\phi+\frac{4\pi}{\lambda'}h)}$. The later part will continue being reflected. Therefore, the total reflected wave in air is

$$U_1 = U_0 e^{j\phi}[r_{12}+t_{12}r_{23}t_{21}e^{j\frac{4\pi}{\lambda'}h}+t_{12}r_{23}t_{21}(r_{23}r_{21})e^{j\frac{4\pi}{\lambda'}h \times 2}$$

$$+ t_{12}r_{23}t_{21}(r_{23}r_{21})^2 e^{j\frac{4\pi}{\lambda'}h \times 3} + \ldots\ldots]$$

$$= U_0 e^{j\phi}[r_{12}+t_{12}r_{23}t_{21}e^{j\frac{4\pi}{\lambda'}h} \sum_{n=0}^{\infty} (r_{23}r_{21}e^{j\frac{4\pi}{\lambda'}h})^n]$$

$$= U_0 e^{j\phi}[r_{12} + \frac{t_{12}r_{23}t_{21}e^{j\frac{4\pi}{\lambda'}h}}{1-r_{23}r_{21}e^{j\frac{4\pi}{\lambda'}h}}] \quad \text{(B-1)}$$

The above equation indicates that when multiple reflections are taken into account, the equivalent reflectivity r_{te} of the top surface is given by

$$r_{te} = r_{12} + \frac{t_{12}r_{23}t_{21}e^{j\frac{4\pi}{\lambda'}h}}{1-r_{23}r_{21}e^{j\frac{4\pi}{\lambda'}h}} \quad \text{(B-2)}$$

where r_{12} actually is r_t in Eq. (4) of the paper. It should be noticed that r_{te} is no longer a real number. Let $r_{te} = r_{te} e^{j\psi}$. By using the complex value of r_{te} in Eq. (A-4). We can obtain an expression for i in a manner similar to the derivation of Eq. (7).

$$i = i_1 - i_2 = 2\alpha p_0 |r_{te}| A\frac{w'}{d} \sin(\frac{4\pi}{\lambda}h-\psi) \quad \text{(B-3)}$$

This is Eq. (11) in the paper.

A new method of 3-D quantitative analysis of holographic interferometry in the applications of solid biomechanics and vibration

Ding Zuquan and Bao Naikeng

Photomechanics Research Div. of Tongji University
Shanghai, China

Abstract

A method of HFR combined with double exposure holography, sandwich holography and stroboscope holography respectively, and its applications of the quantitative analysis of 3-D displacement of human skull; of the change of dentomaxilla and vibration of violin are presented in this paper.

Introduction

As we all know, the quantitative analysis method of holographic interferometry in determining 3-d displacements as well as the applicable apparatus is the key of application of holographic interferometry. A large number of research work have been done.[1]

Here, the holographic fringe reader method (HFR method) used for quantitative analysis of 3-d displacements is proposed by the authors.[2] It bases on the principle of fringe counting method. Only with a single double-exposed hologram, the number of fringe shift and corresponding changes of direction angle at the measured point can be determined by successive changes of the observation directions. By the use of microcomputer, 3-d displacement components can be obtained.

The principle and the method

The holographic fringe reader (HFR) shown in Figure 1. The HFR system and its sensitivity vectors indecated in Figure 2.

Figure 1. The holographic frunge reader (HFR)

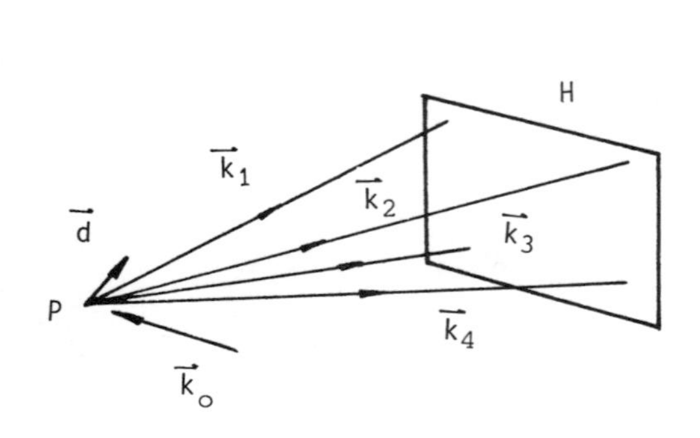

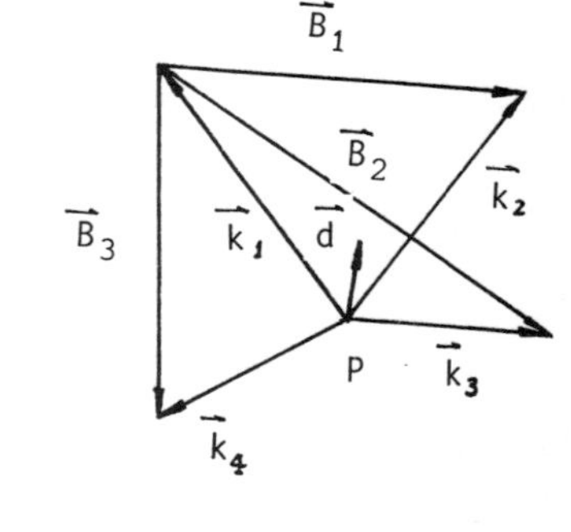

Figure 2. The HFR system and its sensitivity vectors.

If the coordinate is determined, the three components d_x, d_y, d_z of displacement vrctor $\bar{d}$ can be obtained from the equation set expressed by the matrix.

$$\begin{bmatrix} B_{1x} & B_{1y} & B_{1z} \\ B_{2x} & B_{2y} & B_{2z} \\ B_{3x} & B_{3y} & B_{3z} \end{bmatrix} \begin{bmatrix} d_x \\ d_y \\ d_z \end{bmatrix} = 2\pi \begin{bmatrix} N_{21} \\ N_{31} \\ N_{41} \end{bmatrix} \quad (1)$$

Where N_{21} is the number of fringe shift, from $\bar{k}_1$ to $\bar{k}_2$, at the measured point p, N_{31} is also the number of fringe shift, from $\bar{k}_1$ to $\bar{k}_3$, and so is the N_{41} .

In order to enhance the accuracy, the overdeterminate set of equations and the least-square method are used to process a large number of displacement data.

$$
\begin{bmatrix} B_{xj}^2 & B_{yj}B_{xj} & B_{zj}B_{xj} \\ B_{xj}B_{yj} & B_{yj}^2 & B_{zj}B_{yj} \\ B_{xj}B_{zj} & B_{yj}B_{zj} & B_{zj}^2 \end{bmatrix} \begin{bmatrix} d_x \\ d_y \\ d_z \end{bmatrix} = 2\pi \begin{bmatrix} N_{il}B_{xj} \\ N_{il}B_{yj} \\ N_{il}B_{zj} \end{bmatrix} \qquad (2)
$$

where $i = 1, 2, \ldots m.$ $(m > 4)$, $j = i-1$.

With the use of HFR, the techniques of double-exposure holography, multi-step double-exposed holography, sandwich holography and stroboscopic holography can all be used for quantitative analysis in biomechanics and vibration study.

Combined with double-exposure holography, the HFR method has quantitatively determined the deformation of frontal bone and temporal bone of human skul loaded by concentrated force

The damage of human skull is a dangerous common disease, in order to understant its damage rule and to enhance the curative effect, study of the damage mechanism is necessary. Because of the complexity of human skull, the pure theoretical method is not sufficient to solve the problem, and also, the photoelasticity and strain gage method have certain limitation.

In our test, the human skull dry-sample is used as specimen. (see Figure 3.) The special fixture and the self-symmetrical loading apparatus are designed. (see Figure 4.) The dangerous parts of human skull, such as temporal bone and frontal bone, are loaded by concentrated force, and double exposure holograms describing the deformation of the skull are obtained. (see Figure 7, Figure 9.) With the use of HFR method, the 3-d displacements of the measured points (see Figure 6) on the skull are determined quantitatively. (see Figure 8, Figure 10.)

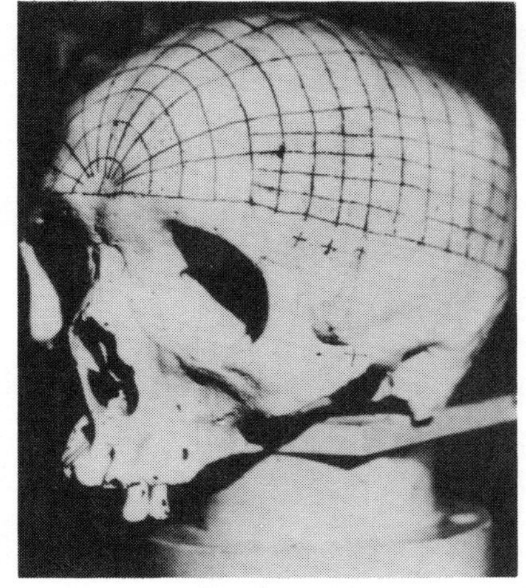

Figure 3 Human skull

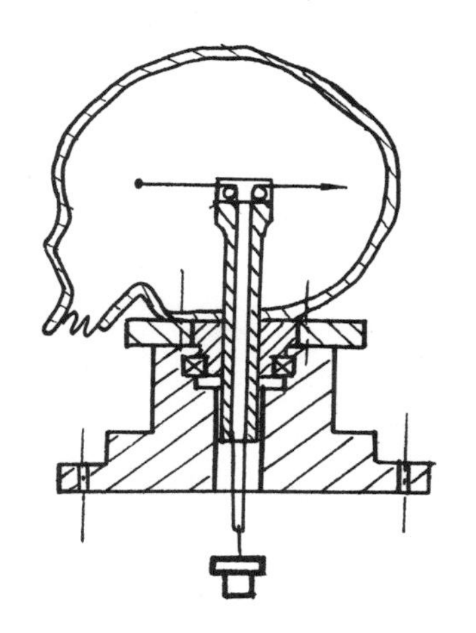

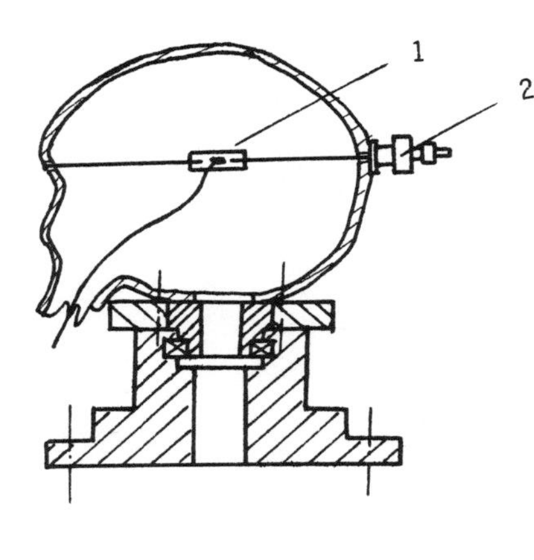

Figure 4. Fixlure and loading apparatus of human skull.
1. mini-strain gauge meter,
2. bolt loading arrangement

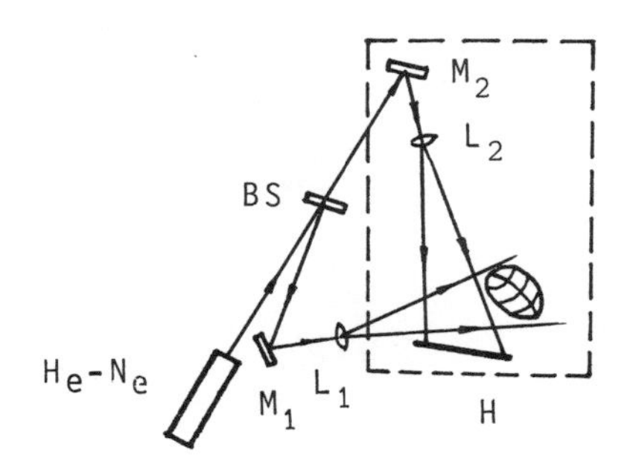

Figure 5. The optical arrangement of human skull.

Figure 6. Measured points of temporal bone.

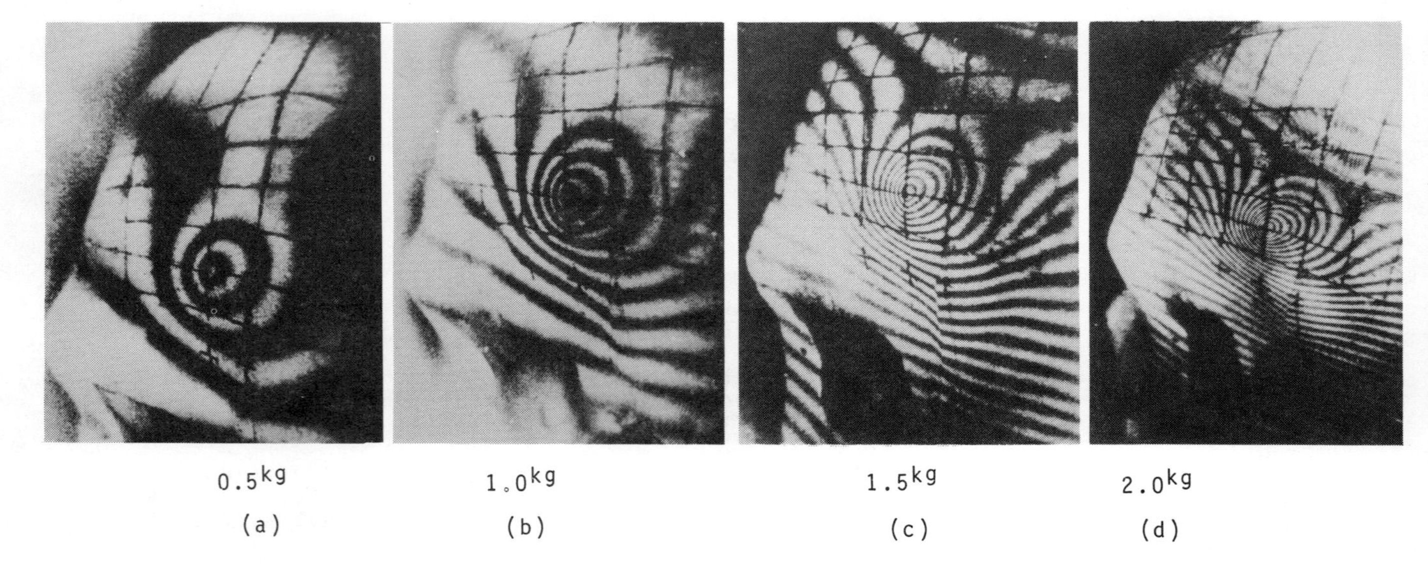

Figure 7. Holographic interferogram of temporal bone.

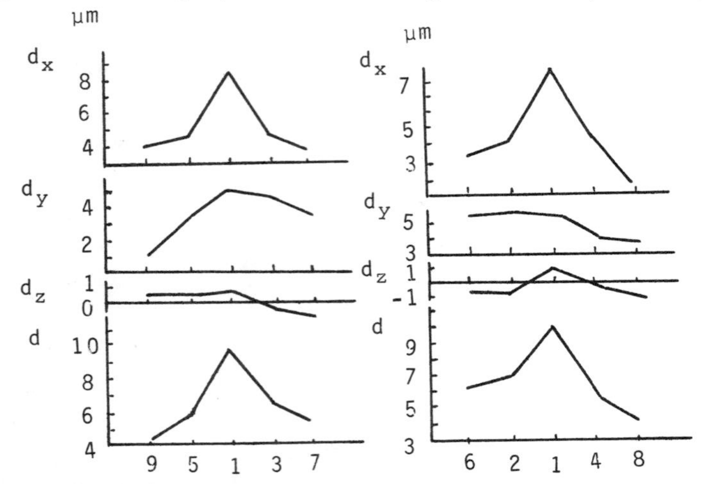

Figure 8 3-d displacements distributions of temporal bone.

Figure 9. Holographic interferogram of frontal bone.

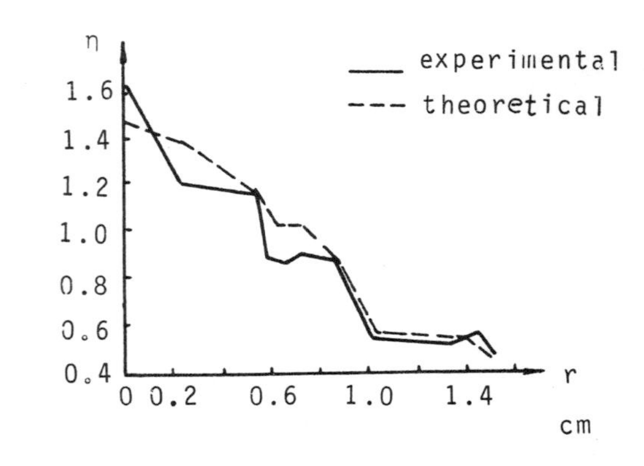

Figure 10. Comparison of relative displacements of frontal bone.

The results show:

(1) To certain loading extent, the human skull material is an elastic material;

(2) The distribution of deformation agrees with the circular cone sunk bone-broken damage rule, the results agree with medical analysis;

(3) Compared with the results calculated from the elastic simplified model, the agreement of the relative displacement is quite well.

Combined with multi-step double exposure holography and sandwich holography, the HFR method is used to quantitatively analyze the change of dentonmaxilla in orthodontic therapy

Rapid Maxillary Expansion (RME) is an orthodontic therapy using a fixed appliance with expansion screws to expand maxillary complex in a short period and so attain therapeutic requirement. Because the RME is easy to control, has a notable therapeutic effect and can be completed within a short time, especially for the adults, the method has been well known by many clinicians. In order to study the mechanism and effect of the RME further, the HFR method is used, combined withreal-time holography, multi-step dojble exposure holography and sandwich hologrpahy, to measure the deformation of dentomaxilla qualitatively and quantitatively.

In the test, child's skull moist-sample is sued as specimen, and is fixed on the holographic bench after the medical treatment and the analogy of periodontal membrance.

The single reference beam, the double illumination beams and multimirror collecting displacement information of object is used in the optical system. (see Figure 11)

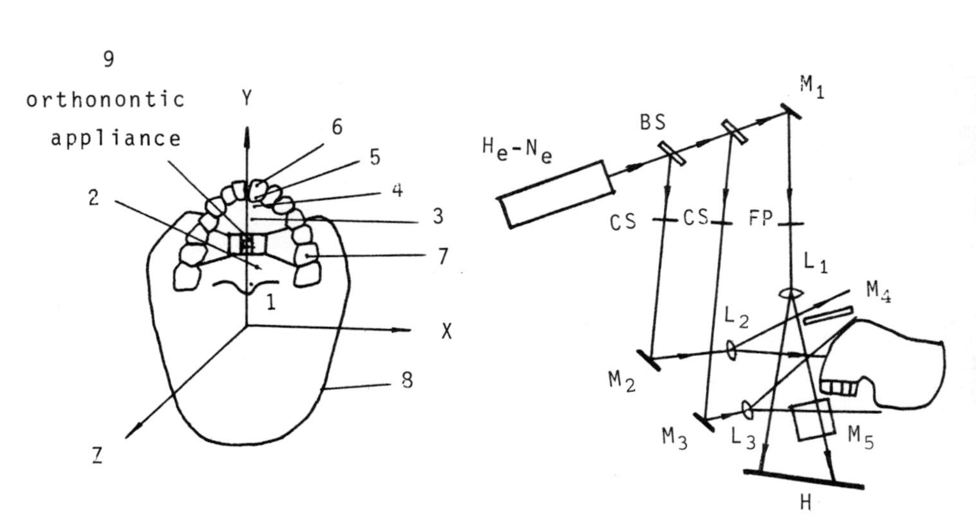

Figure 11. The measured points of maxilla and its optical arrangement.

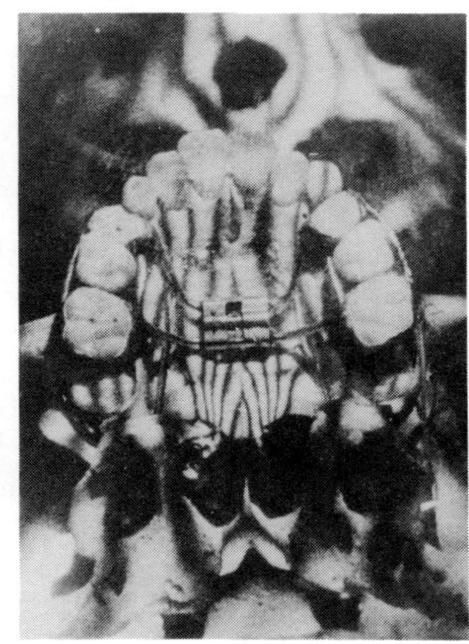

Figure 12. Holographic interferogram of maxilla.

A new hologram replacement holder with high accuracy is designed. After the specimen is loaded by therapeutic force under different conditions, the real-time, sandwich and multistep holograms are obtained respectively. (see Figure 12) The 3-d displacements at the measured points are calculated by the HFR method. (see Figure 13)

The results show:

(1) In REM process the orthopedic force is widely distributed on the maxillary bone as well as the other parts of the skull. The displacements are nonlinearly changed;

(2) holographic patterns and quantitative results provide scientific basis for selecting the arrangement of orthodontic appliance, such as the magnitude, the direction, the loading point and the loading time of orthopedic force;

(3) The experimental methods used in our research work are suitable for large displacement measurement. The scope of the application of holographic interferometry has been widened.

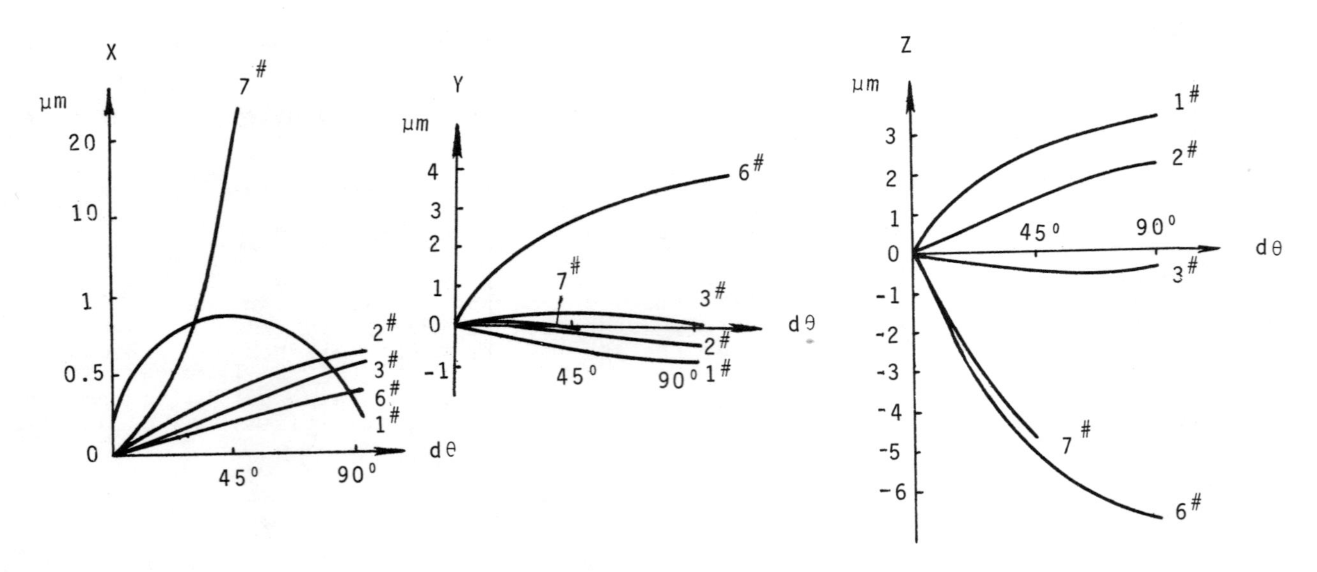

Figure 13. The diagrams of displacements changes of dentomaxilla.

Combined with stroboscopic holography, the HFR method is used to measure the vibration of violin bridge

Because of the agreement on essence of holographic interferometry analysis method between vibration displacement and static displacement, the authors once presented a method of multi-plate time-average holograms to quantitatively analyze steady vibration displacement.[3] As an improvement, the single stroboscopic hologram is used, and the vibration displacement vector is quantitatively analyzed by HFR method.

Because the bridge is the decisive factor of the quality of violin, it is important to quantitatively measure the vibration property of the bridge by the use of optical nondestructive testing techique.

The experimental system indicated in Figure 14. The specimen is violin bridge. Under the sine wave vibration excitation of E string, the stroboscopic hologram of the bridge vibration are recorded. The stroboscopic modulator is the improved EOM which is designed by the author. Stroboscpic exposure is made at two maximum amplitudes in a vibration period T. The width of stroboscopic light pulse ϵ is T/30.

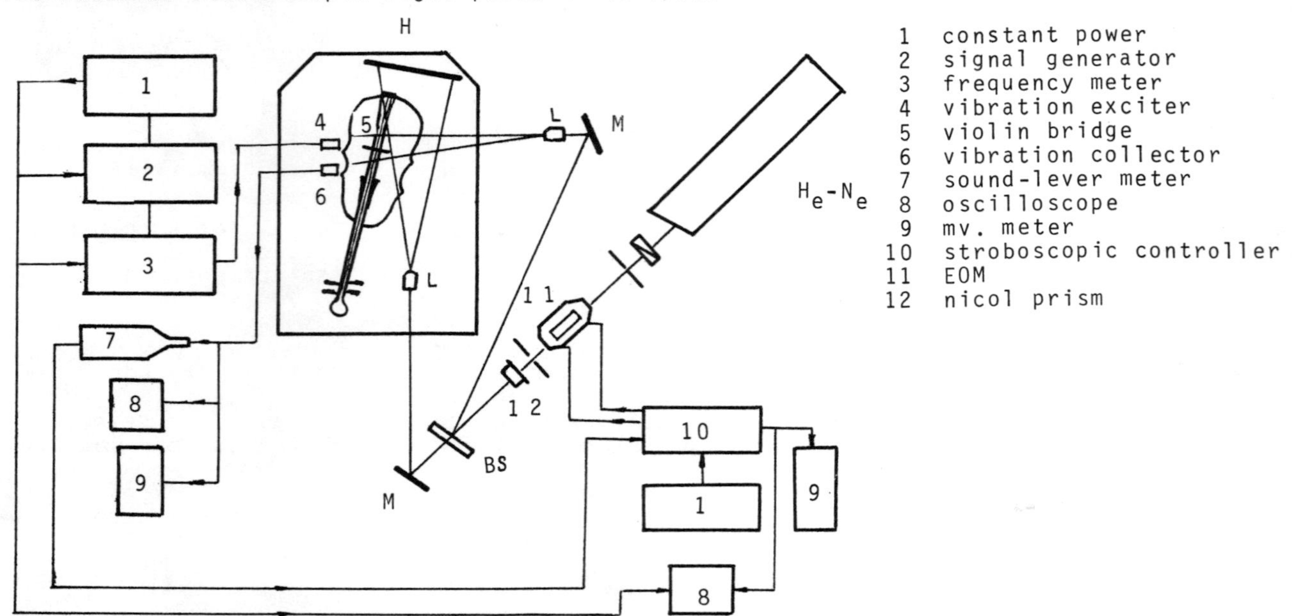

Fig. 14 Experimental arrangement of stroboscopic holography of violin bridge vibration

The stroboscopic holograms (see Figure 15) under different conditions are obtained, the three vibration displacement components calculated by HFR method. (see Table 1.)

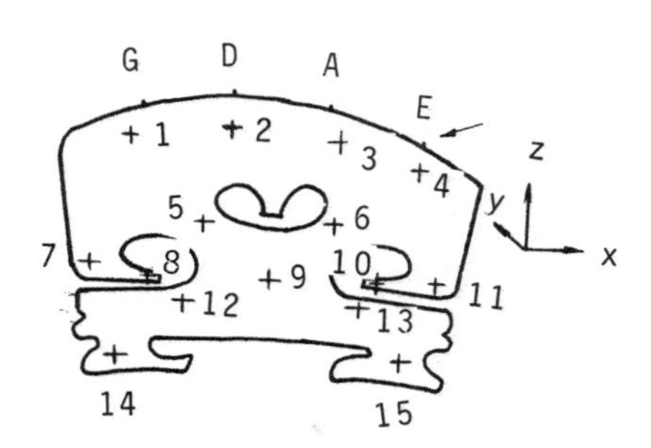

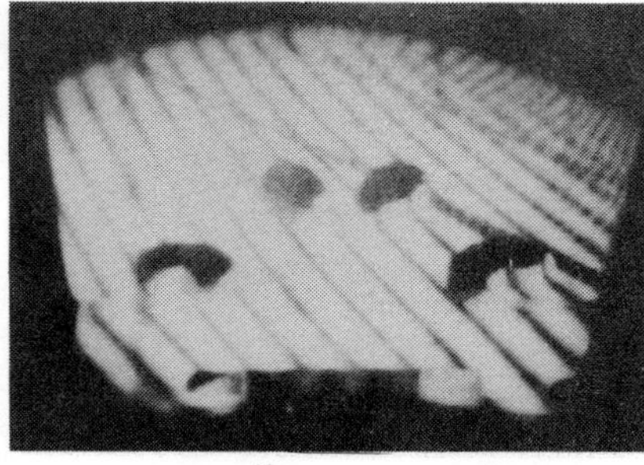

f 544Hz 5.8V

Figure 15. Measured points and stroboscopic holograms of violin bridge.

Table 1. 3-d vibration displacements of violin bridge

No.	d_x	d_y	d_z
1	-2.7960	0.6319	-1.9242
2	-2.9668	-0.1405	-0.3207
3	-2.8562	-0.7234	1.2429
4	-2.5234	-0.4279	2.5279
5	-1.4957	-0.7798	-0.2744
6	-0.8317	-1.6666	2.0371
7	-0.7185	0.8008	-1.7630
8	-0.4676	0.3904	-1.3437
9	-1.1933	-0.6869	0.6465
10	-1.1176	-1.8819	1.9369
11	-0.1609	2.4332	2.8047
12	-0.7054	-2.8559	-0.2355
13	-0.5463	2.2377	1.1327

The results show:

(1) Although the violin bridge is complex, the distribution of vibration displacement components can be obtained efficiently by the method mentioned above;

(2) From the analysis of three displacement components, it is clear that, under the sine wave vibration excitation of E string, the bridge vibrates in a complex movement. The experimental results agree with what the violin experts have analyzed.

(3) If combined with frequency analysis and design technique, if considering the interaction of vibration excitation conditions et al, the method mentioned above can be used to obtain practicable technical results and to guide the violin production and the violin reparation.

(4) The method and experimental system can be used in other structures for vibration measurement.

Conclusions

The results indicate that the application of HFR method in solid biomechanics and complex vibration in more effectife and useful. The applicable range of HFR method has been widely expanded.

Acknowledgments

The authors wishes to thank Xu Genlin of Shanghai Medical Apparatus and Instrument College, Zhou Hongkang of Xi-an Jiao-tong University, Lou Zhaohua and Cai Zhong of Shanghai Institute of Stomatology, Wang Ming and Zhu Shuilin of Tongji University, and engineers of Shanghai Violin Factory for their cooperation throughout the research work.

References

1. Vest, C.M. Holographic Interferometry, John Wiley & Sons, New York, 1979.
2. Bao Naikeng and Ding Zuquan, "Quantitative Determination of 3-D Displacements Using the Method of 3-D Reader Measuring the Fringe Shift From the Hologram", Journal of the Tongji University, No.2, pp. 29-42, 1980.
3. Ding Zuquan and Bao Naikeng, "Quantitative Determination of 3-D Displacements and Vibration Usinf Holographic Interferonetry", Journal of the Tongji University, No.1, pp 49-62, 1980.

Vibration Analysis of Large-sized Structure Using CW Laser Holographic Interferometry

Hajime Yamashita

Technical Research Center, Mazda Motor Corporation
3-1, Shinchi, Fuchu-cho, Aki-gun, Hiroshima, Japan

Abstract

Employing a tuff rock cave as the vibration isolation system made possible the measurement of the vibration mode of large-sized structure such as a fully equipped automotive bodies and producing 2 m by 1 m hologram for display.

Introduction

The reduction of automobile vibration and noise is one of the important tasks for automobile engineers to achieve. While the demand is increasing for reduction in exterior noise and for a quieter passenger compartment, the recent strong emphasis on higher engine output and lighter vehicle weight for improved fuel economy is proving a disadvantage to efforts for cutting noise and vibration.

Holographic interferometry is used as one of the effective methods to identify the vibration characteristics of engines, etc., and is thus a great step toward improving their structures (1), (2), (3).

Continuous wave laser holographic interferometry offers benefits of, among others, showing the vibration amplitudes and nodal lines as interference fringes, for easier interpretation of vibration modes, and permitting real time measurements. It needs, however, a vibration isolation system to accommodate a laser, an optical system, and a test object, imposing an inevitable limitation on the size of the object to be tested to about 1.5 m in diameter--the size of an engine, etc. To resolve this problem, a rock cave laboratory superior in vibration resistance was built as a huge vibration isolation system (4) and was used for vibration measurement of large-sized structures, such as car bodies (5) and producing 2 m by 1 m hologram for display.

Experimental setup

A vibration isolation laboratory was constructed in a cave in a quarry located in an upper layer of a tuff rock (called Ohya rock) about 300 m thick. The laboratory is 7.5 m wide, 13 m long, and 3.5 m high, offering sufficient space for car vibration testing (Fig.1, 2 and 3). The maximum vibration amplitude of the floor is on the order of 0.1 micron at 0.7 Hz in the horizontal direction.

The experimental setup of the optical elements is shown in Fig. 4. The setup is composed mainly of an object beam optical system and a reference beam optical system and, as need arises, a reference beam phase modulation optical system and a Michelson interferometer which monitors vibration and air turbulence are added. The light source used was an argon-ion laser with an output of 4W, and the vibration modes of a test object were photographed mainly by the time average method.

The main problems in measuring the vibrations of a body include:

a) A large object such as a car body requires a long exposure time.

b) The rigidity of a car body itself is low.

c) Rubber, plastic, and other visco-elastic materials are used in automotive structures.

To avoid deformation under its dead weight and due to temperature gradients of the laboratory and vehicle, it was necessary to make the vehicle stationary overnight, i.e.

Fig. 1 Entrance to the tuff rock cave

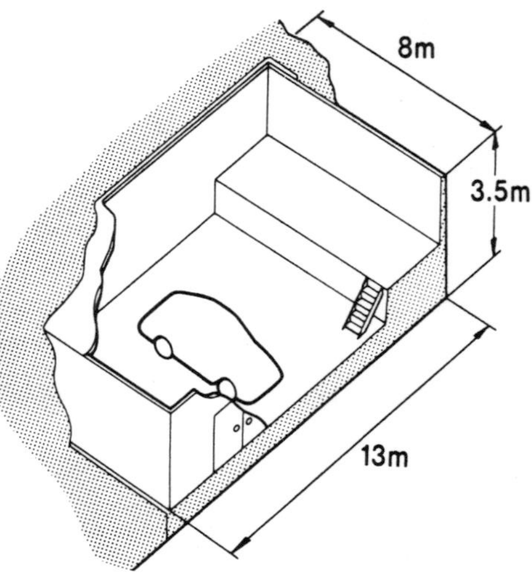

Fig. 2 Laboratory in rock cave

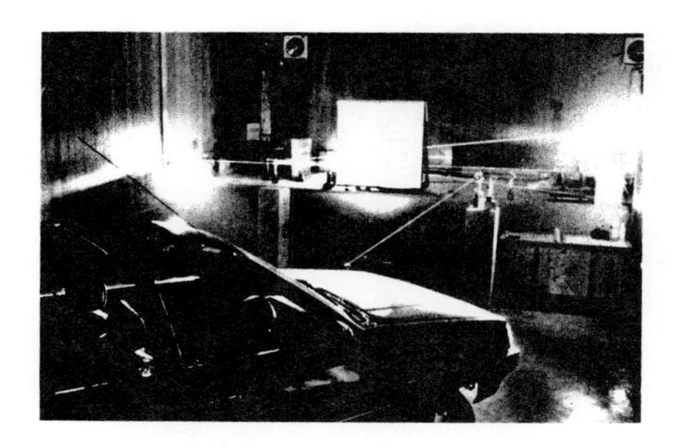

Fig. 3 Inside of the vibration isolated laboratory

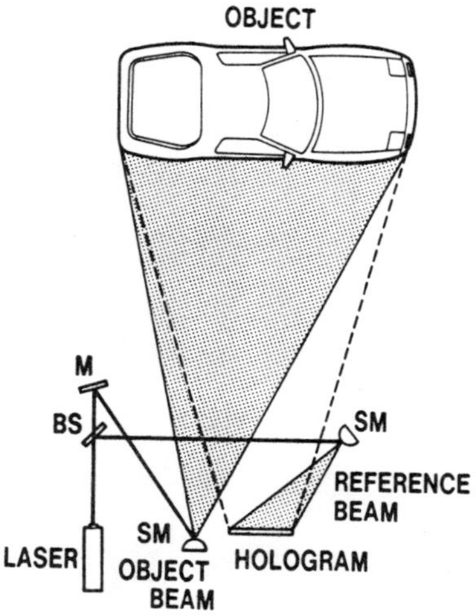

Fig. 4 Experimental setup of object beam to analyze vibration modes of vertical and horizontal components

nearly 12 hours, though varying with exposure time.

The time requirement to achieve the standstill condition can be shortened by supporting the suspension of the car by steel or tuff rock blocks to exclude the influences of the tires. In this case also, little influence is given to the vibration modes.

Fig. 5 shows the exciting equipment. Sine wave signals produced by the function generator are sent through the amplifier to the electromagnetic shaker to excite the vehicle body. The vibration is monitored by a fast Fourier transform analyzer. The vibrations of the vertical panels, such as the rear quarter panels, doors, and fenders, can be measured directly, while the measurements on horizontal components, such as the roof and trunk lid, have to be taken using a reflection mirror. (Fig. 6)

Results

Examples of interferogram with a Mazda RX-7 are shown in Fig. 7. The figures show vibration modes of the quarter panel at frequencies of 187 and 235 Hz. Note that the positions of the loops and nodes of the vibrations can be readily judged, making it easy to take actions to reduce the vibrations. For purposes of reducing vibration amplitude, used were ribs, beads, damping materials, etc. to change the spring constant, mass, and damping of a vibrating component.

Fig. 8 shows the results of modal analysis by an FFT analyzer, which are roughly the same as with holographic interferometry. In modal analysis, the number of measuring points is restricted by the computer's storage capacity, and this imposes great limitations on the magnitude of modes that can be analyzed. By contrast, holographic interferometry is considered capable of handling a virtually infinite magnitude of modes.

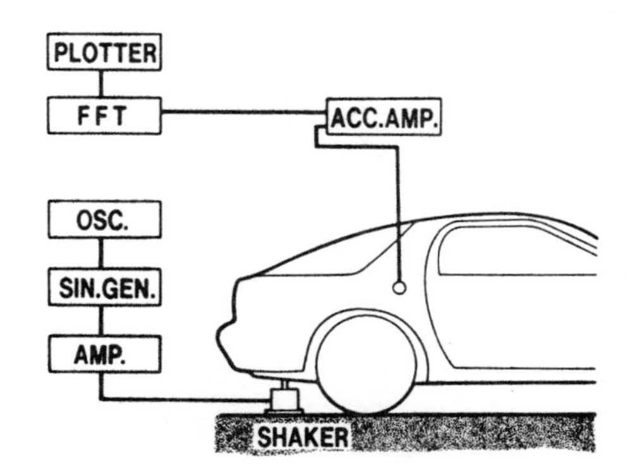

Fig. 5 Experimental setup for exciting automotive body

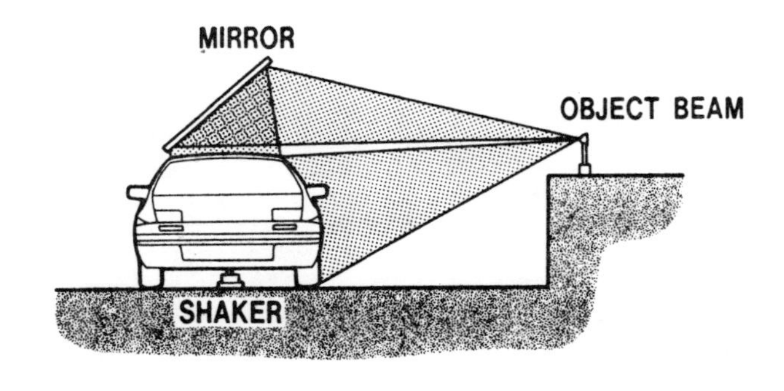

Fig. 6 Experimental setup of object beam to analyze vibration modes of vertical and horizontal components

(a) 187 Hz

(b) 187 Hz

(c) 235 Hz

Fig. 7 Vibration modes of automotive body

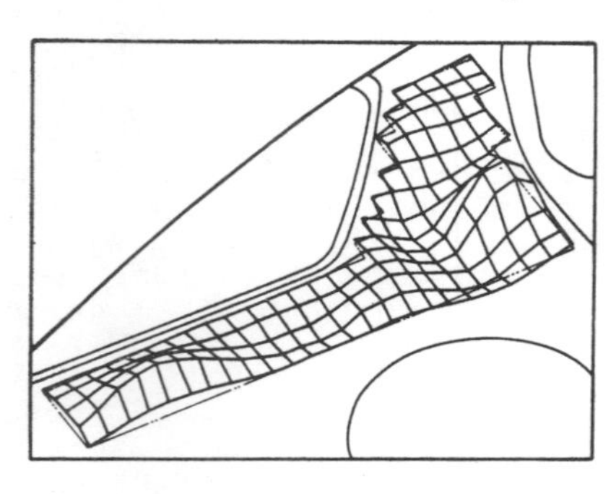

(a) 187 Hz

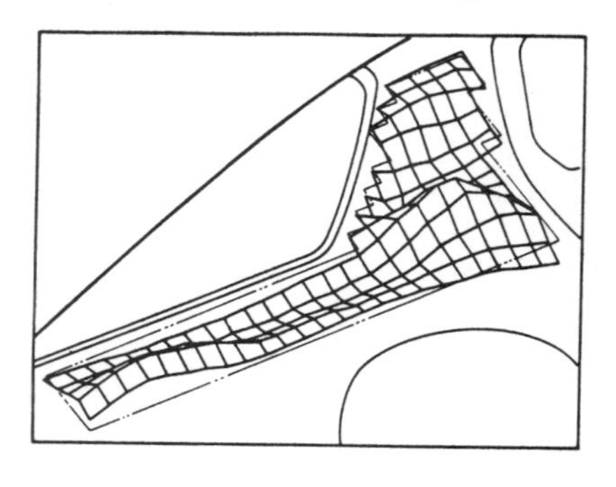

(b) 235 Hz

Fig. 8 Results of modal analysis

Fig. 9 shows a reference beam phase modulation system used in this program and Fig. 10 gives the results of measurement. A small mirror was fixed on the vehicle body and the light reflected from that mirror was used as a reference beam. Compared with Fig. 7, the phase of vibration is easily identified.

As opposed to vertical surfaces such as vehicle sides, horizontal faces like the roof and hood are difficult to measure because of opticals layout, so that a large reflective mirror of 1.4 m by 0.9 m was used. It was necessary in this case to keep the mirror and its frame in a highly stationary condition. Fig. 11 shows measurements on the roof and windshield. Both results indicate well-defined interference fringes.

Conclusion

Employing a tuff rock cave as the vibration isolation system made possible the measurement of the vibration modes of a fully equipped vehicle body by CW laser holographic interferometry. As a result, development of quieter and more comfortable vehicles can now be carried out in a more efficient manner. In addition, it became possible to make large-sized hologram, 2 m by 1 m.

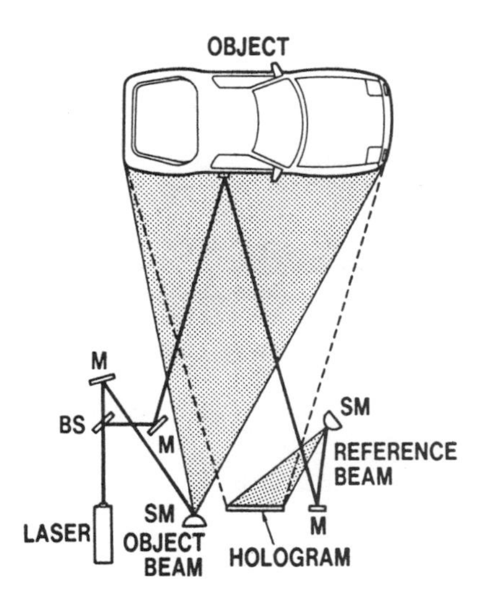

Fig. 9 Experimental setup for reference beam phase modulation method

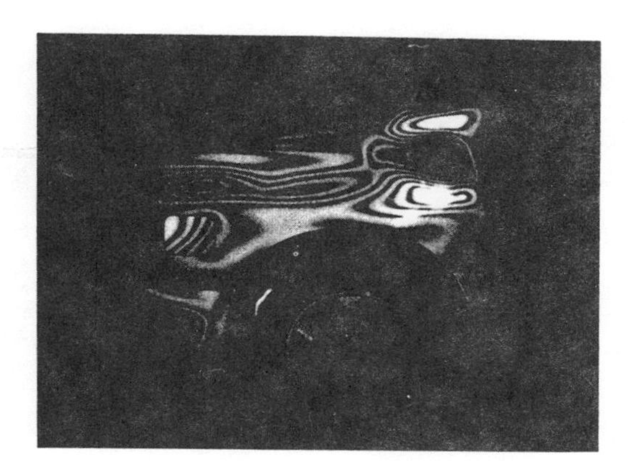

Fig. 10 Vibration mode using reference beam phase modulation method (235 Hz)

(a) Roof

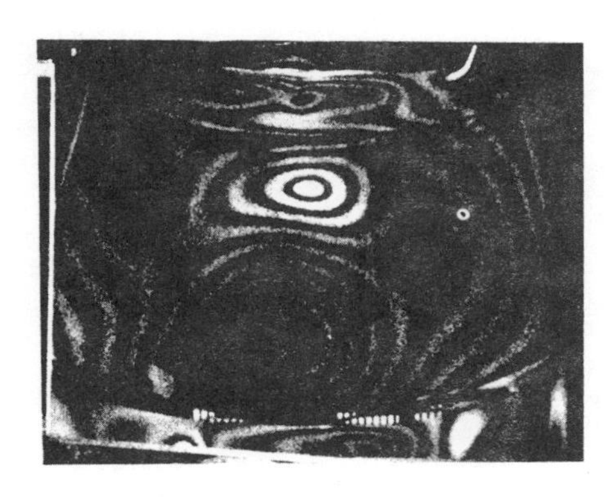

(b) Windshield

Fig. 11 Vibration modes of roof and windshield

Acknowledgements

The author wish to thank Mr. K. Sasanishi and Mr. I. Masamori, Mazda Motor Corp. for their help in this experiment and express their gratitude to Mr. Masane Suzuki and Mr. Takayuki Saito of Fuji Photo Optical Co., Ltd., Mr. Kohei Watanabe and Mr. Akio Hirosawa of Holotec Inc., and Mr. Tsutomu Morie of Shimizu Construction Co., Ltd. who promoted the construction of the laboratory inside the rock cave.

References

(1) A. Felske, A. Happe: Vibration analysis by double pulsed laser holography, SAE Paper No. 780333, 1978.

(2) M. Murata, M. Kuroda: Application of Pulsed-wave Laser Holography to Practical Vibration Study, Mitsubishi Juko Giho, Vol. 20, No. 5, 1983.

(3) G. M. Brown, R. R. Wales: Vibration analysis of automotive structure using holographic interferometry, Proceedings of SPIE, Industrial Applications of Laser Technology, Geneva, Switzerland, 1983.

(4) K. Watanabe, H. Hirosawa, M. Suzuki, T. Morie and H. Yamashita, Large Holography Studio in a Huge Rock Cave, Proceedings of SPIE, International Conference on Holography Applications, Beijing, China, 1986.

(5) H. Yamashita, K. Sasanishi and I. Masamori, Vibration Analysis of Automotive Bodies Using Continuous Laser Holographic Interferometry, Proceedings of SPIE, International Conference on Holography Applications, Beijing, China, 1986.

A numerical method for holographic optical fiber diagnostics

Shun-ichi Himeno*, Masaharu Seki**, and Hitoshi Mochizuki***

* Department of Management, Tokai College of Industry
12-5, Harayama, Okamachi, Okazaki, Aichi Pref., 444 Japan
** Department of Management, Hokkaido Automotive Junior College
2-6-2-1, Nakanoshima, Toyohira-ku, Sapporo, 042 Japan
*** Department of Communication Systems, University of Electro-Communications
1-5-1, Chofugaoka, Chofu, Tokyo, 182 Japan

Abstract

A new method of holographic interferometry and an accurate data processing for determination of refractive index profiles of optical fibers are presented in this paper.

Introduction

Many plasma diagnostics measure line integrals of quantities such as electron density, visible, and X-ray emission. In optical fibers whose cross sectional distribution are circularly symmetrical, the two dimensional distributions corresponding to those line integral projection are routinely recovered by the use of the usual Abel inversion[1].

For more complicated cross sectional distributions having no symmetry such as the ellipse, the local quantities can better be reconstructed by methods of computerized tomography or our asymmetrical elliptical Abel inversion.

In this paper, the 1st section describes our method of holographic interference microscope and its results, and the following section discusses the numerical method and its results.

Principles of holographic interferometry and experimental results

First, for convenience, we will briefly summarize the principles of holographic interferometry. Then we will explain some of the results which have not previously been produced.

As is well known, the reference beam is represented by

$$E_r = A_o \exp(-\alpha x) \tag{1}$$

where x is a coordinate in the hologram plane, and α is simply related to the angle between the reference beam and the hologram plane. The scene beam is represented by

$$E_s = A(x)\exp(i\phi(x)) \tag{2}$$

where $A(x)$ and $\phi(x)$ are the amplitude and the phase of the beam, respectively.

Concerning to holographic interferometry measurement, if the phase of the object wave changes between two successive exposures (1,2) of a single emulsion, the intensity on the hologram is given by

$$I(x) \propto \left| A_o e^{-\alpha x} + A_1(x) e^{j\phi_1(x)} \right|^2 + \left| A_o e^{-\alpha x} + A_2(x) e^{j\phi_2(x)} \right|^2 \tag{3}$$

After reconstruction by the reference beam, the amplitude is modulated as

$$I'(x) \propto A_1(x)^2 + A_2(x)^2 + 2A_1(x)A_2(x)\cos(\Delta\phi(x))$$

with

$$\Delta\phi(x) = \phi_2(x) - \phi_1(x) \tag{4}$$

If the refractive index of the object is n_o before displacement and n after displacement, then the phase difference or fringe shift number between the wavefronts reconstructed from this object is given by

$$F = \frac{1}{\Lambda}\int (n - n_o)\, dl \tag{4}$$

where l and Λ are the pathlength and the wavelength of the probing beam, respectively.

Next, we will describe a new set-up in optical holography and its results.

Figure 1 is the schematic diagram of the experimental arrangement used in the application of optical holography to optical fiber diagnostics. A Mach-Zehnder type holographic interference microscope was constructed. The light source is a 6328 Å He-Ne laser. An optical fiber was immersed vertical to the object beam in matching oil whose refractive index is

close to the cladding. Straight background fringes were added by a change of freon gas pressure to air pressure in the glass hollow wedge between exposures. A diffuser was used to illuminate all points of the hologram plane. Interference occures between the comparison beam arising from the matching oil and the beam arising from the fiber.
Figure 2 is a typical example of a side-on view holographic interferogram of a graded index optical fiber which is almost symmetrical.

Figure 3 is an example of asymmetrical distributions.

Asymmetrical elliptical Abel inversion and discussions

Thirdly, we will explane a new method for obtaining the refractive index profiles of optical fibers.

Let us first compare our asymmetrical elliptical Abel inversion with the usual symmetrical Abel inversion. As is well known, a problem met in fiber diagnostics or plasma diagnostics is the reconstruction of local values from line integrated data.

As shown in Figure 4, in the circular case, the solution of this problem requires the inversion of this integral equation

$$I(y) = 2\int_0^{\sqrt{R^2-y^2}} \varepsilon(r)dx \quad (6)$$

where $I(y)$ is the line integrated data and $\varepsilon(r)$ is the local value. In interferometry, $I(y)$ and $\varepsilon(r)$ correspond to fringe shift number and refractive index distributions, respectively. In an axially symmetrical case, eq.(6) reduces to a one-dimensional problem

$$I(y)=2\int_y^R \frac{\varepsilon(r)rdr}{\sqrt{r^2-y^2}} \quad (7)$$

Inversion of (7) yields the well-known Abel inversion equation

$$\varepsilon(r)=-\frac{1}{\pi}\int_r^R \frac{dI(y)}{dy}\frac{dy}{\sqrt{y^2-r^2}} \quad (8)$$

In general, it is impossible to solve eq.(6). However, if we make assumptions on symmetry, eq.(6) can be solved. One such case occures when asymmetry exists only in the direction perpendicular to the line of observation. A solution of this problem for the circular case was presented by Matoba et al.[1].

In our analysis, the local value distribution is assumed to be represented by

$$\varepsilon'(r,y)=f(y).\varepsilon(r) \quad (9)$$

where $\varepsilon(r)$ and $f(y)$ are the axially symmetrical part corresponding to eq.(8) and the asymmetrical part of the distribution, respectively. Observation takes place in the symmetrical x-direction. The observed data $I'(y)$ is also separated into an even part $I(y)$ corresponding to $\varepsilon(r)$ and the odd part $f(y)$. The function $f(y)$, describing the degree of asymmetry, is defined as

$$f(y)= 1 + \alpha y \quad (10)$$

Combination of eq.(6) and eq.(9) yields

$$I'(y)= 2\int_0^{\sqrt{1-y^2}} \varepsilon'(r,y)dx$$

$$= 2f(y)\int_0^{\sqrt{1-y^2}} \varepsilon(r)dx=f(y).I(y) \quad (11)$$

$f(y)$ can be derived directly from eq.(10), while $\varepsilon(r)$ can be expressed by a symmetrical polynomial

$$I(y)=a_1I_1(y)+a_2I_2(y)+.....+a_nI_n(y) \quad (-1\leqq y\leqq +1) \quad (12)$$

$$I_1(y)= \frac{4}{3}(1-y^2)^{3/2}$$
$$I_n(y)= \frac{4n}{1+2n}((1-y^2)^{3/2}+y^2I_{n-1}) \quad (-1\leqq y\leqq +1) \quad (13)$$

The least square linear Taylor differential correction method[2] was introduced for the determination of $\alpha,a_1,a_2,...,a_n$. This method takes the form, for given initial values $\alpha_1^o,a_1^o,a_2^o,...,a_n^o$ and observed values I' $(i=1,2,...,N)$ as

$$\begin{pmatrix} \sum_i^N \left(\frac{\partial I'(y_i)}{\partial \alpha}\right)^2, & \sum_i^N \frac{\partial I'(y_i)}{\partial \alpha}\cdot\frac{\partial I'(y_i)}{\partial a_1}, \ldots\ldots, & \sum_i^N \frac{\partial I'(y_i)}{\partial \alpha}\cdot\frac{\partial I'(y_i)}{\partial a_n} \\ \sum_i^N \frac{\partial I'(y_i)}{\partial a_1}\cdot\frac{\partial I'(y_i)}{\partial \alpha}, & \sum_i^N \left(\frac{\partial I'(y_i)}{\partial a_1}\right)^2, \ldots\ldots, & \sum_i^N \frac{\partial I'(y_i)}{\partial a_1}\cdot\frac{\partial I'(y_i)}{\partial a_n} \\ \cdots & \cdots & \cdots \\ \sum_i^N \frac{\partial I'(y_i)}{\partial a_n}\cdot\frac{\partial I'(y_i)}{\partial \alpha}, & \sum_i^N \frac{\partial I'(y_i)}{\partial a_n}\cdot\frac{\partial I'(y_i)}{\partial a_1}, \ldots\ldots, & \sum_i^N \left(\frac{\partial I'(y_i)}{\partial a_n}\right)^2 \end{pmatrix} \times \begin{pmatrix} \delta\alpha \\ \delta a_1 \\ \vdots \\ \delta a_n \end{pmatrix}$$

$$= \begin{pmatrix} -\sum_i^N \frac{\partial I'(y_i)}{\partial \alpha}\cdot R_i \\ -\sum_i^N \frac{\partial I'(y_i)}{\partial a_1}\cdot R_i \\ \vdots \\ -\sum_i^N \frac{\partial I'(y_i)}{\partial a_n}\cdot R_i \end{pmatrix} \qquad (i=1,2,\ldots,N) \tag{14}$$

where

$$R_i = I'(y_i;\alpha^o, a_1^o, \ldots\ldots, a_n^o) - I_i' \tag{15}$$

Determining $\delta\alpha, \delta a_1, \delta a_2, \ldots, \delta a_n$ from this simultaneous equations, initial values are corrected according to the equation

$$\begin{aligned} \alpha^1 &= \alpha^o + \delta\alpha \\ a_1^1 &= a_1^o + \delta a_1 \\ &\vdots \\ a_n^1 &= a_n^o + \delta a_n \end{aligned} \tag{16}$$

The local value $\varepsilon'(r,y)$ is obtained corresponding to eq.(12) for the elliptical case[3] as

$$\varepsilon'(r,y) = \frac{b}{a}(1+\alpha y)(a_1(1-r^2)+a_2(1-r^4)+\ldots\ldots+a_n(1-r^{2n})) \tag{17}$$

where a and b represent major axis and minor axis of the ellipse, respectively.

The asymmetrical elliptical Abel inversion was tested using a hypothetical distribution for the circular case without destroying generality. If the local value distribution on the y axis is given by

$$\varepsilon'(r,y) = (1+y)(1-r^2) \qquad (0 \le r \le 1, -1 \le y \le 1) \tag{18}$$

so

$$I'(y) = \frac{4}{3}(1+y)(1-y^2)^{3/2} \tag{19}$$

The local value distribution by eq.(18) is shown in Figure 5, and the corresponding integrated function (19) is shown in Figure 6.

Figure 7 shows the local value result by this inversion method using the same hypothetical input data from (18). This result coincides well with that represented in Figure 5. The order of the polynomial (12) is chosen as 2 in this case because of the following discussions.

The degree of approximation of the polynomial I(y) was determined by the number n. In order to estimate the optimum number of n, the standard deviation[4]

$$\sigma(n) = \left(\frac{\sum_{j=1}^{N}(\Delta\varepsilon_j)^2}{N}\right)^{1/2} \tag{20}$$

was calculated for each n, where $\Delta\varepsilon_j$ is the difference between the inverted local value and the assumed one at the j-th sampling point. The result is shown in Figure 8. The total number of sampling points was 41 in this case. We have chosen n=2 in this paper,because the standard deviation is nearly constant for small n. This new method was applied to determination of refractive index profiles of optical fibers.

A typical example of almost symmetrical fringe shift profile from the optical fiber of Figure 5 is shown in Figure 9. The inverted refractive index profile from Figure 9 is shown in Figure 10.

Another example of extremely asymmetrical fringe shift profile from Figure 6 is shown in Figure 11. The inverted refractive index profile from Figure 11 is shown in Figure 12.

Conclusions

The method for asymmetrical Abel inversion is extended to cases where the distribution is confined within an elliptical boundary with arbitrary ellipticity. This extended method was successfully applied to optical fiber diagnostics within an accuracy of 0.1%.

Acknowledgements

We wish to thank Mr.S.Takahashi and Mr.E.Wakabayashi of Ibaragi College of Vocational Training for their encouragement.

References

1. Matoba T. and Funahashi A.,JAERIM-6239,Rep.Japan Atomic Energy Res.Inst. 1975.
2. McCalla T.R.,Introduction to Numerical Methods and FORTRAN Programing,John Wiley & Sons 1967.
3. Himeno S.,Seki M.,Mochizuki H.,Enoto T. and Sekiguchi T.,J.Phys.Soc.Japan,Vol.54, pp.1737-1742 1985.
4. Cremers C.J. and Birkebak R.C.,Appl.Opt.,Vol.5,pp.1057-1064. 1966

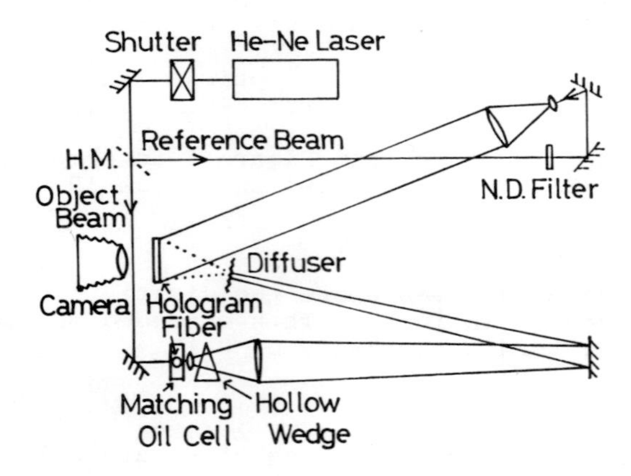

Figure 1. Experimental set-up for the holographic interference microscope.

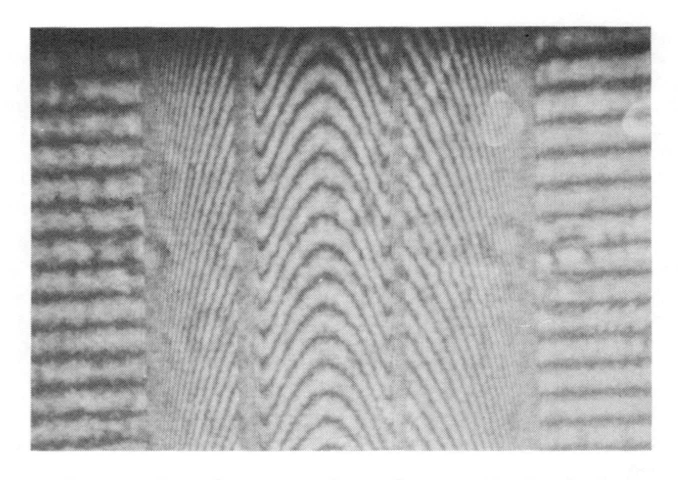

Figure 2. An example of symmetrical interferogram of graded index optical fibers.

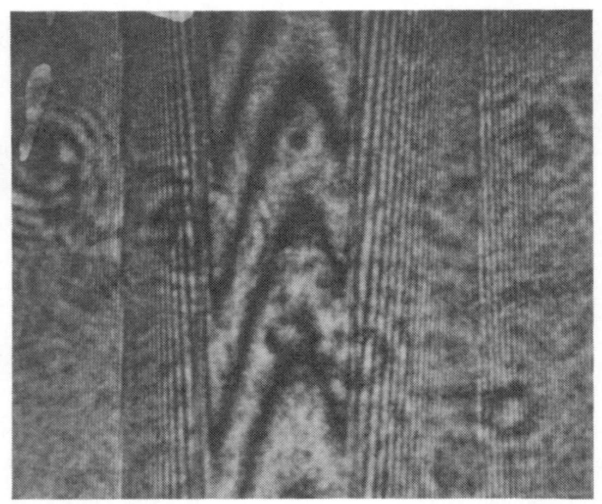

Figure 3. An example of asymmetrical interferogram of graded index optical fibers.

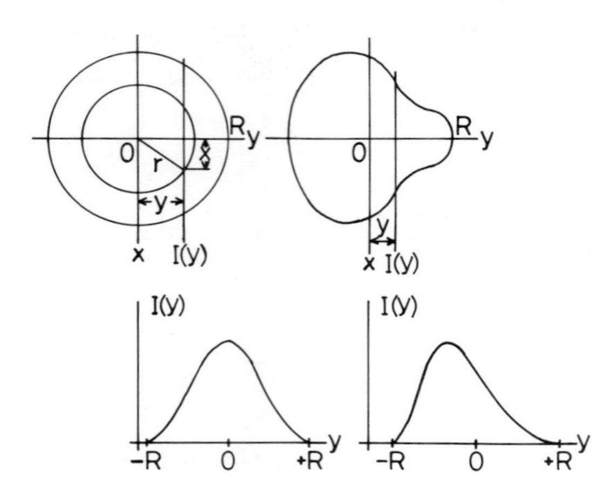

Figure 4. Coordinate system used in Abel inversion.

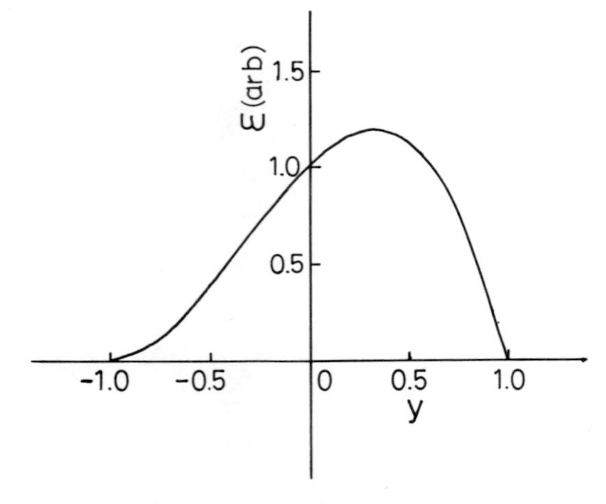

Figure 5. Test data for the asymmetrical Abel inversion.

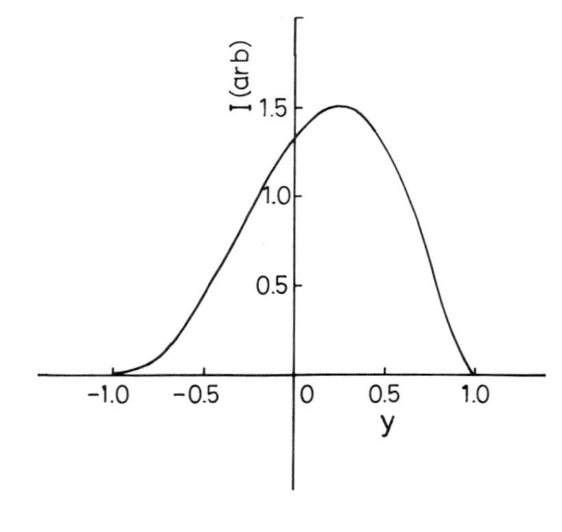

Figure 6. Integrated value corresponding to eq.(18) or Figure 5.

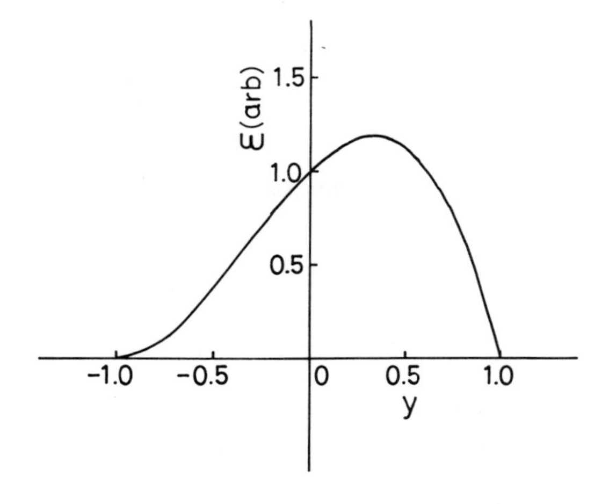

Figure 7. Inverted result by our asymmetrical Abel inversion.

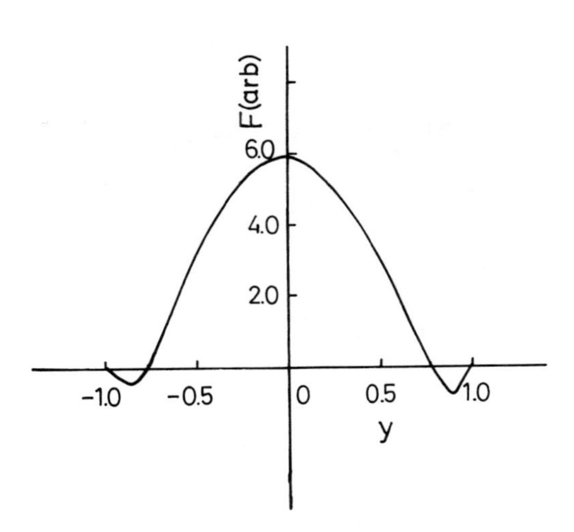

Figure 9. Symmetrical fringe shift profile obtained from Figure 2.

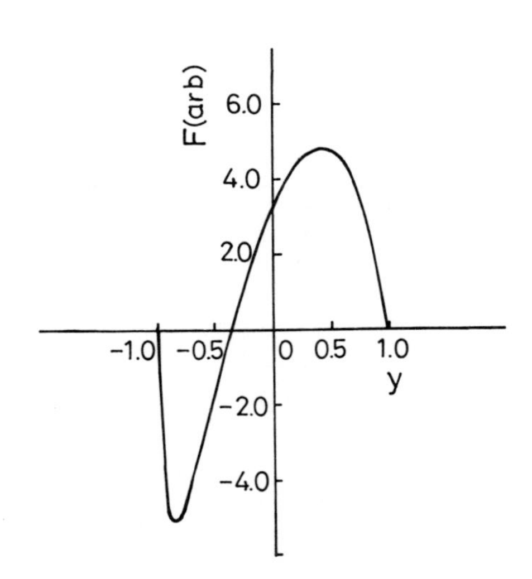

Figure 11. Asymmetrical fringe shift profile obtained from Figure 3.

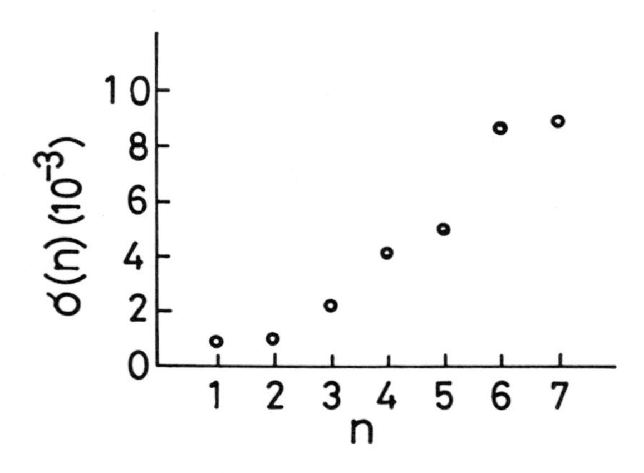

Figure 8. Standard deviation σ versus n in eq.12.

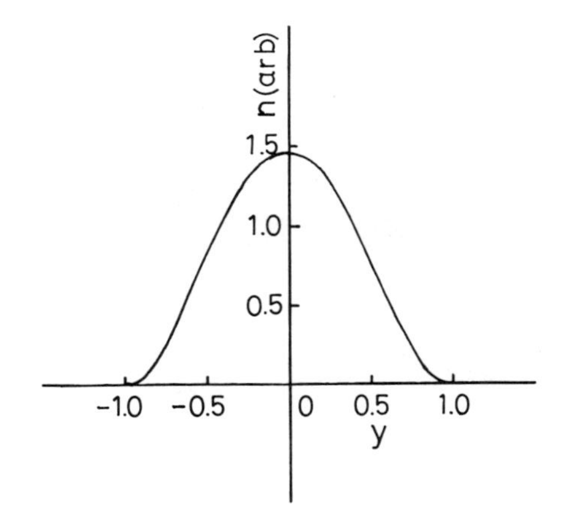

Figure 10. Inverted refractive index profile obtained from Figure 9.

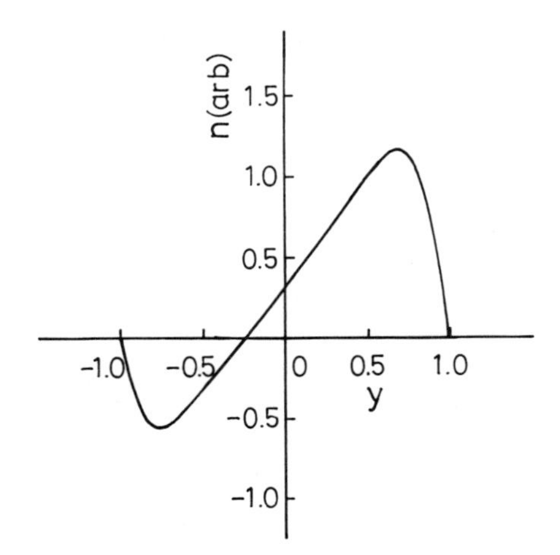

Figure 12. Inverted refractive index profile obtained from Figure 11.

The application of holography to research reason of spurting water of the antique "Water Spurting Basin"

Zhu Hongfan Zhou Genyu

Laboratory of Shanghai Museum, 16 South Henan Road, Shanghai

Abstract

In this paper application of holographic time-average method to research the reason of spurting water of the antique "Water Spurting Basin" is presented.

Introduction

" "Water Spurting Basin" was made in Ming dynasty (1368-1644A.D.), (see figure 1), a vessel for loading water or used as a washing basin in ancient times similar to that used now. It is a shallow basin with wide mouth rim, and a flat bottom, but it is different from modern basin. There are two semi- circular handles symmtrically on the mouth rim, the bottom of the basin is ornamented with four carps spilling wave lines on the inside wall of the basin.

Figure 1. "Water Spurting Basin", Bronze Ming dynasty (1368-1644 A.D.)

Figure 2. Picture of the "Water Spurting Basin" spurtingw water

When the basin is filled with water, a droning sound can be heard as soon as the tops of the handles are rubbed with two clean hands. As the sound becomes louder, the water waves are agitated increasingly at the four potions, and finally, many little drops are spurted upwards, as if they come from the mouths of the carps. The height is about 50 centimetres or even more (see figure 2).

Why the mouth of the carps on the basin can spurt water pearls as the two handles are rubbed with hands ?

Experiment

We think that the spurting water from the mouths of the carps is a physical phenomenon, It is closely related to the vibration frequency of "Water Spurting Basin". For this reason wé designed an experimental device to study the vibration state of the basin.

See the schematic diagram of the excitation and pick-up of the vibration for the time-average laser holography (see figure 3).

We connected the output of audio-frequency oscillator with piezocrystal pieces which are stuck on one side of the rim of the basin, and are used as exciting components. They transfer vibration to the basin and are connected with the input of x-axis of the oscilloscope. The other group of piezocrystal pieces are stuck on the opposite side of the rim, and are used as vibration pick-up. They can transform the vibration into electrical signals and are connected to the input of y-axis of the oscilloscope. Then as adjusting the frequency of the audio-frequency oscillator to the inherent frequency of the basin, the hum, a resonant sound, will occur and we can see a stable Lissajous'pattern on the screen of the oscilloscope. The resonant peak of the basin can be determined, and the frequency can be directly observed from the digital frequency indicator. By continuously changing the frequency of exciting source, a series of resonant vibrations can also be found. Take one picture the time-average holographic interferogram for each from the lowest frequency resonance. (see figure 4)

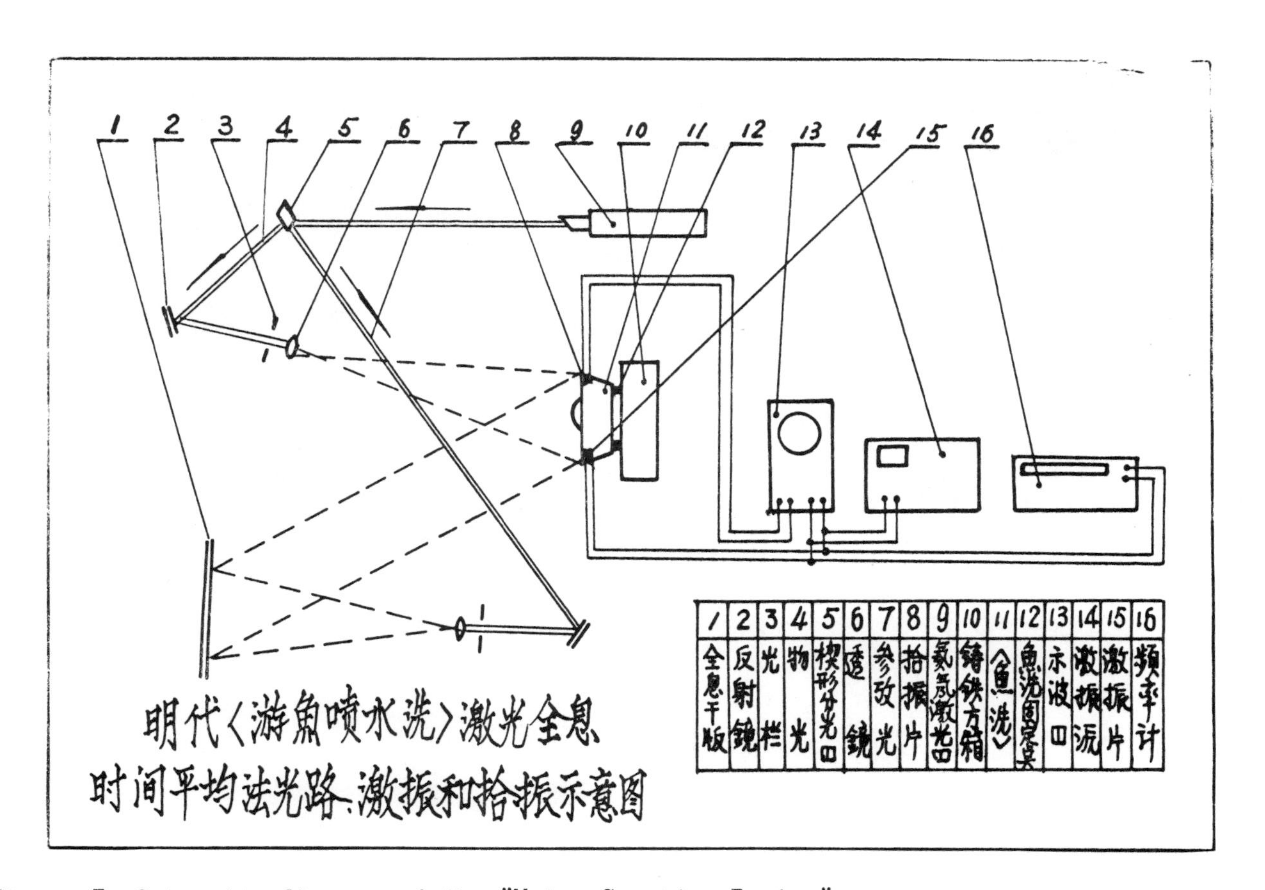

Figure 3. Schematic diagram of the "Water Spurting Basin " for the time-average laser holography.
1. holographic plate; 2. reflector; 3. diaphragm; 4. object beam; 5. spectroscope; 6. lens; 7. reference beam; 8. vibration pick-up pieces; 9. He-Nelaser; 10. a box of cast-iron; 11. "Water Spurting Basin"; 12. fixed points on the basin; 13. oscilloscope; 14. audio-frequency oscillator; 15. vibration exciting pieces; 16. digital frequency indicator.

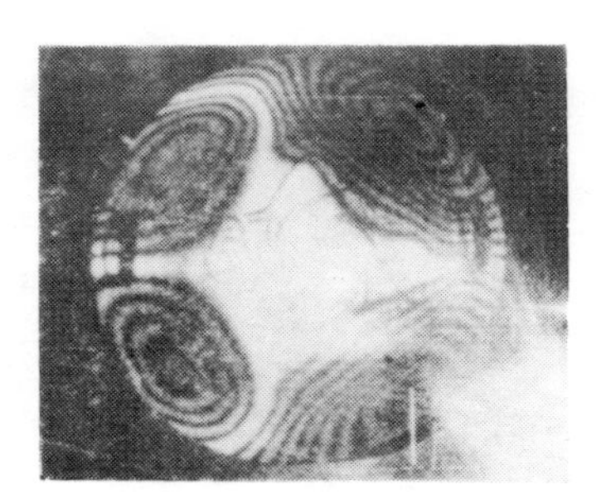

Figure 4. Holographic interferogram for "Water Spurting Basin", 2ooHz

Figure 5. Holographic interferogram for "Water Spurting Basin", 600Hz

Because the hands are in contact with the handles all the while, the nodal points of vibration always located near the handles, and the number of lines are always even number due to the symmetry of the basin. See figure 4 and 5, which are respectively the state of 4 and 6 nodal lines.

Results and Discussions

It has been found that the lowest frequency resonance from the audio-frequency oscillator adjusted to 200Hz produced the same tone as that by rubbing the handles with two hands for spurting water, and a good correspondence between the holographic interferogram and the spurting water phenomenon was also found (see figure 4). There are four groups of interference patterns intervened between four nodes, corresponding to four antinodes existing in the extension of the four carp-mouths. The vibration amplitude is maxium at the centre of the four antinodes and four water spurting centres are formed here.

Furthermore, if a higher frequency is chosen, it may generate another vibration with six or eight and even more antinodes and nodes. Rubbing the handles just with hands, the vibration mode with four antinodes and four nodes is most likely to appear. Therefore, when the basin is filled with water and the handles are rubbed with hands, resonance takes place on the wall of the basin. As a result of simutaneous action of four antinodes, water pearls are highly sprayed from the four directions. This is generally known as standing wave in physics. Because the bottom of the basin is fixed with several strutted points, when we rub the handles of the basin, the incident wave and the reflected wave overlap each other at the wall of basin. And so the standing wave is formed. Therefore take a glance at it, the water pearls look like spurting from the mouths of the four carps. It is actually an ingenious application of resonance principle.

The experiment of holography proved that the "Water Spurting Basin" cleverly linked science and technology with artistic appreciation and imagination in ancient China.

Hologram Interferometer for Film Deposits Evaluation

Diana Tentori and Martin Celaya

Optics Department, Applied Physics Division, CICESE Research Center
P.O. Box 2732, 22830, Ensenada, B. C. Mexico

Abstract

The design and performance of a holographic interferometer for the assessment of thin film deposits made on dielectric substrates is presented. Thickness and uniformity of film deposits thicker than 1 micron can be evaluated with this device. This interferometer can also be used in a reflection mode to test non-dielectric deposits made on opaque substrates.

Introduction

In the development of photosensitive materials like dichromated gelatins and thermoplastics, it is necessary to evaluate the uniformity of the deposited film and its thickness. Usable film thicknesses are within 1 micron to 20 microns and refractive indexes of these materials are very close to that of the glass substrate. In these conditions common interference techniques can not be used, since the optical path difference is high and fringe visibility is poor due to the strong differences in reflectivity between the air-film deposit interface and the film-glass interface. In this work it is presented the design and performance of an optical setup for the assessment of these film deposits based on hologram interferometry of diffuse wavefronts. In the design of this optical system the sensitivity range for longitudinal variations of an in-line object beam hologram interferometer is shifted to the desired interval of measurement. This change of the range of work is made by introducing straight fringes as it is done in common interferometry. Hence small longitudinal changes are detected as deviations from straightness in the fringes produced by a rigid lateral displacement. This optical array can also be used in a reflection mode.

In the proposed setup, the diffuse source of light, the film sample and the optical viewing system are aligned, as it is shown in Figs. 1 and 2. Fig. 1 corresponds to a transmission arrangement used to test dielectric films deposited on dielectric substrates. Fig. 2 shows a reflection set up. It can be used for opaque films deposited on polished substrates. The beamsplitter as well as the surface of the substrates can have a poor optical quality.

For non-destructive uniformity assessment, after the hologram of the plane substrate has been recorded the film deposit is made. By repositioning the substrate with film and introducing a rigid lateral displacement, a fringe pattern is produced. Interference fringes show a level mapping of the equal optical thickness regions. This optical arrangement has the advantage that it can be used with substrates with a poor optical quality. No contribution of the lack of homogeneity and parallelism is observed since it is an in-line arrangement using a diffuser to encode the complex amplitude distribution of the object wavefront. The observed interference fringes are produced just by interference between homologous points of the diffuser. The object is located behind the ground glass plate used as diffuser.

The contribution of the numerical aperture of the optical viewing system is also analyzed. The theoretical analysis of this interferometer shows that the film thickness variation can be determined by measuring the separation between the straight fringes, the shift of fringes from straightness, the refractive index and the wavelenght of the used light. Experimental results are shown.

Theory

The optical setup here presented is shown in Figs. 1 and 2. The reference wave front is a collimated beam of light and the film deposit is also illuminated with a collimated beam of light. This beam transmited through the film and its substrate is projected on a diffuser. The complex amplitude distribution on this diffuser constitutes the object wave front we will record at the hologram. The illuminating source, the film deposit, the diffuser plate, the hologram and the optical viewing system are aligned. Considering this geometry an analysis of the interference fringe pattern obtained for measuring the thickness of the film deposit using the interferometer in a transmission mode has been previously presented (Ref. 1). In this previous work, by means of holography, the wavefront coming from a film deposited on a glass substrate is compared with itself after

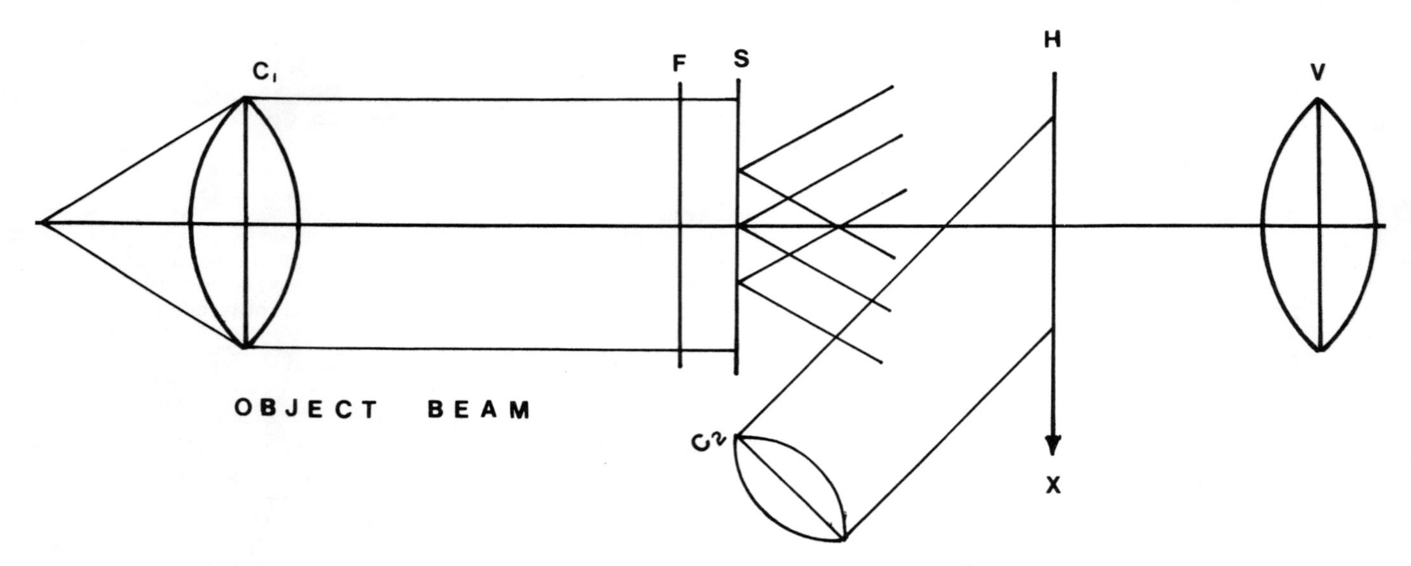

Fig. 1 Optical Arrangement for the assessment of dielectric films deposited on dielectric substrates

part of the film has been removed. Lateral displacement fringes introduced by a transverse rigid displacement of the hologram show a shift of the fringe pattern formed in the region where the film has been removed to that in the region that remained the same. This fringe shift is related to the film thickness through the following equation

$$t = \frac{\lambda(m_2 - m_1)}{n - 1} \quad \frac{x'_1 - x_1}{x_2 - x_1} \tag{1}$$

where λ is the wavelength of light, n the refraction index of the film deposit, m_2 and m_1 are the fringe orders corresponding to the fringes located at positions x_2 and x_1 on the

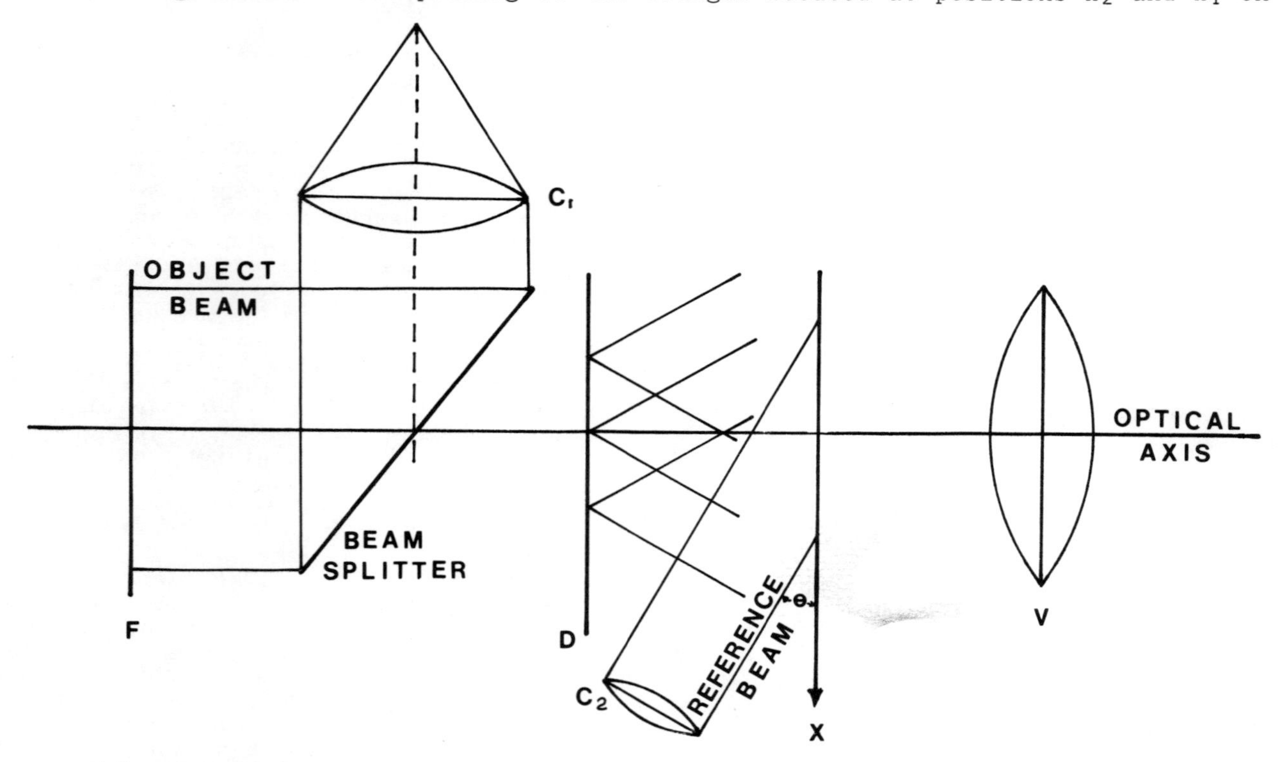

Fig. 2 Optical arrangement for the assessment of opaque deposits substrates made on polished substrates

image plane and x_1' is the new position of the fringe at x_1 when the film deposit has been removed. To obtain this equation a pure lateral displacement along the x axis has been considered.

When the interferometer is used in a reflection mode this equation presents a slight change due to the double pass of the beam of light when light is reflected from the substrate after removing part of the film. In this case the film thickness is given by

$$t = \frac{\lambda(m_2 - m_1)}{2(n-1)} \quad \frac{x_1' - x_1}{x_2 - x_1} \tag{2}$$

to evaluate the film thickness uniformity in transmission mode, we can use the film thickness variation dt,

$$dt = \frac{\lambda}{n-1} \quad \cdot\frac{dx}{\Delta x} \tag{3}$$

where dx is the deviation of the fringe from straightness and Δx is the spacing along the x axis between consecutive fringes. By varying the amount of lateral displacement introduced the position of fringes can be changed. In this way the whole area of the film deposit can be evaluated. When the interferometer is used in reflection mode a $^1/_2$ multiplying factor must be introduced.

Experimental Results

Typical fringe patterns for thickness measurement are shown in Fig. 3. If a sharp step is made on the film deposit the fringe pattern obtained looks as those here shown. With these patterns it is not possible to recognize which fringe in the second pattern corresponds to a given fringe in the first pattern. In order to be able to recognize them, a continuous step must be made. The continuous change of the optical thickness allows us to follow the fringe deviation. Another way of finding out the corresponding fringe is proposed in Ref. 2. As it is shown there, introducing an additional longitudinal displacement and varying the numerical aperture of the viewing system (making it wider), it is possible to identify the shift direction and its value. When no longitudinal displacement is introduced, a wider aperture stop just produces a uniform decrement of fringe contrast over the whole pattern. This result has been demonstrated in Ref. 2 and 3, and can be observed in Fig. 3. Fig. 3a has been obtained using a photographic objective with f/no. = 11 and Fig. 3b changing it to f/no. = 4. In both pictures the expossure energy was the same.

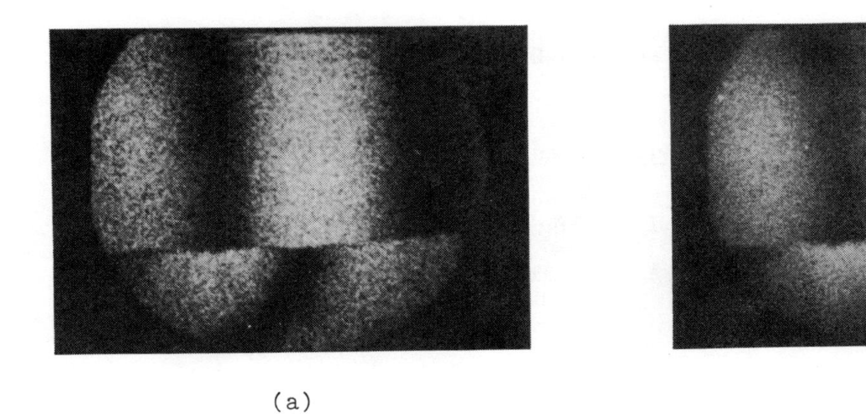

(a) (b)

Fig. 3 Interference patterns obtained introducing a step in the film deposit. a) f/no. = 11, b) f/no. = 4

To demonstrate the results that can be obtained with this technique for the evaluation of thickness uniformity, a hologram of a non uniform deposit has been recorded. Then the film deposit has been removed and lateral displacement fringes were introduced. Fig. 4a shows a fringe pattern obtained when a lateral displacement is introduced and no variation has been made on the film. Fig. 4b shows the fringe pattern obtained when the non uniform film deposit is removed. This figure shows equal thickness contours of the film deposit. From Eq. 2 it can be noticed that for a given thickness variation dt in a certain part of the film, since the wavelength of light and the refractive index keep constant, the ratio of the lateral variation of the fringe from straightness dx to the fringe spacing Δx, must also keep constant. This means that higher sensitivity will be obtained for a small lateral displacement; i.e. a large fringe spacing. Following again the considerations made in Ref. 1, we have that the percent error in this measurement is given, for an interferometer working in a transmission mode, by

$$\frac{\delta t}{t} \times 100\% = \left(\frac{1}{dx} + \frac{1}{\Delta x}\right) 2\delta x \times 100\% \tag{4}$$

where the contribution of the error in the measurement of the refractive index has been neglected and δx is the error in the determination of the positions along the x axis, considered to be the same for all the measurements. To neglect the error contribution due to the measurement of the refrative index, it has been considered that it can be measured to the third decimal place ($\delta n = + 0.001$). Under these conditions its contribution is smaller than 1%. In this case, as in classical interferometry working in a reflection mode we obtain twice the precision than working in a transmission mode; i.e. the factor 2 is dropped in Eq. 4.

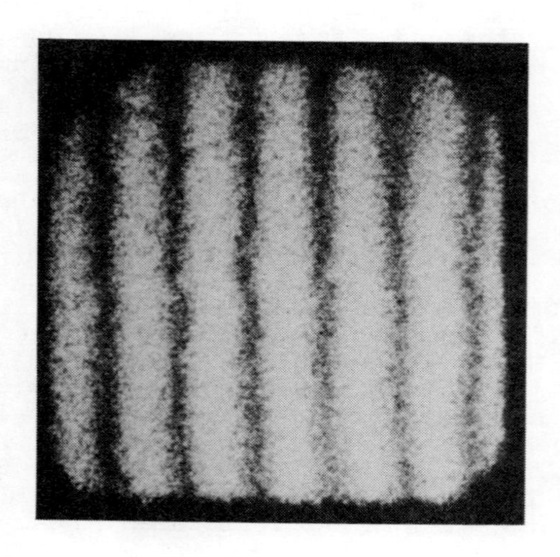

(a)

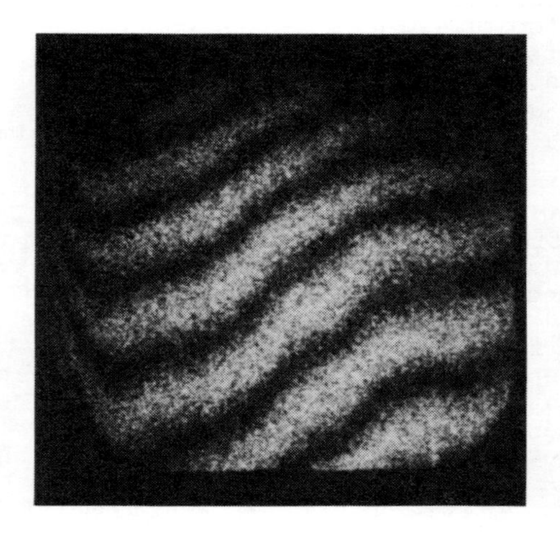

(b)

Fig. 4 Comparison between fringe patterns obtained: a) keeping the non uniform deposit in its place and b) removing the deposit.

Conclusions

In this work a modification of an in line object beam hologram interferometer for film deposist evaluation has been presented. By using a diffuser plate as a screen where the complex amplitude distribution of the object wave front is encoded, it is possible to use this device both in a transmission or in a reflection mode. Its capability for measuring thickness and for the evaluation of film deposits has been demonstrated.

Acknowledgements

This work has been sponsored by the Organization of American States through DAAI-CONACYT. We specially wish to thank Dr. Asdrubal Flores for their support.

References

1.- Diana Tentori and Martin Celaya "Film Deposit Assessment with Hologram Interferometry". Applied Optics 25, No. 16 (1986) (in press).

2.- Martin Celaya, Diana Tentori and Ricardo Villagomez, "Analysis of an In Line Object Beam Hologram Interferometer". SPIE Proceedings, Vol. 673 (1986) (in press).

3.- Martin Celaya and Diana Tentori, "Fringe Contrast Variation in Hologram Interferometry with the Numerical Aperture of the Viewing System". Applied Optics, 25, No. 16 (1986).

"Heat Transfer Studies by Microholographic Interferometry"

Bloisi F.*, Cavaliere P.**, Martellucci S.+, Meucci R.++, Mormile P.++, Pierattini G.++, Quartieri J.++, Vicari L.*

*Facoltà di Ingegneria, Università di Napoli, Italy
**Istituto di Fisica, Università di Palermo, Italy
+Facoltà di Ingegneria, II Università Tor Vergata, Roma, Italy
++Istituto di Cibernetica CNR, Arco Felice, Napoli, Italy

Abstract

A set up for real-time and double-exposure microholographic interferometry has been assembled to determine temperature distribution in very small volumes of transparent mediums traversed by a laser beam. The axisymmetric refractive index distribution was evaluated from experimental data by means of a suitable mathematical model for the interpretation of the microinterferograms. Computer model calculations for the temperature distributions were also developed and compared with the experimental data.

Introduction

The determination of temperature profiles induced by laser irradiation of organic and inorganic materials is of great interest [1,2]. In a previous work [5] we introduced a mathematical model able to evaluate the temperature profiles as a function of the beam power and shape and of the thermal and optical characteristics of the irradiated specimen. The mathematical model used for the computation of the temperature profiles considers the propagation of a laser beam through a slab in which it is both absorbed and scattered. It assumes the cylindrical symmetry of the incoming irradiation, the spherical unpolarized scattering of the light and uses the Lambert-Beer law for the absorption. It solves the heat equation with the suitable boundary conditions by means of a double transformation, i.e. Fourier transform on z and Hankel transform on r. In this communication we show the very good agreement between the theoretically computed temperature profiles and the experimental distributions obtained from microholographic interferometry in a transparent sample traversed by an Argon laser beam (λ=5145 Å). The sample was a solution of red ink in distilled water, of known absorption at this frequency. A He-Ne laser was used for microholographic interferometry of the small volume of the solution heated by the absorption of the Argon laser beam. The actual values of the temperature must be calculated from the interferograms i.e. estimating the variations of the refractive index due to the local heating of the water; for this we have elaborated a model for the interpretation of the holographic interferograms of a transparent specimen with axisymmetric refractive index distribution. This method is based on a numerical approach and is thus well suited to computer processing. It is of a general form and may be useful for many purposes.

Experimental set-up

The optical configuration is shown in fig. 1. The sample is irradiated by an Ar^+ laser (λ =5145 Å); the beam is appropriately reduced in diameter ($\approx$ 100 μm) before it crosses the sample. The microholographic set-up follows a previous system designed by us [6]. The 10 x microscope objective O_2 focuses the enlarged image of the sample on the image plane I, behind the holographic plate H. A red filter is used between the sample and the objective to block the green light of Ar^+ laser from the holographic plate.

During the first phase the Ar^+ laser is off and one records the real-image hologram of the undisturbed sample. The holographic plate is then removed, photochemically processed, bleached and repositioned in the same position as before by means of micro-movements in the plate holder. With a suitable angular displacement of the reference beam it is possible now to produce the background reference fringes in the image plane with the desired orientation and spacing.

In the second phase the Ar^+ laser is turned on and the heating of the sample produces a phase shift in the object (probe) beam. The interference with a wavefront reconstructed from the hologram illuminated by the reference beam produces the deformation of the background fringes in the image plane I. In this way it is possible to follow the temperature rise in real time and to record its evolution with a cinecamera. In fig. 2 we can see a typical interferogram. The specimen was a 1.o/$_{oo}$ aqueous solution of red ink. The main characteristics to be measured were the absorption coefficient, the refractive index and its dependence on temperature.

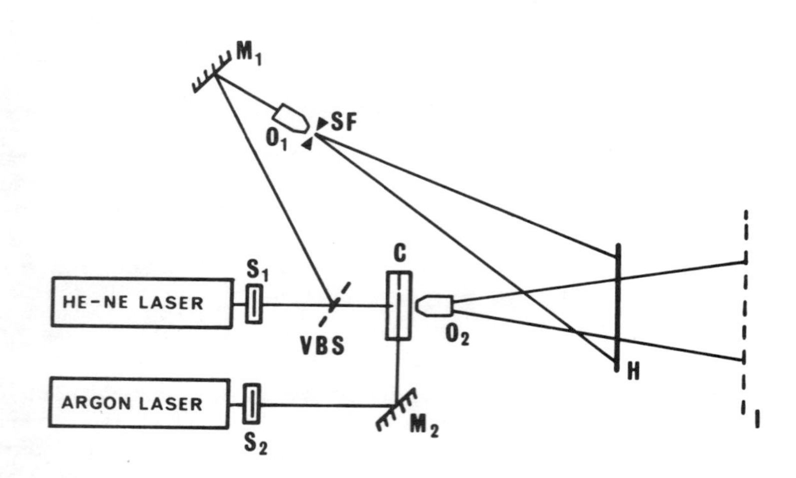

Fig. 1. Microholographic set-up. S_1,S_2, Shutters; M_1,M_2, Mirrors; VBS, Variable Beam-Splitter;O_1,O_2 Microscope Objectives; SF, Spatial Filter; C, Sample; H, Holographic Plate; I, Image plane.

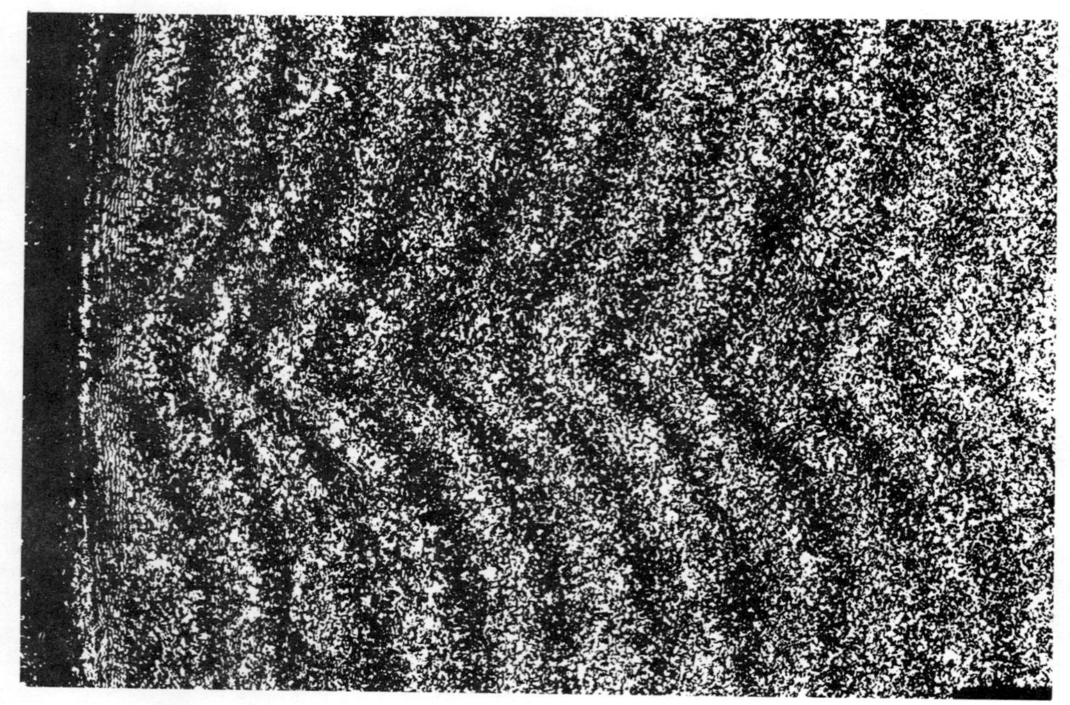

Fig. 2. A typical interferogram

The absorption coefficient was determined measuring the decay of the intensity of the Ar^+ beam passing through various thickness of the specimen. Assuming the validity of the Lawson-Beer law an absorbing coefficient 1.0 cm^{-1} was deduced.

The refractive index was compared to that of distilled water. A hollow prism was filled with distilled water and used to deflect the HeNe laser beam. The ink was then added and the deviation of the beam was observed. The two indices were found to coincide within .1%. The same comparison was made for the deflection of the HeNe laser beam by a vessel of water of 20 cm thickness in which there was a transverse temperature gradient of about 3°C/cm. The observation of the beam deflection allows the direct comparison between the derivatives of the refractive index with respect to the temperature for the distilled water and the ink solution. They were found to coincide, for the red radiation of the HeNe laser, within 1% in the range of temperatures considered (30-40 °C). The problem of determining the temperature rise is reduced to the evaluation of the local refractive index. In fact the function T(n) is well known for distilled water [7,8].

To reduce the noise due to the errors in the reading of the fringes we took for the phase shift the average value on three fringes. Since in our apparatus the distance between two successive fringes was 125 μm the range considered has a length negligible compared to the absorption length.

Experimental results and discussions

Information on the refractive index distribution can be obtained using interferometry; the deviation of the fringes from straight lines (the references fringes) is proportional to the phase shift of the object beam due to the local temperature rise in the sample.

In our case we have an axisymmetric phase object: let us assume the z axis to be the symmetry axis (Ar^+laser beam) and that the probe beam (HeNe laser beam) passes through the object in the x direction (fig. 3).

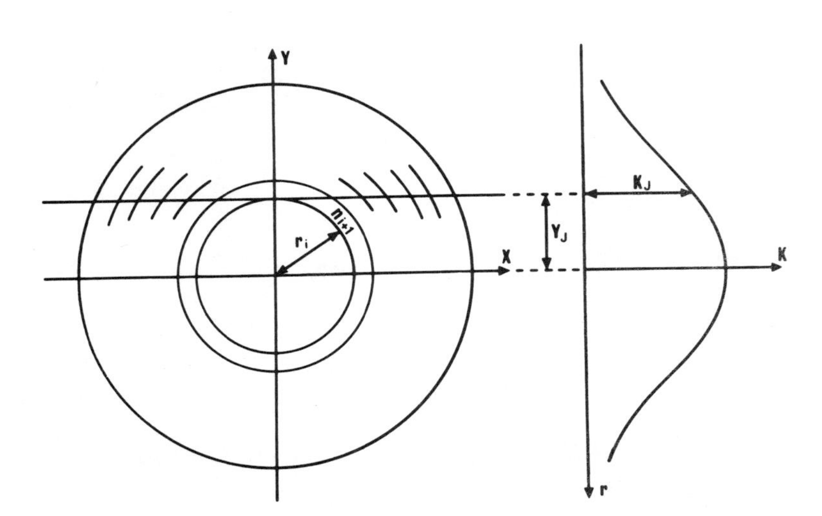

Fig. 3 Cross section of the sample (on the left) and the reconstructed phase shift (on the right).

The zone of disturbance is small compared to the volume of the whole sample. We assume the value of n at $r=\infty$ to be independent of time, and equal to the value of n everywhere at $t=0$. So far:

$$n_f(r,z,0) = n_f(\infty,z,t) = \text{const.} = n_o$$

The fringe shift is

$$k(y,z,t) = \frac{\Delta\phi(y,z,t)}{2\pi} = \frac{2}{\lambda}\int_y^\infty n(r,z,t)\,\frac{r}{\sqrt{r^2-y^2}}\,dr \tag{1}$$

where $r = \sqrt{x^2+y^2}$, $\Delta\phi$ is the phase shift and n is the variation of the refractive index:

$$n(r,z,t) = n_f(r,z,t) - n_o$$

Equation (1) is an Abel integral equation which has the exact solution

$$n(r,z,t) = \frac{\lambda}{\pi}\int_r^\infty \frac{k_y(y,z,t)}{\sqrt{r^2-y^2}}\,dy \tag{2}$$

where k_y is the partial derivative oy k with respect to y.

The distribution $k(y,z,t)$ is, however, not given analytically, so that numerical methods are more convenient [9,10].

Let us divide the sample into m concentric annular zones Z_i of width R/m plus an external zone Z_{ext} for $r>R$, where R is the radius of the zone explored by the probe beam (fig. 3), the zone Z_i being delimited by the circles $r = r_{i-1}$ and $r=r_i$, where $r_i=iR/m$.

When observing along the line $y=y_j=jR/m$ the measured fringe shift is k_j.

Equation (1) can then be written as

$$k_j = \frac{2}{\lambda}\sum_{j+1}^{m}{}_i\left[\bar{n}_{ij}\int_{r_{i-1}}^{r_i}\frac{r\,dr}{\sqrt{r^2-y_j^2}}\right] + \frac{2}{\lambda}\int_R n(r,z,t)\,\frac{r\,dr}{\sqrt{r^2-R^2}} \tag{3}$$

where $\bar{n}_{ij}$ is the average value of $n(r,z,t)$ into the zone Z_i along the line $y=y_j$.

We will now make the following assumptions:

a) Taking into account the radial symmetry of the problem, for a sufficiently small thickness R/m of the annular zones Z_i, we can assume that the average value $\bar{n}_{ij}$, depends only on the zone Z_i and not on the line $y=y_j$, so that

$$\bar{n}_{ij} = \bar{n}_i$$

b) remembering that $n(\infty,z,t) = 0$, and following the results of simulation studies we assume an exponential asymptotic behaviour for $n(r)$

$$n(r) = n^*\exp(-r/\alpha) \qquad \text{(for } r > R) \tag{4}$$

with n^* and α unknown coefficients. The final results will depend very little on the choice of α.

Under such hypotheses, one can find that:

$$n^* = \frac{\tilde{k}_m}{I_m(\alpha)} \tag{5}$$

$$\bar{n}_m = \frac{\tilde{k}_{m-1} - n^* I_{m-1}(\alpha)}{\tilde{d}_{m-1,m}} \tag{6}$$

$$\bar{n}_j = \frac{\tilde{k}_{j-1} - n^* I_{j-1}(\alpha) - \sum_{j+1}^{m} i \; (\bar{n}_i \, \tilde{d}_{j-i,i})}{\tilde{d}_{j-1,j}} \qquad (j = m-1, \ldots .1) \tag{7}$$

introducing the dimensionless quantities:

$$\tilde{d}_{ji} = 2\, d_{ji}\, m/R \qquad \text{where} \qquad d_{ji} = \int_{r_{i-1}}^{r_i} \frac{rdr}{r^2 - y_j^2} = \sqrt{r_i^2 - y_j^2} - \sqrt{r_{i-1}^2 - y_j^2} \qquad \begin{matrix} j=0,\ldots m-1 \\ i=j+1,\ldots m \end{matrix}$$

$$\tilde{k}_j = k_j \, \lambda \, m/R$$

The computer program determines the value of α imposing continuity of $f(r, \alpha)$ and of its derivatives at $r=R$, where for $r>R$ the function $f(r, \alpha)$ is given by equation (4) while for $r>R$ it is the best fit of the points given by equations (6) and (7).

Figs. 4, 5 and 6 show the temperature rises in the sample at various distances from the beam axis for three values of beam intensity. The error is computed propagating the error on the phase shift as derived from the fringe deformation. In the figures the temperature rises are compared to the theoretical predictions deduced by equation (1). As we can see the experimental points are close to the theoretical expectations and the trends are the same.

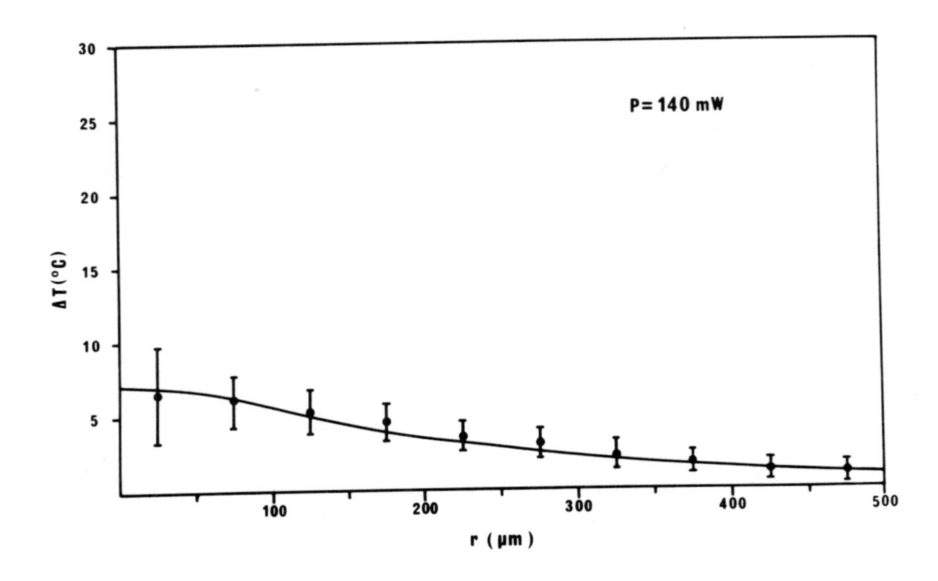

Fig. 4. Temperature profile for 140 mW laser beam

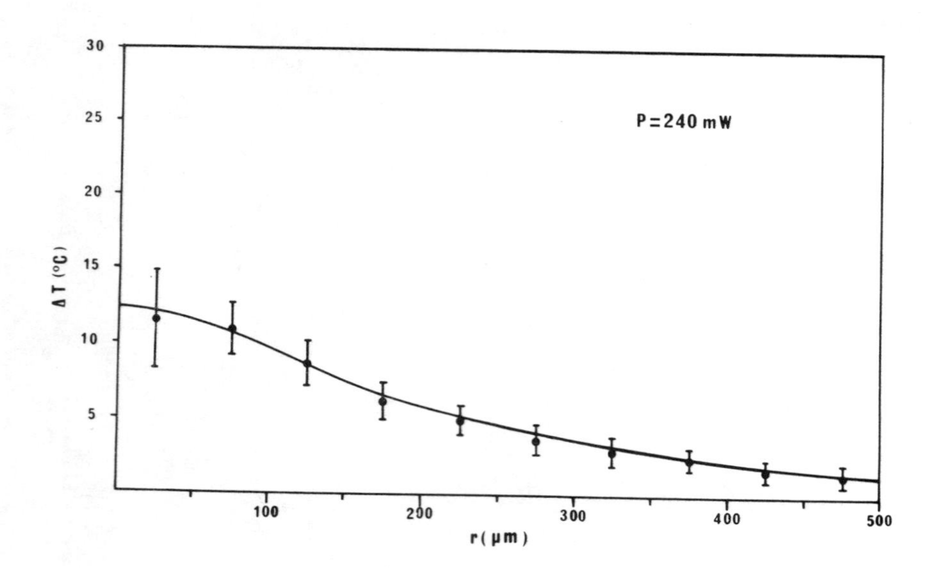

Fig. 5 Temperature profile for 240 mW laser beam

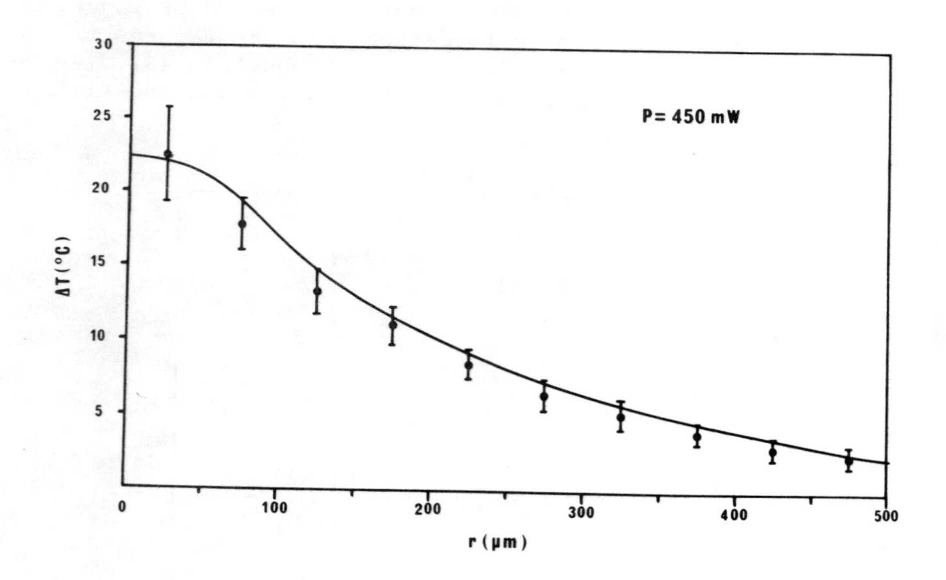

Fig. 6 Temperature profile for 450 mW laser beam.

Acknowledgements

The authors acknowledge the very fruitful help of A. Finizio and E. Casale

References

1. Halldorsson, T. and Langerholc, J., Appl. Opt. 17: 3948 (1978).
2. Gupta, M.C., Hang, S., Gupta, A. and Moacanin, J. Appl. Phys. Lett. 37: 505(1980).
3. Ernst, K. and Hoffman I.J., Phys. Lett. 87A: 133 (1981).
4. Delfino, G., Kemali, M., Martellucci, S. and Quartieri, J., IEEE J. Quantum Electr. QE-20:1489 (1984).
5. Delfino, G., Martellucci, S., Quartieri, J., Quarto E., Lett. Nuovo Cimento, 38: 353(1983).
6. Pierattini, G., Opt. Comm. 5:41 (1972).
7. Tilton, L.W. and Taylor, J.K., J. Research NBS. 20:419 (1938).
8. Dobbins, H.M. and Peck, E.R., J. Opt. Soc. Am. 63: 318 (1972).
9. Vest, C.M.,"Holographic Interferometry",John Wiley & Sons, New York (1979).
10. Ostrovsky, Yu. I., Butusov, M.M., Ostrovskava, G.V.,"Interferometry by holography", Springer, Berlin (1980).

Session 6

Holographic Interferometry II

Chairman
Xin-Gen Zhou
Tsinghua University, China

Laser holographic inspection of solder joints on printed circuit board (PCB)

Hong Jing, Geng Wanzhen, Jiang Lingzhen, Xue Wei,
Ma Jing, Zhao Shijie, Yang Jingfeng

Department of Applied Physics, Harbin Institute of Technology, Harbin, China

Abstract

The principle of inspecting solder joints by holographic interferometric technique is introduced. The criterion of discriminating the flaws is suggested. The results are given and testified by metallographical sectioning.

Introduction

Nowadays, accurate electronic systems are widely used in fields of production and research. Among the operational troubles of electronic installations, many are introduced by the flaws of solder joints of PCB. The existence of flaws constitutes serious imperiling to the reliability of electronic control systems, sometimes may produce tremendous damage. Therefore, seeking a flaw detecting means of high reliability, high accuracy and high degree of automation is an urgent need in recent electronic industry.

There are two aspects in investigating the quality of solder joints of PCB. The first is to increase the solderbility of circuit components, to find out the cause of the presence of flaws, to improve the composition of solder and the technology of soldering. The other aspect is to look for some kind of efficient and convenient means of detecting solder joints with flaws. Domestically many research works have been engaged in solderbility but few have been engaged on the detection of flaws of solder joints.

Domestically and abroad there are several means of detecting flaws in solder joints of PCB[1-6]; visual detection manually or automatically; functional testing; testing by sound waves or ultrasonic waves; method by holographic correlation measurements; testing by X-ray of micro-focus; laser infra-red detection; optical holographic detection. The first means is currently used but is very subjective; the second is hardly applicable to individual solder joint; method by X-ray is of low speed; most of others are rarely applicable or have significance only by principle. Since 1982 Vanzetti Co. USA has produced several models of laser Infra-red Detection Instrument which has advantages of fastness (claimed to be 10 joints/sec.), good repetition and high automation[7]. But its principle is a kind of relative testing. At the start it needs a certain number of similar boards. The thermal signatures (temperature-rise and temprature-decay curves) of a large number of the solder joints are being statistically processed from which the upper and lower limits of the signature peaks for normal solder joints are obtained. Therefore the joints on a board capable of being tested by this technique has to be strictly identical in structure, the technological process of soldering must be under severe control. It needs automatic insertion of leads of components and automatic cleaning.

The holographic testing method here introduced has its distinguished feature. The technological processes of soldering need not be strictly identical. The criterion of existence of flaws depends on whether on the surface of a solder joint there exists any pecularity of interference fringes caused by the heat loading.

Testing process and the related problems

Testing method

The optical diagram of experiment is shown in figure 1 which contains two parts: the holographic camera (including the observation unit--the dotted rectangle) and the thermal loading portion.

The holographic camera consists of a He-Ne laser. The laser's output is split into a reference beam and two illuminating beams. The latter illuminate the joint surface, serve as two object beams. The object beams interfering with the reference beam form two holograms on a single plate. After developing and replacing the hologram plate to the original position, the solder joint is thermally loaded by a YAG laser pulse through an optical fiber. With a strong reconstructing beam from the initial reference beam direction, real-time interferometric phenomenon is observed. Through the hologram, on the reconstructed image, systematic interference fringes are disturbed. After stabilization if any pecularity of systematic fringes on the joint appears, there might exist flaws inside the solder joint, this is due to the fact that a proper thermal loading introduces a slight permanent deformation in a joint with flaw. A double-exposure hologram is taken and is kept for analyzing.

The observation optical system is an amplification system. Between the photographic plate and the solder joints is an objective lens, an eyelens is behind the photogaphic plate. The objective is to amplify the solder joint to make the observation of the joint and the interference fringes easier. The larger the aperture of the lens the smaller the depth of focus of the image. Since the systematic interference fringes and the fringes of pecularity may have different surfaces of localization with the surface of the solder joint, sometimes fringes can hardly be localized on the joint surface.

Fringe control technique

Rotating mirrors M_2 and M_4 makes the fringe control technique possible. It can adjust the localization of the fringes. Besides, it can be used to test whether the fringes on the joints are independent to make sure that they reflect the inside flaws.

The thermal loading allowable to be applied

The thermal loading depends on the power of the thermal source (for convenience the working current is taken instead) and the time of duration. On principle, the quantity of thermal loading should be large enough to reflect the interior flaws to the surface, but not so large as to change the microstructure of the joint surface, or to damage the circuit component thermally.

Inspection by metallographic dissection

Every solder joint after holographic testing is dissected metallographically to confirm the testing. Plane of dissection is chosen to be the horizontal cross section or the vertical cross section of the solder joint. Because of the limitation of the laboratory conditions, every joint is undergone 3-4 planes of dissection. From metallographic dissection the following conclusions can be drawn:

1) After thermal loading and stabilization of systematic interference fringes the fringes on the joint surface appearing just as that at pre-loading reflects a "good" joint.

2) Fringes on the joint of different orientation and spatial frequency or fringes curved compared with systematic fringes on the board reflect flaws between the land and the solder tin.

3) Existence of independent peculiar fringes at the ends of leads refects flaws between leads and solder tin.

4) Curved fringes on the joint reflect cavity inside joint. Sometimes contrast on the joint decreases or fringes disappear.

5) The quantity of cavities inside wave-foldering solder joint is small, the size is also small. Manual solder joints have many cavities.

Results of testing

Typical appearances of peculiar interference fringes in cases of "good" joint, joint with land flaws, lead flaws and cavity flaws are shown in figures 2, 4, 6, 8. The corresponding metallographic dissection photographs are shown in figures 3, 5, 7, 9.

The statistical results of testing are given in Table 1.

The above results indicate that holographic technique applied to flaw detection of PCB is a research work of significance and of promise. Using continuous scanning heat stressing, and applying the technique of pattern recognition to the interferometric fringes, a complete automatic testing unit is feasible by a computer control.

References

1. Y. Nakagwa, Automatic Visual Inspection of Solder Joints on Printed Circuit Board, Pro. SPIE. Int. Soc. Opt. Eng. (USA), Vol. 336, p. 121-7 (1982)
2. Trend Inspection Station for Printed Circuit Board Solder Joints, Final Report, AD-A095971
3. Acoustic Inspection of Solder Joints, Battelle Development Corp, US Patent 4 218 922
4. R. W. Jenkins and M. C. Mellwain, Holographic Analysis of Printed Circuit Board, Mater. Evalua., Sep. (1971)
5. The Ultimate in automatic Inspection, Vanzetti systems Inc., Circuit World (GB), Vol. 8, No.4, pp. 12-17 (1982)
6. J. R. Williams, Holographic Stress Analyzer for Solder Joints, N72-11415
7. Philip J. Klass, Laser Inspection Spots Flawed Solder Joints, Aviation Week & Space Technology, September 3, p. 307, (1984)

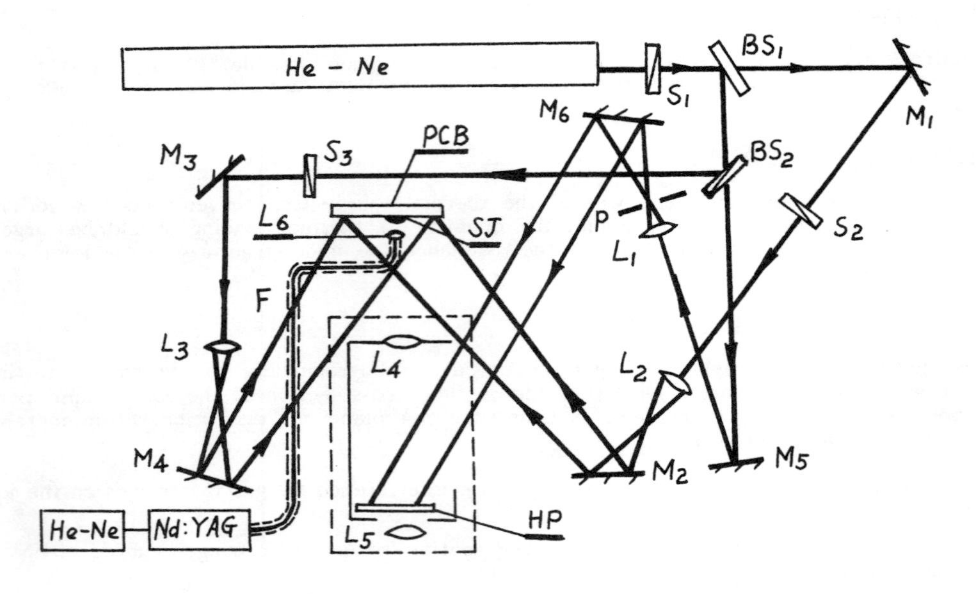

Figure 1. Optical diagram of experiment

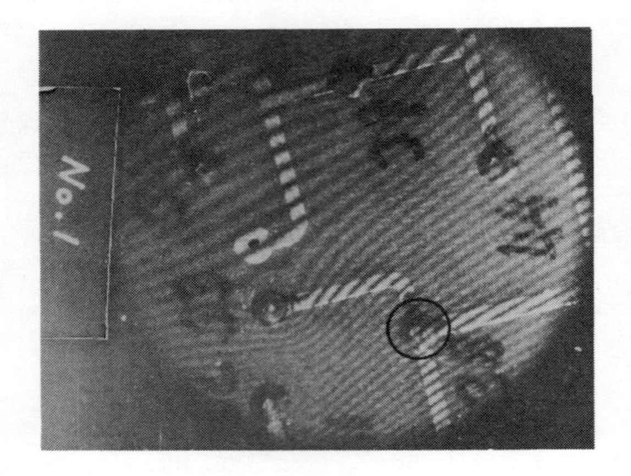

Figure 2.Fringes on joint in case of "good" joint

Figure 3. Corresponding dissection photograph of figure 2

Table 1. Statistical results of testing

Types of detected joints	Overall types of detecting joints	Joints with lead flaws	Joints with land flaws	Joints with cavity flaws		
				d=0.1mm	d=0.2mm	d=0.3mm
Actual no. of joints	387	128	38	209	131	99
No. of joints with correct detection	320	107	35	168	110	87
Rate of successiveness	82.7%	83.6%	92.1%	80.4%	84.0%	87.9%

d: Cavity diameter

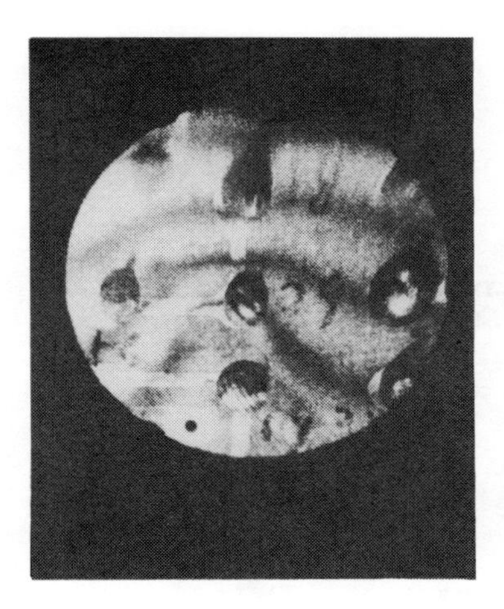

Figure 4. Peculiar fringes in case of land flaws

Figure 5. Corresponding dissection photograph of figure 4

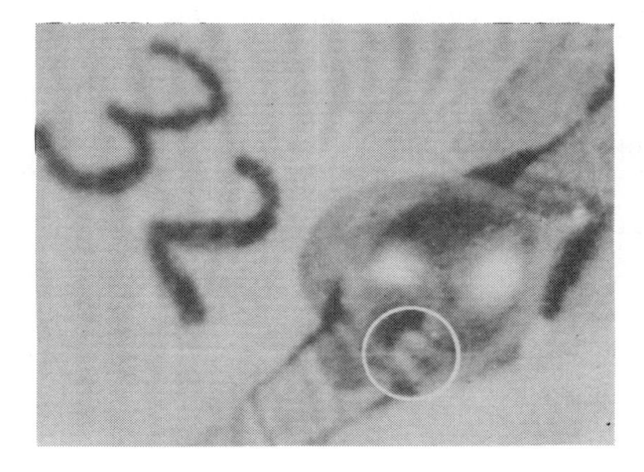

Figure 6. Peculiar fringes in case of lead flaws

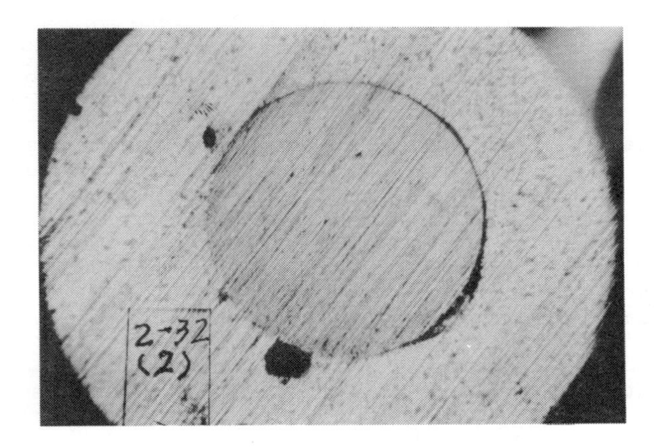

Figure 7. Corresponding dissection photograph of figure 6

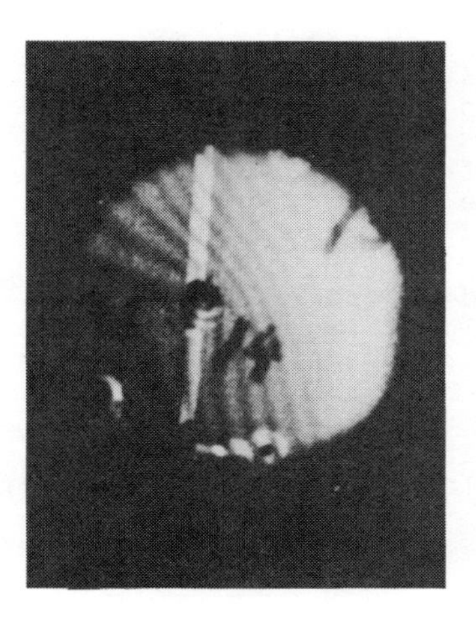

Figure 8. Peculiar fringes in case of cavity flaws

Figure 9. Corresponding dissection photograph of figure 8

Automatic measurements of the small angle variation using a holographic moire interferometry and a computer processing

Yoshiaki Nakano

Department of Physics, Hokkaido Institute of Pharmaceutical Sciences,
7-1 Katsuraoka Otaru 047-02 JAPAN

Abstract

A method for automatic measurement of the small tilt angle variation of an object surface by using a digital holographic moire interferometry[1] based on computer processing is described. The test surface is placed in the middle of a set of two gratings. A monochromatic plane wave illuminates the first grating and undergoes diffraction. After being reflected by the test surface, the first grating produces the magnified Talbot image. The moire fringe is generated by superimposing this Talbot image on the second grating. The inclination angle of the moire fringe is a measure of the tilt angle of the test surface. Fast and automatic measurement has been achieved with improved accuracy and sensitivity. In the experiment, the small tilt angle variation of the plane mirror is measured.

Introduction

Moire interferometry is a complements to conventional holographic interferometry. In a previous publication[2], a method for measuring the small tilt angle variation has been reported by the authors. The method is based on Talbot interferometry[3,4]. This optical system is constructed with two gratings, in which the moire fringe is generated by superimposing the Talbot image g_1' of the first garting g_1 on the second grating g_2. In this paper we present a real time and automatic method for measuring the small tilt angle variation of an object surface using image processor. Image processor is constructed with an image sensor, a computer image memory and a main computer.

In measuring the small tilt angle of a surface, an autocollimator method and moire technique[5] using the autocollimator-optical lever combination are used. It is difficult to measure the local variation of tilt angles of a large object surface because the area of measurement is limited. However, this method can be applied for measuring a large surface having a local tilt variation.

The applications of the Talbot interferometry and the digital Talbot interferometer[6,7] have been reported by the authors[8,9]. The sensitivity of the Talbot interferometry can be easily tuned by varying the crossed angle between the two gratings. Furthermore, the optical system is simple and a precise optical arrangement is not necessary.

Principle

A schematic representation of the optical system is shown in Fig.1. The ξ and x axes are taken in the direction of the periodical amplitude variation of two gratings g_1 and g_2, respectively. The plane of the second grating g_2 is set in the direction perpendicular to the plane of the first grating g_1. The test surface M is placed in the middle of two gratings. The optical path length along the optical axis from the grating g_1 through the test surface M to the grating g_2 is set to be equal with the Talbot length kp^2/λ, where k is an integer, p is the period of the grating g_1, and λ is the wavelength of the incident light.

A monochromatic plane wave illuminated the grating g_1 and undergoes diffraction. After being reflected by the test surface, which is tilted by angle δ from the optical axis, the wave propagates to the grating g_2. therefore, the grating g_1 produces the Talbot image g_1 on the second grating g_2. If the tilt angle of the test surface is $\pi/4$, the pitch of the Talbot image g_1' is the same as grating g_1. However, as the tilt angle is usually not equal to $\pi/4$, we obtain the magnified Talbot image. If the tilt angle of the surface deviates from $\pi/4$ by an amount δ', the pitch p' of the magnified Talbot image g_1' on a screen at the plane x axis is given by

$$p' = \frac{p}{\cos 2\delta'} = \frac{p}{\sin \delta} \quad . \tag{1}$$

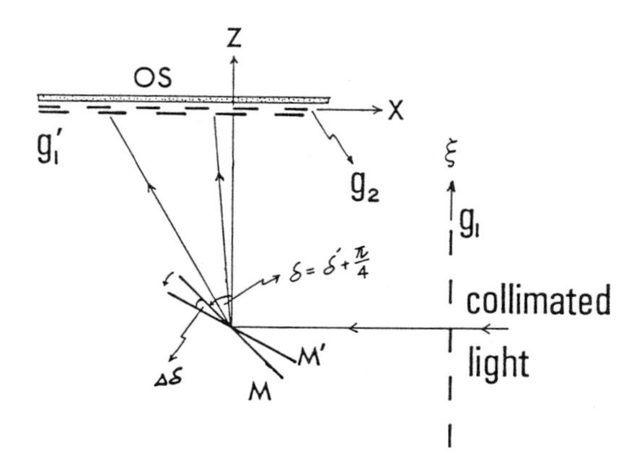

Figure 1. Optical arrangement for measuring the small tilt angle variation of an object surface: g_1, grating of the period p; g_2, second grating of the same period with g_1; M, test surface; OS, observing screen ; g_1', magnified Talbot image g_1' of the period p' on the second grating g_2 (x axis).

Next we obtain the moire fringe pattern by superimposing the Talbot image g_1' on the second grating g_2, which is the same as the first grating g_1 but is crossed by a small angle θ. The resulting moire fringes are observed with an inclination angle α to the x axis. The relation between the pitch p' and p is given by[5]

$$p' = \frac{p\cos\alpha}{\cos(\alpha+\theta)} . \tag{2}$$

Therefore, the tilt angle δ from the optical axis of the test surface is represented by

$$\sin 2\delta = \frac{\cos(\alpha+\theta)}{\cos\alpha} . \tag{3}$$

Hence, if the test surface is slightly tilted further by a small amount $\Delta\delta$, the pitch p" of the Talbot image g_1'' on the plane x axis is given by

$$p'' = \frac{p}{\sin 2(\delta+\Delta\delta)} . \tag{4}$$

Let the inclination angle of the moire fringe in this case be given by α'. The tilt angle $\delta + \Delta\delta$ of the test surface is represented by

$$\sin 2(\delta+\Delta\delta) = \frac{\cos(\alpha'+\theta)}{\cos\alpha'} . \tag{5}$$

Accordingly, from Eqs.(3) and (5) the small angle variation $\Delta\delta$ of the test surface is given by

$$\Delta\delta = \frac{1}{2}\left\{\sin^{-1}\left[\frac{\cos(\alpha'+\theta)}{\cos\alpha'}\right] - \sin^{-1}\left[\frac{\cos(\alpha+\theta)}{\cos\alpha}\right]\right\} . \tag{6}$$

When the crossed angle θ is fixed, the small tilt angle variation $\Delta\delta$ can be obtained by measuring the tilt angle α and α' of the moire fringe.

Experiments and Results

Figure 2 shows a block diagram of the experimental set up. An expanded monochromatic plane wave produced by a He-Ne (λ=632.8nm) and a collimator illuminates the first grating g_1, test surface M, and second grating g_2 successively. The two gratings are the 0.22mm pitch Ronchi type grating. The optical path length along the optical axis from g_1 through M to g_2 is the first Talbot image length ($2p^2/\lambda$ =153mm). The crossed angle θ between the two gratings is set by 1°. The experiments are performed by changing the tilt of the plane mirror surface of 0.16 square meters in area.

An analysis of the moire fringe patterns is tried by a computer image processing system. The system consists of the image sensor, an A-D converter, an image memory, an interface, a microcomputer, and CRT display. The image sensor used is the CCD TV camera. The output of the image sensor is read out sequentially, stored in computer image memory (IDS-98) through the A-D converter and the interface, and processed by a main computer (NEC. PC-9801m).

The lens L_3 and the slit SF for special filtering are set up, because the carrier component is removed.

The experimental results are shown in Fig.3 and Table 1. Figure 3 shows an example of the moire fringe patterns in the experiments. Figure 3(a) shows the fringes with inclination angle α_0 =0.917° for the test surface at arbitrary tilt, and Fig.3(b)-(e) show the fringes observed after the slight tilt variation of the test surface. Using Eq.(6), the small variations of the tilt angle of the test surface are calculated. The results are shown in Table 1. For example, the moire fringes between the photograph (c) and (d), the small variation of the tilt angle of the test surface is given by $\Delta\delta_3$ =0.67°. In this case the shift amount of the inclination angle of moire fringes was $\Delta\alpha_3$ =8.23°, which is about 12 times as much as the small tilt angle variation. Figure 4 shows the theoretical curves giving the relation between the inclination angle α of the moire fringe and the tilt angle δ of the test surface. The appropriate region for measuring variation of tilt angles of the test surface can be found by these curves. For example, for a crossed angle θ =1°, when the tilt angle of the test surface is 40° or 50°, this system is most sensitive to the tilt angle.

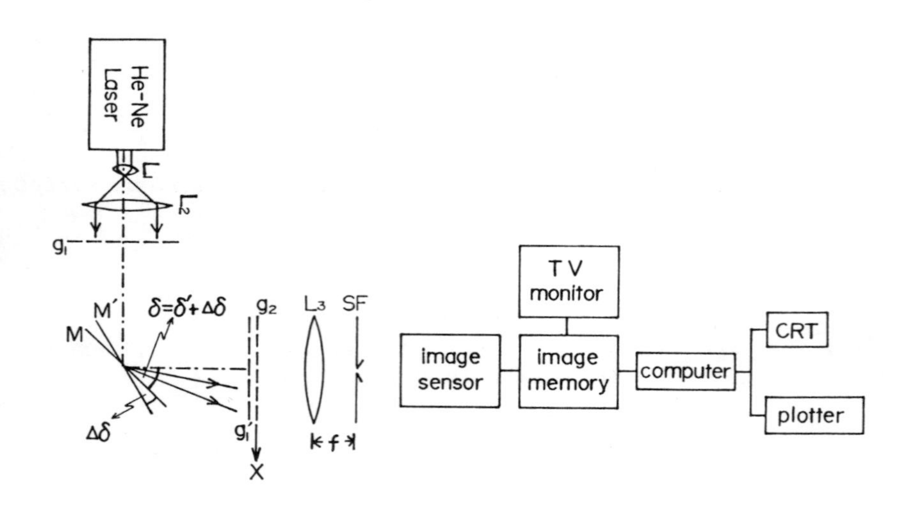

Figure 2. Block diagram of the experimental set up.

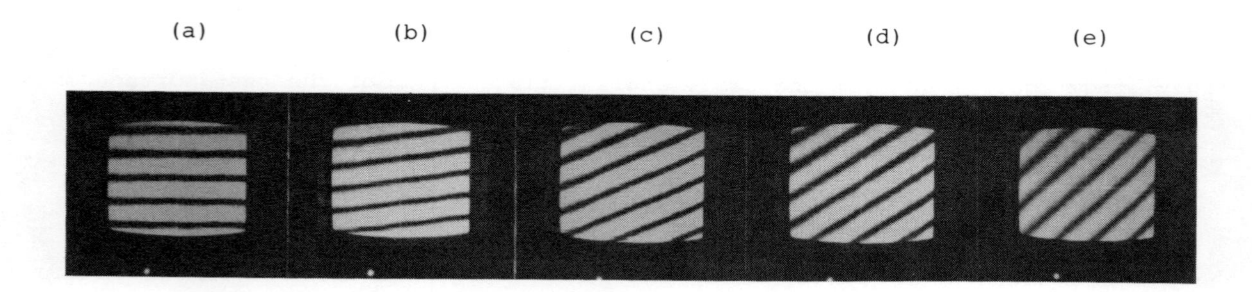

Figure 3. Moire fringe pattern observed through the TV monitor.
(a), α_0=0.917° ; (b), α_1=7.25° ; (c), α_2=23.51° ;
(d), α_3=31.74° ; (e), α_4=42.99°.

Table 1.

	moire fringe inclination angle α		test surface tilt angle variation
a	0.917°		
		$\Delta\alpha_1$ = 6.33°	$\Delta\delta_1$ =1.13°
b	7.25°		
		$\Delta\alpha_2$ =15.26°	$\Delta\delta_2$ =1.59°
c	23.51°		
		$\Delta\alpha_3$ = 8.23°	$\Delta\delta_3$ =0.67°
d	31.74°		
		$\Delta\alpha_4$ =11.25°	$\Delta\delta_4$ =0.95°
e	42.99°		

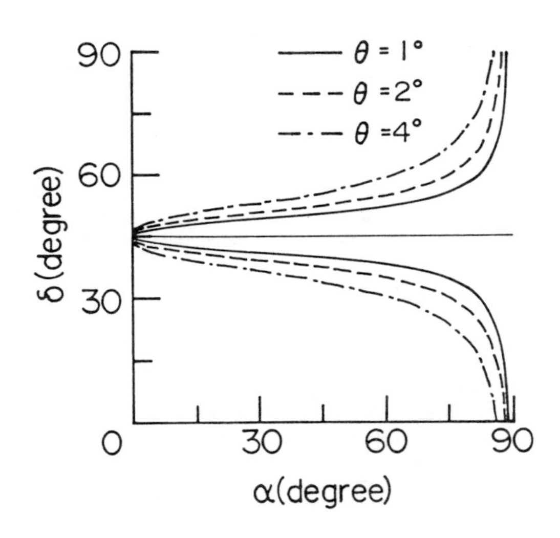

Figure 4. Relation between the tilt angle δ of the test surface and the inclination angle α of the moire fringes.

Conclusion

It has been shown that holographic moire interferometry can be applied for measuring the small variation of the tilt angle of the test surface. Furthermore, by using computer image processing, this method enabled automatic measurements of small tilt angle variation. This method has the following merits:
(1) The optical system for the measurement is simple. A precise optical arrangement is not necessary.
(2) As the second grating is placed in the Talbot image plane of the first grating, the moire fringe patterns are good contrast.
(3) It is especially effective for measuring a large surface having a local tilt variation.

Acknowledgements

The author would like to thank Prof. Kazumi Murata of Hokkaido University for advice and encouragement, and Miss Ryoko Ishioka for valuable assistance.

References

1. Rastogi, P.K., "Comparative holographic moire interferometry in real time," Appl. Opt., Vol.23, pp.924-927.1984.
2. Nakano, Y. and Murata, K., "Talbot interferometry for measuring the small tilt angle variation of an object surface," Appl. Opt., Vol.25, pp.2475-2477.1986.
3. Lohmann, A.W. and Silva, D.E., "An interferometry based on the Talbot effect," Opt. Commun., Vol.2, pp.413-415.1971.
4. Yokozeki, S. and Suzuki, T., "Shearing interferometer using the grating as the beam splitter," Appl. Opt., Vol.10, pp.1575-1580.1971.
5. Jones, R.V. and Richards, J.C.S., "Recording optical lever," J. Sci. Instrum., Vol.36, pp.90-94.1959.
6. Takeda, M. and Kobayashi, S. "Lateral aberration measurements with a digital Talbot interferometer," Appl. opt., Vol.23, pp.1760-1764.1984.
7. Yatagai, T., "Fringe scanning Ronchi test for aspherical surface," Appl. Opt., Vol.23, pp.3676-3679.1984.
8. Nakano, Y. and Murata, K., "Measurements of phase objects using the Talbot effect and moire techniques," Appl. Opt., Vol.23, pp.2296-2299.1984.
9. Nakano, Y. and Murata, K., "Talbot interferometry for measuring the focal length of a lens," Appl. Opt., Vol.24, pp.3162-3166.1985.

A Shearing - holographic - moire for strain patterns

Jing Fang Fulong Dai

Department of Engineering Mechanics, Tsinghua University, Beijing, China

Abstract

A new holographic-moire is presented. The object is illuminated by double beams separately so that two different carriers are introduced. Pure strain patterns are obtained by filtering twice and shearing the plate.

Introduction

It is well known that the holographic interferometry is a useful tool for the determination of the surface displacement of a deformed object. Illuminating with a beam and recording with a holographic plate, one can obtain the component of the displacement vector in the bisector of the angle between the illuminating direction and the viewing direction. Therefore, it is easy to acquire the patterns of the out-of-plane displacements but no possibility to achieve the ones of the in-plane with one beam and one plate. Illuminating with double beams which are symmetric about the normal of the object surface, Boone[1] obtained the moire fringes of the in-plane displacements. However, it is difficult to distinguish those moire fringes from 'the grating lines' because the density of the patterns produced by each beam is low. To solve this problem, Hung and Taylor[2], Sciammarella and Gilbert[3] used different auxiliary system to introduce carriers and achieved the patterns of the in-plane displacements by filtering. Since then, holographic-moire was further developed. Sciammarella and Chawla[4], for instance, introduced image holography to this method so that the fringes are localized in the plane of object. Gilber, Sciammarella and Chawla[5] extended the technique to the object with curved surface. Shearing the object beam and introducing the carriers with a glass edge in the reference beam, Zhong and Dai[6] achieved the patterns of derivatives of the out-of-plane displacements. However, up to now, there is no method in holographic interferometry to achieve directly the in-plane strain patterns of three dimension displacement field. Shearing two same patterns of in-plane displacement, Sciammarella[7] acquired the moire of moire for strain patterns. The same as the superposition of two patterns without carriers, the fringes of derivatives are discernable especially when the subtractive fringes are about parallel to the original fringes. Introducing two carriers of different directions, the authors obtain the in-plane strain patterns of three dimension displacement field. By filtering twice and shearing the plate, not only can the out-of-plane displacements be concealed, but the patterns of the in-plane displacements are eliminated as well.

Basic principle

To eliminate the out-of-plane information of a deformed object, symmetric double beams illumination is essential for the measuring of the in-plane components. (see Figure 1). When the object is unload, object beams O_1 and O_2 produced by beam 1 and beam 2 respectively are recorded simultaneously with reference beam R, and the intensity distribution can be expressed as

$$I_1 = \left| e^{i\phi_1(x,y)} + e^{i\phi_2(x,y)} + e^{i2\pi f_0 x} \right|^2 \tag{1}$$

where $f_0 = \sin\theta_x/\lambda$, which is the spatial frequence of plane wave R. $\phi_1(x, y)$ and $\phi_2(x,y)$ are respectively random phases of each object beam and they are independent each other. After the object is loaded, only beam 1 is used to illuminate and the second exposure is made with R_1 which is in the direction of $\theta_x+\Delta\theta_x$, the intensity is

$$I_2 = \left| e^{i\phi_1'(x, y)} + e^{i2\pi f_1 x} \right|^2 \tag{2}$$

where $f_1 = \sin(\theta_x+\Delta\theta_x)/\lambda$. In the same way, changing the direction of the reference beam to R_2 which is at an angle of $\Delta\theta_y$ with R_1, the third exposure is made by O_2' and R_2 when the object is illuminated only by beam 2, and, the intensity is

$$I_3 = \left| e^{i\phi_2'(x, y)} + e^{i2\pi(f_1 x + f_2 y)} \right|^2 \tag{3}$$

where $f_2 = \sin(\Delta\theta_y)/\lambda$. After the holographic plate with total intensity $I = I_1+I_2+I_3$ is developed, it is reconstructed by reference beam R. The intensity of the object wavefront can be written as

$$I_t = I_o + K\{\cos 2\pi[\alpha_1(x, y) + f_{10}x] + \cos 2\pi[\alpha_2(x, y) + f_{10}x + f_2 y]\} \quad (4)$$

where $\alpha_1(x, y) = [\phi_1(x, y) - \phi_1'(x, y)]/2\pi \qquad \alpha_2(x,y) = [\phi_2(x,y) - \phi_2'(x,y)]/2\pi$

$f_{10} = f_1 - f_o \doteq (\Delta\theta_x)\cos\theta/\lambda \qquad f_2 = \sin(\Delta\theta_y)/\lambda \doteq (\Delta\theta_y)/\lambda$

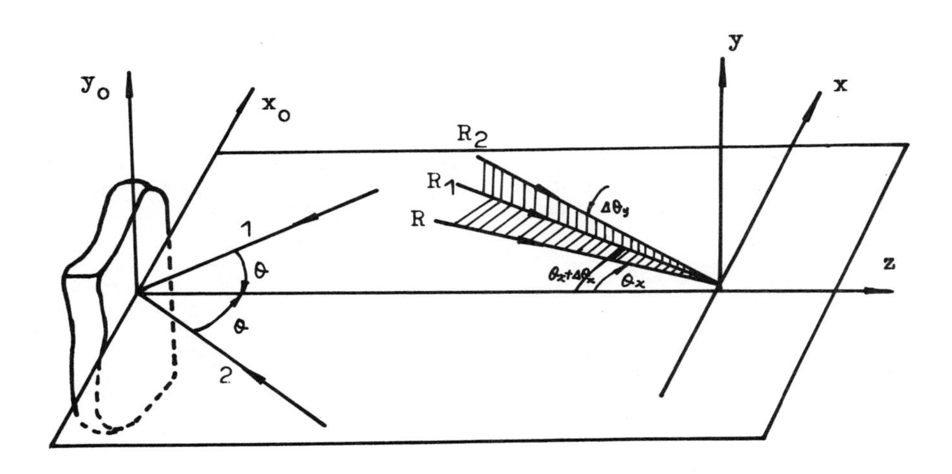

Figure 1, schematic representation of the recording system for three exposures

After recording this intensity distribution with a plate, plate 1, for example, it is processed in a filtering system. (see Figure 2). When the plate 1 is illuminated with a coherent source, a Fourier spectrum of five bright points appears in the back focal plane of the image lens, and, the complex amplitude can be expressed as

$$U(f_x,f_y) = U_o\delta(f_x, f_y) + U_1\{\delta[f_x-(f_{10}+\alpha_1/x), f_y] + \delta[f_x+(f_{10}+\alpha_1/x), f_y] + \delta[f_x-(f_{10}+\alpha_2/x), f_y-f_2] + \delta[f_x+(f_{10}+\alpha_2/x), f_y+f_2]\} \quad (5)$$

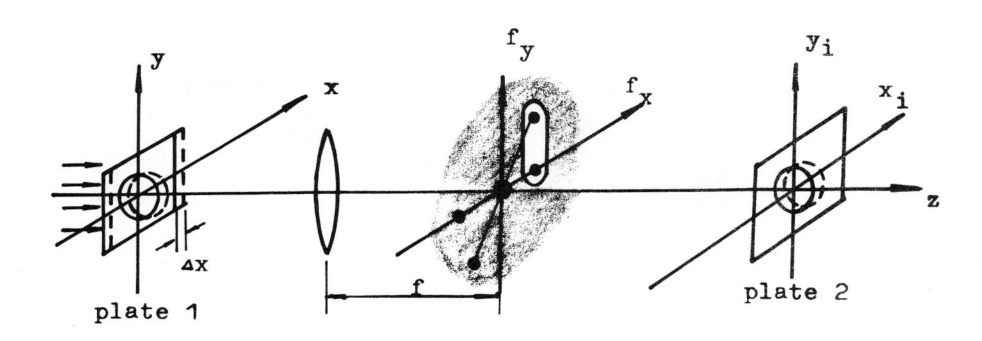

Figure 2. Schematic representation of filtering system

When only two +1 order diffractive waves are allowed to pass through the transform plane, and suppose the magnification is unit, the intensity distribution in the image plane is

$$I_{i1}'(x',y') = K'\left[1+\cos\frac{\alpha_2(x',y')-\alpha_1(x',y')+f_2}{2}\right] \quad (6)$$

After this intensity is recorded by plate 2 which is in the image plane, the plate 1 is moved laterally a small distance, x, for instance, in the x direction. Two +1 order diffractive waves are also recorded by the plate 2

$$I'_{i2}(x',y')=K'\left[1+\cos\frac{\alpha_2(x'+\Delta x',y')-\alpha_1(x'+\Delta x',y')+f_2}{2}\right] \tag{7}$$

and the total intensity is

$$I'_i(x',y')=I'_{i1}+I'_{i2}=2K'\left(1+\cos\frac{\omega\alpha_2-\omega\alpha_1+2f_2}{2}\cos\frac{\Delta\alpha_2-\Delta\alpha_1}{2}\right) \tag{8}$$

where
$$\omega\alpha_2=\alpha_2(x'+\Delta x',y')+\alpha_2(x',y') \qquad \omega\alpha_1=\alpha_1(x'+\Delta x',y')+\alpha_1(x',y')$$
$$\Delta\alpha_2=\alpha_2(x'+\Delta x',y')-\alpha_2(x',y') \qquad \Delta\alpha_1=\alpha_1(x'+\Delta x',y')-\alpha_1(x',y)$$

The function $I'_i(x',y')$ shows that the first cosine term is of higher spacial frequency because the existance of f_2. According to the analysis of optical path before and after the deformation of the object, $\Delta\alpha_2-\Delta\alpha_1$ is related to the in-plane strain component $\partial u/\partial x$ with

$$\Delta\alpha_2-\Delta\alpha_1=\frac{4\pi\sin\theta}{\lambda}\cdot\frac{\partial u}{\partial x}\cdot\Delta x$$

So that, the last term of the cosines is what we are interested. Replacing the plate 1 with plate 2, the second filtering is made. When only +1 order wave is allowed to pass through the back focal plane, the intensity distribution in the image plane is

$$I''_i(x',y')=K''\cos\frac{\Delta\alpha_2-\Delta\alpha_1}{2} \tag{9}$$

and $I''_i=I_{max}$ when $\Delta\alpha_2-\Delta\alpha_1=2n\pi$, or

$$\varepsilon_x=\frac{\partial u}{\partial x}=\frac{n\lambda}{2\sin\theta} \qquad n=0,\pm1,\pm2\ \ldots\ldots, \tag{10}$$

That means, the isoplathes of strain can be observed as the bright fringes. If the plate 1 is translated Δy in y direction, the derivative of displacement $\partial u/\partial y$ will be obtained. In the same way, $\partial v/\partial x$ or $\partial v/\partial y$ can also be achieved by illuminating the object in yoz plane and recording, filtering and shearing as foregoing.

Experimental

The experimental arrangement is shown in Figure 3. Two switches S_2 and S_1 are placed respectively in the optical axies of beam 2 and beam 1, they are closed sequently in the second and the third exposure after the object is loaded. A plane mirror M_r which can rotate about X axis and Y axis is used to change the direction of reference beam, by which, the carriers for purpose are introduced during the recording procedure. The information recorded is processed in the filtering system shown in Fig. 2.

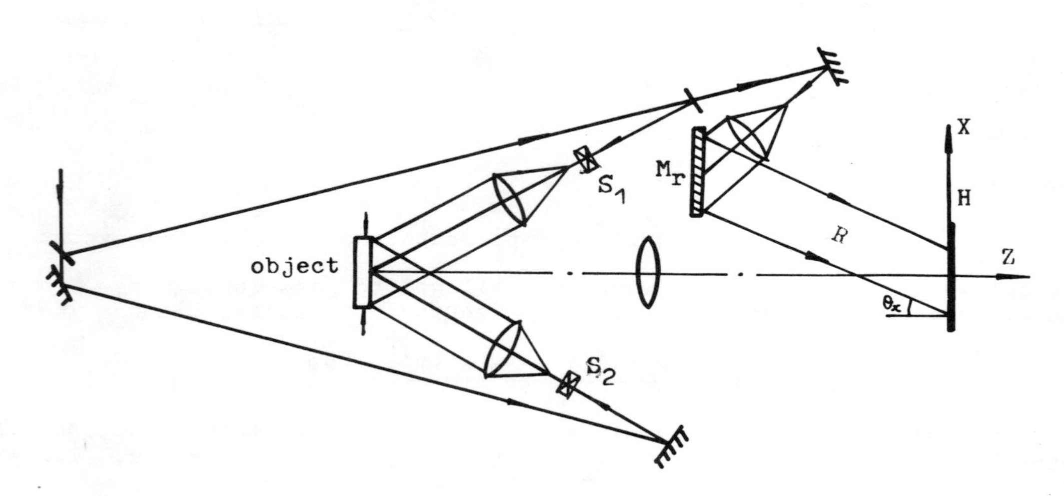

Figure 3. Optical arrangement of recording system

The patterns of derivatives of a disk subjected to the diametral compression are shown in Figure 4. The left one is the pattern of $\partial u/\partial x$ and the right one is the pattern of $\partial u/\partial y$.

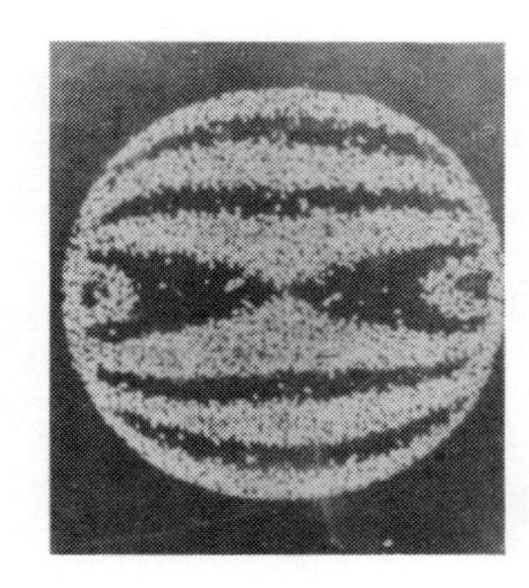
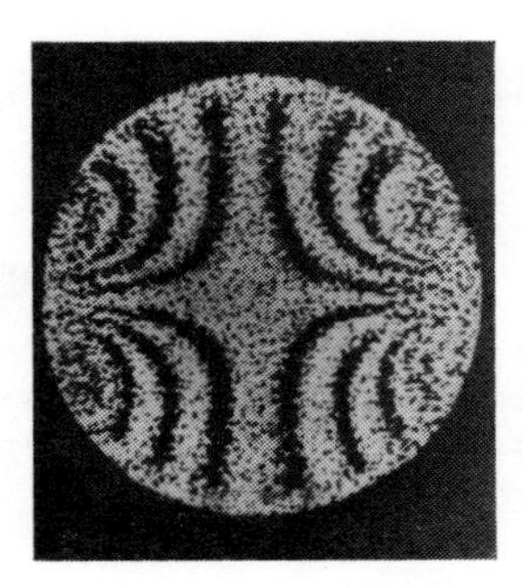

Figure 4. The patterns of derivatives of the in-plane displacements

Conclusion

Holographic-moire is a very useful tool for the solution of the displacements. Developing this method, new technique presented in this paper obtains pure and clear derivatives patterns of the displacements. A procedure of recording with double exposures or in real time is being studied by the authors, and the paper about it will be published soon.

References

1. P.M. Boone, Holographic Determination of In-Plane Deformation, Optics Technology 2, May, 1970.
2. Y.Y. Hung and C.E. Taylor Measurement of Surface-Displacement Normal to Line of Sight by Holo-Moire Interferometry, Journal of Applied Mechanics, 42, 1, 1975.
3. C.A. Sciammarella and J.A. Gilbert, A Holographic-Moire Technique to obtain Separate Patterns for Components of Displacement, Experimental Mechanics 16, 6, 1976
4. C.A. Sciammarella and S.K. Chawla, A Lens Holographic-Moire Technique to obtain Components of Displacements and Derivatives, Experimental Mechanics 18, 10, 1978.
5. J.A. Gilbert, C.A. Sciammarella and S.K. Chawla, Extension to Three Dimensions of Holographic-Moire Technique to Separate Patterns Corresponding to Components of Displacement: Experimental Mechanics, 18, 9, 1978.
6. G.C. Zhong, F.L. Dai, Shearing Holographic-moire, Journal of Experimental Mechanics (in Chinese) 1, 1, 1985.
7. C.A. Sciammarella, Holographic-moire, Optical Methods in Solid Mechanics.

Phase-only Fourier hologram as an optical matched spatial filter

K.Chałasińska-Macukow and T.Nitka

Institute of Geophysics, Warsaw University
Pasteura 7, 02-093 Warsaw, POLAND

Abstract

Phase-only Fourier hologram is recorded optically and applied as matched phase-only spatial filter to the discrimination of characters. The comparison between this filter and the overexposed classical matched spatial filter is presented.

Introduction

The relative role played by phase and amplitude information in the Fourier domain in preserving the accuracy of the restoring image was discussed in many recent papers (see Ref.1). The general result of this discussion in the image processing field is that the phase information is more important than the amplitude information. "Is the same true in the case of a matched filter?" - the question has posed by Horner and Gianino. They presented the computer simulation results and the comparison of the phase-only and the amplitude only filter with classical matched spatial filter.[2,3,4] They also have compared the autocorrelation signal in the presence of noise. The correlation signal and the discrimination capability of the phase-only filter was proved considerably better. They have shown that phase-only matched filtering can provide a higher optical efficiency and a narrower correlation spot compared with classical matched filtering. But the classical matched filter has a higher SNR in the output plane.

Our goal was an optical realisation of the phase-only filter and its application to the discrimination of characters in the optical correlator. In this paper we present qualitative comparison of the discrimination capability of the classical, but overexposed, matched spatial filter with phase-only spatial filter. For the comparison we have chosen overexposed matched spatial filter (i.e. with the saturation of the central part), because of the nonlinearity of materials it is the easiest version for the optical realisation.

Experiment and results

We define the phase-only filter $F_\varphi(u,v)$ as

$$F_\varphi(u,v) = F^*(u,v)\ |F(u,v)|^{-1} = \exp\ |-i\ \varphi(u,v)| \tag{1}$$

where $F(u,v) = |F(u,v)| \exp[+i\,\varphi(u,v)] = \quad f(x,y)$ is the Fourier transform of the object $f(x,y)$. In general case, the Fourier transform is complex with two terms: the amplitude $|F(u,v)|$ and phase $\exp[-i\,\varphi(u,v)]$. $|F(u,v)|^{-1}$ is the inverse of the amplitude and symbol * indicates complex conjugate.

In our experiment the character "A" was chosen as the object. The phase-only Fourier hologram of the character "A" was recorded using optical set-up presented in Fig. 1.[5] This is a typical set-up for the optical realisation of the phase component of the Stroke inverse filter. The filter with the amplitude component $|F(u,v)|^{-1}$ inserted in the first Fourier transform plane was obtained by recording the intensity of the Fourier transform function on the photographic plate Agfa Gevaert 10E75 which was processed with $\gamma = 1$.

The object wave (Fig. 1) passes from the object through the filter proportional to $|F(u,v)|^{-1}$ and in the back focal plane we obtain only the phase term of the Fourier transform of the object i.e. $\exp[+i\,\varphi(u,v)]$. In this plane one recordes the phase-only Fourier hologram adding the plane reference wave $R(u,v)$. The amplitude transmittance $t(u,v)$ of such a hologram is

$$t(u,v) \sim |\exp\ [+i\,\varphi(u,v)] + R(u,v)|^2 = I_0 + R^*(u,v)\ \exp\ [i\,\varphi(u,v)] + \\ + R(u,v)\ \exp\ [-i\,\varphi(u,v)] \tag{2}$$

where I_0 is constant and the last term plays the role of the phase-only filter $F_\varphi(u,v)$.

The overexposed matched spatial filter of the character "A" was recorded as a classical Fourier hologram. Both filters have amplitude recording of the information. They are not bleached.

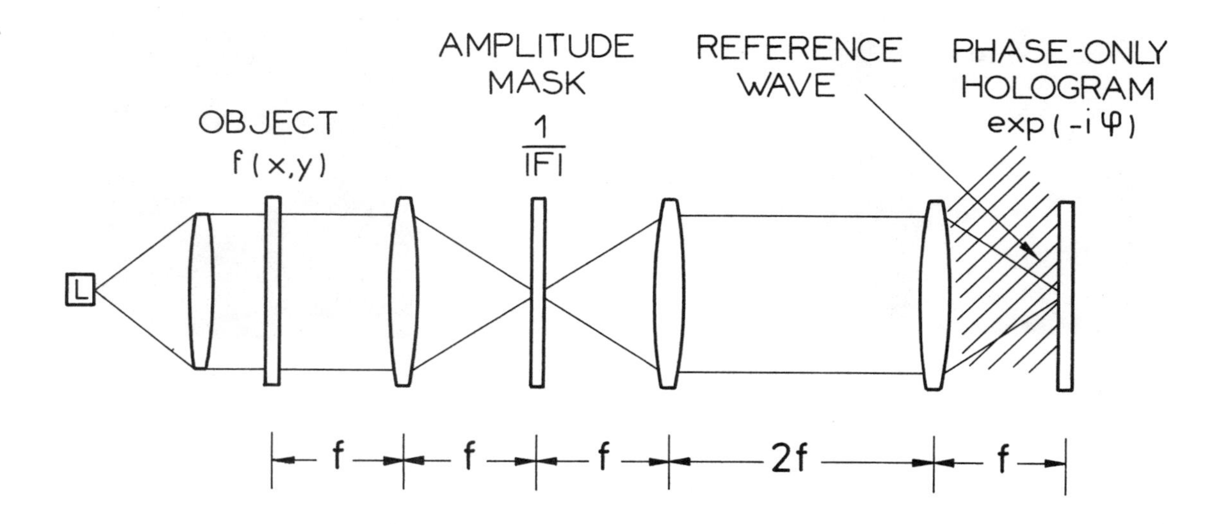

Figure 1. Optical set-up for phase-only filter recording

Figure 2 presents a negative of the overexposed Fourier hologram. The central part of hologram cut off the low frequencies because of the saturation. Figure 3 presents a negative of the phase-only hologram. The distribution of the intensity is more uniforme than in the first one. The amplitude information of the Fourier transform is lost. This is amplitude recording of the phase-only information.

Fig. 2. Negative of the overexposed Fourier hologram

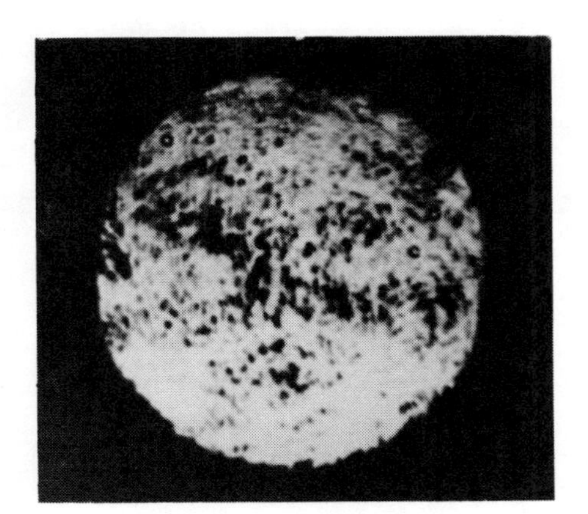

Fig. 3. Negative of the phase-only hologram

Figures 4 and 5 present the impulse response of the overexposed Fourier hologram and of the phase-only hologram respectively. Both holograms preserve the visual intelligibility of the object. The band-pass filter characteristics of both filters are produced by the amplitude term $|F(u,v)|^{-1}$ for the phase-only filter and by the saturation of the central part for the overexposed filter. The aperture of both holograms are limited by the same manner.

Fig. 4. Impulse response of the overexposed Fourier hologram

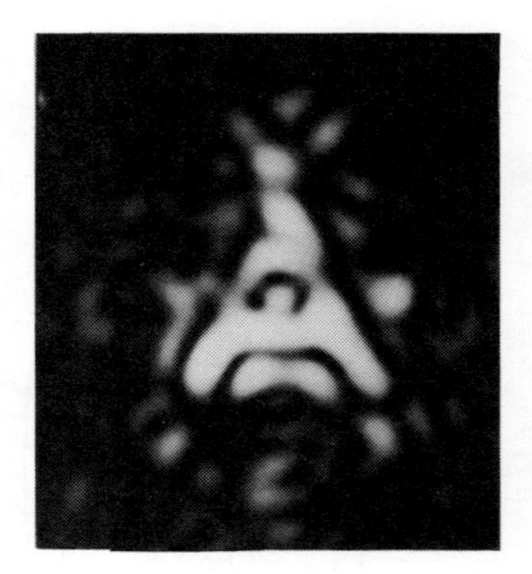

Fig. 5. Impulse response of the phase-only hologram

We used classical optical correlator set-up to discriminate the character "A". The object in the input plane of the correlator was the 3x3 matrix with different characters (Fig. 6).

Figure 7 presents result of the correlation between the input object and the classical overexposed matched spatial filter. Figure 8 presents the similar result but for the phase-only filter. In both cases the character "A" was correctly recognized.

The results presented in this paper are only qualitative. The capability to discrimination of both filters proved to be almost identical.

References

1. Oppenheim, A.V., Lim, J.S., Proc. IEEE 69, 529 (1981).
2. Horner, J.L., Gianino, P.D., Appl. Opt. 23, 812 (1984).
3. Gianino, P.D., Horner, J.L., Opt. Eng. 23, 695 (1984).
4. Horner, J.L., Gianino, P.D., Appl. Opt. 24, 851 (1985).
5. Zetzche, Ch., Appl. Opt. 21, 1077 (1982).

This research was carried on under research project CPBP 01.06.

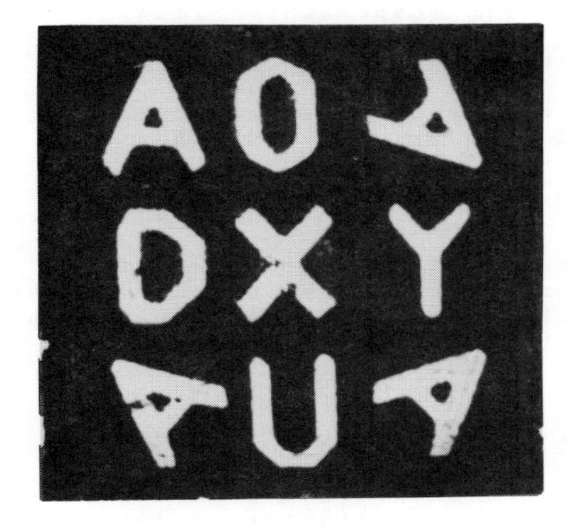

Fig. 6. Input image containing nine characters

Fig. 7. Recognition of the character "A" by the overexposed matched spatial filter

Fig. 8. Recognition of the character "A" by the phase-only filter

Holographic Interferometry Using Digital Image Processing Technique for the Measurement of 3-D Axisymmetric Temperature Field

Han Lei, He Shi-ping and Wu Xiao-ping

Modern Mechanics Department, University of Science and Technology of China, Hefei, Anhui 230029, P. R. CHINA

Abstract

Holographic interferometry combined with the digital image processing technique was firstly used in whole field thermometry. Some discussions on the theoretical basis and operational steps of image processing were given in detail. Using M-75 image system, we easily obtained the 3-D temperature distribution of an alcohol flame within a few minutes.

Introduction

The holographic interferometry is a new optical method for nondestructive testing which has high accuracy and precision. One of its advantages is that it can compensate the faults of the optical devices. Employing it to measure the "phase object"--- a transparent medium with nonhomogeneous refractive index, we could obtain the whole-field refractive index distribution from a double-exposure infinite holographic fringe pattern. For the axisymmetric "object" heated by a small flame, the resulting temperature field may be analyzed and determined according to the state equation of the mixed gases.

However, manual processing and calculating of optical data is a difficult job because this work is time-consuming and easy to fall into error, especially where the physical hypotheses are not same. It is noted that the repeatability of different cross sections is dispensable when we deal with the differences of distinct theoretical analyses. The user is not only faced with the problem of how to convert the optically recorded information to usable engineering data, but also the more detailed analyses.

In recent years, digital image processing techniques have evolved in many fields. We used the image processing combined with the holograhpic interferometry firstly to measure the whole-field temperature distribution of an alcohol flame. For the experiment of this kind, we discussed the theoretical basis of the optical method and the operational steps of image processing. The methods discribed in this paper were found to be more convenient and flexible.

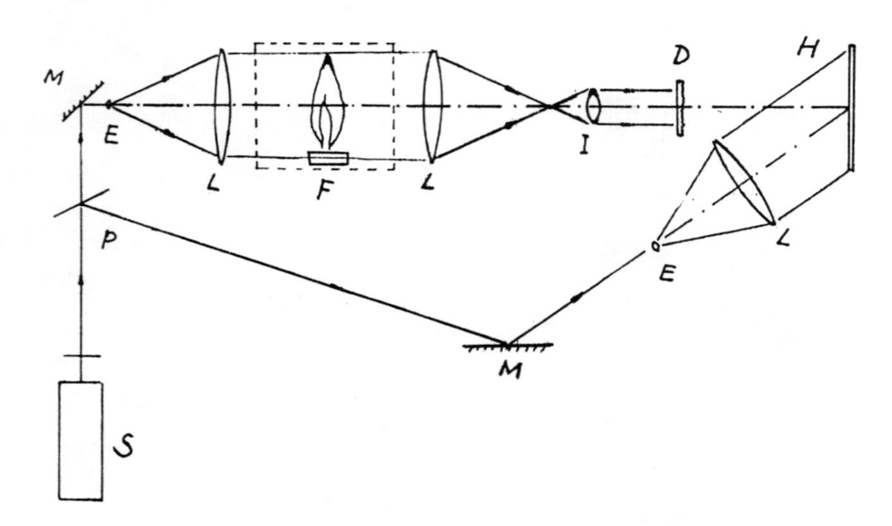

Figure 1.

Holography and Interferometry with Axisymmetric Flame

The thermometric arrangement we used is shown in Figure 1. The system consists of a schlieren device and a reference beam. In addition to being used in holographic interferometry, the system can also be used to measure the temperature gradient directly. Where F-flame, S-laser, B-beamsplitter, L-lenses, H-hologram, D-diffuser, M-mirror, E-beam expander, P-partial mirror, I-imaging lens.

Within the phase object the refractive index distribution is $n0(x,y,z)$ during the initial holographic exposure and $n(x,y,z)$ during the second. The optical pathlengthes of the rays through the medium are:

$$Q1 = \int n0(x,y,z)\, ds \qquad Q2 = \int n(x,y,z)\, ds \tag{1}$$

which are the path integrals of the refracting index along the light ray. When refraction is negligible, rays remain straight lines, and the path integrals become line integrals. If the ray is parallel to the Z axis, we simply replace ds by dz in equation (1):

$$Q1 = \int n0(x,y,z)\, dz \qquad Q2 = \int n(x,y,z)\, dz \tag{2}$$

Here, the refractive errors of the light ray in radially symmetric refractive index distributions have been studied, and found by computer simulation that the refraction errors are generally quite small if the focusing on the center plane containing the axis of symmetry. This is why so arranged as in Figure 1.

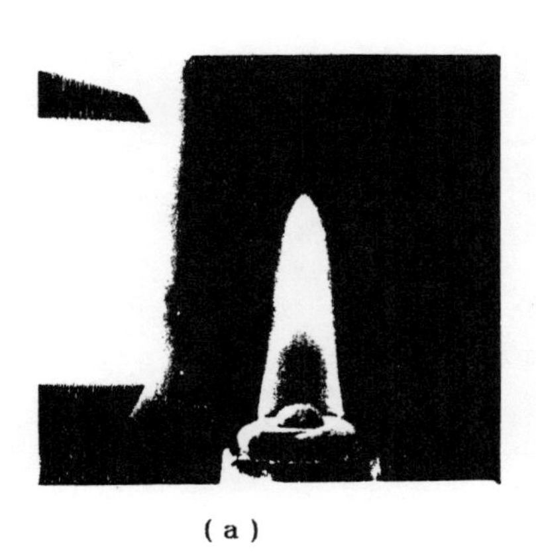

(a)

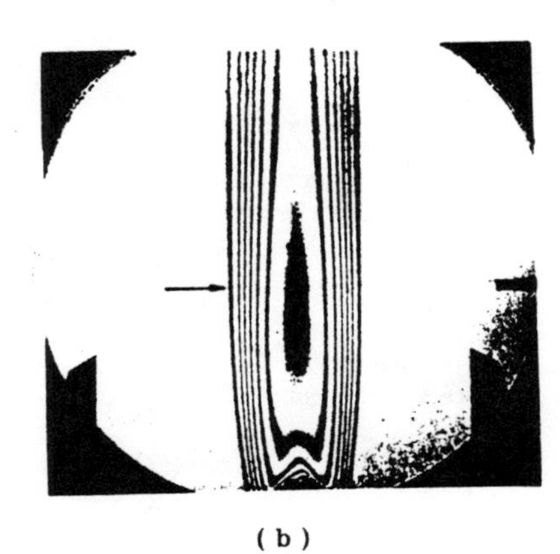

(b)

Figure 2. (a)Flame (b)Infinite Fringe Pattern

Thus, when the hologram is developed and reilluminated by the reference wave, two waves are reconstructed simultaneously. Assuming for simplicity that two waves are uniform and unit amplitudes, we can write this irradiance as:

$$I(x,y) = 2\left(1 + \mathrm{Cos}\frac{2\pi}{\lambda}(Q1(x,y) - Q2(x,y))\right) \tag{3}$$

In our case, the refractive index during the first exposure is uniform and can be denoted by constant $n0$. Then the fringe pattern is

$$I(x,y) = 2\left(1 + \mathrm{Cos}\left(\frac{2\pi}{\lambda}\Delta Q(x,y)\right)\right) \tag{4}$$

so we can define

$$\Delta Q(x,y) = \int (n(x,y) - n0)\, dz = N(x,y)\lambda \tag{5}$$

where

N--number of bright fringe

λ-- wave length of He-Ne laser.

A cross section of the symmetric phase object n(r,y) is shown in Figure 3. The probing plane wave is travelling in the Z direction, as indicated by the typical ray shown in Figure 1. With $dz=(r/\sqrt{r^2-x^2})dr$, the pathlength difference $\Delta Q(x,y)$ for a two-exposure hologram is found

$$\Delta Q(x,y) = 2\int_x^R \frac{n(r,y)-n_0}{\sqrt{r^2-x^2}}\, r\, dr \tag{6}$$

where n(r,y)-n0 = f(r,y). The temperature of flame decay smoothly to room temperature at large radius and has no discontinuities; it is then convenient to write equation (5) and (6) as

$$N(x,y)\lambda = 2\int_x^\infty \frac{1}{\sqrt{r^2-x^2}}\, f(r,y) r\, dr \tag{7}$$

The right-hand side of equation (7) is the Abel transform of f(r,y). The inverse transform is

$$n(r,y)-n0 = f(r,y) = -\frac{\lambda}{\pi}\int_r^\infty \frac{1}{\sqrt{x^2-r^2}}(dN/dx)\, dx \tag{8}$$

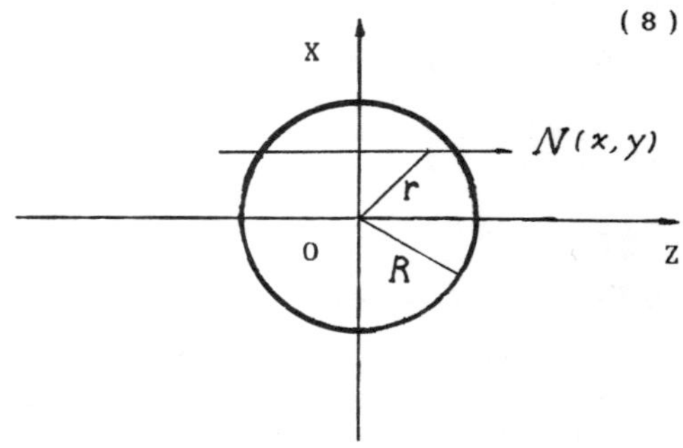

Figure 3.

However, there is a sign ambignity in such interferogram because $+\Delta Q(x,y)$ and $-\Delta Q(x,y)$ yield the same fringe pattern. We made an optical compensator including a photo-elastic plane slab whose thickness and phase shift could be changed by tension loading. Using the compensator, it is easy to know that N is lower and lower when r approches zero. We found that this method has greater precision than the reference fringe method.

Specific volume, denoted by v, is related to the refractive index n of the gas by Gladstone-Dale equation:

$$n-1=K/v \tag{9}$$

where K, the G-D constant, is a property of the gas. It is a weak function of the wave length of light and is nearly independent of temperature and pressure. So we can get temperature T using the ideal gas equation of state:

$$Pv = (RT)/M \tag{10}$$

where P(Pa) is the pressure, R=8.3143 J/mol.K is the universal gas constant, M is the molecular weight of the gas, and T(°K) is the absolute temperature.

In the outer border of flame, M may be the molecular weight of air; in the inner region of flame, this problem become rather complexible. The values of K and M in equation (9) and (10) depend on the spatial distribution of the concentration of each component presenting in the flame. In view of the fact that both the concentration distribution and the temperature distribution are unknown, the actual parameters of a flame are often replaced by that considering the influence of concentration distribution on temperature measurement, more effective method is proposed. For example, in the section A-A, the flame centre (r=0) is in full burning state (at this point, (MK) may be obtained by the condition of full burning), and in the edge of the flame (r=R) is in the air (at this point, (MK) may obtained by air parameters). The (MK) in any point of this section is approximated by linear interpolation. Using this method, the measuring accuracy for temperature may be increased about 6% at the section centre.

Preprocessing of Digital Optical Pattern

We employed the Model-75 Image Processing System finishing this work. The outline of the overall procedure is illustrated in Figure 4. After the optical pattern inputted by TV camera and digitized as a 512*512*8 bits image, the histogram specification of the original image was performed. It is needed not only for the clear image, but also helpful to following operation, that is, thresholding. If the threshold value compared to the mean value of the entire image is too high, then some fringes become indistinguishable from the background. On other hand, if the threshold is too low, then some fringes may get connected to other hearby ones. We found that a satisfactory threshold could be computed by picking the deepest concavity point on the gray level histogram as the threshold value.

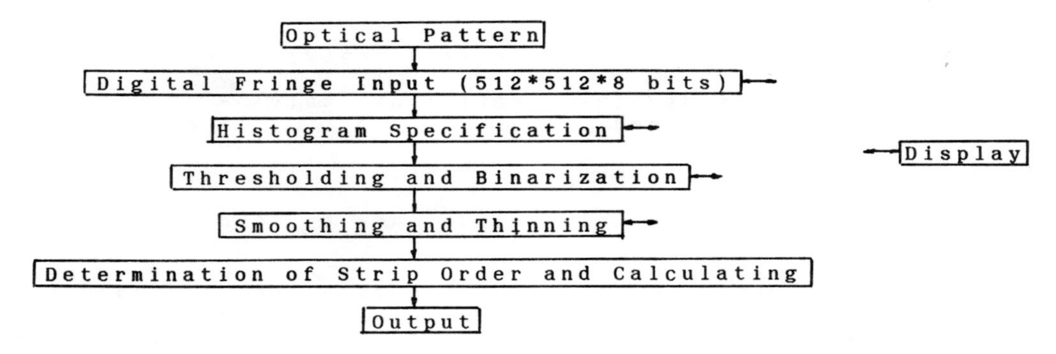

Figure 4.

Thinning reduces fringes region into line-drawing (skeleton) which can then be treated as a graph. This technique is applicable to linelike objects or something where human perception of a shape is in terms of lines. The connecticity of results being desired, our algorithm takes the following steps:

1. NEWSET=R, OLDSET=0
2. IF(NEWSET=OLDSET) STOP
3. OLDSET=NEWSET
4. NEWSET = C(OLDSET) U (OLDSET-IB(OLDSET)) U (OB(C(OLDSET))∩OLDSET)
5. GOTO 2

Where R is the plane region of fringe, IB(P) is the inside boundary of point set P, OB(P) is the outside boundary of P, C(P) denote the region which is 'thin' enough so that no more thinning operation can be performed on it. "U" and "∩" are logical operative symbols "OR" and "AND".

In IB(P) and OB(P) processing, we used Nadler's peephole mask detecting the edges to avoid the speckle noise. They are:

$$M1 = \begin{matrix} + & + & + \\ 0 & 0 & 0 \\ - & - & - \end{matrix} \qquad M2 = \begin{matrix} + & 0 & - \\ + & 0 & - \\ + & 0 & - \end{matrix} \qquad M3 = \begin{matrix} 0 & - & - \\ + & 0 & - \\ + & + & - \end{matrix} \qquad M4 = \begin{matrix} + & + & 0 \\ + & 0 & - \\ 0 & - & - \end{matrix}$$

The procedure is as follows. The center of the template is moved around the image from pixel to pixel. At every position, we multiply every point of the image that is inside the template by the sign or zero indicated in the corresponding entry of the template, and then add the results. Only maximum of their absolute values exceed some thresholding value, we could believe that the central pixel belong to the boundary point set.

The resulting image, containing only fringes and background, is finally smoothed for elimination of discrete noise by 3*3 window operator:

$$\begin{matrix} 1/16 & 1/8 & 1/16 \\ 1/8 & 1/4 & 1/8 \\ 1/16 & 1/8 & 1/16 \end{matrix}$$

Then last binarization should be performed for satisfactory sharpness.

Calculating and Results

After the processings stated above, VAX-11/750 minicomputer automatically recognized the strip orders according to the tendency which had been determined by the optical compensator, and labeled each one. Thus we get the discrete values of N(x,y) and corresponding position x. Considering the symmetric property of flame, we assumed the fringe distribution as an even polynomial function:

$$N(x,y) = a0(y) + a2(y)\,x^2 + a4(y)\,x^4 + a6(y)\,x^6 + a8(y)\,x^8 + a10(y)\,x^{10} \tag{11}$$

It is not difficult to compute the coefficients a0(y) - - - a10(y) by the least squares approximation method. Thus

$$dN/dx = 2a2(y)\,x + 4a4(y)\,x^3 + 6a6(y)\,x^5 + 8a8(y)\,x^7 + 10a10(y)\,x^9 \tag{12}$$

so we could calculate the whole field temperature by equation (8), (9) and (10).

It is noted that the normal equation here is ill-condition and should be carefully treated when using the Gaussian elimination. In addition, proper ratio factor was adopted in computer arithmetic process for satisfactory precision. So all programs of this kind may be solved using the developed FORTRAN or C programs.

The resulting temperature distribution verus radius r at two cross section are shown in Figure 5. Experiment has shown that acceptable subpixel accuracy can be achieved via this method, for the scale 178 pixels/inch. The above curve in Figure 5 shows the temperature in the cross section where the tip of the flame was. The maximum temperature in the ordinary alcohol flame was higher than 1400 -1500 °C, not around 1000 °C (as many scholars had believed). And there was a lower temperature region by incomplete combustion, in the center of the flame.

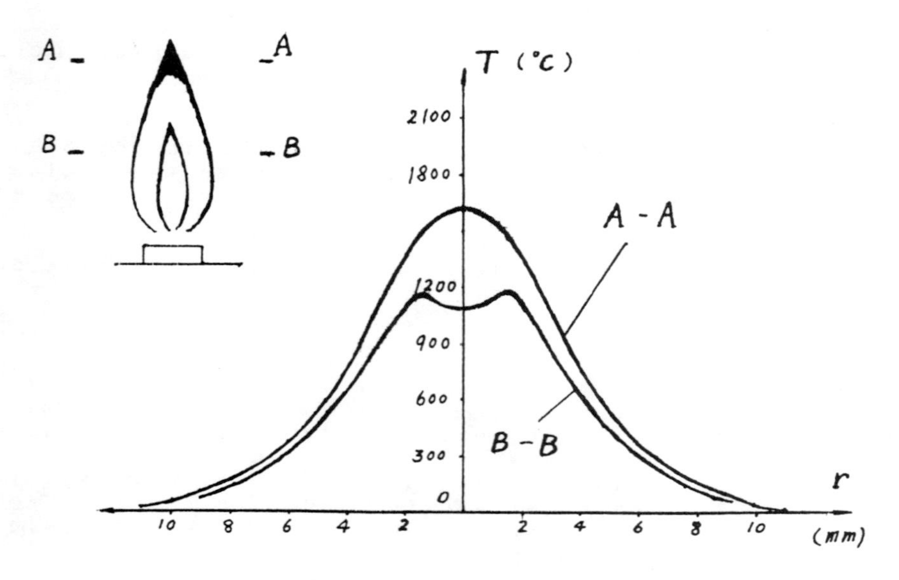

Figure 5.

Conclusions

A new image date acquisition technique combined with holographic interferometry has been presented. It can be employed to provide the full-field visualization of a flame temperature. The optical compensation method is necessary to determine the fringe orders. Experiments have shown that this new technique calculate the results much faster than the ordinary manual processing. The average computation time for a hologram was less than several minutes. Results can be gotten through printer or graph plotter.

References

1. C.M. Vest, *Holographic Interferometry* John Wiley & Sons, New York 1979.

2. He Shi-ping et. al., *J. of Engineering Thermophysics* Vol.3(4), p393. 1984.

MOIRÉ EVALUATION HOLOGRAPHY

Olivério D.D. Soares
Centro de Física, Universidade do Porto, 4000 Porto, Portugal

José F. Fernandez
Applied Physics, E.T.S. Industriales, Universidade de Santiago, Apartado 62, Vigo, Spain

Abstract

The research work on moiré evaluation techniques both in classical and electronic holography, and its practical consequences are reviewed. Advantages of the technique and examples of application are presented.

Introduction

Optical techniques present known advantages, in particular, for metrology and non-destructive testing, of major interest for science, industry and biomedicine, Fig. 1.

Holography a Nobel laureated invention (1971) is regarded as a concept of great technological impact. However, expansion on the use of developed holographic techniques, outside research Laboratory, has been slow and in far less fields and applications than could have been anticipated.

Practical technical difficulties have been indicated as the impediment for a rapid expansion of the use of holography (1) Fig. 2. Some of them can be described in short to illustrate the potentialities of the alternative diversified moiré techniques (2). The inherent high sensitivity shown at the hologram repositioning, and the first order perturbation from uncontrollable environmental parameters fluctuation originates masking effects that make it difficult to use conventional real-time based holographic techniques (3,4), Fig. 3. Moiré evaluation techniques can provide a solution.

Dynamical events require an analysis through different regimes and instants of time. Pulse modulation with convenient synchronization (5) can be combined with moiré evaluation techniques to offer direct data extraction and reducing drastically the number of holograms to be recorded (6).

Holography is usually based on intrinsically long photographic processing resulting on a two step half blind procedure. Moiré evaluation makes it visible, controllable with sensitivity adjustment, and video-recording could be combined with image processing for controlled data extraction (7), being extensive to Fourier analysis as in vibration studies, to quote but an example.

Many other performances can be derived from the use of moiré evaluation techniques and are detailed elsewhere (1-8).

Non-periodic and transient events are a technically difficult topic for holographic techniques. Around 1974, it was asked to Laser manufacturing companies for a triple pulse Laser. Later on such a Laser became available. Meanwhile the concept of moiré evaluation was developed based on electro-optic modulation (amplitude and phase) using CW Lasers. The principle was established, and a variety of techniques analyzed, and later extended to electronic holography or electronic speckle pattern interferometry (4,8).

Principle and Techniques

Holography records the information by a usually complex interference pattern that can be described by its spatial frequency spectrum.

The reconstruction beam decodes this spatial frequency distribution using diffraction and propagation properties of radiation. The entire process relies on spatial frequency spectrum and the metrological codification of the space by the radiation with a dimensional scale related to the radiation wavelength (1).

Moiré concepts are closely related to the various steps of the process so that moiré evaluation techniques may be introduced intentionally with expected advantages.

The foundations of moiré evaluation techniques, Fig. 4 derive from intermodulation of spatial frequency spectra. The carrier spectrum is in principle arbitrary, and represents the metrologically coding of space. It could be the wavefield (as the reference wave in holography), or a transposed spectrum generated intentionally for metrological advantage. Data fringes are extracted by intermodulation and filtering. This gives opportunity to a diversity of the techniques in various methods: change of reference; contouring; differential; integration; derivation in space and time. Previous publications (1-8) were devoted to the matter so that, due to the shortness of space, it will be given here, examples and comments to enlighten the capabilities of the technique.

In time average holography fringe pattern detection at low level of excitation is poor but can be enhanced by the introduction of moiré carrier (carrier frequency fringe shift at every fringe of resonant mode), Fig. 5. Adjustment of carrier frequency (change of reference plane) for optimization of visibility of fringes gives identical results in time-average and stroboscopic mode. Local details can be studies by zooming, and carrier frequency optimization. Electronic filtering would provide a carrier free fringe pattern

OPTICAL TECHNIQUES

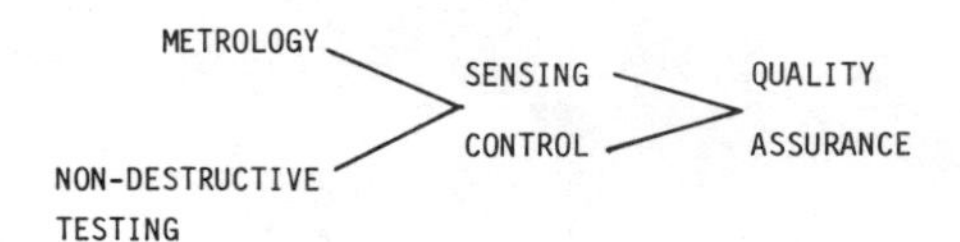

Fig. 1: Expected field of modern industrial application of optical techniques

HOLOGRAPHY

Practical Technical Difficulties

Classical Limitations	Desirable Performances
- Two Steps	- One Step. Real Time
- Photographic Based	- Electronic Based or Self-Processing
- Write-Only Reccording	- Write-Erase Reccording
- High Sensitivity not Adjustable	- Full Adjustment of Sensitivity (globally and locally)
- Masking Effects - mispositioning of hologram - rigid body motion - integrative collective effects - environment fluctuation sensitivity - local transformation	- Direct Data Extraction and Display - large repositioning tolerance - rigid body compensation - adjustable reference for measurement - time differed analysis - insensitivity to local microtransformations
- Coherence and Correlation Range Limitation - Large Amplitudes	- Information Correlation Extended Range - Free Range
- Dynamic Events Limitations - Multitude of holograms - Synchronization Adjustment with Event - Phase Delay Between Exciter and System Response - Frequency Dependence with Phase Delay - Mode Spectrum (amplitude and phase distribution) - Non-Periodic Behaviour (Biomedical random like behaviour) - Transients (fast and slow changes)	- Dynamic Events Full Analysis - Unique Hologram (Rest Position) - Adjustable and Variable Instant of Recording - Controlled Phase Delay Adjustment - Full Frequency Phase Delay Coverage - Mode Spectrum Measurement and Descrimination - Non-Periodic Events Analysis Capabilities - Transients Analysis Capacity

Fig. 2: Sampled examples of classical limitations in holography application for outside research Laboratory employment

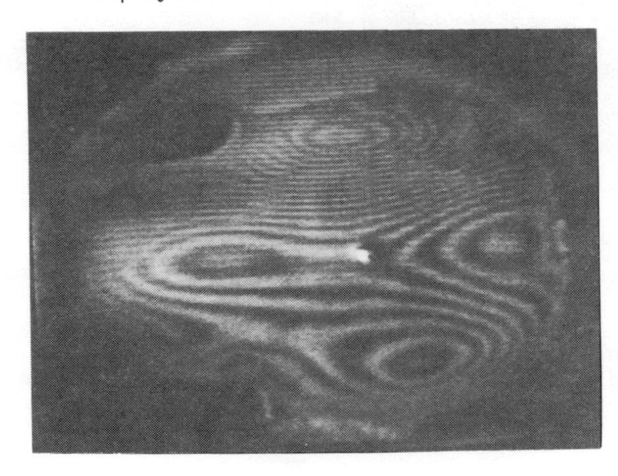

Fig. 3: Environmental and object instabilities in technical objects analysis, and random fluctuactions (biomedical objects) makes it difficult to apply real-time holographic techniques. The back cover of a violin in real-time holography is photographed from video recording showing effect of perturbations by the continuously varying fringe pattern at resting position (9).
Time average observation of resonant modes would require a tight control of perturbations factors and a new hologram recording (see Fig. 12 for clarification)

MOIRE EVALUATION FOUNDATIONS

- The INFORMATION is CODED in the FORM of SPATIAL FREQUENCY SPECTRUM

  Examples: Moiré
  Interferometry (fringes)
  Speckle

- RECORDING of INFORMATION (and TRANSFER) is done by MODULATION with

  CARRIER - HIGH SPATIAL FREQUENCY

  OBJECT or IMAGE } INFORMATION - SPATIAL FREQUENCY SPECTRUM

- DATA EXTRACTION taken from FRINGES in general of

  LOW SPATIAL FREQUENCY

- DATA FRINGES represent the LOCUS of some propriety

  Examples: HEIGHT in contouring
  DEFORMATION on loading

Fig. 4: Principle foundations of moiré evaluation techniques

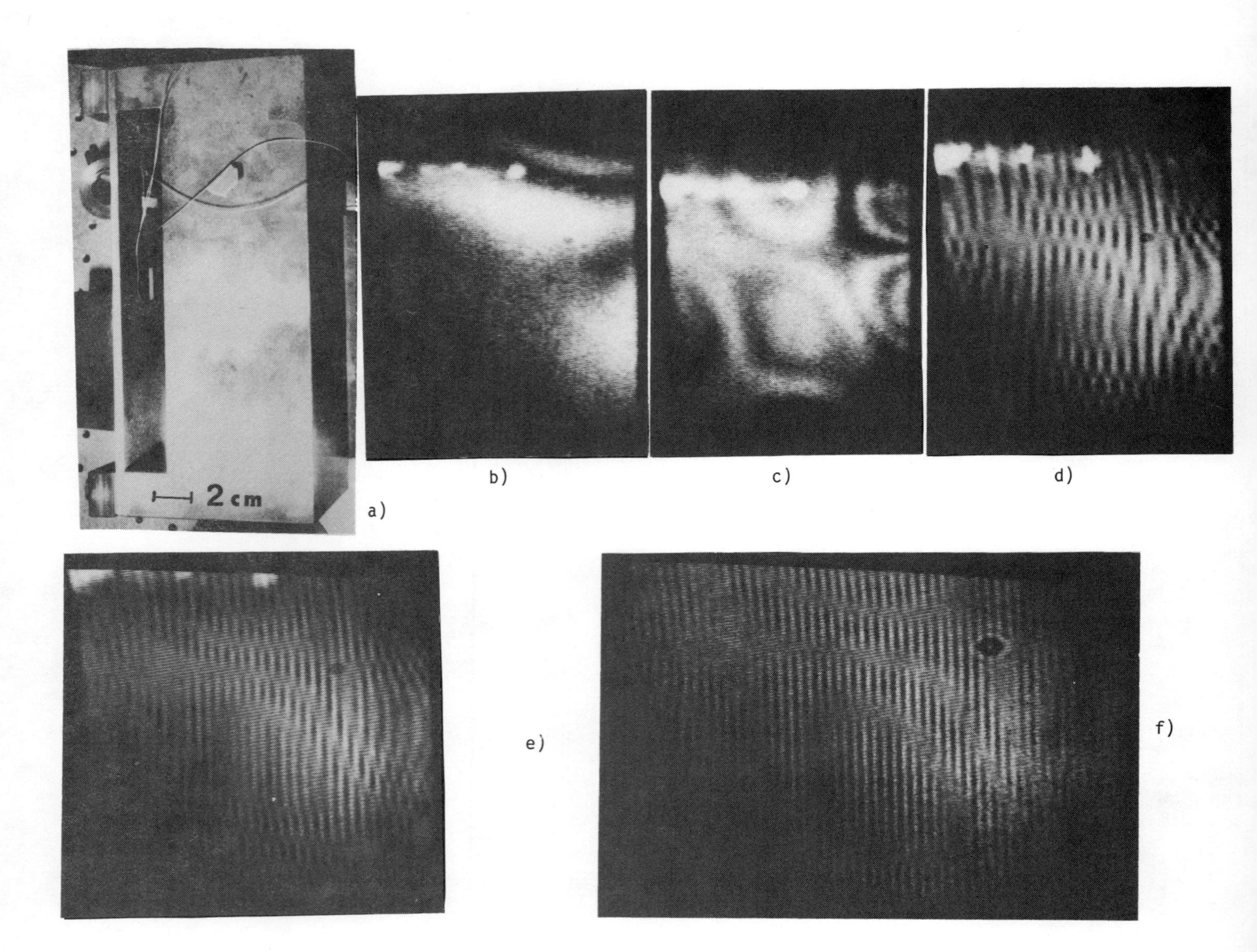

Fig. 5: Time average and pulse holography with a moiré carrier introduced (9).
a) object-steel plate clamped at both ends excited with piezoelectric transducers
b) parasitic fringes at real-time holographic interferometry (two days after exposure)
c) Time-average pattern of a resonant mode (6 KHz)
d) Time-average pattern as in c) with carrier frequency
e) Stroboscopic, one pulse, fringe pattern after adjusting the carrier frequency for optimization of mode pattern visibility
f) local detail analysis by zooming and carrier frequency optimization (stroboscopic mode)

(see section-Technical Advances).

The stroboscopic operation mode becomes necessary as amplitude of vibration increases, Fig. 6, due to the properties of contrast of time-average fringes.

The carrier frequency orientation can be also adjusted, Fig. 7, for metrological advantage and local discrimination. The rotation of carrier fringes combined with the carrier fringe shift at a mode pattern can be used to develop a simpler algorithm in image processing to eliminate the carrier background and produce a filtered mode pattern.

A major practical advantage of the moiré techniques results from the insensitivity to slow varying perturbations such as environmental fluctuations (air stream, temperature, humidity) which introduce phase changes and mode pattern alteration.

The carrier frequency fringes are shifted at the border of every fringe of resonant mode pattern. Interpretation is straightforward for stroboscopic mode as the fringes on the mode pattern represent a step of a phase difference of between wavefronts. The carrier frequency represents then a phase difference scale over the object wavefront. Therefore, phase stepping on the reconstruction beam can be used to produce, in principle, a phase mapping, a possibility being exploited. In time average the modeling is more complex, due to time integration, and complete understanding of mode fringes has not been achieved. However,it seems that the carrier frequency offers information on amplitude and phase.

A peculiar effect is observed when phase modulation is set at the same oscillating frequency of the resonant mode.

These considerations immediately suggest the application of the method in contouring, Fig. 10. When the object deforms during vibration, frozen by stroboscopic illumination, the phase difference between wavefronts is changed and a pattern is formed. Modeling is being pursued for data extraction. The method can also be used for contouring of curved surfaces. It is sufficient to change the reconstruction beam or hologram angular orientation. This constitutes an alternative to the method of projection of interference fringes (10). Too many aspects to fully explore, and understand the capabilities of moiré evaluation techniques, have to be further studied in detail, but present knowledge well supports the believe of the valuable practical interest of the technique.

Technical Advances

Practical applications of the moiré evaluation techniques demand the solution of various technical problems.

Filtering of the carrier background fringes is required. Electronic filtering must discriminate the carrier and the moiré fringes, balancing the contrast, Fig. 11. Resonant modes, Fig. 12, in a violin back cover are observed with differential moiré evaluation, using stroboscopic, two pulses illumination, and a sole hologram.

Elimination of the use of photographic plates suggested the use of photorefractive crystals and phase wave conjugation techniques. Preliminary results confirm capabilities, Fig. 13, but power of the laser must be increased.

Image processing techniques have been introduced but it is still an early stage to see the sought benefits. Meanwhile a dynamical digital memory was conceived, and a prototype was constructed to be tested on the study with transients of slow rate of change (11).

The concepts of moiré evaluation were also extended to electronic speckle pattern interferometry, based on the fact that specklegrams are complex interferograms with amplitude distribution, macroscopically assumed as random but with information contend (phase coded), and therefore producing correlation fringes or moiré like patterns.

Two examples are cited, related to change of reference (4). Fig. 14, relates to a static test with two plane surfaces at an angle (4). Changing the reference plane by reference beam adjustment, sensitivity can be controlled, and direct shape measurement provided. Fig. 15, refers to a case of vibration studies (4). The first specklegram is stored in the E.S.P.I. frame-store-memory for a plate (aluminum, 2 mm thick) at rest. Set-up is realized for out-of-plane displacement measurement. The plate is excited at a resonant mode (0.874 KHz) and correlation fringes are observed with stroboscopic illumination. Acting on the reference beam, the reference plane is changed, producing fringe pattern changes with clear identification of zero order fringe. The signal of the direction of displacement of venters is also resolved. From the change of reference, deformation measurements can be done.

The change of reference plane can be determined by knowing the geometry of the set-up and reference beam direction change. It can also be detected and measured using a background plane surface, stable and properly positioned.

These preliminary results suggested further developments of studies. Optimization of mathematical model and experimental implementation is in progress. It includes other aspects of moiré evaluation techniques in E.S.P.I..

Conclusion

Moiré evaluation techniques were reviewed from previous publications and further work. Advantages and limitations can be identified, Fig. 16. Future developments and theoretical modeling are expected to establish

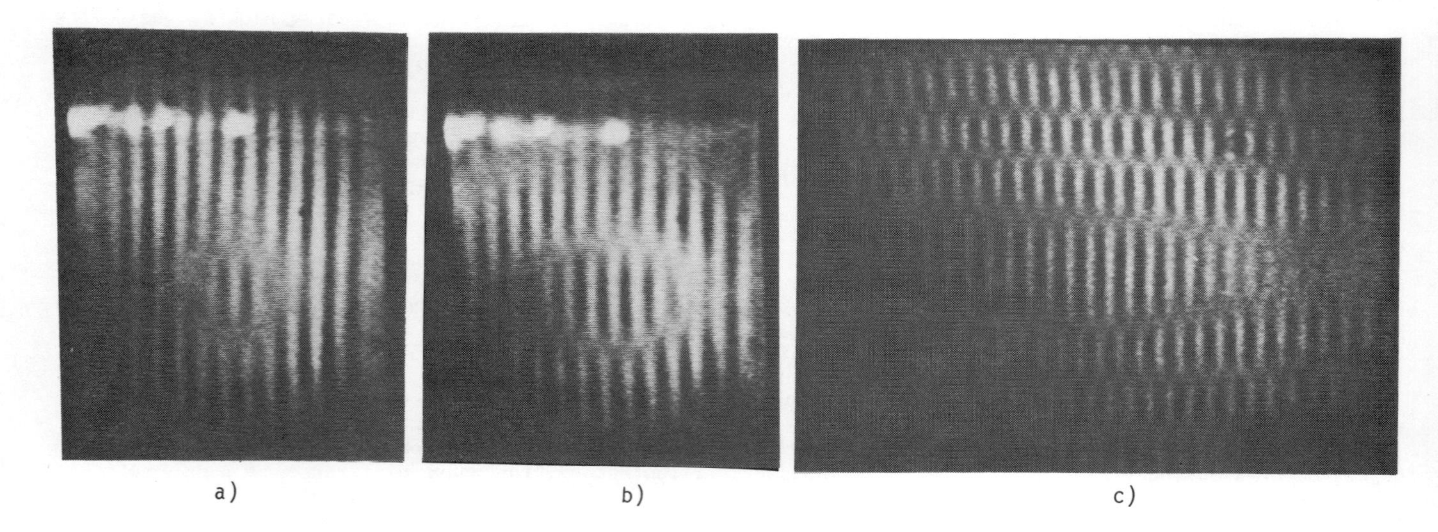

Fig. 6: Effect of amplitude of vibration on fringe visibility of the resonant mode (2,9 KHz) (9).
a) Time-average fringe pattern at low amplitude
b) Same as a) increasing amplitude of vibration
c) Stroboscopic mode pattern for, one illuminating pulse, at maximum deformation

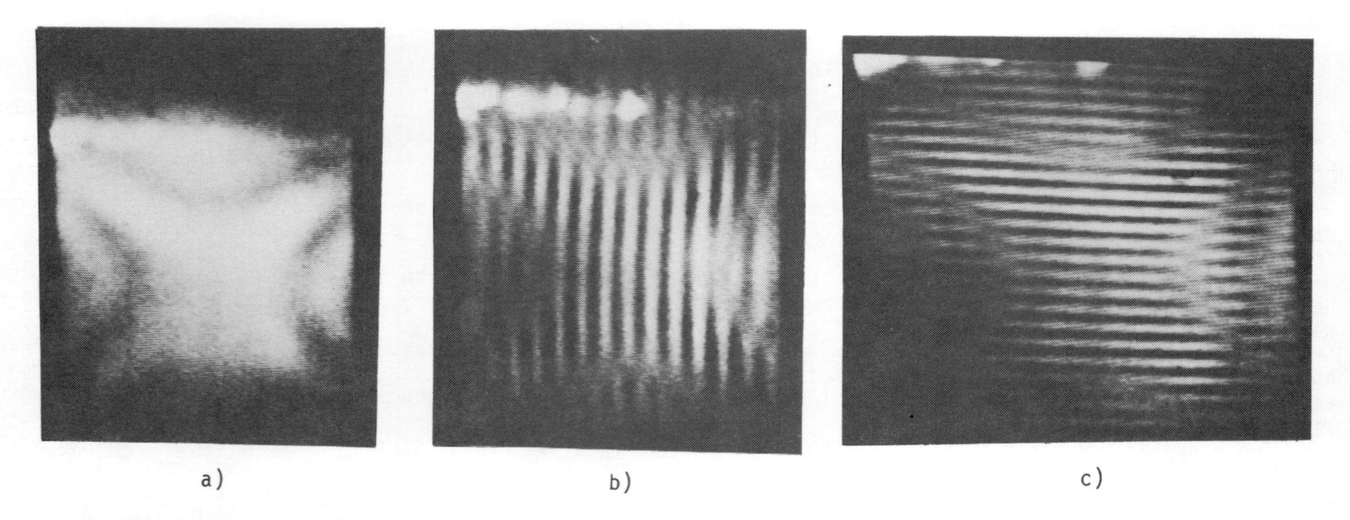

Fig. 7: Resonant mode of plate of Fig. 5 at 3 KHz showing invariance of pattern while changing orientation and frequency of moiré carrier (9).
a) Time-average mode pattern
b) Time-average mode pattern with vertical fringe for carrier
c) Stroboscopic (one pulse) pattern after rotation of carrier fringes and change of periodicity for visibility betterment

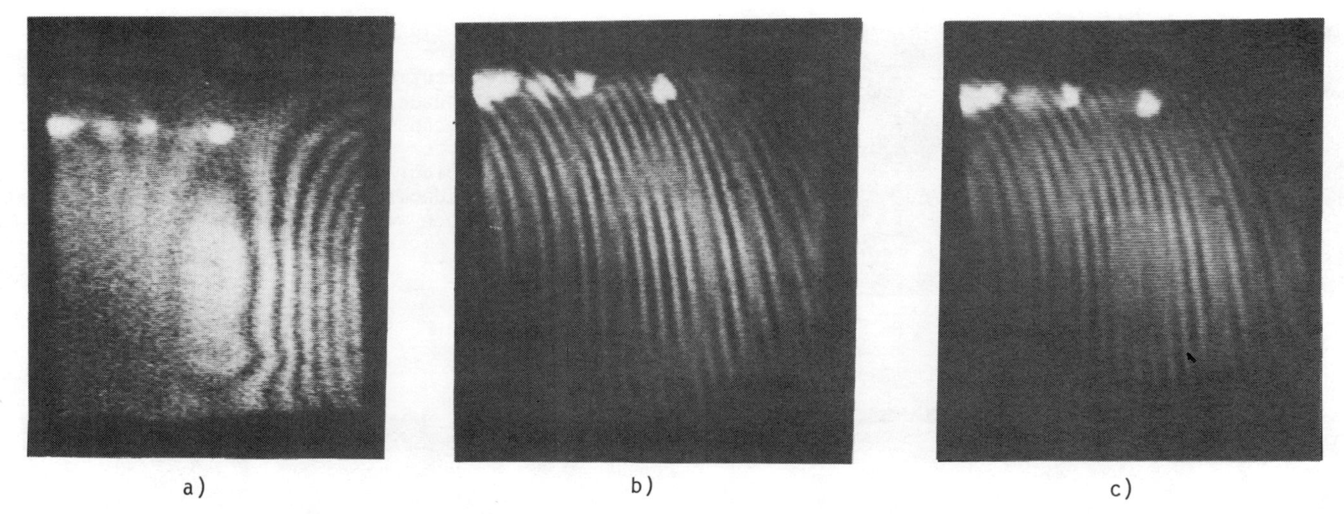

a) b) c)

Fig. 8: Demonstration of mode pattern insensitivity to slow varying fluctuation (9).
- a) Time-average resonant mode pattern (3 KHz) when a temperature gradient is raised by approaching the hand
- b) Stroboscopic (one pulse) resonant mode pattern with same perturbation as in a). Carrier fringes suffer modification but mode pattern remains unchanged
- c) Same as b) but a phase shift of π is introduced in the reconstruction beam as shown in Fig. 9. Carrier fringes are displaced by one fringe and dark and bright moiré fringes interchanged

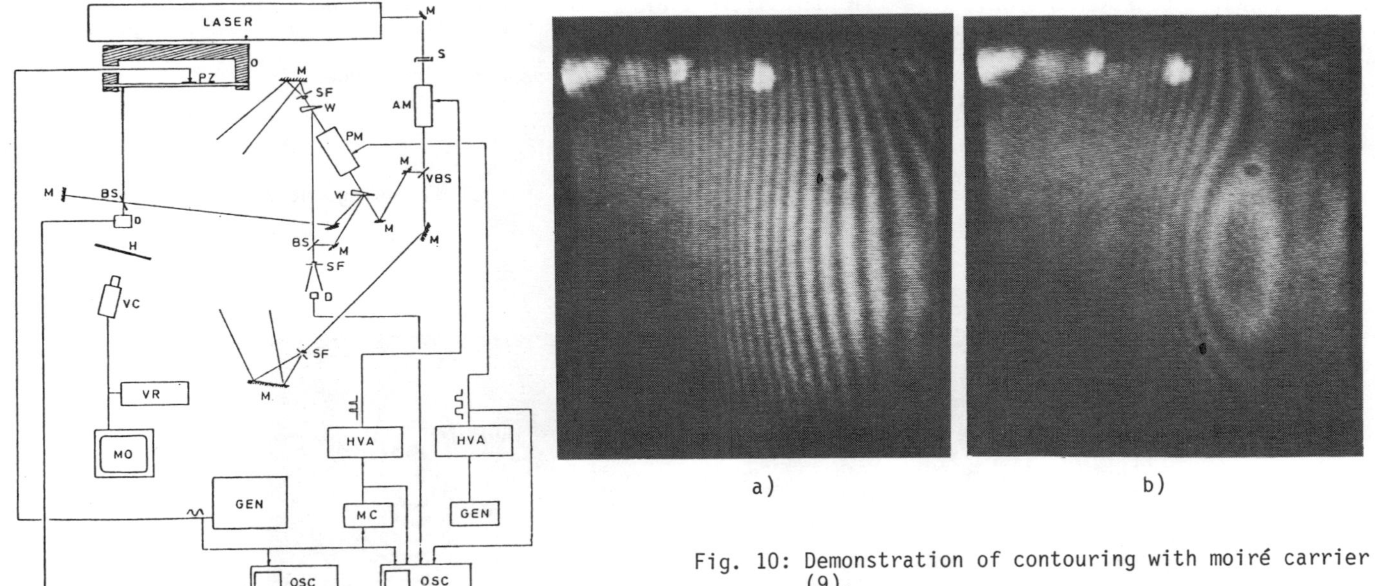

a) b)

Fig. 9: Experimental set-up used to record a video-tape to demonstrate properties and basic performances of moiré evaluation techniques

Fig. 10: Demonstration of contouring with moiré carrier (9).
- a) Stroboscopic recording of carrier fringe pattern of an initial instant of vibration half-cycle.
- b) same as a) at the instant corresponding to a fourth of the vibration period

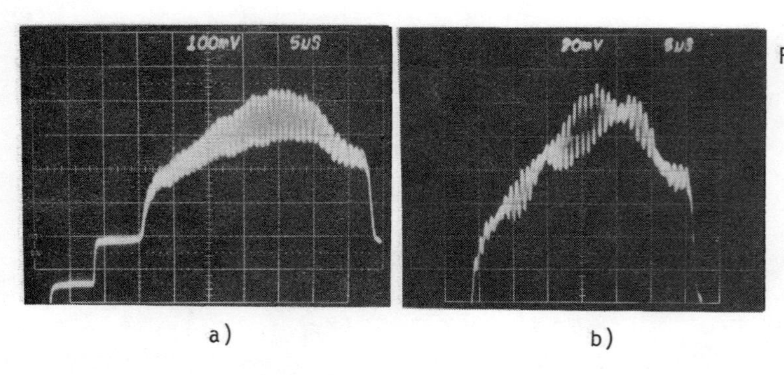

a) b)

Fig. 11: TV-line video signal of hologram restitution with carrier frequency introduced
a) reconstruction with carrier frequency after tilting and translation of the hologram.
b) video signal when recording a resonant mode with carrier frequency shown

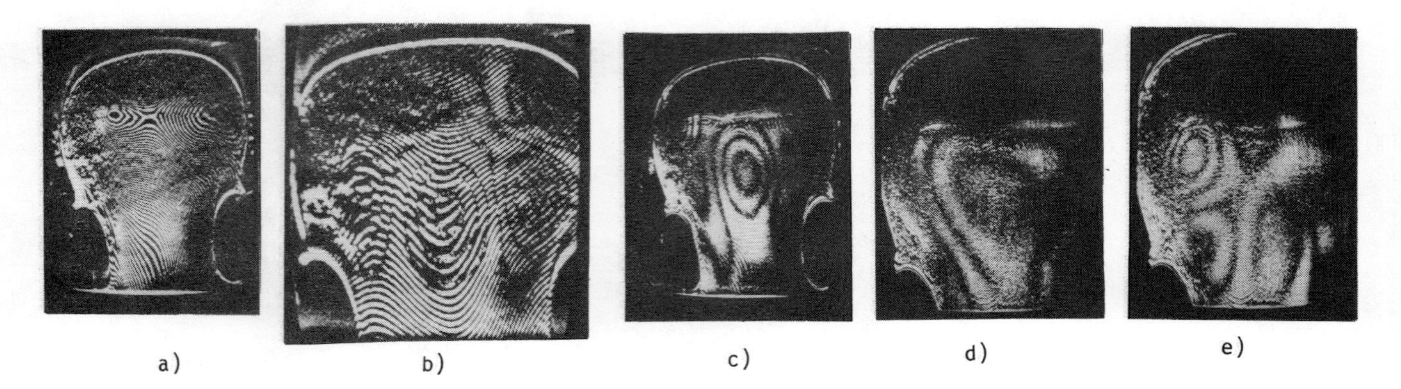

a) b) c) d) e)

Fig. 12: Twin pulse stroboscopic illumination differential moiré evaluation technique of resonant modes of a back cover of a violin. It was used a unique hologram from the violin without excitation (optimum use of illumination intensity).
a) interferogram at an instant of vibration with carrier frequency introduced
b) moiré evaluation of resonant mode at 1.7 KHz
c) Same as b) after carrier frequency filtering
d) Same as c) for resonant mode at 0.8 KHz
e) Same as d) for resonant mode at 2.4 KHz
The resonant modes d) and e) were observed one week later than c) but using the same hologram (1).

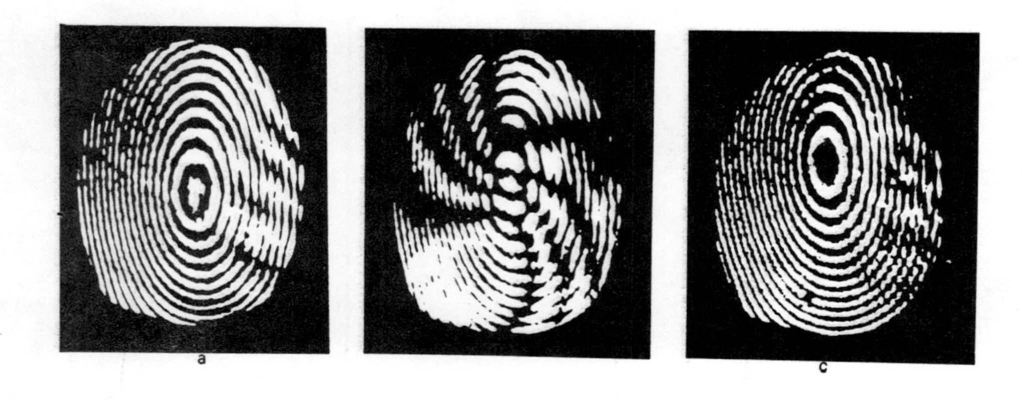

a) b) c)

Fig. 13: Differential moiré evaluation with a photorefractive crystal (BSO) and phase wave conjugation arrangement (1)
a) interferogram of two wavefronts from reconstruction of double exposure hologram corresponding to two stages of deformation of a circular membrane
b) interferogram with moiré pattern seen from reconstruction as in a) but object is in another deformation stage
c) interferogram as a) corresponding to first and last state of deformation
Overlapping of a) and c) results identical to b)

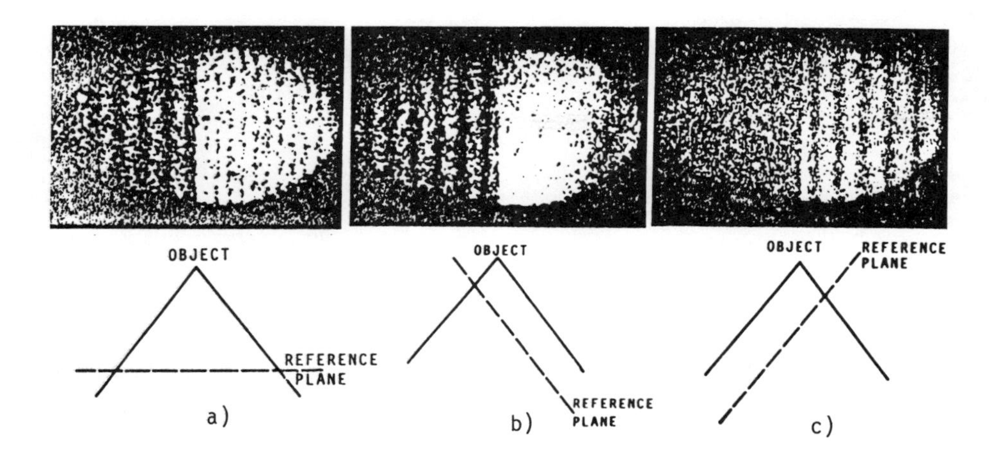

Fig. 14: Change of reference demonstration in contouring for the adjustment of spatial sensitivity in E.S.P.I. (4).
a) observation with sensitivity vector oriented arbitrarily
b) adjustment of reference plane by re-orientation of reference beam
c) same as b) for other reference plane orientation

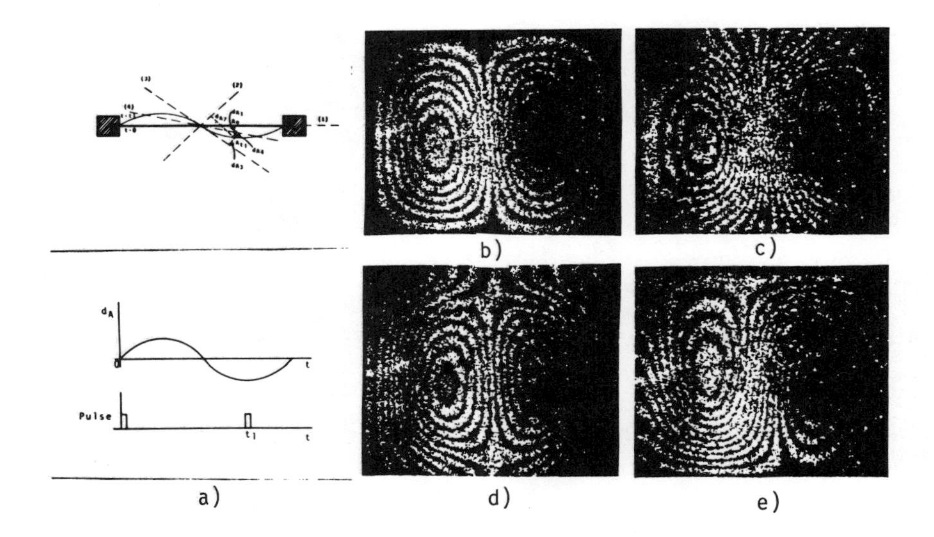

Fig. 15: Vibration analysis with E.S.P.I. at stroboscopic mode (twin pulse technique), and the changing of reference plane (4).
a) geometry and time sequence of vibration and pulse illumination
b) classical pattern of correlation fringes (reference plane (1) in a)).
c) fringe pattern after rotation of reference plane (reference plane (3) in a))
d) use of zero order fringe for direct measurement via rotation of reference beam (reference plane (4) in a))
e) identification of signal on direction of displacement (reference plane (3) in a)).

MOIRE HOLOGRAPHY

ADVANTAGES

- First HOLOGRAM is ARBITRARY
  - repositioning less critical
  - illumination less stringent
  - exposure control easier
  - ambient fluctuations of minor influence
  - tolerances relaxed
- DIFFERENTIAL INFORMATION (e.g. deformation) available between two configurations at discretion
- VIBRATION analysis (complete cycle, and multiple frequencies) with a SOLE HOLOGRAM (taken at rest position)
  Note: correlation, coherence — must be met
- INTRODUCTION of known CARRIER modifications
  Examples: - Tilt and displacement for higher carrier
  - Change of reference plane to screen particular (defects, deformation components, etc)
- TIME DIFFERRED ANALYSIS
- AMBIGUITY ELIMINATION
- ILLUMINATION POWER less stringent
  Alternatively - Larger areas illumination

MOIRÉ EVALUATION

LIMITATIONS

- SAFE GUARD BANDWIDTH
- FILTERING
  - Diffraction
  - Shanon Theorem
- FRINGE VISIBILITY and ENHANCEMENT

Fig. 16: Advantages and limitations on moiré evaluation holography

the potentialities of the moiré evaluation holography towards industrial and biomedical applications.

Advantages should be gained from emerging technologies such as phase stepping digital holographic interferometry.

References

1- O.D.D. Soares, A.L.V.S. Lage, L.M. Bernardo; MOIRE EVALUATION with FRINGES PATTERNS of INTERFEROGRAMS, HOLOGRAMS, MOIREGRAMS and SPECKLEGRAMS, in Optical Metrology, Martinus Nijhoff (1985)

2- O.D.D. Soares, A.L.V.S. Lage; METODO de ANALISE DIFFERENCIAL e INTEGRATIVA em HOLOGRAFIA INTERFEROMETRICA por TECNICA MOIRE, Patent N. 80333, INPI (1985)

3- O.D.D. Soares, A.L.V.S. Lage; MOIRE HOLOGRAMETRY, 13th Congress ICO Sapporo (1984)

4- O.D.D. Soares, A.L.V.S. Lage; REAL-TIME MOIRE HOLOGRAPHY, Proc. SPIE 615 (1986)

5- O.D.D. Soares, A.L.V.S. Lage; CONTROLLABLE SYNCHRONIZED MULTIPULSE ILLUMINATION SYSTEM for E.S.P.I. and HOLOGRAPHY, Proc. SPIE 427 (1983)

6- O.D.D. Soares, A.L.V.S. Lage, L.M. Bernardo; MOIRE EVALUATION of PULSE ILLUMINATED INTERFEROGRAMS by SYNCHRONIZED VIDEO RECORDING , Proc. SPIE 491 (1984)

7- O.D.D. Soares, A.L.V.S. Lage, L.M. Bernardo; PROGRESS on DYNAMICAL STUDIES by MOIRE EVALUATION in HOLOGRAPHY and IMPROVEMENTS on ELECTRONIC SPECKLE PATTERN INTERFEROMETRY, Acta Polytechnica, Scandinavica 150 (1985), 253

8- O.D.D. Soares; ELECTRONIC SPECKLE PATTERN INTERFEROMETRY (E.S.P.I.) for DYNAMICAL STUDIES, , Proc. SPIE (1986)

9- Photographs taken from a video-tape recording to demonstrate moiré evaluation techniques performances. A unique hologram was used. Recording was taken distributed over three days for testing capabilities.

10- O.D.D. Soares; NON OPTICAL SURFACE TOPOGRAPHY by PROJECTED INTERFERENCE FRINGES, Portugaliae Physica 13 (1982), 217-231
- O.D.D. Soares, S.P. Almeida; PROJECTION INTERFERENCE FRINGE MICROSCOPE, Proc. SPIE 429 (1983)
- R.W. Wyant, S.P. Almeida, L.M. Bernardo, O.D.D. Soares; ANALYSIS of MICROSCOPIC SURFACE STRUCTURE by PROJECTION INTERFERENCE FRINGE, OSA Annual Meeting, Seatle (1986)

11- O.D.D. Soares, A.L.V.S. Lage, A.O.S. Gomes; DYNAMICAL DIGITAL MEMORY for HOLOGRAPHY, MOIRE and E.S.P.I., Optical Metrology, Martinus Nijhoff (1985)

Some aspects of application of a double frequency interferometer for distance measurements

J.J. Galiński, H.Z. Kowalski, J. Sanecki

Institute of Geodesy and Cartography, 00-950 Warsaw, Jasna 2/4, Poland

Field vector of laser light with sufficient approximation can be circumscribed by scalar function of the form:

$$E(r,t) = E_r(r,t) + iE_i(r,t) \tag{1}$$

If the function (1) satisfies Gabor postulates, its real and imaginary parts makes the pair of Hilbert transformations - than the real part $E_r(r,t)$ can be circumscribed with the help of analytic signal of the form:

$$E_r(r,t) = A(t)\exp[i\omega t + i\varphi(r,t)] \tag{2}$$

In equation (2) the function $\exp[i\varphi(r,t)]$ represents informative signal which modulates carrier wave - laser light - expressed with the function $A(t)\exp(i\omega t)$.

Measuring information included in the signal (2) can not be identified in a simple way because of the laser light frequency band lies nearby 10^{14} Hz frequency. Informative signal detection is realized by photodetectors, which in the frequency band of the light have square -law characteristic and as low-pass filter satisfies dependences:

$$I_f(t) = [E_r(r,t)]^2 \; ; \; I_f(t) = 0; \text{for} \, \omega > \omega_{gr} \cong 10^9 \text{ Hz} \tag{3}$$

It predisposes them to work in system with optical heterodyne action.

Interferometer structure is based on space - time modulation of light wave and mixing two light beams in a heterodyne system.

Practically the problem is solved in a way that two light beams are inserted onto photodetector input. Informative beam E_i and reference beam E_o are circumscribed by dependences:

$$E_i = A_i \exp\{i[\omega_i t + \varphi(r,t) + \varphi_i]\} \; ; \; E_o = A_o \exp[i(\omega_o t + \varphi_o)] \tag{4}$$

Structure and functional diagram of the interferometer is presented on Fig. 1.

Measuring signal of the arrangement presented on Fig 1 after simplifications is circumscribed by the dependence:

$$I_f = A_i^2 + A_o^2 + A_i A_o \exp\{i[(\omega_i - \omega_o)t + \varphi_o - \varphi_i - \varphi(r,t)]\} + \\ + A_i \exp\{-i[(\omega_i - \omega_o)t + \varphi_o - \varphi_i - \varphi(r,t)]\} \tag{4a}$$

It is possible to transfer measuring information when the arrangement satisfies the condition:

$$\omega_i - \omega_o = \text{const} \tag{5}$$

for sufficiently long time interval.
Practically the condition (5) is satisfied for two cases:

$$\text{a)}\; \omega_i - \omega_o = 0 \qquad \text{b)}\; \omega_i - \omega_o \neq 0 \tag{6}$$

Conditions (5) and (6a) are satisfied by single-frequency interferometer class, with stationary image of interference fringes. Conditions (5) and (6b) are satisfied by double-frequency interferometers class with movable image of fringes. Furthermore for metrological purposes the arrangement should fulfil additional conditions:

$$\omega_i = \text{const, and } \varphi_i = \text{const, and } \varphi_o = \text{const} \tag{7}$$

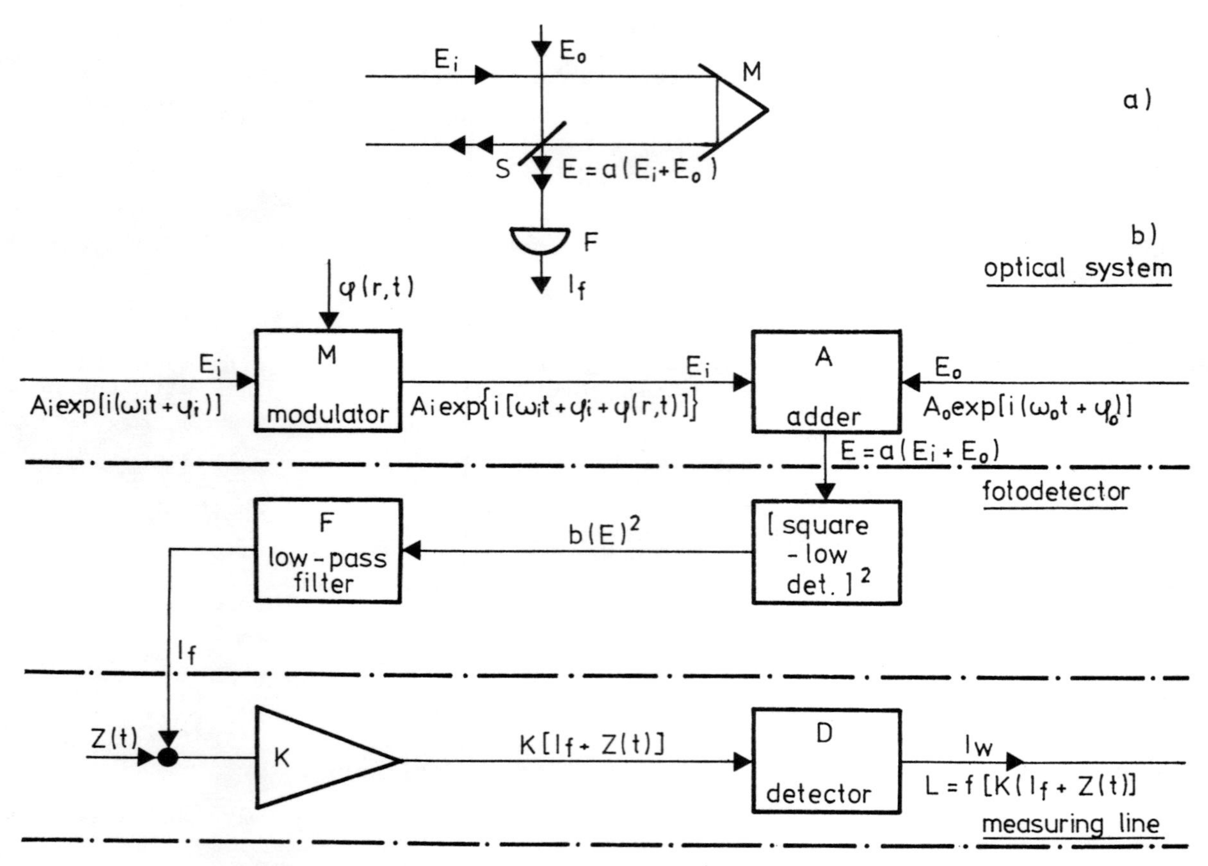

Figure 1. Structure and functional diagram of the interferometer

Measuring signal I_f generally occurs together with the disturbance signal z(t) and the sum of them is an input for the measuring line(measuring circuit). Avoiding noise structure investigation, we can assume that z(t) has the character of narrow-band white noise which spectrum ranges in frequency band $0 \leqslant \omega_z < \omega_{gr}$, which is the same as the informative signal spectrum $\varphi(r,t)$ ranging in the band $0 \leqslant \omega \leqslant \omega_i$.

Resolution of interferometres - being the metrological parameter - is a function of the ratio of mean noise power of the measuring signal S, and mean noise power N related to input of the measuring line (measuring circuit).

In the class of interferometres satysfying conditions (5) and (6a) and because of accurate overlaping of frequency bands of functions $\varphi(r,t)$ and z(t), there is impossible to separate these spectra placing separating filter into measuring line. Because of that, arrangements in which measuring lines have ratio $(S/N) \leqslant 20$ dB with resolution $R \leqslant \lambda \cdot 10^{-1}$ and detector D has characteristic of amplitude discriminator, are constructed. Output signal is circumescribed by dependence:

$$I_{ws} = \frac{1}{2}(A_i^{\ 2} + A_o^{\ 2}) + A_i A_o \cos\left[\varphi_o + \varphi_i - \varphi(r,t) + z(t)\right] \tag{8}$$

Resolution increasement of the arrangement is achieved through stabilization of optical system characteristics:a = const; photodetector characteristics b = const and measuring line characteristics K = const (Fig. 1).

In a class of interferometres satysfying conditions (5)and (6b) it is a possible to separate signals $\varphi(r,t)$ and z(t) inspite of overlapping their frequency bands.

The ratio value $(S/N) \leqslant 50$ dB with resolution $R \leqslant \lambda \cdot 10^{-2}$ is attained through design arrangement fulfilling dependence:

$$(\omega_i - \omega_o) \pm \omega_i \geqslant \omega_{gr} \tag{9}$$

and by implementation of the band filter fulfilling condition (9) on input of measuring line and using detector having frequency discriminator characteristic.

Than output signal is described with the dependency:

$$I_{wn} = A_i A_o \cos\left[(\omega_i - \omega_o)t + \varphi_o - \varphi_i - \varphi(r,t)\right] + z'(t) \qquad (10)$$

and after passing through band filter $z'(t) \ll z(t)$
It must be noticed that resolution increasement of the arrangement is achieved on electrical way.

Requirements of length metrology are conditioned by attributes of metrological lasers for which parameter ω_i = const is satisfied with inacuracy of the order 10^{-12}. That is reason of necessity for constructing interferometers having resolution of the order minimum $\lambda \cdot 10^{-5}$ and (S/N) $\leqslant$ 100 dB. The know interferometer construction do not satisfy these postulates.

The analysis based on methods accepted in information theory leads to increasement of (S/N) ratio through detection of measuring signal in the line with synchronous detector. Interferometer arrangement satysfying mentioned above postulates has been built in the team led by the authors.

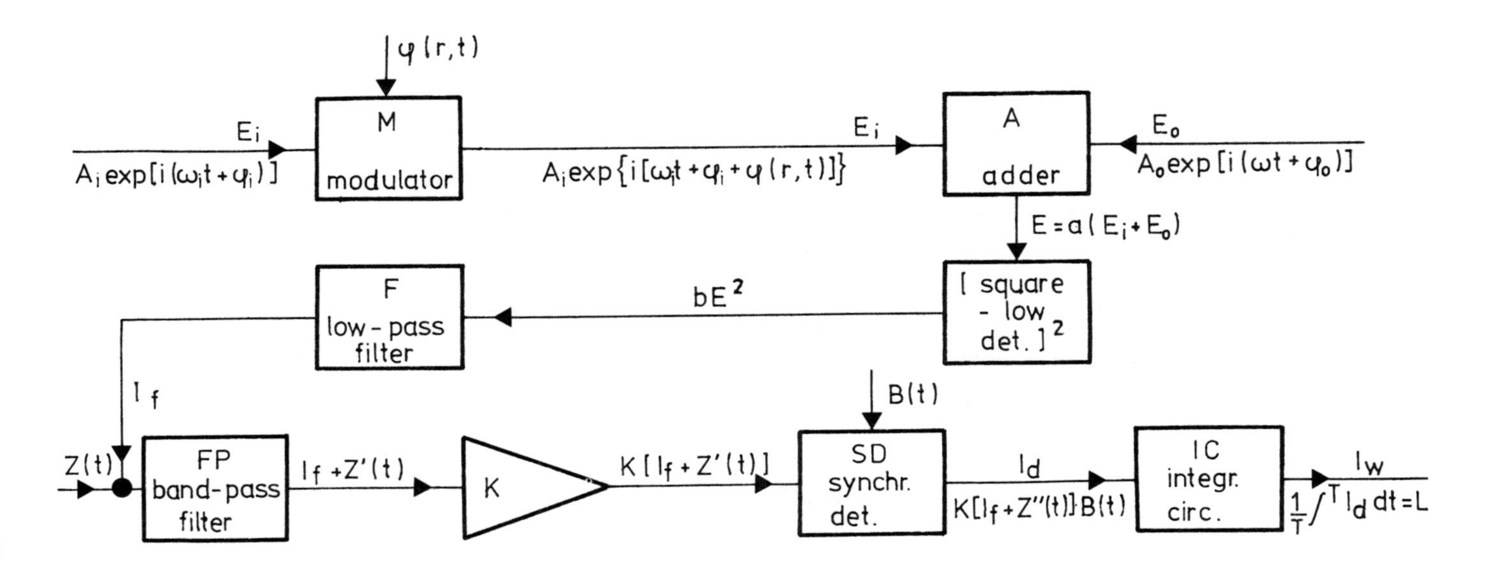

Figure 2. Functional diagram of the double heterodyne interferometer.

On Fig. 2 the interferometer functional diagram with double heterodyne action is presented: on optical waves in optical systems and on radio waves in measuring line (measuring circuit). This interferometer arrangement satisfies conditions (5) and (6b).

The measuring line of the arrangement is provided with band filter matched to dependency (4) at the input, and with synchronous detector controlled by electrical signal $B(t) = B_i \cos(\Omega t + \varphi_b)$, at the same time the condition $\Omega = (\omega_i - \omega_o)$ is satisfied, and on the output - with integrating circuit. The output signal of arrangement from Fig. 2 is described with dependence:

$$I_{wh} = C \sin\left[\varphi_o - \varphi_i - \varphi(r,t)\right] + \frac{B}{T}\int_0^T z''(t)\cos(\Omega t + \varphi_b)\,dt \qquad (11)$$

The first factor of the relation (11) contains undisturbed measuring signal $\varphi(r,t)$. The second factor under integral is the product of filtered noise signal $z''(t)$ and harmonic function. For sufficiently long T it satisfies the dependence:

$$\frac{B}{T}\int_{0}^{T} z''(t)\cos(\Omega t+\varphi_b)\,dt \simeq 0 \quad (12)$$

Technical specifications allow building arrangements according to the scheme from Fig. 2 with $(S/N) \leqslant 100$ dB value and resolution $R \leqslant 10^{-5}$.

Basing on metrological properties of the arrangement with double heterodyne action the authors have tried to solve some problems of length and angle metrology.

1. The interferometer of resolution better then $\lambda \cdot 10^{-3}$ and measuring range up to 100 m is worked out.
   Interferometer arrangement according to diagram from Fig. 2 satisfies the dependence (10) and informative signal is circumscribed with expression:

$$\varphi(r,t) = \frac{4\pi}{\lambda_i}\int_{0}^{T} V(t)\,dt = L \quad (13)$$

   where $v(t)$ - displacement velocity of reflecting prism - modulator.

2. The arrangement for distance measurements of interference fringes of two interfering laser beams, according to dependence:

$$D = \frac{\lambda_i}{2\sin(\frac{A}{2})} \quad (14)$$

   has been worked out in arrangement satysfying the dependence (10); the informative signal satisfies conditions:

$$\varphi(r,t) = 0 \quad (15)$$

   and density of lines D is the function of angle A between electromagnetic field vectors of two interfering laser beams (Fig. 3). With quantity $D \gg 1{,}000$ lines per mm the arrangement enables the measurement with inaccurancy of the order 10^{-7}.
   Basing on the described arrangement one has built interference range - finder with measurement range up to 1,000 m and resolution of 0.1 mm basing on the parallactic triangle rule.

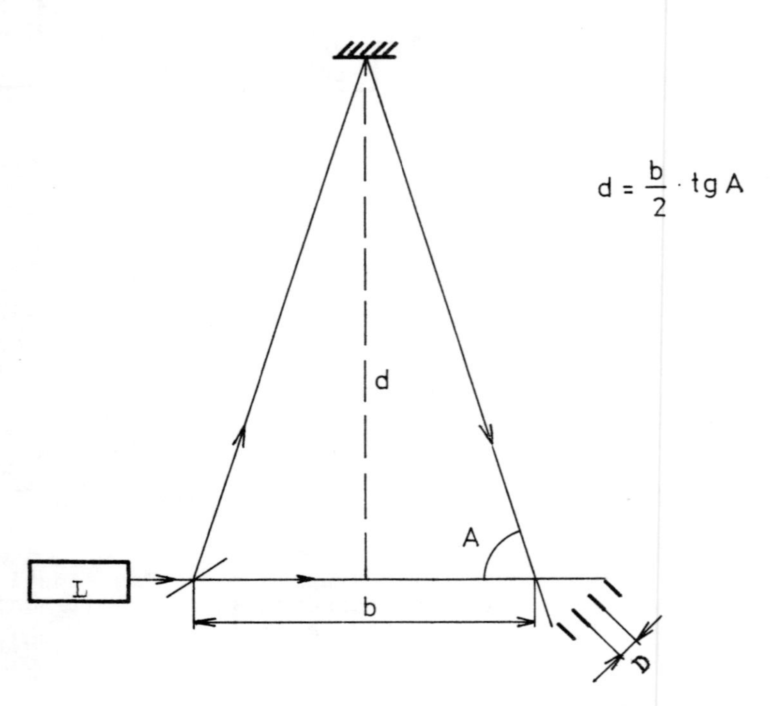

Figure 3. Range measurement using interference range-finder.

Recapitulation

Basing on analysis of general form interferometer, the authors worked out the interferometer arrangement with nonstationary image od fringes with double heterodyne action.

The arrangement assures achieving of ratio $(S/N) \leqslant 100$ dB and building:
a/ interferometer with the range to 100 m and resolution $\lambda \cdot 10^{-3}$,
b/ interference range - finder with the range to 1,000 m and resolution 0.1 mm.

Application of holography in the distribution measurement of fuel spraying field in diesel engines

He Wan Xiang, Li Zhi Xiong, et al

Tianjin University

Abstract

The distribution of fuel spraying field in the combustion chamber is an important factor which influences the performance of diesel engines. Precise data for those major parameters of the spraying field distribution are difficult to obtain using conventional ways of measurement, so its effects on the combustion process cannot be controlled. The laser holographic measurement is used and many researches have been made on the injecting nozzles used in diesel engines Series 95, 100 and 130.

These researches show that clear spraying field hologram can be taken with an "IC Engine Laser Holography System". By rendition and data processing, droplet size, amount and their space distribution in the spraying; the spraying range, cone angle and other dependable data can be obtained. Therefore, the spraying quality of an injecting nozzle can be precisely determined, which provides reliable basis for the improvement of diesel engines' functions.

Introduction

More than half of the power mankind uses now is generated by internal combustion engines. In some developed countries, such as the United States of America, the petroleum consumed by IC engines takes over 50% of the petroleum production. In China, this figure is over 35%. At present, the world is in the energy crisis condition, so it is of great economical value and socialogical meaning to improve the performance of IC engines, reduce their energy consumption as well as the pollution of harmful gases.[1]

The fundamental way to enhance the engine properties is to improve the combustion process. The matching among the droplet distribution in the combustion chamber, the air motion, the size and configuration of the chamber is the key factor determining the combustion state of the diesel engine, where the droplet distribution of the fuel spraying field is the most important parameter. In order to insure that the fuel can be evaporated, dispersed, mixed and combusted rapidly, and that the engine can reach high specifications, different combustion systems must have different spraying characteristics.

Selection of the research method and fundamental principles of spraying field holography

Analysis and selection of research methods used in the spraying property measurement

Since the fuel spraying field distribution is the major reason for the combustion performance of engines, the distribution rule must be given detailed and reliable measurement and study so that it can be controlled.

Normally, the distribution of fuel spraying field is specified with spraying parameters of the injecting nozzle, that is, droplet fineness and evenness in the spraying field, cone angle and penetrability of the spraying,etc.. Many measurements and researches were made in the past using methods such as impression, sedimentation, solidification, laer condensed scattering and high-speed photography,etc.[9] Many useful results were obtained. However, these methods have shortcomings in their principles and in their structures which are difficult to overcome, so the spraying field measured is often disturbed by various reasons. The space distribution of the spraying field is damaged and the shape of the fuel droplet changed, or because the sampling range is so small that it's difficult to exactly measure the size, the amount and space distribution of fuel droplets. The results thus obtained cannot reflect the true content of the spraying field, therefore, they cannot provide a dependable basis for the improvement of combustion process.

The development of laser holography provides a reliable means for the measurement and research of corpuscle field changing instantaneously in the space. Holography uses laser with steady amplitude, phase and good coherence as the light source, which is able to record and render all information (amplitude and phase) of the object light-wave. Especially when pulse laser is used as the light source, the shape, size, amount, space distribution and motion speed of corpuscle groups changing instantaneously can be recorded and rendered. Since the coaxial holography was used by B.J. Thompson in 1964 [2] to measure atmospheric fog field, many countries such as USA, West Germany, Great Britain, Japan and China [3-9] have done researches in this field. In order to measure the combustion process of the IC engine (including fuel spraying field), we started research on the application of laser holography in IC engines from 1977. Now a set of methods for the measurement of fuel spraying field distribution and a "High-speed Instantaneous Laser Holography System" have been developed.

Basic principles of spraying field holograph

Although off-axial holograph has the advantages like strong 3-dimensional feeling, no influence from conjugate virtual image, it still has the disadvantage like high requirement for light source coherence, complex light circuit arragement, high in cost, and difficult in data processing. Coaxial holograph is weak in 3-dimensional feeling and influenced by conjugate virtual image when observing the real image. But what we are studying now are those tiny fuel droplets flying in the space. Those we want to measure are the size, amount, and space distribution of the droplet outline, not the details of the droplet surface. So it is more practical for us to use coaxial holograph with the advantages of simple light circuit, low requirements for light source, low in cost and easy in data processing.

The holographic principle of the fuel spraying field distribution was specially stated before [6]. In order to discuss the question, the recording and rendition of coaxial hologram of fuel droplets in the spraying field is briefly analyzed here.

Recording of droplet hologram . Let's assume that all fuel droplets in the spraying field are plates with even permeability, when lighted up with plane wave laser (wavelength is λ) and if the Fraunhofer distant field condition can be met ($(\xi^2+\eta^2)_{max}/\lambda Z \ll 1$, according to Huygens principle, the light intensity distribution of the light field on the recording plane is as follows:

$$U(R) = \exp(jKZ)\left[1 - \frac{jK}{Z}\left(\frac{jKR^2}{2Z}\right)\tilde{A}\left(\frac{R}{\lambda Z}\right)\right] \quad (1)$$

$$I(R) = 1 + \frac{2Ka^2}{Z}\sin\left(\frac{KR^2}{2Z}\right)\left[\frac{J_1\left(\frac{KaR}{Z}\right)}{\frac{KaR}{Z}}\right] + \frac{K^2a^4}{Z^2}\left[\frac{J_1\left(\frac{KaR}{Z}\right)}{\frac{KaR}{Z}}\right]^2 \quad (2)$$

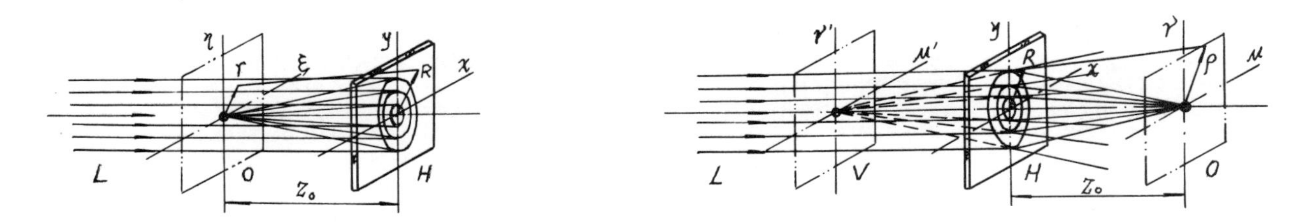

Figure 1. Recording diagram of coaxial light circuit droplet hologram .

Figure 2. Rendition diagram of coaxial light circuit droplet hologram .

If holographic plate is put on the recording plane, a spraying field hologram can be recorded after being exposed, where every fuel droplet hologram is a diffraction ring bright alternating with shade from stage 0 to high stage. Its light intensity attenuates rapidly from stage 0 to high stage.

Rendition of fuel droplet hologram . When plane wave laser (wavelength is λ) is used to light up the fuel droplet hologram , on the viewing plane Zo distance far from the hologram , the recorded droplet light wave can be diffrated to form the real image of the droplet. Assuming the amplitude permeability of the hologram is I(R), and Z=Zo, according to Huygens principle, the light field distribution of the light wave rendered on the viewing plane should be:

$$U(\rho) = \exp(jKZ)\left\{1 - \mathrm{CirC}\left(\frac{\rho}{a}\right) + \exp\left[j\left(\frac{K^2}{4Z} - \frac{\pi}{2}\right)\right]\left[\frac{J_1\left(\frac{Ka}{Z}\right)}{\frac{Ka\rho}{Z}}\right] + \frac{K^2a^4}{Z^2}\left[\frac{J_1\left(\frac{Ka}{Z}\right)}{\frac{Ka\rho}{Z}}\right]^2\right\} \quad (3)$$

The first term in the above equation represents the background light, the last term represents the Airy disk, while the central term represents the light field distribution which includes all droplet information.

Using suitable ways of measurement and mathematical treatment, all kinds of tested data for spraying parameters can be obtained from the rendered spraying field hologram .

Test instruments and test conditions

Test instruments

Test instruments consist of three parts, that is, the hologram recording system, testing device, hologram rendition and data processing system as shown in Figure 3 (a) and (b):

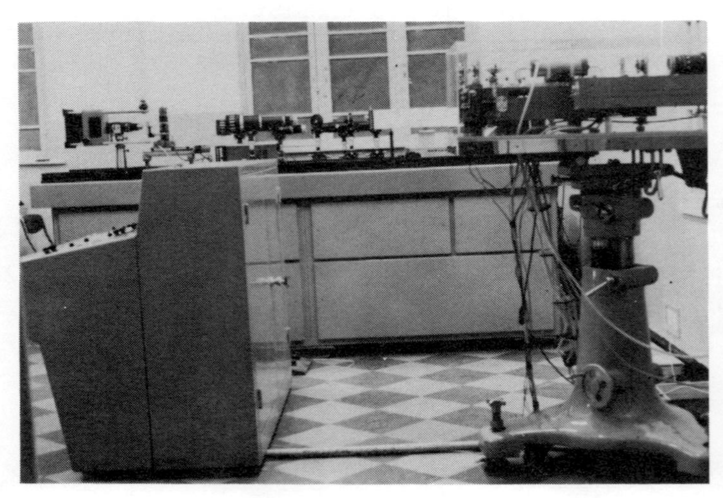

(a)

(b)

Figure 3. Hologram recording system, testing device, hologram rendition and data processing system.

The hologram recording system is composed of the followings: Pulse ruby laser, laser power source, switching power source, synchro-controlling console, syncsignal generator, intaking light circuit, intaking device and universal-purpose optical shock-absorbing test stand. The testing device is composed of the followings: Test model (or test engine), fuel supply and primary injection control system, high pressure gas cylinder. The hologram rendition and data processing system consists of a He-Ne laser, extender collimating device, 3-dimension movable rack, microscopic arbiter and shock-absorbing rendition table. etc. The detailed content of the test instrument was specially described before.[8]

Test conditions

Hologram recording light source. The light source used to intake the spraying field hologram is a pulse Q-adjusted ruby laser. The size of the ruby rod ø8 x 120, luminous energy is 0.2 joule, pulse width is 30-50nS, coherent length is about 1 m.

Intaking light circuit. After the laser beam generated from the ruby laser is extended it becomes a quasi-parralel light with ø44 light field and lights up the spraying field measured. From the analysis of light intensity distribution in the fuel droplet hologram recordings, we know that light intensity I(R) is proportional to the square of the droplet radius a, and is inversely proportional to the distance Z between fuel droplet and the recording plane. J.D.Trolinger connected a, Z, λ with the concept of distant field number N[5].

$$N = \frac{Z}{(2a)^2/\lambda} \qquad (4)$$

Tests show that when N is greater than 1 and less than 100, a clear corpuscle hologram can be taken. When N is greater than 100, since value Z is too large or value a is too small, and the light intensity of the object light diffraction ring is too weak, the object information is covered by light noises and cannot be recorded clearly. When N is equal to 100, if 5 μm droplet is to be recorded, value Z is only 3.57mm, the depth of the spraying field is generally several centimeters, and together with the limit of test instruments and intaking devices, the distance between the holographic plate and the spraying field intaken is over 100 mm. In order to solve the problem, an extender 4f optical system is used, as shown in Figure 4, with which all droplets in the light field's Fresnel diffraction zone and part of larger droplets outside the zone appeared in the exposure are recorded. The advantages in using this system are as follows: a). The droplet image can be properly amplified to compensate the low resolution of the holographic plate and make it possible to take pictures for all droplets at any space layer in the spraying field; b). The distance between the spraying field and the holographic plate can be lengthened, which make it possible to directly take pictures for the fuel spraying field inside the test model or the combustion chamber of the test engine.

Test model. A high pressure simulation device specially designed for taking pictures for the spraying field of the diesel engine measured. The inside diameter of cylinders is 130mm, 300mm high with transparent glass windows (80 x 120mm^2) at both sides. An etalon 3-dimension adjusting mechanism is mounted in the cylinder. Either nitrogen or air may be filled in and its pressure may be adjusted between 1 and 30 standard atmosphere. This system can be used to measure spraying properties of any fuel injection in any models of fuel pumps or injector.

Fuel is injected into the air moving with high temperature, high pressure and high speed in the cylinders of diesel engines. Its motion, evaporation, diffusion, mixing and combustion is a very complex physicochemical process, and it is very difficult to know everything about it overnight. Therefore, it is necessary to make certain simplifications and some simulations beforehand. If the resistance the fuel droplet meets during its movement in the cylinder is caused by collision of air moleculars, then the resistance of the droplet is a function of Reynolds number Re of the air flow, air density and its flow speed v, that is, $F = f(Re. \rho .v)$. Although before compression stops, the gas inside the combustion chamber is making high speed rotating movement, and since the fuel spraying direction is almost perpendicular to that of the air flow and the speed of the air flow in the fuel spraying direction is very low, so the influence of v and Re is small. The resistance borne by the fuel droplet depends mainly on ρ. Let's temperarily not consider the effects of high temperature, evaporation, diffusion and other factors, but, based on the principles that gas densities are equal, we can use the air state equation to convert the simulated back-pressure which should be added in the test model.

$$P_M = \frac{T_o}{T_c} P_c \quad (Pa) \tag{5}$$

Where: P_M Simulated back-pressure(Pa) added in the test model;
P_c Internal pressure (Pa) of the combustion chamber when the compression of the diesel engine stops;
T_o Laboratory temperature (K);
T_c Internal temperature (K) of the combustion chamber when the compression of the diesel engine stops.

Hole diameter of the intaking light field and the intaking position. The laser beam of the intaking light source after being extended and collimated lights up a light field with a hole diameter of ∅44 after extended by the 4f system, the hole diameter of the hologram light field is ∅75 that has been amplified for 1.7 times enough to record a complete spraying hologram used for measuring exactly various parameters of the spraying property.

High pressure fuel pump and test injectors. The following types of fuel pumps and injectors were used in the test:

a). Fuel pump No I and injectors used in the reformed diesel engine Series 95 under normal temperature and normal pressure;

b). Fuel pump No I and three types of injectors (A, B - made in China, C - made by Bosch of West Germany,) used in diesel engines Series 100 under normal temperature and 1.5 MPa simulated back-pressure;

c). Fuel pump No III and injectors used in diesel engines Series 130 under normal temperature and 1.55 MPa simulated back-pressure. The measurement of spraying field under rated operating condition and max. torque condition are also made.

Test results and their analysis

Identification, counting and scaling methods of fuel droplet rendition image

The identification, counting and scaling of fuel droplet in the spraying field should be solved before data processing. According to holographic principles, only those droplets that their distance to the holographic dry plate is $-100\, d^2/\lambda < Z < 100\, d^2/\lambda$ can be recorded clearly. Light diffraction patterns of those on the focal plane are Fresnel diffraction rings, with even light intensity of rendition and clear boundary as shown in the Index 1 of Figure 5. The fuel droplets located much further than the focal plane at the time when Z is greater than $100(d^2/\lambda)$ belong to the distant field Fraunhofer diffraction. Besides the background light and object light in light intensity distribution of the diffraction ring, there are also Airy disks in the centre of images as shown in the Index 2 of Figure 5; The diffraction light field of those located nearer than the focal plane has a virtual image with blurred boundary as shown in the Index 3 of Figure. At present condition of our optical instruments, fuel droplets with diameter of 5μm can be measured

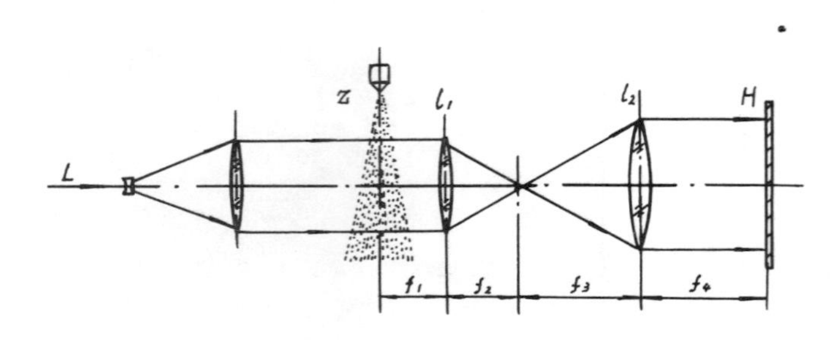

Figure 4. Extender 4f intaking light circuit.

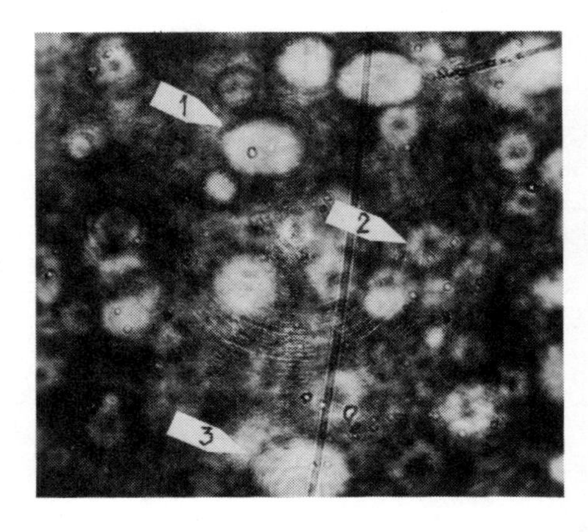

Figure 5. Rendition image of fuel droplets in spraying field.

Sizes of the fuel droplet are scaled out using the material object scaling method.Put an etalon with known diameter exactly on the focal plane and take picture for them together with the spraying field. Then the amplification ratio of the rendition based on the etalon can be used as the measuring means for fuel droplets.

<u>Data processing method for the rendered spray image</u>

The light field diameter of the spraying field hologram is ø75. If all data of the light field are to be processed, the counting of the droplet dimension and amount will be enormous. In order to assure a high degree of accuracy and to reduce the amount of work, the interpolation function developed us is used in the data processing, that is, the hologram is divided into limited element. The number and diameter of fuel droplets in each element are measured. The average diameter of fuel droplets, the SMD diameter and other spraying field parameters are also calculated. The result will then be computed with the interpolation function to find the distribution law of the spraying field (which was described before).

<u>Test results and their analysis</u>

<u>Droplet fineness in the fuel spraying field</u>. Fuel droplet fineness in the spraying field is an important parameter in defining the quality of fuel spraying, which reflects the evaporation speed of fuel under same condition in the combustion chamber, and the availability of air. The frequency spectrum distribution of the spraying field droplets (particle spectrum and diameter spectrum distribution) is used in description. Let's take the direct spraying space mixing type injector used in diesel engines Series 100 for example. Figure 6 shows the changing curve of droplet number N with different diameters and droplets diameter d based on the measurement in the processed areas, that is, the particle spectrum distribution. Figure 7 shows the changing curve of droplet percentage n with different diameters in the total amount of fuel droplets and the droplet diameter d, that is, diameter spectrum distribution. From these curves we know that under rated working condition and the maximum torque working condition, the frequency spectrum distribution curves of the above mentioned three types of injectors used in engines Series 100 rise sharply, their peak value is in the range of 5 to 15μm, where 90% of the total amount of droplets are in this range. This shows the droplet fineness in the spraying field of these injectors are good, in which the frequency spectrum distribution curve of the injector C has the highest peak value, the maximum number of fuel droplets and the highest percentage of fine droplets in the total amount of them, that is to say it has the best quality of droplet fineness. The injector A is the second best while the injector B is the worse one.

<u>The space-time distribution evenness of fuel droplets in the spraying field</u>. This is another important parameter of the spraying property, which reflects the space-time distribution status of the fuel and gas in the combustion chamber, and directly relates to the distribution of the mixed gas content in sprays. Limited by the measuring means in the past, no reliable data could be obtained directly. However, with holographic method instantanous

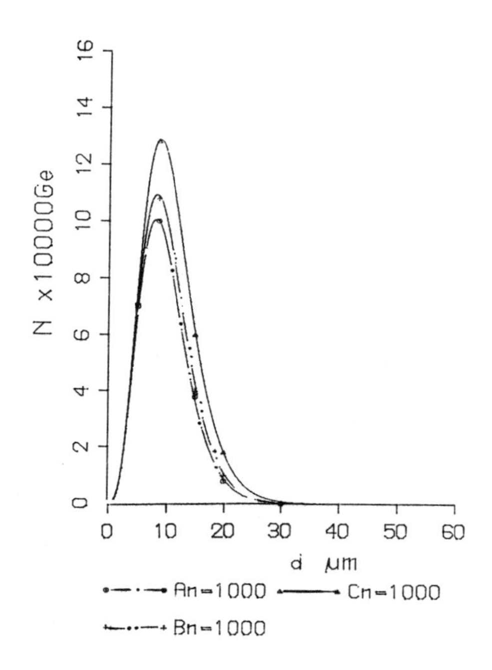

Figure 6. N-d particle spectrum distribution.

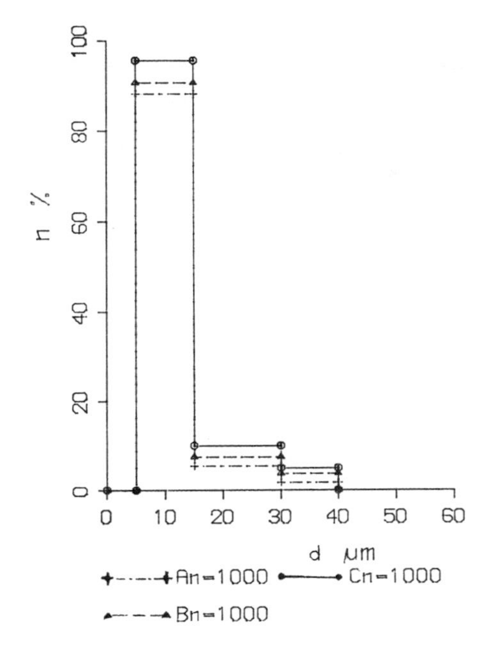

Figure 7. n-d diameter spectrum distribution.

status of droplet distribution in the space and with time during the whole spraying process can be exactly recorded, so the space distribution and the time distribution can be used to precisely describe the space-time distribution evenness of fuel droplets in the spraying field. Note: In the figures 6 - 17, Ge-piece/cm^3.

a). Space distribution: The unit volume droplet number EN, the droplet number in different diameter range N, the average diameter of droplets AD and the change of Sauter average diameter SMD with the spraying radial coordinate R in the spraying field are called the droplet space distribution. Figures 8 to 12 show the space distribution curves of 3 Series 100 injectors in the rated working conditions, from which the general trend can be found: There are much more EN and fine droplets with their d are both smaller than 15μm and greater than 5μm in the spraying centre than that at the spraying edges, as for larger droplets (d is greater than 15μm) there are more in the centre and less in the edge. From centre to the edge AD changes gradually and there are more SMD in the centre than at the edge. Here, injector C has the max. EN (1.35×10^5 piece/cm^3), the highest percentage (96.6%) of fine droplets (d is both geater than 5 m and smaller than 15 m), the smallest AD and SMD (AD=10.74μm, SMD=13.48μm) and its differential value between the max. value and the average value is the smallest, so injector C has the best space distribution evenness of spraying droplets.

b). Time distribution: Figures 13 to 17 show the changing of droplet dimension and their number distribution in the spraying field with time (calculated based on the rotating angle of the fuel pump's cam shaft) which is called the time distribution. The curve group has certain randomness because they come from different cycles, but it still reflects the distribution rule, i.e. more EN and fine droplets with d being greater than 5μm and smaller than 15μm at the edge than that in the centre, while more large droplets with d being greater than 15 m in the centre than that at the edge. The injector C at the max. torque working condition: AD=10 - 12μm, SMD=11.5 - 14μm, which are same as the values of the space distribution rule. So the space-time distribution evenness of the injector C is the best among the tested injectors, while injector A and B are much equivalent.

c). Penetrability and cone angle of the spraying: Figure 18 is a fuel spraying field hologram taken at different times (calculated based on the rotating angle of the fuel pump's cam shaft). The changing of spraying penetrability (indicated with the spraying range) and the cone angle of various injectors at rated working condition are processed. From drawings attached we can see that injector C's spraying range is short (L=89.6mm), its cone angle is big ($\alpha=25^\circ$). This shows the covered volume in the combustion chamber is large and wall-touching ratio of the fuel during spraying is small, so the air utilization rate is high.

Conclusions

The following conclusions are made after studying all of the measurement results:

a). Comparing with the injectors Series 95, 130 already measured, the spraying property of three types of injectors used in diesel engines Series 100 is better. Their parameters and indexes are as follows: More unit volume droplets, EN=0.84--1.35 x 10^5(piece/cm^3), 89.4%--96.6% Of 5<d<15μm fine droplets in total amount of fuel droplets, calculated average diameter (AD=10.74--11.92μm) and Sauter average diameter (SMD_{max} =15.24--19.69μm, $\overline{SMD}$= 13.47--16.6μm) are much smaller, spraying range L=79.3--98.2mm, cone angle α =23.6--25.3°. The spraying quality of the injector C is the best in the three. It has the maximum EN (1.03--1.35 x 10^5 piece/cm^3), d=the highest percentage (96.6%) of 5<d<15μm fine droplets, min. droplet average diameter (AD=10.74--10.91μm, SMD_{max}=15.24--16.29μm, $\overline{SMD}$=13.47--15.08μm), the shortest spraying range (L=79.3--89.6mm) and the largest cone angle (α=23.6--25°). The injector A is the second best and the injector B is the worse one.

b). The average diameter AD, SMD of spraying field droplets of three types of injectors used in Series 100 are smaller(SMD for Series 95 is 31.2μm,for Series 130 is 26.8μm), the differential value for the average diameter in centre area and that in border area is small, and the spraying field distributions at different time are same as the space distribution rule. These show the space-time distribution evenness of this three types of injectors are better, particularly, the injector C. Its AD, SMD are smallest, average diameter unevenness (δ =(Max. Value - Average Value)/Average value x %) is small, L value is small, value is large, so during fuel spraying, the wall-touching amount is small and the coverage of fuel spraying in the combustion chamber is large, and the distribution is reasonable. This can lower the requirement for inlet air flow and will be good for rapid evaporation, mixing, and combustion of fuel in diesel engines with direct spraying space mixing type combustion system. Therefore, it is estimated that higher economy and driving power will be obtained.

The above analysis have been verified by the property measurement of diesel engines Series 100.

Concluding Remarks

From the previous research works we have realized that:

1. With laser holography, we can micro-quantitatively measure the size, amount, spraying range and cone angle of fuel droplets in the spraying field of diesel engines, we can also observe the real shape of droplet while flying in the space, especially the space-time distribution evenness of the fuel droplet can be measured, which is unable to be measured with any other measuring means including high-speed photography. The measuring results can be used as the reliable basis in determining the spraying quality of injectors in order to improve the combustion process of the diesel engine and to obtain better properties.

2. The test result shows that the spraying field holographic research method and the high-speed instantaneous laser holographic system we found and developed can wholly meet the requirements for the measurement of fuel spraying field distribution. With China-made components and instruments, we have obtained spraying field distribution hologram for down to 5μm fuel droplets. The system is especially characterized by its intaking light circuit, synchro-system, droplet scaling, shock-absorption, testing devices and ways of data processing.

Besides that the system can be used in the measurement of IC engines, it can also be used, after a certain modification is made,in the microscopic measurement of corpuscle field, flow-rate field, combustion field, dynamic stress field, mechanical vibration, multi-phase reaction and other high-speed instantaneous changes in steam turbines,chemical reaction, powder-spray drying and agricultural equipment, etc.

Notes

The following are the participants of this research: Wang Bao Lin, Zhao Yi, Liu Gui Lian, Tan Cong Min, Li Guo Shun, Gao Yan Lai, Zhang Ying Chun, Wang Hong Sheng, Hou Jian Wen,et al

References

1. Shi Shao Xi,"On the Problem of Energy Conservation in Internal Combustion Engines", J. Internal Combustion Engines, Vol.1, No 1, pp 1-8. 1983.
2. Thompson, B.J., "Fraunhofer Diffraction with Coherent Background", J. Society Photographic-Optic Instru. Engineering, Vol.2, No 43, 1964.
3. Thompson, B.J., Parrent, G.B., Jusfh, B.J., and Ward, J.H., "Image Reconstruction with Fraunhofer Hologram", J. Optical Society of America, Vol.56, No 4, pp423. 1966.
4. Thompson, B.J., Ward, J.H., and Zinky, W.R., "Application of Hologram Techniques for

Particle Size Analysis", Applied Optics, Vol.6, No 3, pp 519-526. 1967.
5. Trolinger, J.D., "Particle Field Holography", Optical Engineering, Vol.14, No , pp 383. 1975.
6. He Wan Xiang, Li Zhi Xiong, and Ma Shi Ning, "Theoretical Investigation on Application of Laser Holography in the Measurement of Fuel Spraying Field of Diesel Engines", J. Internal Combustion Engines, Vol.2, No 3, pp 215-230. 1984.
7. He Wan Xiang, Li Zhi Xiong, and Qian Qi Yun, "The Application of Laser Holography in the Measurement and Study of Fuel Spraying Field Distribution of Diesel Engines", J. Internal Combustion Engines, Vol.3, No 1, 1985.
8. He Wan Xiang, Li Zhi Xiong, and Qian Qi Yun, On the Research of Laser Holographic Camera System of IC Engines , Intern. Conf. on Combustion of IC Engines, 855012.
9. Dong Yao Qing, "Present States of Diesel Engines Spraying Features Study Abroad", Foreign Fuel Pumps &. Injectors, No 3, 1985.

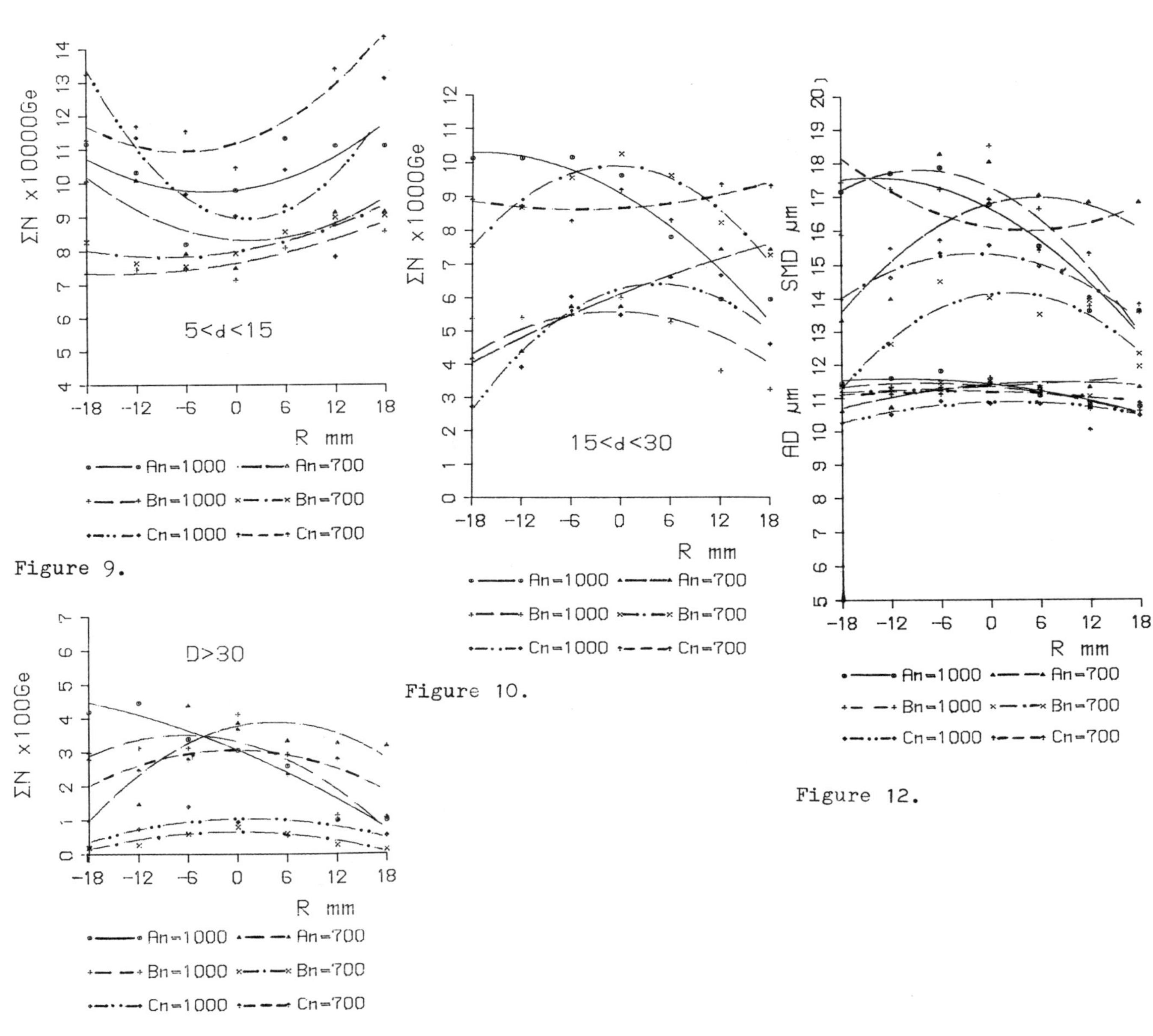

Figure 9.

Figure 10.

Figure 11.

Figure 12.

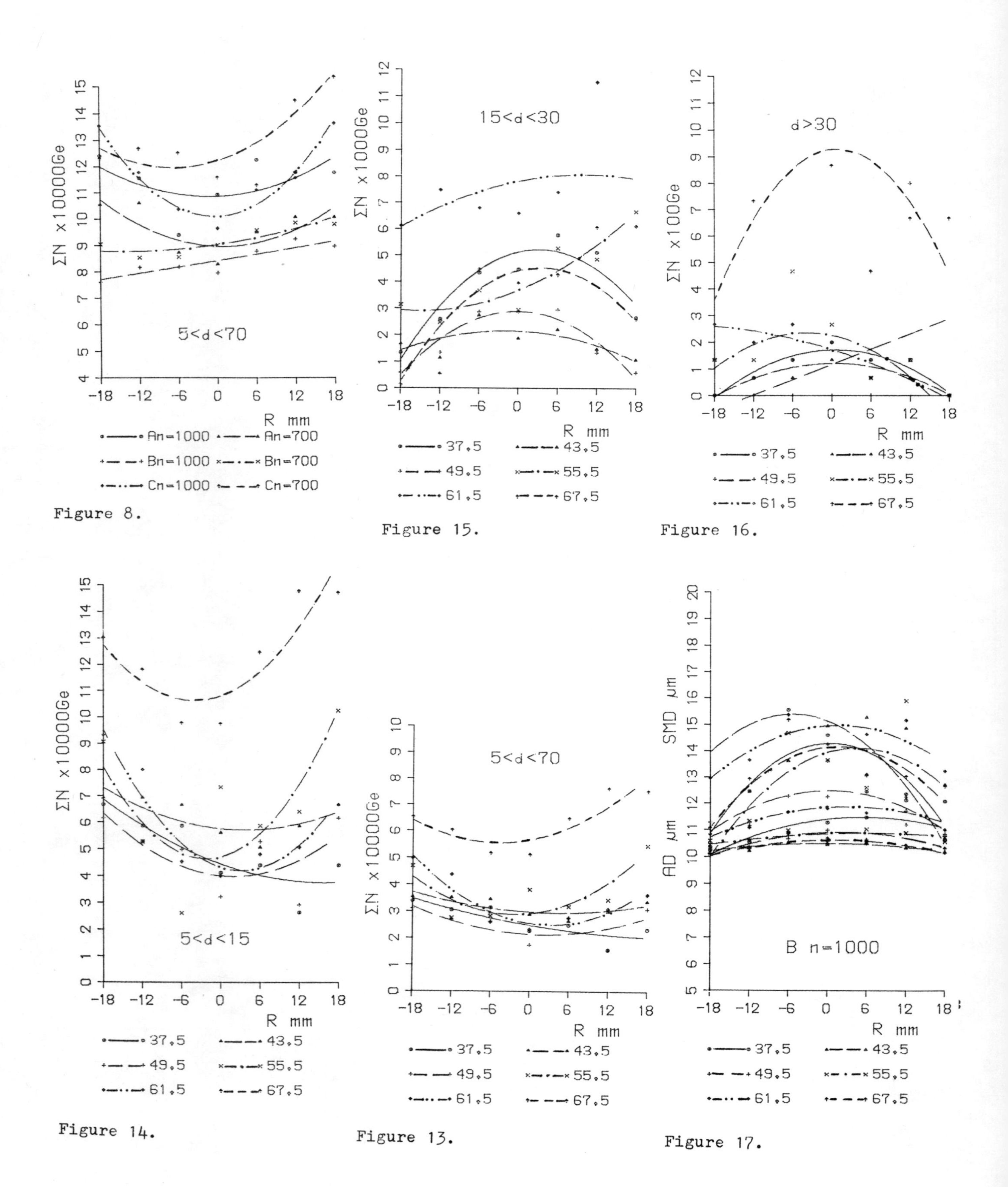

Figure 8.

Figure 15.

Figure 16.

Figure 14.

Figure 13.

Figure 17.

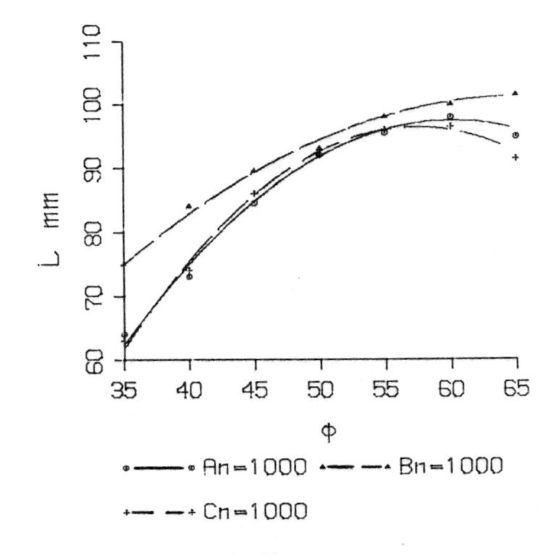
L mm
60
70
80
90
100
110
35
40
45
50
55
60
65
φ
An=1000
Bn=1000
Cn=1000

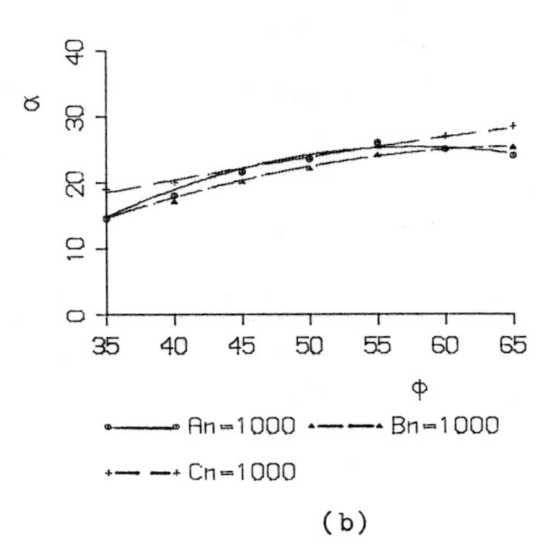
α
0
10
20
30
40
35
40
45
50
55
60
65
φ
An=1000
Bn=1000
Cn=1000

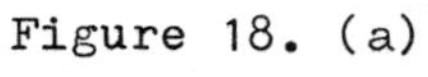
Figure 18. (a) (b)

White-Light Color Holography and Its Applications

F. T. S. Yu
Electrical Engineering Department
The Pennsylvania State University
University Park, PA 16802

Abstract

In this paper, several techniques of white-light holography and some of their applications are discussed. These techniques include the reflection hologram, rainbow hologram, spatial encoding, coherent and white-light speckle encodings, double-aperture encoding, dual-beam encoding, incoherent spatial encoding, source encoding, and white-light Fourier holography. Since one of the interesting applications of white-light optical signal processing is psuedocolor encoding, we shall demonstrate some of those applications to holographic interferometry and metrological studies. Techniques of producing pseudocolor encoded holographic fringe patterns and misfocused speckle interferograms are provided. These techniques allow the viewing of a multiset of encoded interferograms simultaneously.

I. Introduction

The technique of wave front reconstruction, or rather the holographic process, is now over three decades old, having been proposed by Dennis Gabor in 1948 as a possible means of improving the resolving power of the electron microscope.[1] It was not until in the early 1960s, when Leith and Upatnieks produced the first high-quality holographic image using a strong coherent source that wide attention was given to holography.[2,3]

Although holography has provided a great step towards practical three-dimensional imagery, its acceptance for commercial and educational uses has been slow for a number of reasons. The high cost of the holographic process, low hologram image luminance, and the necessity of special illuminators for high quality images are among the primary causes for its rejection. In this paper, we will discuss several techniques of producing brighter and more efficient holographic images. The types of color holographic images that we will mostly address are observed through the transmitted light field, which are called white-light transmission color holograms, to distinguish them from Denisyuk's white-light reflection holograms.[4] We will also in this paper demonstrate a few of its applications.

II. Reflection Color Holography

The best-known white-light color holographic process must be that due to Denisyuk.[4] In 1962, he reported a technique in which the process of holography was combined with a form of color photography invented[5] by the French physicist Gabriel Lippmann in 1891. Thus the Denisyuk hologram can be viewed with a white-light source. In other words, Denisyuk's work is one of the cornerstones of white-light holography, combining the work of Lippmann and Gabor by using coherent light for hologram construction and a white-light for hologram image reconstruction. In his method, a coherent polychromatic wave field, with the primary colors of coherent light, passes through a recording plate, falls on a diffused color object and then is reflected back to the recording plate, traversing it a second time but in the opposite direction, as shown in Fig. 1a. As in the Lippmann color photographic process, interferometric fringes are formed throughout the depth of the emulsion. The result is a color hologram that has the characteristics of the Lippmann process and can be viewed with a white-light source of limited spatial extent. An example is an ordinary high intensity desk lamp or slide projector, as illustrated in Fig. 1b. Such a color holographic image viewable by a white-light source was subsequently demonstrated by several investigators in the past.[6-8] However, the reflection color hologram does possess several drawbacks which prevent wide spread practical applications. Two of those drawbacks are: 1. An elaborate film processing technique is required to prevent the emulsion shrinkage; 2. The hologram image diffraction efficiency is relatively low. Nevertheless, the reflection hologram image can be viewed by direct white-light illumination, and it is convenient for the decorative displaying purposes.

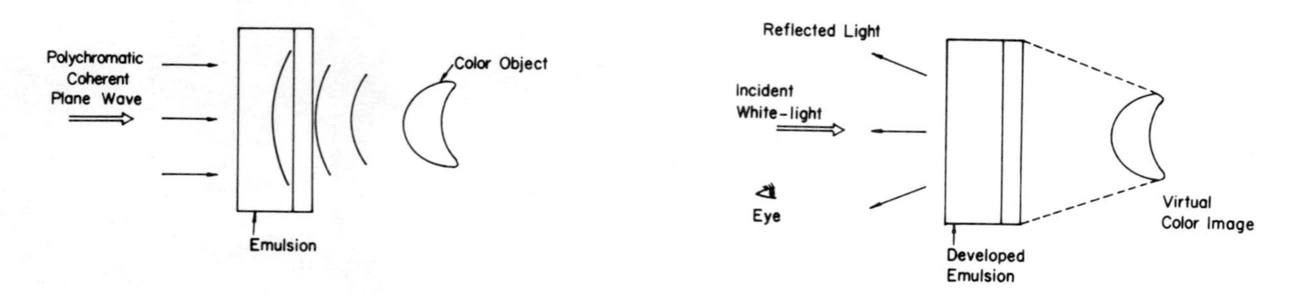

Fig. 1a Construction process

Fig. 1b Reconstruction process

Figure 1. Reflection Color Holography

III. Color Rainbow Holography

In 1969 another type of white-light hologram was reported by Benton.[9] This type of hologram is also known as white-light transmission hologram but it is best known as rainbow hologram. A Benton-type rainbow hologram involves a two-step recording technique. First a hologram (called the primary hologram) is formed with the conventional off-axis technique. Next a second hologram is made, using the real image from the first primary hologram as the object beam. In the second step of the holographic construction process, a narrow horizontal slit is placed at the primary hologram. Since each point of the primary hologram produces the entire holographic image but from only one viewpoint, the narrow slit aperture eliminates all the parallax in the vertical direction. Thus the second hologram, under a conjugate illumination, forms a holographic image (either real or virtual) of the original object and a real slit image behind the hologram. To view the object image, the viewer must place his eye at the slit image. If the viewer moves up and down transversely, the object image disappears; thus the vertical parallax is lost. The object image, perceived by the viewer, can be quite bright since the real slit image is convergently reconstructed and the light gain can be several hundred fold.

There is however an interesting phenomenon of the Benton-type hologram: If the hologram is illuminated by two different wavelengths of coherent sources, the lateral dispersion of the diffracted light results in two vertically displaced reconstructed slit images, one for each color. The viewer can observe the object image in one color if he places his eyes at one of the slit images and in the other color if he moves his eyes to the other slit image. By extrapolating the displacement of the slit image due to wavelength, it is rather straightforward to visualize that slit image would smear into a continuous spectrum, if the hologram is illuminated by a white-light source. To view the holographic image, the viewer would simply place his eyes at the smeared slit image where he would see the image in one color. It is apparent that the image color changes if one views from the top to the bottom of the smeared slit image. Thus this type of hologram is known as rainbow hologram.

Although color holographic images have been generated by a two-step rainbow holographic technique by Hariharan et. al.[10], the process of constructing a true color rainbow hologram is rather cumbersome and time consuming. First, three primary holograms have to be constructed with three primary color coherent sources. Then the projected real images of these three primary holograms are multiplexed onto a fourth hologram sequentially, again with three color coherent read out. These three primary holograms must be aligned very carefully to assure that their reconstructions would be exactly overlapped. Thus this technique for generating a color holographic image is quite complicated and not very easy to implement. We shall now describe a technique of color rainbow hologram that can be constructed with one-step.[11] The optical arrangement is illustrated in Fig. 2a. A He-Ne laser is used to provide the red light (6328Å), and an argon laser is used to provide the green and blue lights (5145Å and 4765Å). The illuminated object is imaged through an achromatic lens to a plane just in front of the hologram. A narrow slit of about 1.5 mm is placed between the object and the focal plane of the imaging lens. A collimated reference beam is used to insure the carrier frequency would be the same across the hologram. The intensities of the three lights are measured independently, and the exposure time for each is calculated. The hologram is first exposed to the red light of the He-Ne laser, then the green light (5145Å) of the argon laser. The argon laser is then tuned to the blue line (4765Å), and a third exposure is made.

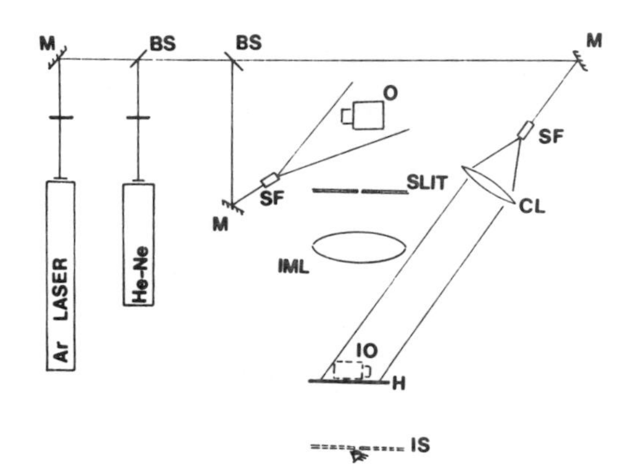

Fig. 2a Construction process

Fig. 2b A black-and-white photograph of a color holographic image

Figure 2. Color Rainbow Holography

A Kodak 649F plate is used because of its relatively flat spectral response. After the plate is developed, a rainbow hologram is formed. When the hologram is viewed with a polychromatic point source, a very bright color image is reconstructed. It should be noted that with this arrangement, the same reference angle is used in the reconstruction instead of the conjugate angle as with the Benton-type rainbow hologram. Similar to the color rainbow holograms generated with the two-step Benton method using three primary holograms, a holographic image with exact color reproduction is observed when the hologram is viewed in the correct plane. If the viewer moves off this plane, different shades of color can still be seen, but the color would be different from that of the original object. This is not a serious drawback since three-dimensionality is preserved along only one axis in rainbow holograms. The greatest disadvantage of this technique is that the field of view is restricted by the lens aperture. In order to achieve a large field of view, an imaging lens with large aperture and small focal length is required, which tends to be expensive. However, this technique is so much simpler than other techniques in producing color holographic images, that it should be a useful addition despite its drawback. In Fig. 2b, we have a color image generated by the method we described. The image is very bright and sharp, and the color reproduction is very distinctive and quite faithful. There is however a marginal resolution loss along the axis perpendicular to the slit[12] In addition, the smeared slit images also cause some degree of color blur in the object image.[12-15]

IV. Color Holographic Imaging with White-Light Processing

One of the major advantages of optical signal processing[16] must be the color image processing. Since the white-light source emanates all the visible wavelengths, it is particularly suitable for color holographic imaging. In the following we will describe a few techniques of color holographic imaging utilizing a simple white-light processor.

4.1 Spatial Holographic Encoding

We shall now describe a method of generating color holographic images that utilizes a white-light processing technique.[16] We have previously shown that multiple holographic interferometric images can be encoded onto a holographic plate and reconstructed separately or simultaneously using a white-light optical processor.[17] We shall use the same technique to generate a color hologram. Since there is no slit involved in the construction process, in principle, this technique will eliminate those disadvantages posed by the rainbow holographic technique.

In this technique, the color hologram is constructed by sequentially illuminating the object using three primary color coherent sources and three corresponding reference beams each oriented at equi-angular positions about the object beam axis, as illustrated in Fig. 3a. By making three primary color exposures, a multiplexed image plane hologram is encoded. The encoded hologram is placed at the input plane of a white-light processor diagramed in Fig. 3b. Spatial filtering is performed on the smeared Fourier spectra, using narrow band spatial filters placed in the appropriate locations corresponding to the three primary colors. The filtered light then recombines to form a color image at the output plane. To enable all of the diffracted light emanating from the multiplex hologram to pass through the achromatic transform lens during reconstruction, a very narrow angle is required between the one object beam and the individual reference beams used in constructing the hologram. A broader angle may be used if a large aperture and short focal length lens can be found. For experimental demonstration, Fig. 3c shows a color holographic image obtained by this white-light processing technique. This technique allows color holographic images to be reconstructed using a white-light source, while, in principle, eliminating the marginal resolution loss and blue color found in the rainbow holographic system. However, some of the previous problems found in processing the holographic image still exist. These include an elaborate recording scheme and the undesirable speckle effect inherent to coherent construction which cannot be totally eliminated, although it is somewhat reduced during white-light reconstruction. This technique also has a limited range of object-to-reference beam angles available for holographic construction. During the recording process the three complex holographic gratings that are formed will also introduce moire fringe patterns. But, by an ingenious design of the holographic encoding process it may be possible to eliminate these fringe patterns.

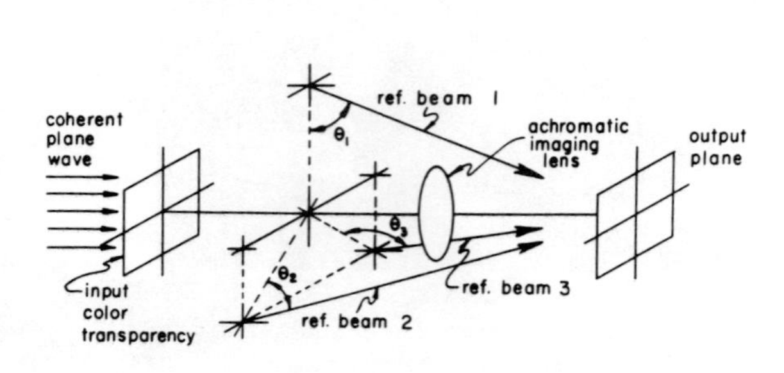

Fig. 3a Construction process

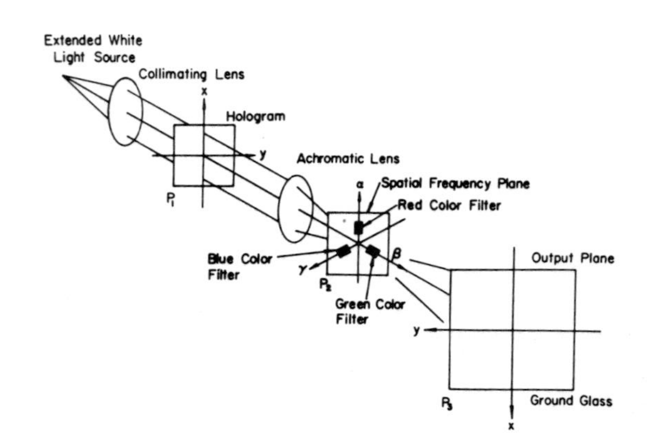

Fig. 3b Color holographic imaging

Fig. 3c A black-and-white photograph of a color holographic image

Figure 3. Color Holographic Imaging by Holographic Encoding

4.2 Coherent and White-Light Speckle Encoding

We shall first demonstrate a simple technique that utilizes coherent speckles to encode color images onto black-and-white film.[18] In encoding, a diffuse color object illuminated by coherent light is imaged onto a photographic plate through a narrow slit by an imaging lens as shown in Fig. 4a. Let us assume that the recording was sequentially performed using only red and green coherent illumination with the slit oriented at 0° and 90°, respectively, thereby, giving us red and green encoded color image multiplexed onto one photographic film. Due to the different orientations of the slit aperture, the red encoded specklegram will have speckles elongated in one direction, while the green encoding has speckles elongated in the other direction (90° apart). Thus an encoded monochrome multiplex specklegram or speckle hologram can be recorded. Decoding the color image, we insert the multiplexed specklegram in the input plane P_1 of a white-light optical processor, as depicted in Fig. 4b. Since the elongated speckles of each specklegram are oriented about 90° apart, the corresponding Fourier spectra would be distributed in confined directions perpendicular to these elongations in the spatial frequency plane P_2. It is therefore apparent by spectrally filtering these Fourier spectra, a full color holographic image can be reconstructed at the output image plane P_3.

For experimental demonstration, Fig. 4c shows a multicolor hologram image obtained with this specklegraphic technique. Although the resolution of the reconstructed color image suffers, the color reproduction is relatively faithful. With the use of a finer diffuser and appropriate slit size in the construction process, an optimum color holographic image reproduction may be obtained. Finally, we also note that color holographic imaging can also be obtained by white-light speckle encoding.[16] With the same white-light optical processing technique, a true color holographic image can be viewed, as shown in Fig. 4d. However this technique also possesses the same drawbacks as the coherent speckles, except the simplicity of white-light illumination.

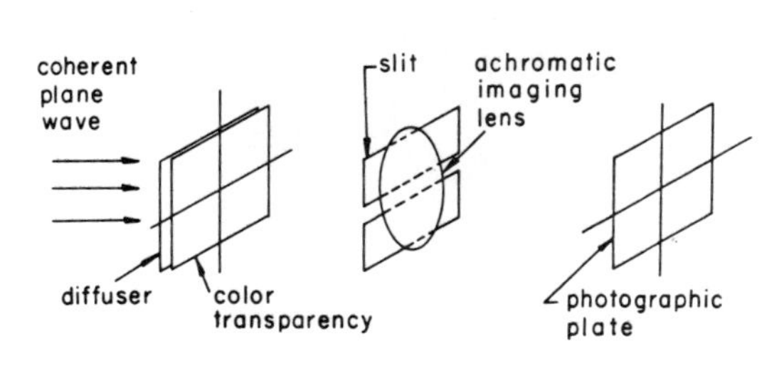

Fig. 4a Construction process

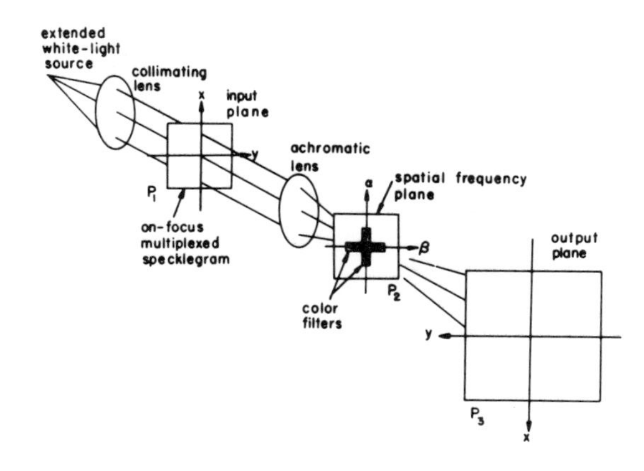

Fig. 4b Color holographic imaging with a white-light processor

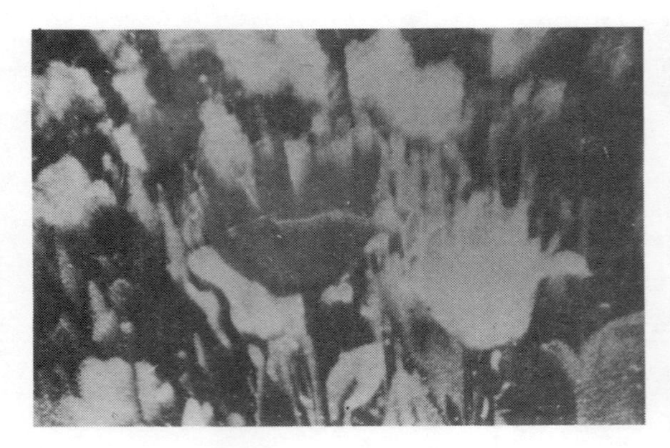

Fig. 4c A black-and-white photograph of a coherent speckle color holographic image

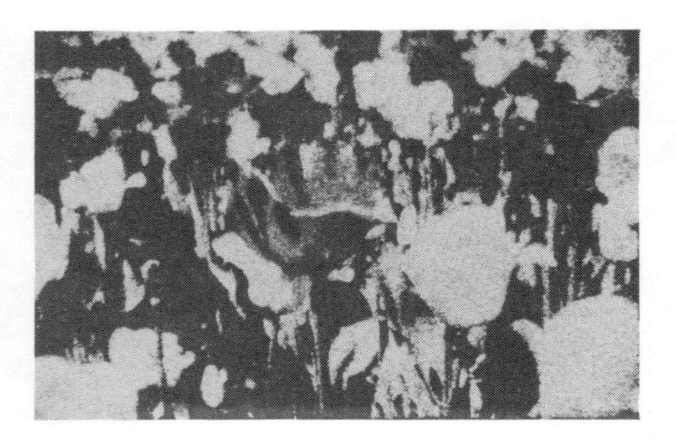

Fig. 4d A black-and-white photograph of a white-light speckle color holographic image

Figure 4. Coherent and White-Light Speckle Encoding.

4.3 Double-Aperture Encoding

We shall now show a technique by Gerhart and Ruterbush[20] for producing a color holograhic image utilizing fringe modulation. Although multiple aperture speckle interferometry has been used previously, the application to color holograhic imaging is quite new.

A process for constructing a fringe modulated speckle pattern is illustrated in Fig. 5a. The diffusing element produces wide band spatial frequency noise with a sufficient bandwidth for the light from each object point to pass through both apertures. The irradiance function at the recording plane generates a complicated speckle pattern by the random spatial frequency carriers. The two apertures are separated by a distance 2b which is greater than the diameter 2a of each individual aperture. These two light beams will coherently superimpose to produce a single image. Since each pair of light rays is mutually coherent, it will interfere to produce a random speckle pattern which is modulated by parallel fringes. These interference fringes are perpendicular to a line joining the aperture centers, which can be observed by placing an optical microscope in the image plane. Figure 5b shows a typical speckle fringe pattern with a spatial frequency of 90 cycles/mm and a speckle diameter of 55μm. The lower spatial frequency content of the input object and the random speckle pattern is modulated onto these holographic fringes which act as a periodic spatial frequency carrier. Thus a hologram encoding can be recorded at the image plane, and the holographic image can be retrieved by a white-light processor, as shown in Fig. 5c.

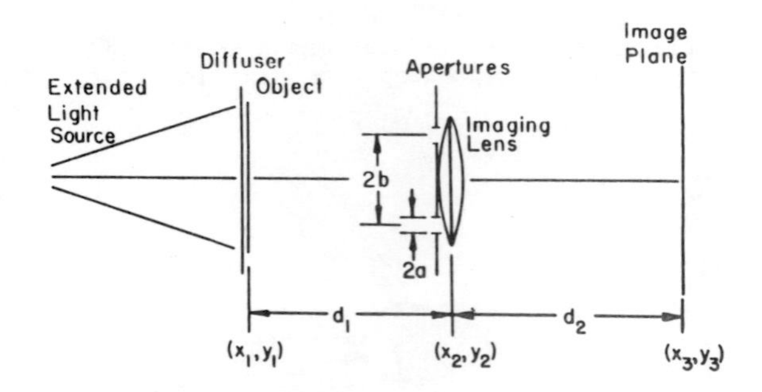

Fig. 5a Construction process

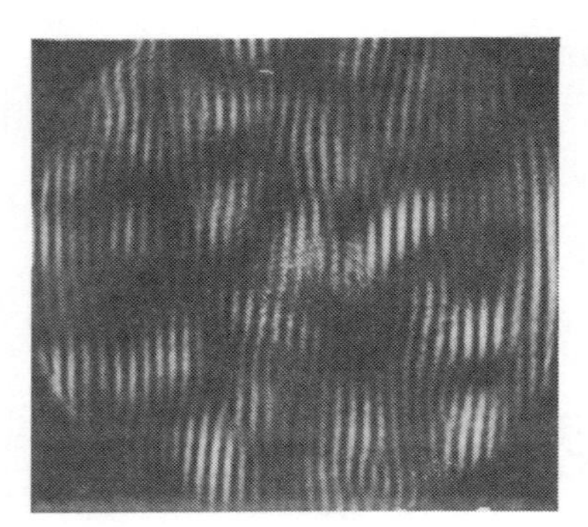

Fig. 5b Typical modulated speckle pattern

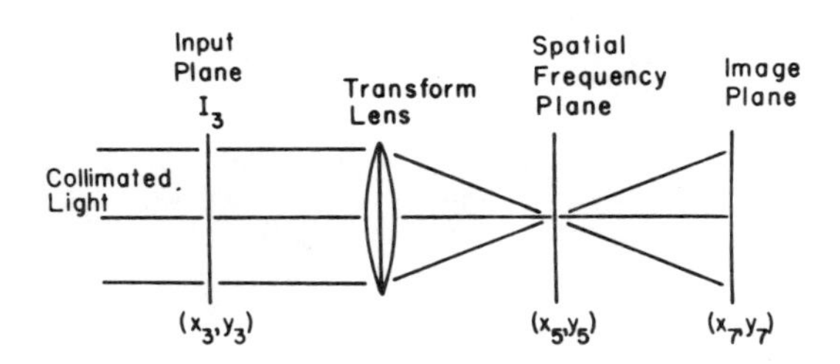

Fig. 5c Holographic image reconstruction by a white-light processor

Figure 5. Double-Aperture Encoding.

The procedure for a three-dimensional color hologram encoding is similar of the 2-D case where variations in the axial positions of the object are encoded by variations in the speckle fringe spacings on the corresponding image points, as shown in Fig. 6a. A Rubik cube with six distinct colors is illuminated by red, green and blue coherent sources. The holographic encoding can take place either sequentially or simultaneously with the coherent illuminations. Three pairs of apertures containing red, blue, and green filters are positioned in front of the imaging lens. The various diffraction orders are recorded onto a black-and-white holographic film. If the recorded color hologram is placed at the input plane of a white-light processor, as shown in Fig. 6b, a multicolor holograhic image can be viewed at the ouput image plane as shown in Fig. 6c. Although this technique suffers some degree of resolution loss, the holographic encoding has a much less stringent temporal and spatial coherence requirement, which means a white-light source can be used. In addition, the simplicity of this technique may offer a wider dimension in color holographic imaging applications.

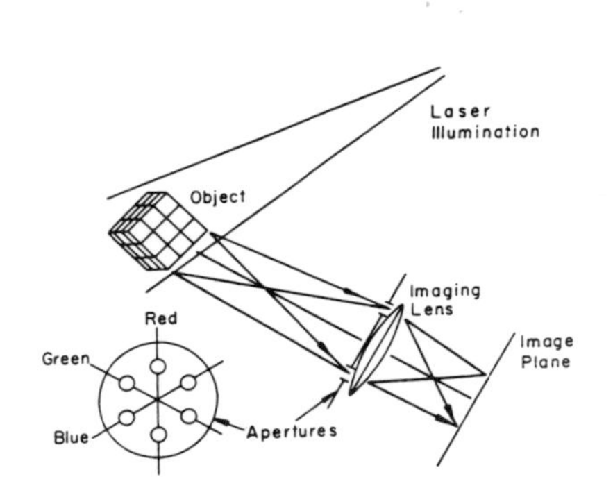

Fig. 6a Construction process

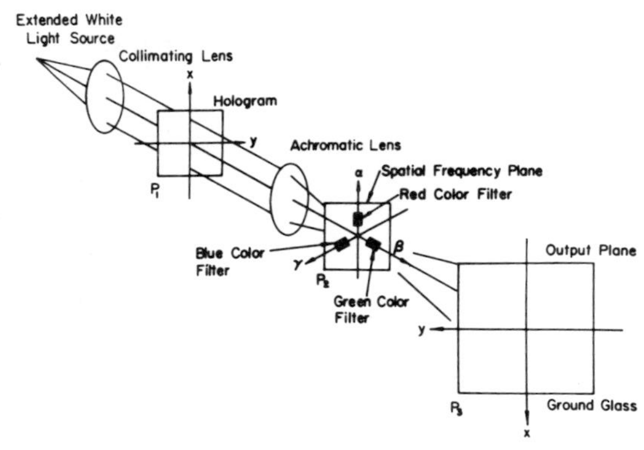

Fig. 6b Color holographic imaging with a white-light processor

Fig. 6c A black-and-white photograph of a color holographic image

Figure 6. Multi-Aperture Encoding

4.4 Dual-Beam Encoding

Although the color holographic image encoding can be easily obtained by the double-aperture speckle modulation technique, the technique suffers one major drawback, namely the holographic image resolution is limited by the size of the apertures. We will describe a technique of dual-beam encoding to alleviate this disadvantage.[21] We have in a previous paper[22] proposed a real holographic image contouring method utilizing dual-beam coherent illumination. We shall utilize this dual-beam technique for a color-hologram construction, as illustrated in Fig. 7a.

With reference to this figure, the color holographic encodings can be sequentially or simultaneously taken by dual beam coherent illuminations, each with a specific spatial sampling direction. In other words, the encoding can be performed by spatial sampling of the color object with the red, green, and blue coherent lights. Since these primary color coherent lights are mutually incoherent, a spatially encoded color image can be recorded on a black-and-white photograhic plate to form a multiplexed color hologram. If this recorded hologram is inserted at the input plane of a white-light processor shown in Fig. 7b, then a true color holographic image with high quality can be observed at the output image plane of the processor, as shown in Fig. 7c. The advantage of this technique, in principle, is that it can produce a high resolution color holographic image limited only by the imaging lens. This technique is suitable for three-dimensional holographic image reconstruction. In addition this technique eliminates the use of a reference beam, which simplifies the hologram construction procedure. However, with use of coherent illumination for the construction process, the inherent coherent noise can not be eliminated.

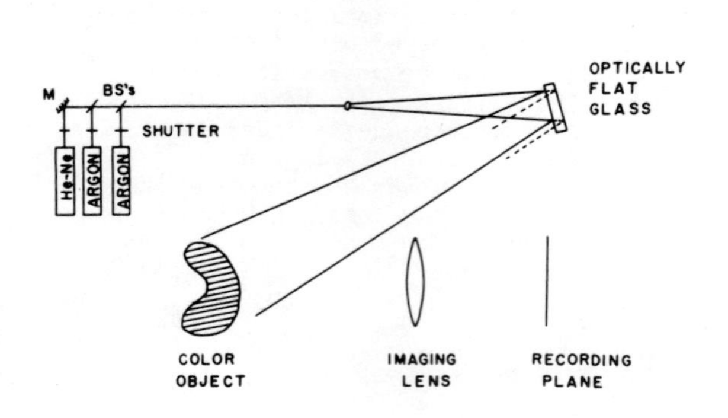

Fig. 7a Construction process

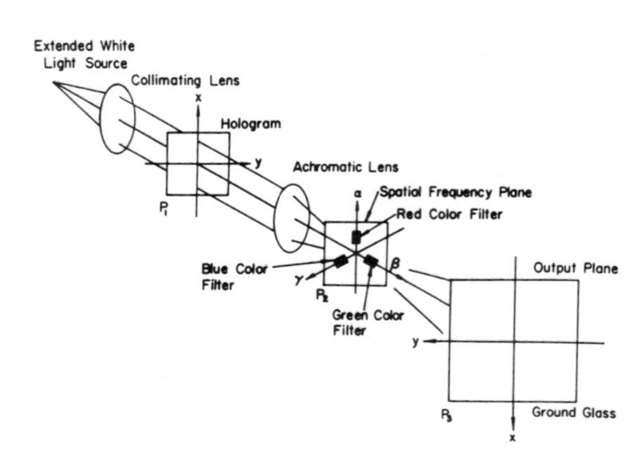

Fig. 7b Color holographic imaging with a white-light processor

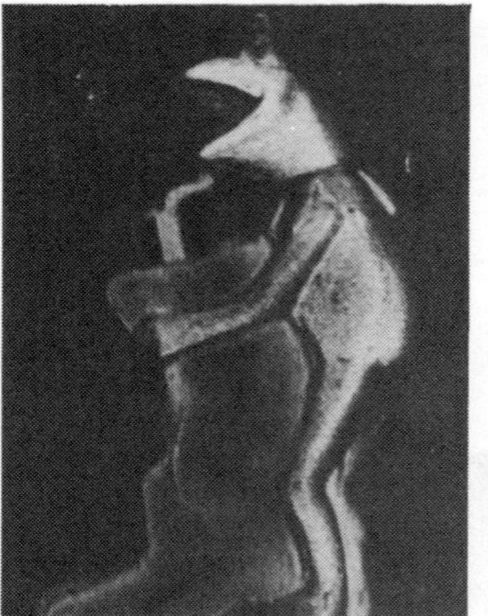

Fig. 7c A black-and-white photograph of a color holographic image

Figure 7. Dual-Beam Encoding

4.5 Incoherent Spatial Encoding

We shall now illustrate a tricolor sampling technique for white-light color hologram construction, as depicted in Fig. 8a. The use of a monochrome transparency to retrieve a color image by diffraction was first reported by Ives[23] as early as in 1906. However, the use of a tricolor grid for image sampling may be first due to Mueller[24] in 1969. In other words, an image sampling hologram may be obtained by spatially sampling the primary colors of a color image with three spatial directions onto a black-and-white film. For example, the red color image of the object is sampled in one direction, while the green color and the blue color images of the object are sampled at 60° and 120° respectively to the original direction. Thus the three primary color images of the object can be spatially multiplexed onto a black-and-white film to form a multiplexed color hologram. However this encoded in-focus hologram is a negative image hologram, which is not suitable for color holographic image reconstruction by a white-light processing technique.[25] To alleviate this drawback, the encoded hologram is bleached by a R-10 formula[26] to convert it into a surface relief phase object hologram. If this surface relief hologram is inserted in the input plane of a white-light processor shown in Fig. 8b, then a true color holographic image can be viewed at the output plane of the processor, as shown in Fig. 8c. For simplicity, the color holographic image was obtained with a red-and-green color encoding grating. The advantage of this technique is that the color multiplexed hologram can be constructed by simple white-light illumination. However the color hologram image loses the three dimensionality, due to the incoherent construction. Nevertheless, with the use of a white-light source, this technique offers a high quality holographic color image with virtually no coherent artifact noise. In addition, the system is more economical and simple to maintain as compared with the coherent color holographic processor.

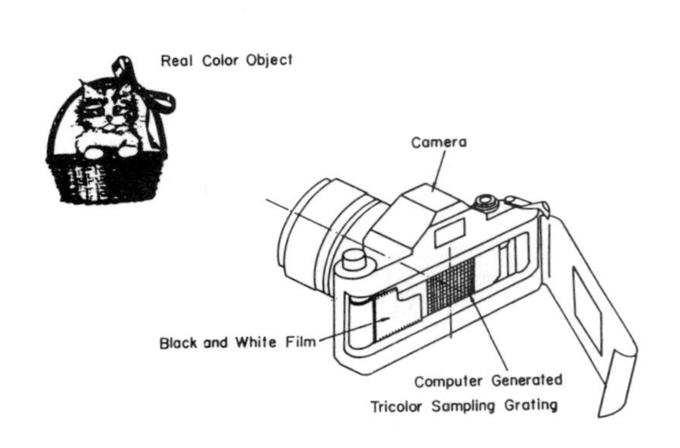

Fig. 8a Construction process

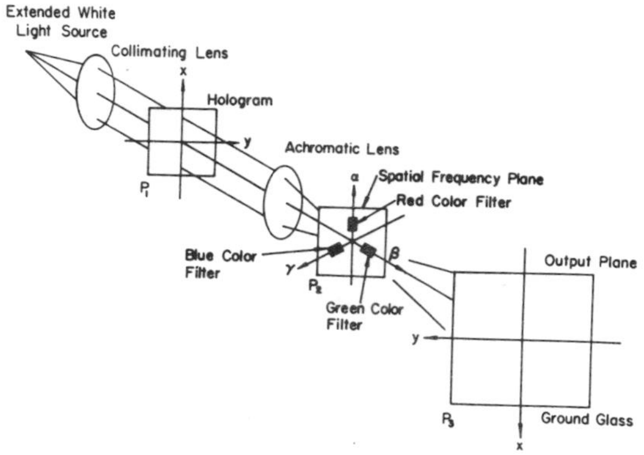

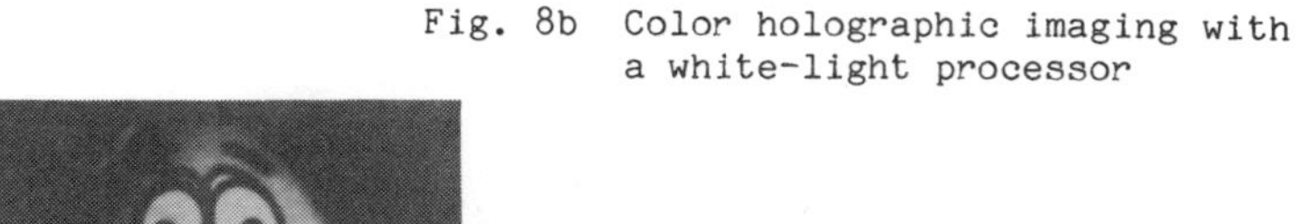

Fig. 8b Color holographic imaging with a white-light processor

Fig. 8c A black-and-white photograph of a color holographic image

Figure 8. Incoherent Spatial Encoding

4.6 Incoherent Source Encoding

We have in a previous paper described a color image subtraction technique utilizing encoded extended incoherent sources.[27] We have shown with an appropriate source encoding[28], a point pair coherence function can be obtained at the input object plane of an optical processor, as shown in Fig. 9a. We shall utilize this source encoding concept to establish the coherence requirement for a color hologram construction. A color holographic image can then be reconstructed with a white-light processor.

Again, for simplicity, we shall use only two primary color light sources for the color hologram encoding. This source encoding system of Fig. 9a is similar to the system for the color image subtraction[27], except the diffraction gratings are placed at a distance ℓ from the Fourier plane P_2. With reference to this figure, one sees that if a color object transparency is inserted in one of the input open apertures, say $O_1(x,y)$, then the light field from the open aperture O_2 would act as a reference beam for the hologram construction.

Since the two beams of light, one from the object transparency and others from the open aperture, are coherently added at the output plane, a multiplexed color hologram is formed at the output plane. The carrier spatial frequencies of the hologram are dependent on the distance ℓ and the wavelength λ of the light sources.

If we insert this encoded hologram at the input plane of a white-light processor, as shown in Fig. 9b, two sets of smeared Fourier spectra can be seen in the Fourier plane. By a pre-implementation of the spatial frequency carriers of red and green color images during the holograhic construction, it can be shown that the red and green spectra of the smeared Fourier spectra appear at the same band in the Fourier plane. By using a narrow slit filtering, a true color holographic image can be reconstructed, as shown in Fig. 9. This technique can also be extended for three color holographic imaging. We see that by properly adjusting the distances of the diffraction gratings, it is possible to make the three spectral bands of the smeared Fourier spectra to be diffracted at the same band in the Fourier plane.

Although the color hologram is constructed by incoherent sources, this technique possesses a severe drawback since it is limited to two-dimensional objects. Furthermore, in view of the color holographic image of Fig. 9c, the image quality is rather poor as compared with the other techniques.

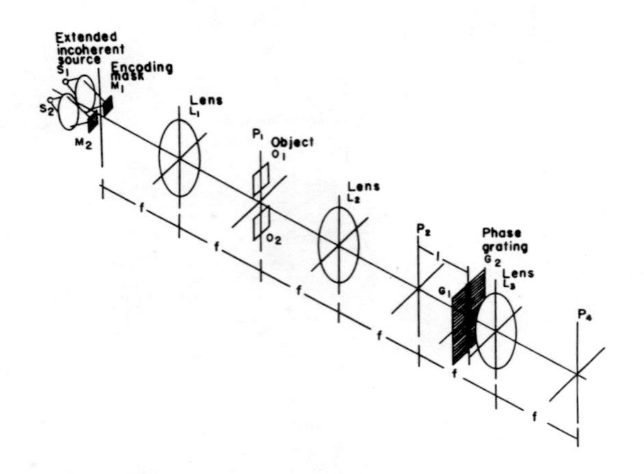

Fig. 9a Construction process

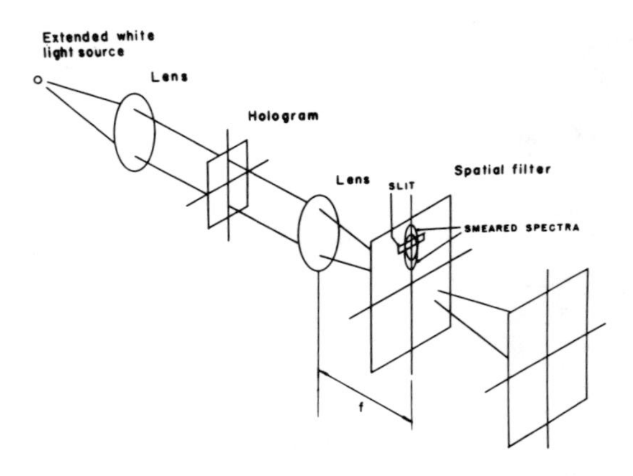

Fig. 9b Color holographic imaging with a white-light processor

Fig. 9c A black-and-white photograph of a color holographic image

Figure 9. Incoherent Source Encoding

4.7. White-Light Color Fourier Hologram

We shall now describe a technique of generating a broad spectral band Fourier hologram with an encoded white-light source[29], as schematized in Fig. 10a. This technique utilizes a high-efficiency dispersive grating and a source encoding mask to obtain the required temporal and spatial coherence for broadband Fourier hologram construction. In other words a sequence of Fourier sub-holograms for a broad range of wavelengths can be constructed in the Fourier plane. If the encoded broadband Fourier hologram (or rather sub-hologram) is inserted back at the Fourier plane of the white-light processor shown in Fig. 10b, it can be seen that twin color holographic images would be constructed at the output image plane.

Figure 10c shows a twin color Fourier holographic image obtained with this technique. Although the image quality is still somewhat poor as compared with the original color object transparency, the color reproduction is quite faithful. Nevertheless, this result may be the first white-light color Fourier holographic image ever obtained.

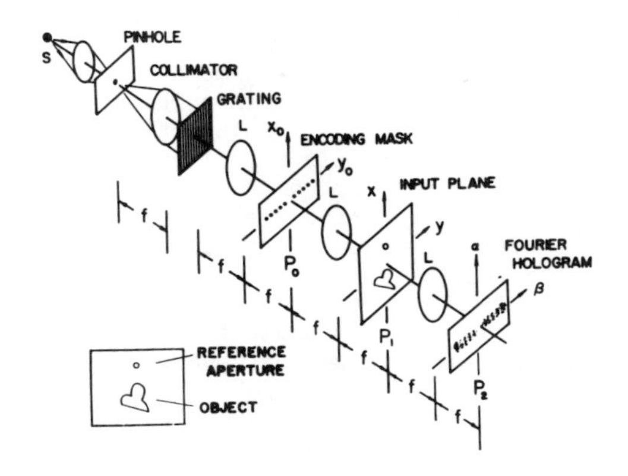

Fig. 10a Construction process

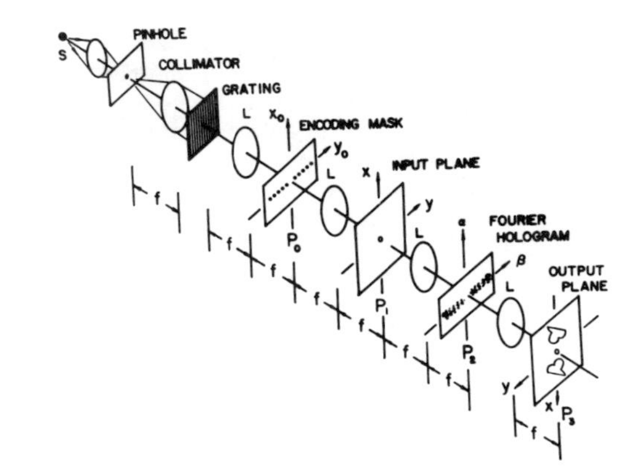

Fig. 10b Color holographic imaging with a white-light processor

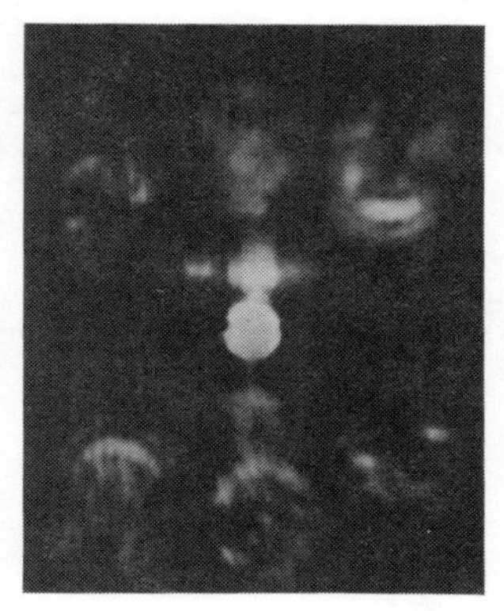

Fig. 10c A black-and-white photograph of a twin color Fourier holographic image

Figure 10. White-Light Fourier Hologram

V. Applications

A wide variety of colorful applications for white-light optical signal processing have been recently reported by Yu[30]. Since the white-light optical signal processor is capable of processing the signals in complex data, it is suitable for the applications to interferometric studies. We shall now demonstrate some of the applications to holographic and misfocused speckle interferometric studies.

5.1 Color Encoding of Multiplexed Holographic Interferometric Fringe Patterns

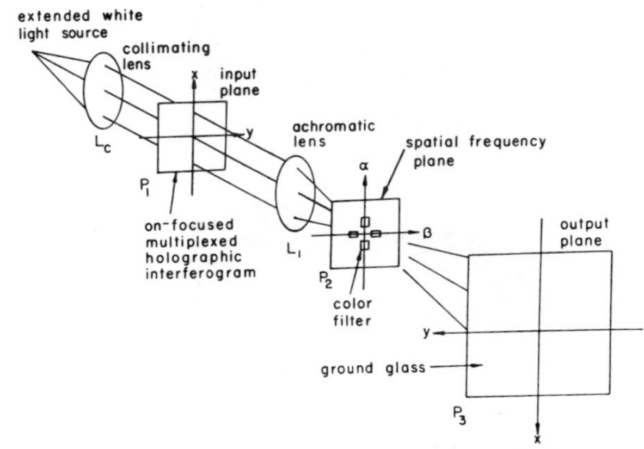

Figure 11. White-Light Optical Color Encoding of Holographic Fringe Patterns

Holographic interferometry has been shown to be a valuable means of studying the deformation of an object[31]. Recently a technique of white-light reconstruction of a multiplexed on-focus holographic interferogram was reported by Gerhart et al[32]. They have shown that the holographic fringe patterns, resulting from different physical effects, can be color-coded and reconstructed one at a time or simultaneously with a white-light optical processor. For example, a set of N on-focused interferograms can be multiplexed onto a single holographic plate with different angular reference beams. This set of interferograms can then be reconstructed one at a time or simultaneously with a simple white-light optical signal processor, as shown in Fig. 11. Since the multiplexed on-focus interferograms are illuminated by a white-light source, we would expect to observe N set of rainbow colors Fourier spectra displaying over the Fourier plane, P_2. Thus by simple spectral band filterings of these smeared Fourier spectra, a color coded holographic fringe data can be obtained at this output plane P_3, of the white-light processor.

Figure 12 shows a set of results obtained with this technique. The object is a small 8 Ω, 0.2W speaker, which was driven by a 2 kHz sinusoidal generator. A pair of multiplexed time-average on-focus interferograms were recorded onto a holographic plate, for different applied voltages (e.g.) 2V P-P and 4V P-P respectively, to the speaker. Thus two sets of color smeared Fourier spectra, about 90° apart, can be observed in the spatial frequency plane P_2. If a red and a green filter were used over the smeared Fourier spectra, a color coded interferogram can be seen at the output plane P_3.

Figure 12(a) shows a broader interferometric fringe pattern, which is color encoded in green. This holographic interferogram is produced with applied voltage to the object speaker during this holographic recording. The finer red fringe pattern of Fig. 12(b) is the reconstructed interferogram corresponding to a 4V P-P applied voltage. Figure 12(c) shows combination of Figs. 12(a) and 12(b) obtained at the output end of the white-light processor. If the individual fringe patterns have nearly identical intensity functions, the superposition of two bright fringes shows a yellow hue. The brightest fringes on a time-average interferogram are nodal regions where the vibration amplitude is zero. This phenomenon is described mathematically by an intensity distribution which is proportional to $(J_o)^2$, where J_o is a zero-order Bessel function. The less intense yellow regions correspond to a superimposed pair of higher-order bright fringes. The red and green fringes similarly correspond to a superposition of one dark and one light fringe. This approach to holographic fringe analysis shows great promise for comparing two or more fringe patterns to identify nodal or zero-motion regions on the vibrating test object.

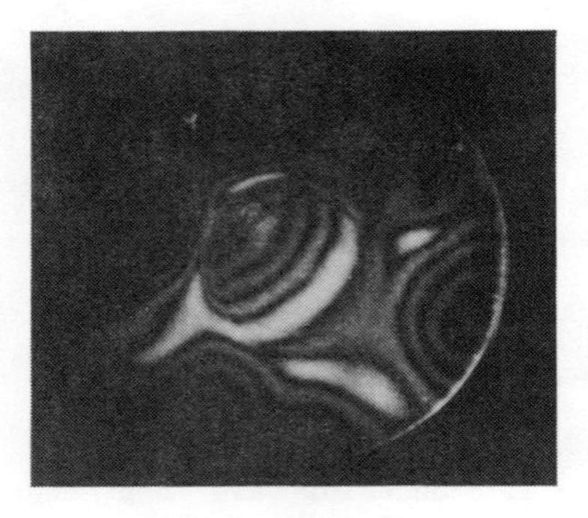

Fig. 12a Green color-encoded interferogram of a speaker oscillating at 2 KHz with an input voltage of 2 V P-P

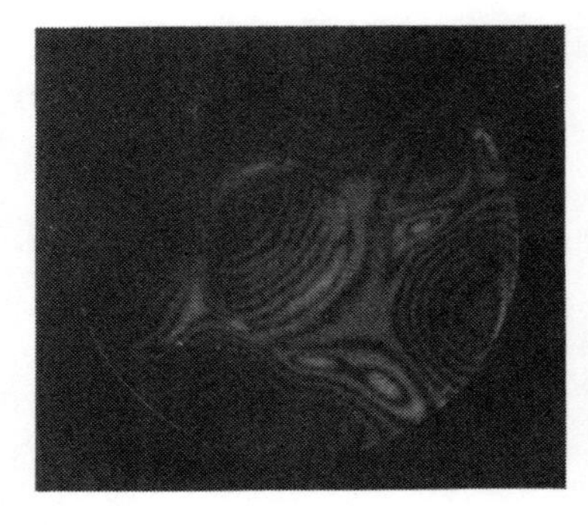

Fig. 12b Red color-encoded interferogram of a speaker oscillating at 2Khz with an input voltage of 4 V P-P

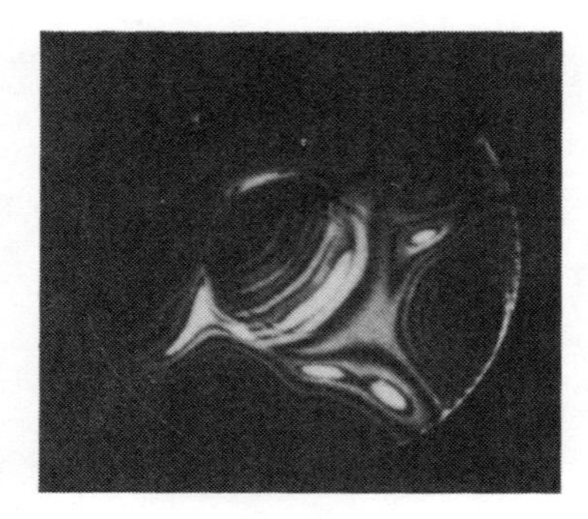

Fig. 12c A color coded multiplexed interferogram of (a) and (b)

Figure 12. A Black-and-White Picutre of Color-Encoded Time-Average Interferograms of a Vibrating Speaker

5.2 Color Encoding of Multiplexed Speckle and Holographic Interferometric Fringe Patterns

One of the interesting applications of speckle interferometry may be the speckle shearing metrology[33]. The advantage of the speckle shearing technique is to allow a direct measure of displacement derivatives. The multiplexed speckle and holographic interferometry with color encoding by white-light processing has been reported by Ruterbusch et al[34]. Holographic interferometry records out-of-plane displacement on the order of several micrometers, while speckle interferometry records in-plane displacements on the order of several tenths to hundredths of a micrometer. The defocused speckle interferometric techniques[35] were chosen to best illustrate the advantage of multiplexing with color encoding. A vibrating object is illuminated with a coherent source of light. The image of the vibrating object is defocused from the image plane by a small distance Δz. Here lies the tradeoff between the holographic and speckle fringes; the holographic fringes focus at $\Delta x = 0$, while the speckle fringes are optimized at $\Delta z = 3$ cm. A compromised defocus distance $\Delta z = 1$ cm was chosen to optimize the multiplexed image.

After the photographic processing, the multiplexed interferogram is placed at the input plane of a white-light optical processor, as shown in Fig. 13. A collimated beam of white light emanating from an extended xenon arc source is used to illuminate the multiplexed interferometric transparency. Spatial filtering and color encoding are then performed in the spatial frequency plane. The result is a set of color-coded interferometric fringe data that can be observed at the output image plane.

Figure 13 shows the reconstruction scheme for a multiplexed interferogram. We note that the speckle interferometric fringes are available in the halo surrounding the optical axis of the Fourier plane. The directional derivative of the object motion fringe patterns can be reconstructed by placing a small aperture in the halo[35]. This aperture can be rotated to show the various directional derivatives of the object motion. We also note that the holographic interferometric fringe pattern is available in the higher spatial frequency space. Color encoding can be accomplished by placing appropriate color filters over the sampling apertures. The intensity of the fringe pattern can be adjusted for proper color encoding. With this technique, it is possible to view either the speckle or holographic interferometric fringe patterns separately or simultaneously in a multiplexed manner at the output image plane of the white-light optical processor. These color-coded fringe patterns are suitable for photographing or recording with a color video system.

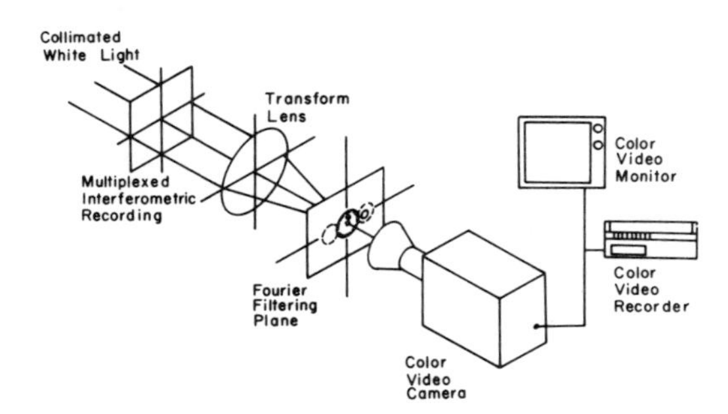

Figure 13. White-Light Optical Processor for Reconstruction of Multiplexed Time-Averaged Interferograms

We present an experimental demonstration in Fig. 4 which shows some of the interferometric fringe patterns that can be obtained with a time-averaged interferometric technique. The center photograph, Fig. 14 (b), shows the holographic interferometric fringes obtained from a 64 mm diameter speaker vibrating at 4 kHz with a 2.2Vp-p sinusoidal input signal. The photographs above and traveling in a clockwise direction to 5 o'clock [Fig. 14(a)] are the speckle interferometric fringe pattern. From 6 o'clock to 11 o'clock [Fig. 14(c)], the photographs show the multiplexed holographic and speckle interferometric patterns. Both types of interferometric fringe patterns were multiplexed onto the recording film during a single construction exposure. All of these fringe patterns can be directly viewed by a color video monitor, photographed, or studied by the human eye. Color encoding of the separate fringe patterns allows easy access for interpretation of the overlapped interferometric fringe data. Although the black-and-white photographs do show the basic advantages involved, in reality the color-coded fringe patterns do enhance the observability of the interferometric properties. The diagrams beside the photographs show the position and orientation of the sampling apertures as they were placed in the spatial frequency plane to obtain the respective interferometric fringe patterns.

Perhaps the most suitable use for this technique is in finding the location of the J_o order holographic fringe. Since speckle-shearing interferometry measures the derivative of the object motion, the fringe pattern contains no information where the object has been stationary. This dark area corresponds to the J_o order Bessel function in the holographic interferometric fringe pattern. In the color-encoded multiplexed interferogram, the areas of pure holographic color are the J_o order fringes. (Holographic color=the color filter used in encoding the holograhic information.) This is very beneficial when the J_o term is not the brightest fringe.

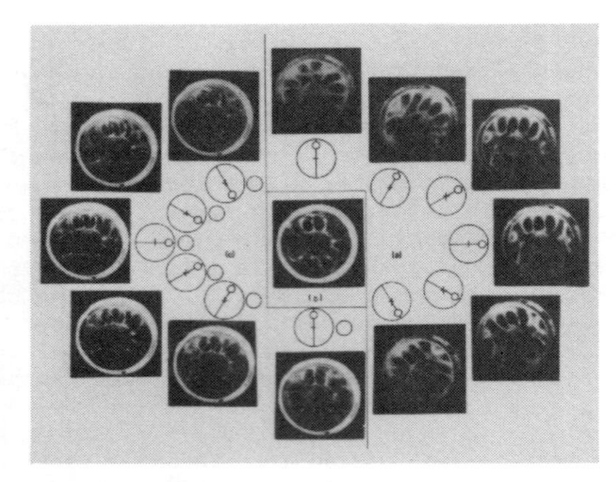

Figure 14. Black-and-White Pictures of Color-Coded Time-Averaged Interferogram

(a) 12 o'clock to 5 o'clock showing speckle interferometric fringe patterns; (b) Center photograph interferometric fringe patterns; and (c) 6 o'clock to 11 o'clock showing multiplexed interferograms

VI. Concluding Remarks

We have briefly discussed several techniques of color holographic image reconstruction, which can be obtained with reduced coherence, and some of its applications. Although color holographic images can be easily generated by a one-step rainbow holographic process, this technique introduces poor image resolution, color blur, and loss of parallax. To alleviate some of these constraints, we have shown a few other techniques in color hologram generation. However, those technique also possess some other severe drawbacks.

It would be exhaustive to discuss all the possible techniques of generating color holographic images. In this connection, we would refer the interested readers to a review article of color-holography by Hariharan[36]. Some of the applications of rainbow holography can also be found in an article published by Yu, Tai and Chen[37].

VII. References

1. D. Gabor, Nature, 161, 777 (1948).
2. E.N. Leith and J. Upatnieks, J. Opt. Soc. Am., 53, 1377 (1963).
3. E.N. Leith and J. Upatnieks, J. Opt. Soc. Am., 54, 1295 (1964).
4. Y.N. Denisyk, Sov. Phys. Doklay, 7, 543 (1962).
5. M.G. Lippmann, Compt. Rend., 112, 274 (1891).
6. G.W. Stoke and A.E. Labeyrie, Phys. Letters, 20, 368 (1966).
7. L.H. Lin et al., Bell System Tech. J., 45, 659 (1966).
8. E.N. Leith et al., Appl. Opt. 5, 1303 (1966).
9. S.A. Benton, J. Opt. Soc. Am., 1545 (1969).
10. P. Hariharan, W. H. Steel and Z.S. Hegedue, Opt. Lett., 1, 8 (1977).
11. H. Chen, A. Tai and F.T.S. Yu, Appl. Opt., 17, 1490 (1978).
12. S.L. Zhuang, P.H. Ruterbusch, Y.Z. Zhang, and F.T.S. Yu, Appl. Opt 20, 872 (1981).
13. J.C. Wyant, Opt. Lett., 1, 130 (1977).
14. P.N. Tamura, SPIE, 126, 59 (1977).
15. H. Chen, Appl. Opt., 17, 3290 (1978).
16. F.T.S. Yu, Optical Information Processing, Wiley-Interscience, N.Y., 1983.
17. G. Gerhart, P.H. Ruterbusch and F.T.S. Yu, Appl. Opt., 20, 3084 (1981).
18. F.T.S. Yu and P.H. Ruterbusch, Appl. Opt., 21. 2300 (1982).
19. P.H. Ruterbusch and F.T.S. Yu, Opt. Eng.. 21. 798 (1982).
20. G. Gerhart and P.H. Ruterbusch, Opt. Lett., to be published.
21. F.T.S. Yu, J.A. Tome and F.K. Hsu, Opt. Commun., 46, 274 (1983).
22. F.T.S. Yu and A. Tai, Japanese J. Appl. Phy., 14, 213 (1975).
23. H.E. Ives, Br. J. Photog., 609 (1906).
24. P.F. Mueller, Appl. Opt., 8, 2051 (1969).
25. F.T.S. Yu, Appl. Opt., 19, 2457 (1980).
26. B.J. Chang and K. Winiek, SPIE, 215, 172 (1980).
27. F.T.S. Yu and S.T. Wu, J. Opt., 13, 183 (1982).
28. S.T. Wu and F.T.S. Yu, Appl. Opt., 20, 4082 (1981).
29. F.T.S. Yu and F.K. Hsu, Opt. Commun., 15, 384 (1985).
30. F.T.S. Yu, White-Light Optical Signal Processing (Wiley-Interscience, New York, 1985).
31. C.M. Vest, Holographic Interferometry (Wiley-Interscience, New York, 1985).
32. G. Gerhart, P.H. Ruterbucsh and F.T.S. Yu, App. Opt. 20, 3085 (1981).
33. R.K. Erf, Speckle Metrology (Academic Press, New York, 1978).
34. P.H. Ruterbusch, J.A. Tome, F.T.S. Yu, Opt. Eng. 22, 501 (1983).
35. Y.Y. Hung, I.M. Daniel and R.E. Rowlands, Exp. Mech. 18, 56 (1978).
36. P. Hariharan, "Colour Holography" in press in optics, by E. Wolf, 22, 256 (1983).
37. F.T.S. Yu, A. Tai and H. Chen, Opt. Eng., 19, 666 (1980).

Session 7

Holographic Interferometry III

Chairmen

O. D. D. Soares
Universidade do Porto, Portugal
Erzhen Sheng
Beijing Institute of Optoelectronics, China

Automatic processing of holographic interference fringes to analyze the deflection of a thin plate

S.Toyooka, Y.Iwaasa, and H.Nishida

Faculty of Engineering, Saitama University
255 Shimo-ohkubo, Urawa, Saitama, JAPAN

Abstract

Holographic interference fringes of deflection of a plate are automatically analyzed by spatial phase detection technique. The first and the second derivatives of deflection which are essential to stress analysis can be numerically derived from resultant phase distribution.

Introduction

Holographic interferometry is a powerful technique for measuring deformation of a rough surface object. In the theory of small deflection of a thin plate, the bending moments per unit length are proportional to the second derivatives of deflection. It is usually not easy to derive the derivatives of deflection by numerical differentiation from the holographic contour map in usual fringe readout, since the operation of differentiation leads significant errors. Purely optical method was presented[1], but it required troublesome procedure and was time-consuming and the accuracy was not sufficient. To overcome these problems, computer-based techniques for fringe analysis should be well available. In this paper, we apply spatial phase detection technique(SPD)[2,3] to analyze holographic interference fringes. In this method, interference fringes including carrier term superimposed on the modulation term are compared to sinusoidal reference functions with the same frequency as that of the carrier term of fringes. Two-dimensional phase distribution or deflection can be precisely determined by simple calculations. The first and the second derivatives of deflection are approximated by numerical difference operation.

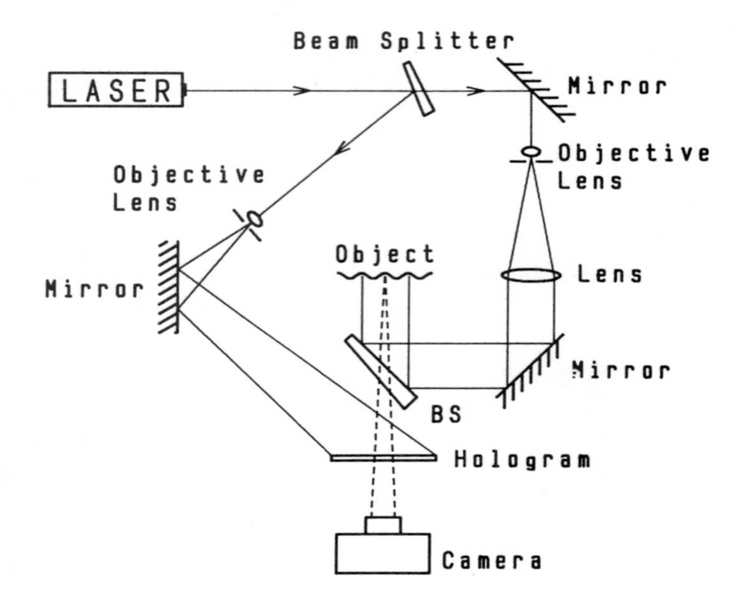

Fig.1 Optical system for making and reconstructing a hologram

Experimental setup

A double-exposure hologram was made by an usually used holographic setup shown in Fig.1, in which a normal illumination and viewing system were used. An object being tested was a rectangular steel plate of the dimension shown in Fig.2. It was clamped on one edge and concentrically loaded. The plate was viewed from the opposite surface in the figure.

The schimatic illustration of an analyzing system is shown in Fig.3. The reconstructed fringe pattern or it's printed pattern is viewed by a TV camera which scans along the y-axis with 1024 pixels and 8 bit gray scale. Adequate number of scanning lines are selected and transfered to a microcomputer through a camera controller. Two-dimensional phase distributions due to the deflection of the plate are calculated by SPD. The first and the second derivatives of deflection are numerically calculated from the original phase distribution. These results are displayed on an x-y plotter or color CRT.

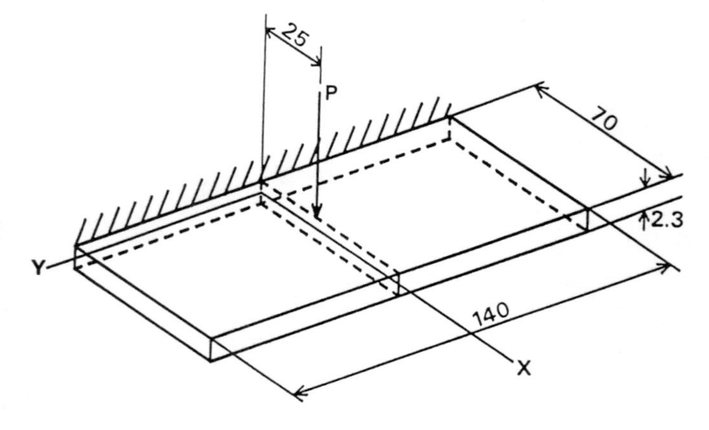

Fig.2 Dimension of a steel plate under test

Argorithm of spatial phase detection

Figure 4 shows the reconstructed fringes of the central part of the area of 36X47 mm^2 which includes a loading point. In the present experimental model, deflection of the plate monotonousely increases as an observation point departs from the clamped edge. In other words, the fringes consist of carrier term and modulation term. Intensity distribution is given by Eq.(1) in which w(x,y) is deflection of the plate and λ is the wavelength of the light source. Coefficient a(x,y) and b(x,y) are bias intensity and contrast of fringes, respectively.

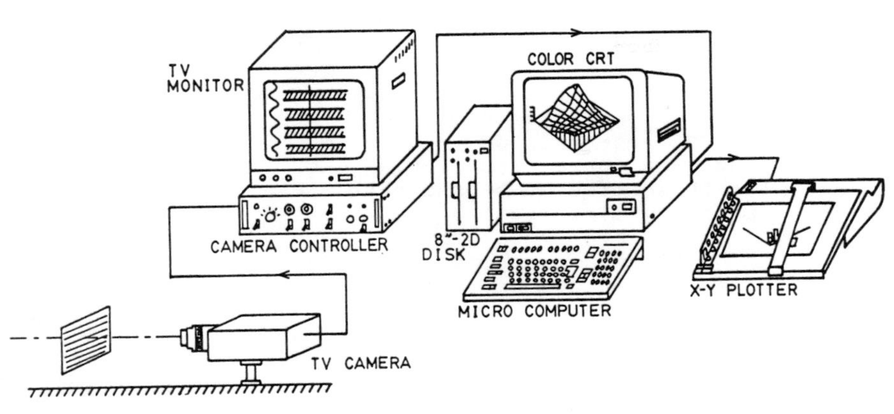

Fig.3 Illustration of an analyzing system

$$I(x,y) = a(x,y) + b(x,y)\cos(4\pi/\lambda)w(x,y)$$
$$= a(x,y) + b(x,y)\cos[2\pi f_0 x + \varphi(x,y)]. \qquad (1)$$

In the bracket of the second equation, the first term is the carrier term in which f_0 is the spatial frequency, and the second term, $\varphi(x,y)$, is the modulation term. If deflection does not vary monotonously or has a peak, we can make carrier fringes by tilting the object illuminating beam while making a double-exposure hologram. It is sufficient to know only the modulation term, because the important term is not deflection itself but its derivatives which are independent on the carrier term. In the agrorithm of SPD, acquired irradiance data along scanning lines are equally devided into small part of the interval of each carrier fringes. If the carrier fringes are sufficiently fine or f_0 is sufficiently large, a(x,y), b(x,y) and $\varphi(x,y)$ become constant in each truncated interval of carrier fringes. They can be replace a_i, b_i, and φ_i respectively, in the i-th interval of carrier. The acquired data in each interval of carrier fringes are multiplied by two sinusoidal function, $\sin(2\pi f_0 x)$ and $\cos(2\pi f_0 x)$ which are generated in the microcomputer and integrated over one period of carrier fringes. Phase values are obtained by the calculation of arctangent of the ratio of both integrated values as follows.

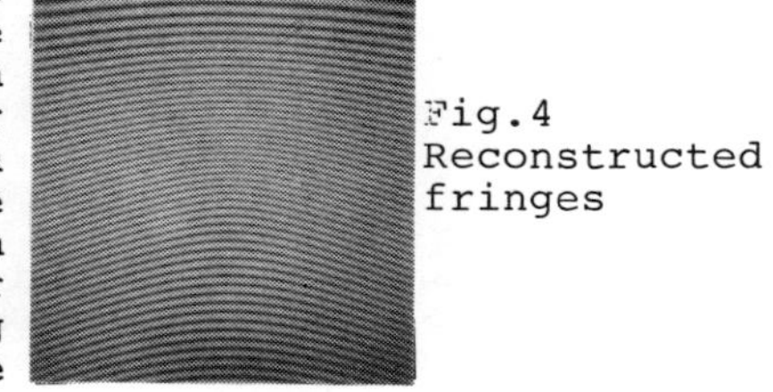

Fig.4 Reconstructed fringes

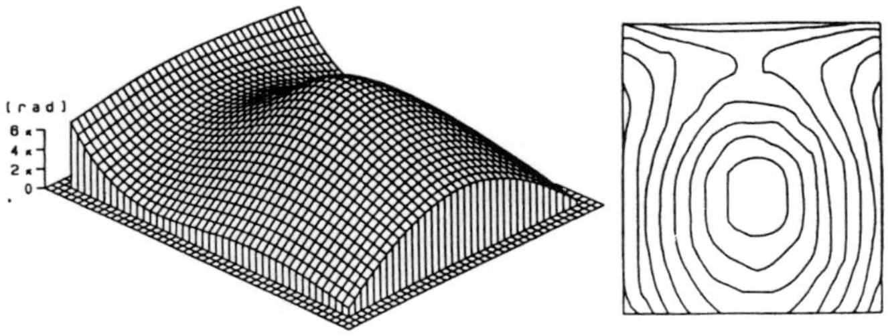

Fig.5 Distribution of deflection w.

$$\varphi_i = -\tan^{-1}\left(\frac{\int_i I_i(x)\sin(2\pi f_0 x)dx}{\int_i I_i(x)\cos(2\pi f_0 x)dx}\right) \qquad (2)$$

In such a way, the phase over every interval of the carrier fringes is successively determined along each scanning line. In our present experiment shown later, rms errors are estimated to be less than $2\pi/200$. The first and the second derivatives of deflection are approximated from the first and the second differences of the resultant phase distribution. They are simply written as for only the variable x

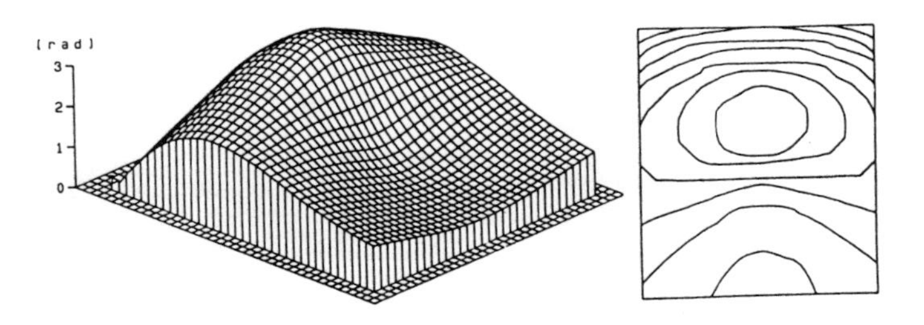

Fig.6 Distribution of slope $\partial w/\partial x$.

$$\Delta\varphi(x) = \varphi(x + \Delta x) - \varphi(x)$$

$$= (4\pi/\lambda)[w(x+\Delta x) - w(x)] - 2\pi f_0 \Delta x$$

$$\approx (4\pi/\lambda)(\partial w/\partial x)\Delta x - 2\pi f_0 \Delta x \qquad (3)$$

where Δx is the amount of shear. In order to get good approximation to the true differentiation, the amount of shear is prefarable as small as possible. We choose it to be equal to the minimum interval of the data location. Each phase value is subtracted by the next one in the x- or y-direction. Resultant difference data are smoothed in each small square of 5x5 data points. The second differences can be obtained by repeating twice the same procedures.

Experimental results

Phase distribution analyzed from the interference fringes are shown in Fig.5 as a perspective plot and a coutour type plot. The analyzed area is a rectangle which is 43 mm along the scanning lines or the x-axis and 36 mm perpendiculat to them. Figures 6 and 7 show the distributions of the first differences with respect to x and y in which the amount of shear is 1mm. Figures 8 and 9 show the second differences with respect to x and y. They are, in other words, the distributions of curvature or normal strain along the x-direction and y-direction. Figure 10 shows the second difference with respect to x and y. This is the distribution of twist or shearing strain. According to the Fook's law, the distributions of normal stresses along the x and y direction can be obtained by different linear combinations of Figs.8 and 9 and the distribution of shearing stresse is proportional to Fig.10.

Conclusions

Holographic interference fringes of deflection of a plate were analyzed by spatial phase detection technique. Two-dimensional optical phase distribution can easily be obtained with sufficient accuracy to get the first and the second derivatives of deflection which are approximated by numerical difference operation. Resultant distributions were displayed as perspective plots and contour lines. Color contours are also available.

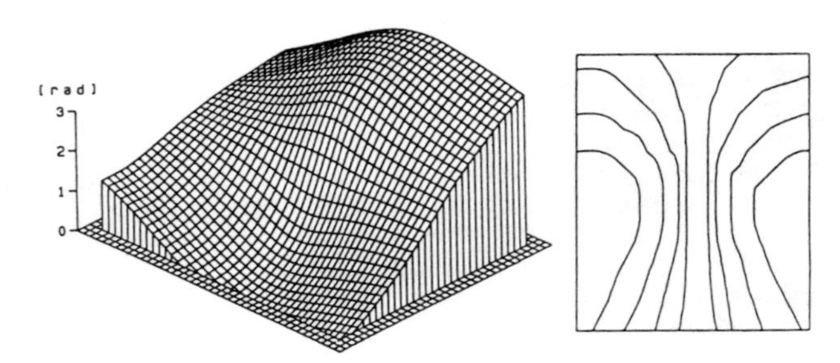

Fig.7 Distribution of slope $\partial w/\partial y$

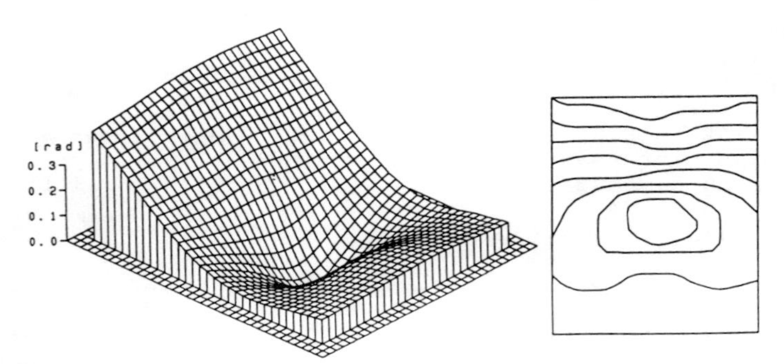

Fig.8 Distribution of curvature $\partial^2 w/\partial x^2$

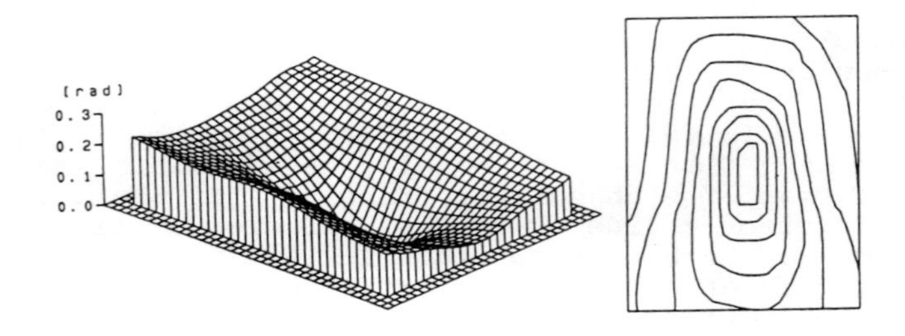

Fig.9 Distribution of curvature $\partial^2 w/\partial y^2$

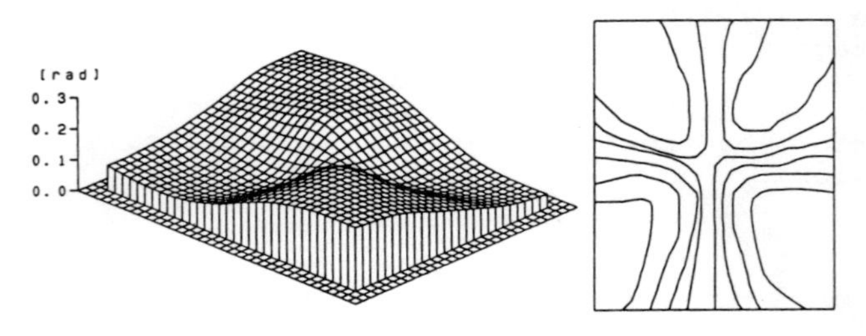

Fig.10 Distribution of twist $\partial^2 w/\partial x \partial y$

References

1. Toyooka, S., "Contour mapping of the first and the second derivatives of plate deflection by using modulated diffraction gratings made by double-exposure holography," Opt.Acta, Vol.25, pp.991-1000. 1978.
2. Toyooka, S. and Tominaga, M., "Spatial Fringe Scanning for Optical Phase Measurement," Opt. Commun., Vol.51, pp.68-70. 1984.
3. Toyooka, S. and Iwaasa, Y., "Automatic profilometry of 3-D diffuse objects by spatial phase detection," Appl. Opt., Vol.25, pp.1630-1633. 1986.

A dynamic process study by preset light pulse method in holographic system with Faraday Rotator

Wu Xing, Ren Guoquan, Zhang Zunlin
Ye Zhisheng, Wang Shengqing

Department of Precision Instruments Engineering,
Tianjin University, Tianjin, China.

Abstract

This paper introduces a new method using preset light pulse technique to study dynamic process in holographic system with Faraday Rotator. The feasibility of the method is analysed theoretically. A controlling system is designed. The experimental results are successfully obtained.

Introduction

A accurate timing is requred when holographic interference is used to study dynamic process. The experimental results make sence only when each dynamic process is recorded precisely at any moment. Otherwise, they do not make sence or they produce serious error. Therefore, one always concentrates on studies of both recording system (including light source) and controlling system in the holographic interference measurement. But in ordinary holographic interference measurement, it is extremely difficult to record precisely a certain instantaneous moment, such as 5us or 10us from the dynamic trigger.

To obtain exactly a requred moment of dynamic process, often one has to experiment repeatedly. Even though, the results exist error which is in a range of 2us to 10us observed by our experiment. And this error is stochastic. If the dynamic process requred is also in the range of 2us to 10us, the results are never ideally reached.

Used the method presented in this paper, arbitrarily requred dynamic moment can be directly preset according to practical requrement. Thus the dynamic exposure preset function can come into effect. And the requred state of dynamic moment operation can be precisely achieved. The present paper analyses theoretically a feasibility of preset light pulse technique in holographic system with Faraday Rotator. On the basis of dynamic holographic interference principle, the a controlling system is designed based on double-instruction principle(static and dynamic). It was systematically experimented successfully to use the method to study the photoelastic materials (PSM-1, 6*63*65mm) dynamic process under free drop steel ball shock. The separate dynamic isochromatics and isopachics at different moment were recorded simultaneously. From the results, one can get accurate information for the stress analysis of materials. As a new mehod, preset light pulse technique in holographic system can be extensively used in analysis of various instantaneous process.

Basic theoretical analysis

In 1986, Fourney and Hovanesain used holography in the studies of stress. Then, Lallemand and Lagarde used pulse ruby laser holographic system and Faraday Rotator in the experimental studies of shocked photoelastic materials. On the basis of holographic interference theory, the expression of Double exposure is as follows:

$H_1 = O_1 + R$ for the first exposure
$H_2 = O_2 + R$ for the second exposure

where R is reference beam, O_1 and O_2 and object beams. After two time exposures, field on dry plate is

$$E = |H_1|^2 + |H_2|^2 = |O_1|^2 + O_1R^* + |R|^2 + RO_1^* + O_2R^* + RO_2^* + |R|^2 + |O_2|^2 \quad (1)$$

If reconstruction beam is identical with reference beam, regarding recording material works in its linear range, amplitude transmissivity is directly proportional to exposure, the transmitted field from hologram is

$$H' = RE \quad (2)$$

From equation (1) and (2), it can be obtained

$$H' = R|O_1|^2 + R^2 O_1^* + |R|^2 O_1 + R|R|^2 + R|O_2|^2 + R^2 O_2^* + R|R|^2 + |R|^2 O_2 \quad (3)$$

seen from (3), object-information field is

$$H'' = |R|^2 O_1 + |R|^2 O_2 \quad (4)$$

and intensity of the field is given by

$$I = |H''|^2 \quad (5)$$

In holographic system with Faraday Rotator, isochromatics and isopachics are obtained by double exposed holographic interference, as shown in figure 1.

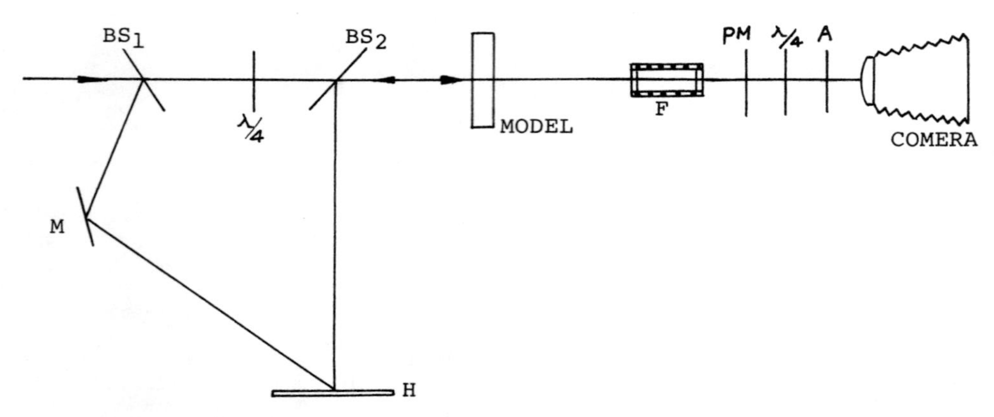

Figure 1. The scheme of recording isochromatics and isopachics

When the model is not shocked, the first exposure is produced. At a certain moment after the model is shocked, the second exposure is produced. From figure 1, in the first exposure, change in phase of object beam after two time passing through the model is given by

$$\varphi_0 = 2K(n_o - 1)d \quad (6)$$

Where $K=2\pi/\lambda$, n_o is index of non-shocked model, d is thickness of the model. Therefore, the object wave-front is expressed as

$$O_1 = A\, e^{2ik(n_o - 1)d} \quad (7)$$

When the model is shocked, double refraction of the model appears. The changing phase of object beam passing through the model for the first time in the direction of stress are:

$$\varphi_1 = K(n_1 - 1)d$$
$$\varphi_2 = K(n_2 - 1)d$$

Where n_1, n_2 are index of refraction in two main stress directions, respectively. The polarization plane of object beam first time transmitted through Faraday Rotator is rotated by 45°. It is rotated also by 45° for second time transimission. Therefore the polarization plane is rotated by 90° in comparison with the polarization plane of original object beam. Thus after two time transimission through the shocked model, the direction of fast-axis and low-axis change position with each other. In that case, the total change in phases of two polarization planes are equal:

$$\varphi_1' = \varphi_2' = \varphi_1 + \varphi_2 = K(n_1 + n_2 - 2)d$$

So object wave-front is:

$$O_2 = B\, e^{ik(n_1+n_2-2)d} \quad (8)$$

Substituting (4), (7), and (8) into (5) gives

$$I = |R|^4 \left[A^2 + B^2 + AB\, e^{ikd[(n_1-n_o)+(n_2-n_o)]} + AB\, e^{-ikd[(n_1-n_o) + (n_2-n_o)]}\right] \quad (9)$$

set
$$\psi_1 = Kd(n_1-n_o)$$
$$\psi_2 = Kd(n_2-n_o)$$

(9) can be expressed in cosinform as

$$I = |R|^4 \, [A^2 + B^2 + 2AB\cos(\psi_1 + \psi_2)] \qquad (10)$$

From (10), visibility of fringes can be got:

$$V = \frac{2AB}{A^2 + B^2} \qquad (11)$$

From (11), A^2 and B^2 are intensities of two laser pulses. If letting A^2 be 1, the value of V changes with B^2 as follows:

B^2	1	0.9	0.8	0.7	0.6	0.5	0.4	0.3	0.2	0.1
V	1	0.999	0.994	0.984	0.968	0.943	0.924	0.843	0.745	0.576

Obviously, in holographic system with Faraday Rotator, visibility does not requre exact equality of light intensity of both pulses. Our experiment has still got distinct fringes under the condition of 50% difference between two pulses. It is just because of the fact that we have designed preset light pulse technique to realize the function of dynamic preset exposure.

Controlling Principle

In ordinary double-exposure holographic measurement, lasers used are single instruction. Under the trigger of a single instruction, laser produces two fixed-interval light pulses. But shocking process between the two laser pulses is not accurate in time, for example, aifgun, pendulum, free-drop steel ball are very discrete when they shock the model. So it is difficult using the second pulse to record exactly the dynamic process of the model at a certain moment after it is shocked.

To solve the problem. We have designed the Double-instruction Holographic system as shown in Figure 2:

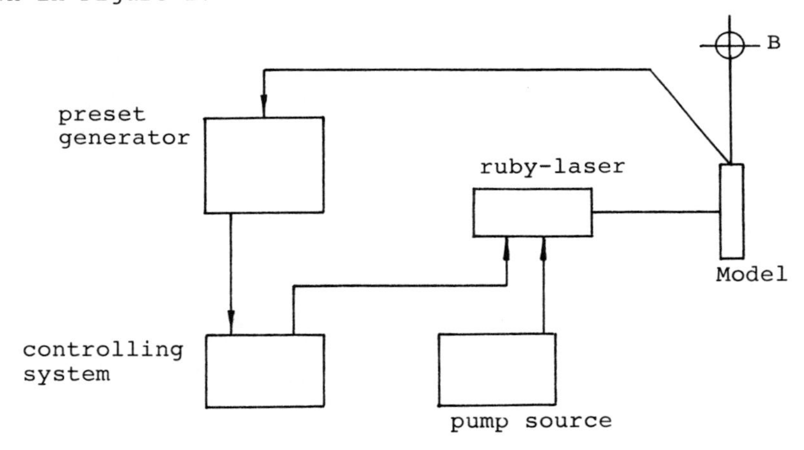

Figure 2. The scheme of controlling principle

At the moment when controlling system produces the first instruction signal to open Q-switch of laser system, the first laser pulse is emitted to finish the first exposure of non-shocked model. At the same time when the model is shocked, the shock produces a signal to control Preset Signal Generator to give the second instruction signal to finish to second exposure of shocked model. If we record a certain state of the dynamic process of shocked model, the second instruction signal can be preset at the moment requred after the shock happened. Thus the preset exposure process is realized. In the process of operation, the second preset laser pulse has more change in intensity compered with the first pulse because at different preset moment laser has different stored energy. But by the basis analysis above and experiment result, the difference in intensity between two pulses is allowed.

Experiment and Its Result

Experimental device is as shown in figure 3:

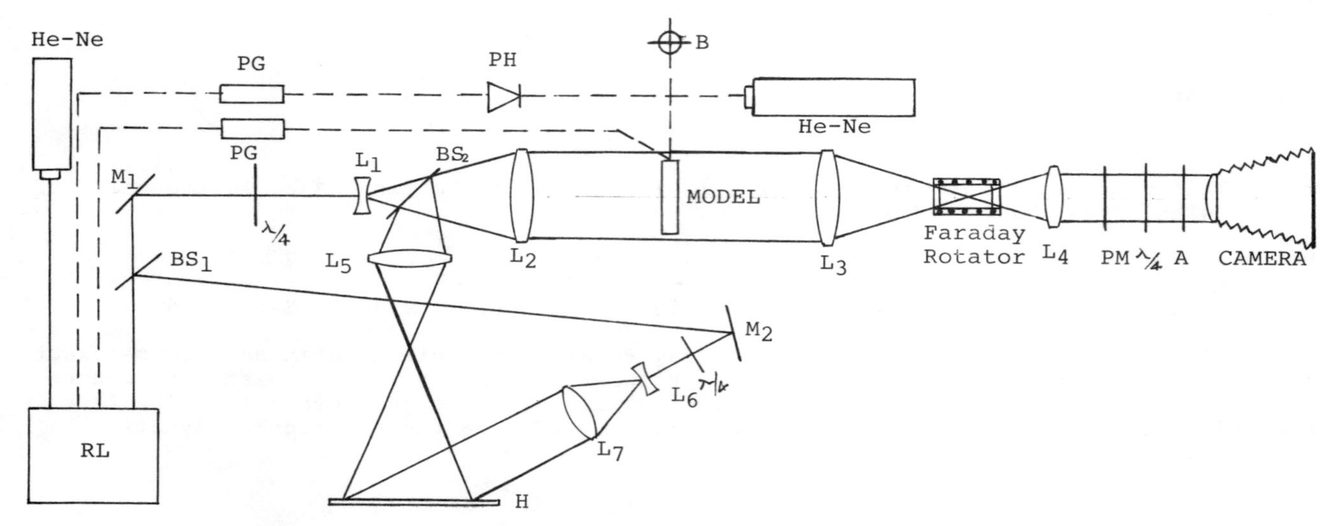

Figure 3. Experimental setup to simultaneously obtain separete isochromatics and isopachics.

Its main technical specifications:

1. Energy of two pulses: more than 200 mj.
2. Coherence lenghth: more than 1 m,
3. Preset time range: 0——99 us,

isochromatics and isopachics are shown in figure 4:

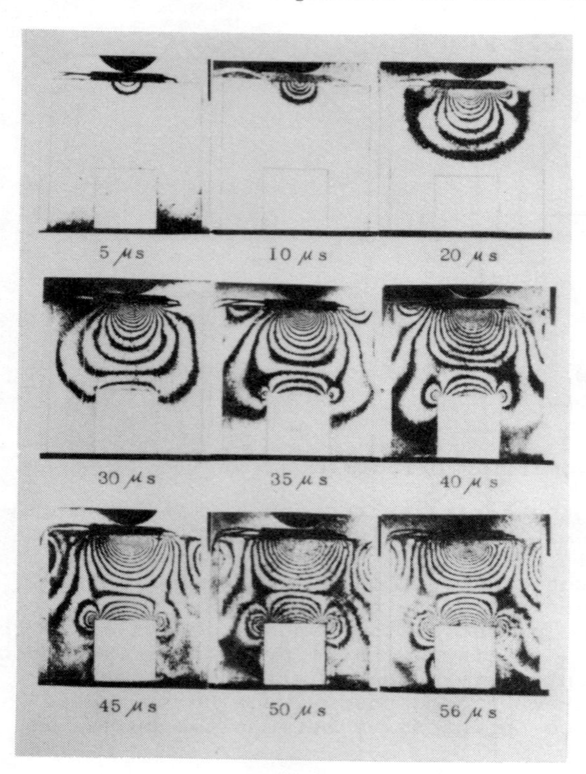

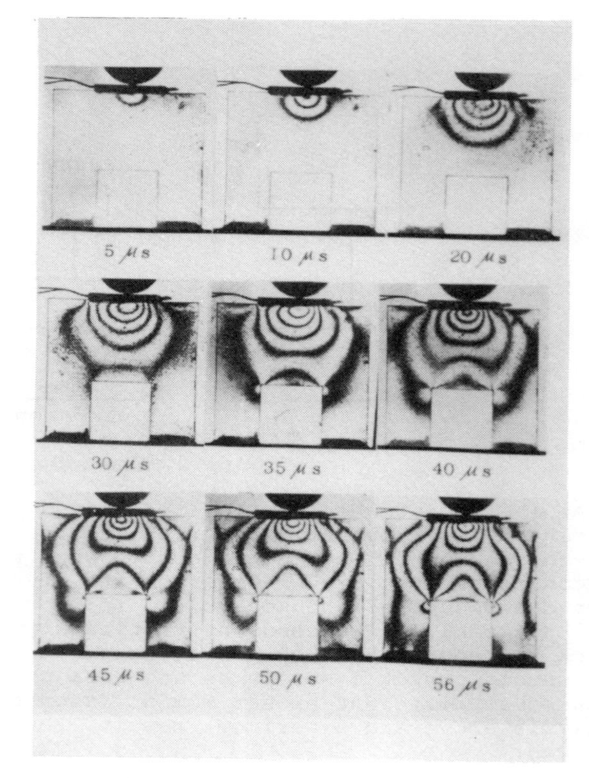

Figure 4. Experimental results

References

1, J.P.Lallemand, and Lagarde, Exp. Mech. 22(5), 1982, PP174-179.
2, Jean-Paul Lallemand, Exp. Mech. 12, 1981, PP477-480.

Application of monowavelength pulsed laser holometry in the measurement of arc plasma

Li Junyue, Liu Jinhe, Lian Jinrui, Li Yishan

Department of Mechanical Engineering, Tianjin University
Tianjin, China

Abstract

In this paper a new demodulating procedure for monowavelength holometry of arc plasma is advanced. The method of measuring arc plasma by monowavelength pulsed laser holometry has been investigated and realized. The temperature field and the density fields of electron, ion and atom of an argon arc plasma have been measured out by this method. It can be, in principle, extended to measuring other arc plasma. It has many advantages and wide prospects for application.

Introduction

It is well known that the measurement of arc plasma by holometry has many advantages of the high measuring velosity and sensibility, and of the broad temperature range measured, of obtaining the information on the whole field at one time of measurement and without disturbing to measured flow field etc.. The arc plasma is, however, different from the common gases. It has very high temperature and complicated component. Especially its refractive index is influenced not only by atoms, but also by electron and ions. Therefore, in demodulating holographic interferogram there appears two unknowns for one equation, which is the problem in solving the equation. Up to new in order to overcome the difficulty the biwavelength laser interferometry has to be adopted. It needs the complicated devices, must be made with two times of measurement and calculation, and can not obtain continuosly the physical parameters below 6000K. Consequently, its extension and application is limited. In this paper a new demodulating procedure for monowavelength holometry of arc plasma is advanced. The method of measuring arc plasma by means of the monowavelength pulsed laser holometry has been investigated and realized. It has many advantages and wide prospects for application.

Demodulating procedure

In this paper the premises for the demodulating are as follows:

(1) the arc plasma is a phase object and consists in the axial symmetry state, and satisfies the local thermodynamics equilibrium.

(2) the pressure at each point in arc plasma is all equal to 1 atm.

According to the theories relative to the plasma physics and optics the demodulating procedure for monowavelength holometry of arc plasma is put forward in Figure 1.

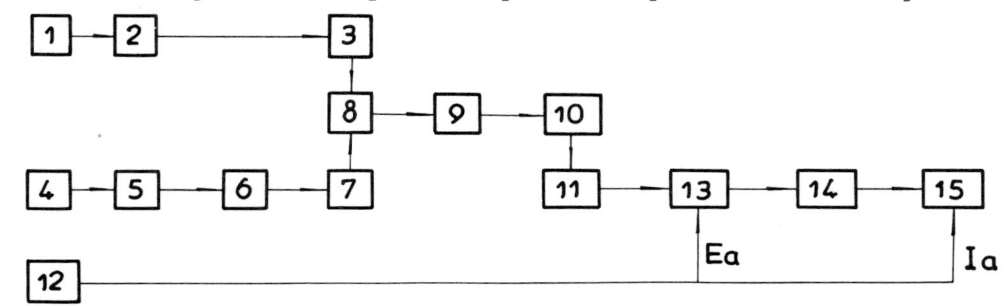

Figure 1. Square scheme of demodulating procedure for monowavelength holometry of arc plasma.

1 - writing out the system of character equations of arc plasma.
2 - calculating the relationship between the number densities of particles and temperature of arc plasma $N_k=f(T)$.
3 - calculating the relationship between the refractive index and temperature of arc plasma $n=f(T)$.
4 - monowavelength holographic interferogram of arc plasma.
5 - measuring interference fringe desplacement $F(x)$.
6 - Abelian transformation.
7 - obtained radial distribution of refractive index of arc plasma $n=f(r)$.
8 - comparing.

9-obtained radial distribution of temperature of arc plasma T=f(r).
10-obtained radial distribution of various particle number densities of arc plasma N_k=f(r).
11-obtained radial distribution of specific conductivity of arc plasma σ=f(r)
12-macro-measuring intensity of electric field in the column E_a and current I_a of arc.
13-obtained the radial distribution of current density of arc plasma j=f(r).
14-obtained arc current I_a' by integral.
15-current verification of measuring results.

Call the demodulating procedure a comparison method.

For an argon arc plasma, we have following system of character equations.

$$p = N_t KT \tag{1}$$

$$N_t = N_a + 2N_1 + 3N_2 \tag{2}$$

$$N_e = N_1 + 2N_2 \tag{3}$$

$$\frac{N_1 N_e}{N_a} = \frac{2Z_1}{Z_a}\left(\frac{2\pi m_e KT}{h^2}\right)^{3/2} e^{\frac{-\mathcal{E}_{i1}}{KT}} \tag{4}$$

$$\frac{N_2 N_e}{N_1} = \frac{2Z_2}{Z_1}\left(\frac{2\pi m_e KT}{h^2}\right)^{3/2} e^{\frac{-\mathcal{E}_{i2}}{KT}} \tag{5}$$

Where, N_a, N_1, N_2 - number density of the atom, first postive ion, second positive ion of the argon respectively; N_e - number density of electron; N_t - total number density of particle of argon arc plasma; Z_a, Z_1, Z_2 - internal partition function of the atom, first positive ion, second positive ion of the argon respectively; m_e - mass of electron; p - pressure of argon arc plasma; T - temperature of argon arc plasma; K - Boltzmann constant; h - planck's constant; $\mathcal{E}_{i1}$ - first ionization energy of the argon atom; $\mathcal{E}_{i2}$ - second ionization energy of the argon atom.

The above system of equations is composed of five equations including seven unknowns, i.e. N_a, N_1, N_2, N_e, N_t, p and T. Suppose p=760 mmHg, and if T is given, the system of equations is solvable, the N_a, N_1, N_2, N_e and N_t can be calculated out.

According to the theory of dispersion, the refractive index of arc plasma (n-1) is a function of wavelength of the incident light and number density of particles. It can be expressed by

$$n-1 = \sum_i K_i N_i \tag{6}$$

where, N_i is the number density of i-th particle of arc plasma; K_i is the specific refractivity of i-th particle. When the wavelength of incident light λ is given (in this paper, λ = 0.6943 µm), K_i can be calculated out in accordance with references[2.3].

From above, if T is given, the number density of various particles N_i and refractive index (n-1) can be calculated out. Figure 2 is the relationship calculated between the refractive index and temperature of the argon arc plasma.

The information that is obtained directly from the holographic interferometer is variation of refractive index in measured region. This variation is recorded by means of holographic interferogram. Therefore, analysing holographic interferogram the field of refractive index of measured region can be obtained.

Suppose that the incidance of the probing laser beam and coordinate system are shown in Figure 3. A cross-section perpendicular to the axis of arc column (i.e. a section of Z=const.) is given in Figure 4. After the laser pass through the arc plasma (See Figure 4), the variation of the optical path can be expressed by

$$L = \int_{y_0}^{y_1} [n(x,y) - n_0]dy = 2\int_0^{y_1} [n(x,y) - n_0]dy \tag{7}$$

where, n is refractive index of the medium at each point inside the disturbance region; n_0 is refractive index of the medium outside the disturbance region, which can be found from reference[4]; y_0 and y_1 are places where incident light pass into and out disturbance region respectively.

The variation of the optical path results in displacement of interference fringe. Its number F(x) is as follows

$$F(x) = \frac{2}{\lambda} \int_{0}^{y_1} [n(x,y) - n_0]dy \tag{8}$$

F(x) can be measured from the holographic interferogram. By Abelian transformation, we obtain

$$n(r) - 1 = -\frac{\lambda}{\pi} \int_{r}^{r_0} \frac{[dF(x)/dx]}{\sqrt{x^2-r^2}} dx + (n_0 - 1) \tag{9}$$

Therefore, the radial distribution of refractive index of arc plasma [n(r)-1] can be calculated from the holographic interferogram.

Set-up and method of experiment

Figure 5 is the scheme of the holographic interferometer with ruby pulsed laser for this experiment. Where, 1 is a He-Ne laser for the adjustment of optical path; 2 is a ruby laser for the light source of the holography; M_3 is a separating mirror to separate laser beam into two laser beams, one of them is for reference beam, the other for object beam; M_4, M_5, M_6 are full mirror; L_1, L_2 and L_1', L_2' are lens for collimated beam. M_4 is adjustable in angle of dip to get reference interference fringe.

The double exposure method for holometry is adopted for the experiment (Figure 6). The first exposure (Figure 6, a) goes at no arc to get reference beam r. the second (Figure 6, b) goes at arc burning to get object beam 0. At reappearing with beam R (Figure 6, c) the reference beam r and the object beam 0 will reappear simultaneously and interfere each orther. So, the holographic interferogram of arc can be obtrained. If M_4 remains stationary between the double exposures we will obtain the infinite space holographic interferogram (Figure 7). If M_4 is adjusted in angle of dip slightly between the double exposures we will obtain the finite space holographic interferogram (Figure 8), which has reference fringe, and it is convenient to demodulating of interferogram.

Experimental results and verification

The holographic interferogram to be measured of an argon arc is shown in Figure 9. The numbers of fringe displacement F(x) is measured by a tool microscope with x10 and with tolerance of 0.01mm.

The numbers of fringe displacement F(x) measured on the section at a distance of Z=7.6mm to the cathode peak are traced out in Figure 10. The obtained radial distributions of the refractive index, temperature, number density of particles, specific conductiveity and current density on the same section are given in the Figure 12, Figure 13, Figure 14 and Figure 15 respectively. The temperature field of the argon arc has also been obtained and is given in the Figure 16.

The integral formula for electric current

$$I_c = 2\pi \int_{0}^{R} \sigma E r dr \tag{10}$$

is used for verifying the measured results. Where

I_c - integral current;
σ - specific conductivity which values can be looked up from reference[5] according to the obtained temperature at given point;
E - intensity of electric field in the arc column, determined by macro-measurement;
R - electric radius equal to the distance from the centre of arc column to the points where the temperature is 5000K.

The integral results of current for ten sections is compared respectively with macro-measured arc current I_a and the maximal relative error is not greater than 4%. Therefore, the measured results are reliable.

Conclusions

1. The demodulating procedure for monowavelength holometry of arc plasma proposed in this paper can be used for quantitative measurement. It has overcome the principal difficulty in the demodulation.
2. The method of measuring arc plasma by monowavelength pulsed laser holometry can measure temperature up to 16000K or even higher. The number densities of atoms, ions and electron and that other parameters can be obtained simultaneously.

3. This method has advantages of higher measuring accuracy, wide measuring range, obtaining information of whole field at one time of measuring, very short time of exposure and simpler device, etc.

4. this holometry is a differential interferometry (adopting two times of exposure), and therefore it is suitable particularly for study of non-normal pressure arc.

5. This method can be, in principle, extended to measurement of other arc plasma.

Acknowledgements

This work has been supported by the Science Foundation of the Academy of Sciences of China.

References

1. Dushin, L.A., et al. Applied Spectroscopy Journal, Vol.25, No.3, pp.379-407, 1976 (in Russian).
2. Martelluci, S., Nuoro Cimento, Vol.5, No.1-4, pp642-679, 1967.
3. Allen, C.W., Astrophysical quantities, University of london, the Athlone Press, 1963.
4. Shkarofky, L.P., et al., Planetary and Space Science, No.6, p 24, 1961.
5. Devoto, R.S., Phys. Fluids, No.2, P 354, 1967.

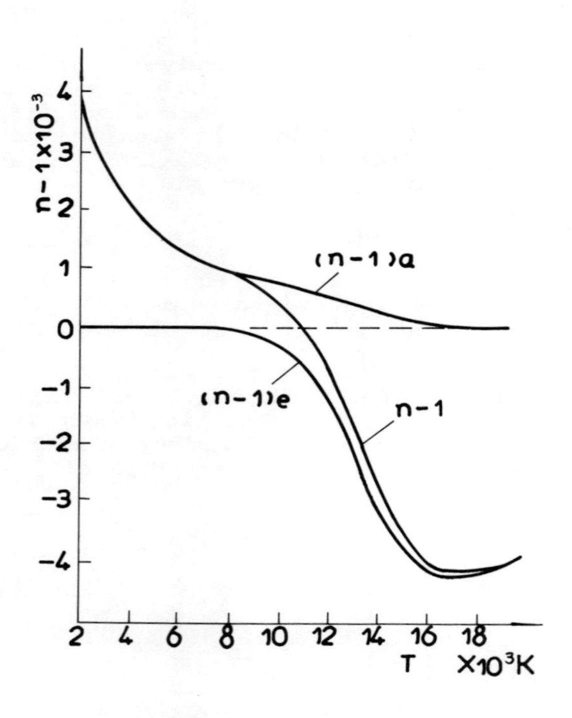

Figure 2. Calculated relationship between the refractive index n-1 and temperature T of the argon arc plasma.
λ= 0.6943 μm;
$(n-1)_a$-refractive index of argon atoms;
$(n-1)_e$-refractive index of electrons;
(n-1)-refractive index of argon plasma.

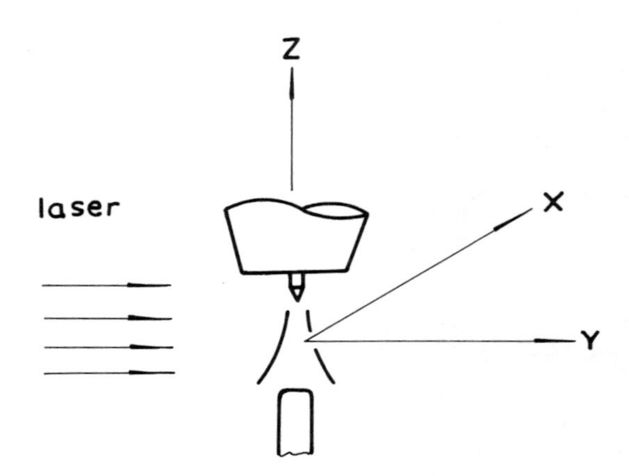

Figure 3. Scheme of incidence of probing laser beam.

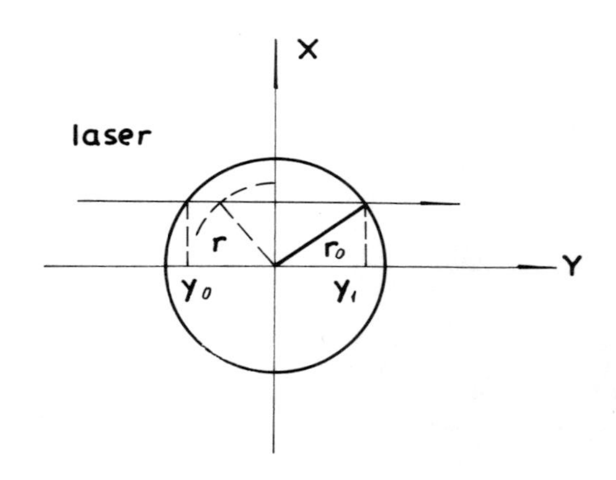

Figure 4. A section of Z=const. of arc plasma.

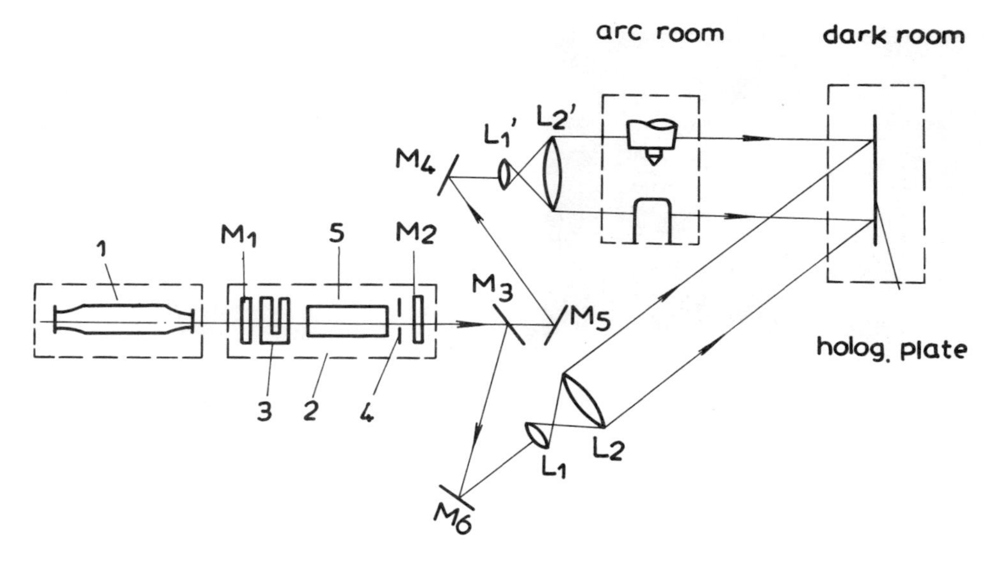

Figure 5. Scheme of the holographic interferometer for this experiment.

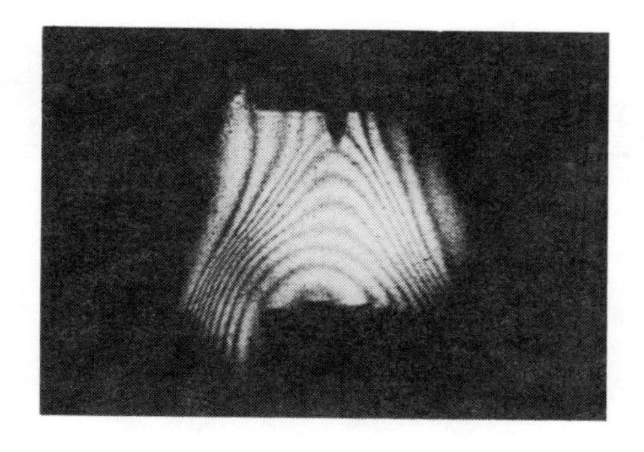

Figure 7. Infinite space holographic interferogram of an argon arc taken by double exposure method.

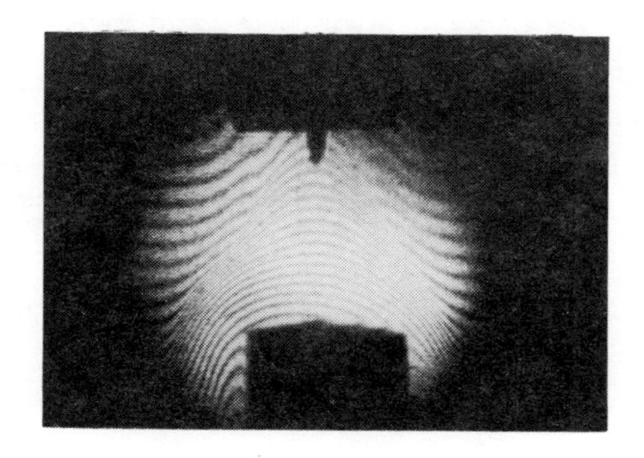

Figure 8. Finite space holographic interferogram of an argon arc taken by double exposure method.

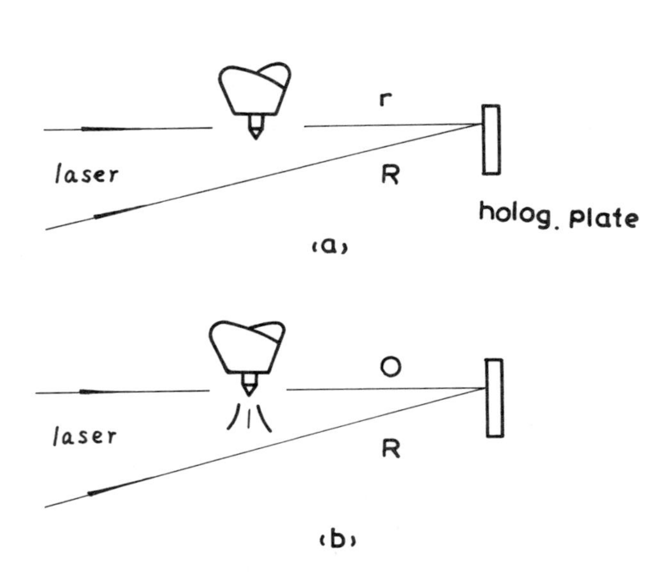

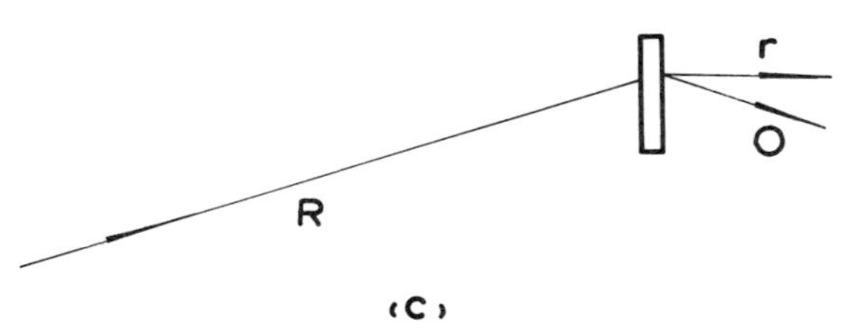

Figure 6. Principled schema of the double exposure method for holometry.

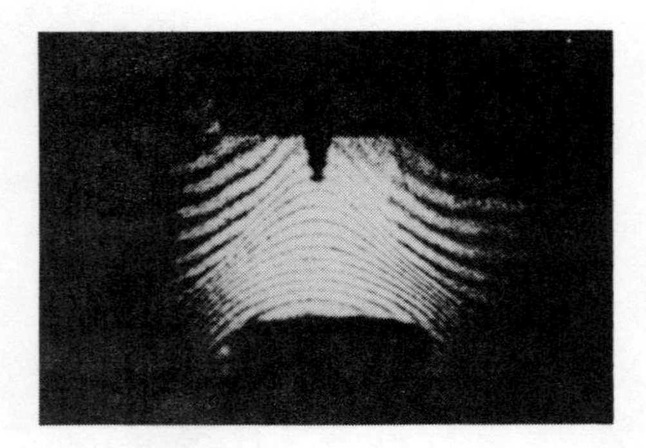

Figure 9. Holographic interferogram to be measured of an argon arc.
arc current: 40 A,
arc voltage: 20 V,
arc length: 12.8 mm,
volume of argon flow: 400 L/h,
electrodes: W(-), Cu(+)(water),
atmospheric pressure: 760 mmHg,
ambient temperature: 15°C.

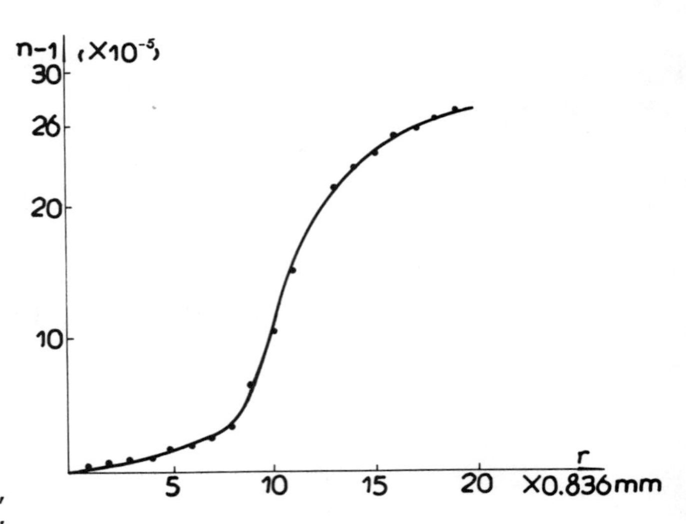

Figure 11. Radial distribution of refractive index on the section of Z=7.6 mm.

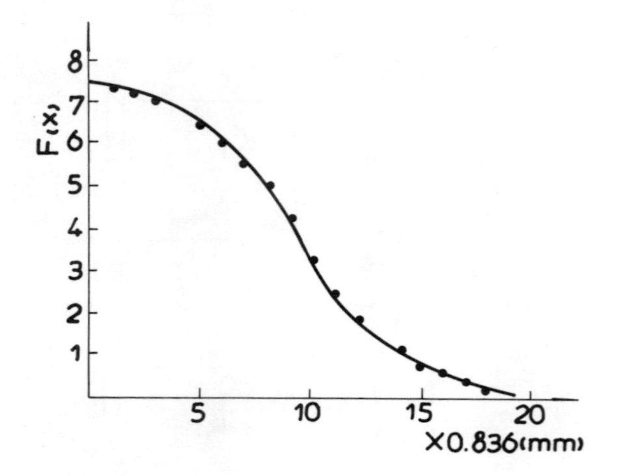

Figure 10. Numbers of fringe displacement F(x) measured on the section of Z=7.6 mm.

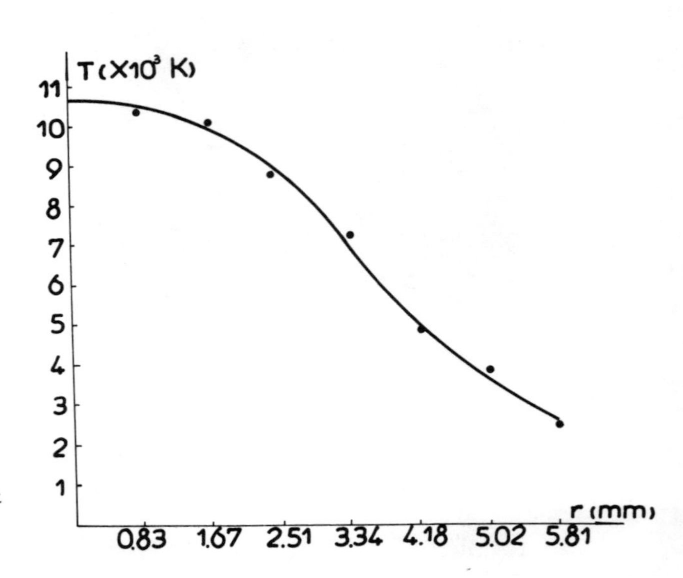

Figure 12. Radial distribution of temperature on the section of Z=7.6 mm.

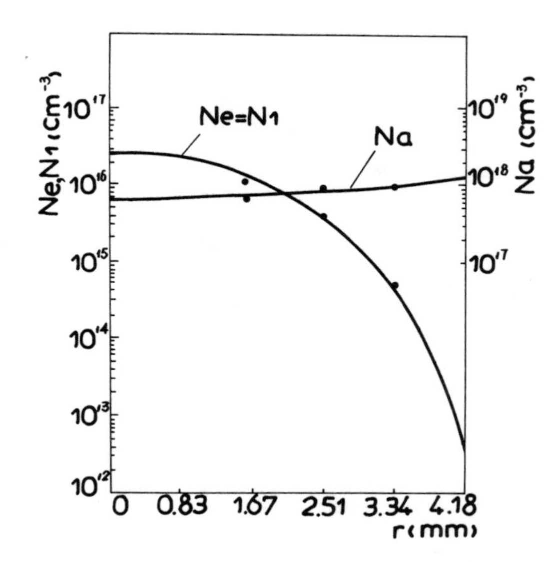

Figure 13. Radial distribution of number density of particles on the section of Z=7.6 mm.
N_e-number density of electron;
N_1-number density of argon first ion Ar^+;
N_a-number density of argon atom Ar.

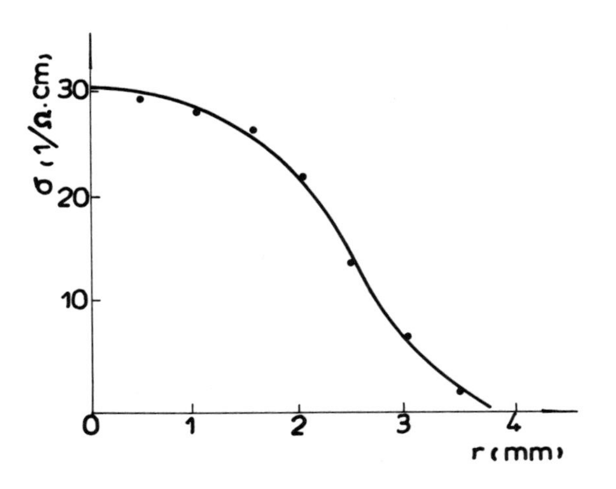

Figure 14. Radial distribution of specific conductivity on the section of Z=7.6 mm.

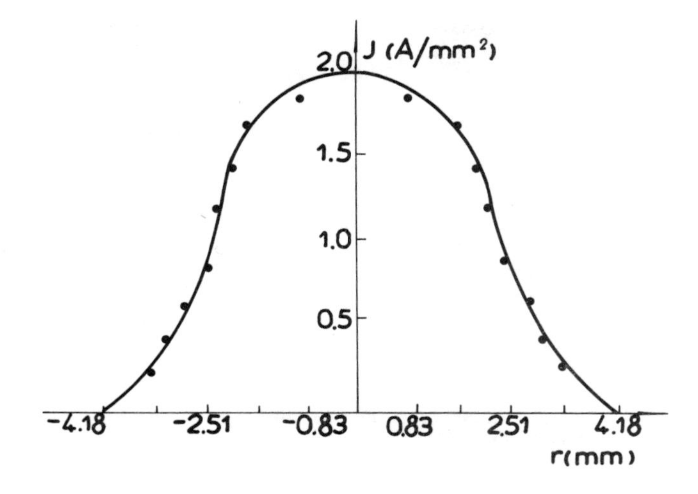

Figure 15. Radial distribution of current density on the section of Z=7.6 mm.

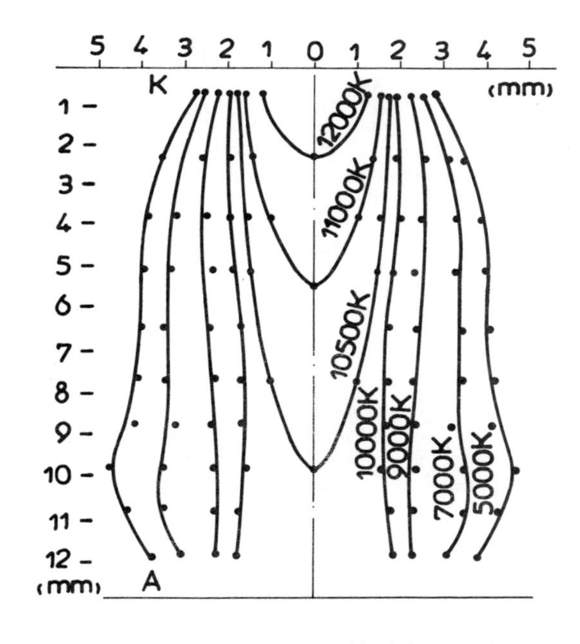

Figure 16. Temperature field of the argon arc.

COMPUTER AIDED FRINGE ANALYSIS

Ryszard J. Pryputniewicz

Center for Holographic Studies and Laser Technology
Department of Mechanical Engineering
Worcester Polytechnic Institute, Worcester, MA 01609

ABSTRACT

Computer aided methods for quantitative fringe analysis are presented. In these presentations, special emphasis is placed on automated data reduction from holographic interferograms.

1. INTRODUCTION

To effectively use laser hologram interferometry, one must get involved with quantitative fringe analysis. This involvement, however, is not straightforward, because of a number of different methods that are available today[1-12]. These methods are, in general, based on multiple observations of holographic images. Although the methods based on multiple observations of the images work well when true fringe order is known, their particular utility is in the cases when the zero-order fringes are not identifiable within the hologram reconstruction. In such cases, fringe shifts, that is, number of fringes "crossing" the points of interest on the object, as the observation perspective is changed, are determined and, together with parameters defining illumination and observation geometry, are used in computations. Measurement of system coordinates specifying the illumination and observation geometry is straightforward. However, accurate determination of fringe shifts, or more specifically, local phase changes, is not easy. To overcome this difficulty, a number of automated, computer aided methods were developed to facilitate measurement of the fringe shifts.

Today's methods for automated interpretation of holographic interferograms are usually based on one of the following approaches[12-18]: (i) vide digitization, (ii) heterodyne readout, (iii) phase step readout, or (iv) direct electronic readout and processing of holograms. Each of these approaches has certain advantages over the others, depending on the particular application. Some of the characteristics of these methods are discussed in Section 2.

2. AUTOMATED INTERPRETATION OF HOLOGRAMS

In this section, brief descriptions of the four, most frequently used, representative approaches to automated interpretation of holograms are given. First, the video digitization is discussed in Section 2.1, then it is followed with discussions of the heterodyne and phase step readouts in Sections 2.2 and 2.3, respectively, while in Section 2.4, electro-optic holography is outlined.

2.1. Video digitization

A typical system for automated interpretation of holograms, involving scanning of images reconstructed from holograms with a computer compatible video digitizer, is shown in Fig. 1. In this system, the digitizer, in addition to converting the scene being observed into a composite video signal, which is viewed on a monitor, produces a digital signal that is transmitted directly to a computer. The computer, in turn, reads the electronic signal corresponding to the video image being digitized. It also processes the digitized data, producing plots of intensity distribution within the image plane. Data characterizing these intensity distributions, together with other pertinent parameters, are used in quantitative interpretation of holographic images. These results can be obtained for any region within the reconstructed image by simply instructing the computer to perform calculations for a point, or a number of points, at specified coordinates.

A system such as that shown in Fig. 1 provides a unique capability for quantitative interpretation of holograms by allowing automated determination of displacements at any point on the surface of the studied objects. This is obtained by recognizing that the intensity distributions measured across the object relate directly to the fringe-locus function, constant values of which define fringe loci on the object's image.

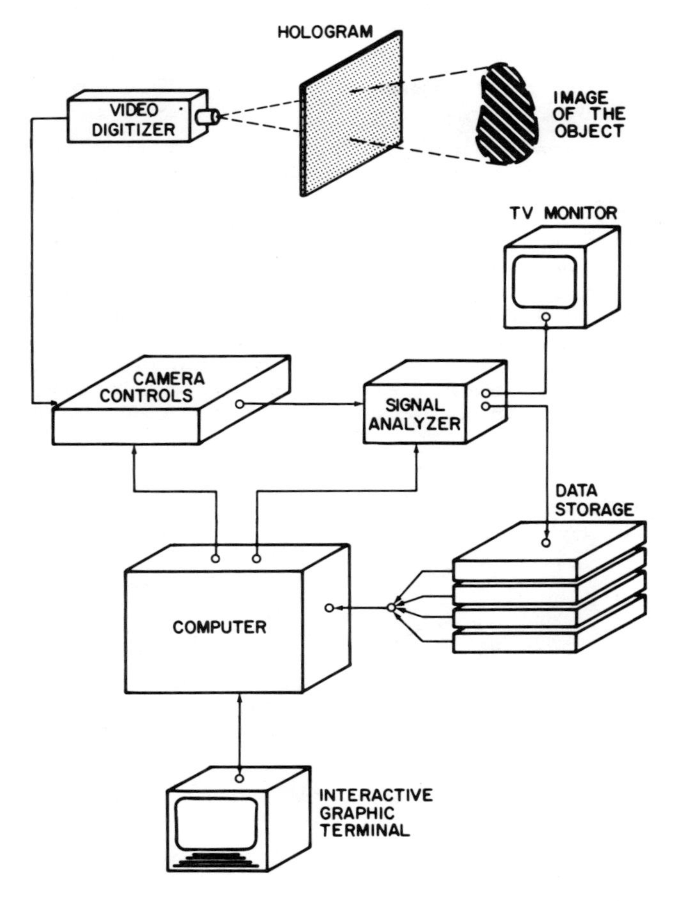

Fig. 1. Video digitizer for automated interpretation of holograms.

The fringe-locus function Ω can be related to the unknown displacement vector **L** and the sensitivity vector **K** via the equation

$$\mathbf{K}(\mathbf{R}) \cdot \mathbf{L}(\mathbf{R}) = \Omega(\mathbf{R}) \quad , \tag{1}$$

where **R** is the position vector defining points on the object's surface, and

$$\mathbf{K} = \mathbf{K}_2 - \mathbf{K}_1 \tag{2}$$

with $\mathbf{K}_1$ and K_2 being vectors specifying average directions of illumination and observation, respectively.

Following the procedures of multiple observations of the holographically produced images[19], displacement vector and strain-rotation matrix can be computed from

$$\mathbf{L} = \underset{\sim}{K}^{-1} \mathbf{\Omega} \tag{3}$$

$$\underset{\sim}{f} = \left[\underset{\sim}{K}^T \underset{\sim}{K}\right]^{-1} \left[\underset{\sim}{K}^T \underset{\sim}{K}_{fc}\right] \quad , \tag{4}$$

respectively. In Eqs 3 and 4, $\underset{\sim}{K}$ represents a matrix of sensitivity vectors, corresponding to the set of observations used in the observation of the holographic images, while $\underset{\sim}{K}_{fc}$ is the matrix of the fringe vectors corrected for perspective[20]. Decomposition of the matrix

$\underset{\sim}{f}$, computed from Eq. 4, into symmetric and antisymmetric parts, yields strains and rotations, respectively, at the particular point on the object[1].

2.2. Heterodyne readout

A heterodyne readout is characterized by the fact that each of the two object fields (in the case of a double-exposure hologram) is recorded with a different reference beam[9]. The reference beams are set up in such a way that they can be reconstructed independently, during the hologram reconstruction process, with an introduction of a known frequency shift between them. As a result of this, the two reconstructed and interfering light fields are intensity modulated at the frequency equal to the frequency shift between the reference beams. The intensity modulation takes place at all points within the interference pattern produced during the reconstruction of the hologram. Now, the optical path differences, corresponding to the displacement and/or deformation recorded within the hologram being reconstructed, are converted into phase of the beat frequency of the two interfering light fields. This phase is, in turn, interpolated opto-electronically, resulting in determination of fringe shifts with an accuracy of 1/1000 of one fringe.

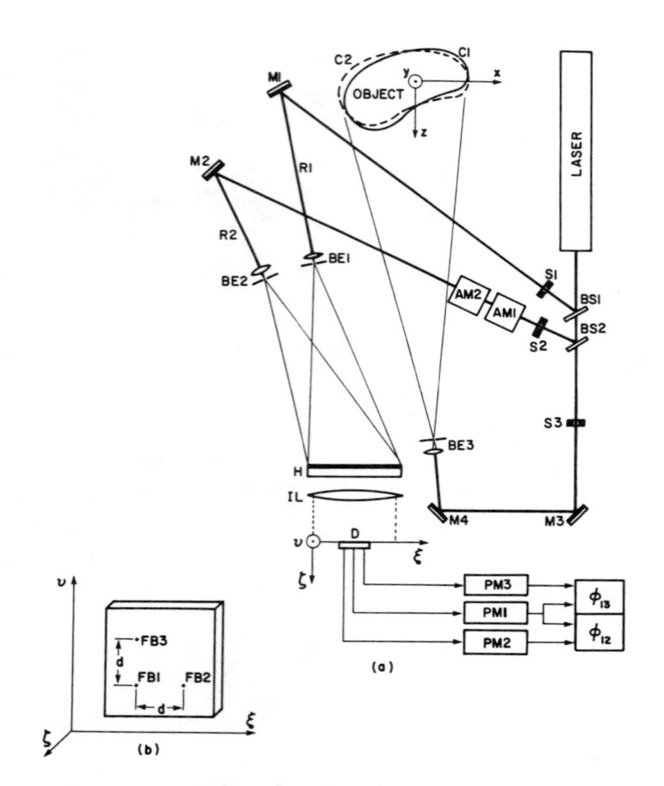

Fig. 2. Heterodyne system: (a) system setup, (b) fiber optic detector head.

A typical setup used in heterodyne hologram interferometry is shown in Fig. 2. In this arrangement, laser output is divided into three beams by means of beamsplitters BS1 and BS2. The first beam, reflected from BS1, is directed by mirror M1, via beam expander BE1, toward recording medium H; this is reference beam R1. The second beam, reflected from BS2, is reference beam R2. The third beam, directed by mirrors M3 and M4, is expanded by BE3 to illuminate an object being studied. This beam, modulated by a reflection from the object, at two different configurations C1 and C2 corresponding to the object's two different states of stress, is recorded against R1 and R2, one at a time, with the exposure being controlled by shutters S1 and S2, respectively.

The desired frequency shift is conveniently produced using acousto-optical modulators AM1 and AM2 which can, for example, be cascaded in R2, Fig. 2.

To measure interference phase variations, during reconstruction of the holograms, at least two photodetectors D are required to be within the image space produced by an imaging lens IL. One of these detectors can be fixed and the other can scan the image, or both of the detectors can be coupled and can simultaneously scan the image; in fact, it is often advantageous to use more than two coupled detectors, as detailed below. In the case when a fixed and a moving detectors are used, cumulative phase difference is measured with reference to the fixed detector. Such a measurement is needed when, for example, determining cumulative displacements. In the case when coupled detectors are used, local phase changes are measured, based on which, for example, one computes local derivatives of displacement, which, in turn, relate to strains.

More specifically, Fig. 2b shows a detector head consisting of three fiber optic bundles, FB1, FB2, and FB3, arranged in an orthogonal pattern with the center to center distance d. In this way, local phase changes in two mutually perpendicular directions are measured simultaneously. The signal from the detector is transmitted, via the fiber optic bundles, to photomultipliers PM1, PM2, and PM3; it should be noted that other suitable optical detectors can also be used in lieu of the photomultipliers. The photomultiplier signal is, in turn, fed into the differential phase meters, which output the optical phase differences ϕ_{12} and ϕ_{13} between the light beams sensed by the corresponding pairs of the detectors.

In quantitative deformation analysis, the first and the second derivatives, L' and L", respectively, of the displacement vector **L** are computed. In general, for the n-th point, these derivatives can be approximated numerically as

$$L'_{i_n} = \frac{L_{i_{n+1}} - L_{i_{n-1}}}{d} \tag{5}$$

and

$$L''_{i_n} = \frac{L_{i_{n+2}} - 2L_{i_n} + L_{i_{n-2}}}{d^2} \quad , \tag{6}$$

where the subscript i denotes Cartesian coordinates normal to the direction of observation, while $L_{i_{n-2}}$, $L_{i_{n-1}}$, L_{i_n}, etc., represent displacement components at points n-2, n-1, n, etc., respectively, which are separated by the distance d/2. If the distance d, appearing in denominators of Eqs 5 and 6, is equal to the center to center separation between pairs of the fiber optic bundles in the detector head of the heterodyne system, Fig. 2b, then the first derivatives of the displacement vector can be related to the corresponding phase differences as

$$L'_{\xi_n} = \frac{L_{\xi_{n+1}} - L_{\xi_{n-1}}}{d} = \frac{1}{d}\left(\frac{\phi_{n+1}}{2\pi}\lambda - \frac{\phi_{n-1}}{2\pi}\lambda\right) = \frac{\lambda}{2\pi d}\left(\phi_2 - \phi_1\right) = \frac{\lambda}{2\pi d}\phi_{12} \tag{7}$$

and

$$L'_{\upsilon_n} = \frac{\lambda}{2\pi d}\phi_{13} \quad . \tag{8}$$

In Eqs 7 and 8, λ is the wavelength of the laser light used in recording and reconstruction of the hologram, d is the center to center distance between the pairs of the fiber optic detectors, ϕ_{12} and ϕ_{13} are the phase differences between the signals detected by FB1-FB2 and FB1-FB3, respectively, while ξ and ν are Cartesian coordinates defining the detector space.

2.3. Step phase readout

In step phase (also known as quasi-heterodyne) hologram interferometry, like in heterodyne hologram interferometry, each of the two object fields (in the case of a double-exposure hologram) is recorded with a different reference beam[12-14]. The reference beams are set up in such a way that they can be reproduced independently, during the hologram reconstruction process, Fig. 3. As a result of such a setup, a known phase step can be imposed on R2 relative to R1 by a well controlled adjustment of the signal driving the piezoelectric mount of mirror M2. Now, the fringe pattern, corresponding to the displacement and/or strain recorded within the hologram being reconstructed, is shifted in accordance with the magnitude of the phase step imposed on R2. The intensity variation, corresponding to the given fringe pattern is, in turn, detected and stored for subsequent computations.

Figure 3 shows the setup for recording of step phase holograms where the two reference sources are set up on the same side of the object with a mutual separation larger than the

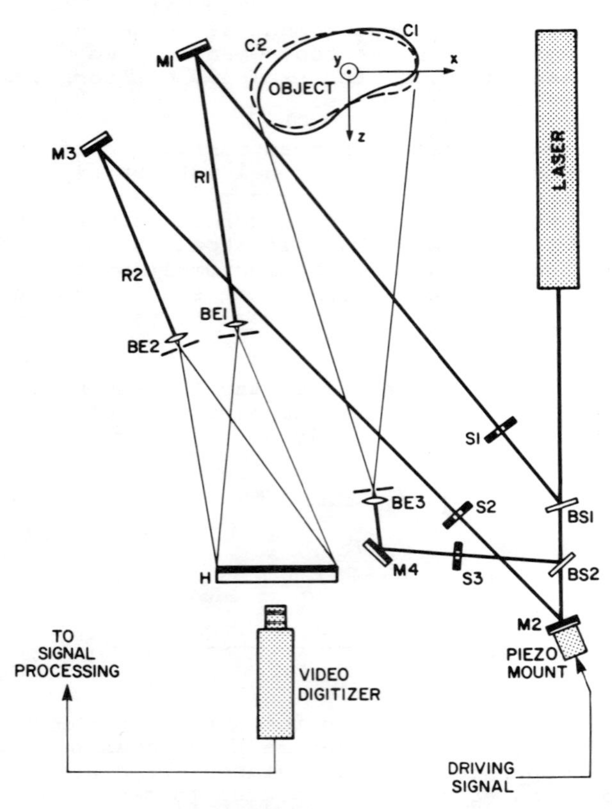

Fig. 3. Setup for step phase hologram interferometry with well separated reference sources.

angular size of the object, to avoid overlapping of the different reconstructions. However, large separation of the reference sources leads to high sensitivity to hologram repositioning errors. To overcome this difficulty, a compromise is made. That is, at the expense of overlapping of the reconstructing images which causes decrease in image contrast, the reference sources are moved close together, Fig. 4. The required minimum angular separation between the two reference sources is given by the resolution imposed by the aperture of the imaging lens.

Phase step hologram interferometry allows electronic scanning of the image by photodiode arrays or TV cameras and use of microprocessor controlled digital phase evaluation[12-14]. In this method, a single interferogram is analyzed at least three times. For each of these analyses, phase of one of the reference beams is changed in a step-wise fashion by a well controlled phase shift, while data acquisition and processing are carried out with a video system and a computer. Using such a system, successive hologram reconstructions, with different relative phases of the reference beams, are stored digitally.

In the case when three different phase steps are used to reconstruct the hologram, the interference phase can be represented mathematically as

$$\phi(\mathbf{R}) = \arctan\left[\frac{I_1(\mathbf{R})(\cos\gamma - \cos\beta) - I_2(\mathbf{R})(\cos\gamma - \cos\alpha) - I_3(\mathbf{R})(\cos\alpha - \cos\beta)}{I_1(\mathbf{R})(\sin\gamma - \sin\beta) - I_2(\mathbf{R})(\sin\gamma - \sin\alpha) - I_3(\mathbf{R})(\sin\alpha - \sin\beta)}\right], \quad (9)$$

where $I_1(\mathbf{R})$, $I_2(\mathbf{R})$, and $I_3(\mathbf{R})$ are local intensities, corresponding to the three phase steps α, β, and γ, respectively; it should be noted that these intensities are measured at the specific point on the object specified by the position vector $\mathbf{R}$.

Equation 9 can be significantly simplified if the specific values of the three phase shifts are used. For example, in the case of typical phase step values of $\alpha = 0^\circ$, $\beta = 120^\circ$, and $\gamma = 240^\circ$, Eq. 9 reduces to

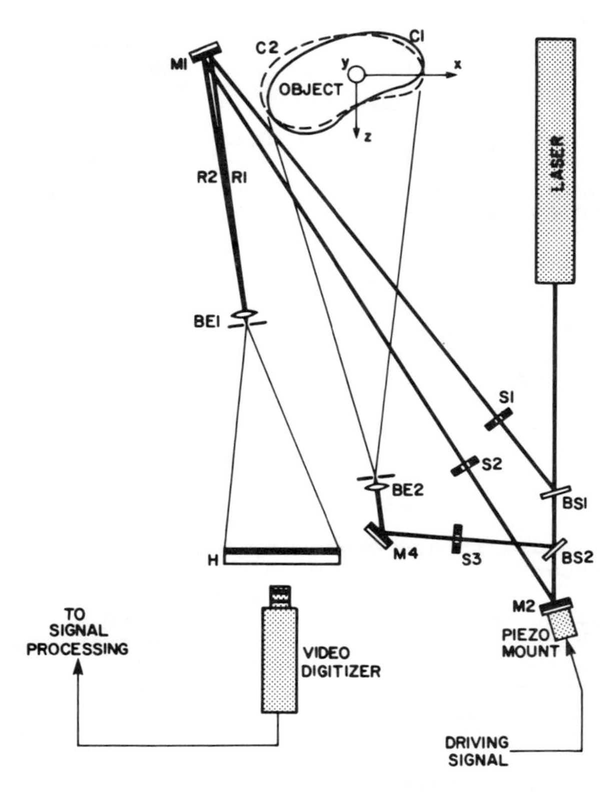

Fig. 4. Setup for step phase hologram interferometry with reference sources close together.

$$\phi(\mathbf{R}) = \arctan\left\{\frac{\sqrt{3}\,[I_3(\mathbf{R}) - I_2(\mathbf{R})]}{2I_1(\mathbf{R}) - I_2(\mathbf{R}) - I_3(\mathbf{R})}\right\} \quad . \tag{10}$$

The interference phase obtained from Eq. 10 can, in turn, be used to compute the fringe-locus function, that is,

$$\Omega(\mathbf{R}) = 2\pi\,\frac{\phi(\mathbf{R})}{360} \quad . \tag{11}$$

Once the fringe-locus function is determined, it can be employed in computations of the displacement vector and/or strains using, for example, Eqs 1 through 4.

2.4. Electro-optic holography

Electro-optic holography utilizes advances in heterodyne and phase step methods to directly record holograms and to transmit them by television systems for automated measurement of object displacements and/or deformations[12,16,21]. The system does not require photographic recording of holograms, but instead does it electronically.

One system proposed for electro-optic holography consists of a modified speckle interferometer, which produces speckles large enough to be resolved by the TV camera[21]. The output of the TV camera is fed to a computer that calculates and stores the magnitude and phase (relative to the reference beam) of each picture element in the image of the illuminated object. The process is repeated for each increment in the phase, with the number and magnitude increments depending on the system's characteristics.

In the electro-optic holography system, the phase of the reference beam is modulated between TV frames. Then, the corresponding pixels are compared, in a number of successive frames, to detect irradiance modulation. In this approach, the aim is to display the interference fringes that will be present in the amplitude of the holographic reconstruction of the object. Ideally, this should be done by performing computations based on the pixel values from the successive frames. However, because of long computation times preventing real-time operation, thresholding of the camera output was employed to utilize high speed logic operations[21].

Any one of the usual phenomena that generate characteristic fringes in hologram interferometry, such as vibration, beam modulation, linear or random motion, will also work with the electro-optic holography. As such, electro-optic holography may be used in recording of time-average holograms, conventional double-exposure holograms, as well as in the nondestructive testing.

Initial estimates of the computation time for the electro-optic holography indicate that images can be recorded and displayed in approximately 2 seconds[21]. Although these times are much shorter than for the conventional holography, they are considerably longer than the corresponding times for the electronic speckle pattern interferometry (ESPI), which can operate at video rates. However, the electro-optic holography could be employed with specular fields that have no speckle patterns and any spatial fluctuations of the reference amplitudes should have lesser effect on the electro-optic holography than on ESPI[21].

3. CONCLUSIONS

The methods for computer aided fringe analysis, discussed in this paper, allow rapid, remote, non-invasive measurement of object dimensions, displacements, and strains. These measurements can be made with high accuracy. In fact, accuracies on the order of 1/1000 of one fringe are obtainable using the methods of heterodyne hologram interferometry, whereas the methods of phase step hologram interferometry lead to fringe measurements to within 1/100 of one fringe. Furthermore, the electro-optic hologram interferometry allows direct, electronic recording of holograms. This, in turn, permits processing and display of the information at rates much faster than those currently possible with holographic systems based on photographic processing of holograms.

Use of computers and video digitizers for data acquisition, processing, and display, offers the possibility to perform all analysis in real-time. Since results obtained are quantitative, it is hoped that as such, the computer aided methods of hologram interferometry will find applications in fields where analysis must be done in three dimensional space to provide information for automated process monitoring and engineering control of operations.

4. REFERENCES

1. K. A. Stetson, "Homogeneous deformations: determination by fringe vectors in hologram interferometry," Appl. Opt., 14:2256 (1975).
2. R. Dändliker and B. Ineichen, "Strain measurement through holographic interferometry," Proc. SPIE, 99:90 (1976).
3. R. J. Pryputniewicz and K. A.Stetson, "Holographic strain analysis: extension of fringe-vector method to include perspective," Appl. Opt., 15:725 (1976).
4. R. J. Pryputniewicz, "Determination of the sensitivity vectors directly from holograms," J. Opt. Soc. Am., 67:1351 (1977).
5. C. M. Vest, Holographic Interferometry, Wiley, New York (1978).
6. W. Schuman and M. *Dubas, Holographic Interferometry, Springer-Verlag, Berlin (1979).
7. R. J. Pryputniewicz, "State-of-the-art in hologrammetry and related fields," Internat. Arch. Photogram., 23:630 (1980).
8. K. A. Stetson, "The use of projection matrices in hologram interferometry," J. Opt. Soc. Am., 71:1248 (1981).
9. R. J. Pryputniewicz, "High precision hologrammetry," Internat. Arch. Photogram., 24:377 (1982).
10. P. Hariharan, B. F. Oreb, and N. Brown, "Real-time holographic interferometry: a microcomputer system for the measurement of vector displacements," Appl. Opt., 22:876 (1983).
11. W. Schuman, J. P. Zürcher, and D. Cuche, Holography and Deformation Analysis, Springer-Verlag, Berlin (1985).
12. R. J. Pryputniewicz, "Quantification of holographic interferograms: state of the art methods," Technical Digest, OSA, Washington, DC (1986).
13. P. Hariharan, "Quasi-heterodyne hologram interferometry," Opt. Engrg., 24:632 (1985).
14. R. Dändliker and R. Thalmann, "Heterodyne and quasi-heterodyne holographic interferometry," Opt. Engrg., 24:824 (1985).
15. J. D. Trolinger, "Automated data reduction in holographic interferometry," Opt. Engrg., 24:840 (1985).

16. K. A. Stetson and W. R. Brohinsky, "An electro-optic system for vibration analysis and nondestructive testing," Proc. SPIE, 746:15 (1987).
17. J. D. Trolinger, J. E. Craig, and H. Tan, "On-line recording and data reduction for holographic flow diagnostics," Proc. SPIE, 746:000 (1987).
18. R. J. Pryputniewicz, "Real-time automated hologrammetry," in press.
19. R. J. Pryputniewicz, "Quantitative interpretation of holograms and specklegrams," Proc. SPIE, 669:000 (1987).
20. R. J. Pryputniewicz, "Holographic strain analysis:an experimental implementation of the fringe-vector theory," Appl. Opt., 17:3613 (1978).
21. K. A. Stetson and W. R. Brohinsky, "Electro-optic holography and its applications to hologram interferometry," Appl. Opt., 24:3631 (1985).

FATIGUE DETECTION BASED ON THE CHANGE OF LASER-PRODUCED DIFFRACTION PATTERNS

Xu Boqin Li Li Wu Xiaoping

Department of Mechanics, University of Science & Technology of China
Hefei, Anhui, PRC

Intruduction

Fatigue induced damage and crack initiation is certainly one of the most important problems in solid mechanics. Safety consideration for structures ranging from aircraft to nuclear reactor all hinge upon , to a large extent ,the early detection of fatigue induced damages. Thus, it is paramount that effective means be developed that can detect ,or better yet, predict the onset of a fatigue crack. Optical techniques being non-contact methods are ideally suited for such a tast. Indeed a number of techniques aiready exist, e.g. the methods of holographic correlation, speckle pattern correlation, speckle spectrum pattern correlation, speckle spectrum core energy variation, etc . In this paper a new method is given to detect fatigue damage and crack initiation and judge the shape , position and damage level of the damage zone, based on the change of laser-produced diffraction pattern.

Theory

When leser light reflects from a being fatigued two phenomena will be observed . First, owing to the change of positions of scattering units, reflective speckle field will move according to the law of speckle movement[1]. It is evident that this change of the speckle pattern should be related to the strain state of scattering units. It makes the reflective light pattern spread out , draw in or revolve, but not change the whole shape . For example, not from a band-shaped distribution to curcular-shaped. We call this phenomon the grating effect. Next, when the damge starts on the specimen surface the permanent deformation accumlates resulting in surface texture change, which becomes rougher. From this, the reflective light pattern will change ,but this change is completely different from aforesaid .Plastic deformation's taking place in a metal specimen is as a result of lattice slip inside the specimen.The slips arise only along the special slip planes and slip directions. For polycrystal metal the crystal lattice orientations in various grains are different, thus the possible slip directions have nothing in common with each other. Therefore , the probablity with which the slips happen in each direction ,thus the level to be rougher is equal approximatly. It is clear that the tendency of specimen surface to be rougher should only depend on metallographic properties of material and the level of permanent deformation not its strain state. Since the polycrystal metal specimen becomes rougher in each direction evenly , after plastic deformation occurs, the light intensity of the reflective speckle field will increase curcularly symetrically.In other words,the reflective diffraction halo will tend towars symetric , gradually , with the increasing of the level of plastic deformation. We call this phenomenon isotropic rough effect[2].

From the above ,when a thin beam of laser light illuminates the surface of specimen which is worked in the same direction (grinded or polished in single-direction),the reflective halo is a bright band.As soon as fatigue induce damage on the surface of specimen,becouse of its becoming rougher evenly, the reflective light pattern tends isotropic, that is,band-shaped halo widens by degrees and tends towars circular-spot-shaped . We can consider this change of reflective light pattern to be a criterion and measurement. It is emphasized that the change of the grating effect also make the reflective halo move , but always remain the band-shaped distribution. In order to get rid of the influence of grating effect we have designed a spetial band-pass aperture of the receiver , as shown in Fig.1. If we make the oringinal diffraction in middle part of the

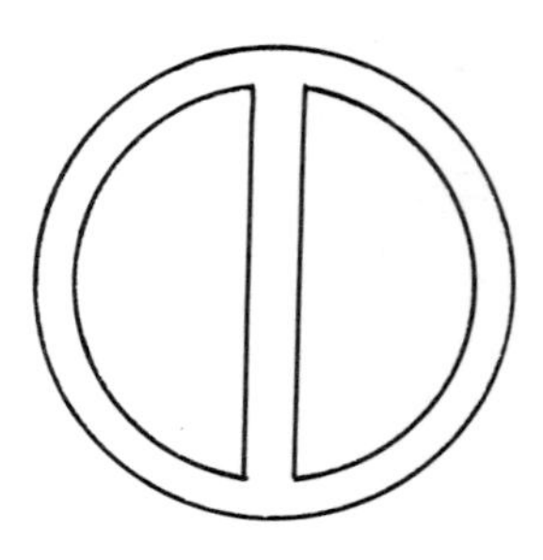

Fig.1

aperture our detection can't be effected by the change of positions of scattering units. Only if the plastic deformation starts, the surface of specimen becomes rougher, a part of light will flow out of the mask band of the aperture and be received . Thus, we can get the information about damage initiation and development.

Experimental Research

(A) Detection of the fatigue damage and the shape of damage zone ----pointwise

(1) **Optical System**: as shown in Fig.2. A thin beam of laser light reflects from the surface of specimen. A photodetector detects the intensity of reflective light at a fixed point in the space. The specimen may move in the horizontal plane, whose position can be measuremented with a displacement sensor.The relationship between the signals from the photodetector and the sensor can be obtained by a X-Y recorder.

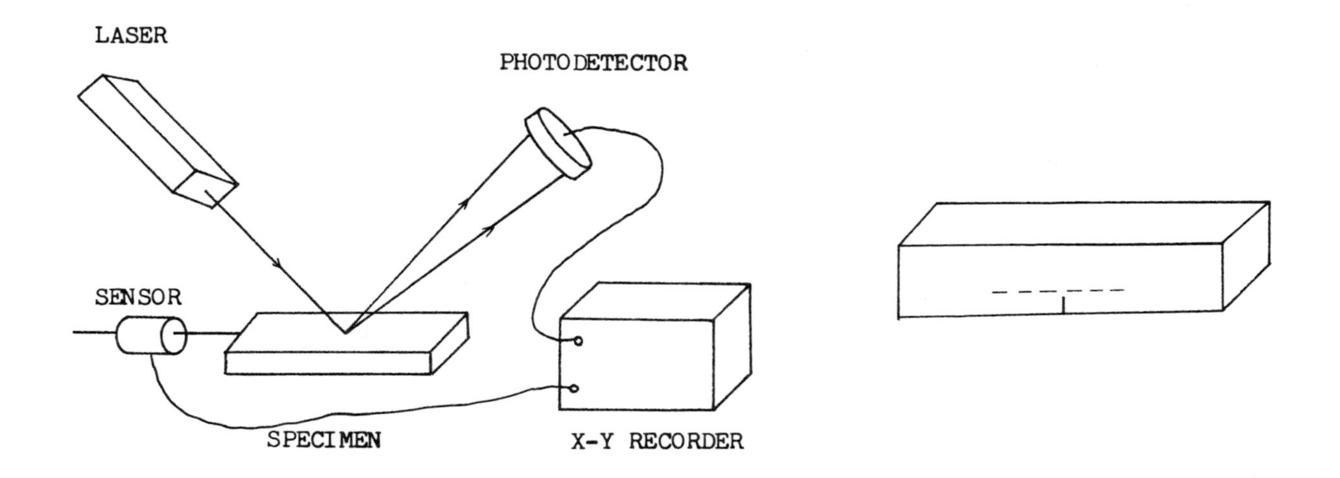

Fig.2 Fig.3

(2) **Specimen**: as shown in Fig.3. This is a three-point-bend steel specimen whose surface is ground on up to Rz =0.5 μm. There is an original crack whose witdth is 0.2 mm approximately.

(3) **Operation**: Move the specimen to make laser light scan the intrested area on its surface. Detecting the position X with the displacement sensor and the light intensity I at a fixed point with a photodetector we can obtain the function curve of X-Y with a X-Y recorder. Scanning the same area before and after loading respectively, we can get the information about fatigue damage initiation by comparing the two curves. Correspondent parts of the curves with the damaged area will be completely different ,but the other parts of the curves will be similar.

(4) **Result**: Exsample 1 refers to Fig.4. The beam-scan-line is 0.5 mm above the crack tip. The dead load P_o=800 kg, the amplitude of cyclic load p=600 kg. The cycle number N=22.5*10^3 .After fatigue test we haven't seen any change on the surface of specimen with the naked eye and microscope.Compearing the oringinal curve I-X (curve 1) and the curve obtained after testing I'-X (curve 2) we can see a pair of evident uncorrelated areas .. In order to determine the uncorrelation degree, we can define a non-demensional parameter

$$R = \frac{I'(X)-I(X)}{\langle I'-I \rangle} \qquad (1)$$

Curve 3 in Fig.4 is curve R-X, where <I'-I> is the average value of the quantity (I'-I). From this figue we can see a obvious peak around the crack tip at point A . From this figue we can judge the dimension of the area in which the material is damaged and the damage level.

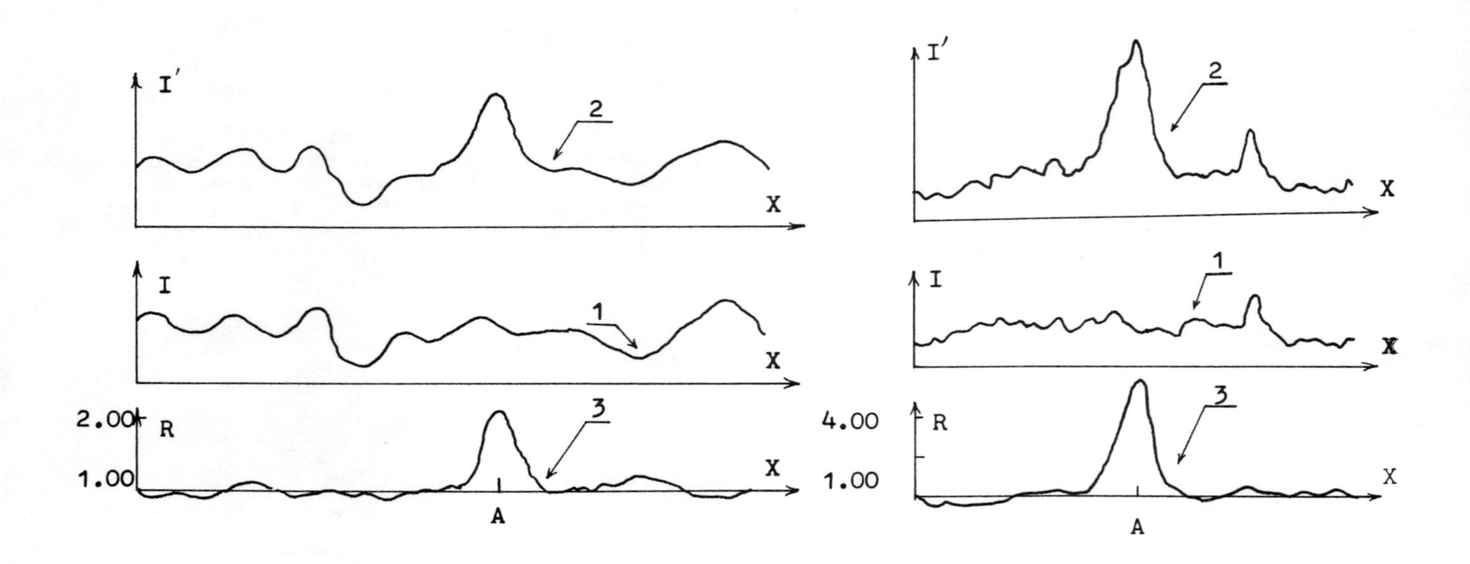

Fig.4 Fig.5

Exsample 2 is shown in figure 5. The scanning line is 0.2mm above the crack tip , which is parallel to the upper end plane. Po=1500 kg, p=800 kg. Cycle number $N=10*10^3$. After testing we can see a unusually reflective area by the naked eye and a macroscopic crack 1.1 mm long with the microscope.In Fig.6 the implications the curves stand for are the same as example 1. From this figue we can realize that , with accumulate and development of the damage, the uncorrelation degree drops down greatly.

(B) **Real-time detection of the initiation and development of fatigue damage**

(1) **Optical System**: The principle is the same as the above.

(2) **Specimen**: It is idem.

(3) **Operation**: We illuminate the interested area with a thin beam of laser light and detect the intensity at a fixed point by a photodetector with a aperture shaped as shown in Fig.1. During fatigue test we keep watch on the change of light intensity I. Befor the surface is damaged the intensity I will not change basiclly .As soon as the surface is damaged, from the above, the reflective light pattern must widen , consequntly ,a part of reflective light will flow out of the mask band of the aperture , which will change the intensity I received by the photodetector. From this special change of curve I-N we can detect the initiation and development of the fatigue crack.

(4) **Result**: Exsample 3 is shown in figue 6. The material of specimen is Steel A3. Po=500 kg, p=250 kg. From the curves in Fig.6 we can see that after $5*10^3$ cycles the slope of curve I-N has changed. When N=6.4 *10 we stop testing ,in this time, and see a unusual reflective area 0.5 mm long near the crack tip, but don't see any macroscopic crack. Thus we can judge the initiation of the fatigue damage.

Exsample 4 is shown in Fig.7. The material of specimen is Steel No45. Po=700 kg, p=500 kg. From the curves we can notice that after 150 $*10^3$ cycles approximately curve I-N make a sudden change. Up to $N=290*10^3$, we stop testing . At this time we have seen a fatigue crack 0.15 mm long near the crack tip by a microscope.Thus, observing the change of curve I-N we can inspect the development of fatigue damage.

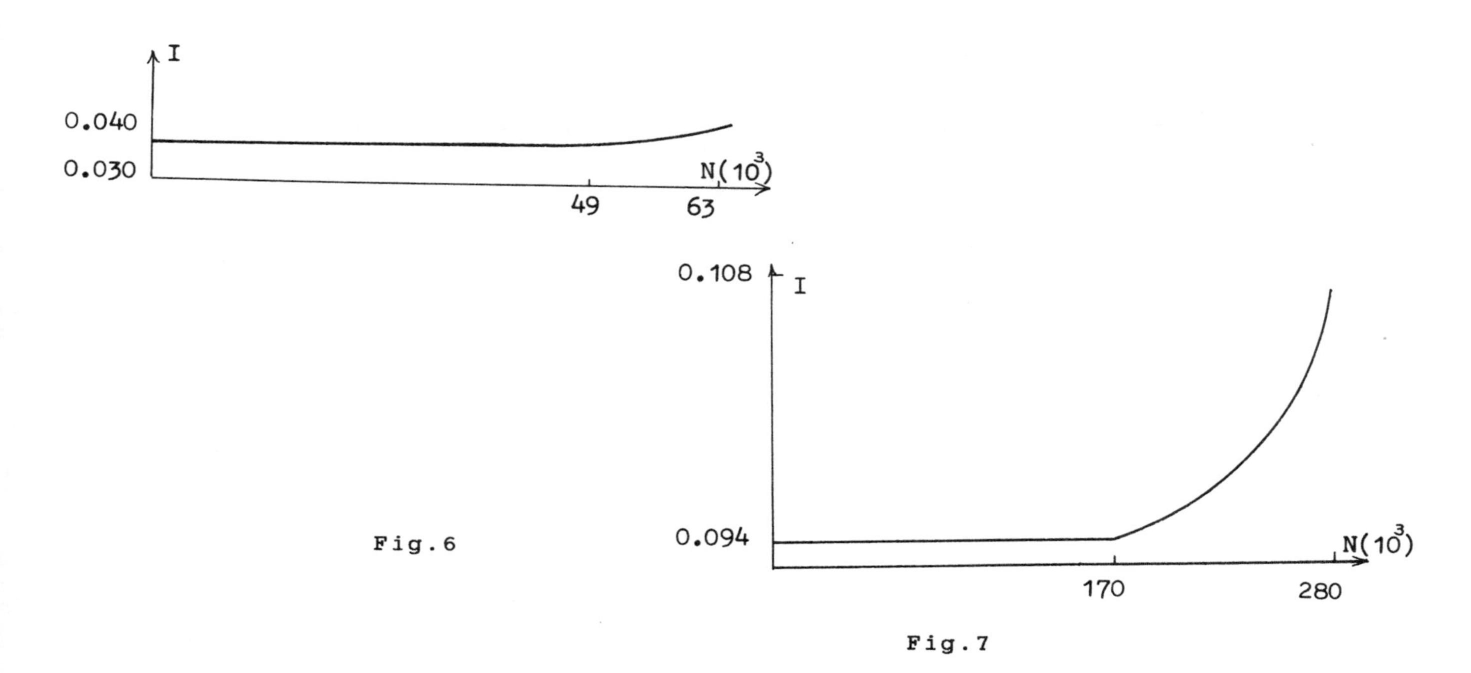

Fig.6

Fig.7

(C) **Detection of the shape of damage zone ---- whole field analysis**

(1) **Optical System**: The set-up is shown in Fig.8 which is a 4f system with filter. The band-pass filter can be use to delet the grating effect and extract the information of isotropic rough effect which relates to the fatigue damage.

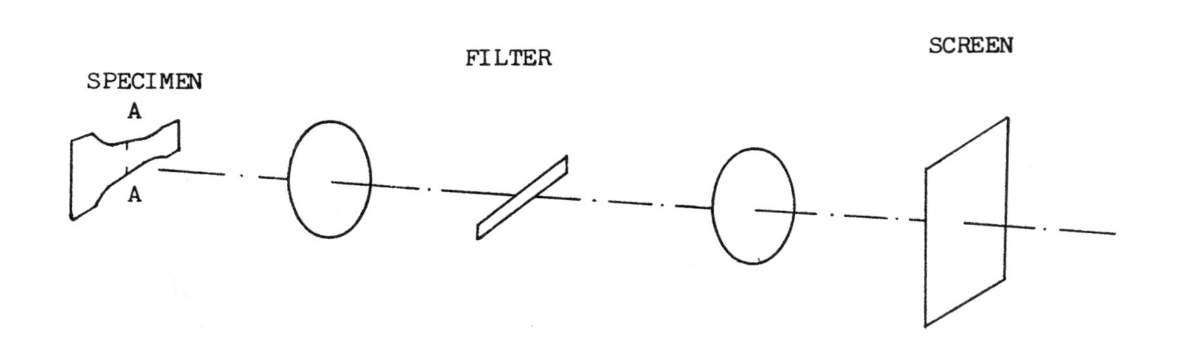

Fig.8

(2) **Data Collecting and Processing System** : The intensity changing at a point of image plane (screen) is the flow-out energy at spectrum plane from the conjugate point of specimen.The flow-out energy is caused by the variation of surface roughness of the specimen and can be used to describe the fatigue damage. A digital camera is a detector. The detected area , 100mm on specimen ,includes 256 elements. The array processor deposites the intensity distributions befor and after fatigue and caculates the relative variation of the intensities.

(3) **Monitor**: Computer shows the digital results and draws the isothetics of the relative change of intensity I . We define

$$\tilde{I} = \frac{I(X)-I(X)}{I(X)} \tag{2}$$

Fig. 9 is a set of isothetics of fatigue damage after 3500 cycles. From them we can find the shape of fatigue damage zone.

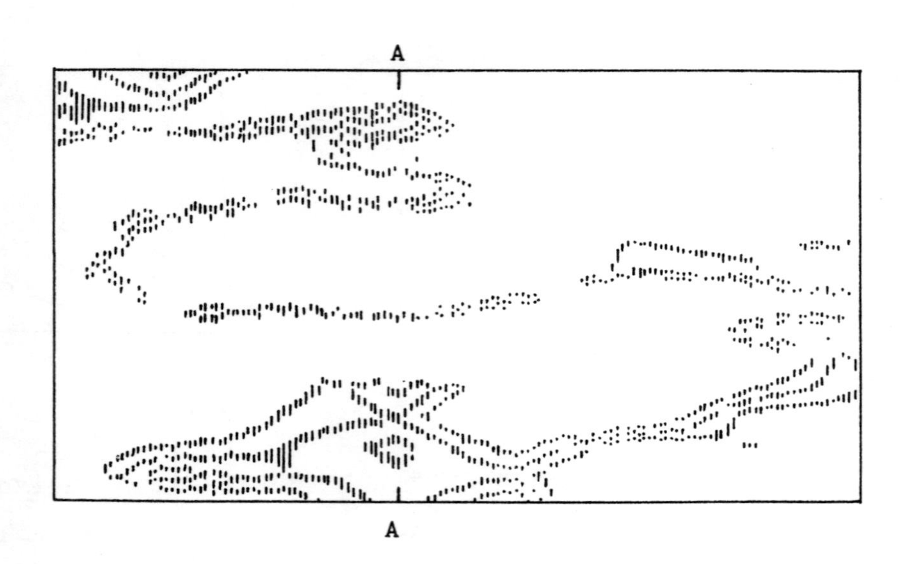

Fig.9

Reference

1. Wu Xiaoping, He Shiping, Li Zhichao , Acta Physica Sinica, Vol. 29, No.9 pp.1142, 1980
2. Xu Boqin, Wu Xiaoping, Acta Mechanica Sinica, Vol.16, No.2, pp.1751, 1984

Polychromatic speckle interference for edge enhancement of image

S.Q.Shen*, X.P.Wu**, F.P.Chiang***

* Beijing Institute of Posts & Telecommunications, Beijing,China
** University of Science & Technology of China, Hefei,China
*** Laboratory for Experimental Mechanics Research
SUNY at Stony Brook, Stony Brook, NY.11794 USA.

Abstract

A new method for edge enhancement of image using polychromatic speckle is presented.

By using the polychromatic speckle pattern as the carrier of an object function,the edge enhancement of image has been realized. This technique is similar to that proposed by M.Francon[1].

The generalization of polychromatic speckle has been discussed by many authors[2,3]. in this article, we just want to point out that when a rough object is illuminated by broad band-limited light or even white light, a field of coloured speckle will exist near the surface of the object, or if an optical system is used, the coloured speckle will appear on the image plane.

The system for recording is shown in Fig.1.

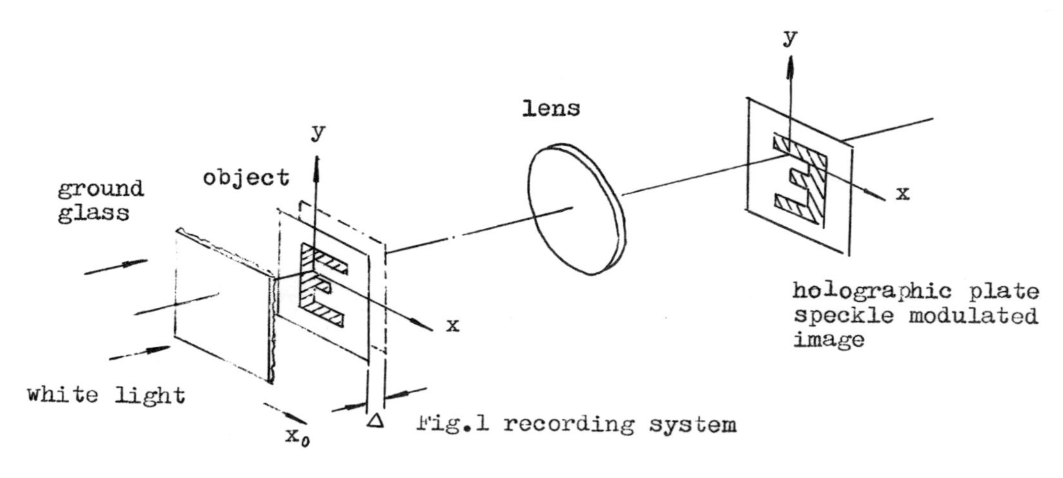

Fig.1 recording system

The holographic plate on the image plane is recorded with the exact image A which is modulated by the image plane polychromatic speckle pattern D. Then the ground glass is displaced transversely by an amount x_0, and the object is moved an amount Δ along the axis at the same time.

During the second exposure, a defocused image A' is recorded on the same plate. The two images are overlapped, both modulated by the image-plane speckle patterns which are separated from each other by x_0. After development, the intensity transmittance of the plate is given by

$$t = a-b\left\{ AD*\left[\delta(x+x_0/2\,,y)+\delta(x-x_0/2\,,y)\right] -(A-A')D*\delta(x-x_0/2\,,y)\right\} \qquad (1)$$

Where A and D are the intensity distribution of the image and the speckle pattern in the image plane, respectively, and " * " represents the convolution operation. A-A' is the difference between the focused image and defocused image.

Fig.2 shows a white light optical system for extracting the difference C=A-A'. In the back focal plane of lens O_1, the spectrum of the negative plate is given by the Fourier transform of the intensity t:

$$\tilde{t}(u,v) = a\,\delta(u,v)-b\left\{\left[\tilde{A}(u,v)*\tilde{D}(u,v)\right]\cos\frac{\pi u x_0}{\bar{\lambda}} -\left[\tilde{C}(u,v)*\tilde{D}(u,v)\right]\exp(-j\frac{\pi u x_0}{\bar{\lambda}})\right\} \qquad (2)$$

where u, v are spatial frequencies, " ~ " is the symbol for Fourier transform, $\bar{\lambda}$ is the average wavelength of the illuminating white light.

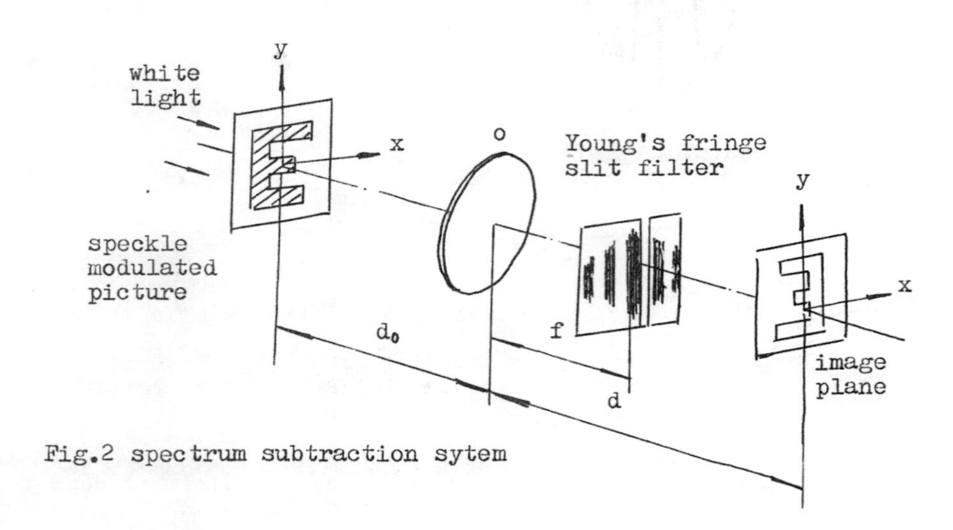

Fig.2 spectrum subtraction sytem

If an opaque screen with a narrow slit is put on the Fourier transform plane in such a way that the slit coincides with a minmum of the Young's fringes: the term

$$(A * D) \cos \frac{\pi u x_0}{\lambda} \tag{3}$$

will be blanked out, but the slit allows only

$$(C * D) \exp \left(-j \frac{\pi u x_0}{\lambda}\right) \tag{4}$$

to go through to the image plane. Only the difference C=A-A' can be seen in the image plane, which is the differentiated image of the original one.
The key point of this experiment is based on the subtraction of the two spectrum $\tilde{A}$ and $\hat{A}'$.
We know the incoherent transfer function(OTF) is

$$I(u,v) = I_t(u,v)H(u,v) \tag{5}$$

where I and I_t are the spectrums of the image intensity and object intensity, respectively. H is the OTF function. Therefore the difference between the focused and defocused spectrum can be written as:

$$A-A' = H(u,v)A_0 - H'(u,v)A_0 = (H-H')A_0 \tag{6}$$

where $\tilde{A}_0$ is the spectrum of the object intensity, H' is the OTF with defocused aberration. After filtering, only the difference $\tilde{A}-\tilde{A}'$ can pass through the slit. So, we come to the conclusion that as a result of the subtraction between the two spectrums, the higher frequency components are raised, as is shown in Fig.3.
Fig.2 shows an edge-enhanced image. When changing the width of the slit filter, we can get different ratio of contrast between the intensity of the edge and the body of the image. If changing the position of the slit filter, we can also get different coloured image, but the edge of it keeps white.

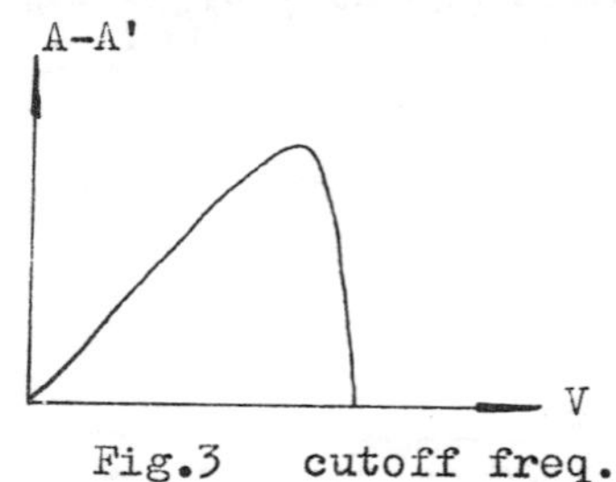

Fig.3 cutoff freq.

Fig.4 gives the experimental results.

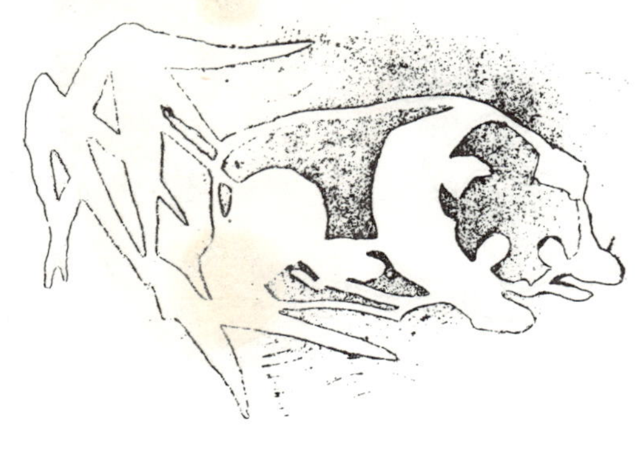

Fig.4 the edge-enhanced image (a Chinese penda)

References

1. M.Francon. "Laser Speckle and Its Applications in Optics"
2. M.M. Pedersen, Optics Acta, 22, 523 (1975)
3. N.George, Appl. Phys., 4, 201 (1974)

Session 8

Holographic Interferometry IV

Chairmen
Y. Y. Hung
Oakland University, USA
F. T. S. Yu
The Pennsylvania State University, USA

An interferometric method for measuring optical spherical surfaces using holographic phase conjugate compensation

Zou Zhenshu, Liao Jianghong, Gu Yunan, and Gu Quwu

Changchun Institute of Optics and Fine Mechanics, Academia Sinica
P.O.Box 1024, Changchun Jilin China

Abstract

In this paper a new interferometric method for measuring errors of optical surfaces and their radii using a holographic phase conjugate compensation is proposed. It has the advantages such as good contrast of the fringes, low quality requirement of the main lens of the system and self-examination. It can be used to test a large size of convex spherical surfaces. The fundamental principle of the method is discribed in detail and the experimental results are given followed by the discussion of the features of this method.

Introduction

In the optical manufacturing the surface errors and radii of the optical spherical surfaces are generally measured by a standard spherical surface and a spherometer respectively. These methods have some disadvantages that the surfaces of optical elements are easily damaged due to the contact measurements and many different standard spherical surfaces are required for different radii. The method of non-contact interferometric measurement has been used in order to overcome these disadvantages. But it requires a group of high quality lenses of which the last surface is a standard spherical surface. Generally, it is difficult to make, especially for a large aperture of the group of the lenses used to examine the large size of a spherical convex surface. The method of interferogram compensation which reduces the requirement for the standard spherical surface is proposed in Ref.(1). However, the method still requires standard spherical surfaces to make the different compensation plates for measurement of surface errors and different radii. The holographic method for measuring the errors of spherical convex surfaces based on synthetic hologram and wavefront reversal, which removes a standard lens group, is described in Ref.(2). But the method is only used to measure a special spherical surface for each hologram because the hologram depends on the chosen surface.

The new interferometric method discribed here can be used to measure the errors of optical spherical surfaces and their radii using holographic phase conjugate compensation which is suitable for both concave and convex spherical surfaces with large aperture. this method is similar to the method shown in Ref.(3), but the arrangement is more convenient to adjust and the hologram is easier to replace in our arrangement.

Fundamental principle

The optical scheme of the interferometric measurement and the hologram recording is shown in Fig.(1).

Hologram recording

The light emitted from a laser is divided into two beams, the object and the reference beam , by the splitter BS. The object beam is then converted into a standard spherical wave by a microscope objective O and a pinhole P after reflection by a mirror M_1 and then is converted into a plane wave by a positive lens of large aperture and a small negative lens. It is evident that the object wave has aberrations due to passing through the lenses AL and L_1. The object beam illuminates the holo-plate together with the reference beam after reflection at M_2 and M_3 via a beam expander E and the hologram is formed after processing and should be replaced.

Measurement

The interferometric measuring system is formed when the mirror M_2 is removed and the standard plane SS is inserted as shown in Fig.(1). In this case the reference beam is transfered into its conjugate by mirror M_4. Adjust the tested surface UTL (suppose it is a convex surface) until its center of curvature is coincident with the point P. The conjugate reference beam is incident on the H and is diffracted into zero order and first order . The zero order is returned by the standard plane SS and diffracted again by the H into the first order which contains the aberration of lenses AL and L_1, expressed by (0,1). On the other hand, the first order becomes a standard spherical wave after passing through L_1 and AL and is returned towards H along the original path after it is reflected from the UTL. After that, it is diffracted again by H into the zero order, expressed by (1,0). Both (0,1) and (1,0) interfere to form the interferometric pattern on the screen S. The field on the S is straight and equal-spaced fringes, or homogeneous bright or dark, if the UTL is a standard spherical surface. If not, the fringe pattern will show the aberration of the UTL.

When the UTL is moved to the location of the point P the interferometric pattern of the system errors can be seen on the screen S, called apex interferometric pattern of the system. The distance between the UTL and the point P is the radius of the UTL.

Experiment and results

An experimental arrangement is shown in Fig.(1). Both surfaces of the large lens are symmetric aspherical surface and the aperture is 270mm with 0.2mm of Airy disk. The point source is combined by a 40x objective lens with 0.65 numerical aperture and a pinhole with 4μ diameter. The distance is 1060mm between large lens AL and the pinhole P. The surface error of the standard optical plane SS is $\lambda/20$ and its aperture is 20mm. The holo-plate (Model I . Tianjin) is mounted in a liquid gate with a refractive index of 1.5. In general case, when object beam and reference beam simultaneously illuminate the hologram, straight equal-spaced Moire fringes are observed because of interferometric compensation. When bright sence occurs in the interferometric field by adjusting the location of the hologram the hologram is completely replaced. Fig.(2) shows the apex interferometric pattern of the system. The deviation from the standard spherical wavefront is 0.09λ peak-to-valley and 0.02λ rms. Fig.(3) shows the interferogram of a convex spherical surface. The result is 0.11λ peak-to-valley and 0.03λ rms. An experimental result shows that the accuracy for measuring radius is +2μ.

Features of this method

There are several features of this method.

A group of standard lenses is not needed and accuracy is increased.

As described above, general spherical interferometer must have a lenses group with a standard spherical surface on the last one. In fact, it is difficult to make such a lens group. The more its size, the more difficult to make. For the measurement of a large convex surface it is not avoidable to require a large standard lenses group, if general spherical interferometric method is used. On the other hand, because the light beam from the lenses group to the tested surface is not a good spherical wave, the apex interferometric pattern is not good and the measuring accuracy is low. In our method, the microscope objective lens and the pinhole are used instead of a standard lenses group to produce a standard spherical wave. Of couse, the measuring accuracy is high, the objective lens and the pinhole are made easilier than a standard lenses group.

We have a standard spherical wave if the diameter of the pinhole satisfies the following formula

$$d = 2.44\frac{L\lambda}{D} \tag{1}$$

where L: distance between the point source P and the lens AL, D: the aperture of the large lens AL, λ: wavelength, d: the diameter of the pinhole.

In our method the hologram is both a beam splitter and an element of wavefront compensation. The conjugate reference beam passes through the hologram and is diffracted into zero order and first order. After both the beams are reflected, they illuminate the hologram again and interfere with each other. It is like a beam splitter. On the other hand, both the beams (0,1) and (1,0) have the same aberration produced by AL and L_1, so that the interferogram does not reveal the aberration. It is called effect of wavefront compensation. Therefore, the single large lens can be chosen as a main element of the system whose requirement of the quality may be much lower than that of the standard lenses group. This property is very useful to make a large interferometer to check large optical convex surfaces.

The high contrast of interferometric fringes

It is necessary to make the contrast of fringes as high as possible in interferometric measurement. In our method the theoretical contrast is 1, because the intensities of the beams (0,1) and (1,0) are equal. Suppose that transmittance of the hologram is T , and its diffraction efficiency is η , and that the reflectance of the standard plane SS is R. For beam (0,1) , the intensity is

$$I_{(0.1)} = I_o TR\eta \quad (2)$$

Similarly, if the reflectance of the tested surface UTL is also R, for beam (1,0), the intensity is

$$I_{(1.0)} = I_o RT\eta \quad (3)$$

where I : intensity of the conjugate reference light. So

$$I_{(0.1)} = I_{(1.0)} \quad (4)$$

Self-examination of the optical system

In our optical system, a good apex interferogram can be obtained if the pinhole and the standard plane have high quality . Therefore, the quality of the optical system can be determined by the apex interferogram. It is easy to remove the system error by means of a computer in the process of the measurement.

Acknowledgements

The authors would like to thank Mr. Li Xizeng for tis help in processing interferometric patterns by computer, Mr. Lu Zhenwu has taken part in a part of this work.

References

1. Yang Li, " Laser Holograohic Spherical Surface Interferometer ", Laser (Chinese) Vol. 6, pp. 51. 1979.
2. A.A.Gorodeslskii, N.P.Larionov, A. V. Lukin, and K. S. Muslafir " Holographic Testing of Convex Surfaces Employing Wavefront Reversal," Soviet J. Opt. Technol. Vol. 50, pp. 787-789, 1983.
3. A. B. Zenzinov, A. A. Shchetnikov, " Application of the Holographic Wave-front Correction Method in Interferometry to Moniter the Shape of Reflecting Surfaces," Opt. Spectros., Vol. 56, pp. 435, 1984.

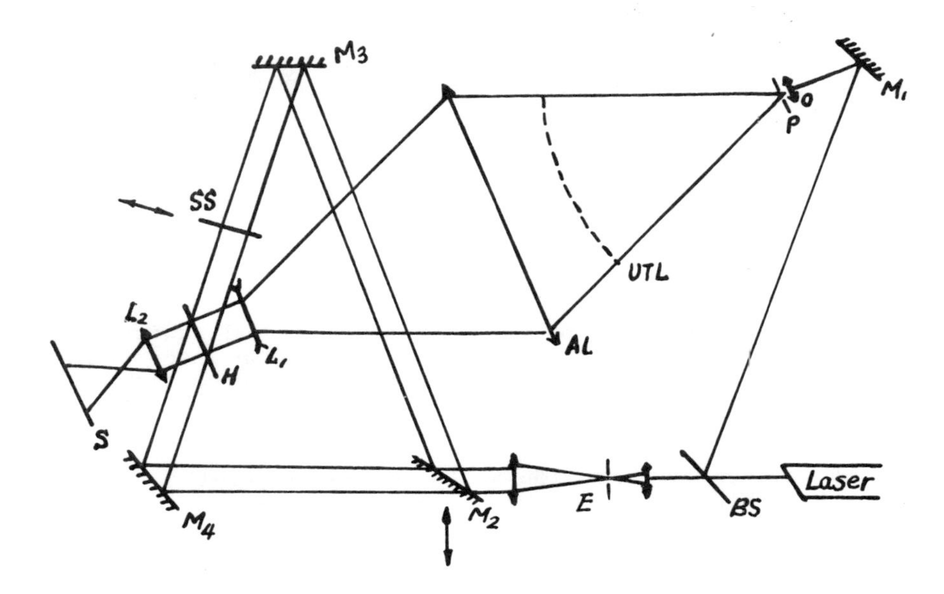

Fig.(1) Scheme of the interferometric measuring system and hologram recording

Fig.(2) Apex interferometric pattern of the system

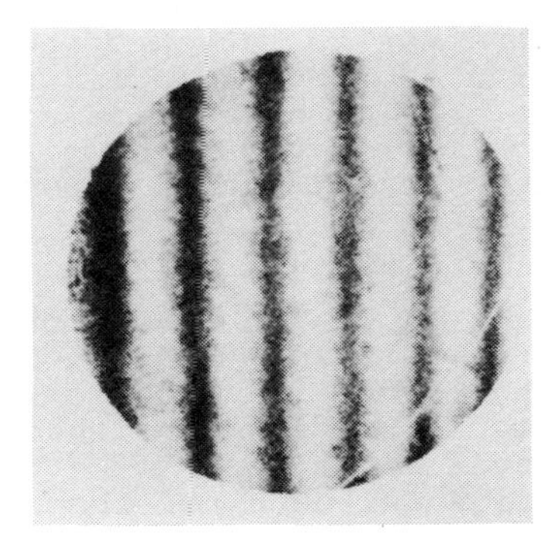

Fig.(3) Interferometric pattern of the tested convex spherical surface

Vibration Analysis of Automotive Bodies
Using Continuous Wave Laser Holographic Interferometry

Hajime Yamashita, Kazuto Sasanishi and Ichiro Masamori

Technical Research Center, Mazda Motor Corporation
3-1, Shinchi, Fuchu-cho, Aki-gun, Hiroshima, Japan

Abstract

Vibration modes from 30 Hz to 1 kHz were measured in full equipped automotive body, door, roof and windshield in a rock cave laboratory using continuous wave laser holographic interferometry and analyzed to reduce noise and vibration.

Introduction

The reduction of automobile vibration and noise is one of the important tasks for automobile engineers to achieve. While the demand is increasing for reduction in exterior noise and for a quieter passenger compartment, the recent strong emphasis on higher engine output and lighter vehicle weight for imporved fuel economy is proving a disadvantage to efforts for cutting noise and vibration.

Holographic interferometry is used as one of the effective methods to identify the vibration characteristics of engines, etc., and is thus a great step toward improving their structures (1), (2), (3).

Continuous wave laser holographic interferometry offers benefits of, among others, showing the vibration amplitudes and nodal lines as interference fringes, for easier interpretation of vibration modes, and permitting real time measurements. It needs, however, a vibration isolation system to accommodate a laser, an optical system, and a test object, imposing an inevitable limitation on the size of the object to be tested to about 1.5 m in diameter--the size of an engine, etc. To resolve this problem, a rock cave laboratory superior in vibration resistance was built as a huge vibration isolation system (4) and was used for vibration measurement of large-sized structures, such as car bodies.

Experimental setup

A vibration isolation laboratory was constructed in a cave in a quarry located in an upper layer of a tuff rock (called Ohya rock) about 300 m thick. The laboratory is 7.5 m wide, 13 m long, and 3.5 m high, offering sufficient space for car vibration testing (Fig. 1). The maximum vibration amplitude of the floor is on the order of 0.1 micron at 0.7 Hz in the horizontal direction.

The experimental setup of the optical elements is shown in Fig. 2. The setup is composed mainly of an object beam optical system and a reference beam optical system, as need arises, a reference beam phase modulation optical system and a Michelson interferometer which monitors vibration and air turbulence are added. The light source used was an argon-ion laser with an output of 4W, and the vibration modes of a test object were photographed mainly by the time average method.

The main problems in measuring the vibrations of a car body include:

a) A large object such as a car body requires a long exposure time.
b) The rigidity of a car body itself is low.
c) Rubber, plastic, and other visco-elastic materials are used in automotive structures.

To avoid deformation under its dead weight and due to temperature gradients of the laboratory and vehicle, it was necessary to make the vehicle stationary overnight, i.e. nearly 12 hours, though varying with exposure time. The time requirement to achieve the standstill condition can be shortened by supporting the suspension of the car by steel or tuff rock blocks to exclude the influences of the tires. In this case also, little influence is given to the vibration modes.

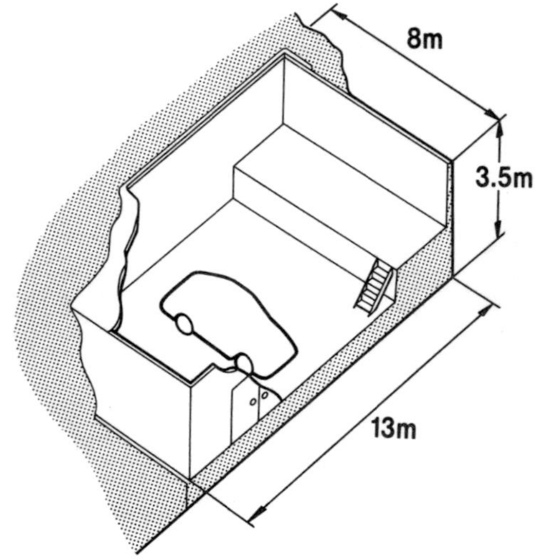

Fig. 1 Laboratory in rock cave

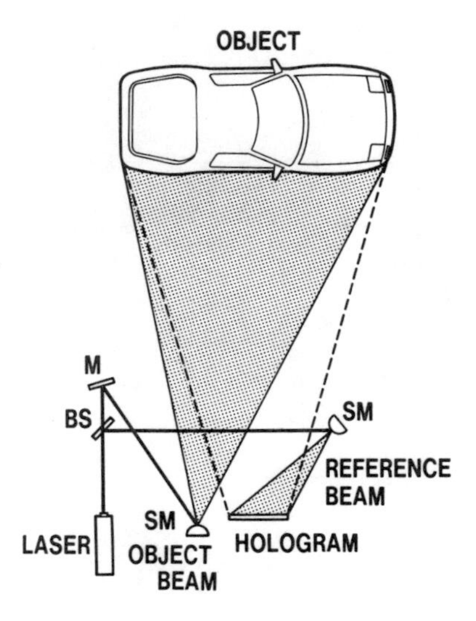

Fig. 2 Experimental setup for vibration analysis using CW laser holographic interferometry

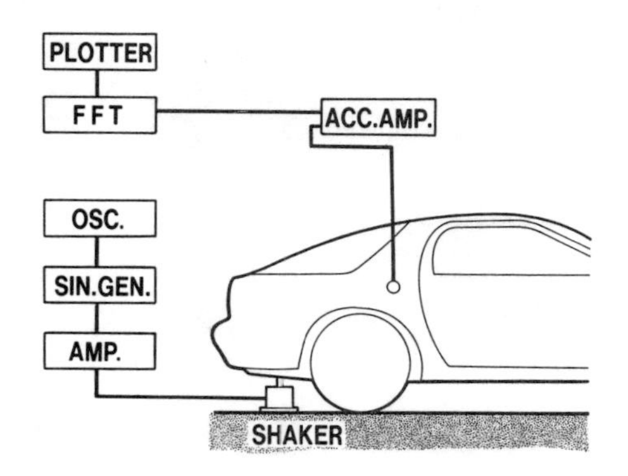

Fig. 3 Experimental setup for exciting automotive body

Fig. 3 shows the exciting equipment. Sine wave signals produced by the function generator are sent through the amplifier to the electromagnetic shaker to excite the vehicle body. The vibration is monitored by a fast Fourier transform analyzer. The vibrations of the vertical panels, such as the rear quarter panels, doors, and fenders, can be measured directly, while the measurements on horizontal components, such as the roof and trunk lid, have to be taken using a reflection mirror. (Fig. 4)

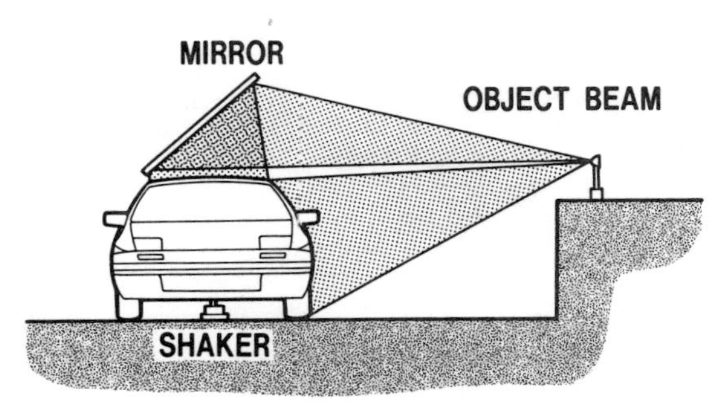

Fig. 4 Experimental setup for vibration analysis using CW laser holographic interferometry

Results

Examples of interferogram with a Mazda RX-7 are shown in Fig. 5. The figures show vibration modes of the quarter panel at frequencies of 187 and 235 Hz. Note that the positions of the loops and nodes of the vibrations can be readily judged, making it easy to take actions to reduce the vibrations. For purposes of reducing vibration amplitude, used were ribs, beads, damping materials, etc. to change the spring constant, mass, and damping of a vibrating component.

Fig. 6 shows the results of modal analysis by an FFT analyzer, which are roughly the same as with holographic interferometry. In modal analysis, the number of measuring points is restricted by the computer's storage capacity, and this imposes great limitations on the magnitude of modes that can be analyzed. By contrast, holographic interferometry is considered capable of handling a virtually infinite magnitude of modes.

Fig. 7 shows a reference beam phase modulation system used in this program and Fig. 8 gives the results of measurement. A small mirror was fixed on the vehicle body and the light reflected from that mirror was used as a reference beam. Compared with Fig. 5, the phase of vibration is easily identified.

As opposed to vertical surfaces such as vehicle sides, horizontal faces like the roof and hood are difficult to measure because of opticals layout, so that a large reflective mirror of 1.4 m by 0.9 m was used. It was necessary in this case to keep the mirror and its frame in a highly stationary condition. Fig. 9 shows measurements on the roof and windshield. Both results indicate well-defined interference fringes.

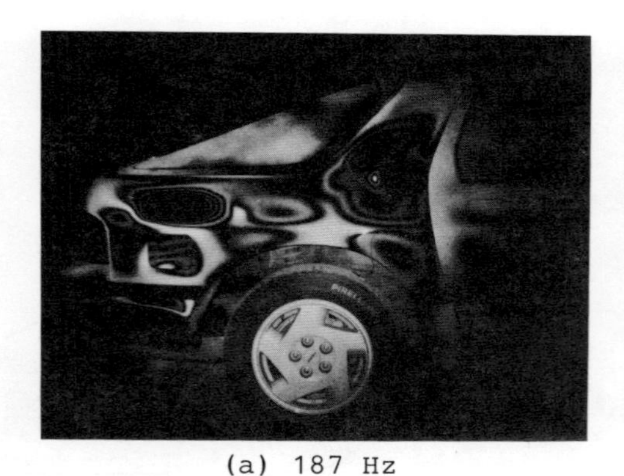

(a) 187 Hz

(b) 235 Hz

Fig. 5 Vibration modes of automotive body

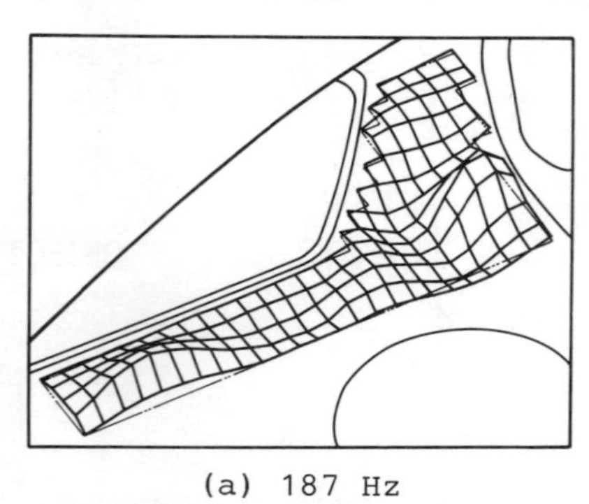

(a) 187 Hz

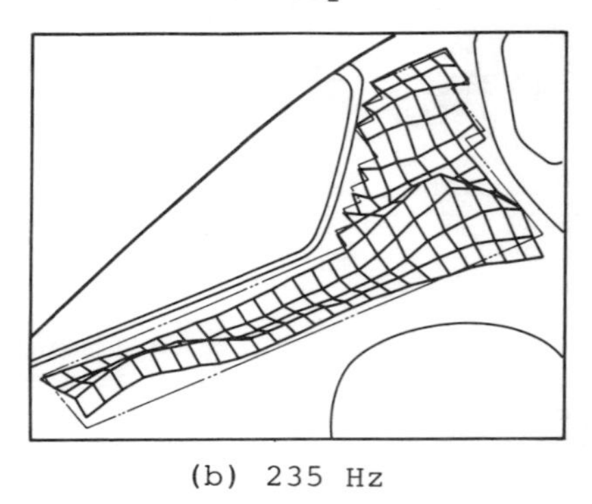

(b) 235 Hz

Fig. 6 Results of modal analysis

Conclusion

Employing a tuff rock cave as the vibration isolation system made possible the measurement of the vibration modes of a fully equipped vehicle body by CW laser holographic interferometry. As a result, development of quieter and more comfortable vehicles can now be carried out in a more efficient manner.

This shows that the present method is not only applicable to vibration measurement but also has potential for measuring static and thermal deformations of larger structures.

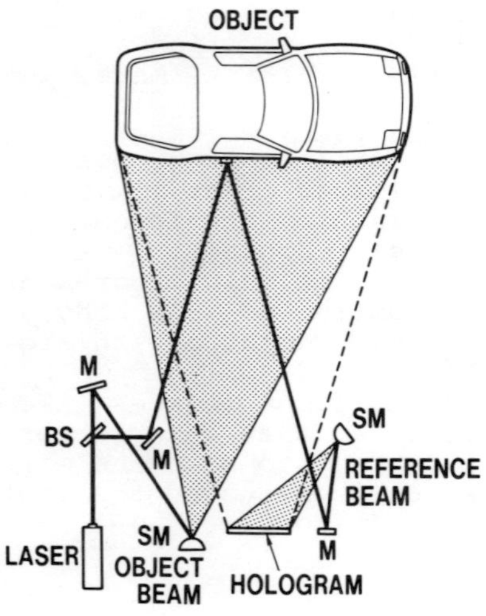

Fig. 7 Experimental setup for reference beam phase modulation method

Fig. 8 Vibration mode using reference beam phase modulation method (235 Hz)

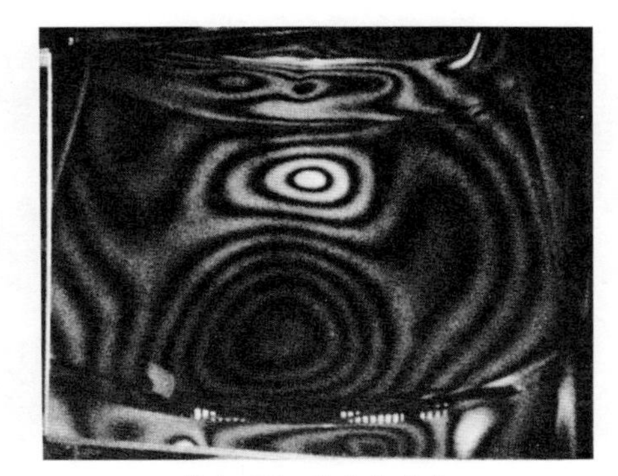

(a) Roof (b) Windshield

Fig. 9 Vibration modes of roof and windshield

Acknowledgements

The authors wish to express their gratitude to Mr. Masane Suzuki and Mr. Takayuki Saito of Fuji Photo Optical Co., Ltd., Mr. Kohei Watanabe and Mr. Akio Hirosawa of Holotec Inc., and Mr. Tsutomu Morie of Shimizu Construction Co., Ltd. who promoted the construction of the laboratory inside the rock cave.

References

(1) A. Felske, A. Happe: Vibration analysis by double pulsed laser holography, SAE Paper No. 780333, 1978.

(2) M. Murata, M. Kuroda: Application of Pulsed-wave Laser Holography to Practical Vibration Study, Mitsubishi Juko Giho, Vol. 20, No. 5, 1983.

(3) G. M. Brown, R. R. Wales: Vibration analysis of automotive structure using holographic interferometry, Proceedings of SPIE, Industrial Applications of Laser Technology, Geneva, Switzerland, 1983.

(4) K. Watanabe, H. Hirosawa, M. Suzuki, T. Morie and H. Yamashita, Large Holography Studio in a Huge Rock Cave, Proceedings of SPIE, International Conference on Holography Applications, Beijing, China, 1986.

Real time grating shearing interferometry applied to investigating of evaporative convection in liquid drops

Xu Youren

East China Technical University of Water Resources, Nanjing, China

Zhang Nengli, B. X. Wang

Tsinghua University, Beijing, China

Abstract

Two holographic gratings are used as beam splitting elements to form a special shearing interferometer through which time changing profiles of liquid drops evaporating on transparent flat plates are visualized and measured in real time. Binary liquids with various mixture ratios have been tested. Characteristics of thermal stability in binary droplet vaporization are revealed from the sequential changes of interference patterns. An interfacial flow map for binary drops is constructed that defines the domains of stable and unstable type evaporations.

Introduction

The physical process of liquid droplet vaporization on solid surfaces is of great importance in various areas of science and technology. Since there are several forces interacting in an evaporating droplet, considerable attention has been attracted to the interfacial instability phenomena in liquid drops.[1,2,3]

Several optical methods have been employed by the authors to investigate the droplet morphology during the process of evaporation. The dynamic behavior of minute liquid drops evaporating on flat plates were investigated cinematographically by using methods of direct photography and laser shadowgraphy.[4] A quantitative technique based on holographic and shearing interferometry was then developed and used in studying of the detailed profile of liquid droplets.[5] For investigating the complicated physical phenomena in binary droplet vaporization, real time shearing interferometry has been conducted in visualizing and measuring the time history of evaporating droplet profiles.[6]

In this article, we report the recent progress in the project. A special designed shearing interferometer using two holographic diffraction gratings as the beam splitting elements is constructed which makes an improvement upon the procedure used in work[6]. An analysis to the function of the shearing elements is carried out and new experimental results are presented.

Holographic gratings used as shearing elements

In paper[6], optical arrangement shown in Fig. 1 was adopted in the real time shearing interferometry. After expanded and collimated a He-Ne laser beam was divided by a beam splitter into two parts, which were reflected by several mirrors and symmetrically passed through the droplet with an angle 2θ between them. The two beams interfered with each other and formed a set of interference fringes at the emulsion plane of a holographic plate, which had been exposed to the same two beams, developed, and replaced exactly as it was before the droplet was placed on the test plate. There were two sets of interference fringes, with and without the distortion caused by the liquid droplet, forming a moiré fringe pattern at the emulsion plane, which contained necessary information for calculating the droplet profile and showed the profile changes in real time.

This optical arrangement has the same disadvantages as commonly seen in two beam interferometers with shallow intersection angle, in which the interference fringe pattern is very sensitive to minute environmental perturbation. Any mechanical disturbances may cause the beam splitter and mirrors to vibrate relative to each other, which produces random fluctuations in the intersection angle 2θ. Air currents in the passage of the two beams may also affect the optical path length difference. When the intersection angle 2θ becomes very small, the moiré fringe pattern will suffer from a strong jitter and distortion due to these disturbances.

To avoid the problem mentioned above, we adopt a compact optical arrangement similar to that proposed by G. Laufer in paper[7]. Two holographic diffraction gratings are used to form an integrated beam splitter, which split the incident laser beam into two beams with an intersection angle 2θ, and immediately intersect them. The perturbation from mechanical vibrations and air currents affects almost equally to the two laser beams, therefore compensating the fluctuations in angle 2θ and eliminating jitters and distortions in the moiré fringe pattern. Furthermore, in this compact optical system, the intersection angle 2θ can be adjusted continuously without affecting the optical arrangement, and it is possible to get stable moiré interference fringes while the intersection angle 2θ becomes very small.

The integrated beam splitter is built of two holographic diffraction gratings G_1 and G_2. Both of the gratings have the same spatial frequency f=1000 1/mm and are arranged in tendem in the xy plane as shown in Fig. 2. The gratings are rotated around the normal to the xy plane through a small angle $\pm\delta$ respectively. The incident collimated laser beam is lain in the xz plane with an incident angle ϕ. For monochromatic illumination the incident light beam can be described by its complex amplitude

$$u_0 = \exp\left\{j(2\pi/\lambda)(x \sin\phi + z \cos\phi)\right\} \tag{1}$$

where we have droped the constant amplitude for simplicity, λ is the wave length of laser light. The two holographic gratings in the xy plane can be represented by their amplitude transmittance functions

$$T_{1,2} = 1/2 + (m/2)\cos 2\pi f\ (x \cos\delta \pm y \sin\delta) \tag{2}$$

where we have assumed that the holographic grating can be represented by a sinusoidal amplitude grating with parameter m representing the peak-to-peak change of amplitude transmittance across the grating. T_1 and T_2 correspond to grating G_1 and G_2 respectively.

According to Eq.(1) and Eq.(2), light beams transmitted through the gratings can be derived explicitly, in which terms corresponding to the first term in the right side of Eq.(2) represent zero order transmitted beams, whereas terms corresponding to the second term in the right side of Eq.(2) represent first order diffraction beams. Thus when drop the constant factor, the complex amplitude of first order diffraction beams can be expressed as

$$u_{1,2} = \exp\left\{j(2\pi/\lambda)[x(\sin\phi - \lambda f \cos\delta) \mp y\ (\lambda f \sin\delta) + z \cos\phi]\right\} \tag{3}$$

where f denotes the spatial frequency of the gratings. We impose the following condition on the incident angle ϕ

$$\sin\phi = \lambda f/2 \tag{4}$$

thus the direction cosines of these two first order diffraction beams are

$$\begin{aligned} \cos\alpha_{1,2} &= \sin\phi(1-2\cos\delta) \\ \cos\beta_{1,2} &= \mp 2 \sin\phi \sin\delta \\ \cos\gamma_{1,2} &= \cos\phi \end{aligned} \tag{5}$$

It is easy to derive the intersection angle 2θ between beams u_1 and u_2 by use of the following relation

$$\cos(2\theta) = \cos\alpha_1 \cos\alpha_2 + \cos\beta_1 \cos\beta_2 + \cos\gamma_1 \cos\gamma_2 \tag{6}$$

When the angle 2δ is very small we have

$$2\theta = 2\sqrt{3}\delta \sin\phi \tag{7}$$

Therefore, two diffraction gratings arranged as in Fig.2 can be used to split a laser beam into two beams with an intersection angle 2θ, which can be used as a substitute for the beam splitter and mirrors in Fig.1. From Eq.(4) and (7) we have

$$2\theta = \sqrt{3}\ \lambda f \delta \tag{8}$$

For He-Ne laser $\lambda = 0.6328\mu m$, spatial frequency of the grating $f = 1\ 1/\mu m$, then

$$2\theta = 1.096\,\delta \tag{9}$$

which shows that the intersection angle 2θ is of the same order of magnitude as angle δ. It follows that the angle 2θ can be conveniently adjusted by changing the angle 2δ, in order to meet different sensitivity requirements on the shearing interferometry for a certain reason.

Physically, the two diffraction beams u_1 and u_2 can be understood as following. When a collimated laser beam u_0 passes through the holographic gratings under the condition of

Eq.(4), it is split into several modes. Light beams u_1 and u_2 are among those beams. Beam u_1 is the one transmitted through the first grating G_1 and then diffracted by the second grating G_2 as its first order diffraction beam. Beam u_2 is the one diffracted by the first grating G_1 as its first order diffraction mode and then transmitted through the second grating G_2 as its zero order mode. When the rotated angle 2δ is very small, the incident beam u_0 can still be thought in matching with both gratings G_1 and G_2. The first order diffraction beams are lain in the planes perpendicular to their grating lines respectively and their diffraction angles are equal to 2Φ. The relationship among beams u_0 , u_1 and u_2 is illustrated as in Fig. 2.

Experiment apparatus and procedure

Fig. 3 illustrates the experiment set-up of the real time grating shearing interferometry. A type HN-3, 50 mw, helium-neon laser is used as the light source. The laser beam is spatially filtered, expanded, collimated and reflected by a mirror with an incident angle Φ impinging onto the holographic gratings G_1 and G_2. The orientation of the gratings is adjusted so that the two first order diffraction beams u_1 and u_2 illuminate the test substrate T symmetrically. The substrate T is a high quality glass plate, which is put on the horizontal emulsion plane H of a high resolution photographic plate (tianjin type-1). The rotated angle 2δ between the two gratings is chosen to produce an appropriate intersection angle 2θ between the two diffraction beams. Before the droplet is added, the photographic plate is exposed to the light beams u_1 and u_2, and removed for developing and fixing, which is then returned carefully to its oringinal position. A uniform field should be formed under the illumination of u_1 and u_2, if the developed plate is replaced properly.

The droplet being tested is placed on the glass substrate by means of a 50 μl microsyringe. The needle tip of the microsyringe is slightly touching the substrate surface before and during the injection to permit the formation of a calm unsplashing test droplet. The substrate plate can be moved and replaced during the experiment with no affection on the test results. The substrate plate must be cleaned and left overnight before use for the purpose of maintaining consistency and reproduction of the test results.

When a droplet is placed on the substrate surface, the two beams u_1 and u_2 pass through the droplet and the transparent substrate impinging on two regions AA' and BB' in the H plane respectively as shown in Fig. 4. In regions AB and A'B', wave fronts distorted due to passing through the droplet interact with wave fronts reconstructed from the developed photographic plate forming real time interferometry. In region BA' two wave fronts coming from $\pm\theta$ directions interfere with each other producing a set of shearing interference fringes. The spatial frequency of these fringes is too high to be resolved, but they interact with the fringes which had been recorded in the photographic plate before the droplet was placed on the substrate forming a visible moiré fringe pattern in H plane. While a droplet is evaporating on the substrate, a time changing moiré fringe pattern can be seen in the H plane, which can be recorded by a video recorder or a movie camera for later analysis.

A technique was presented in paper[8] for calculating the volume-time history of the evaporating droplet. From the recorded moiré fringes, the calculations of drop profiles at any moment can be made by an iterative ray tracing procedure using the following equations:

$$Y_N^2(n_1-n\sin\theta\sin\beta_2)/\cos\beta_2 - Y_N^1(n_1 - n\sin\theta\sin\beta_1)/\cos\beta_1 +$$

$$+d((n_2 - n\sin\theta\sin\beta_2')/\cos\beta_2' - (n_2 - n\sin\theta\sin\beta_1')/\cos\beta_1') - n(Y_N^2 - Y_N^1)\cos\theta = N\lambda \tag{10}$$

$$x_N^1 = x_N - d\,tg\beta_1' - Y_N^1\,tg\beta_1 \tag{11}$$

$$x_N^2 = x_N + d\,tg\beta_2' + Y_N^2\,tg\beta_2 \tag{12}$$

$$\beta_1 = \sin^{-1}(n\sin(\theta - Y_N^{1\prime})/n_1) - Y_N^{1\prime} \tag{13}$$

$$\beta_1' = \sin^{-1}(n_1\sin\beta_1/n_2) \tag{14}$$

$$\beta_2 = \sin^{-1}(n\sin(\theta + Y_N^{2\prime})/n_1) - Y_N^{2\prime} \tag{15}$$

$$\beta_2' = \sin^{-1}(n_1\sin\beta_2/n_2) \tag{16}$$

where Y_N^2 and Y_N^1 are the height of the drop surface at both sides of the N-th order fringe respectively; x_N^2 and x_N^1 are the x-coordinates corresponding to Y_N^2 and Y_N^1 respectively; x_N is the position of the N-th fringe; d is thickness of the glass plate; n, n_1, n_2 are the refractive indexes of air, liquid drop and the substrate respectively; and $Y_N^{1}{}'$ and $Y_N^{2}{}'$ are the slopes of the drop surface at positions of x_N^1 and x_N^2 respectively. In the calculations the refractive index n_1 should be considered as variable with the compositions of the drop. Here a first approximation was taken: the more volatile component in the binary drop would reduce linearly and then its refractive index would change linearly during the evaporation process.

Experiment results and discussion

Many analytically pure organic liquids and a lot of binary liquids which consisted of pure liquids with various mixture ratios have been studied. The recorded moiré fringes and their changes for different drops showed that there existed two different evaporation modes: "stable" and "unstable" type evaporations. For stable type evaporation, the moiré fringes exhibited a perfect symmetrical pattern and kept a similar pattern from beginning to end. In Fig. 5, we present two typical examples of this type evaporation. Fig. 5(a) indicate three stages in the evaporation process of an analytically pure cyclohexane droplet. Three photos in Fig. 5(b) show the sequential changes of moiré fringe pattern for a binary droplet. The test sample was a mixture of 50% volume methylene chloride and 50% volume carbon tetrachloride. The drops were shown to have a spherical segment with smooth surface and retained it during the entire evaporation process, only their size changed according to time. All the drops of pure liquids and many of the binary drops possess this type of evaporation features which was classified as "stable-interface type evaporation". On the other hand, there were many binary drops exhibiting interfacial turbulent evaporation processes. Fig. 6 shows an example. It was a binary drop of 50% volume chloroform and 50% volume ethyl ether. The moiré fringes changed from a symmetrical pattern to an irregular one and then finally recovered. The sequential changes of moiré fringes indicated that the droplet surface was varying from smooth to rippled and then restored to smooth. This type of evaporation was then called "unstable-interface type evaporation".

Forty-one binary liquids have been tested, in which twenty-four of them exhibited features of stable-interface type evaporation and seventeen showed typical evaporation processes of unstable-interface type. By correlating the evaporation modes with their physico-chemical characteristics we have found that the dimensionless excessive free-energy of surface in binary liquids

$$N_E = F_e/\sigma_m \quad (17)$$

and the crispation number of the binary droplets

$$N_R = \mu_m \alpha_m/(\sigma_m d_e) \quad (18)$$

are two important parameters which affect the interfacial instability for binary drops. F_e represents the excessive free-energy of surface in binary liquids due to the mixture

$$F_e = (\Delta\sigma)^2 x_a x_b \bar{A}/(2RT) \quad (19)$$

where $\Delta\sigma$ expresses the surface tension difference of the pure liquid components in the binary liquids; x_a and x_b, the mole fractions of components a and b respectively; and $\bar{A}$, the averaged mole surface area; R, the universal gas constant; T, the absolute temperature; μ_m, α_m and σ_m are the dynamic viscosity, thermal diffusivity and surface tension of the binary liquid; and d_e, the equivalent diameter of the droplet on the test substrate at zero time.

Based on the experiment results, an interfacial flow map for binary drops is constructed, in which drops are classified according to their N_E and N_R numbers. Sign "x" indicates the unstable type droplet and sign "o" indicates the stable type droplet. It is found that all the unstable type evaporation modes are located in the regime with high N_E and N_R numbers, and the stable type corresponding to low N_E and N_R. Between these two regimes a line describing quasi-stable mode can be revealed. The experiment results are shown in Fig. 7.

Fig. 8 illustrates typical calculated results of droplet profiles at several moments. By using the profiles at sequential moments the volume-time history can be made very easily. It is observed that the evaporation rate and droplet life-time are strongly affected by the evaporation modes. The evaporation rate for unstable type is larger than the stable type. It is worth noting that droplet evaporation mode of some binary liquids may change from one to another when the N_E and N_R numbers go across the quasi-stable line in the flow map due to the concentration effect, which may cause the evaporation rate to change rapidly. Fig. 9 is an experiment result of concentration effect on binary drop lifetime. The test liquid is a mixture of ethyl ether and chloroform. The result is plotted as the drop lifetime versus % volume ethyl ether, which shows that the lifetime sharply deviates from the linear change and crosses the linear line at the concentration of about 25%. This coincides with the fact that the droplet exhibits stable evaporation mode when the concentration of ethyl ether is less than 20% and changes to unstable mode when the concentration is larger than 25%, but droplet of pure ethyl ether possesses typical stable type evaporation features, whose lifetime is larger than the binary drops with concentration in the regime from 65% to 95%, which may be useful for the purpose of enhancing the droplet vaporization.

Conclusions

Recent progress in the study of evaporative convection in liquid drops by means of a holographic technique is reported. Two holographic gratings are used to construct a special beam splitting element which makes it possible to simplify the shearing interferometer, increase the stability of measurement, eliminate the jitter problem in the interference pattern, and change the sensitivity of the shearing interferometry by simply adjusting the intersection angle between the two gratings. The technique has been used in investigating the droplet evaporation process, which discloses two evaporation modes existing in binary drops, and two important parameters N_E and N_R which are closely related to the interfacial flow instability in evaporating drops. The experiment result gives an inspiration that the interfacial flow instability may be useful in enhancing the droplet evaporation rate.

Acknowledgment

This research work is supported by a grant from the Science Fund of the Chinese Academy of Sciences, Beijing (Grant No.TS 840220).

References

1. Yang, W. J., "Theory on vaporization and combustion of liquid drops of pure substances and binary mixtures on heated surfaces", ISAS Report No.535, Institute of Space and Aeronatical Science, University of Tokyo, Japan, Vol. 40, No. 15, 1975
2. Zhang, N. and Yang, W. J., "Evaporation and explosion of liquid drops on a heated surface", Experiments in Fluids, Vol. 1, pp. 101-111, 1983
3. Mann, R. F. and Walker, W. W., "The vaporization of small binary drops on a flat plate at maximum heat flux", The Canadian Journal of Chem. Eng., Vol. 53, pp. 487-493, 1975
4. Zhang, N. and Yang, W. J., "Natural convection in evaporating minute drops", J. Heat Transfer, Vol. 104, pp. 656-662, 1982
5. Xu, Y., Zhang, N., Yang, W. J. and Vest, C. M., "Optical measurement of profile and contact angle of liquids on transparent substrates", Experiments in Fluids, Vol. 2, pp. 142-144, 1984
6. Xu, Y., Zhang, N., "Recording evaporation time history of droplet on flat plate by laser real-time interferometry", International Symposium on Heat Transfer, Beijing, China (in print), 1985
7. Laufer, G., "Instrument for velocity and size measurement of large particles", Appl. Opt., Vol. 23, No. 8, pp. 1284-1288, 1984
8. Xu, Y., and Zhang, N., "An optical method for measuring surface shape of a liquid-drop on base piece", Mechanics and Practice, Vol. 7, No. 6, pp. 27-29, 1985
9. Zhang, N., Xu, Y. and Yang, W. J., "Thermal stability in binary droplet vaporization on a flat plate by real-time holographic interferometry", (To be apear in Proceedings of Eighth International Heat Transfer Conference, San Francisco, U.S.A., 1986)
10. Tong, J. S. and Li, J., The Calculations of Thermal Properties of Fluids, Tsinghua University Press, Beijing, China, 1982 (in chinese)

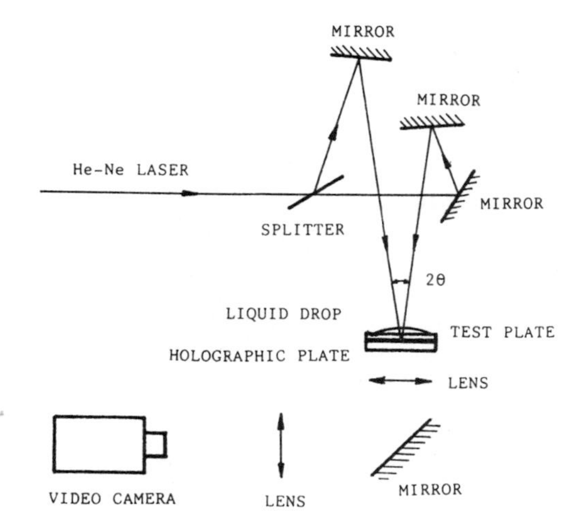

Fig. 1. Real-time shearing interferometry

Fig. 2. Beam splitter by two holographic gratings

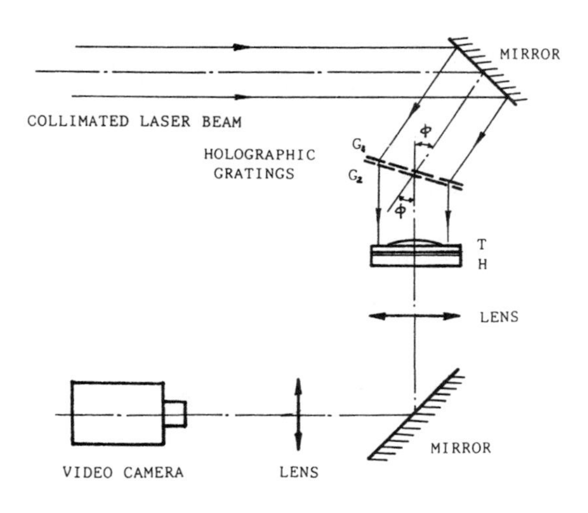

Fig. 3. Scheme of experimental set-up

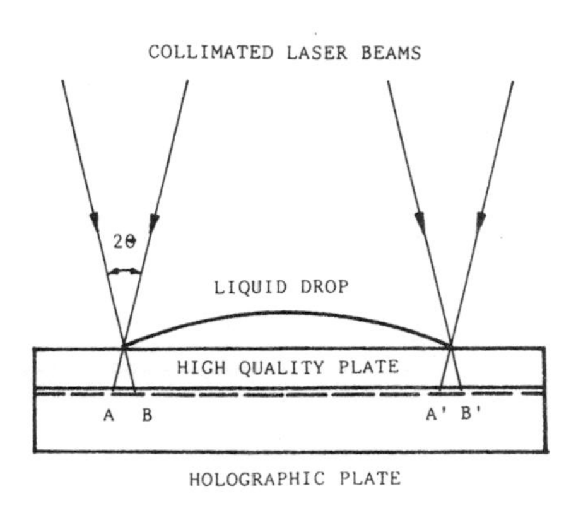

Fig. 4. Schematic diagram of test appratus

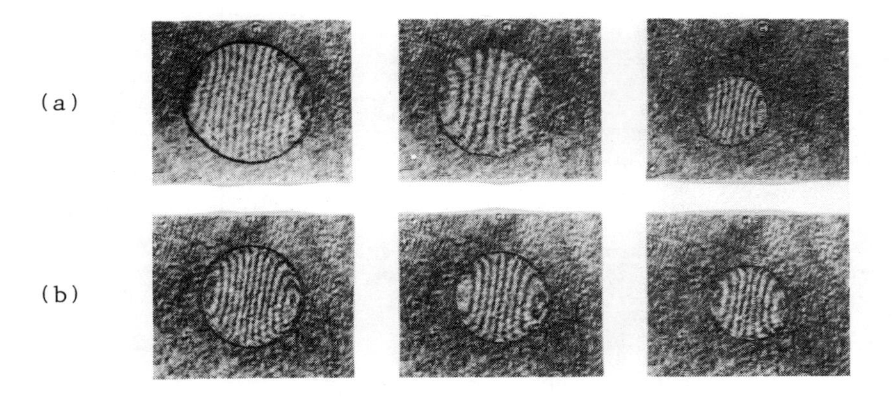

Fig. 5. Stable-interface type evaporation. (a) pure drop. (b) binary drop

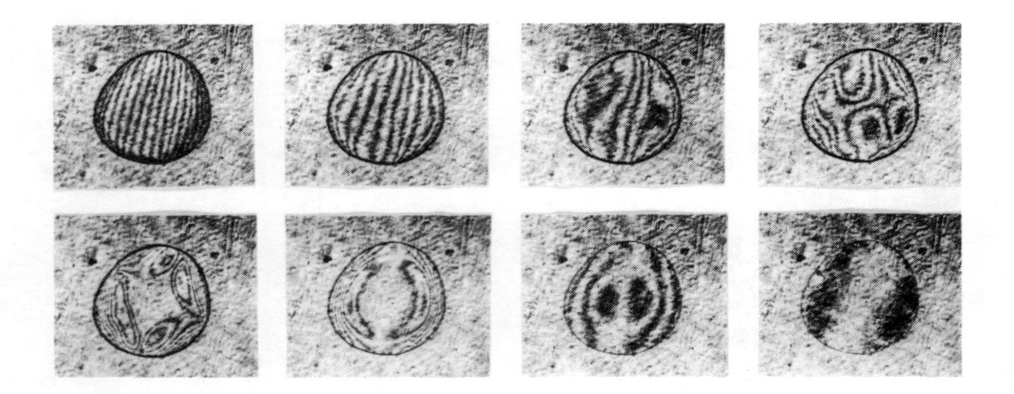

Fig. 6. Unstable-interface type evaporation

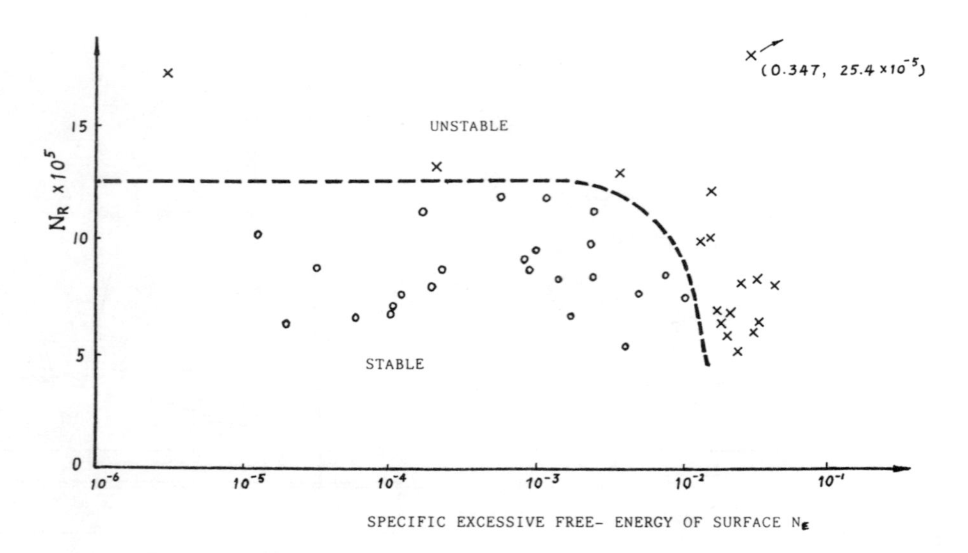

Fig. 7. Interfacial flow map of binary drops

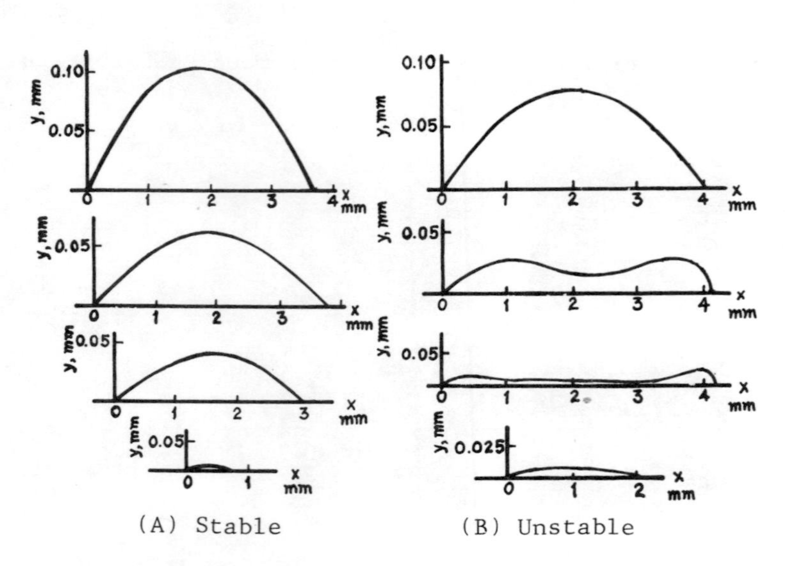

Fig. 8. Typical changes of binary drop profiles

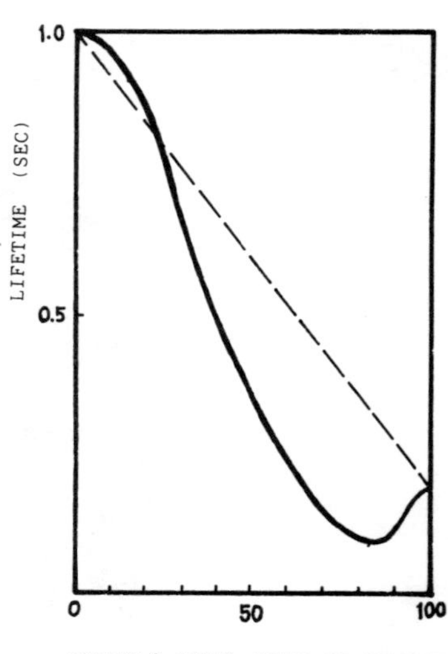

Fig. 9. Concentration effect on binary drop lifetime

A holographic testing of gear tooth surface

D. Y. Yu*, T. Kondo, N. Ohyama, T. Honda and J. Tsujiuchi

Imaging Science and Engineering Laboratory, Tokyo Institute of Technology
Nagatsuta, Midori-ku, Yokohama 227, Japan

*On leave of absence from Department of Precision Instrument, Tianjin University, Tianjin, China.

Abstract

Reflected light with a grazing angle from a gear tooth surface is put into a hologram recording a standard surface. Deviation of reconstructed wavefront from the standard one gives a surface error, and is detected by a simple photodetector with a small pinhole and by the fringe scanning technique.

Introduction

For the inspection of surface shape of precisely made machine parts, i.e. a tooth surface of gear, the stylus scanning method is usually used. This method, however, takes a long time in measuring one-surface and sometimes gives scratches on the surface measured. The mass production of precision machine elements requires a new method free from these difficulties. This paper proposes a holographic method in which the non-contact measurement of gear tooth surface is made by illuminating the surface with a grazing angle of incidence and by measuring the reflected light with the fringe scanning interferometry. Experiments are performed for a gear with shaving tooth surface.

Principles of experiment

1. Principles of experiment with photo-detector

Principles of the present experiment are shown schematically in Figure 1. A hologram of an object beam reflected from a standard tooth surface of a gear is made with the aid of a collimated reference beam. In order to decrease the influence of the roughness of the surface finish a grazing angle of incidence is required. To avoid the repositionning error of the hologram after development, it is recommendable to record the hologram in a thermoplastic photoconductor material whose development can be done in situ with a high diffraction efficiency. Then, a tooth surface to be measured is set in position, light reflected from the surface is put into the above mentioned hologram, and the reconstructed beam from the hologram become the subject to measure. For that purpose, a lens is placed behind the hologram, and the light intensity through a pinhole located at the focus of the lens is measured by a photo-detector. If the tooth surface has a shape error with regard to the standard surface, the light intensity through the pinhole will decrease, and the amount of decrease gives a shape error of the surface. Figure 2 shows results of four tooth surfaces

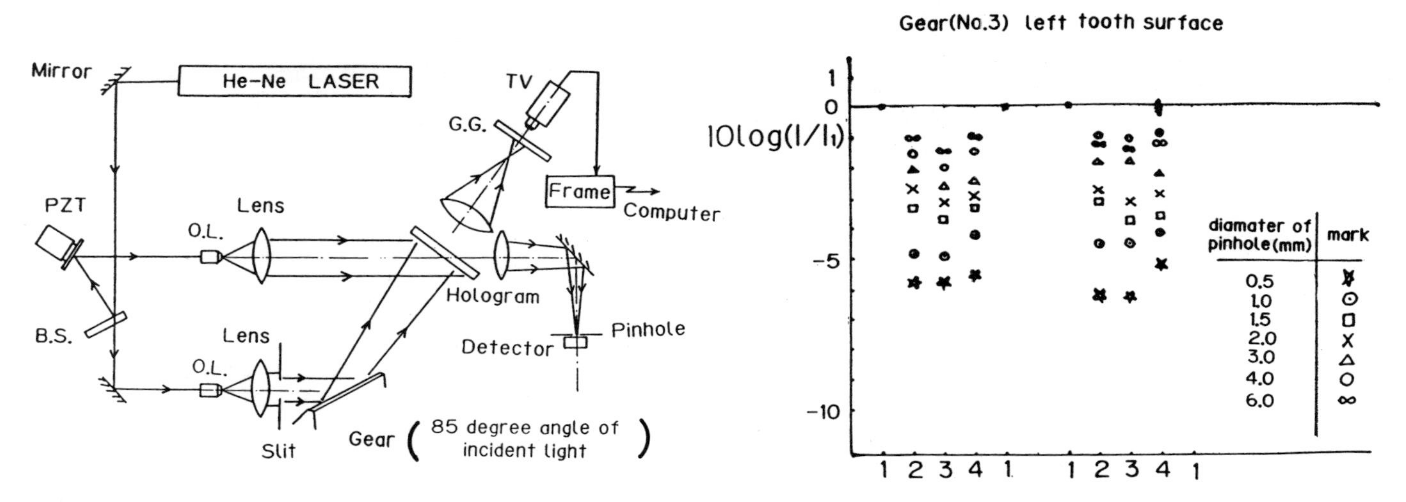

Figure 1. Optical system of experiment.

Figure 2. Measured data by photodetector.

of a gear, and different marks indicate data measured with pinholes of different diameters. Figure 3 shows the corresponding results given by the stylus method on comparing two results, it can be concluded that the result obtained by a small pinhole well correspond to the shape error of the tooth surface. But, it is not yet clear where the error takes place in the surface, and such a difficulty can be relieved by adopting the fringe scanning interferometry.

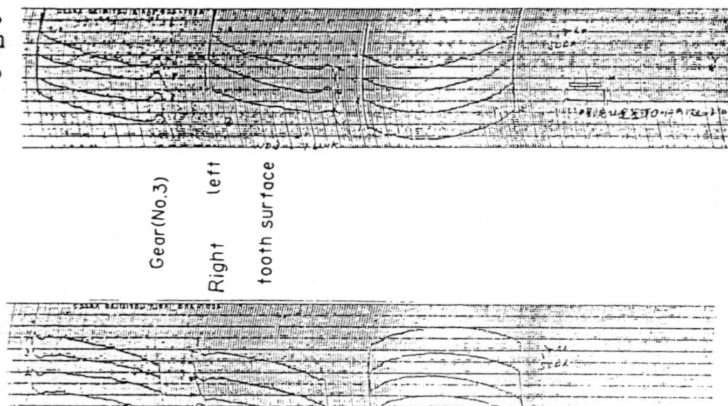

Figure 3. Measured results by stylus method.

2. Fringe scanning interferometry

An optical system for fringe scanning interferometry is also shown in Figure 1. A light reflected from the standard surface is reconstructed from the hologram and is interfered with a beam reflected from the surface to be measured. If there is an error in the surface, the resultant interference fringes will give the informations. Due to the large incident angle of object light, sensitivity of surface error measurement becomes low. For increasing the sensitivity, the fringe scanning method is used. A plane mirror attached to PZT piezo-electric element is inserted into the path of reference beam, and the optical path of reference beam can be changed with equal step by adding an equal voltage to the PZT. So, interference patterns are recorded in a frame memory for every change of optical path of equal interval, and sent to a computer. Thus, the shape errors distribution in the surface can be obtained at all sampling points on the surface. The principles are as follows:

Provided that the shape to be measured is h(x, y), the wavefront reflected from the surface can be written as

$$Wo(X,Y)=A\exp\left[i2k\ h(X,Y)\right] \quad (1)$$

and provided that the phase variance of reference beam is δ, the reference wavefront can be written as

$$Wr(X,Y,\delta)=B\exp(ik\delta) \quad (2)$$

Thus, the intensity of the interferogram is obtained as

$$I(X,Y,\delta)=\alpha+\beta\cos k\left[2h(X,Y)-\delta\right] \quad (3)$$

where $\alpha=A^2+B^2$ $\beta=2AB$

expanding formula (3) for δ by Fourier series, the following equation is obtained.

$$I(X,Y,\delta)=Ao/2+\sum Ar\cos rk\delta+\sum Br\sin rk\delta \quad (4)$$

If the phase $k\delta$ of reference beam is changed in one period of the fringe with P equal intervals, and interferograms are measured in P steps, by making use of the orthogonal properties of the trigonometric function, the following formula are given,

$$Ar=(2/P)\sum_{j=1}^{p} I_j \cos r\ _j$$

$$Br=(2/P)\sum_{j=1}^{p} I_j \sin rk\delta_j \quad (5)$$

$$(j=1,\ 2,\ \ldots\ldots\ P)$$

Taking into consideration of the first harmonics only, the following formula are obtained from (3) and (4),

$$A_1=\beta\cos 2kh(X,\ Y)$$

$$B_1=\beta\sin 2kh(X,\ Y) \quad (6)$$

then, we have

$$h(X, Y)=1/2K \text{ arc tan } (B_1/A_1)$$

$$=1/2K \text{ arc tan } \frac{\sum_{j=1}^{p} I_j \sin(2\pi j/P)}{\sum_{j=1}^{p} I_j \cos(2\pi j/P)} \tag{7}$$

i.e. the optical path of reference beam is changed by an equal interval of λ/P and P interferograms are measured. Then, in order to achieve precisely shape error of tooth surface, intensity distributions of interference patterns are measured in P steps, h(x, y) may be obtained from formula (7) at every sampling point.

Least-squares fitting

Figure 4 shows relations between intensity at a point in the interferogram and voltage added to the PZT element, and the result is not along an exact sine curve. The reason is the nonlinearity of the PZT element, the noise in the optical system and the roughness of tooth surface etc.. So, it is impossible to obtain an exact phase difference directly from these results. A least square method is used to get an optimum sine curve fitting from the experimental result. At the bottom of Figure 4, a sine curve thus obtained is shown, and the shape error of tooth surface can be calculated from the fitted data.

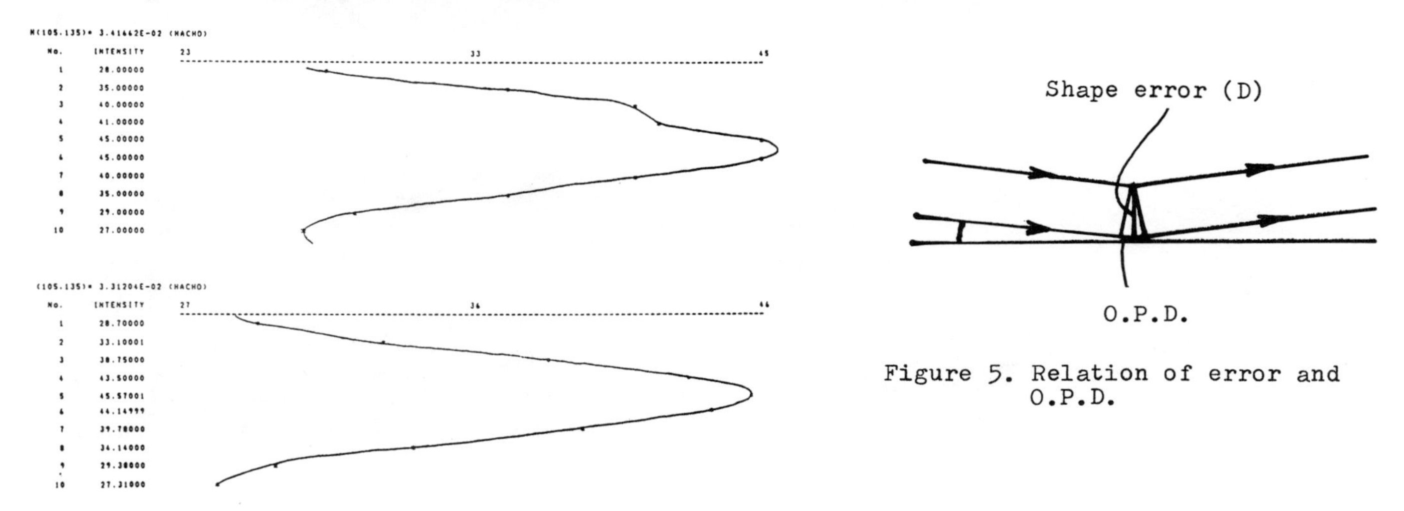

Figure 5. Relation of error and O.P.D.

Figure 4. Cosine curve of data.

Shape error experimental results

Figure 5 shows the relation between the shape error D and the optical path difference O.P. obtained from interferograms. We have then

$$O.P.=D\cos\theta+D\cos\theta$$

$$=2D\cos\theta$$

$$\frac{O.P.}{D}=2\cos\theta=\frac{\lambda_o}{\lambda_{eff}}$$

$$\lambda_{eff}=\frac{\lambda_o}{2\cos\theta}=\frac{o}{2\cos 85^{\circ}} \tag{8}$$

where λ_o is the wavelength of a He-Ne laser, 85° is the incident angle of light at the center point of tooth surface, and λ_{eff} is the effective wavelength for evaluating the shape error from the interferogram.

Figure 6,(a) is the standard surface, and a small tilt of the surface is found. (b), (c) and (d) are results of surfaces No. 2, 3 and 4 respectively. The error of No. 2 and 3 are large and that of No. 4 is small in total but has a fairly large roughness. These results

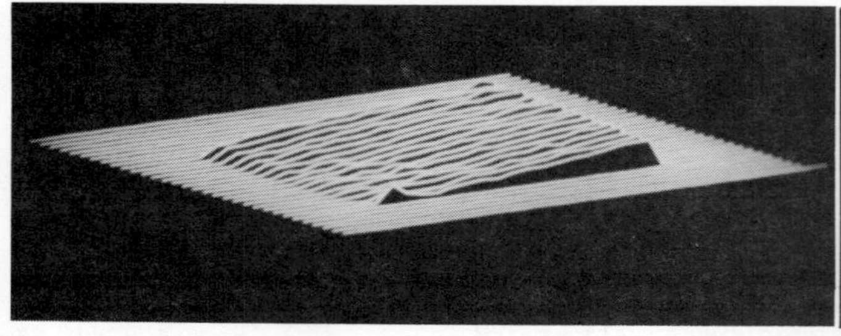

(a). Standard surface.

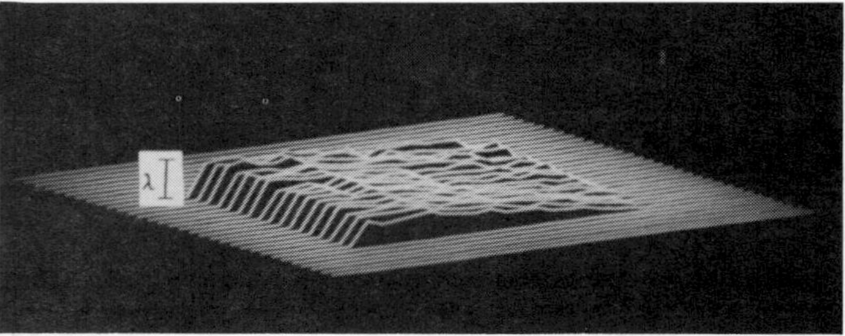

(b). Tooth surface of No. 2.

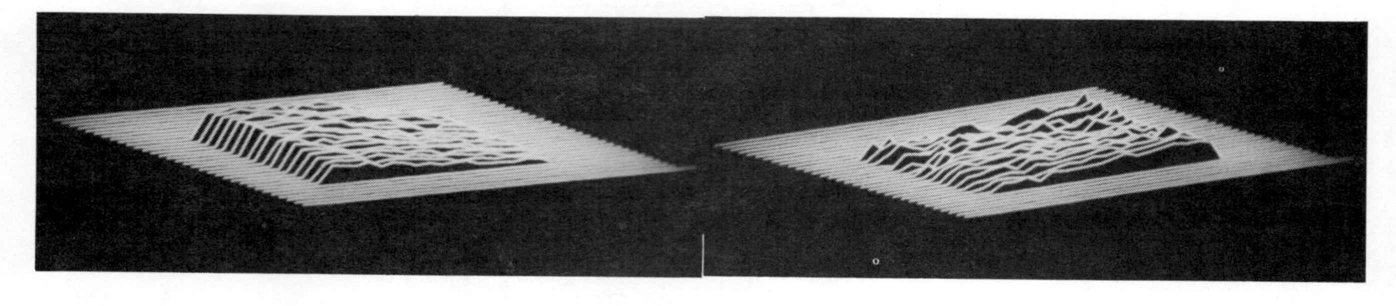

(c). Tooth surface of No. 3.

(d). Tooth surface of No. 4.

Figure 6. Shape error of gear tooth surface by fringe scanning method.

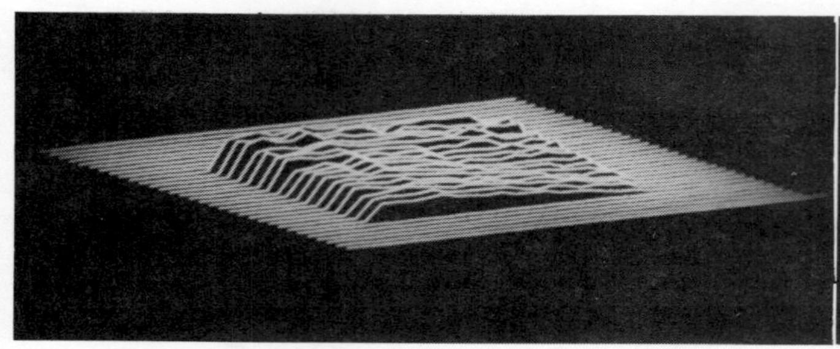

(a). Surface shape of No. 2.

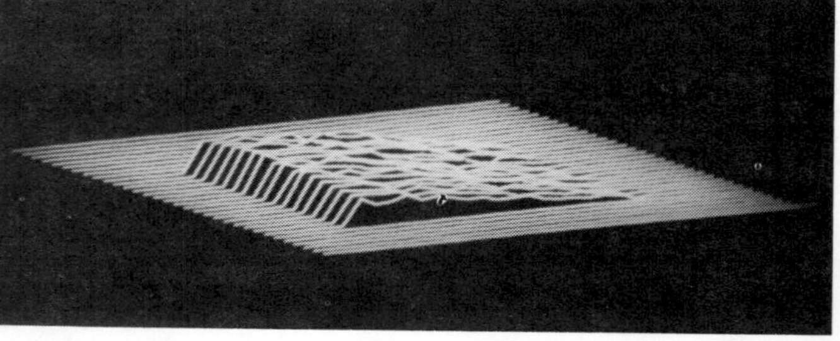

(b). Surface shape of No. 3.

Figure 7. Subtracted results from shape error of standard surface.

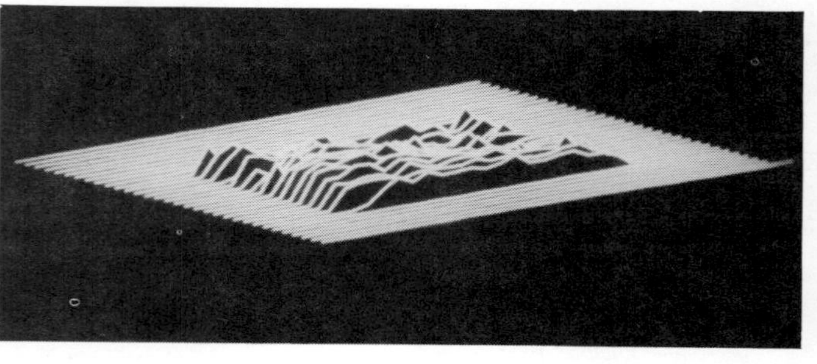

(c). Surface shape of No. 4.

correspond well to those of Figure 2, and this means that we can obtain the shape error of tooth surface from this experiment. As for the standard surface, no error should be caused in principle, but a small tilt like Figure 6 (a) appears, because a small lateral shift of the reference beam takes place while changing the optical path of reference beam, and this phenomenon is due to the fact that the reference beam enters the PZT mirror not perpendicularly. So, the same tilt error as the standard surface equal error is superposed in all the other results, and a correction of such an error should be made. Figure 7 gives the results of surfaces No. 2, 3 and 4 after the correction of this error.

Analysis of experimental results

Measured errors of tooth surfaces in the experiment are errors obtained by the optical beam reflected from the surfaces in question, and λ eff, the effective wavelength, is a calculated value at the center of the tooth surface.

But, the error calculated with λ_{eff} defined at the center of the surface is not correct in all points of surface due to the change of incident angle at different positions. It is recommendable to use the exact incident angle at every point where the surface error is measured. Since it is difficult to measure the exact incident angle at all points, the largest error is calculated in the surface. As shown in Figure 8, the largest and smallest incident angles are about 85°, 83° respectively, we have

$$\lambda_{eff} = \frac{\lambda_o}{2\cos} \tag{9}$$

$$\lambda_{prac} = \frac{\lambda_o}{2\cos\alpha'} \tag{10}$$

where α' is the practical incident angle at a point of tooth surface, and λ_{prac} is the effective wavelength at that point.

Therefore, we have

$$\lambda_{eff} - \lambda_{prac} = \frac{\lambda_o}{2\cos\alpha} - \frac{\lambda_o}{2\cos\alpha'}$$

$$= \frac{\lambda_o(\cos\alpha' - \cos\alpha)}{2\cos\alpha \, \cos\alpha'} \tag{11}$$

$$\frac{\lambda_{eff} - \lambda_{prac}}{\lambda_{eff}} = \frac{(\cos\alpha' - \cos\alpha)}{\cos\alpha'}$$

$$= 1 - \frac{\cos\alpha}{\cos\alpha'} = \frac{\Delta D}{D} \tag{12}$$

where ΔD is the fluctuation of D caused in the calculating procedure, D is the shape error evaluated by λ_{eff}. We may estimate from eq.(12) the accuracy of D at every point on the tooth surface, and especially, the maximum amount of fluctuation by estimating the largest and the smallest incident angles on the surface.

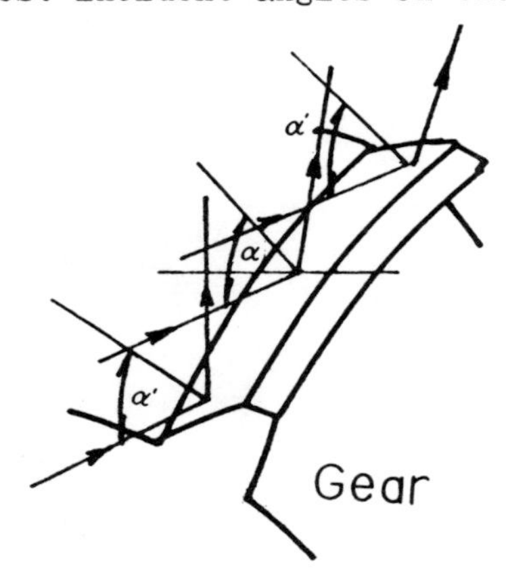

Figure 8. Angles of incidence in the tooth surface.

Conclusion

A method for measuring surface shape errors is developed and is applied to the measurement of gear tooth surface. Properties of this method are as follows:

1. A method for measuring output intensity through a pinhole at the focus of a lens behind the hologram is suited to the inspection in mass production.

2. The shape errors of tooth surface can be managed by using the real-time interferometry with fringe scanning method.

References

1. Bruning, J.H., Herriott, D.R., Gallagher, J.E., Rosenfeld, D.P., White, A.D. and

Brangaccio, D.J., Appl. Opt., 13, pp.2693. 1974.

2. Bruning, J.H., Optical Shop Testing, Malscara, D. ed., John Wyley and Sons 1978, pp.409.

3. Hariharan, P., Oreb, B.F. and Brown, N., Opt. Commu., 41, pp.393. 1982.

4. Ming Chang, Ching-Piao Hu, Philip Lam and Wyant, J.C., Appl. Opt., 24, pp.3780. 1985.

Optical Testing Using a Point Diffraction Holographic Interferometer

Zhou Wanzhi, Lu Zhenwu

Changchun Institute of Optics and Fine Mechanics
P.O.Box 1024, Changchun, Jilin, P.R. China

Abstract

A new interferometer, the Point Diffraction Holographic Interferometer (PDHI), is described in this paper, which covers its principle, structure, applications and error analysis. It is shown that this interferometer has some advantages over the conventional Point Diffraction Interferometer (PDI) such as easy alignment and adjustable visibility. It can be used for routine optical testing in optical shops or laboratories.

Introduction

Interferometry has been widely used in optical testing since the introduction of the laser in the early 1960s. Its main development are concentrated on two aspects, in which one is to develop new interference techniques and the other is to develop more suitable techniques for handling the data from interferogram. As far as interference techniques are concerned, there have been many kinds of interferometers besides the traditional Twyman-Green interferometer, such as lateral and radial shearing interferometers[1,2,3]. Like all common path interferometers, they have the advantages such as low sensitivity to vibration and temperature variations. However, they also have the drawback that the interferograms from lateral or radial shearing interferometers are not so convenient to interpret. Perhaps the simplest interferometer is the Point Diffraction Interferometer which was first discussed by Linnik and put into use by Smartt[4]. In fact, it is just a glass plate on which is coated a highly absorbing layer with a very small hole in the layer. The wavefront to be tested is focused at the hole so that there are two kinds wavefronts passing through the plate, of which one is a uniform spherical wavefront diffracted by the hole and the other is the test wavefront directly transmitted by the plate. The interferogram thus formed, like the Twyman interferogram, directly shows the contours of the tested wavefront. However, the PDI has the drawbacks that the alignment is difficult because it is not easy to find such a small hole by the naked eye, and also the visibility of the fringes is affected by the transmitance of the layer, the aberrations of the tested wavefront and the size of the hole. Therefore a number of plates are needed for choosing one in order to get optimum visibility.

These drawbacks can be avoided by the Point Diffraction Holographic Interferometer that we have developed recently. In this paper we will discuss the principle of the PDHI followed by an error analysis.

The principle of the PDHI

The fundamental element of the PDHI is a point diffraction hologram which should be made before testing. As shown in Fig.1, a plane wave from a He-Ne laser via a beam expander BE is divided into two waves, reference and object waves, at a beam splitter BS. The object wave is then reflected by a mirror M_1 towards the tested lens L. The emerging wave from L is focused at its focal point, where a pinhole is placed. The size of the pinhole is smaller than that of the Airy Disk of the tested lens so that a uniform spherical wave is formed due to its spatial filtering action. This wave passes through the lens L_1 and illuminates a holographic plate H together with the plane reference wave reflected by M_2. The role of L_1 is to convert the spherical wave into a plane wave so that a grating-type hologram is formed, thus reducing the sensitivity of its replacement, also the interferogram formed has the same dimensions within the interfering region. The pinhole is used to convert the aberrated wave into a spherical wave and thus the tested lens can be used to make a point diffraction hologram without the need of another aberration-free lens. After processing of the plate H it is replaced at the original position. In order to reduce the error of replacement, in situ processing is suggested.

Let O and R represent the object and reference waves respectively , then the transmitance t of the hologram after processing is expressed as

$$t \propto C + RO^* + R^*O \tag{1}$$

where C is a constant and * expresses the complex conjugate. A Point Diffraction Holographic Interferometer is formed after replacement of the holographic plate and taking away the pinhole, as shown in Fig.2. For testing lens L the hologram is illuminated by both the reference and unfiltered object waves. If the aberration of the unfiltered wave is expressed as W, then its complex amplitude can be written as $O\exp(i2\pi W/\lambda)$ and the amplitude distribution A behind H is written as

$$\begin{aligned} A &\propto (O\exp(i2\pi W/\lambda) + R)t \\ &= CR + (R)^2O^* + O + OC\exp(i2\pi W/\lambda) \\ &\quad + R\exp(i2\pi W/\lambda) + (O)^2R^*\exp(i2\pi W/\lambda) \end{aligned} \tag{2}$$

Thus the amplitude distributions A_o and A_r in both object and reference beams are expressed as

$$\begin{aligned} A_o &\propto O + OC\exp(i2\pi W/\lambda) \\ A_r &\propto CR + R\exp(i2\pi W/\lambda) \end{aligned} \tag{3}$$

And the corresponding intensities are

$$\begin{aligned} I_o &\propto 1 + \mu\cos(2\pi W/\lambda) \\ I_r &\propto 1 + \mu\cos(2\pi W/\lambda) \end{aligned} \tag{4}$$

where μ is a constant which depends on the intensity ratio of the reconstructed waves to the original waves. Unit visibility is obtained when μ is equal to one by suitable filtering the original waves. It is evident from Eq.(4) that two interferograms are formed, one in the object beam and the other in the reference beam with the same intensity distributions. The principle of this interferometer is almost the same as that of the PDI except that point diffraction plate is replaced by a point diffraction hologram in the PDHI. The interference in the PDI takes place between the diffracted and transmitted wavefronts from the same wavefront, but in the PDHI it takes place between the original and reconstructed wavefronts. The advantages of this interferometer are that the hologram is easier to make compared with making a point diffraction plate and the alignment of the PDHI is simpler with in situ processing of the hologram. Fig.3 shows two interferograms of a tested lens, in which (a) is shown with a pinhole at the focal point of the tested lens and (b) without the pinhole.

In practice it is troublesome to make a hologram for each test. The best way is to use a versatile hologram which should be made using a lens with a large enough numerical aperture and a pinhole at its focal point. In this case only replacement and alignment of the tested lens is needed in order to ensure that its focal point is coincident with the image of the pinhole reconstructed by the hologram.

Error analysis

Two kinds of errors, system and alignment errors, should be considered.

System errors

It is well known that the requirements on optical components are much reduced in normal holographic interferometry because of common path interference. But it is a different case in the PDHI since the pinhole is removed from object beam during the test and thus the state of the interferometer is changed. In fact, the pinhole not only filters the aberations of the tested lens, but also filters the errors produced by the surface irragularities of the beam splitter BS and mirror M_1. Therefore W in Eq.(2) involves the aberrations of the tested lens plus the errors produced by BS and M_1. In order to obtain satisfactory accuracy these errors should be too small to be significant. For visual detection it is acceptable if they are less than $\lambda/10$.

Alignment errors

The errors concerned here are those of orientation of the tested lens when it is replaced for testing using a point diffraction hologram made with another lens. They can be divided into longitudinal and lateral shifts of the focal point of the tested lens and the tilt related to the axis of the interferometer.

Longitudinal shift. It occurs when the image of the reconstructed pinhole has a longitudinal shift from the focal point of the tested lens. The OPD related to this shift is expressed as

$$OPD_1 = \frac{1}{2}\delta_1\sin^2\alpha \tag{5}$$

where δ_1 is the longitudinal shift and α is the angle between the axis and the emerging ray from the tested lens. This interferogram is just like Newton's Rings.

Lateral shift. The extra OPD caused by lateral shift is expressed as

$$OPD_2 = \delta_2 \frac{x}{d} \tag{6}$$

where d is the distance between the pinhole and lens L_1, δ_2 is the lateral shift and x is the coordinate of the interferogram in the shift direction. It is clear from Eq.(6) that Yang's fringes are formed in the interference field.

Tilt of the tested lens. Unlike lateral and longitudinal shift which introduce extra errors, tilt of the tested lens introduces oblique aberrations which can be tested if needed. If tilt is not required it can be gradually reduced to zero by adjusting the lens and observing the shape of the interferogram until only spherical aberration is shown on the pattern.[5]

The lateral and longitudinal shifts also can be adjusted using the same means. In fact, one needs definite amout of these shifts in order to balance spherical aberration and to interpret interferogram with ease.

Conclusion

We have discussed the Point Diffraction Holographic Interferometer, including the principle and error analysis. It is an example of a combination of interferometry with holography, by which the drawbacks of PDI are avoided. Moreover it has a simple structure and is flexible in operation. The hologram used is easy to produce and can be used for a long period. Therefore this interferometer is useful for routine optical testing.

References

1. J. W. Bates, "A Wavefront Shearing Interferometer,"Proc. Phys. Soc., Vol. 59, p . 940. 1947.
2. P. Hariharan and D. Sen, "Radial Shear Interferometer," J. Sci. Instrum., Vol. 38, p. 428, 1961b.
3. Zhou Wanzhi, "Reflecting Radial-shear Interferometers with an Air-spaced System," Optics Commun., Vol. 53, p.74. 1985.
4. R. N. Smartt and W. H. Steel, "Theory and Application of Point-Diffraction Interferometers," Jpn. J. Appl. Phys., Vol. 14, Suppl. 14-1, p. 351. 1975.
5. D. Malacara, Optical Shop Testing (New York, John Wiley), p.71.

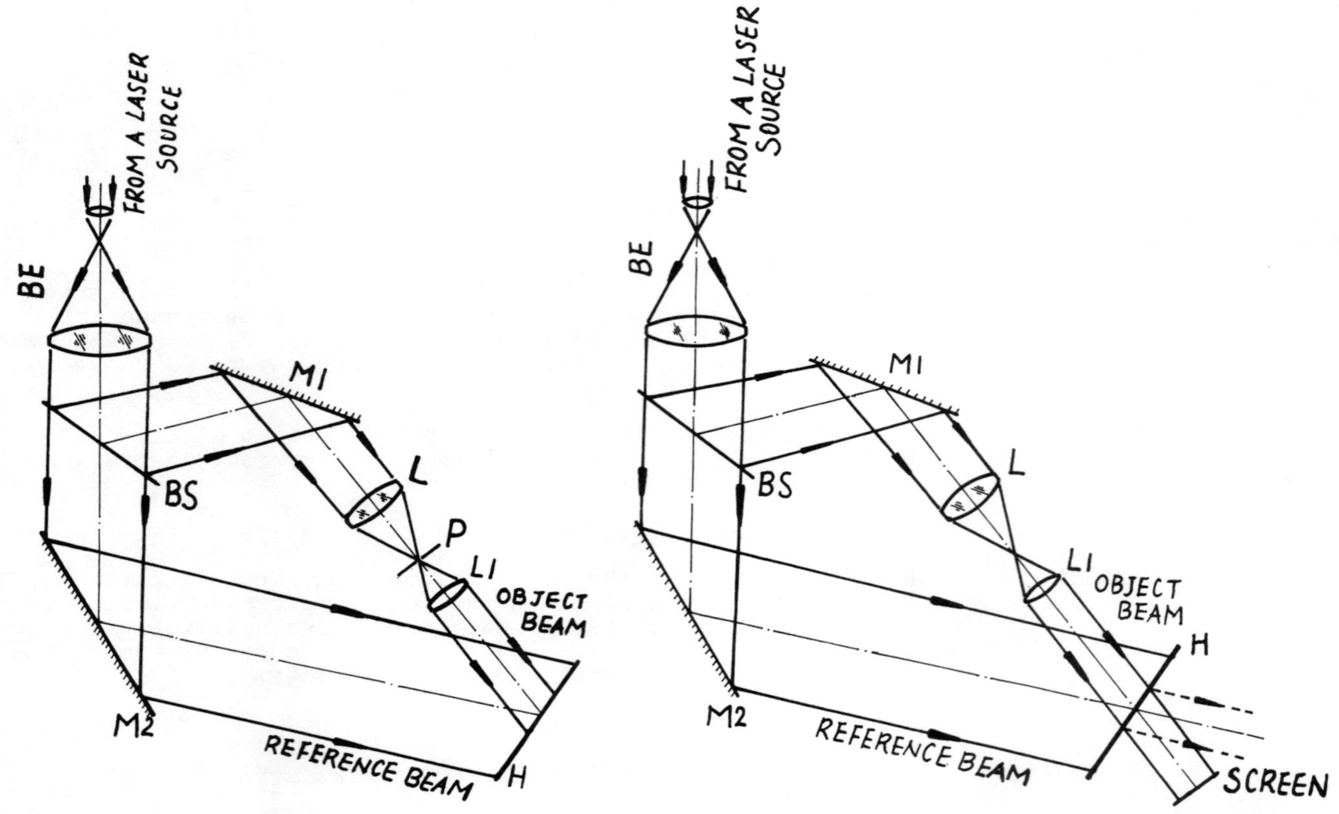

Figure 1. Schematic of the optical system for making a point diffraction hologram.

Figure 2. The point diffraction holographic interferometer.

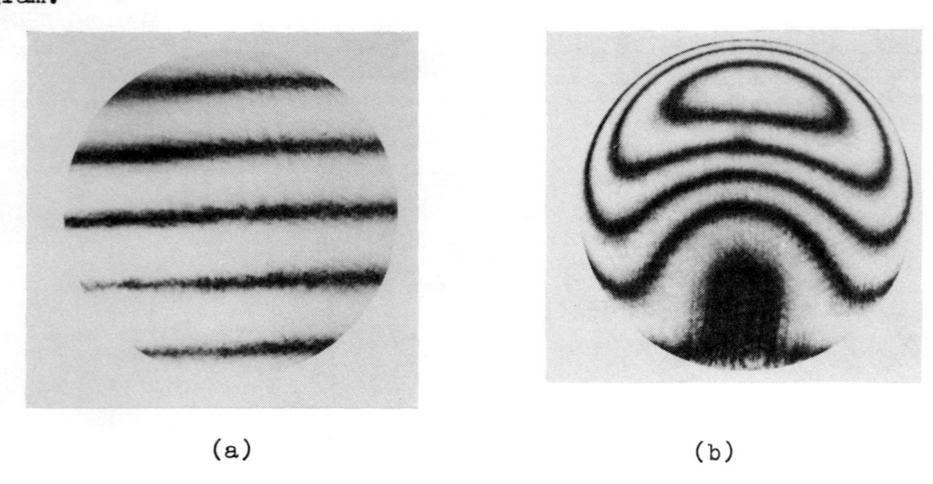

(a) (b)

Figure 3. The interferograms by the PDHI.

A method for measuring strain directly by laser objective speckles

Zhou Xingeng Niu lin

Department of Engineering Mechanics, Tsinghua University, Beijing, China

Abstract

A laser beam impings on a small region of object surface to be measured and objective speckles are recorded by a photoplate. Strains at measured point on object surface determined from a double-exposure specklegram. The equations were derived and two experiments were done.

Introduction

Most of the laser speckle method for strain analysis developed so far utilize subjective speckles in that they are recorded through a imaging system. The objective speckles are those produced intrinsically from a scattering surface. Only few works exist where objective laser speckles are utilized for strain analysis. P.M.Boone developed a displecement measuring method with close range objective speckles.[1] In this method a photoplate is placed very close to the scattering surface to receive directly the scattering wavelets. F.P. Chinag and C.C.Kin presented a method for strain determination on curved surfaces using far-field objective laser speckles which are produced by a ground glass illuminated by an expanded laser beam.[2] I. Yamaguchi utilized digital image processing technique.[3] In this paper we present a method for measuring strain directly by laser objective speckles.

Description of the method

As schematically shown in Fig 1, a laser beam impinges vertically on a small region (0.5-2mm) at a measured point on the object surface and a photoplate is also set at a right angle to the laser beam. The surface element illuminated by the laser beam can be regarded as a random diffraction grating of continuously varying pitch and orientation. The speckle field is produced by diffraction from this random grating. The configuration of those speckles in air space are ellipsoidal.[4,5] The speckles will move due to any displacement, rotation and deformation of the surface element. As shown in Fig.2, a recording plane is parallel to the surface element with a distance D. $Q_1(-a,0,D)$ and $Q_2(a,0,D)$ are a pair of points on the recording plane symmetrical about the nomal of the surface element. x direction translation of speckles at Q_1, Q_2 are only influenced by the surface element displacement component u along x direction, w along z direction, rotation Ω_y about axis y and strain ε_x. When the surface element moves with a distance u, all speckles will move along direction x with a same distance, namely $U_1^u=U_2^u=u$. The contribution from rotation Ω_y can be obtained by differentiating the grating equation. $d\theta=(1+1/\cos\theta)\Omega_y$. The displacement components at Q_1 and Q_2 which are induced by Ω_y are equal. $U_1^{\Omega_y}=U_2^{\Omega_y}$. One can consider the fact that ε_x equals db/b, where b is the random grating pitch and db is the change of the pitch, when the surface element is deformed. Also by differentiating the grating equation one can obtained that $d\theta=-\tan\theta\, db/b=-\varepsilon_x\tan\theta$. Thus the speckles at Q_1 and Q_2 on the recording plane should move with $U_1^{\varepsilon_x}$ and $U_2^{\varepsilon_x}$ respectively, where $U_1^{\varepsilon_x}=\varepsilon_x D\tan\theta/\cos^2\theta$, $U_2^{\varepsilon_x}=-\varepsilon_x D\tan\theta/\cos^2\theta$. The contributions from the surface element displacement component w are U_1^w, U_2^w, where $U_1^w=w\tan\theta$ and $U_2^w=-w\tan\theta$.

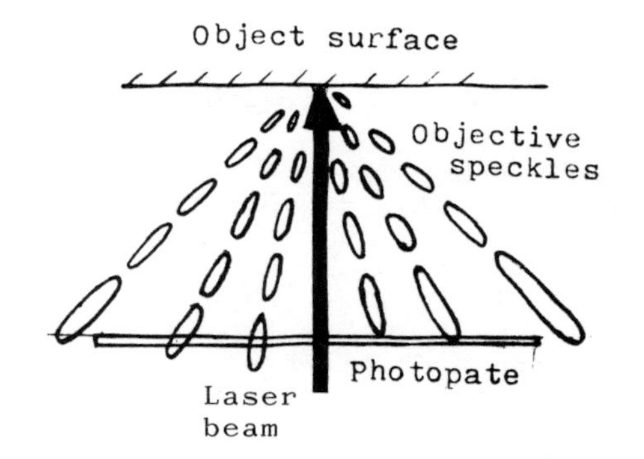

Figure 1. Experimental arrangement

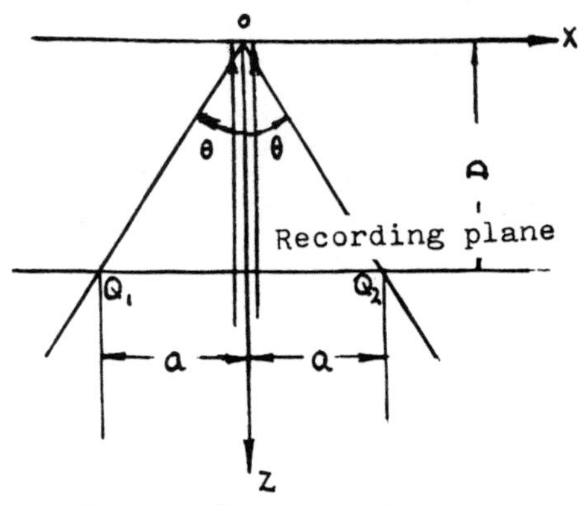

Figure 2. Geometry

A double-exposure specklegram can be taken before and after loading. One can obtained U_1 and U_2 from fringe patterns by point wise filtering.[6] According to previous analysis, the difference between U_1 and U_2 only related to ε_x and w. It is given by

$$\Delta U = \frac{2D\tan\theta}{\cos^2\theta}\varepsilon_x + 2w\tan\theta \qquad (1)$$

In the case that $w \ll D\varepsilon_x/\cos^2\theta$,

$$\varepsilon_x = \frac{\cos^2\theta}{2D\tan\theta}\Delta U \qquad (2)$$

Thus one can obtain ε_x from a pair of measured points on the double-exposure specklegram. If the effect of w can not be ignored, one can obtain ε_x and w from two pairs of measured points on a double-exposure specklegram by resolving simultaneous equations.

Thus

$$\varepsilon_x = \frac{\Delta U_I/\tan\theta_I - \Delta U_{II}/\tan\theta_{II}}{2D(1/\cos^2\theta_I - 1/\cos^2\theta_{II})} \qquad (3)$$

and

$$w = \frac{\Delta U_I\cos^2\theta_I/\tan\theta_I - \Delta U_{II}\cos^2\theta_{II}/\tan\theta}{2(\cos^2\theta_I - \cos^2\theta_{II})} \qquad (4)$$

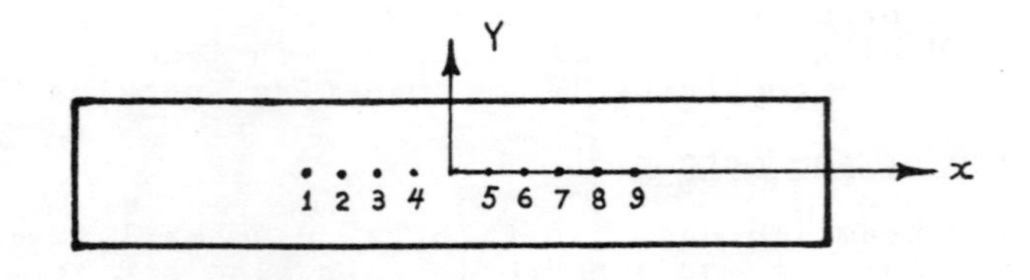

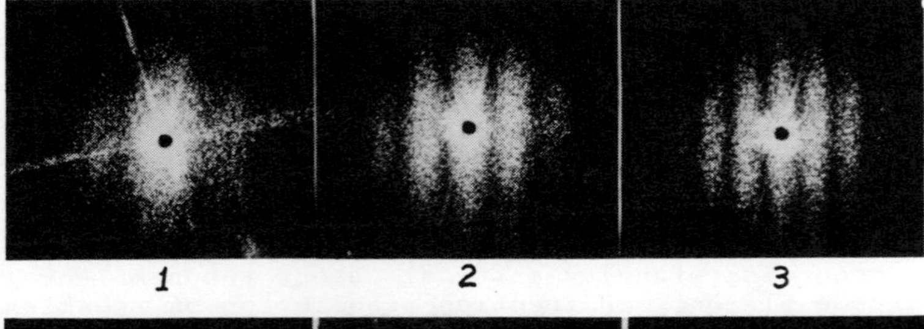

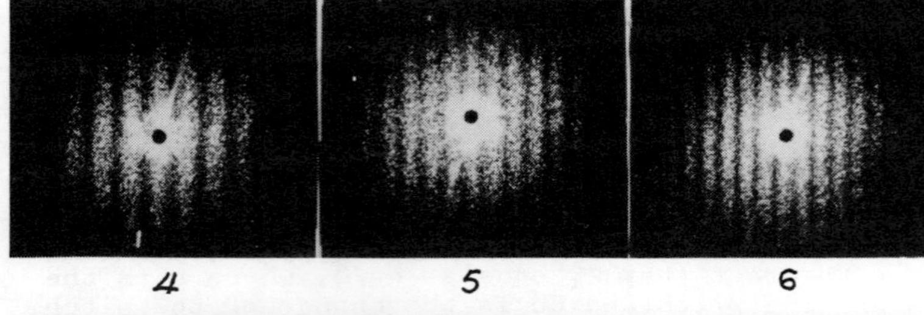

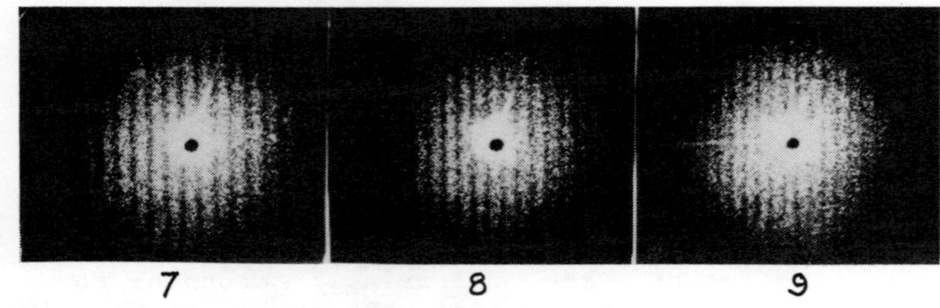

Fig.3 Young's fringe patterns at selected points on a double-exposure specklegram

Two experiments was done. In the first experiment, a thin cylindrical shell was compressed with axial load. In this case w can be ignored. Three pairs of ponts were measured on speckegram The deviations of measured value from average value are smaller than 2%. The second experimental sample is a straight bar has a rectangular cross section under compression. In this case, the effect of w can not be ignored. Three sets simutaneous equations were taken. The deviations of measured value are not over than 5%.

In Fig.3, a series of the fring patterns taken from different ponts on a double-exposure specklegram, in which case w can be ignored, the change in these Yung's fringe patterns can demonstrates distinctly the effect of the strain.

Of course one can obtained strain component along any direction from the same double-exposure specklegram with the same procedure. Only difference is that the point pairs on the specklegram should be selected along a direction corresponding to the strain component.

References

1. D.M. Boone, Use of close range objective speckles for displacement measurement, Optical Engineering Vol. 21 No.3 pp. 407-410 1982.
2. F.P. Chiang and C.C. Kin, Strain determination on curved surfaces using far-field objective laser speckles, Optical Engineering Vol.21 No.3 pp. 441-446. 1982.
3. I. Yamaguchi, simplified laser-speckle strain gauge, Optical Engineering Vol. 21 No.3, pp. 436-440 1982.
4. B. Eliasson and F.M. Mottier, Determination of the Granular radiance distribution of a diffuser and its use for vibration analysis , J. Opt. Soc. Am. Vol. 61 No.5, pp. 559-565. 1971.
5. B. Colombeau, C.Froehly, M. Vampouille, Fourier description of the axial structure of speckle, J. Optics (Paris), Vol. 10 No.2, pp.65-69 1979.
6. F.P. Chiang. J. Adachi, R.Anastasi, J.Beatty, Subjective laser speckle method and its application to solid mechanics problems, Optical Engineering Vol. 21 No.3, pp.379-390. 1982.

ADVANCED HOLOGRAPHIC SCANNINGS AND APPLICATIONS*

C. S. Ih and L . Q. Xiang

Department of Electrical Engineering
University of Delaware
Newark, DE 19716

ABSTRACT

Holographic scanners with advanced features are described. Novel and synergetic scannings that can be created by holographic scanners and in particular those with an aspherical reflector are presented.

INTRODUCTION

Holographic scanners have many unique features and offer flexibilities and versatilities unparalleled with other scanners. Their speed and resolution can exceed those of polygonal scanners. The rotating holographic scanner was first described by Cindrich [1] and subsequently by McMahon and *et al.*[2]. Based on the same concept of a rotating polygon, individual mirrored facets are replaced with either simple plane holograms (or gratings) or complex holograms.

The advantages of holographic scanners are many. The smooth surfaces and simple geometry of holographic scanners result in a much lower windage losses and lower distortions at high rotation speeds. They are inherently better for two- and three-dimensional scannings [2,3]. They can be manufactured at low cost by either contact (holographic) printing or embossing. As an input device, holographic scanners have the unique ability to selectively collect the scattered light back from the scanning position [4]. Also for a transmissive holographic scanner, the scanning beam can be designed to be almost completely insensitive to bearing wobble or substrate surface tilt [5].

For bar code readers, such as the supermarket UPC readers, the capabilities of operating the 3-D and complex scannings with holographic scanners offer enormous advantages over conventional techniques. For more demanding applications, such as laser printers and document readers (scanners), more accurate and linear scan lines must be maintained. We will describe holographic scanners which not only can meet these severe requirements but also in many instance can automatically compensate the mechanical errors which cause line scan irregularities. Several interesting and potentially very useful scan patterns are presented.

HOLOGRAPHIC SCANNERS

As indicated by Bryngdahl and Lee, the scanning motion in the holographic scanners is achieved by both the rotational and linear (translation) motion of the hologram [3]. The former is similar to the operation of the rotating polygons. The latter resembles the translation of a lens along the direction normal to the optical axis. Cindrich made a hologram with an off-axis point source and a parallel reference beam. He demonstrated a circular scan pattern by rotating the hologram with a stationary and collimated reconstruction beam [1]. McMahon[2] presented a two-dimensional scanning system with approximately 2×10^3 (50×40) total resolution elements and curved scan lines by rotating a set of holograms recorded on the circumference of a disc. These holograms are formed by two parallel beams with the incident angle of object beam being increased by a given amount for each hologram in succession. The focal points of the scanning beams can be moved by changing the convergence of the reconstruction beam. The scan lines in these systems are curved. Holographic scanners for laser printers and laser facsimile systems require a flat-field and straight scan lines. Various configurations of holographic scanners and technologies have been proposed for meeting these requirements.

---

* Supported in part by a National Science Foundation grant, ECS-8209184.

Pole and Wollenmann [4] made a variable spatial frequency hologram on a rotating cylinder surface for their scanner. The reconstruction beam comes from a point source located on the axis of rotation. When the cylinder is rotated, an one-dimensional scanning is realized. The scan line is straight but its scan range is limited due to defocusing and aberrations. Furthermore, a cylindrical shaped scanner is more difficult to duplicate.

The Computer Generated Hologram (CGH) offers a unique way to provide novel beam deflections as demonstrated by Bryngdahl and Lee [6] and at the same time to correct the scanning line curvatures. Lee [7] made a CGH to correct the wavefront of the reference beam in making the Interferometric-Zone-Plate (IZP). A cylinder-shaped scanner with this IZP can correct the scan line aberrations that exists in the conventional IZP scanner. The scan range without aberrations can be extended to $\pm 12^{o}$. Ishii and Murata [8] designed a scanner using reflective dichromated gelatin holograms. An object beam with an optimized aspherical wave is formed from a CGH, and its phase, ϕ, is given by,

$$\phi(x,y) = \frac{2\pi}{\lambda}(\sqrt{F^2 + x^2 + y^2} - F + C_{20}x^2 + C_{40}x^4 + C_{60}x^6 + C_{80}x^8 + C_{02}y^2 + C_{04}y^4 + C_{06}y^6 + C_{08}y^8 - C_{22}x^2y^2 - C_{44}x^4y^4)$$

where F is the focal length of the hologram and $C_{20}, \ldots, C_{44}$ are unknown coefficients to be determined by the ray tracing to obtain a linearized scan with a flat-field.

Hologram scanners using a flat or a slightly concave disk are attractive from a practical point of view. They are easy to make and to duplicate and create a minimum wind resistance in high speed operation. Kramer [9] demonstrated a flat disc scanner with planar gratings, which can be made holographically or otherwise. When properly designed, the planar grating scanner has a minimal line bow and is insensitive to mechanical wobble and eccentricity. However, because of its lack of the built-in focusing power, the flexibility and cost reduction are limited.

Ono and Nishida proposed a generalized holographic zone plates (GHZP) to correct the scan aberrations in IZP scanner by using a plural coherent spherical waves [10]. When N spherical waves are incorporated in the GHZP, the scan length can be extended to N^2 times. They also found another method to correct the scan line bow by matching the rotation radius of the hologram to the center of the GHZP [11].

Iwaoka and Shiozawa also developed an aberration-free linear holographic scanner for use in a diode-laser printer [12]. A computer program was used to design and to optimize system. The wavelengths of lasers used in recording and in reconstruction are different.

Ih demonstrated a holographic scanner using an Auxiliary Reflector (AR) [13]. The hologram was recorded along the circumference of a disk. The AR can be convex or concave and has a rotational symmetry about the disc rotating axis. The holograms can be either transmissive or reflective. After reflecting from the AR, the reconstructed principal ray will be normal to the scanned plane. The aberration of the object beam after reflecting from the AR during the recording is completely canceled. The symmetry of the AR and the scanning with straight lines make this type of scanner suitable for high resolution applications. Such scanners can be extended to a raster scanner [14]. The resolution of these scanners can be increased using multiplexing techniques, and a total resolution of 10^7 and a temporal frequency of $100MHz$ are achievable [14].

Scanners with a concave AR provides an effective enlarged aperture and it is thus possible to achieve high resolution without a lens. Operating in the reflective mode, these scanners are also suitable for microwave and infrared imaging systems [15]. The characteristics of these holographic scanners with a spherical AR are well understood [16], however, because of the large aberrations of the spherical reflector at a large field angle, the theoretical resolution as predicated by the analyses cannot be achieved. If the AR assumes an aspherical shape, high resolution can be obtained [17].

The holographic scanner with an aspherical reflector has been investigated and a scanning system with high resolution and without a lens has been demonstrated [18]. This novel system can perform 1-D, 2-D, 3-D, crossed and selective scans and thus has the potential for broad applications for either object identification or input/output devices for computers. The quality of the scan is comparable to that of the polygonal and planar grating scanners. The influence of the eccentricity and wobble on the scanning can be minimized by proper design. Because computer-generated holograms can be used in making the holograms, we can make scanning systems practically for any wavelengths, from mm wave to IR and to acoustical waves [15,16]. For these longer wavelengths applications, the reflective mode operation is preferred, since it would put less restriction on substrate materials. In this paper, we will describe several systems using this configuration.

SCAN CONFIGURATIONS

One of the most important advantages in using a holographic scanner is the ability to easily generate different and complicated scanning patterns. For instance, for UPC code reading, 20 to 30 scans all with different scan lengths, deflection angles and focal lengths can be easily generated by a holographic scanner [19]. This task, however, would be very difficult to perform using conventional techniques.

It is still a challenge to design a scanning system which combines flexibility and high resolution while maintaining high scan qualities, such as, good straightness, good linearities, and insensitivity to wobble and eccentricity. A high quality and high resolution 1-D, and 2-D scan can be obtained by a planar grating scanner. But since a focusing lens is still needed, its flexibility is restricted. Similarly many proposed holographic scanners cannot satisfy these requirements simultaneously. The high resolution holographic scanners utilizing an auxiliary reflector discussed above is capable of performing 1-D, 2-D, 3-D, crossed and selective scan and at the same time maintaining high qualities. We will elaborate on these scanners and suggest some possible applications.

The basic holographic scanner with an aspherical reflector is shown in Figure 1. In making the holograms, the reference beam is convergent onto the rotating axis and the object reaches the hologram recording area after reflecting from the aspherical surface. During the reconstruction, a straight scan line is generated when the hologram is rotated. If all the holograms are recorded identically, it is then a conventional (1-D) scanner. If the positions of the point object source are different from each other, the system will perform a 2-D scan. Since the number of holograms that can be put on the disc is limited, so is the number of scan lines. A high resolution 2-D scanner can be arranged by combining a 1-D scanner with a galvanometer as shown Figure 2. If the focal lengths of each hologram are also made different, a 3-D scanning is performed.

A multi-hologram 2-D scanner can be easily arranged into a crossed scanning and it is shown in Figure 3. The laser beam is split into two parts. One of them is used for the normal reconstruction. The reconstruction beam is incident on the hologram disc after reflection from beam splitter. The reconstructed beam is focused on the scanned plane after reflection from aspherical reflector and mirror M3. The other reconstruction beam is first shifted in y-direction, and reflected from mirror M2 to reach the hologram. The reconstructed beam, after being reflected by the aspherical reflector and mirror M4, is focused on the scanned plane. Since diffraction efficiency of a relief phase is sensitive to the polarization of the laser beam, care must be taken to assure that the polarization of each reconstruction beam is properly oriented.

Another interesting and useful scanning pattern is the selective scan[20]. It can be used

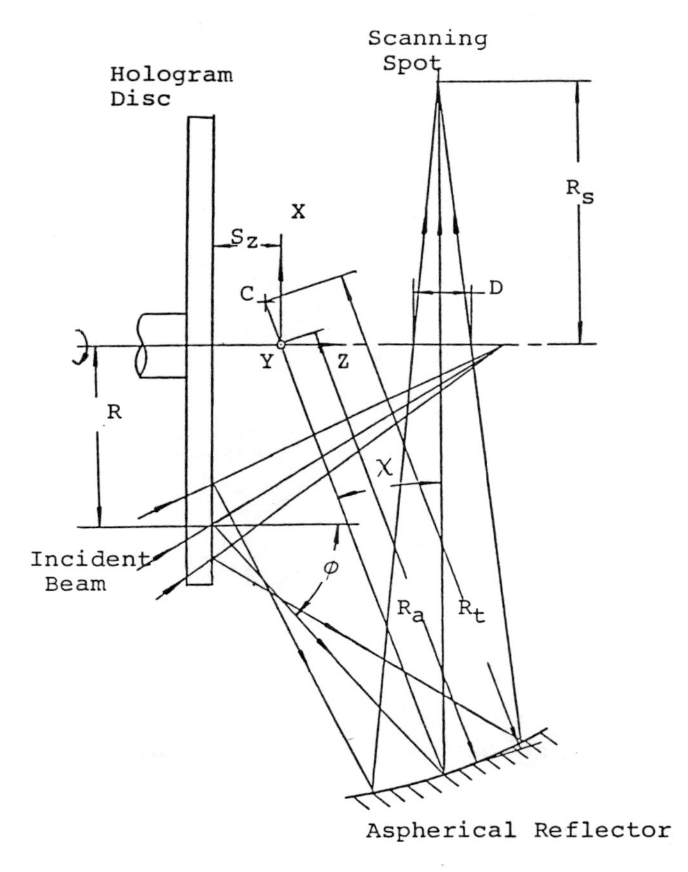

Figure 1 Holographic scanner with an aspherical reflector.

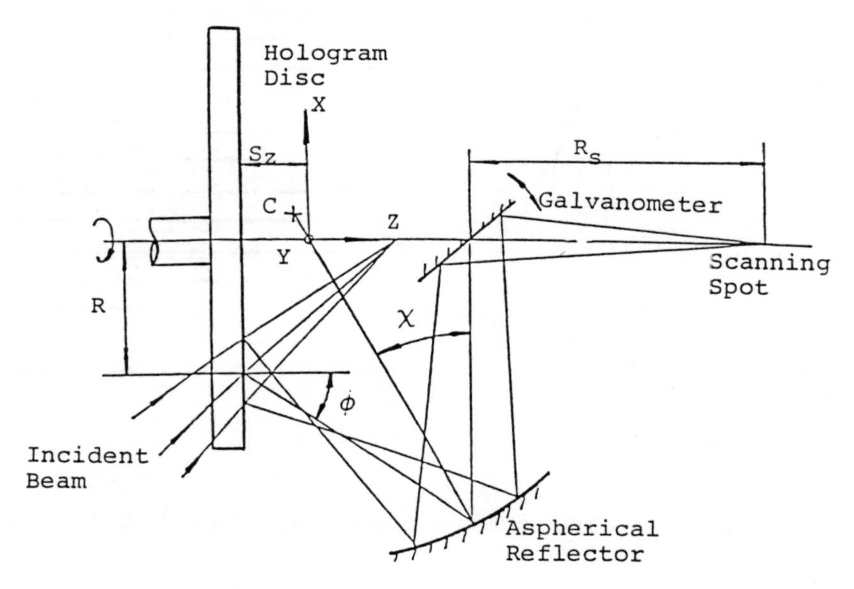

Figure 2 2-D holographic scanner with a galvanometer.

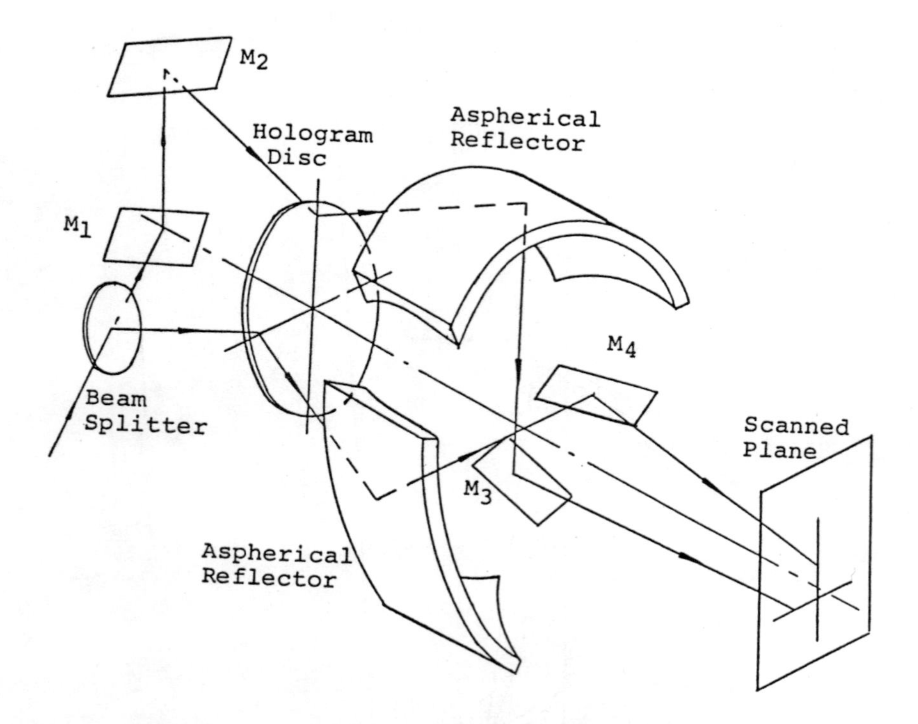

Figure 3 A cross scanner based on a 2-D holographic scanner.

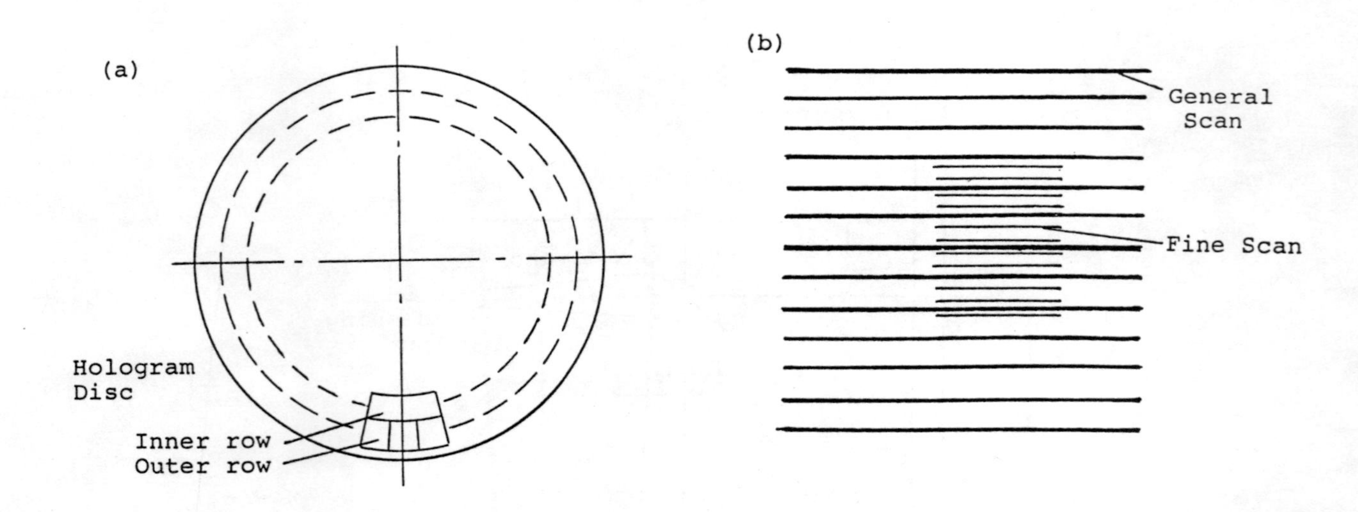

Figure 4 (a) Hologram arrangement in a selective scanner, (b) the scan pattern.

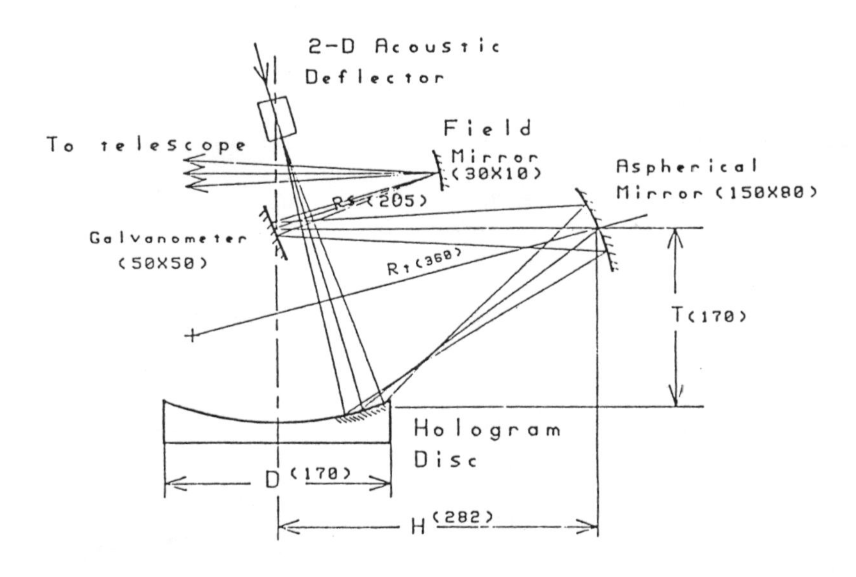

Figure 5 A selective scanner.

for robot or machine vision systems. As shown in Figure 4a, the holograms are distributed on two concentric rings. There are more holograms on the outer than the inner ring. The outer holograms thus scan a smaller angle than the inner holograms. The scanner is designed such that the scanning from the outer holograms has a higher resolution than the inner ones, i.e., the scanning spot is smaller. Therefore the inner holograms generate a larger scan with lower resolution and the outer holograms scan a smaller raster but with a higher resolution. The scan pattern is shown in Figure 4b. As shown in Figure 5, the laser beam can be switched between the two rings by a mirror or an acousto-optical deflector or using two switchable beams. The small scanning pattern can be moved around within the larger raster by changing the reconstruction beam position in the tangential direction and by adjusting the angular position and scan speed of the mirror.

COMMENTS

Holographic scanners utilizing an aspherical reflector can generate many intriguing and useful scan patterns. To extend the application of holographic scanners to longer wavelengths and ultrasonic waves, the technique of computer-generated hologram (CGH) is important. The use of CGH alleviates the need for making the hologram at the operating wavelength. With the help of the CGH technique, these scanners can be made and used at any wavelengths, including long IR and even ultrasonics. These holographic scanners can be used for machine and robot visions and imaging lidars. With further improvements in wavelength compensation and in stabilizing the wobble and eccentricity, these scanners can also be used for laser printers and image scanners.

REFERENCES

1. I. Cindrich. Image scanning by rotation of a hologram. *Appl. Opt.*, Vol.6, pp1531-1534, Sept. 1967.
2. D. H. McMahon, A. R. Flanklin, and J. B. Thaxter. Light beam deflection using holographic scanning techniques. *Appl. Opt.*, Vol.8, pp.399-402, Feb. 1969.
3. O. Bryngdahl and W. H. Lee. Laser beam scanning using computer generated holograms. *Appl. Opt.*, Vol.15, No.1, pp.183-194, 1976.
4. R. V. Pole and H. W. Werlich. Holographic opaque document scanner. U. S. Patent 4,113,343, Sept. 12, 1978.
5. C. J. Kramer. Optical scanner using plane linear diffraction grating on a rotating spinner. U. S. Patent 4,289,371, Sept. 15, 1981.
6. O. Bryngdahl and W. H. Lee. Laser beam scanning using computer-generated holograms. *Appl. Opt.*, Vol.15, No.1, pp.183-194, Jan. 1976.
7. W. H. Lee. Holographic grating scanners with aberration corrections. *Appl. Opt.*, Vol.16, No.5, pp.1392-1399, May 1977.
8. Y. Ishii and K. Murata. Flat-field linearized scans with reflection dichromated gelatin holographic grating. *Appl. Opt.*, Vol.23, pp.1999-2006, No.12, June 1984.
9. C. J. Kramer. Holographic laser scanners for nonimpact printing. *Laser Focus*, pp70-82, June 1981.
10. Y. Ono and N. Nishida. Holographic laser scanners using generalized zone plates.*Appl. Opt.*, Vol.21, No.24, pp.4542-4548, Dec. 1982.
11. Y. Ono and N. Nishida. Holographic disk scanners for bow-free scanning. *Appl. Opt.*, Vol.22, No.14, pp.2132-2136, July. 1983.
12. H. Iwaoka and T. Shiozawa. Aberration-free linear holographic scanner and its application to a diode-laser printer.*Appl. Opt.*, Vol.25, No.1, Jan. 1986.
13. C. S. Ih. Holographic laser beam scanners utilizing an auxiliary reflector. *Appl. Opt.*, Vol.16, No.8, pp.2137-2145, Aug. 1977.
14. C. S. Ih. Design considerations of 2-D holographic scanners. *Appl. Opt.*, Vol.17, No.5, pp.748-754, March 1978.
15. C. S. Ih, N. S. Kopeika, and E. LeDet. Characteristics of active and passive 2-D holographic scanner imaging systems for the middle infrared. *Appl. Opt.*, Vol.19, No.12, pp.2041-2045, June 1980.
16. C. S. Ih, E. G. LeDet, and N. S. Kopeika. Characteristics of holographic scanners utilizing a concave auxiliary reflector. *Appl. Opt.*, Vol.20, No.9, pp.1656-1662, May, 1981.
17. C. S. Ih. Holographic scanners with an aspherical auxiliary reflector. *SPIE Vol. 299, Advances in Laser Scanning Technology (1981)*, pp.169-174.
18. L. Q. Xiang. *Investigation in Holographic Scanners*, Dissertation for Ph.D. degree, 1986, University of Delaware.
19. Leroy D. Dickson, Glenn T. Sincerbox. Optics and holography in IBM supermarket scanner. *SPIE vol. 299, Advances in Laser Scanning Technology (1981).*
20. C. S. Ih and L. Q. Xiang, Compound holographic scanners, *SPIE Vol. 498, Laser scanning and recording (1984)*, pp.191-198.

Industrial Holography combined with Image Processing

J. Schörner, H. Rottenkolber, W. Roid,
Rottenkolber Holo-System GmbH, Henschelring 15, 8011 Kirchheim/München, FRG

K. Hinsch
University of Oldenburg, FRG

Abstract

Holographic test methods have gained to become a valuable tool for the engineer in research and development. But also in the field of non-destructive quality control holographic test equipment is now accepted for tests within the production line. The producer of aircraft tyres e. g. are using holographic tests to prove the guarantee of their tyres. Together with image processing the whole test cycle is automatisized. The defects within the tyre are found automatically and are listed on an outprint. The power engine industry is using holographic vibration tests for the optimization of their constructions. In the plastics industry tanks, wheels, seats and fans are tested holographically to find the optimum of shape. The automotive industry makes holography a tool for noise reduction. Instant holography and image processing techniques for quantitative analysis have led to an economic application of holographic test methods. New developments of holographic units in combination with image processing are presented.

Introduction

In 1963 the first holograms were produced by Leith and Upatnieks. At that time only weak laser-radiation was available, therefore the exposure times were very long. The optical setups were very sensitive to vibration. Holograms only could be produced at night and in the cellar. Nowadays, holographic testing equipment is installed in factories near producing machines.
This change was the result of three main improvements:
1. the higher output and reliability of lasers,
2. the higher quality of vibration isolating systems,
3. the production of instant recording and computer analyzing techniques.

Therefore now holographic interferometry is not only used as a method within research and development but also as a method for quality control in series production, for example, in the tyre and aircraft industry.

There exist three main fields of application of holographic non-destructive testing methods: quality control, optimization of constructions and vibration analysis. Examples for application out of these fields are presented within this paper. Apart from that, the state of art of holographic industrial equipment is shown.

Holographic Construction Optimization

The presuppositions for application of holography in an industrial environment are the following ones:
1. provision of suitable antivibration isolation
2. instant development of the test results
3. short exposure times by means of high laser power
4. automatic interpretation of the interference hologram.

These requirements are realized in the HOLO-ANALYZER HT-80, a holographic interferometer, which can be placed in an industrial environment (Fig. 1). Rottenkolber Holo-System have developed special active, pneumatic air cushions, which are characterized by their high load capacity and their very low resonance frequency. At an operating pressure of 8 bars, the load may be as high as 1000 kilograms per shock absorber. The natural resonance in the vertical direction is about 1.3 cps, and about 3.2 cps in the horizontal direction. Particularly when holography is used in mechanical engineering, where cast-iron T-grooved disks are used to clamp the heavy test bars, this pneumatic suspension system must be designed for high loads. The Argon-Ion-Laser, which is mounted aside the experimental plate, has an holographically usable output of 1.5 watt. Therefore short exposure times, 100 to 400 ms typically can be realized. The beam is directed within a pipe to avoid laser hazards to the HOLO-RECORDER or the HOLO-INSTANT-CAMERA. Both camera systems store the hologram as a phase information on a thermoplast film. This thermoplast film, called HF-85, is usually used in this method. It is a panchromatic material with a high sensitivity and a high diffraction efficiency (Fig. 2). The optimum exposure energy ranges at 10erg/cm^2, but even with 2 erg/cm^2 good results can be obtained. Hence the sensitivity of this thermoplast material comes under the same range as that of the usually employed holographic silver film materials (for example: Agfa Holotest 10 E 75 has 20 erg/cm^2). There is no other thermoplast recording material with the same sensitivity. Moreover, the photographic thermoplast material allows a wide range in exposure time. Remarkable variations of the diffraction efficiency were established in the range from 0.1 to 20 concerning the intensity ration between the object beam and the reference beam.

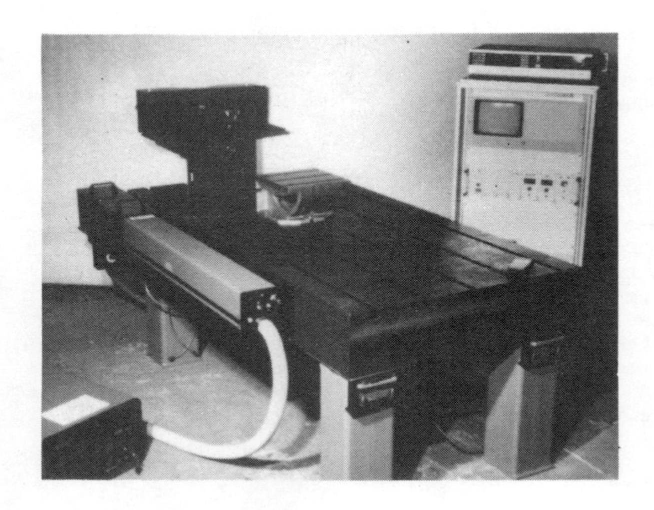

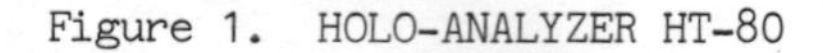

Figure 1. HOLO-ANALYZER HT-80

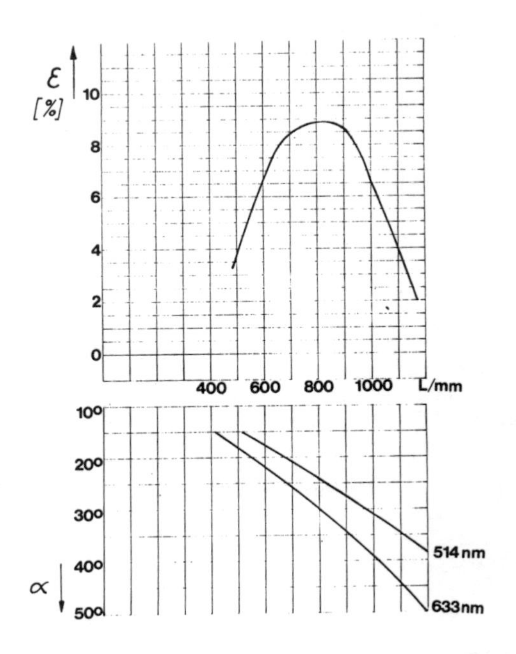

Figure 2. Diffraction efficiency of thermoplast film HF-85

The HRC-110 hologram recorder of Rottenkolber Holo-System is a unit which includes both the optical system required for illumination and the reference beam generation in hologram production as well as the instant development on a thermoplast film. When a hologram is shot, the interim object image produced by an image forming objective is recorded on the thermoplastic film. By thesemeans the information to be stored is restricted to the area of optimum diffraction efficiency. Thus , object angle areas as wide as 110° can be covered and yet instant holography is possible. Fig. 3 shows the scheme of the structure of this HRC-110 HOLO-RECORDER. In this hologram recorder it is only necessary to introduce through reflectors a laser beam expanded to 10 mm. The optical illumination system is a component of this unit. Simple change of the objectives or the use of a zoom objective allows adaption of the field of view to the respective requirements in each case.

The developed hologram is registered by a TV camera and displayed on a monitor. A video recorder may be used for documentation purposes, or the image may be transferred onto a 35 mm film via an additional optical system. Fig. 4 shows how this hologram recorder is constituted by various units e. g. within the HOLO-TYRE-ANALYZER. This hologram recorder is universal as far as its application is concerned. It is suited for use with both continuous-wave laser and pulsed-laser holography. It allows simple realization of all holographic methods such as double exposure, real-time and time average methods. The packaged design allows the HOLO-RECORDER to be used also for testing in areas which involve three-dimensional problems. This hologram camera, or the HOLO-INSTANT-CAMERA (without optics) is an inseparable component of all of our industrial holographic testing equipments.

In 1985 this HOLO-ANALYZER was combined with an image processing system. Together with the Academy of Sciences, Berlin, a special software program was developed for quantitative analysis of the interference hologram. On the basis of static evaluation the automatic calculation of the amplitude of deformation is possible. The resolution of the system is 64 x 64 pixel. The procedure is the following: The displacement of one point of the pointmatrix is known and all points of the matrix are calculated. The projection of the displacement vector to the sensitivity vector of the interferometer for each point is measured. If the vector of the displacement has to be calculated, three holograms have to be used with illumination of the object from three different directions. The result of this calculation can be shown in a pseudo 3-d-plot.

Typical applications of the HOLO-ANALYZER HT-80 are deformation or strain studies for design optimization, quality control of products of series fabrication and vibration analyses according to the time average method. Fig. 5 (BASF, Ludwigshafen) shows the control of adhesiveness of a sport shoe. The circular line pattern marks the defect. In Fig. 6 the holographic testing of a tube made out of glass fiber enforced material is to be seen. The tube at the top of the picture is the good one, the tube beneath shows defects caused by incorrect winding. Fig. 7 (BASF) shows the deformation behaviour of a cover plate of a plastic tank. How perfect construction optimization has been done in nature is shown in Fig. 8 and 9 (BASF): An egg pressed by a force of 20 N. The first one having a crack, the second one without crack.

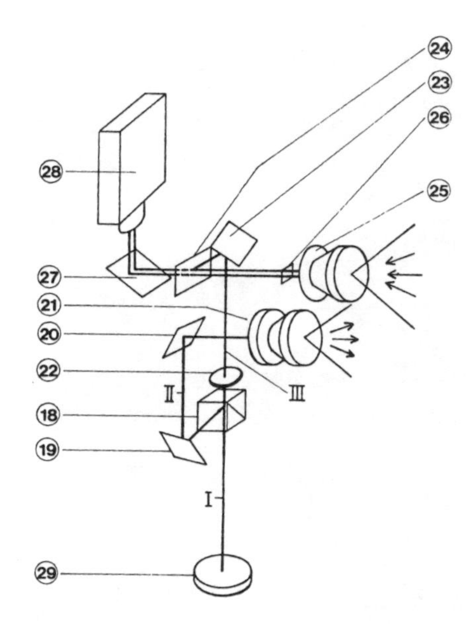

Figure 3. Schematic setup of HOLO-RECORDER HRC-110, I. main beam, II. illumination, III. reference beam , 26. first image, 24. hologram

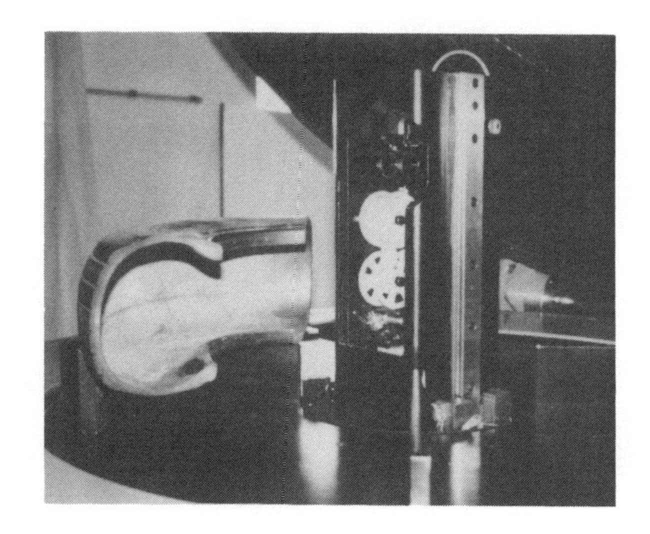

Figure 4. HOLO-RECORDER HRC-110 for automatic instant holography

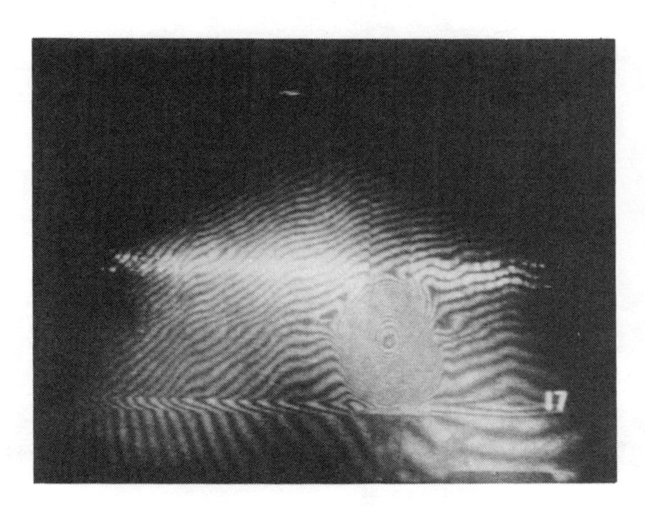

Figure 5. Holographic control of adhesiveness of a sport shoe (BASF)

Figure 6. Quality control of a glass fibre tube; top: the good one; bottom: tube with defects

Holography at Transparent Media

There is a special application of holography for the study of thermal processes, wind channel research and flow pattern research of gases and liquids. In the past often Mach-Zehnder interferometers were used in these fields. The great advantage of holographic interferometers is: they do not need cuvettes and windtunnel windows of high quality in flatness. System errors or optical component defects are compensated by the HOLO-INTERFEROMETER HIF-12, which is a specail construction made by Rottenkolber Holo-System. The optical setup of this equipment is shown in Fig. 10. The HOLO-INTERFEROMETER itself with control cabinet and video-monitor are to be seen in Fig. 11. Two examples for application: Fig. 12 shows the temperature field around a candle, and around a metal ball (Fig. 13). This HOLO-INTERFEROMETER HIF-12 was firstly shown at the exhibition "Laser and Opto-Electronics 1985" and then delivered to the Department of Thermal Engineering of Quenghua University in Peking.

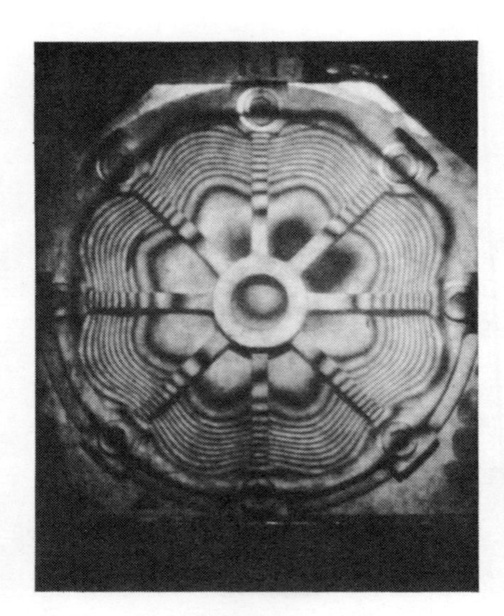

Figure 7. Deformation of the cover plate of a plastic tank (BASF)

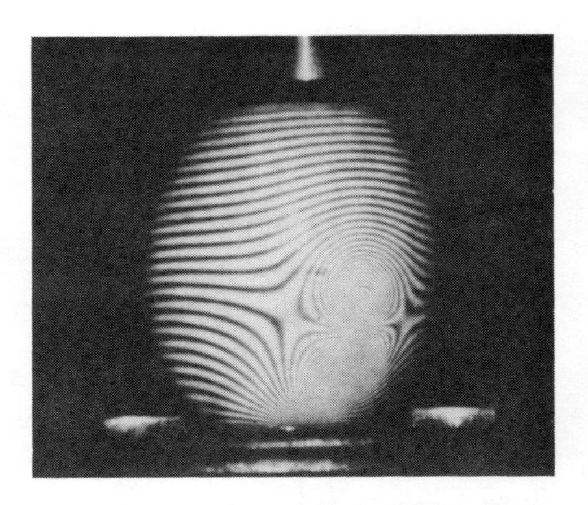

Figure 8. Egg pressed by a force of 20 N Interference pattern shows crack (BASF)

Figure 9. Interference pattern shows uniform distribution of force (BASF)

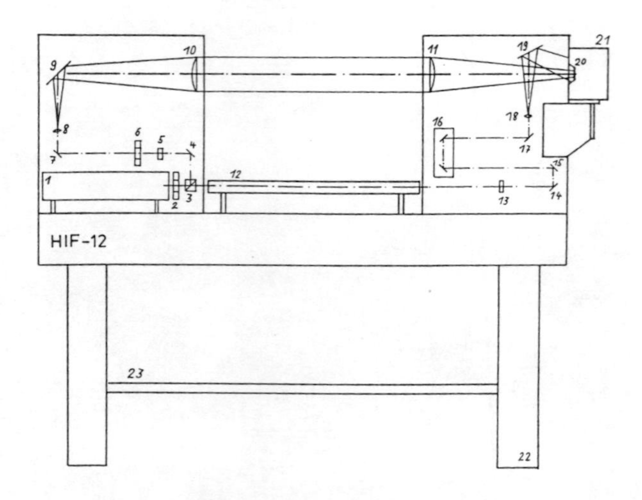

Figure 10. Optical setup of HOLO-INTERFEROMETER HIF-12

Holographic Vibration Analysis

The vibration analysis both according to the time-average method and using the double-pulse holography is a vast field of industrial application of instant holography. Both methods are employed by leading companies in turbine and automobile industry with great success, with application focussed on development and design.

Time Average Method

Let me give ycu some examples to demonstrate this point. Fig. 14 (KWU; Mühlheim/Ruhr) shows the vibration of a turbine blade. The course of the interference lines distinctly shows the nodal lines and the locations of maximum vibrational behaviour with the respective resonance frequencies of a structural element or component, as well as the localization where the design has to be changed. A holographic experimental setup to test vibrational behaviour of a railway wheel and the time average hologram is shown in Fig. 15 and Fig. 16 (MAN). The knowledge of such vibration diagrams is extremely important in order to determine the localization of maximum sound emission. DAIMLER, for instance, utilize one of our systems for vibration analysis with an 8-cylinder motor block.

Double-Pulse-Holography

This method offers the advantage of vibration analysis being carried out under operational conditions. First successful tests of this kind were made by the VOLKSWAGENWERK, by Dr. Felske and his colleagues. In the meantime BMW, DAIMLER, FORD, KWU and MAN have adopted the double-pulse holographic method in Germany as well.

Fig. 17 shows the HOLO-VIBRO-ANALYZER PHK-100, which has been developed to meet these requirements. It comprises a camera carriage supporting a scissor tyre table which carries the pulse laser and the HOLO-RECORDER. The control cabinet accomodates the laser power supply, a vibration-processing system, a specific trigger system, the HOLO-RECORDER-Controller and a TV monitor for reproduction of the interference hologram. Two HOLO-VIBRO-ANALYZERS were delivered to China up to now: one to Aero Engine Factory in Xian and the other to Machine Tool Research Institute in Miyun near Bejjing. Fig. 16 (MAN) illustrates an example of holographic vibration analysis at a large Diesel engine. The PHK-100 system can be combined with a derotator to study rotating objects. This system was applied to analyse the vibrations at a turbine periphery at 10.000 revolutions per minute (Fig. 19, BIAS a. TU Hannover). Also for the evaluation of double pulse interference holograms the image processing method can be applied. The software is similar to that shown in combination with the HOLO-ANALYZER HT-80.

Figure 11. HOLO-INTERFEROMETER HIF-12

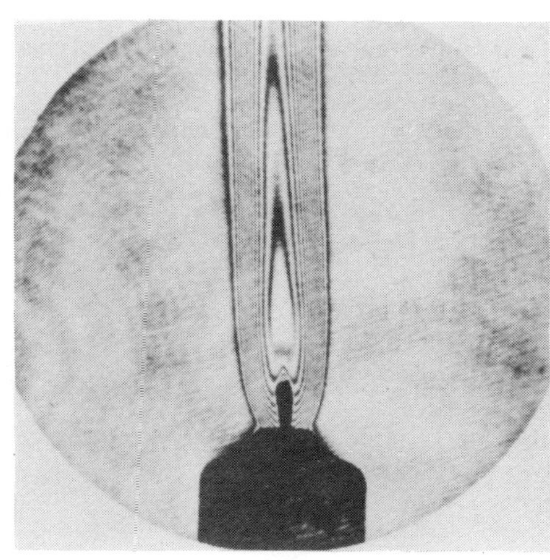

Figure 12. Temperature field around a candle

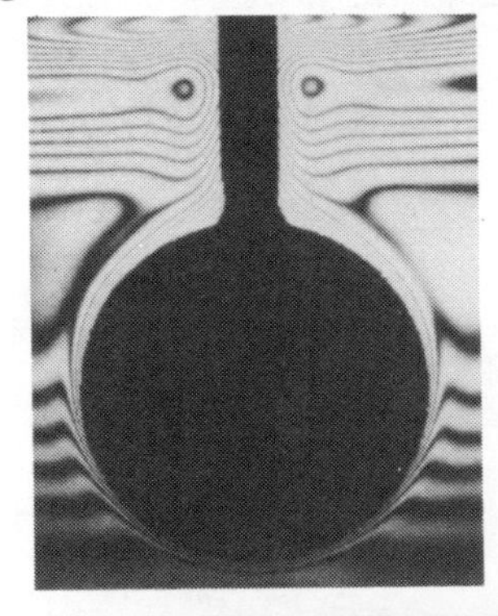

Figure 13. Temperature field around a metal ball

Figure 14. Time average hologram of turbine blade (KWU)

Figure 15. Holographic setup for testing vibration behaviour of a railway wheel (MAN)

Holographic Tyre Testing

A typical example for holography in industry is the HOLO-TYRE-ANALYZER HRT-220. This equipment is constituted by the essential elements of the HOLO-ANALYZER such as shock-absorbing system, hologram recorder, and Argon-Ion continuous-wave laser, which are arranged in a testing chamber. A great deal of experience collected in industrial application has been incorporated in the latest development of this holographic tyre tester, which was done successfully in cooperation with DUNLOP, Birmingham. Every retreated aircraft tyre is checked there with this HOLO-TYRE-ANALYZER.

A prominent feature is the degree of automation in this system, which can be used to inspect tyres for passenger-cars, trucks and airplanes for separations, damage and structural weakness. The tyre size to be tested can be set at the controller, and the complete holographic test of both the side walls and the tread can be started by pushing a button, without it being necessary to turn the tyre again. Fig. 20 and 21 show the general arrangement of the system, the control cabinet and the hologram at the monitor. The tyre periphery is checked in fcur sectors by the HOLO-RECORDER with the illumination and registration objectives scanning an angle of approximately 100° each. A 4 watt Argon-Ion laser is used to produce the hologram (i. e. a holographical output ranging between 500 and 1000 mwatt). One exposure at normal pressure and one exposure after pressure reduction in the chamber are realized per tyre segment. As has been explained in the description of the HOLO-RECORDER, the hologram is instantly developed and is immediately available for display at the monitor. The test result can then be documented on a video tape or directed to an image processing system.

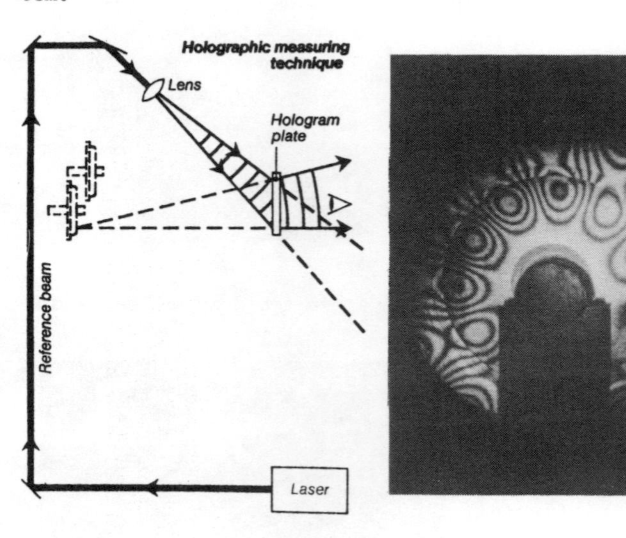

Figure 16. Optical scheme for holographic setup and interference hologram o railway wheel (MAN)

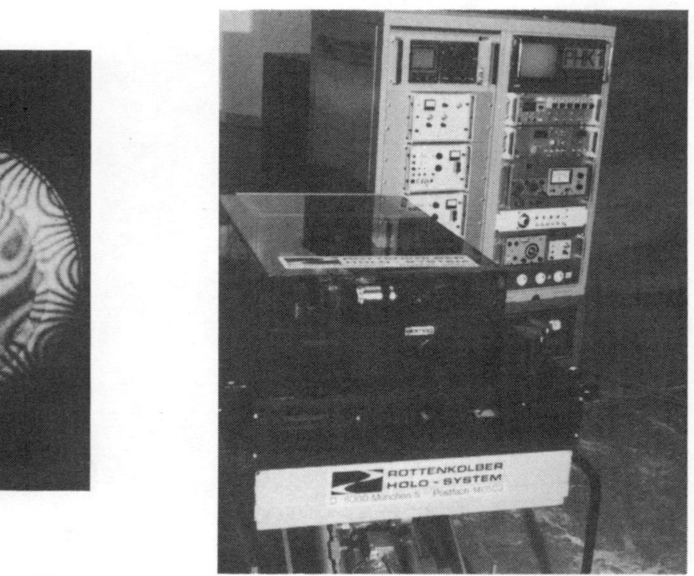

Figure 17. HOLO-VIBRO-ANALYZER PHK-100

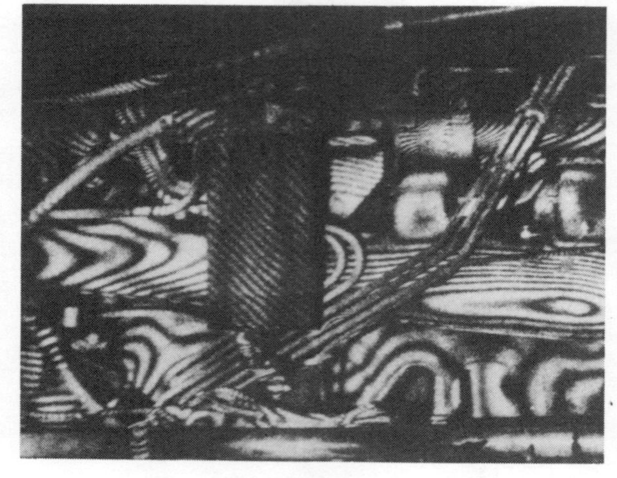

Figure 18. Holographic vibration analysis of Diesel engine (MAN)

Figure 19. Vibration study of a turbine rotor (BIAS/TU Hannover)

Figure 20. HOLO-TYRE-ANALYZER HRT-220

Figure 21. Tyre hologram at control unit of HRT-220

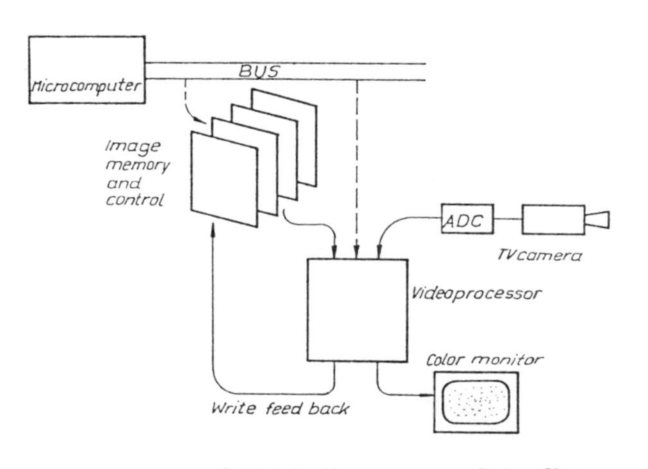

------ Control flow; ——— Data flow

Figure 22. Scheme of image processing system

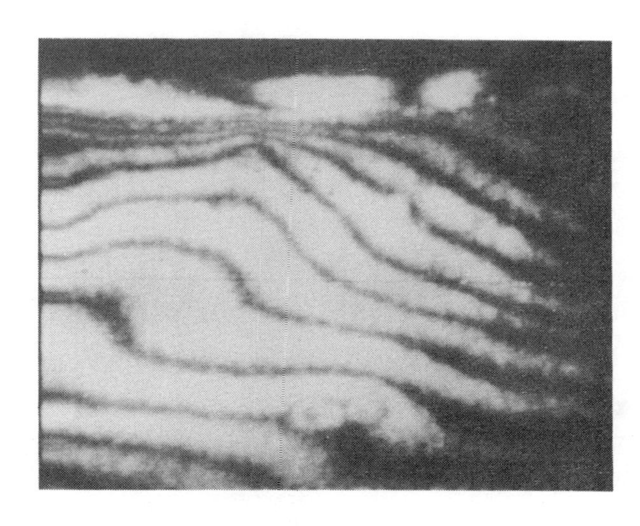

Figure 23. Tyre hologram

Figure 24. Tyre hologram after image processing with marked defects

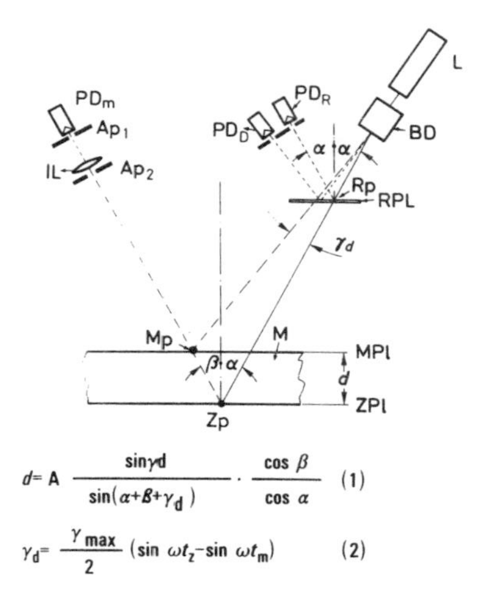

Figure 25. Principle of LASERMETER LMS

Digital Analysis of Tyre Holograms

The advantage of automatic interpretation of interference holograms are: objectivity, saving time, no necessity for special trained people for interpretation. In cooperation with the Academie of Sciences, Berlin, Rottenkolber Holo-System has developed an image processing system, which informs the user whether the tyre is faulty, conditionally faulty or in an acceptable condition. The system is interactive and the thresholds for acceptance of the tyre can be fixed by the user. In Fig. 22 the principle of hardware is shown. The essential components of the system are listed below:

1. A very high-speed analog/digital converter is used to digitize the video image in video real-time (40 ms).
2. Four image memories are available to store the images; their storage capacities 512 x 512 pixel, each 8 bites in depth; this allows a maximum of 256 gray steps per pixel.
3. A 16 bit micro-computer with 248 kbyte memory is provided for process control.
4. The actual nucleus of the image processor is a very high speed pipeline processor with a processing speed of 12.5 mbytes/sec. this unit includes also all the other subassemblies required for fault analysis, e. g. the components for histogram production, minimum/maximum recognition, input and output function generators and output channels for multi-colour display.
5. The individual processing steps can be monitored on the supervisor colour monitor.
6. A terminal is provided for interactive use.
7. A printer is provided for the output of the test result with the necessary information about the tested tyre.

The software is written to recognize typical line patterns in correlation to essential kinds of defects. These are elliptical patterns or sequences of them, sharp curved and in most cases Z-shaped fringes, which, for example, indicate a separation between layers of the tyre or breaks of the tyre cord. Fig. 23 shows the original tyre interference hologram and Fig. 24 the hologram at the end of image processing with the fault marked on the image.

Figure 26. LASERMETER LMS measuring the profile of a tyre press

Prospects and Conclusions

Computerized optical systems are more and more developed, not only in the combination shown: holography together with image processing. One example is the distance, profile and thickness measurement with laser-light and personal computer (developed by Dr. Bodlaj, SIEMENS, Munich). Rottenkolber Holo-System is producing this LASERMETER; which is working on the basis of time-measurement of an optical modulated laser beam. Fig. 25 shows the principle and Fig. 26 shows application, where the profile of a tyre press is measured. A printed output from the computer shows the engineer the profile contours and their values.

The development and production of holographic test equipment with a high degree of automatization and with instant display of the test result has brought a successful and effective progress in the application of technical holography to the industry. The instant holography was a presumption for this development. CW-laser- and double-pulse-laser-holography deliver valuable results for non-destructive testing, quality control, optimization of construction and vibration analysis, as well. The combination computer and optical test system will open the industrial market for a wide field of effective applications.

Session 9

Holographic Applications in Medicine

Chairmen

P. Greguss
Technical University of Budapest, Hungary
Zhi-Min Qu
Shanghai Institute of Laser Technology, China

MULTIPLEX HOLOGRAMS AND THEIR APPLICATIONS IN MEDICINE

Jumpei Tsujiuchi

Imaging Science and Engineering Laboratory, Tokyo Institute of Technology
4259 Nagatsuta, Midori-ku, Yokohama 227, Japan

Abstract

Fundamental properties of reconstructed images from a multiplex hologram are studied, and conditions for compensating distortions and for designing a reconstructing source are proposed. Applications of multiplex hologram to medical objects are reviewed, and a computer-aided hologram synthesizing system is proposed for obtaining better images and wider applications. An example of multiplex holograms synthesized from a series of CT images is also presented.

Introduction

A multiplex hologram is a cylindrical holographic stereogram with white light reconstruction, and is a very attractive medium to display 3D (three dimensional) images because it is synthesized from a series of ordinary photographs of an object taken from different directions in the horizontal plane [1,2]. This means that it can achieve 3D display of an object whose hologram is very difficult or impossible to realize in a conventional method. Also, the reconstructed image is very bright and has wide viewing angle to be observable by many people in the same time. These advantages are very useful for many applications, especially for medical purposes.

In this paper, a brief interpretation of multiplex holograms is given together with fundamental properties of reconstructed images, and some applications in medical field are also described.

Synthesis of Multiplex Holograms

To synthesize a multiplex hologram, first a series of photographs of an object should be taken. Ordinarily, an object to be recorded in a hologram is put on a horizontal turn table, and the image is taken by a fixed movie camera with the horizontal optical axis. So, images with successively different lines of sight are recorded in a movie film, and this film is used as an original film for synthesizing a multiplex hologram.

Fig. 1 shows an optical system for synthesizing holograms [3]. The image in a frame of the original film OF is projected through a lens L by using a laser beam onto the pupil of a large aperture lens system composed of a spherical lens SL and a cylindrical lens CL. The optical system has two foci, the vertical focus S and the horizontal focus M, and a narrow strip hologram of the projected image is recorded on a film H at the vertical focus S by using a point reference source SR. This strip hologram with a width of 0.5 mm becomes an elementary hologram of the multiplex hologram, and the entire hologram can be obtained by successively transforming all the frames of the original film into the respective elementary holograms.

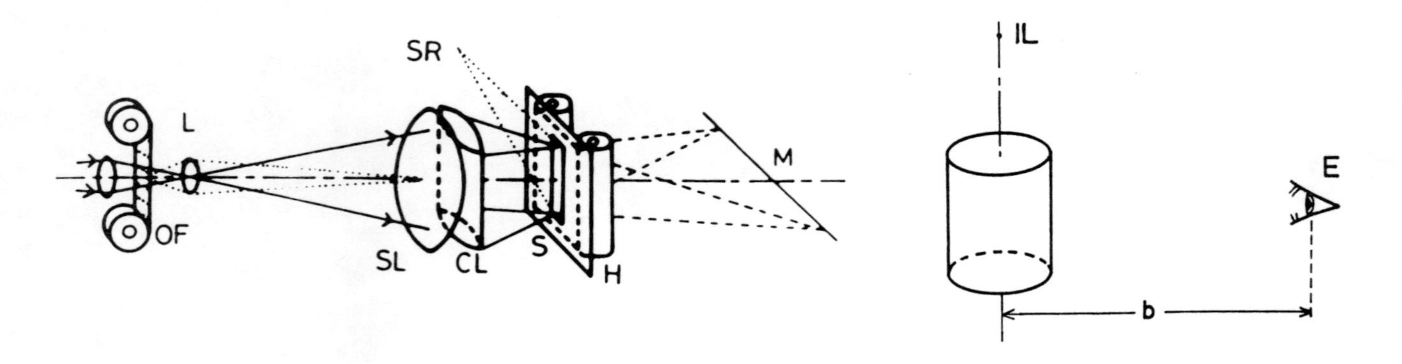

Fig. 1 Optical system for synthesizing multiplex holograms.

Fig. 2 Reconstruction of multiplex hologram.

After completing the entire hologram, it is shaped into a cylinder as shown in Fig. 2 and is illuminated by a small white light source IL located on the vertical axis of the cylinder, then the reconstructed image can be observed inside the cylinder. If you put your eyes at E, a position b apart from the axis, where b is the distance between CL and M in Fig. 1, you can observe the reconstructed 3D image with a uniform color.

Fundamental Properties of Reconstructed Images

The reconstructed image is composed of many narrow vertical segments reconstructed from the elementary holograms, and the reconstructed image shows 3D effect because your eyes detect the parallax of reconstructed images in the horizontal direction. In the vertical direction, however, the original images are recorded as an image hologram similarly to a rainbow hologram, and have no parallax.

So, imaging processes of the reconstruction in the horizontal and vertical directions are different, and the reconstructed image has in general a distortion called static distortion. This is due to the change of line of sight of reconstructed image from that of taking the original image. In fact, certain conditions exist to make the reconstructed image distortion-free separately in vertical and horizontal directions, but these conditions are not compatible in both directions except for the case of holographic stereogram with laser reconstruction [4]. As for multiplex holograms with white light reconstruction, it is impossible to obtain distortion-free images in ordinary processing [5]. Several methods have been proposed for compensating the static distortion [6,7], and two methods we are using are as follows:

One is an optimum conditions of taking the original images and reconstructing the hologram, and the condition $a = 2r/m$ gives the minimum distortion, where a is the distance between the object (center of rotation) and the camera lens, r is the radius of hologram cylinder, and m is the magnification of the reconstructed image relative to the object. So, if a hologram is made under the above-mentioned condition and is observed from a point E in Fig. 2, you can observe the reconstructed image with the minimum distortion but not distortion-free [3,5]. If you come closer to the hologram from E the distortion increases considerably, but even if you go further from E the increase of distortion is of small importance.

The other method is the rearrangement of the original film [8]. If each frame of the original film is divided into many narrow vertical segments, and then the original film is recomposed by arranging segments of different frames so as to compensate the change of line of sight between recording and reconstructing the hologram. This processing can be made by computer, and the hologram synthesized from the rearranged original film can reconstruct a distortion-free image.

Another sort of distortion called dynamic distortion appears if the object motion is too rapid during taking the original film. If the object motion is sufficiently slow, the distortion becomes negligible and the reconstructed image moves smoothly as the hologram rotates slowly around the axis. However, if the object moves rapidly, the reconstructed image becomes discontinuous and the dynamic distortion takes place. The compensation of this distortion is also possible by rearranging the original film in a computer so as to reconstruct an image at an instant from a series of elementary holograms [8]. This proces-

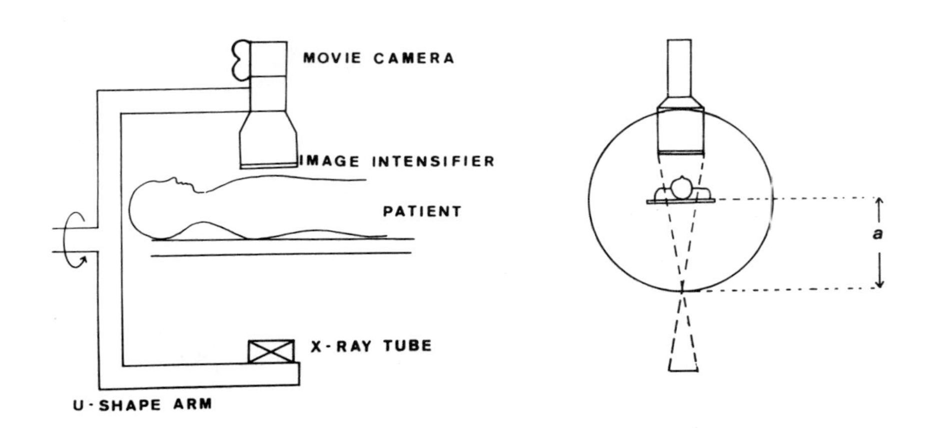

Fig. 3 An equipment for taking X-ray original images of medical objects.

sing is very similar to the previous one, and both the static and dynamic distortions can be eliminated simultaneously if needed.

Resolution is also an important property of reconstructed images [9]. If the original film has sufficiently high resolution, the resolution of reconstructed image is mainly determined in the reconstruction process, and the size of reconstructing source, the diameter of the pupil of your eyes, and the position of reconstructed image are important parameters to determine the resolution. The size of reconstructing source makes blurs in both vertical and horizontal directions, but the pupil of your eyes stops down the vertical blur, and the vertical resolution will be kept constant by the pupil of your eyes. In the horizontal direction, however, no effect of stopping by your eyes exists, and the horizontal resolution depends on the horizontal width of the source, and if the reconstructed image is apart from the direction from your eyes toward the axis of hologram the horizontal resolution depends not only on the horizontal width of the source but also on the vertical width of the source and the diameter of the pupil of your eyes. From these points of view, the size of reconstructing source should be as small as possible, and a typical example of the source for reconstructing a 40 cm diameter hologram would be an incandescent lamp with a vertical linear filament smaller than 1 mm wide and 3 mm long.

Medical Applications of Multiplex Holograms

One of the most interesting applications of multiplex holograms would be found in medicine. The original film in medical application is a series of X-ray images taken by a special equipment, an example of which is shown in Fig. 3 [10]. This equipment has a U-shaped arm with a mechanism to turn around a patient, and the arm has a point X-ray source at one end and a 35 mm movie camera behind an image intensifier at the other. The distance between the X-ray source and the center of rotation corresponds to the object distance a in the ordinary case. Multiplex holograms made of such an original film have an advantage for making better understanding of 3D structures of bones and internal organs of the patient, and also provide a possibility of measuring 3D coordinates of particular points of the objects [11].

On the other hand, since medical X-ray images are in general of low contrast, and reconstructed image from such original images cannot be expected to be of good quality. Computer processing of original images such as enhancement and distortion correction is very effective to obtain better images. For that purpose, a schematic diagram of multiplex hologram synthesizing system for medical applications is shown in Fig. 4.

There are many possibilities of synthesizing holograms from various kinds of medical images as shown in Fig. 4. The simplest is to use X-ray images recorded by a movie camera with the aid of a U-shaped arm equipment, and thus obtained movie film is directly sent to the hologram synthesizer without any processing, so reconstructed images thus obtained may have some distortions and poor image quality. To apply computer processing, images must be scanned by an appropriate scanner such as drum scanner and the digital data of the image are sent to a computer, but scanning many frames seems very troublesome and not suitable for practical use. Another possibility is to take original X-ray images by a TV camera, preferably of high definition type. in a video-tape by using the U-shaped arm equipment, and the digital data are sent to a computer through a frame memory. A digital X-ray

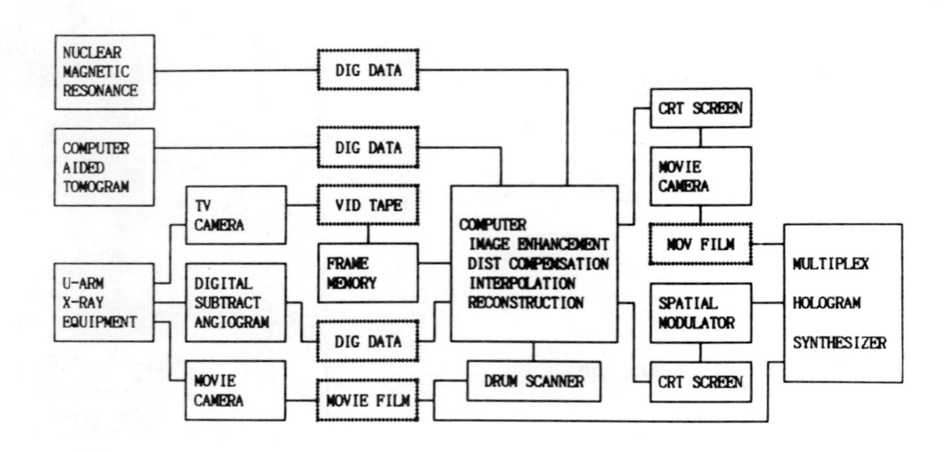

Fig. 4 A computer-aided hologram synthesizing system.

imaging system such as DSA (digital subtraction angiography) can also be used for taking X-ray original images, and the digital data will be obtained directly from the equipment.

Image processing in a computer includes image enhancement, compensation of both static and dynamic distortions, and synthesis of intermediate images if the total number of original images is limited because of the dose of X-ray to the patient. After processing, two methods for sending the processed image to the hologram synthesizer are to be considered: one is to display processed images in a CRT screen and take these images in a movie film by using a movie camera controlled by computer. The other one is to use a coherent spatial light modulator [12], and processed images displayed in a CRT screen can be directly connected to the hologram synthesizer.

This computer-aided hologram synthesizing system brings a possibility of using the other kinds of medical images such as CT (computer-aided tomogram) images or NMR (nuclear magnetic resonance) images. These images are given as sectional or volume images of a patient, and original films for making holograms can be made by using a computer. An example of holograms synthesized from CT images is shown in the next section.

Synthesis of Multiplex Holograms from CT Images

CT images are sectional images along the body axis, and they are very suitable for making holograms because they are already enhanced by computer and obtained in digital data. However, these image are different from ordinary original images for hologram synthesis mentioned above, and images should be transformed into the ordinary original images.

A series of 93 CT images of the head of a patient with an interval of 2 mm along the body axis is given as original images. The object of interest is the anterior half of the skull of the patient, and only the skull is extracted by level slicing in the front half of tomograms. Then, projected images of the skull are calculated for every 0.86 degree of arc around the body axis with a shadow casting and interpolation in the interval space of tomograms [13]. After calculating 419 projected images for 360 degrees, these images are successively displayed in a CRT screen and recorded in a movie film by a movie camera controlled by computer. This movie film is used as an original film for synthesizing a hologram. Fig. 5 shows an example of reconstructed images of the hologram made in this way.

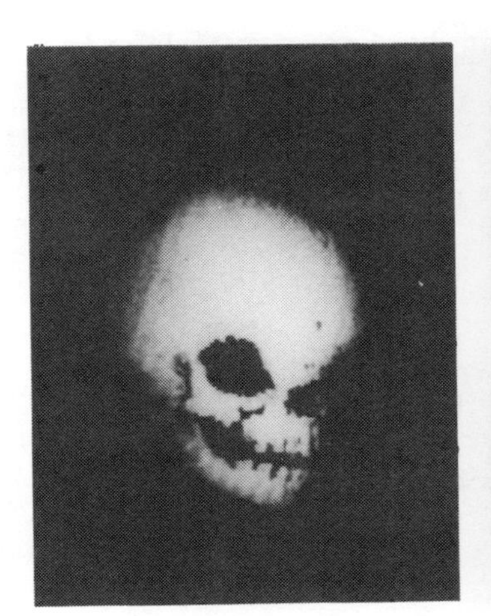

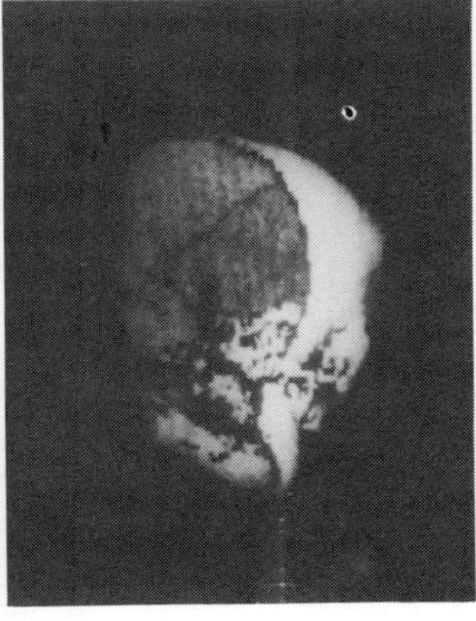

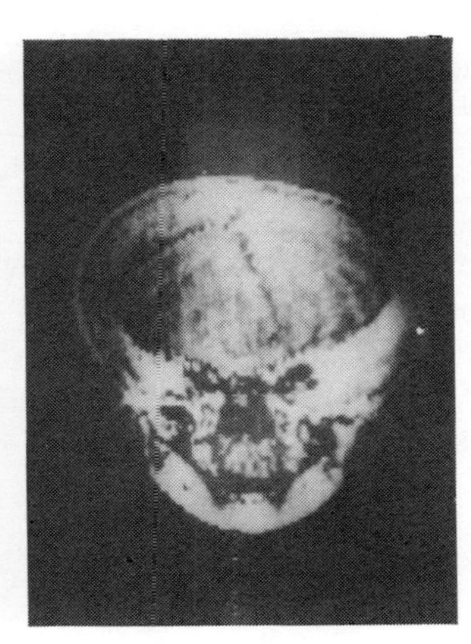

Fig. 5 Reconstructed images of a skull: from left to right, front view, side view and rear view.

Conclusion

Fundamental properties of the reconstructed image of a multiplex hologram, in particular distortions and resolution are reviewed, and methods for compensating distortions are proposed together with a typical condition of choosing the illuminating source for reconstruction.

Medical applications of multiplex hologram are very promising, and a computer-aided synthesizing technique is proposed to obtain better images and wider utilization. As an example of applications, the synthesis of multiplex hologram of a living skull from CT images is shown. Multiplex holograms made by such a system will be very useful to medical doctors for making better understanding and easy measurement of 3D structure of bones and internal organs and are possible to apply for the purpose of medical diagnosis, planning surgical operations, collecting samples of bones such as a skull, and education in medical school.

These studies have been continued in these several years in our laboratory in cooperation with Fuji Photo Optical Co. and Toppan Printing Co.. The author would like to express his thanks to Dr. M. Fujioka, M.D., Saitama Children's Medical Center, for providing CT images and for kind suggestions and advises from medical points of view. His thanks are also due to members of Japan Medical Holographic Display Society, presided by Prof. S. Ikeda, M.D., National Cancer Center, and Prof. S. Hashimoto, M.D., Keio University School of Medicine, for their kind cooperations and encouragements.

References

1. Benton, S. A., "Holographic Displays - A Review", Opt. Eng. vol. 14 (1975) 402
2. Huff, L. and Fusek, R., "Color Holographic Stereograms", Opt. Eng. vol. 19 (1980) 691
3. Tsujiuchi, J., Honda, T., Okada, K., Suzuki, M., Saito, T. and Iwata, F., "Conditions for Making and Reconstructing Multiplex Holograms", AIP Conf. Proc. No. 65 - Optics in Four Dimensions (1981) 594
4. Honda, T., Okada, K. and Tsujiuchi, J., "3-D Distortion of Observed Images Reconstructed from a Cylindrical Holographic Stereogram. (1) Laser Light Reconstruction Type", Opt. Commun. vol. 36 (1981) 11
5. Okada, K., Honda, T. and Tsujiuchi, J., "3-D Distortion of Observed Images Reconstructed from a Cylindrical Holographic Stereogram. (2) White Light Reconstruction Type", Opt. Commun. vol. 36 (1981) 17
6. Benton, S. A., "Distortions in Cylindrical Holographic Stereogram Images", J. Opt. Soc. Am. vol. 68 (1978) 1440A
7. Huff. L. and Loomis, J. A., "Three Dimensional Imaging with Holographic Stereograms", Proc. SPIE vol. 402 - Three Dimensional Imaging (1983) 38
8. Okada, K., Honda, T. and Tsujiuchi, J., "A Method of Distortion Compensation of Multiplex Holograms", Opt. Commun. vol. 48 (1983) 167
9. Okada, K., Honda, T. and Tsujiuchi, J., "Image Blur of Multiplex Holograms", Opt. Commun. vol. 41 (1982) 397
10. Tsujiuchi, J., Honda, T., Suzuki, M., Saito, T. and Iwata, F., "Synthesis of Multiplex Holograms and Their Application to Medical Objects", Proc. SPIE vol. 523 - Applications of Holography (1985) 33
11. Okada, K., Honda, T. and Tsujiuchi, J., "3-D Measurement by Using a Multiplex Hologram", Opt. Commun. vol. 45 (1983) 320
12. For example: Casasent, D., "Recyclable Input Devices and Spatial Filter Materials for Coherent Optical Processing", Laser Applications, vol. 3, Academic Press (1977) pp. 43
13. Ohyama, N., Minami, Y., Watanabe, A., Tsujiuchi, J. and Honda, T., "Multiplex Hologram of Skull Made of CT Images", Opt. Commun. (to be published)

Dynamic laser speckles and refractive measurements of the eye

Guojian Chen, Jiajun Lui*, Hong Tang

Department of Applied Physics, Hefei Polytechnical University, Anhui, China.

Abstract

A method for determining ametropia of the eye by using dynamic laser speckles is described in this paper, including its theoretical basis,optical system and results of clinical applications.

Introduction

When the highly coherent laser illuminates a rough translucent screen moving in a plane with constant velocity, the dynamic speckles are produced in the space behind the screen. the moving states of speckle spots are different to the different observers with various visual refractive defects of eyes when they view the speckle field.The refraction of the eye may be determined quantitatively according to the moving behavior of the speckles[1].

Theoretical basis

As a diffuse object, a model of the random phase screen is employed here. Dynamic speckles are investigated in the diffraction field with Fresnel diffraction approximation and in the image field with lens law for the image formation[2].

The moving behavior of speckles in the diffraction field

Figure 1 shows the coordinate system for describing the moving behavior of dynamic speckles produced in the diffraction field by the moving diffuse screen. When the diffuse screen illuminated by laser light is moving at constant velocity V in the object plane, the translation velocity Vt of dynamic speckles in the observation plane is given by[3]

$$Vt = V\left(1+\frac{A}{R}\right) \tag{1}$$

where A stands for the distance between the light source S and the diffuse screen P and R for the distance between the diffuse screen P and the observation plane Q.

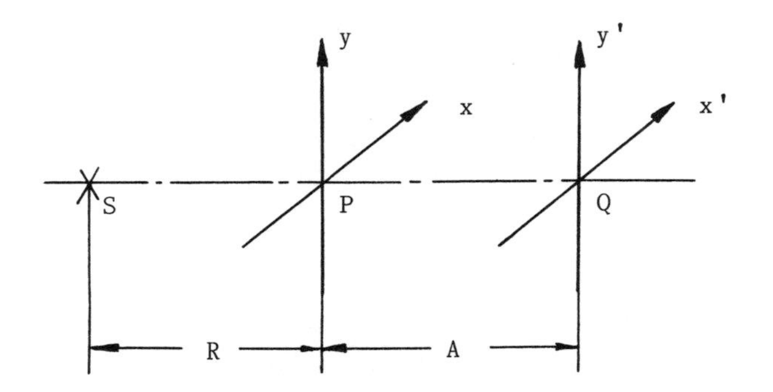

Figure 1. Coordinate system describing speckle movement in the diffraction field.

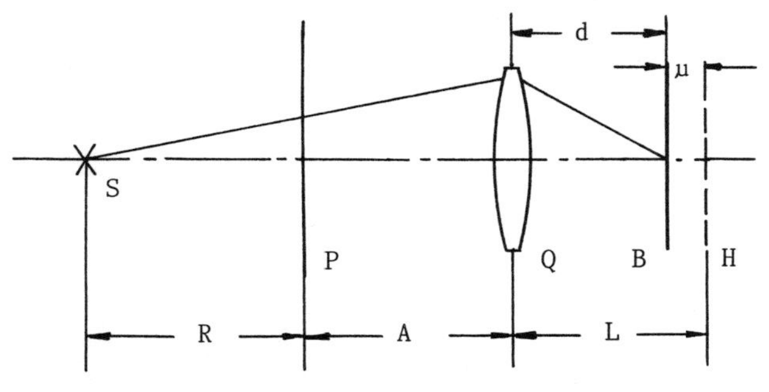

Figure 2. Schematical diagram describing speckle movement in the image field.

The moving behavior of apeckles in the image field

As shown in Fig. 2, dynamic speckles in the image field are investigated if at the position of the plane Q a lens or just the lens of the subjective eye in question is placed. The position H is the retina of the eye where the distance between the retina and the pupil of the eye is denoted by L. As the result of the analysis[4], one can conclude that the translation velocity Vt' of dynamic speckles on the retina surface is given by

$$Vt' = \left(1+\frac{A}{R}\right)\frac{\mu}{d}\,V \tag{2}$$

*Jiajun Lui is with Anhui Enginerring Institute.

where $L=\mu+d$, d is the distance from the pupil to the image plane B which is conjugate to the light source, and μ is the distance between the image B and the retina H. The significations of the signs A, R and V are the same as the above.

It is seen clearly from equation (2) that several extreme cases of interest which is the basis of this measurement should be noted.

1. In equation (2), if $\mu=0$, i.e. if the retina H is optically conjugate with the light source S, $Vt'=0$, thereby the speckles are regarded as being stationary. This situation is equevalent to the normal vision or the case in which the proper correction of the eye-sight has been obtained, as illustrated in Fig. 3.

2. When a divergin light illuminates the diffuse screen, namely, $R>0$, the translation velocity Vt' of speckles on the retina is determined by the equation(2). If $\mu<0$, that means if the image B of the light source is made in front of the retina H', the movement of speckles Vt' is in the opposite direction to the movement of the diffuser V. This situation is equivalent to the near-sighted state (with extra-refracting power); If $\mu>0$, this, if the image B of the light source is made behind the retina H", the direction of the speckle movement Vt' is the same as V. this situation is equivalent to the far-sighted state (with weak-refracting power).

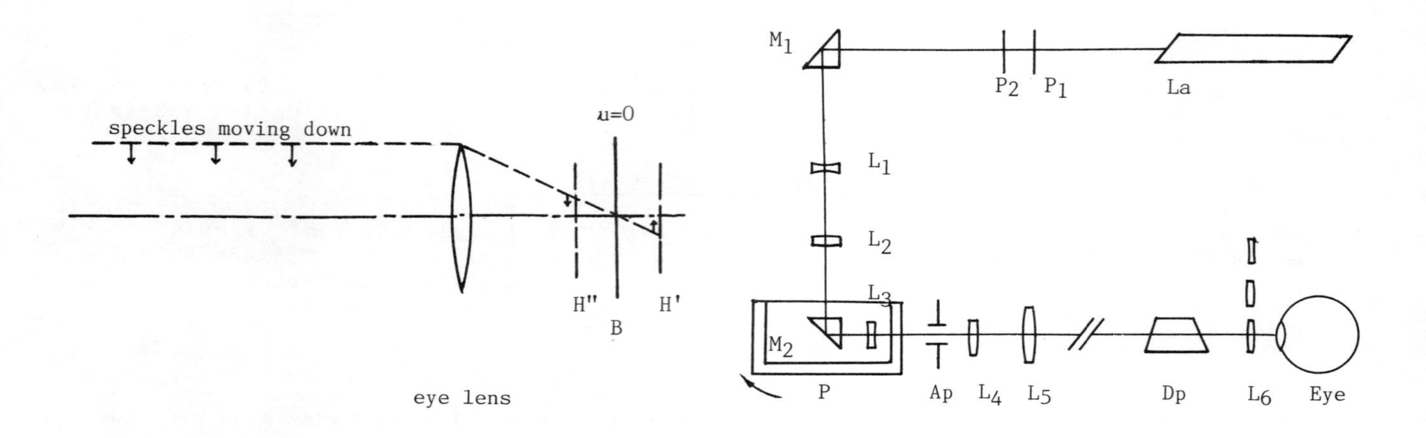

Figure 3. Diagram of the correspondence between speckle movement and ametropia of eyes.

Figure 4. Diagram of optical system of the speckle optometer.

It is noted that we must recall that the brain automatically interprets an inverted image on our retina to be an upright object. The same type reversal is also acomplished for a motion across the retina.

Optical system

The basic design of the speckle optometer is shown in Fig.4. La is the He-Ne laser with 0.5-1.0mW power output. P_1 and P_2 are the polarizers for changing the intensity of the laser. M_1 and M_2 are the reflecting prisms. The system of expanding the light beam consistes of lenses L_1, L_2 and L_3 in order to change the diverging angle of laser. The size of the light beam projected on the diffuse screen with homogeneous luminance is about 30 mm. The diffuse object is a 50 mm high frosted-glass cylinder with a 180 mm diameter which is driven by motor and rotated slowly. Ap is an aperture for constricting the range of the speckle pattern. The amplification system consistes of lenses L_4 and L_5, which has 2.2 ratio of amplification and 14° visual field angle. L_6 is a corrective group of lenses. The subject selects the proper lens in front of the eye while viewing dynamic speckles for correcting the ametropia itself. A rotating Dove prism Dp is placed also in front of the eye in order to determine the astigmatism, with which the movement of speckles may be made along the transverse axes at 0°, 60°, 120° and 90°.

Method and results

Method

1. For the measurement, a subject is positioned 3-5m away from the cylinder, and the luminance of the laser beam is maintained at a constant level under the threshold of the eye

safety. then, as the subject is looking at dynamic speckles, the corrective lenses with different focal lengths are sequentially put in front of the eye untile the apparent motion along the transverse axis at 0° or 90° spots or reverses. In other words, the required corrective power of the eye has been obtained when the speckle pattern is consisdered to stationary by the subject.

2. For the astigmatinsm, as suggested by Malacara[5], three measurements along the transverse axes at 0°,60°, and 120° are made in order to determine the following three parameters: a) spherical power, b) cylindrical power and c) cylindrical axis orientation. Let P_0, P_{60} and P_{120} denote spherical powers at ϕ=0°, 60° and 120° respectively as shown in Fig. 5. Then spherical power Ps, cylinder power Pc and cylindrical axis orientation θ can be calculated by means of three equations below:

$$\cos 2\theta = 1.155\left(\frac{P_{60}-P_0}{P_{120}-P_{60}}\right)+0.577 \quad (3)$$

$$P_c = 1.155\left(\frac{P_{120}-P_{60}}{\sin 2\theta}\right) \quad (4)$$

$$P_s = P_o - \left(\frac{1-\cos 2\theta}{2}\right)P_c \quad (5)$$

This calculating procedure should be performed by the special micro-computer developed by us to avoid a very tedious labor to do numerical calculations every time the measurements are made. Now, having P_0, P_{60} and P_{120} input, the parameters of the astigmatism Ps,Pc and θ are printed out by the computer.

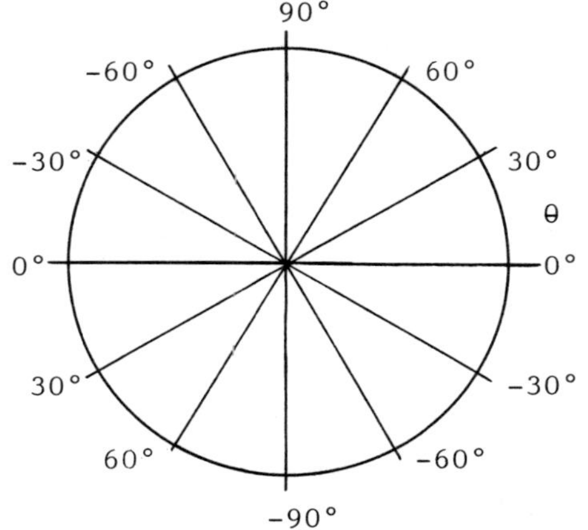

Figure 5. Possible cylinder axis orientation.

Results

According to the previous principle and design the instrument has been developed and experimently applied in some hospitals and opticians. Some of the refraction measurement results performed by ophthalmology of Anhui People's Hospitat are reported by comparing with the traditional optometry.

a) Spherical lenses. We define the deviation Ds=Psl-Pst where Psl denotes the measurement value of the spherical lens obtained by using the laser speckle method and Pst denotes the value obtained from traditional optometry to be called as objective method in which the eye muscle is in the state of relaxation by means of some medicine method. 37 eyes with ametropia are tested. The deviations Ds range as follows:
0—±0.13 diopters, for 19eyes(19/37≐48.7% of the total);
0—±0.25 diopters, for 9 eyes (66.7%)(including the above 19 cases);
0.25—0.5 diopters, for 12 eyes (30%).
Beside, one case is highly myopic (12 diopters). The deviation value comes up to 1 diopter.

b) Cylindrical lenses. Similarly, we have Dc=Pcl-Pct about the cylindrical power. 21 eyes with astigmatism have been tested. The Dc are presented in the following:
0—±0.13 diopters, for 9 eyes (9/21≐42.9%);
0—±0.25 diopters, for 14 eyes (66.7%)(including the above 9 cases);
0.25—0.73 diopter, for 7 eyes (33.3%).

c) Cylindrical axis orientation. In the same way, the deviation of the cylindrical axis orientation may be written as D_θ=θl-θt. The deviations D_θ of the same as above 21 eyes with astigmatism are given by :
0°—5° for 15 eyes (71.4%);
5°—14° for 5 eyes (24%).
Besides, in one case the deviation D_θ could be as high as 52° because of the mis-understanding of the subject.

Discussion

Although the optometry of the laser speckle is a subjective test, but its accuracy is higher than traditional sudjective test and resembles the objective.Because the speckle pattern has no definite plane of position. That means it exists at every point in space between the diffuse screen and observer's eye. The eye muscle quickly goes into an involuntary state of relaxation. In this state the eye is checked under ideal test condition. And,it is suitable not only for measuring the refractive errors of eyes in hospitals and in opticians, and also makes the mass vision monitoring possible.

References

1. Mohon, V., et al., "Laser Speckle for Determining Ametropia and Accommodation Respons of the Eye", Appl. Opt., Vol. 12, pp. 783-787. 1973.
2. Francon, M., Laser Speckle and Application in Optics, New York, 140. 1979.
3. Asakura, T., et al., "Dynamic Laser Speckles and Their Application to Velocity Measurements of the Diffuse Object", Appl. Phys., Vol. 25, pp. 179-194. 1981.
4. Ohzu, H., "The Application of Laser in Ophthalmology and Vision Research", Opt. Acta, Vol. 26, pp. 1089-1101. 1976.
5. Malacara, D., "Measurement of Visual Refractive Defects with a Gas Laser", Am. J. Optom. and Physiol., Vol. 51 pp. 15-23. 1974.

Mechanical reaction of human skull bones to external load examined by holographic interferometry

Halina Podbielska
Institute of Physics, Technical University of Wroclaw,
Wybrzeze Wyspianskiego 27, and
Medical Academy, Surg. Clinic, Lab. Experimental Surgery
and Biomaterials, Poniatowskiego 2, Wroclaw (Poland)

Gert von Bally
Medical Acoustics and Biophysics Laboratory, Ear, Nose and Throat Clinic,
University of Münster, Kardinal-von-Galen-Ring 10, D-4400 Münster,
(Federal Republic of Germany)

Henryk Kasprzak
Institute of Physics, Technical University of Wroclaw,
Wybrzeze Wyspianskiego 27, Wroclaw (Poland)

Abstract

Holographic interferometry was used to study the mechanical properties of human calvaria samples taken from macerated skulls. Deformations caused by static loading as well as their derivative functions are calculated from the recorded holographic interferograms. Elliptically shaped interference fringe patterns indicate the.generation of a shallow funnel around the point of force introduction. The lowest point of this impression is shifted sagittally in anterior direction with increased load. The load deformation curve is nonlinear before entering a linear range, demonstrating that in this experiment the calvaria can be regarded as a so called Kelvin body.

Introduction

Different methods of stress-strain analysis are commonly used in biomechanical research to study the mechanical properties of bones. These techniques show drawbacks such as restrictions to investigations on models, when using photoelastic techniques [1], to pointwise analysis, when using extensometers or strain gauges [2], or to structural models of stress distribution, when applying the finite element analysis method [3].

Holographic interferometry can be used to measure the vector displacement of each point of an opaque, diffusely reflecting object surface which undergoes a small deformation [4]. This technique allows a non-contactive, high-resolving deformation analysis on embalmed as well as fresh, wet or dry whole specimens. Examples of holographic applications in biomechanics and experimental orthopedics are presented in [5], including a preliminary holographic deformation analysis of the human skull base [6]. Results of further holographic interferometric investigations on the deformation of human calvaria samples under static load are presented in this paper.

Material and methods

For the investigations described here samples of calvaria were taken from five dry, macerated human skulls (Fig. 1).

As shown by Evans [7], differences in the mechanical properties and biomechanical behaviour of living and dead bones are neglegible. Minor changes in tensile and compressive strength characteristics as well as in modulus of elasticity and hardness of bones occur with increased drying. At low stress values these differences are also expected not to be significant.

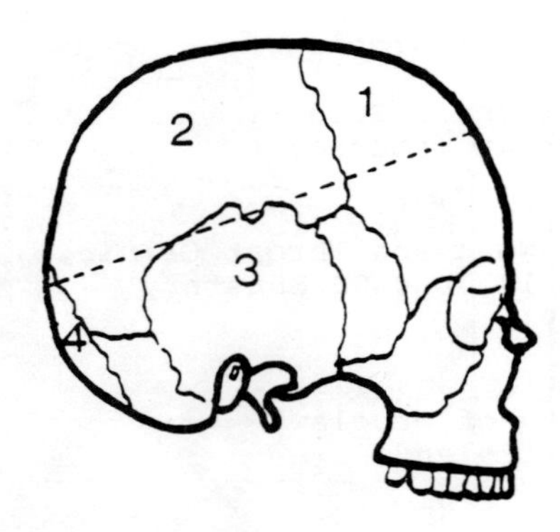

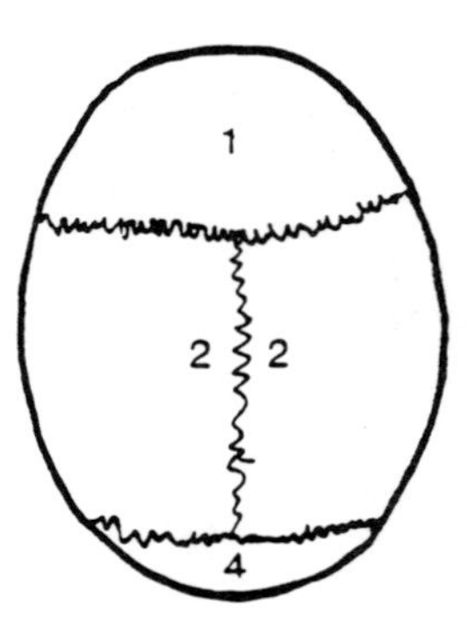

Figure 1. Schematic drawing of the human skull; (left) lateral view: (1) os frontale, (2) os parietale, (3) os temporale, (4) os occipitale; the calvaria samples were resected along the dotted line; (right) vertical view.

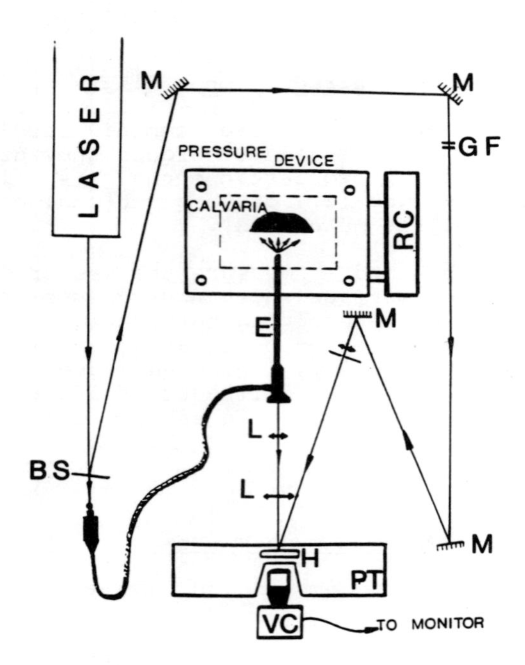

Figure 2. Experimental set-up: BS - beam splitter, E - endoscope, L - field and imaging lenses, H - hologramplane, PT - photothermoplastic film camera, VC - video camera, M - mirrors, GF - gray filter, RC - remote control system.

The calvaria samples were placed in a pressure device like a turned-over cup and fixed with their bases in a specially designed metal plate in order to ensure maximal stability of the object. Load was applied by a pneumatic piston to the highest point of the calvaria samples. The force was introduced perpendicular to the plane of the skull base. An electronic remote control device in combination with a piezo crystal pick-up allowed digital measurement of the actually acting force.

Double-exposure as well as real-time holographic interferometry was used to record the deformation W of the calvaria samples resulting from differential loads ΔF, ranging from 2 to 4 kp, after applying preloading forces F in the range of 0 to 120 kp. The optical set-up is outlined in Fig. 2. A 50 mW He-Ne laser (λ = 633 nm) was used as coherent light source. The inner surface of the calvaria was illuminated and imaged via an endoscopic system (laryngoscope with 90° viewing direction). The use of fiber optics and Hopkins optical systems for holographic endoscopy is described e.g. in [8, 9]. Holographic interferograms were recorded on a photothermoplastic film and displayed by a video system.

Experimental results

As an example Fig. 3 shows a holographic interferogram of the deformed inner surface of a human calveria sample. Deformation curves at different ΔF and F values were determined graphically a posteriori from video records of holographic interferograms (Figs. 4 and 5). First derivative functions of selected deformation curves are presented in Figs. 6 and 7. Such diagrams give informations about the symmetry of the mechanical reaction of the specimen under study to the applied force [10].

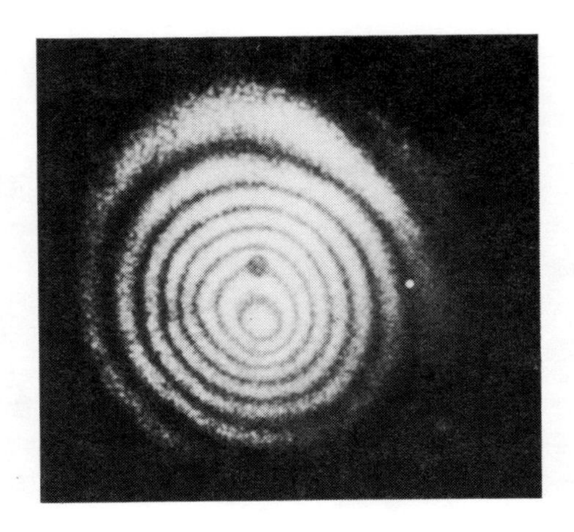

Figure 3. Sample of a holographic interferogram.

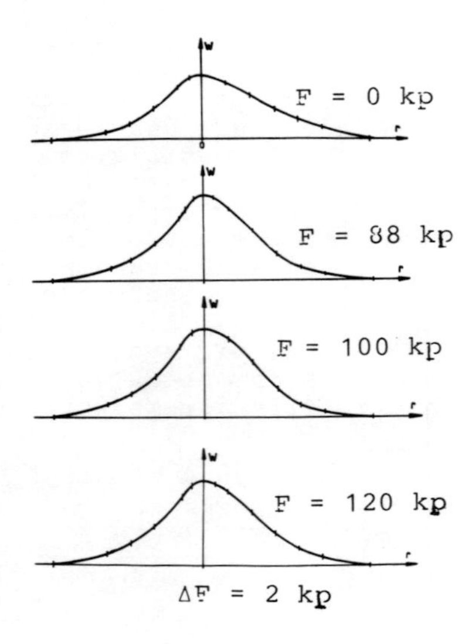

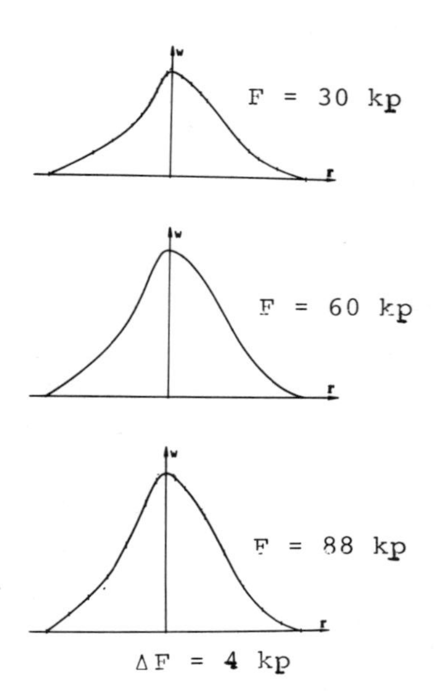

Figure 4. Deformation W as a function of location r along the medial saggital axis, for a differential load ΔF=2 kp at different preloading levels F.

Figure 5. Same sample as in Fig. 5, but here for a differential load ΔF=4 kp.

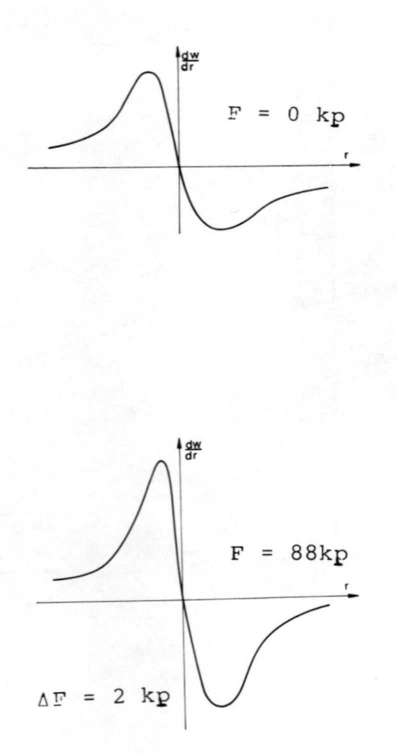

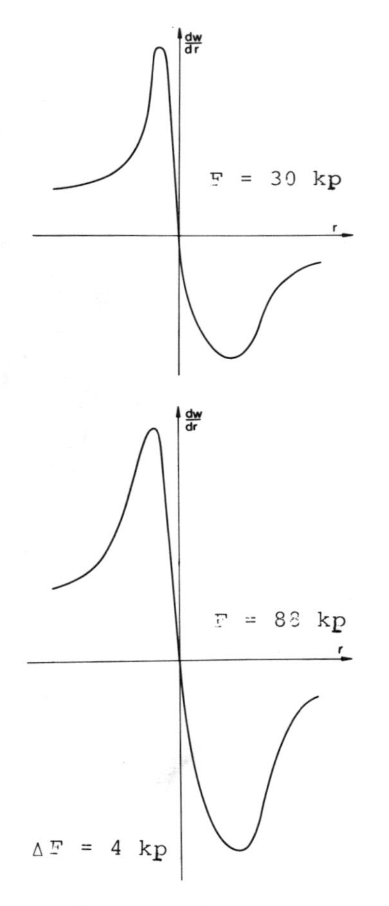

Figure 6. First derivatives of some selected deformation curves from Fig. 5.

Figure 7. First derivatives of some selected deformation curves from Fig. 6.

Conclusions

Holographic fringes caused by deformation of the inner surface follow the elliptical shape of the calvaria sample. These fringes are not concentric. Thus, the deformation functions are not symmetric which becomes especially evident in the presented derivative curves. This unsymmetry decreases with increasing force. These findings are similar to results of holographic investigations on the outer surface of human calvaria samples [11]. Near the area where the load is applied, a shallow funnel is formed and its minimum point is shifted with increasing load sagittally in anterior direction. Figure 8 demonstrates the dependence of the relative maximal deformation difference per 4 kp differential load $\Delta W/\Delta F$ on the applied preloading force F. This function is nonlinear up to about 80 kp, and approaches a constant value asymptotically with increasing preloading force.

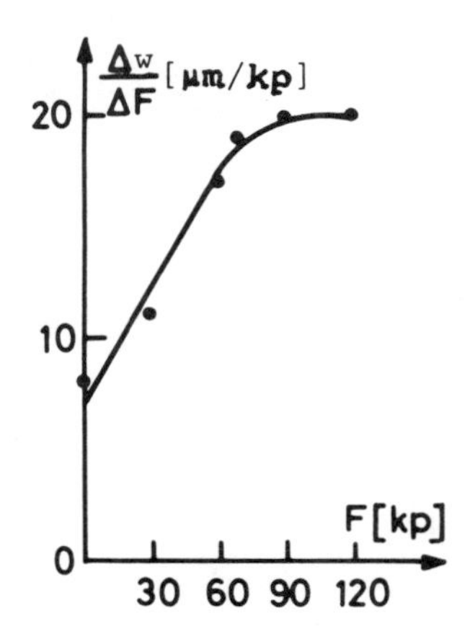

Figure 8. Dependence of the maximal deformation per 4 kp differential load $\Delta W/\Delta F$ on the applied preloading force F.

Figure 9 shows the preloading force vs. relative deformation curve as calculated qualitatively from the function in Fig. 8. The curve in Fig. 9 demonstrates a nonlinear relation between the preloading force F and the relative deformation W as long as the preloading force is kept within the linear range of the function shown in Fig. 8. For higher force values the functional dependence in Fig. 9 becomes linear. Thus, for the point of maximal deformation the response of the calvaria is that of a Kelvin body [12].

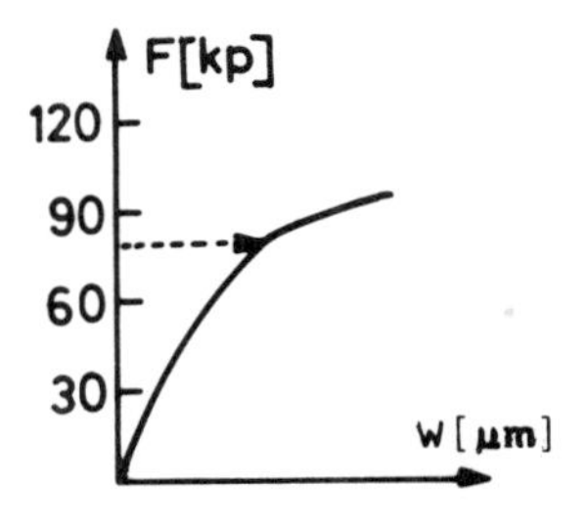

Figure 9. Preloading force F vs. relative deformation W curve, calculated from the function in Fig. 8; the arrow indicates the end of the nonlinear range.

References:

1. Kummer, B., Photoelastic studies on the functional structure of bone, Fol. Biotheor. 6, 31. 1966.
2. Pellcev, R., Saha, S., Stress wave propagetion in bone, J. of Biomech. 16, 481. 1983.
3. Huiskes, R., Chao, E.Y.J., A survey of finite element analysis in orthopedic biomechanics, J. of Biomech. 16, 385. 1983.
4. Vest, Ch., Holographic interferometry, John Wiley and Sohn. 1979.
5. v. Bally, G. (ed.), Holography in Medicine and Biology, Springer-Series in Optical Sciences Vol. 18, Springer-Verlag, Heidelberg, Berlin, New York. 1979.
6. v. Bally, G., Holography in otology, in: [5], p. 183. 1979.
7. Evans, F.G., Mechanical properties of bone, Charles Thomes, Publisher, Springfield. 1971.
8. v. Bally, G., Otoscopic investigation by holographic interferometry, in: G. v. Bally and P. Greguss (eds.): Optics in Biomedical Sciences, Springer-Series in Optical Sciences Vol. 31, Springer-Verlag Heidelberg, Berlin, New York, p. 110. 1982.
9. v. Bally, G., Brune, E., Mette, W., Holographic endoscopy with gradient-index optical imaging systems and optical fibers, Appl. Opt. (in press). Oct. 1986.
10. Kasprzak, H., Podbielska, H., Sultanova, N., Mechanical features of the human thigh bone investigated by means of holographic interferometry, Acta Polytechn. Scandinavia, Proc. Conf. Image Science ´85, 2, 198. 1985.
11. Podbielska, H., Kasprzak, H., Investigation of the human calvaria using the two-exposure holographic interferomtry, Abstracts European Photonics Workshop No. 9, Holopro ´85, 12. 1985.
12. Frenkel, V., Burstein, A., Orthopedic biomechanics, Lea and Febiger, Philadelphia. 1970.

HOLOGRAPHY IN BIOMEDICAL SCIENCES

G. von Bally
Laboratory of Biophysics, Institute of Experimental Audiology,
University of Muenster, Kardinal-von-Galen-Ring 10, D-4400 Muenster,
Federal Republic of Germany

Introduction

Today not only physicists and engineers but also biological and medical scientists are exploring the potentials of holographic methods in their special field of work. Most of the underlying physical principles such as coherence, interference, diffraction and polarization as well as general features of holography e.g. storage and retrieval of amplitude and phase of a wavefront, 3-d-imaging, large field of depth, redundant storage of information, spatial filtering, high-resolving, non-contactive, 3-d form and motion analysis are explained in detail in other contributions to this book. Therefore, this article is confined to the applications of holography in biomedical sciences. Because of the great number of contributions and the variety of applications [1,2,3,4,5,6,7,8] in this review the investigations can only be mentioned briefly and the survey has to be confined to some examples.

As in all fields of optics and laser metrology, a review of biomedical applications of holography would be incomplete if military developments and their utilization are not mentioned. As will be demonstrated by selected examples the increasing interlacing of science with the military does not stop at domains that traditionally are regarded as exclusively oriented to human welfare like biomedical research [9]. This fact is actually characterized and stressed by the expression "Star Wars Medicine", which becomes increasingly common as popular description for laser applications (including holography) in medicine [10]. Thus, the consequence - even in such highly specialized fields like biomedical applications of holography - have to be discussed.

Three-Dimensional Imaging

Biomedical sciences are involved in research on biological processes and their interactions in living organisms, so that the knowledge of the spatial structure - in particular the microstructure - of the object under consideration is of importance. Since three dimensional imaging and the additional advantage of a large field of depth are basic features of holography, this technique has been applied in ophthalmological research (fundus holography) [11,12,13,14] with the advantage that - contrary to conventional fundus photography - any layer of interest within the eye may be investigated in the reconstructed image. The possibility of white light reconstruction does not only provide an easy display e.g. in biological or medical teaching, but in combination with image plane holography and by using limiting apertures inherent in the investigated medical object a considerable speckle reduction as achievable [15].

A general problem in biomedical applications of holographic methods, particularly in clinical applications in-vivo without anesthesia, is the requirement of interferometric stability of the experimental set-up and the object during exposure. This can be solved satisfactorily only by utilizing a Q-switched ruby laser [16]. Using this technique safety requirements have to be followed carefully, especially in ophthalmological applications [11,12]. Due to the short exposure times the use of Q-switched ruby lasers allows holographic imaging of fast moving objects. Therefore, it can be applied e.g. to investigations of chemical reactions, the distribution and size of droplets, jets or the function of spray nozzles. Thus, in-line holographic arrangements have been used for the assessment of aerosols and pollution [17]. Such investigations are especially important e.g. for the development of sprays which should optimally penetrate to the lung like anti-asthmatic sprays or for optimizing the distribution of insecticides and pesticides in agriculture. But there have been also proposals for applications in the defoliation-actions in Vietnam [18].

Holographic Endoscopy

In order to investigate biological processes which take place in the interior of living organisms special optical devices so called endoscopes, have been developed. Conventional endoscopic photography suffers from the limited field of depth. Holographic methods may help to solve this problem. In developing a "holoendoscope" it has to be taken into account that external recording of holograms requires the reflected object wave to be led from the object space to the photographic plate by lenses and mirrors. This results particularly in phase aberrations and a small entrance pupil limiting the parallax. Therefore, an endoscope has been constructed including a holographic recording device inserted in the head of the instrument [19]. The use of mirrors and lenses within the shaft of the endoscope results necessarily in a rigid instrument. Thus, the introduction of fiber optics has considerably improved the performance of various endoscopes. In holographic endoscopy multimode fiber bundles can only be used for object beam guidance due to mode and phase distortions. Effects of image transmission in such a fiber bundle and movements of the object and the fiber bundle itself on the image quality, especially the interference fringe visibility (ref. chapter "Double Exposure Holography") have been investigated [20], demonstrating that the use of pulsed ruby lasers is possible. Using a multimode fiber bundle for the illuminating object beam and a monomode fiber for reference beam guidance, holographic endoscopy is feasible in spite of the limited power transmission capability of a single monomode fiber. Thus, the development of an easy to handle and flexible set-up for a hand-held holoendoscopic camera became possible [21], which proved to be usable for clinical applications (ref. chapter "Double Exposure Holography"). Investigations using a singlemode fiber in both the reference as well as object beam path and a multimode fiber bundle for image transmission are described in [22]. A multilayer fiber using a single mode core for reference beam transmission and a multimode cladding for object illumination can provide a very small holographic set-up in the tip of a flexible holographic endoscope [23]. Reduction of speckle noise caused by the small entrance pupil of endoscopic optics has been gained by using a gradient index rod lens as illumination and imaging guide, simultaneously. Parasitic light reflexes from the endfaces of the gradient index rod lens could be suppressed in the reconstructed image by appropriate shaping and holographic subtraction [24], even when delivering illuminating object and reference beam via multimode or single mode fibers, resp. [25].

Holography with Non-Visible waves

In addition to three-dimensional imaging of the surface of a biological object, 3-d display of its internal structure is also of interest. Due to the high absorption coefficient of biological materials, recording of holograms of internal structures using electromagnetical waves within the visible spectral region is usually not possible. On the other hand, certain non-visible waves are commonly used in biology and medicine because of their potentiality to penetrate tissue. Non-visible waves which can be generated coherently are e.g. micro-, infrared-, acoustical-waves and, recently, X-rays. Thus, there is the possibility of recording in-line and off-axis holograms, but with difficulties and unsolved problems in detection and transformation into the visible spectrum at reconstruction.

Microwave Holography

The capability of microwaves to penetrate optically opaque dielectrics can be used to locate internal anomalies by holographic methods [26]. Experimental results such as the detection of metallic objects like concealed weapons through clothes [27] may indicate potential biomedical applications.

Infrared Holography

As to the author's knowledge at the time there are only suggestions for the use of coherent infrared waves for holographic three-dimensional imaging of cancer of skin and breast [28]. The advantage would be contrast enhancement based on the different absorption of infrared radiation by normal and cancerous tissue.

Acoustical Holography

As far as biomedical applications of holography with non-visible waves are concerned most extensive investigations have been carried out in the field of ultrasonic holography. The interest in the use of ultrasonic waves is based on its capability to image soft tissue structures without - contrary to e.g. X-ray exposure - the risk of radiation damage. It is beyond the scope of this short survey to describe the different techniques for holographic ultrasonic imaging. Those particularly interested in biomedical applications of acoustical holography may consult comprehensive literature [28,29,30,31].

X-Ray Holography

Two-dimensional X-ray imaging is the most commonly used non-visible wave technique in medical diagnostic of internal structures within the human body. Thus, already for a long time it was the hope of researchers in the biomedical field to develop a three-dimensional X-ray imaging technique using holography. Such a method is expected to render possible recording of 3-d images of molecules, pinpointing cancer cells, and developing a safer and more effective X-ray therapy.

Basic ideas to overcome problems of finding an appropriate high resolving recording medium and a coherent X-ray source were proposed long ago [32]. However, an (expensive) technical realization was reached only after military interest was raised. Today X-ray lasers emitting at short wavelengths are developed for military purposes but information is not sufficiently available from the scientific literature [33], although detailed considerations on technical realization and expected results of e.g. X-ray biomicroholography are published [34,35].

Holographic Multiplexing

Holographic multiplexing can be a solution to the problem of three-dimensional imaging without a coherent X-ray source by combining holographically several two-dimensional X-ray pictures recorded from different views [36,37]. The advantage compared to conventional tomography is the possibility of analysing any layer of interest in the image reconstructed from the synthetic hologram, although it has been generated only by a limited number of radiographs. The type of recording process of the two-dimensional images used for holographic multiplexing is obviously of no importance. It could be an ultrasonic-B-scan record [28], electron micrograph [28], or simply two-dimensional photographs [38]. A semiautomatic combination of tomographic X-ray recording and multiple hologram display with the possibility of white light reconstruction is presently under study [39].

Holographic Microscopy

In studying microorganisms or microstructures of biological specimens microscopy is an important tool in biomedical sciences. Some properties of holography can be used with advantage in microscopic imaging, in particular the large field of depth and the two step principle of recording and reconstructing. This enables microscopic investigations to be carried out without preparing sections [40], or without focusing to a certain layer during recording e.g. in exobiology [41], or the method of analysis to be chosen afterwards (dark field, phase contrast, or interference microscopy) [42]. Basically there are two methods for holographic microscopic imaging. Firstly, there are so called holographic lensless techniques, which, however, lead to considerable aberrations caused by the enlargement of the expanding object and reference beam. Yet, pulsed ruby lasers can be applied since no cemented lens systems are used. Secondly, the already microscopically enlarged image can be used as object for holographic recording [28]. In this case the use of a groundglass leads to diffuse but speckled object illumination, while the use of a point source results in a non-uniform illumination. A good compromise seems to be found by using four light beams, entering the optical system with the aperture angle [43]. A regular interference pattern that remains superimposed on the reconstructed image may be eliminated e.g. by holographic spatial filtering techniques.

Holographic Spatial Filtering

The possibility of "a posteriori" image deblurring of photographs unintentionally blurred by motion, improper focusing, imperfect instruments etc., using a "holographic Fourier-transform division filter" [44] has also led to image improvements in electron microscopy [45,46]. A well known result of this application is an electron microscopic picture of the double-helical structure of a fd virus [47]. Holographic spatial filtering techniques can also be applied to non-coherent waves e.g. X- and Gamma-rays as used in radiology and nuclear medicine. Instead of taking serial two-dimensional radiographs for three-dimensional imaging by holographic multiplexing, the object can be projected simultaneously from different views, thus producing a coded image e.g. on a X-ray film.

In these methods - known as coded source and coded aperture techniques [36,48] - on-axis and off-axis Fresnel-zone plates, or discrete point distributions may be used for the coding process. Such coded aperture devices can be produced in form of holographic optical elements (HOE) [35], the first well known practical application of which was the use as "Head-Up Display" for supersonic fighter aircraft pilots [49,50]. If discrete point distributions are used, these may be an array of holes (Gamma-ray imaging), or a distribution of radiation sources (X-ray imaging). Using a non-redundant distribution of the sources to optimize the signal-to-noise ratio, decoding can be provided during reconstruction in laser light e.g. by means of a Fourier transform hologram of this distribution [51,52]. This latter technique ("flashing tomosynthesis") is under study with the aim of displaying layers of moving objects like the pulsating heart or fast flowing contrast media, among other objectives.

Pattern Recognition

The possibility to "recognize" wavefronts using holographic spatial filtering techniques gives rise to a variety of suggestions for biomedical applications of this feature of holography. Cell identification - especially differentiation between normal and cancerous cells - may be an important example for clinical use of holographic pattern recognition techniques [1,53]. Automatic recognition and counting of diatoms (algae) as a measure of water pollution is under study, using averaged filters [54]. Coherent optical filtering by means of light optical diffraction is used for pattern recognition and image evaluation of electronmicrographs of biological specimen [55]. Because of the great interindividual variety of the shape of biological specimens as well as the problem of orientation and size variance the generation of appropriate filters is sophisticated. New optical transforms particulary suited for scale, positional, and rotational invariant correlations without loss in signal-to-noise ratio can be applied advantageously to holographic pattern recognition [56].

Spatial filters can also be generated by digital computers [57]. Biomedical applications of this technique e.g. in image enhancement and pattern recognition are mentioned in [58,59].

Another important domain in biomedical research is the study of alterations of form and structure as well as movements of biological objects. Utilizing holographic methods this can be realized either by succesive recording of single holograms of the process under consideration (cineholography) or by means of holographic interferometry.

Cineholography

Cineholography has been used for microscopic investigations of living marine plankton organisms [60]. Holograms have been taken by stroboscopic illumination using a pulsed Argon laser synchronized to the recording sequence of a camera. Because of the large field of depth it was possible to investigate the microscopic subjects as they moved freely in the object space. A similar technique has been applied to studies of Bends decompression sickness of deep sea combat divers, which is caused by bubbles forming in the blood vessels [61,62].

Holographic Interferometry

As known by its applications to non-destructive testing, holographic interferometry provides the possibility of a three-dimensional, non-contactive, high resolving analysis of alterations either in shape, structure, and position of the object under test. These changes are characterized by interference fringes macroscopically visible in the reconstructed image. In the following, examples of biomedical applications are presented, classified by the most commonly used techniques of holographic interferometry.

Time-Averaged Holography

In order to generate a macroscopically visible interference fringe pattern by time-averaged holography the object has to move periodically. Therefore, this technique is used for vibration analysis. A biological object with periodic movements within the range of displacement resolution of holographic interferometry, is the tympanic membrane.

Thus, time-averaged holography has been used for the analysis of the vibration pattern of tympanic membranes in cats and human temporal bones to determine the role of the tympanic membrane in sound transmission by the middle ear [63,64]. Vibration analysis of the round window in cats [65] and of the human ossicular chain [66,67] has been carried out using this technique. Contrast enhancement of the interference fringes was achieved in the latter experiments by means of a phase modulated reference wave to shift high fringe orders to lower ones. In this way a small tilting movement of the stapes could be detected, besides the expected piston like oscillation. Phase modulation of the reference beam also renders possible phase mapping and increase in amplitude resolution to the order of 1 nm when using time-averaged holography which is otherwise insensitive to the phase of vibration.

Real-Time Holography

Biomedical applications of real-time holography are complicated by the requirement of precise repositioning of the reference hologram and the rapid changes of biological specimens, even in in-vitro experiments [68]. Therefore these investigations are restricted to experiments on models or objects, e.g. teeth or macerated bones, not suffering from uncontrollable alterations. Since there are possibly adverse influences on osteosynthesis by unphysiological load, e.g. after implantation of prostheses or fixation by plates after fractures, the mechanical properties of bones have to be known. Comparative holographic investigations have been carried out in real-time on the human femur in-vitro before and after implantation to optimize hip joint prostheses [69,70], as well as on the human tibia after fracture fixation by compression plates [71]. Real-time holography has been used to study the function of the human ankle joint and the leg-foot complex [72], as well as the thermal expansion of human teeth and dental materials [73,74]. Biological membranes may vibrate unsymmetrically around the resting position. Time-averaged holography cannot detect such unsymmetry of oscillations because it does not provide (vibration-)phase information. Real-time holography combined with synchronized stroboscopic illumination proved to be capable of investigating arbitrary vibrations by Fourier analysis and synthesis [75] by damping mechanically one half-wave of the oscillation on models of the tympanic membrane.

Double-Exposure Holography

Concerning its basic principle double-exposure holography can be regarded as part of holographic multiplexing, a technique used, among other things, for three-dimensional imaging, as mentioned previously. Thus, there is a capability of this technique, known as "contour mapping", related to three-dimensional imaging rather than to the analysis of vibrations, deformations, or structure changes. Isocontour lines, generated by superposition of two holograms of the same object on one photographic plate, can be used to measure the three-dimensional contour of an object. For that purpose each hologram has to be recorded e.g. with a different wavelength or slightly changed angle of the reference or object beam. The first method has been used to generate depth contour lines of the eye [11,12] and it is suggested to determine in this way the curvature of the sclera or front corneal surface for the production of well fitting contact lenses [68]. A combination of contour mapping and real-time holography is described to measure the wear of knee prostheses [76]. Similar techniques have been used to investigate the wear of hip joint prostheses, dental materials, and prosthetic mitral heart valves [77]. For the analysis of vibration and deformation double-exposure holograms are taken by time selective holographic recording of two phases of vibration or states of deformation, resp., on the same photographic plate. Concerning the use of CW-lasers the domain of biomedical applications is similar to that of real-time holography, e.g. in-vitro experiments on the biomechanics of the locomotor system, particularly the investigation of the function of the tibia/fibula system [72,78], the pelvis [79], studies on deformations under load of hip joint prostheses/femur systems [80], and human vertebrae [81,82,83]. A review on orthopedic applications of holographic interferometry can be found in [84]. Basic investigations of the mechanism of muscle fiber contraction have been carried out using double-exposure holography [85]. This holographic technique has also been used in experimental dentistry to investigate deformations of teeth, jaws, prosthodontic appliances, and skulls [86,87,88,89]. Dental applications of holographic interferometry are comprehensively reviewed e.g. in [90,91]. The possibility to measure the growth of seeds and plants by means of double-exposure holography has also been considered [92]. Using an extremely simple, easy to handle, and inexpensive set-up such an application of double-exposure holography can become helpful in on spot experiments for the enhancement of food production in third world countries [93]. Using a Q-switched ruby laser for the recording of double-exposure holograms - besides periodic vibrations - fast, non-periodic processes can be studied. This possibility was discovered firstly during military oriented investigations of bullets in flight using a ruby laser which accidentally emitted two short pulses. The reconstructed images showed fringe patterns according to the propagation of the shock wave [94]. An interesting application of this method, which has been used in biomedical research, is the study of transient processes. Thus, movements of tympanic membranes subjected to acoustic impulses have been investigated in in-vitro experiments on guinea pigs by superposition of a hologram recorded at rest and a second hologram taken on the same holographic plate at a certain time after the acoustic event [95]. The laser pulses were separated by a time interval of about one minute for technical reasons, which prevents in-vivo applications. The aim of these experiments was the study of generation of lesions of the tympanic membrane caused by acoustic impulses such as bursts emanating from weapons [96]. Releasing two laser pulses within one flashlamp pulse of a Q-switched ruby laser system the same technique has been applied to the same object but for the development of clinical diagnostics in audiology. As pathological changes of the mechanical properties of the middle ear have an influence on the vibratory pattern of the tympanic membrane, a vibration analysis may provide the possibility of a differential diagnosis of dysfunctions without opening the tympanic cavity. After model and in-vitro experiments, results of which have demonstrated the capability of double-pulsed holography to detect unsymmetric oscillations, a special closed acoustic system for simultaneous application of sound and holographic recording of the tympanic membrane vibration through the intact outer ear canal has been used on patients [97]. For clinical routine applications it would be desirable to have an easy to handle, flexible holographic-endoscopic arrangement. Thus, a small hand held holootoscopic camera has been developed using fiber optics [21] (ref. chapter "Holographic Endoscopy"). Successful in-vivo experiments to investigate motions of teeth and bridge-work to optimize the design of prosthodontic appliances have been carried out on patients, releasing the laser pulses at certain masticatory force levels [98,99]. Human chest motions have been investigated in-vivo during inhalation with the aim of lung diagnosis [100], as well as by triggering the laser pulses in relation to heart action to test the possibility for detection of heart diseases by double-pulsed holography [101]. In order to study the function of the human vocal organ holographic vibration analysis has been carried out in-vivo on the frontal part of the human neck [102].

Electronic Speckle Pattern Interferometry

Speckles usually regarded as an inevitable disadvantage in coherent optics can be used for holographic interferometrical purposes, e.g. in combination with videotechnical means, as in electronic speckle pattern interferometry (ESPI). Basically, this is an in-line, image plane holographic technique using a video target as recording medium, which renders possible a quasi real-time display of speckle interferograms according to the TV frame rate [103]. Examples of biomedical applications of this technique are vibration analysis of the human tympanic membrane [104], ossicular chain [66], basilar membrane [105], and of the human skull [106]. A review on this technique can be found in [107].

Conclusions

Although only some of the numerous examples have been mentioned, this review demonstrates that already early after the development of lasers, by which holography became a practical tool, the different holographic techniques have been used extensively in biomedical sciences. Today holography has established its place in biomedical research. Unfortunately, clinical applications are still rare and none of the holographic techniques has been used really routinely in clinical diagnostics, up to now. Thus, intensive interdisciplinary research activities between physicists, engineers, and biological and medical experts are still necessary in order to make available the advantages and potentials of holography for all day clinical practice.

On the other hand, the given examples of military applications demonstrate that the problems caused by an increasing interlacing of science with military interests are present even in such specialized fields like biomedical applications of holography, although this is commonly not anticipated. We see that nearly all techniques can be used as well in the civil as in the military domain. Not only the techniques themselves but also the results of their applications even in research fields commonly regarded as purely civil like biomedical sciences can be used for military purposes. In spite of a need for human welfare and medical care, which may have been expressed first, many technical developments - if not most - have been or will be used first in the military domain or exclusively developed for it. The unbalanced increase of military vs. civil research budgets results in the preference of developments for military use. In turn, the preference of military demands and military budgets leads to situations, in which scientists - although originally working with different intentions - propose already by themselves military applications of their ideas, in order to get the funds for its realization. Concentration of financial means and personel within giant military programs like SDI intensify such developments and lead to dependencies and constraints of science and scientists. Alarming signs for the threat of freedom of our science are the open political censorship on the presentation of non-classified papers in scientific meetings with the argument of "necessary" military security restrictions [108,109,110]. Unfortunately, we have to confess that we as individual scientists as well as members of our scientific societies have our share in this development by not clearly opposing these trends in principal and from the very beginning [9,111].

It is obvious that the results of scientific and engineering research are at least one important cause of the arms race. This competition for destructive power does not only distract scientific and engineering resources from contributing to the solution of the major social issues like hunger, overpopulation, insufficient education and medical care, but rather increases the threat, mankind has to face [112]. Thus, no scientist or engineer can deny his share of responsibility, including myself, since, as demonstrated in this review, even results of holographic investigations in the biomedical field can be and are used for military purposes.

I guess, it is high time to demonstrate clearly that we are willing and - as shown by many examples in this review - are able to work on what should be selfevidently the goal of science i.e. to form the basis for understanding of nature and for social development of mankind.

References

[1] Felleppa, E.J.: Biomedical applications of holography, Physics Today 22, 25 (1969).

[2] Leith, E.N. et al.: Reconstructed wavefronts and communication theory, J. Opt. Soc. Amer. 52, 1123 (1962).

[3] Greguss, P. (ed.): Holography in Medicine, IPC Science and Technol. Press (1975).

[4] Hoke, M. and G. von Bally (eds.): Proc. Symp. 1976 Spec. Res. Area 88 and Int. Conf. on Electrocochleography and Holography in Medicine, Muenster (1976).

[5] Marom, E., Friesem, A.A. and Wiener, E. (eds.): Proc. Int. Conf. Appl. Hol. and Opt. Data Process., Pergamon Press (1977).

[6] von Bally, G. (ed.): Holography in Medicine and Biology, Springer-Series in Optical Sciences, Springer-Verlag, Heidelberg, Berlin, New York, Vol. 18 (1979).

[7] Shankar, P.M. et. al.: Applications of Coherent Optics and Holography in Biomedical Engineering, IEEE Transactions on Biomedical Engineering 29, 8-15 (1982).

[8] von Bally, G. and P. Greguss (eds.): Optics in Biomedical Sciences, Springer-Series in Optical Sciences, Springer-Verlag, Heidelberg, Berlin, New York, Vol. 31 (1982).

[9] von Bally, G.: Remarks of the chairman: scientists, scientific societies, and military research, in: D. Vukicevic (ed.): Holographic Data Nondestructive Testing, SPIE 370, 26 (1983).

[10] Andrus, W.S.: Fibers and lasers: the "Star Wars" medical team, Photonics Spectra 71 (December 1985).

[11] Vaughan, K.D. et al.: Holography of the eye: a critical review, in: M.L. Wolbarsht (ed.): Laser applications in medicine and biology, Plenum Press (1974), 77 pp.

[12] Calkins, J.L.: Fundus camera holography, see [3], 85 pp.

[13] Tokuda, A.R. et al.: Development of a holocamera for 3-D microscopy of the unanesthetized human eye, J. Opt. Soc. Am. 68, 1382 (1978).

[14] Ohzu, H. et al.: Application of holography in opthalmology, see [6], pp. 133.

[15] Martin, C. et al.: Application of rainbow holography for speckle reduction in tympanic membrane interferometry, see [8], pp. 121.

[16] Ansley, D.A.: Techniques for pulsed laser holography of people, Appl. Opt., 9, 815 (1970).

[17] Bexon, R. et al.: In-line holography and the assessment of aerosols, Optics and Laser Technol. 8, 161 (1976).

[18] Bals, E.J.: The principles of and new developments in ultra low volume spraying, Proc. 5th Br. Insectic. Fungic. Conf. 189 (1969).

[19] Hadbawnik, D.: Holographische Endoskopie, Optik 45, 21 (1976).

[20] Yonemura, M. et al.: Endoscopic hologram interferometry using fiber optics, Appl. Opt. 20, 1664 (1981).

[21] von Bally, G.: Otoscopic investigations by holographic interferometry: a fiber endoscopic approach using a pulsed ruby laser system, see [8], pp. 110.

[22] Dudderar, T.D. et al.: Remote vibration measurement by time averaged holographic interferometry, Proc. Vth Int. Cong. Exp. Mech., Montral, 362 (1984).

[23] Raviv, G. et al.: Fiber optic beam delivery for endoscopic holography,

Opt. Com. 55, 261 (1985).
[24] von Bally, G. et al.: Gradient-index optical systems in holographic endoscopy, Appl. Opt. 23, 1725 (1984).
[25] von Bally, G. et al.: Holographic endoscopy with gradient-index optical imaging systems and optical fibers, Appl. Opt. 25, 3425 (1986).
[26] Tricoles, G. et al.: Microwave holography: applications and techniques, Proc. IEEE. 65, 108 (1977).
[27] Farhat, N.H. et al.: Millimeter wave imaging of concealed weapons, Proc. IEEE 59, 1383 (1971).
[28] Greguss, P.: Thoughts on the future of holography in biology and medicine, Optics and Laser Technol. 253 (1975).
[29] Proc. Int. Symp. Acoust. Holography, Plenum Press (1967) et seq.
[30] Greguss, P.: Optical evaluation of ultrasonic scattering in animal tissue, Ann. New York Acad. Sci. 267, 312 (1976).
[31] Hildebrand, B.P. et al.: An introduction to acoustical holography, Plenum Press (1972).
[32] Caulfield, H. et al.: The applications of holography Wiley-Interscience, New York (1970).
[33] New harmonic technique opens up extreme UV, Laser and Applications 40 (1983).
[34] Solem, J.G.: X-ray biomicroholography, Opt. Eng. 23, 193 (1984).
[35] Howells, M. et al.: Applications of holography to X-ray imaging, SPIE 523, 347 (1985).
[36] Groh, G.: Tomosynthesis and coded aperture imaging: new approaches to three-dimensional imaging in diagnostic radiography, Proc.R.Soc. Lond. B. 195, 299 (1977).
[37] Sugimura, K. et al.: Clinical application of multiplex holography, SPIE 370, 20 (1983).
[38] Tsujiuchi, J.: Holographic stereograms as a tool of nondestructive testing, SPIE 370, 17 (1983).
[39] Tsujiuchi, J.: Synthesis of multiplex holograms and their application to medical objects, SPIE 523, 33 (1985).
[40] Greguss, P.: Laser as a probe in biomedical research, in: Waidelich, W. (ed.): Laser 75 Optoelectronics Conference Prob., Munich (1975), pp. 155.
[41] van Ligten, R.F.: Holographic microscopy in exobiology, see [3], 44 pp.
[42] Ellis, G.: Holomicrography: transformation of image during reconstruction a posteriori, Science 154, 1195 (1966).
[43] Haendler, E. et al.: Contribution to experimental holographic microscopy, see [3], pp. 51.
[44] Stroke, G.W. et al.: Image improvement and three-dimensional reconstruction using holographic image processing, Proc. IEEE 65, 39 (1977).
[45] Stroke, G.W. et al.: Image improvement in high-resolution electron microscopy using holographic image deconvolution, Optik 41, 319 (1974).
[46] Reuber, E. et al.: Use of synthetic holograms in coherent image processing for high resolution micrographs of a conventional transmission electron microscope (CTEM), this issue (1987).
[47] Stroke, G.W.: Optical computing, IEEE Spec. 9, 24 (1972).
[48] Barett, H.H. et al.: Fresnel zone plate imaging in radiology and nuclear medicine, Opt. Eng. 12, 8 (1973).
[49] Close, D.H.: Hologram optics in head-up displays, Proc. Int. Symp. Information Display (SID), 58 (1974).
[50] Close, D.H.: Holographic optical elements, Opt. Eng. 14, 408 (1975).
[51] Weiss, H. et al.: Coded aperture imaging with X-rays (flashing tomosynthesis), Opt. Acta 24, 305 (1977).
[52] Jiang, Y.G. et al.: 3-d X-ray imaging using a new inverse filter technique - optical tomography, this issue (1987).
[53] Caulfield, H.J.: The applications of coherent optical image processing in medicine and biology, see [3], pp. 39.
[54] Almeida, S. et al.: Water pollution monitoring using matched spatial filtering, Appl. Opt. 15, 510 (1976).
[55] Boseck, S. et al.: Some progress in image evaluation by light optical diffraction and computer analysis for electron microscopy of biological Specimen, specially in virology, Proc. XVth Czech. Slov. Conf. Electr. Micr., Prague, 450 (1977).
[56] Casasent, D. et al.: New optical transforms for pattern recognition. Proc. IEEE 65, 77 (1977).
[57] Lohmann, A.W. et al.: Computer generated spatial filters for coherent optical data processing, Appl. Opt. 7, 651 (1968).
[58] Stroke, G.W. et al.: Holographic image restoration using Fourier spectrum analysis of blurred photographs in computer-aided synthesis of Wiener filters, Phys. Lett. 51A, 383 (1975).
[59] Huang, Th.S.: Computer holography and its possible applications to medical diagnosis, see [3], pp. 36.

[60] Knox, G. et al.: Holographic motion picture microscopy, Proc. Roy. Soc. Lond.B. 174, 115 (1969).
[61] van der Haagen, G.A.: Ein Mikroskop mit holographischer 16-mm-Filmaufzeichnung, Laser 2 (1970).
[62] Buckles, R.G. et al.: Holographic study of bubble dissolution in human plasma, SPIE 236, 185 (1981).
[63] Khanna, S.M. et al.: Tympanic membrane vibrations in cats studied by time-averaged holography, J. Acoust. Soc. Amer. 51, 1904 (1972).
[64] Tonndorf, J. et al.: Tympanic membrane vibrations in human cadaver ears studied by time-averaged holography, J. Acoust. Soc. Amer. 52, 1221 (1972).
[65] Khanna, S.M. et al.: The vibratory pattern of the round window in cats, J. Acoust. Soc. Amer. 50, 1475 (1971).
[66] Gundersen, T. et al.: Holographic vibration analysis of the ossicular chain, Acta Otolaryngol. 82, 16 (1976).
[67] Hogmoen, K. et al.: Holographic investigation of stapes foot plate measurements, Acustica 37, 198 (1977).
[68] Greguss, P.: Holographic interferometry in biomedical sciences, Optics and Laser Technol. 8, 153 (1976).
[69] Haeusler, G. et al.: Holograhische Deformationsmessungen zur Optimierung von Hueftgelenksimplantaten, see [4], pp. 349.
[70] Hanser, U.: Anwendung der holographischen Interferometrie in der experimentellen Orthopaedie, see [4], pp. 343.
[71] Hardinge, K. et al.: A preliminary study of fracture fixation using holographic interferometry, see [4], pp. 307.
[72] Vukicevic, D. et al.: Holographic investigation of mechanical characteristics of the complex leg-foot in conditions of lesion and reconstruction, see [6], pp. 34.
[73] Kinder, J. et al.: Holographische Untersuchungen des thermischen Verhaltens von Schmelz, Dentin und ausgewaehlten Dentalstoffen, see [4], pp. 301.
[74] Zuquan, D. et al.: A new method of 3-d quantitative analysis of holographic interferometry in applications of solid biomechanics and vibration, this issue (1987).
[75] Sieger, C. et al.: Measurement of vibration waveforms using temporally modulated holography, see [6], pp. 247.
[76] Atkinson, J.T. et al.: Measurement of the area of real contact between, and wear of, articulating surfaces using holographic interferometry, see [4], pp. 289.
[77] Lalor, M. et al.: Holographic studies of wear in implant materials and devices, see [6], pp. 20.
[78] Wagner, J. et al.: Application de l interferometrie holographique a letude du complexe tibio-peronier charge axialement, Acta Orthop. Belgica 41, 24 (1975).
[79] Vukicevic, D. et al.: Holographic investigations of the human pelvis, see [8], pp. 138.
[80] Hanser, U.: Quantitative evaluation of holographic deformation investigations in experimental orthopedics, see [6], pp. 27.
[81] Wesendahl, Th. et al.: Untersuchung des Verformungsverhaltens menschlicher Wirbelkoerper mittels holographischer Interferometrie, Laser u. Elektrooptik 1, 37 (1977).
[82] Piwernetz, K. et al.: Elastomechanical properties of trabecular bone from the human vertebral body, see [6], pp. 15.
[83] Matusomoto et al.: Deformation measurement of lumbar vertebra by holographic interferometry, this issue (1987).
[84] Piwernetz, K. et al.: Holography in orthopedics, see [6], pp. 7.
[85] Sharnoff, M. et al.: Holographic study of contraction in striated muscle fibers, Proc. 35th Ann. Conf. Engineering in Medicine and Biology 24, 252 (1982).
[86] Pryputniewicz, R. et al.: Determination of arbitrary tooth displacements, J. Dent. Res. 57, 663 (1978).
[87] Dirtoft, I.: Holographic measurement of deformation in complete upper dentures - clinical application, see [8], pp. 100.
[88] Pavlin, P. et al.: Strain distribution in the facial skeleton arising from orthodontic appliance activity, see [6], pp. 177.
[89] Podbielska, H. et al.: Mechanical reaction of human skull bones to external load examined by holographic interferometry, this issue (1987).
[90] Bjelkhagen, H.: Holography in dentistry, see [6], pp. 157.
[91] Dirtoft, I.: Dental Holography, SPIE 370, 108 (1983).
[92] Hinsch, K.: Coherent optics in enviromental monitoring, this issue (1987).
[93] Lunazzi, J. et al.: A simple set-up for using holographic interferometry in studies on seeds, see [6], pp. 77.
[94] Brooks, R.E. et al.: (9A9) Pulsed laser holograms, IEEE QE-2, 275 (1966).
[95] Dancer, A.L. et al.: Holographic interferometry applied to the investigation of tympanic-membrane displacements in guinea pig ears subjected to acoustic impulses, J. Acoust. Soc. Amer. 58, 223 (1975).
[96] Smigielski, P. et al.: Application de l interferometrie holographique a l etude des deformations du tympan du cobay sous l effet de bruits de duree breve,

Nouv. Rev. Optique 6, 49 (1975).
[97] von Bally, G.: Otological investigations in living man using holographic interferometry, see [6], pp. 198.
[98] Wedendal, P. et al.: Holography in dentistry, in: M.L. Wolbarsht (ed.), Laser applications in medicine and biology, Plenum Press, (1977) pp. 221.
[99] Pryputniewicz, R.: Holographic determination of rigid body motions, and application of the method to orthodontics, Appl. Opt. 18, 1442 (1979).
[100] Zivi, S.M. et al.: Chest motion visualized by holographic interferometry, Med. Res. Eng. 9, 5 (1970).
[101] Bjelkhagen, H.: Development of hologram interferometry, in particular pulsed sandwich holography, for engineering uses as well as applications within medicine and odontology, Dissertation, Stockholm (1978).
[102] Pawluczyk, R. et al.: Holographic vibration analysis of the frontal part of the human neck during singing, see [8], pp. 131.
[103] Løkberg, O.: Speckle techniques for use in biology and medicine, see [8], pp. 144.
[104] Løkberg, O. et al.: Use of ESPI to measure the vibration of the human eardrum in-vivo and other biological movements, see [6], pp. 212.
[105] Løkberg, O. et al.: Bio-medical applications of ESPI, see [8], pp. 154.
[106] von Bally, G. et al.: Potentials of holographic vibration analysis of the human skull, Arch. Otorhinolaryngol. Suppl. II, 133 (1984).
[107] Løkberg, O.: Electronic speckle pattern interferometry, this issue (1987).
[108] Marshak, R.E.: The peril of curbing scientific freedom, Physics Today, 192 (Jan. 1984).
[109] Hecht, J.: The American laser scene, Laser und Optoelektronik 2, 84 (1985).
[110] Lytle, D.: Barring entry: the societies say no, Photonics Spectra, 50 (Oct. 1985).
[111] von Bally, G.: Holography and the freedom of science - a welcome address to ICHA 86 -, this issue (1987)
[112] Statement of European Physicists, Europhysics News 13, 2 (1982).

HOLOGRAPHY AND THE FREEDOM OF SCIENCE
- A Welcome Address to ICHA´86 -

Gert von Bally
Medical Acoustics and Biophysics Laboratory,
Ear-Nose-Throat-Clinic,
University of Münster, Kardinal-von-Galen-Ring 10,
D-4400 Münster, Federal Republic of Germany

It was a great honour for me to assist in preparing this International Conference on Holographic Applications for that part of the world where the idea of holography was created by Dennis Gabor in the fourties - that is Europe. Although my function as a chairman of the regional European program committee was combined with some additional work, it was always a pleasure for me to contribute to an effort to bring us "holographers" together in a country most of us, who are foreign here, have not yet been before, and in this way spreading the exchange of experience and knowledge to our mutual advantage also in this part of the world.

As scientists we have the great chance to overcome borders and restrictions of ideology and other constraints, and to meet openly and with friendship. The name of the place in Beijing where ICHA´86 was hold - Friendship Hotel - may act as a symbol in this sense. But on the other hand this puts a burden on our shoulders, that is to work on keeping this independance and freedom of science.

If we look on our scientific child or - for the younger ones among us - already our scientific mother - I mean holography, we cannot avoid recognizing that we did not only always use ways to keep this independance and freedom. I mean the deep envolvement of holographic technolgies in the - to my feelings - darkest point of human thinking - that is the development of mutual military deterrence, which threatens mankind and separates people, including ourselves.

We have to confess that this is true from the very beginning of practical applications of holography. Who takes the time to read not only the scientific text of publications in our field but also pays attention to where and what for these investigations were undertaken and by whom they were paid, can easily see that this is still an increasingly growing development.

Also in this meeting, if we want to or not, we have to realize that the ambiguity of applications of the most modern developments in our field is demonstrated.

Thus, I feel we are approaching a cross road, where we have to decide what the future way of our science will be. I hope, we will demonstrate with this conference clearly, what great potentials our work can offer to the benefit and social development of mankind. Thus, I also hope, that our decision for the future of holographic applications will go clearly in a non-destructive direction.

In this sense I wish - not only the organizers of this conference - but all of us success.

The study of optometry apparatus of laser speckles

Wang Bao-cheng, Yao Kun, Wu Xiu-qing

Department of Radio & Electronics, Department of Physics of University of Science and Technology of China

Long Chang-ying, Shi Jia-qi, Shi Shi-zhong

Department of Physics of Anhui University

Abstract

Based on the regularity of laser speckles movenent the method of exam the uncorrected eyes is determined. The apparatus with micro-computer and optical transformation is made. Its practical function is excellent.

Introduction

It has been twenty years past since the first attempt to utilize the laser speckle pattern in eye refraction measurement by Knoll. Over about ten years a compact automated fefractometer had been constructed by H.Ohzu and K.Ukai. Because the problem is interesting we analyzed the dynamic behaviour of speckles in the human eyes and succeeded to make an useful optometry apparatus of laser speckles with micro-computer. Having tested the apparatus, we confirmed its actual function is excellent.

Principle

Many investigators have described the regulations of the dynamic behaviour of laser speckles in the human eyes. According to the space speckle movement theory we can get the following formula:

$$v = \left(1 + \frac{A - L}{R} \right) V$$

Where v is the apparent velocity of the laser speckle pattern, V is the velocity of a moving rough surface. R, A and L are the distances as shown in Figure.1. S, S' and P, P' are two pairs of conjugate points.

Thereby we learn:

1. When A = L i.e. the eye is focused onto rough surface, the observer sees the speckles movement just like the movement of the rough surface with the same velocity i.e. v = V.
2. When A = L the observer sees the speckles movement is different from the movement of the diffuser surface. The ratio of v to V is changed.
3. When P and S are superpised with each other, even though the diffuser is moving the speckles pattern appears to be stationary. This stationary state is called a 'boiling' state. i.e. v = 0.

Design of the apparatus

The fundamental design is shown in Figure.2. The diverging coherent beam from the point source S_1 is focused onto the point S through the lens L_1. G is a ground glass cylinder rotating slowly by motor T_1. The W is adjustable and used to convert the grain speckles into strip-speckles. The P_1 is a conbinative prism. It is used to divibed the strip speckles into two parts moving in opposite directions. The mirrors M_1 and M_2 are used to return the light and shorten the length of the instrument. The mirrors M_3 and M_4 can be moved on a slideway in a distance about 800 mm so that the conjugate point S_2 of the point S_1 can move along the optical axis. The lens L_2 and lens L_3 are used to change the exiting beam's deverging power. When the two mirrors M_3 and M_4 are forced to move to where the speckles seem to be stationary by the patient, the eye refraction can be calculated by mirrors' position. The P_2 is another Dove prism used to test an astigmatic eye by rotating its range by a gear wheel and motor T_2. We use three methods to determine an astigmatic eye. These are two, three and six-meridian methods, as shown in Figure.3.

Clinical experiments

In the clinical use of this instrument, a subject would be press the switch to move the mirrors and rotate the Dove prism until he judges the speckles pattern to be stationary. Moving speed and the range of the prism and all the procedures including the rests output are controlled automatically by the micro-computer.

The apparatus with computer used practical test are named as Model JSB-1. We have made about 300 trials. It was proved that the accuracy of the measurement was councide well with the method of paralysing eye strings. The indexes of this instrument as shown in Table.1.

Table.1. The indexes of the optometer JSB-1

	Range	Accuracy	The minimum resolved accuracy
Sphrer	-20D +20D	0.25D	0.0213D
Cylinder	-8D +8D	0.25D	0.0213D
Axis	0 180	5	1

Special characteristic of this instrument

1. The near point, far point and the range of accommodation are determined by this apparatus, so we can give the diopter of an old eye or the diopter of correction of a youth myopia. The far point is a position of mirrors M and M when the subject observes the strip-like-speckles moving toward left just turn to be 'boiling' state (see the lower part of the speckles pattern). The near point is a position of mirrors when the one observes the strip just moving toward right. The distance between the two points is the range of accommodation. We found the false myopia, the accommodation convulsion and the decay of accommodation can be indicated by this instrument.

2. The method of this apparatus is a subjective method which can be used in the situation that the objective method is not capable. For example the patient eye have nebulas or opcity.

Acknowledgements

This work was supported by Anhui Committee of Science & Technology. The authors wish to thank Z.J.Wang, X.P.Wu and J.P.Xie for valuble help and discussions during the alignment and test.

References

1. H.Ohzu & K.Ukai, "Dynamic laser speckle pattern used to determine eye refraction," Proc. of ICC-11 Conf, Madrid, Spain, 1978.
2. X.P.Wu, et al., "Movement of space speckle," Anta. Phys. Sinica, Vol. 29, 1142.1979.

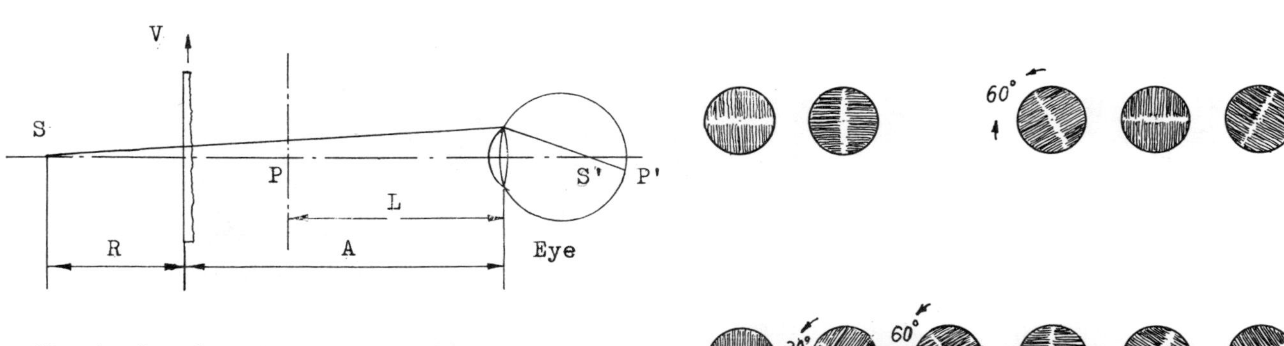

Fig.1. The brief principle diagram

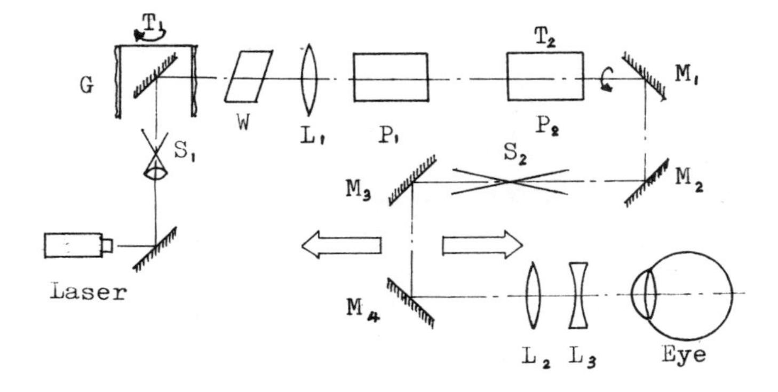

Fig.2. The fundamentals design

Fig.3. The methods to exam an astigmatic eye.

Deformation measurement of lumbar vertebra by holographic interferometry

Toshiro Matsumoto*, Arata Kojima**, Ryoukei Ogawa**, Koichi Iwata***, Ryo Nagata***

* Department of Mechanical Engineering, Osaka Prefectural Technical College
26-12, Saiwai-cho, Neyagawa-shi, Osaka 572 Japan

** Department of Orthopedic Surgery, Kansai Medical University
1, Humizono-cho, Moriguchi-shi, Osaka 570 Japan

*** Department of Mechanical Engineering, College of Engineering, University of Osaka Prefecture
Mozuume-cho, Sakai-shi, Osaka 591 Japan

Abstract

The mechanical properties of normal lumbar vertebra and one with the interarticular part cut off to simulate hemi-spondylolysis were measured by the double exposure holographic interferometry. In the normal lumbar vertebra, displacement due to the load applied to the inferior articular process was greater than that of superior articular process under the same load. The interarticular part was subjected to the high stress. From these points, one of the valuable data to consider the cause of spondylolysis was obtained.

Introduction

Since the vertebral column is very important to support the trunk and protect the neuraxis in the human musculoskeletal system, a quantitative consideration of its mechanics should be useful in understanding clinical problems. In clinical orthopedics, it is often found that the interarticular part of vertebra is fractured in young men playing sports such as weight lifting and football. This condition is known as spondylolysis.[1,2] However, the process by which this part is fractured has not been clarified.[3]

Holographic interferometry[4] has been shown to be useful for measuring the deformations related to the hard tissues of human body.[5-7] The purpose of this study was to examine the mechanical behavior of human lumbar vertebra by double exposure holographic interferometry. First, measurements of deformation of the dried human normal lumbar vertebra and the bone with interarticular part cut off to simulate hemi-spondylolysis subjected to the different loads were carried out. Next, the second derivative of displacement was calculated to examine the cause of spondylolysis. Last, the deformations of dried bone were compared with those of cadaverous bone.

Experiment

Specimens used in the experiments were dried human normal lumbar vertebrae and those with the right interarticular part cut off, as observed in the vertebra with spondylolysis. In addition, the cadaverous bone preserved in formalin was used for comparison with the deformations of the dry bones. The anterior part of lumbar vertebra was embedded in the super hard plaster to fix the specimen. The plaster was held tightly in a precision vice.

The methods for applying the lateral and posterior loads on the ends of superior or inferior articular processes are shown in Fig.1(a) and (b), respectively. One end of a string was fixed on the end of articular process. The other end was passed through the pulley and was pulled by the weight as shown in the figure.

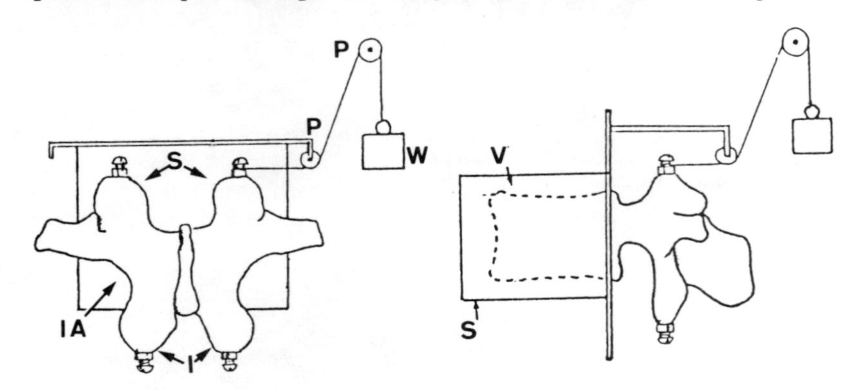

(a)Posterior view (b)Left side view

Figure 1. Method to apply the load to the end of lumbar vertebra. (a) and (b) show the lateral and posterior loads, respectively. S and I:superior and inferior articular processes, respectively. W:weight, V:vertebral body, S:super hard plaster, P:pulley, IA:interarticular part.

Figure 2 shows schematically the optical system used in the experiment. The light source is a 50 mw He-Ne laser. The light is divided into reflected and transparent waves by a half mirror. The former illuminated directly the holographic plate as the reference wave. After the latter was collimated by the lens system, it illuminated the posterior part of the specimen. The light wave reflected on the surface of the object arrives at the plate placed at the focal plane of a Fourier transform lens. Adjacent interference fringes obtained in the system correspond to the displacement to the out of plane of 0.3 μm. The specimen was loaded at predetermined place and direction between the first and second exposures. After the holographic plate was developed, it was returned to its original position. The reconstruction image with interference fringes could be observed in the image plane through the hologram illuminated by the reference beam.

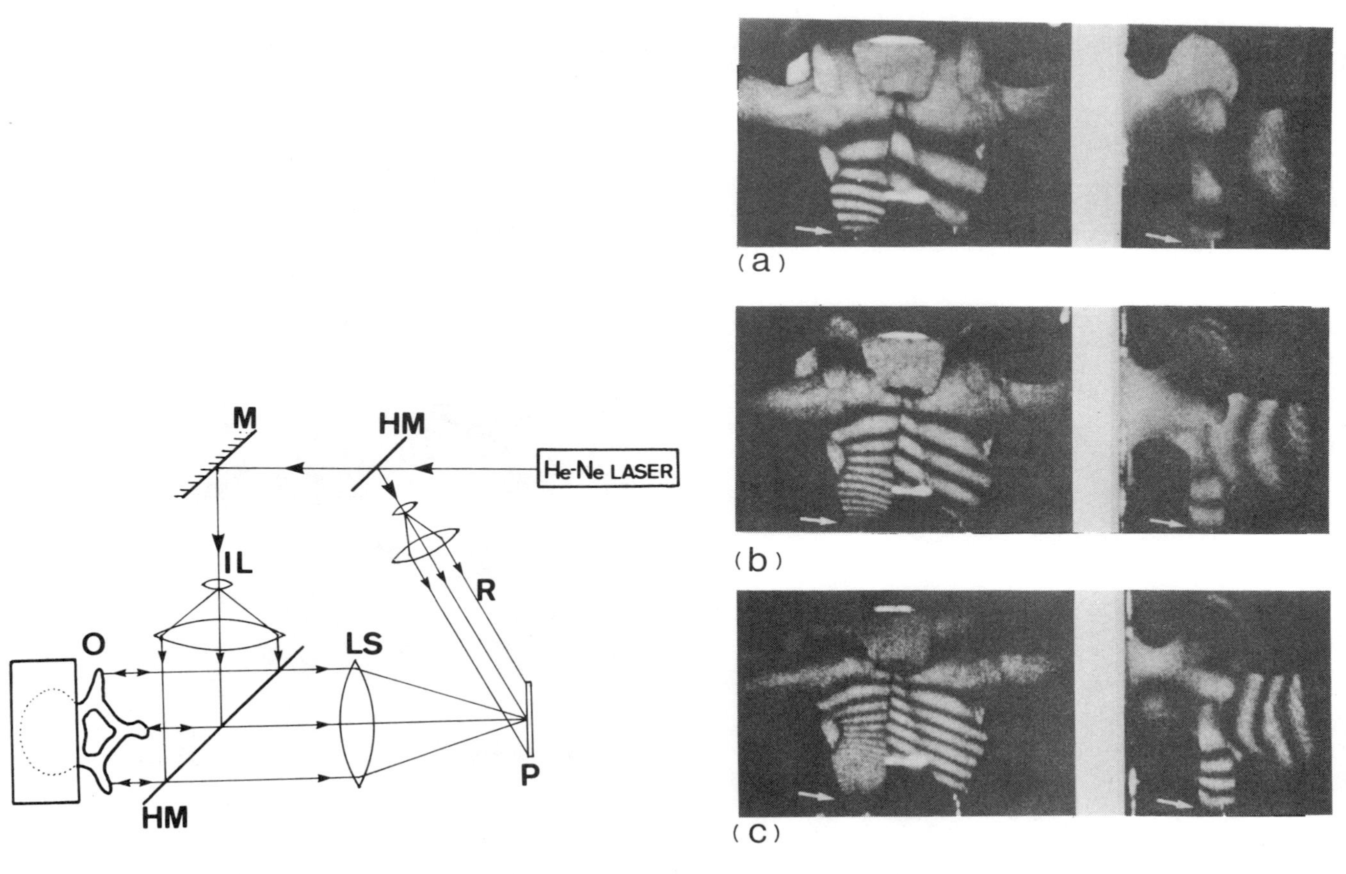

O:posterior portion of lumbar vertebra,
M:mirror, P:photographic plate
IL:illuminating wave, R:reference wave
HM:half mirror, LS:lens

Figure 2. Schematic diagram of the optical system.

Posterior view left lateral side view

Figure 3. Reconstruction images obtained when the load applied to the end of left inferior articular process(as shown with arrow) toward the posterior direction. (a)50gf-->100gf, (b)50gf-->150gf, (c)50gf-->200gf.

Results and discussions

Figure 3 shows the reconstruction images obtained for the normal lumbar vertebra. The magnitude of preloading was 50gf. Before the second exposure was carried out (Fig. 3(a)), a 100gf weight was loaded to the end of left inferior articular process of the lumbar vertebra in the posterior direction. It is found that the density of fringes on the end of left inferior articular process subjected to the load was greater than that on the end of left superior articular process and displacement of the end of articular process was proportional to the increased load. Both ends of superior articular process were displaced slightly in the anterior direction. In the reconstruction images of left lateral side view, the end of inferior articular process was subjected to the load in the posterior direction. However, as shown in Fig. 3, the posterior part was displaced slightly at right angles to the posterior directions.

To consider how the left interarticular part is displaced in the posterior direction for the bone in Fig.3(c), we express the length between superior and inferior articular processes by X and the displacement by Y in Fig.4. The curve-fitting operation was carried out by a personal computer using the measured results, and a good approximation was obtained as a eighth-degree polynomial, Y(x) as shown in the figure. As indicated in the figure, the displacement increased gradually in a nonlinear pattern with the increase in length. To determine where W(x) has a maximum or minimum value, we calculated the second derivative of Y(x) with respect to X. We found that the maximums of the second derivative occur at the interarticular part and the end of inferior articular process. Therefore,these places are under high stresses and weak mechanically. The results obtained in the former part are of value in considering the factors involved in spondylolysis and those obtained in the latter part are reasonable because the end is tapered.

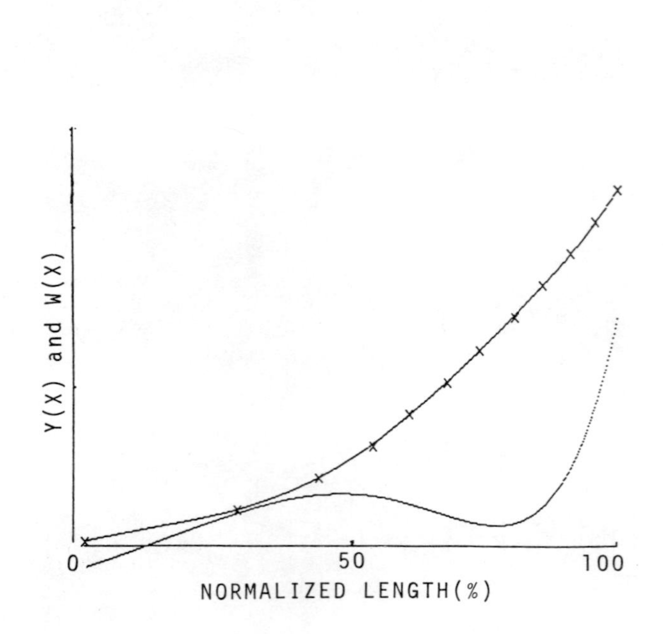

Figure 4. Measuring result of displacement obtained for the reconstruction image of Fig.3(c) and its second derivative. X:measured value, -X-:curve obtained by least-squares method, ---:curve represented by second derivative.

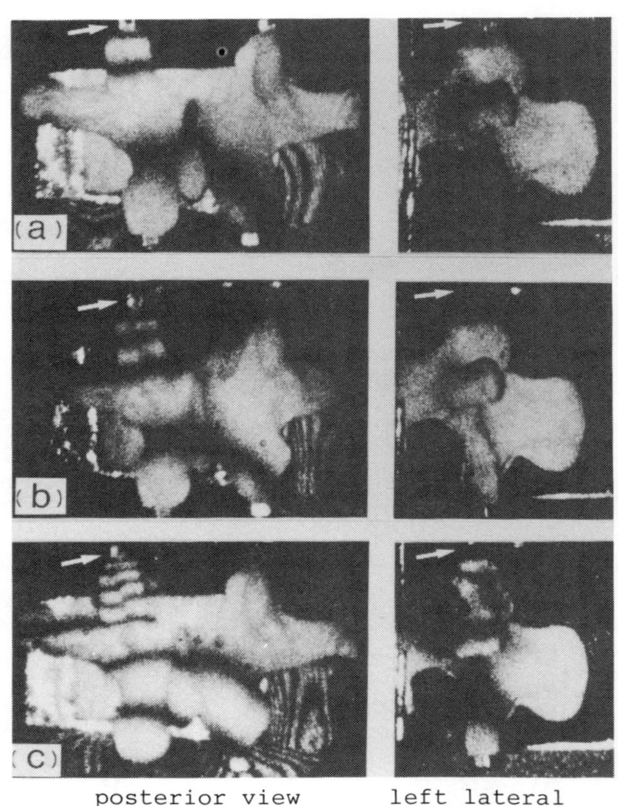

Figure 5. Reconstruction image obtained when the load applied to the end of left superior articular process (as shown with arrow) toward posterior direction. (a)50gf--> 100gf, (b)50gf-->150gf, (c)50gf -->200gf.

In the reconstruction images of Fig.5, the load was applied to the end of the left superior articular process in the posterior direction as represented by the arrow. The magnitudes of the load were 50gf, 100gf and 150gf. The density of fringes on the end of the left superior articular process was greater than that on the inferior articular process. In left side view, the displacements of the specimen against the out of plane rarely occurred.

Two reconstruction images under the different loading conditions is shown in Fig.6. The magnitude of the load is 100gf. The points of application of the load in Fig.6(a) and (b) were the ends of the inferior and superior articular processes, respectively. The number of fringes in Fig.6(b) was fewer than that in Fig.6(a). The moment is the product of the force and the moment arm. The distance from the superior articular process to the pedicle is shorter than that from the inferior articular process to the pedicle. If the magnitude of the load is the same, the moment in Fig.6(a) is greater than that of the right image.

We compared the deformations of the normal vertebra with those of one cut off as shown in Fig.7. The specimens were subjected to a load perpendicular to the plane of the

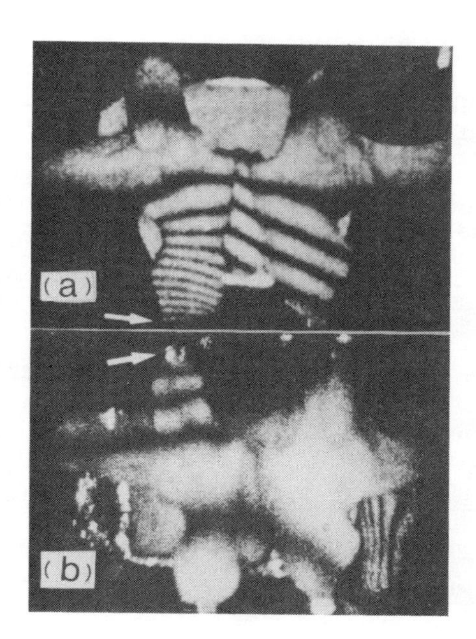

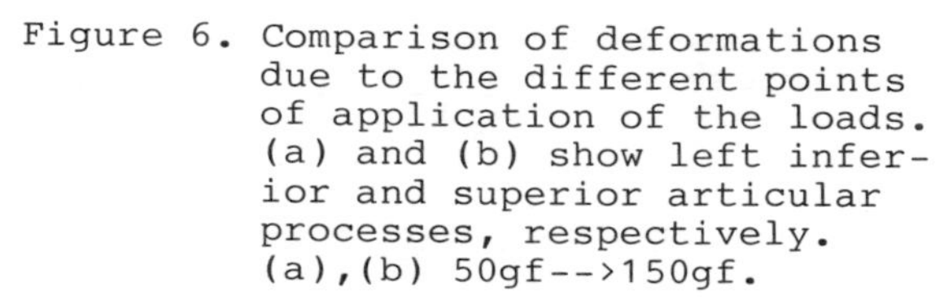
Figure 6. Comparison of deformations due to the different points of application of the loads. (a) and (b) show left inferior and superior articular processes, respectively. (a),(b) 50gf-->150gf.

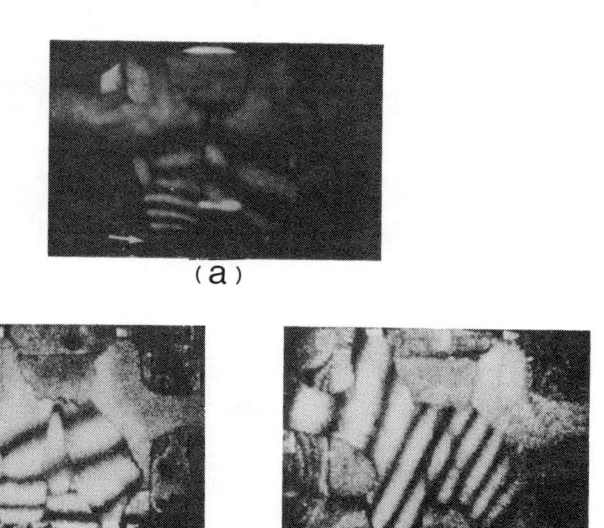

Figure 7. Comparison of deformations obtained for the specimens with and without the right interarticular part cut off artificially. (a)normal lumbar vertebra, (b) and (C) are specimens with the interarticular part cut off artificially. (a)50gf -->100gf, (b)50gf-->70gf, (c)50gf-->60gf

drawing at the end of inferior articular process. The fringes on the right superior articular process bounded by the cutting line could not be observed in Fig 7. This indicates that the posterior part of vertebra with spondilolysis becomes remarkably weak with weight loading.

The same experiments were carried out on dry bone. For the cadaverous bone preserved in formalin , the experiment was carried out as shown in Fig.8. Incremental loads were the same for normal bones(a) and (b) and them cut off (c) and (d). Comparisons of reconstruction images of the normal lumbar vertebrae (a) and (b) and the lumbar vertebrae (c) and (d) cut off revealed the same tendency for the numbers and directions of fringes for each case. The results obtained for dry bone were similar to those obtained for cadaverous bone.

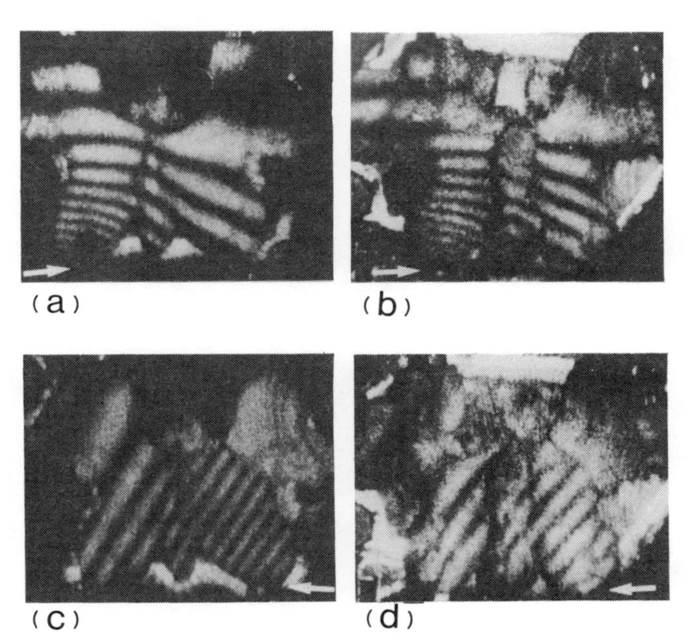

Figure 8. Comparison of deformations between dry(a),(c) and cadaverous(b),(d) lumbar vertebrae. The upper two patterns (a),(b) were obtained for normal lumbar vertebrae. The lower two patterns (c),(d) were obtained for the bones cut off artificially.

Conclusion

We applied holographic interferometry to measure the deformation of lumbar vertebrae, and found that displacement of the normal lumbar vertebra caused by the load applied to the inferior articular process was greater than that due to the same magnitude of the load applied to superior articular process. The amount of displacement at the point of application of the load was proportional to the increase of the load for the same vertebra. However, the lumbar vertebra with interarticular part cut off deformed considerably with small load compared with the normal lumbar vertebra. Displacement distributions of the dry bone were similar to those of cadaverous bone under the same loading conditions. We also showed for the normal vertebra that the interarticular part was subjected to high stress. These findings may be useful in investigating the cause of spondylolysis.

This work was supported in part by a Grant in aid for Scientific Research from the Ministry of Education, Science and Culture(No. 61480324).

References

1. Frankel, V. H. and Nordin, M., Basic Biomechanics of the Skeletal System ,Lea & Febiger, Philadelphia, 1980.
2. Kapandji, I. A., The Physiology of the Joints, Churchill Livingstone, London and New York 1974.
3. Wiltse, L. L., "The Etiology of Spondylolysthesis," J. Bone and Joint Surg., Vol. 44-A, p. 539. 1962.
4. Vest, C. M., Holographic Interferometry, Jone Wiley & Sons, New York 1979.
5. Matsumoto, T., Fujita, T., Nagata, R., Sugimura, T., and Kakudo, Y., "Measurements of Deformations of Teeth and Mandibles due to Occlusal Forces", In Holography in Medicine and Biology, ed. G. von Bally, Springer Series in Optical Sciences, Vol.18(Springer, Berlin, Heidelberg, New York 1979)p. 170.
6. Matsumoto, T., Fujita, T., Nagata, R., Iwata, K., Sugimura, T., and Kakudo, Y., "Holographic Investigation of Tooth Deformations", In Optics in Biomedical Sciences, eds. by G. von Bally and P. Greguss, Springer Series in Optical Sciences, Vol.31(Springer, Berlin, Heidelberg, New York 1982)p. 105.
7. Kojima, A., Uketa, S., Ogawa, R., Matsumoto, T., Iwata, K., and Nagata, R.,"Strength and deformation Measurement of Plate applied on Osteosynthesis by using Double Exposure Holographic Interferometry"(in Japanese), Cent. Jpn. J. Orthop. Traumat., Vol.29,p.362. 1986.

Session 10

Speckle Techniques

Chairmen
O. J. Løkberg
Norwegian Institute of Technology, Norway
Tongshu Lian
Beijing Institute of Technology, China

Electronic speckle pattern interferometry

Ole J. Lokberg

Physics Department, The Norwegian Institute of Technology
Trondheim, Norway

Abstract

The basic principles of electronic speckle pattern interferometry (ESPI) are described, stressing its close similarity to hologram interferometry. The technique's applications for vibration and deformation testing within industrial and medical research are outlined. Future developments are discussed.

Introduction

The ability of hologram interferometry to provide global pictures of mechanical vibrations and deformations has proved extremely useful for testing material and structure behavior. The introduction of thermoplastic materials has made the holographic recording process considerably faster, which makes the technique attractive for industrial testing. The time delay between exposure and reconstruction is, however, still too long for on-line inspection purposes. The process is also too slow to allow for measurements of changes in the object that take place so fast that nearly instantaneous updating of the interferometric reference state is necessary. For such purposes we are waiting for the instant reusable holographic film to appear (BSO crystals and similar materials are promising candidates).

Electronic speckle pattern interferometry (ESPI) represents an interesting alternative to conventional hologram interferometry. The ESPI system is based on direct video recordings of holograms with subsequently electronic filtering and video display and acts at the speed of the video system (25 Hz, European video standard; 30 Hz, American standard). This increased temporal resolution is exchanged for a limited spatial resolution compared to ordinary film-hologram interferometry. However, like video imaging, it is essential for industrial inspection and surveillance, and ordinary photography is used for high-quality pictorial documentation; the same situation will probably apply for ESPI and hologram interferometry.

The ESPI system

The ESPI principle can be explained using speckle theory, or by comparison to Moire effects, or by holographic analogy. For the most commonly used ESPI system where a uniform reference wave is used, the holographic analogy is rather obvious. The hologram interferometry analogy also explains ESPI systems where a speckle reference is used, but historically these are usually categorized as speckle correlation techniques (where strictly speaking also hologram interferometry belongs). Since the optomechanical construction of an ESPI system is very similar to conventional hologram interferometry setups, we will comment only briefly on the various blocks of the ESPI flow diagram depicted on Figure 1. Readers who are interested in a more detailed description should consult the literature.[1,2]

As laser light sources, HeNe, Ar, and Rb are often used. Note that the recording medium, the TV target, has a broader response in the infrared than conventional films. Therefore we are less restricted in the choice of lasers than in hologram interferometry. For

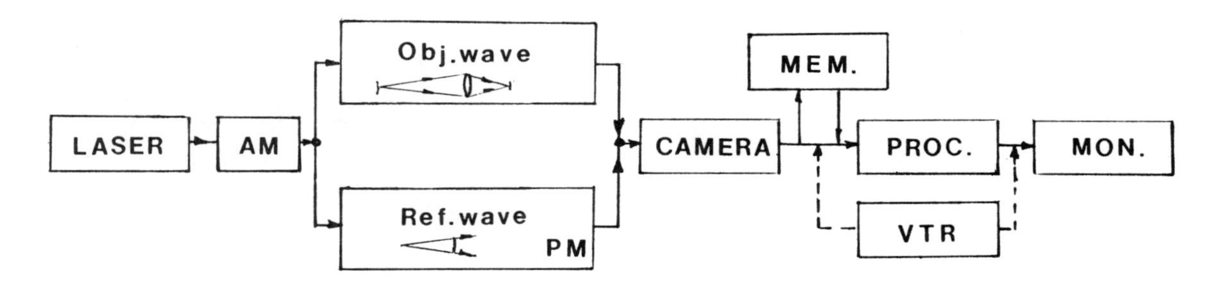

Figure. 1. The elements of an ESPI system.

example, the 1.06-μm light from the YAG laser might be used without frequency doubling while the wavelengths of most laser diodes coincide with the most sensitive part of the silicon cameras. The AM modulating block in Figure 1 simply means that the system can be run stroboscopically, double exposed within a video frame, or used at shortened exposures. Similar to other interferometers the light is split into two branches - the object and the reference branch. In the object branch the laser-illuminated test object is imaged by a lens system onto the light-sensitive plate of the video camera or the target. The lens system is stopped down to about f/30. The reference branch provides a reference wave of uniform quality which impinges on the target from the same direction as the object wave. The two waves are directionally combined by an optical element which may be a simple wedge beamsplitter or more complex optical elements.[1] The stopped-down lens and the in-line combination of the two waves keep the spatial frequencies in the hologram within the resolution of the video system. As shown in Figure 1, phase modulation (PM) might be impressed upon the reference wave. In this conjunction, PM indicates that the phase of the reference wave can be varied either by reflection from a moving mirror or by transmission through a polarizing crystal. The video camera records the image hologram created by the combination of the object and reference waves by transforming the spatial content of the primary hologram into an equivalent video signal. The video camera should have good sensitivity, large dynamical range, and high resolving power. These requirements are met to a very high degree by the modern multi-alkali tubes (like Chalnicon and Newicons) mounted in almost any respectable camera housing. However, these conventional video cameras will soon be replaced by charge-coupled-device (CCD) cameras. For the time being the resolution of the commercially available CCDs is not quite good enough for the highest quality ESPI recordings. The video signal from the camera is fed into an electronic processing unit, which essentially performs a bandpass filtering and a rectification on the signal. This process can be shown to be entirely equivalent to the reconstruction process in ordinary hologram interferometry. For certain applications we want to store a video frame to compare or subtract it from the frames coming from the video camera. The video frame store is today exclusively based on digital registration and storage.

The modified signal is finally fed into the TV monitor, where it is converted into a TV image that can be considered a reconstructed hologram. This hologram lacks of course the spectacular three-dimensional effects of ordinary holography. However, as in ordinary hologram interferometry, the image is interferometrically sensitive to any movement of the object. The interpretation of ESPI fringes is similar to that of ordinary hologram interferometry. The ESPI fringes have a coarser speckle structure because of the limited resolution of the video camera compared to holographic film. However, the ease and speed by which these ESPI interferograms are produced are the strengths of the technique, and cannot be compared to hologram interferometry.

Measurements of vibrations

The ESPI system is well suited for the investigations of vibrating or dynamically excited objects. For example, if we hit the test object with a sharp blow, we see the resulting damped oscillations, provided the damping is not too strong. Analysis of harmonically vibrating objects is particular easy by ESPI. We put the test object in the illumination, excite it, and observe at once its resonance patterns on the video monitor. The fringe pattern represents time-averaged fringes where the mathematical relationship between the amplitude $u_0(x,y)$ and the fringe intensity $I_0(x,y)$ is given by

$$I_0(x,y) = J_0^2 \, [2\pi/\lambda \cdot g(\theta) \cdot u_0(x,y)] \qquad (1)$$

where J_0 is a Bessel function of zero order and first kind, λ represents the laser wavelength, and $g(\theta)$ is the angle factor associated with hologram interferometry.

These fringes are easily interpreted because they represent contours of constant amplitude at near-integral fractions of the wavelength. The zero-order fringes, $J_0^2(0)$, represent nodal lines or points. This important information is easily spotted on the monitor image since the zero-order fringe is six times brighter than the next-order fringe. The resonances of the object are found by scanning through the frequency range of interest.

The actual strength of each resonance might be inferred by measuring the excitation necessary to observe the first dark fringe. Or, what might be more relevant, we find the Q factor of each resonance by measuring the frequency width as we slowly scan the spectral resonances. Time-average ESPI as described above is very easy to use, but does not provide information about the vibration phase and has a limited measuring range. These shortcomings are effectively remedied by using phase modulation usually introduced in the reference branch as indicated in Figure 1. Using phase modulation we obtain effectively the same fringe function, but its argument is now dependent on the vectorial difference between the two movements. Phase modulation can be used to expand drastically the measuring range and, in addition, provide for separate phase and amplitude distributions.[3-5] The information gained by these phase modulation techniques is often essential for a meaningful vibration analysis. For example, the movement of a loudspeaker is nearly impossible to deduce from the amplitude distribution alone. The relative phase distribution across the vibrating surface must also be known. The ESPI technique can be combined with stroboscopy, whereby we can observe the amplitude distribution in any part of the vibration cycle.[6] The fringe distribution $I_S(x,y)$ will in this case be given by

$$I_S(x,y) = \cos^2[(2\pi/\lambda \cdot g(\theta) u_0(x,y)] \, , \qquad (2)$$

where $u_0(x,y)$ now represents the amplitude excursions between two arbitrary parts of the vibration cycle (usually the rest position and the maximum excursion). The stroboscopic techniques (and to these we also include the double exposure within a frame)[7] are particular useful when analyzing vibrations that are not pure harmonical. However, because they usually demand chopping of the light, more laser power is necessary than for time-average methods. Before we leave vibration analysis, we point out the remarkable tolerance of the ESPI techniques with regard to object instabilities and hostile surroundings in this mode of operation. We have looked at the vibrations of objects held freely in our hand. We have observed the vibrations of objects heated to above 1000°C. We have looked at objects in bright sunshine located at distances over 100 m from the instrument. We should also note that the quality of documentary photographs of vibration fringe patterns can be greatly improved using the speckle-averaging technique.[8] An example of this is shown in Figure 2, where we have used the commercial system RETRA 1000, which contains a patented speckle-averaging system.

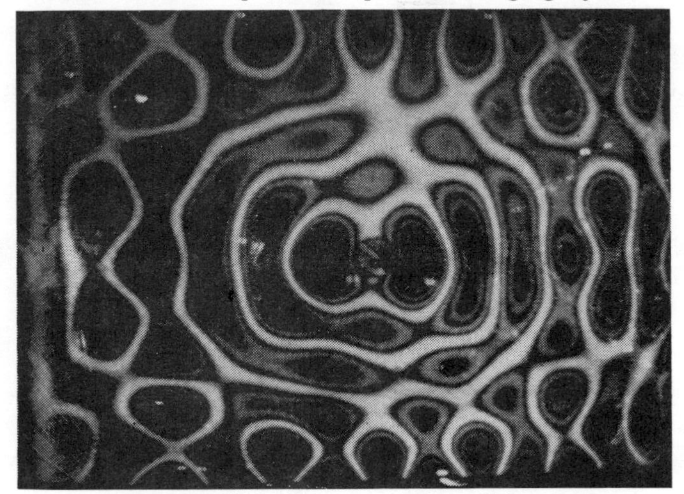

Figure 2. ESPI recording of a vibrating plate.

Deformations

By deformation we mean changes of the object that take place slowly or are induced two-step movements separating several video frames temporally. In such cases we cannot any longer use the storage property of the video camera, but have to provide for a separate video store. As already mentioned, digital frame stores now are exclusively used for this purpose. In practice we record the reference state, subtract it from the incoming frames of the video camera, and have the processed interferogram displayed on the video monitor instantly. The interferogram represents the displacement of the object between the frames and the interpreting function will be yielded by Eq. (2). Figure 3 is an example of ESPI recordings of an approximately 1-cm-square pressure transducer.

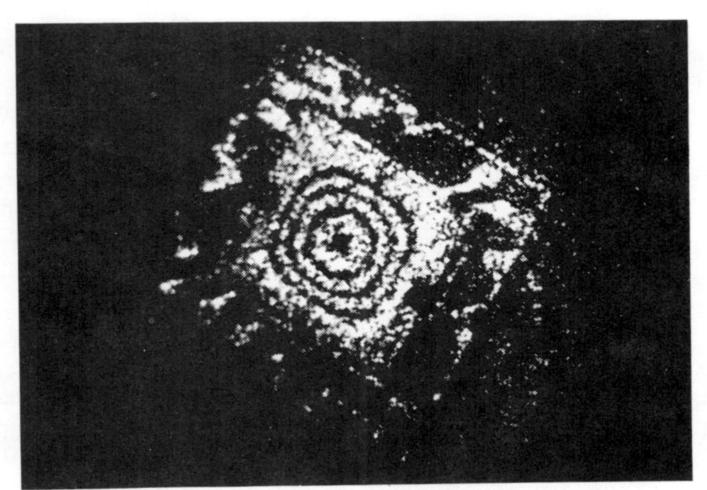

Figure 3. Deformations of a pressure transducer.

The rapid updating allows us to follow rapidly deforming objects from their initial deformation to their final crackdown with high accuracy. For such experiments it is essential to record the entire sequence on a video tape recorder (VTR) for later replay and analysis, either in slow motion or by single frames. If we have access to a VTR of sufficient resolution, it is not even necessary to make separate recordings of the reference states. We simply make a straight recording of all the frames coming from the video camera during the experiment. Afterwards, the videostore is used to compare any frame recordings. In this way, Hologram 234 might be subtracted from Hologram 239 or Hologram 3635 and so forth.

Miscellaneous

In this part of the paper we discuss topics which are either common to each modes of operation and/or general questions regarding the technique.

Comparative ESPI

Using ESPI, the movement or displacement of one object relative to another object attributable to the subtraction process can be directly compared. This can be very useful for rapid inspection of components, where the behavior of an incoming component is compared directly to that of a standard component. This process can even be performed for vibrating objects, although the phase relation makes the results more difficult to interpret.[9]

Displacement direction

ESPI is essentially an image-hologram interferometry process, which means that we observe the object from one direction only and thus can measure only one component of the object's displacement vector. In vibration analysis, it is easy to find the remaining components necessary to construct the vector. Simply rotating the vibrating object in the illumination provides for different fringe patterns from which we can deduce the actual vibration vector at each point of the object.[10] For a deformation study we have no guarantee that the deformation can be repeated with a sufficiently high degree of accuracy. Work is now in progress to use three different illumination directions combined with extended video memories for storing the different patterns.[11]

Remote inspection

The video camera effectively transforms the hologram into a corresponding electronic signal. The consequences of this feature have so far not been really exploited. In practice, the optical head can be placed far from the observation station. In fact, by using video transmission, the optical head could be placed anywhere in the universe while we could sit here on earth and observe and measure, for example, the expansion of lunar rocks. Less far reaching, but potentially more useful, would be to use remote ESPI to inspect dangerous places like in atomic power plants. For inspection of hard-to-reach places like cavities, fiber-optic imaging and illumination works very well in conjunction with ESPI.[12]

Object size

The size of an object that can be recorded simultaneously depends on two different aspects: how strong the available laser and how small the details we want to observe. The necessary laser power is again a function of object reflectance. If a 5-mW laser is used, we normally would say that a $30x30\text{-cm}^2$ white matte surface would represent a maximum area for recording vibration fringes (deformation fringes are much less critical because the intensity of the higher-order fringes is constant). If retroreflective coating can be used on the object, the observed area can be at least 1-m-square. The resolution of the system is most critical for deformation analysis. Here we cannot use speckle-averaging techniques and the digital resolution is at present only 512x512 pixels. If we assume 20x20 pixels to be a reasonable number for positive visual detection of a fringe anomaly (for example a delamination), we find that for a 1-m square object, defects down to about 4 cm (linear size) can be found. If we want to detect 0.5-cm defects, we need to reduce the observed area to about 12x12 cm^2, and so forth. These numbers will vary with the actual setup, the object, and the type of defect, and may also be improved by computer-assisted readout.

Automated readout

Computer-assisted read out and automatic fringe analysis are of great interest for industrial purposes such as robotics. It is also of interest to be able to improve the fringe accuracy for strain calculations. There have lately been published several papers on this subject[13,14] and more are expected to come. It is reasonable to expect an accuracy about $\lambda/200$ for deformation measurements and possibly $\lambda/500$ for vibrations. (In principle lower values for vibration measurements have already been reached,[5] however, these values have been reached by spot measurements.) In general, the computer can be relatively easily incorporated into the ESPI system, for example, to write experimental data on the video frames for future references.

Applications

In principle we can perform the same measurements by ESPI as we do by hologram interferometry. However, the speed and stability of the system allow us to tackle experimental difficulties that would either have been impossible by hologram interferometry or would have called for a pulsed laser. So far vibration analysis has been the most common application of the technique and where most impressive results have been achieved. We have used the technique for medical and industrial purposes. Within the medical field we have measured the vibrations of the human eardrum both in vitro and in vivo.[15] We have even measured the vibrations of the basilar membrane in vitro with a high degree of success,[16] showing among other things that the standard Mössbauer technique for measuring vibrations often provides wrong answers. However, since in vitro results probably are not very representative for the live conditions, the experiments have been postponed. However, there are no doubts that if global interferometric methods ever will progress from the experimental medical stage to the practical clinical usages, then ESPI will play an important role. Within industrial research and inspection we have found a great interest for vibration measurement by ESPI. A typical example of a practical application area in this field is the testing of gas turbine parts. We constructed an ESPI setup for a Norwegian gas turbine manufacturer ten years ago to be used on site to check the resonances of each blade.[17] The rather primitive setup has been used by ordinary mechanics full days in critical periods during production, and the system was claimed to have repaid its cost within the first two months. The first prototype version of the new compact version of the ESPI instrument (RETRA 1000) has now been in use at the same plant for a year without any problems. Analysis of loudspeakers represents another industrial area of interest. The main problem here is that the phase information is essential for an understanding of the movement, and this information is far easier acquired by ESPI than by other methods. The vibrations of car components is another potential field of interest. The left ESPI recording in Figure 4 shows a car door (Mazda 818) vibrating because of external sound pressure. The door was recorded by means of a RETRA 1000 with the built-in 5-mW laser. Note the peculiar vibration pattern on the lower part of the door. The nodal areas have degenerated to nodal points which are typical for combination modes, with resulting traveling waves across the object. This complex movement could easily be analyzed utilizing the phase sensitivity of ESPI. On the right ESPI recording in Figure 4 we have changed the magnification of the imaging lens of the instrument and we are now looking at the area around the key lock, which at this frequency was found to resonate quite strongly with an antinode centered in the doorlock.

Deformation testing by ESPI has been mainly confined to industrial research, where one area of interest is the detection and monitoring of cracks in various materials as the loading increases.

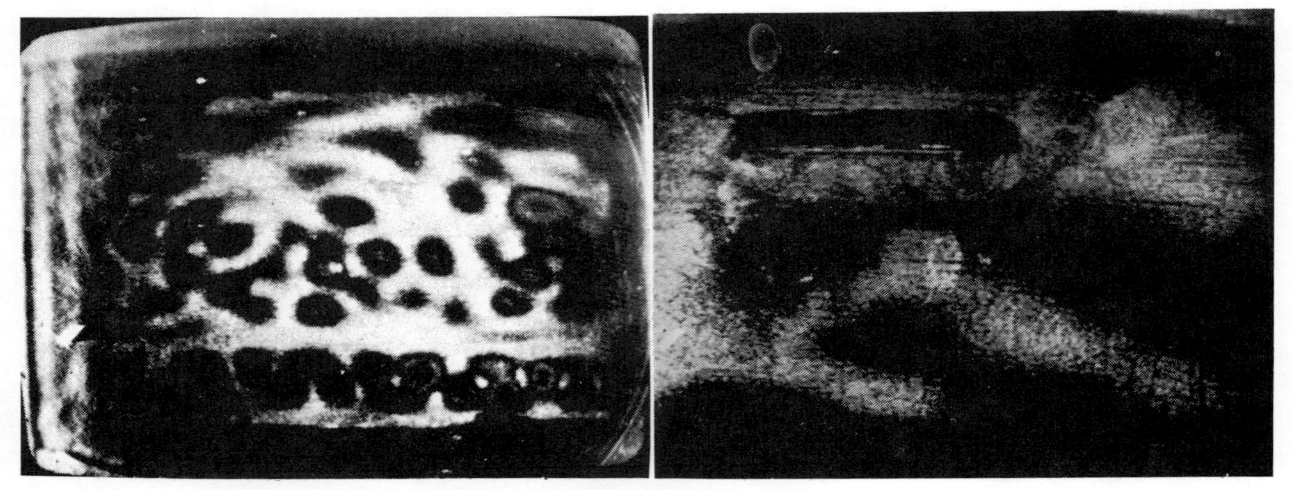

Figure 4. Vibrations of car door.

We have used the instrument at an industrial test laboratory where the strength of materials was tested. The particular material was to be used for melting electrodes. The surface was black with a very rough texture. Looking at the object in the loading rig we were able to predict at a very modest load (about 50 kg) the crack pattern of the surface. The object disintegrated at about 11,000 kg exactly along the lines where we had predicted. Similar experiments have also been performed on concrete samples where dramatic changes in the material behavior have been observed as the loading increased. Information found in this way is very important for determining the long-time strength of concrete. This type of deformation testing is also of particular industrial interest for ceramic materials. Ceramic materials are used at very high temperatures and we plan to do measurements at temperatures above 2000°C. We have already made successful recordings[18] at surface temperatures over 1500°C and going higher is just a matter of suppression of the background light and having a sufficiently good heater. We have also looked at composite materials that seem to be well suited for defect testing by ESPI. In honeycomb test samples, all the interior defects could be found. We have also inspected test samples of laminated panels and were able to detect most faults by various heating procedures. The quality of the fault revelation is not as good as with acoustic imaging with subsequent computer treatment. However, ESPI may also provide information about the relative strength of the bondings.

Concluding remarks

We have presented the ESPI technique and some of its applications. At present the impact of the technique upon industrial testing is small compared to hologram interferometry. However, the system's speed, ease of handling, and ability to work in hostile environments should make it suitable for many industrial inspection purposes. The future instrument will be extremely compact, battery powered, and with computing facilities built in. This instrument will provide a new freedom in holographic testing.

References

1. Jones, R., and Wykes, C., *Holographic and Speckle Interferometry* Cambridge U.P. London 1983.
2. Lokberg, O. J., and Slettemoen, G., "Electronic Speckle Pattern Interferometry," to be published in *Appl. Opt.* and *Opt. Eng.*
3. Lokberg, O. J., and Hogmoen, K., "Use of Modulated Reference Wave in Electronic Speckle Pattern Interferometry," *Journal of Physics E.*, Vol. 9, pp. 847-851. 1976.

4. Lokberg, O. J., and Hogmoen, K., "Vibration Phase Mapping Using Electronic Speckle Interferometry," *Appl. Opt.*, Vol. 15, pp. 2701-2704. 1976.

5. Hogmoen, K., and Lokberg, O. J., "Detection and Measurement of Small Vibrations Using Electronic Speckle Pattern Interferometry," *Appl. Opt.*, Vol. 16, pp. 1869-1875. 1977.

6. Pedersen, H. M., Lokberg, O. J., and Forre, B. M., "Holographic Vibration Measurement Using a TV-Speckle Interferometer with Silicon Target Vidicon," *Opt. Comm.*, Vol. 12, pp. 421-426. 1974.

7. Lokberg, O. J., "Use of Chopped Laser Light in Electronic Speckle Pattern Interferometry," *Appl. Opt.*, Vol. 18, pp. 2377-2384. 1979.

8. Lokberg, O. J., and Slettemoen, G., "Improved Fringe Definition by Speckle Averaging in ESPI," *Proceedings of ICO-13*, pp. 116-117. 1984.

9. Lokberg, O. J., and Slettemoen, G., "Interferometric Comparison of Displacement by Electronic Speckle Pattern Interferometry," *Appl. Opt.*, Vol. 20, pp. 2630-2634. 1981.

10. Lokberg, O. J., "Mapping of In-Plane Vibrations Modes by Electronic Speckle Pattern Interferometry," *Opt. Eng.*, Vol. 24, pp. 356-359. 1985.

11. Winther, S., *Proc. Symposium on Techniques and Applications of Automatic Fringe Analysis*, Loughborough, England, Nov. 4-5. 1986.

12. Lokberg, O. J., and Krakhella, K., "Electronic Speckle Pattern Interferometry Using Optical Fibers," *Opt. Comm.*, Vol. 38, pp. 155-158. 1981.

13. Creath, K., "Phase Shifting Speckle Interferometry," *Appl. Opt.*, Vol. 24, pp. 3053-3058. 1985.

14. Robinson, D. W., and Williams, D. C., "Digital Phase Stepping Speckle Interferometry," *Opt. Commun.*, Vol. 57, pp. 26-30. 1986.

15. Lokberg, O. J., Hogmoen, K., and Holje, O. M., "Vibration Measurement on the Human Ear Drum in Vivo," *Appl. Opt.*, Vol. 18, pp. 763-765. 1979.

16. Neisswander, P., and Slettemoen, G., "Electronic Speckle Pattern Interferometric Measurements of the Basilar Membrane in the Inner Ear," *Appl. Opt.*, Vol. 20, pp. 4271-4276. 1981.

17. Lokberg, O. J., and Svenke, P., "Design and Use of an Electronic Speckle Pattern Interferometer for Testing of Turbine Parts," *Opt. and Laser Eng.*, Vol. 2, pp. 1-12. 1981.

18. Lokberg, O. J., Malmo, J. T., and Slettemoen, G., "Interferometric Measurements of High Temperature Objects by Electronic Speckle Pattern Interferometry," *Appl. Opt.*, Vol. 24, pp. 3167-3172. 1985.

Measurement for dynamic deformation by mismatch white speckle method

Cao Zhengyuan Chen Fang Fang Ruhua Chen Pingping

Division for Photomechanics Research, Department of Engineering Mechanics,
Tongji University, Shanghai, China

Abstract

The method put forward in the paper can be used to measure not only the large dynamic deformation, but the small dynamic deformation as well. If the photoelastic material is taken as the test specimen, the isocromate and displacement patterns are obtained at the same time under the same load. So the principal stresses are separated.

Introduction

Under the loads of impact, explosion and so on, the stress and strain analysis of structure is a very important problem in scientific research and real engineering projects. But all dynamic photoelastic devices, in general, at present can only record the isochromate patterns. So it is necessary to find some other complement condition to separate a principal stress.

Moreover, for inertia of medium, relativity of it to strain ratio and the dynamic similarity of the test specimen to the real problems, it is very important for the dynamic research to research for one small dynamic deformation measurement method which does not depend on the properties of the material measured.

The method put forward in the paper can be used not only to measure large dynamic deformation, but small dynamic deformation as well. If the photoelastic material is taken as the test specimen, the isochromate and displacement patterns are obtained at the same time under the same load with the same test specimen. So the principal stresses are separated or are acquired from the strains. The dynamic stress analysis of the object is realized.

Experimental Principle

Dynamic white light speckle photography

The speckled face of the structure is illuminated by incoherent light. The film is exposured at different times before and after deformation of a body under dynamic loading respectively. The double exposure speckle patterns are obtained, that is, the displacements of a body at different times are got. The high speed photographic system of model WZDD-1 dynamic photoelastic instrument is used. The double exposure speckle patterns are obtained by this system. The sixteen speckle patterns at different times can be obtained during the action of the load.

For the double exposure speckle patterns, the displacement of any point in the specimen is obtained by the pointwise filter analysis method and is given by the equation (1):

$$d = \frac{\lambda Z}{M X} \tag{1}$$

where λ is the wavelength. Z is the distance between the speckle pattern and the screen. M is the magnification. X is the interval of Young's fringe.

When the speckle patterns are put into Fourier translating system in Figure 1, the fringe patterns of the contour of the component displacement of the whole field are obtained.

Mismatch method

If the epoxyresin photoelastic material is used as the specimen which is a high elastic model material and deforms very small, lens of high resolving power are required. But the resolving power of the lens of the dynamic photoelastic instrument, multi-spark model WZDD-1, is only 10 lines per millimetre. The displacement which is smaller than 100 microns cannot be measured by the lens. So mismatch method is used. In order to make the interval between two speckles so large that it can be distinguished, the rigid displacement is added before second exposure; or two single exposure speckle patterns recorded before and after the deformation of the body are mismatched (sandwich speckle). Suppose the deformation displacement is $\vec{d}$, then "speckle pair" in the double exposure cannot form the "double aperture". If horizontal displacement is added for an object $\delta_x \geqslant$ reciprocal of the resolving power of the resolving power of the lens, because $\vec{d}_1 = \vec{d} + \vec{d}_x$, "speckle pair" will form "double aperture" at the time. Franhofer defraction forms interferomatry fringe when the speckle pattern is illuminated by the laser beam. Thus a small deformation is measured when the resolving power of the lens is lower, that is, the sensitivity of the white light speckle photography is increased.

Dynamic white light speckle photography combined with dynamic photoelasticity to obtain the separated principaa stresses

The optical set-up is shown in Figure 2. The square of sixteen spark dischargers is used as illuminating source and is on the left side of the set-up (only four are drawn in the Figure 2). The square of sixteen cameras is on the right side of the set-up. F is the optical colour filter. L_1 and L_2 are the visual field lenses. K is the spectroscope. The spark square plane is located in front of the focal plane. Those cameras correspond to spark sources respectively. H_1 camera square images for the specimen to record isochromate. H_2 camera square images for the specimen to record speckle patterns at the same time.

The equation (2) can be obtained from the isochromate formulas of photoelasticity and linear Hook's law.

$$\sigma_1 = \frac{E}{2(1-\mu)}(\varepsilon_x + \varepsilon_y) + \frac{nf_c}{2t}$$
$$\sigma_2 = \frac{E}{1(1-\mu)}(\varepsilon_x + \varepsilon_y) - \frac{nf_c}{2t} \quad (2)$$

where
- n: isochromatic fringe order
- t: thickness of the specimen
- E: dynamic elastic model
- μ: dynamic poisson's ratio
- f_c: dynamic material fringe value
- ε_x: strain component in x direction
- ε_y: strain component in y direction

Obviously, the separated principal stresses will be acquired from both isochromate and speckle patterns.

Experiments and Results

The specimen is a rectangular plate made from epoxyresin with a round hole in the center as shown in Figure 3. It is fixed by one edge and is subjected to an impact load.

Experiment 1.

The speckles of the specimen are recorded by the film which is closely in touch with the whole roughly polished surface of the specimen. After the spark device is operated for two exposures, the speckle fringe patterns with the information about deformation are got. Figure 4 shows the speckle fringe patterns at the time 580 μs after impacting.

Experiment 2.

In the second experiment, the specimen is put in the optical set-up shown in Figure 2. After the measures of the mismatch are taken, the speckle fringe patterns obtained from different points in the specimen at the time 540 μs after impacting are shown in Figure 5.

The two kinds of method, both objective speckle and dynamic photoelasticity, are compared. The experimental results are in Figure 6. From Hook's law

$$\sigma_1 - \sigma_2 = \frac{E}{1+\mu}\left((\varepsilon_x - \varepsilon_y)^2 + \gamma_{xy}^2\right)^{\frac{1}{2}} \quad (3)$$

and dynamic isochromate

$$\sigma_1 - \sigma_2 = \frac{nf_c}{t} \quad (4)$$

the following equations are obtained

$$(\sigma_1 - \sigma_2)_i = \frac{E}{1+\mu}\left((\varepsilon_x - \varepsilon_y)^2 + \gamma_{xy}^2\right)_i^{\frac{1}{2}} \quad (3)$$

$$(\sigma_1 - \sigma_2)_i = \frac{n_i f_c}{t} \qquad (4)'$$

where i is the number of measured points. For the different points j and i, we have

$$\frac{E}{1+\mu}\left((\varepsilon_x - \varepsilon_y)^2 + \gamma_{xy}^2\right)_i^{\frac{1}{2}} = \frac{n_i f_c}{t} \qquad (5)$$

$$\frac{E}{1+\mu}\left((\varepsilon_x - \varepsilon_y)^2 + \gamma_{xy}\right)_j^{\frac{1}{2}} = \frac{n_j f_c}{t} \qquad (6)$$

Eliminating the constant of the material property from equations (5) and (6) yields

$$\frac{n_i}{n_j} = \frac{\left((\varepsilon_x - \varepsilon_y)^2 + \gamma_{xy}^2\right)_i^{\frac{1}{2}}}{\left((\varepsilon_x - \varepsilon_y)^2 + \gamma_{xy}^2\right)_j^{\frac{1}{2}}} = \frac{A_i}{A_j}$$

On section A, the points whose isochromate fringe orders are 1.5, 2.5, 3.5 are calculated. The results are shown in Table 1.

Table 1.

Number of measures points	j	2	3
n_i/n_j		0.60	0.43
A_i/A_j		0.64	0.47
relative error		6.6%	9.4%

When the equation (2) is used, the principal stresses will be acquired.

Conclusion and Discussion

The method put forward in the paper can be used not only to measure large dynamic deformation, but small dynamic deformation as well. If photoelastic material is taken as the test specimen, the isochromate and speckle patterns are obtained at the same time and under the same load. So the dynamic stress analysis of the object is acquired. Furthermore, the dynamic property of the material can be obtained as well.

Acknowledgements

The authors gratefully acknowledge the help of the colleagues of photomechanic Research Division, Tongji University. The authors also thank Mrs. Li Gongyu for typing the manuscript and Mrs. Zhang Linchun for drawing the picture.

References

1. Ching F.P., "White Light Speckle Method of Experimental Strain Analysis", Applied Optics, Vol. 18, No. 4 (1979).
2. Wang X.L. and Guo Q.H., "Measuring Instantanous Deformation by White Speckle Method", Proc. of ICEM Beijing (1985).

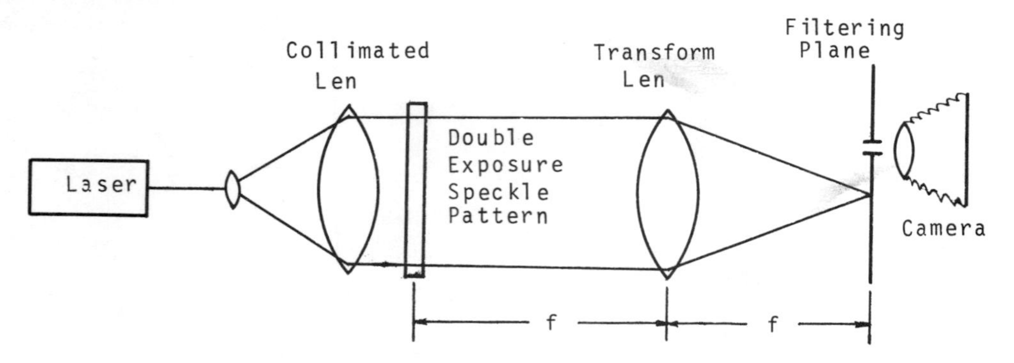

Figure 1. Fourier Filtering system

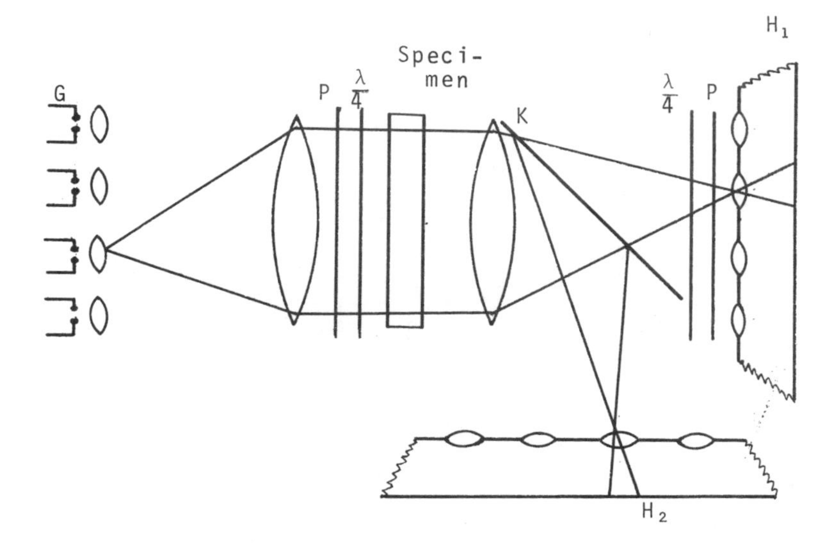

Figure 2. Schematic recording the isochromate and speckle patterns at the same time.

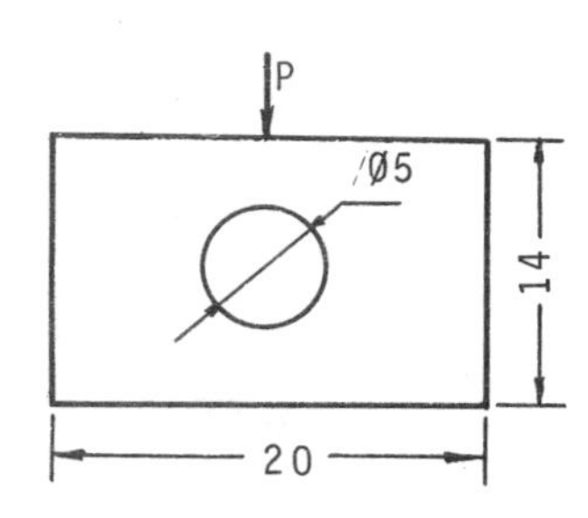

Figure 3. Size of the specimen

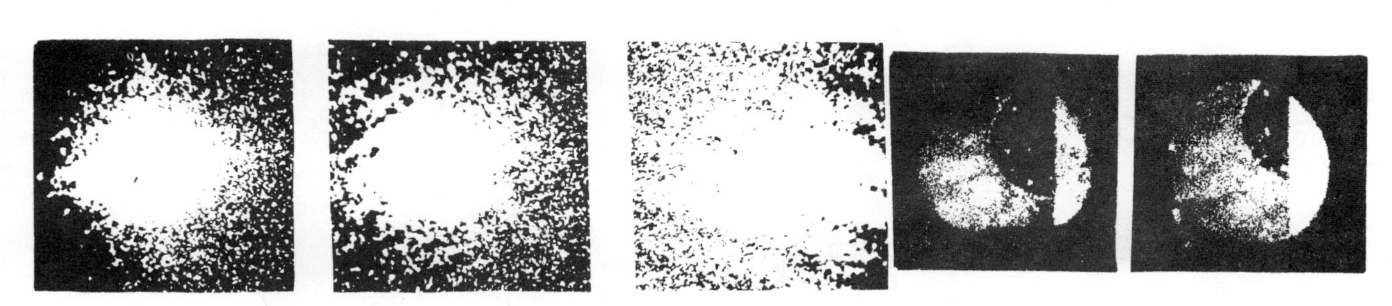

Figure 4. Objective speckle patterns

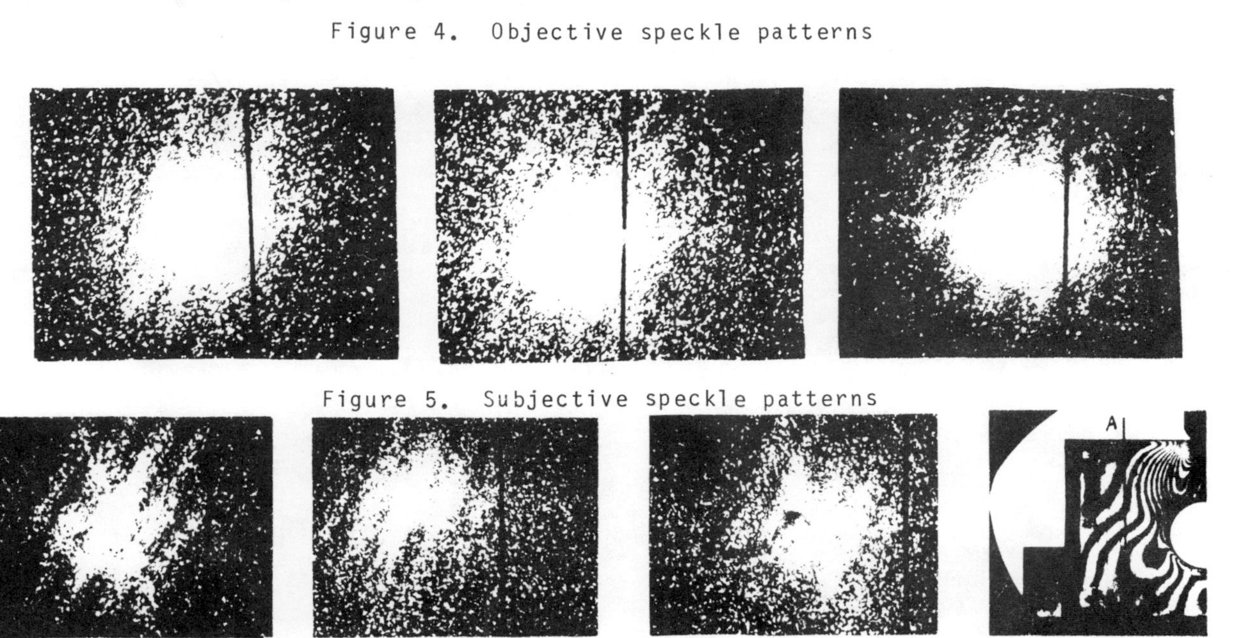

Figure 5. Subjective speckle patterns

Figure 6. Dynamic Isochromate and speckle patterns

Speckle velocimetry applied to wake flows

I Grant and G H Smith

Department of Offshore Engineering, Heriot-Watt University,
Riccarton, Edinburgh EH14 4AS

Abstract

This paper describes the application of particle image velocimetry to wake flows. As an example the vortex shedding from a circular cylinder is considered and a 'captured' instantaneous flow field presented. The current technique of analysis is described and alternative methods for the presentation of vortex wakes demonstrated.

Introduction

The flow around a bluff obstacle, such as a structural member, is generally divided into two distinct regions. The wake, formed approximately in the aerodynamic shadow of the body, is characterised by low pressure, reverse flow and vortex shedding. Outside the wake the fluid behaves as if it were inviscid and this region can be readily modelled by potential flow techniques. The wake, partly because of its complexity and partly because of its dominance in inducing bluff body loads, constitutes a region of continuing research interest. This paper considers the example of vortex shedding from a circular cylinder at a Reynolds number of approximately 200.

The wake regions associated with many industrial, aeronautical or offshore environments frequently show features with a single or multiple periodicity. A major difficulty in investigating such flows is the volume of data which must be collected, marshalled, analysed and displayed. The use of computers in all these processes is now well established. The recent extensions in processing power and the availability of dedicated VDU facilities has allowed a rapid development in the graphical techniques which enable the time-domain analysis of a flow. This approach has advantages over conventional frequency-domain techniques in that visual interpretation of the flow behaviour is considerably improved.

In providing the data base for such a flow reconstruction velocity measurements must be obtained at many points in the flow field. Simultaneous measurements at a number of field points are possible using a number of sensors. There is a limit on the extension of this approach depending on the budget available for such work. In periodic flows a conditional sampling method may be used to provide a flow mapping where a single fixed sensor acts as a strobe on a second mobile sensor, but this is a laborious and computationally expensive process.

A further difficulty in multi-probe wake investigation is that the net disturbance of the many probes perturbs the flow significantly and reduces the credibility of the data. The popular solution to this difficulty until recently was to measure flow velocities non-intrusively using the laser-doppler anemometer. This device provides measurements at a volume commonly of the order of $1mm^3$ or less usually referred to as a "point" measurement. Progressing from the basic one (velocity) component LDA to two or three components escalates the cost and multiplexity of the processing hardware. Flow field mapping using this type of anemometer is time consuming since the scanning of extended regions by the transmitting and receiving optical components usually relies on the availability of substantial mechanical traversing mechanisms. The need to maintain alignment during such procedures leads to the need for frequent and frustrating adjustments during tests. For these reasons, where large wind tunnel or wave tanks are used in experimental investigations, LDA has proved an unacceptable means of quickly obtaining multi-component velocity measurements over an extended field.

There has been considerable interest in the last ten years on alternative methods which would enable instantaneous, multiple component, whole field mapping. Holographic techniques[1] for compressible flows have been pioneered although interpretation of flow holograms is still difficult. Speckle velocimetry which enables the instantaneous recording of two components of fluid velocity over a sheet or plane of moving fluid currently shows more promise[2,3,4,5,6].

The term "speckle" refers to the "pin prick" bright spots produced by the constructive interference of coherent light produced by multiple scattering. For this classical speckle to occur in a fluid, whether moving or at rest, a large concentration of scattering centres is required to ensure that the light scattered by different centres overlaps on the film plane. When low concentrations of seeding are used the method, still often referred to as speckle velocimetry, becomes a particle image technique.

In order to obtain the light intensities required to obtain an instantaneous and extended illumination of the flow field, pulsed or chopped laser beams are generally used. Most previously reported work using this technique has been limited by the laser powers available to the investigators. This paper reports the first experiences in pulsed laser velocimetry (PLV) using a 10 joule Q-switched pulsed ruby laser. The term PLV could be used to describe "speckle" or "particle image" methods and here refers to the latter.

Experimental arrangements

The experiments were conducted in an open circuit wind tunnel having a working section of height 70mm and length 600mm. The width increased over the working length from 225mm upstream to 240mm downstream in order to partially compensate for blockage caused by the increasing boundary layer thickness along the tunnel walls. The working section was constructed from plate glass.

A 1mm diameter, smooth, brass cylinder of length approximately 230mm was fitted centrally at the upstream end of the working section.

An air flow of approximately $3ms^{-1}$ was used in the experiments giving a Reynolds number of 207. The flow was seeded with oil droplets produced by an aerosol generator which gave particles of diameters 2μm to 5μm. The small scale of working was determined by the limited space within the laboratory rather than the available laser energy. As mentioned elsewhere our current work is the application of the technique to larger model test facilities.

The JK Lasers 10 joule Q-switched pulsed ruby laser is fitted in the Heriot-Watt University's Offshore Engineering N°1 Laser Laboratory. The laser head delivers two 30ns pulses of 5 joules separated by a time delay variable between 1 and 800 microseconds. This means that the rate of delivery of energy to the flow during the delivery of either pulse is 166.5 megawatts. The repitition rate is currently 2 double ppm. The laser control units will be upgraded shortly to enable 4-pulse operation with a corresponding improvement in derived fringe quality.

The beam delivered from the laser head is approximately 20mm in diameter. As it passes through an internal spatial filter it is expanding. Previously reported work has used a spherical convex lens followed by a cylindrical convex lens to produce a sheet of light suitable for experiment. Our initial experiments established this was unacceptable for the current energy since focussing a 30ns 5 joule pulse causes air ionisation and damage to the cylindrical lens is highly probable. The arrangement we adopted therefore was a convex cylindrical lens followed by a concave spherical lens (Figure 1). The strengths of the lenses were determined by the sheet dimensions required. The scene was recorded by a camera with its principal axes orthogonal to the object plane (laser sheet).

Although this paper reports only initial tests conducted within the N°1 Laser Laboratory we have also made arrangements to "pipe" the beam to the main fluids laboratory where tests will be conducted in a closed return wind tunnel at a "realistic" experimental scale.

The double image is recorded on Kodak Technical Pan film developed with HC110 at a dilution B (1:30 with water) for 12 minutes to give high contrast and fine grain images obtained at an acceptable film speed. The currently reported tests were recorded on 35mm film. A 5" x 4" format camera is also used where increased spatial resolution is required.

Details of the PLV method

Photographic recordings are made of seeded fluid flow illuminated by a pulsed laser light source. Double or multiple exposures are made using the "pulsed" illumination from the laser. The resulting plates or film therefore contain double or multiple images, respectively, of all the seeding particles.

The local spatial periodicity of the transmittance of the developed film is found by passing an unexpanded laser beam, normally He-Ne, through the exposed film (Figure 2). The convex lens produces an intensity distribution at its focal plane which corresponds to the 2D Fourier transform of the spatial distribution at that point on the negative.

A broad diffuse background is observed, the diffraction halo, within which Young's fringes can be seen (Plate 1). The central bright spot is the undiffracted light from the interrogating beam. The fringe spacing is related to the particle displacement on the film. Hence, knowing the time delay between exposures, we obtain the particle velocity from

$$v_x = \frac{\lambda f_e}{M\lambda_x T} \quad (1)$$

and

$$v_y = \frac{\lambda f_e}{M\lambda_y T} \quad (2)$$

where v_x = velocity in x direction
v_y = velocity in y direction
λ = wavelength of interrogating beam
f_e = focal length of converging lens used in producing the fringes
M = magnification of the camera optics
λ_x = fringe spacing along the x-axis
λ_y = fringe spacing along the y-axis
T = time between laser pulses.

The particle velocity vector is perpendicular to the orientation of the fringes. The instantaneous velocity field is thus found by traversing the photographic negative and finding the fringe spacing and inclination at each point.

In practice it is necessary to interrogate the speckle negative at a regular grid of points. Depending on the particular requirements of the experiment this may contain 10^3 or 10^4 points. Clearly an automated analysis procedure is required.

In the present case the photographic negative was placed in a precision x-y traversing jig driven by computer controlled stepper motors. The low powered He-Ne beam passed through the negative producing fringes which were captured by a vidicon digitising camera and framestore. The fringes were analysed using in-house software based on the FFT technique to provide the necessary two components of fringe periodicity from which the velocity components were calculated.

The micro-computer was programmed in a simple loop to enable an automated analysis to proceed on a regular grid.

Speckle exposure parameters

The condition for obtaining good quality fringe (high visibility) is that the separation of displaced particle images be less that some fraction of the analysing beam diameter, D_ℓ, in practice about $D_\ell/2$, thus

$$Mv_{max}T = 0.5D_\ell \quad (3)$$

defines one limit of the dynamic range of the velocimeter by fixing the maximum velocity, v_{max}.

The exposure time is chosen to ensure that the particle moves less than some fraction of its diameter. Lourenco and Wiffen[5] suggest

$$t = \frac{0.5D_\ell}{AMv_{max}} \quad (4)$$

where $A = 20$.

The minimum velocity is determined by the need for the particle image to be separated by at least one particle image diameter, d_i, so

$$v_{min} = \frac{d_i}{T} = \frac{2d_i Mv_{max}}{D_\ell} \quad (5)$$

Now

$$d_i = d_p M + 2.44\lambda_R(1+M)F\# + Mvt \quad (6)$$

in general, where d_p = diameter of (object) particle
λ_R = wavelength of pulsed laser source
$F\#$ = recording optics F number.

The first term is the image particle diameter. The second allows for the diffraction limited response of the optics while the third is due to the particle movement during the exposure time.

The dynamic range is thus

$$v_{max} - v_{min} = v_{max}(1 - \frac{2d_i M}{D_\ell}) \quad (7)$$

The use of smaller seeding particles increased the dynamic range. This is a well established feature in LDA studies. The scattering efficiency of the particles are, however, reduced with reducing diameter. Adrian and Yao (7) have shown that 5μm diameter in air and 10μm in water are optimum sizes.

In the present investigation the exposure times were fixed at 30ns. Typical parameter values were $M = 1$, $d_p = 5\mu m$, $F\# = 8$. The effect of particle movement during an exposure is therefore negligible. The time between pulses could be varied from 1μs to 800μs and determines the upper and lower measurable velocities respectively.

The upper velocity measurable by the present arrangement is from equation (3) while the lower is determined by equation (5).

At 800μs pulse separation

$$v_{min} = 3.9 \times 10^{-2} ms^{-1}$$

$$v_{max} = 0.6 ms^{-1}$$

at 1μs pulse separation

$$v_{min} = 31 ms^{-1}$$

$$v_{max} = 500 ms^{-1}$$

Examples of wake flow data capture and analysis

The type of wake visualisations obtained using the speckle velocimetry technique are illustrated in Plate 2 where the double images of the seeding particles are clearly evident. An example velocity vector grid obtained at points of interest in Plate 2 is seen in Figure 3.

Enhancing the computed flow field image

In presenting vortex flow field measurements it is helpful to remove the mean convection speed, where this exists. The individual vortices are then easily picked out from the vector representation. An example from an earlier investigation is shown in Figure 4.

The software used in the analysis procedure is now being enhanced by incorporating routines to calculate vorticity. Since vorticity is defined by

$$w = \frac{\partial v}{\partial x} - \frac{\partial u}{\partial y} \quad (8)$$

which becomes

$$w(i,j) = \Delta v(i,j)/\Delta x - \Delta u(i,j)/\Delta y \quad (9)$$

where $\Delta v(i,j) = \frac{1}{2}\{v(i+1,j) - v(i,j) + v(i+1,j+1) - v(i,j+1)\}$
and $\Delta u(i,j) = \frac{1}{2}\{u(i,j) - u(i,j+1) + u(i+1,j) - u(i+1,j+1)\}$

Examples of this grid calculation applied to a typical portion of a vortex wake is shown in Figure 5.

There is an immediate advantage in displaying vortex wakes as contours of vorticity, namely that any convection velocity of the vortices is eliminated by the differencing algorithm. Consequently the isolated eddies are clearly identified and in regions where the translation speed is not easily identified, no error is introduced into the interpretation of the data from an assumed, global, uniform, convection velocity.

Conclusions

The particle image velocimetry (PIV) method ("speckle velocimetry") has been shown to be particularly suitable for the complex and unsteady wake regions behind bluff bodies. The instantaneous whole-field mapping achievable with PIV is desirable where the periodic flow is accompanied by a random turbulence.

References

(1) Lauterborn W and Vogel A: "Modern Optical Techniques in Fluid Mechanics", Ann Rev Fluid Mech 16, pp223-244, 1984.

(2) Simpkins P G and Dudderar T D: "Laser Speckle Measurements of Transient Benard Convection", J Fluid Mech 89, pp665-671, 1978.

(3) Meynart R: "Equal Velocity Fringes in a Rayleigh-Benard Flow by Speckle Velocimetry", Applied Optics, 19, pp1385-1386, 1980.

(4) Meynart R: "Speckle Velocimetry Study of Vortex Pairing in a Low-Re Unexcited Jet", Phys Fluid 26, pp2074-2079, 1983.

(5) Lourenco and Whiffen: "Laser Speckle Methods in Fluid Dynamic Applications", 2nd Intl Symp on Applications of Laser Anemometry to Fluid Mechanics, 6.3, 1985.

(6) Grant I, Smith G and Greated C A: "Velocity and Vorticity Speckle Pattern Simulation on a DAP Computer", 6th Int Conf on Photon Correlation and Other Optical Techniques in Fluid Mechanics, Cambridge University, July 1985.

(7) Adrian R J and Yao C S: "Application of Pulsed Laser Technique to Liquid and Gaseous Flows and the Scattering Power of Seed Materials", Applied Optics 29,1,pp 44-59,1985.

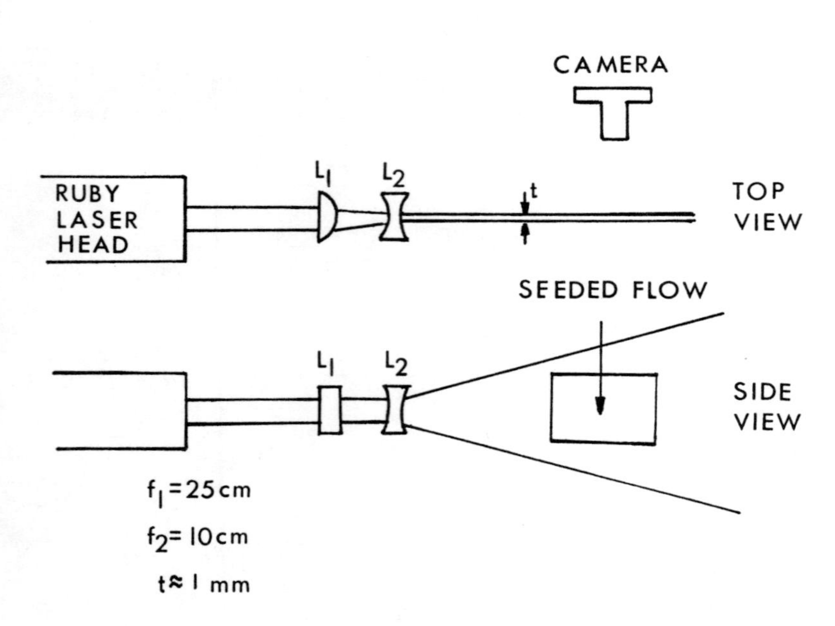

Figure 1: Production of the Laser Sheet.

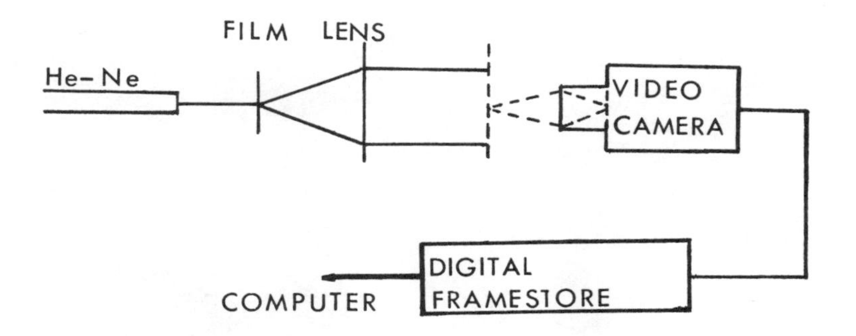

Figure 2 : Production of Fringes and Data Capture Hardware.

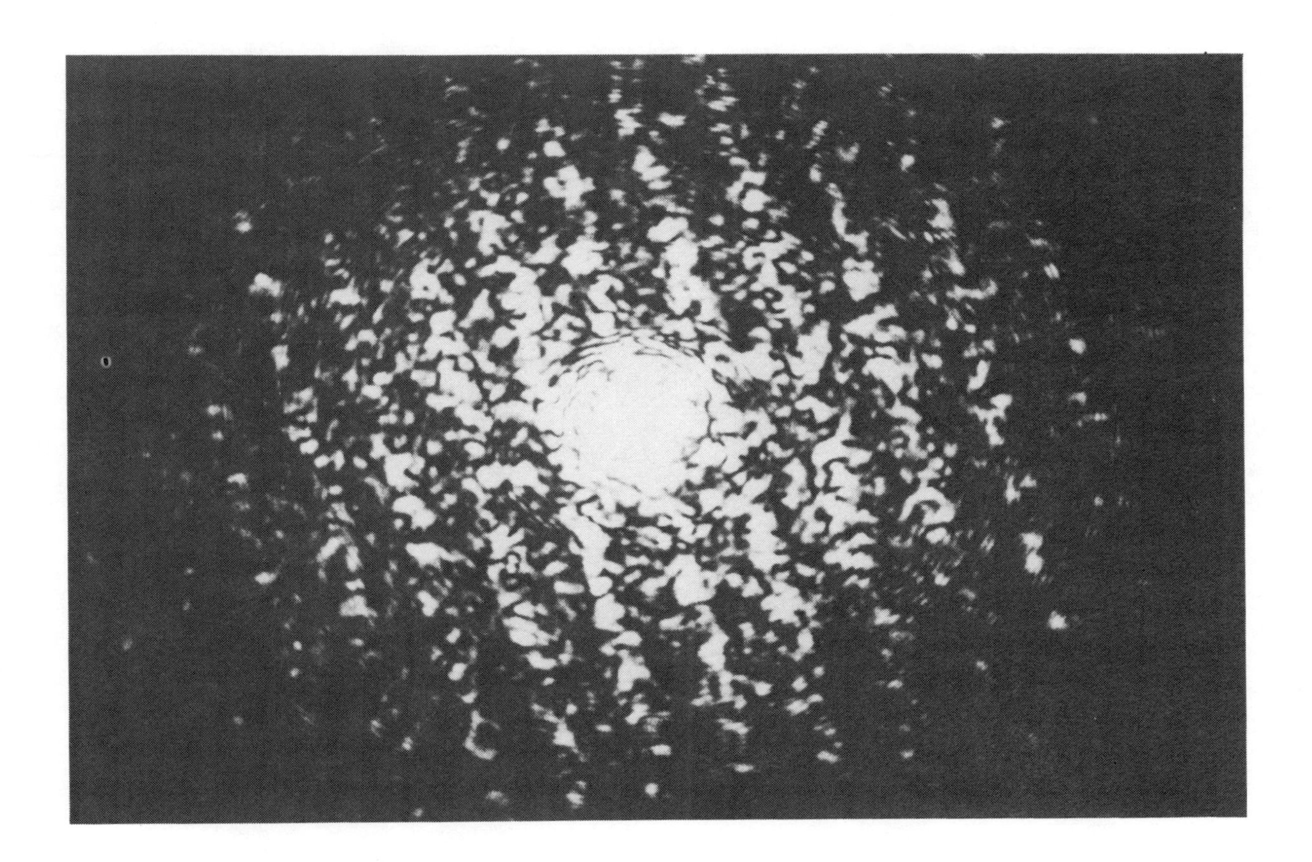

Plate 1: Typical Fringes Obtained From a Negative.

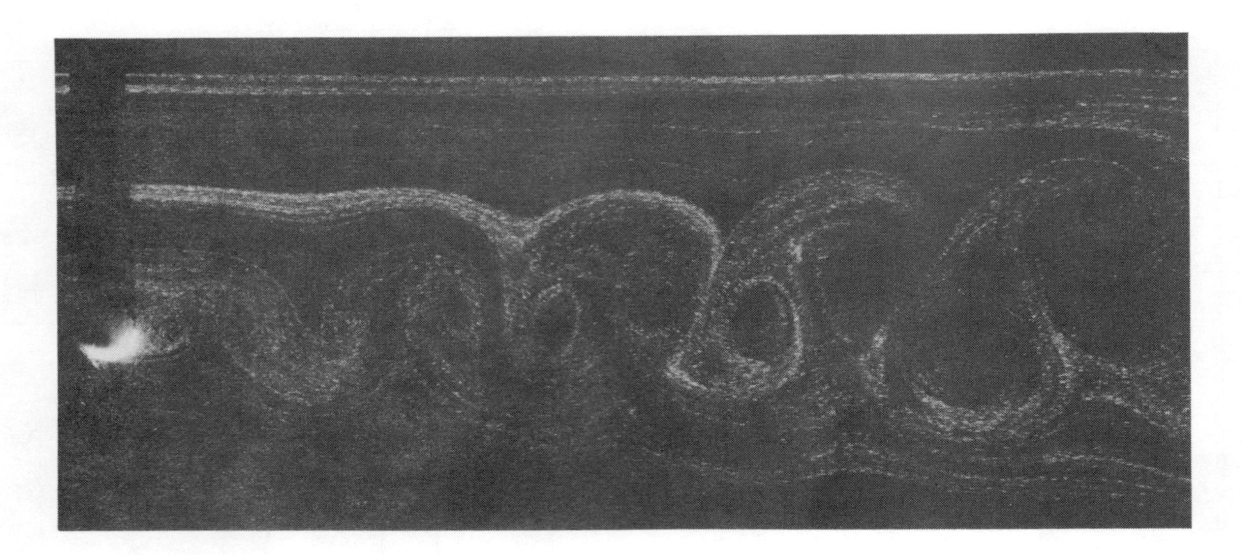

Plate 2: Flow Visualisation Obtained From Double Pulsed Laser Sheet.

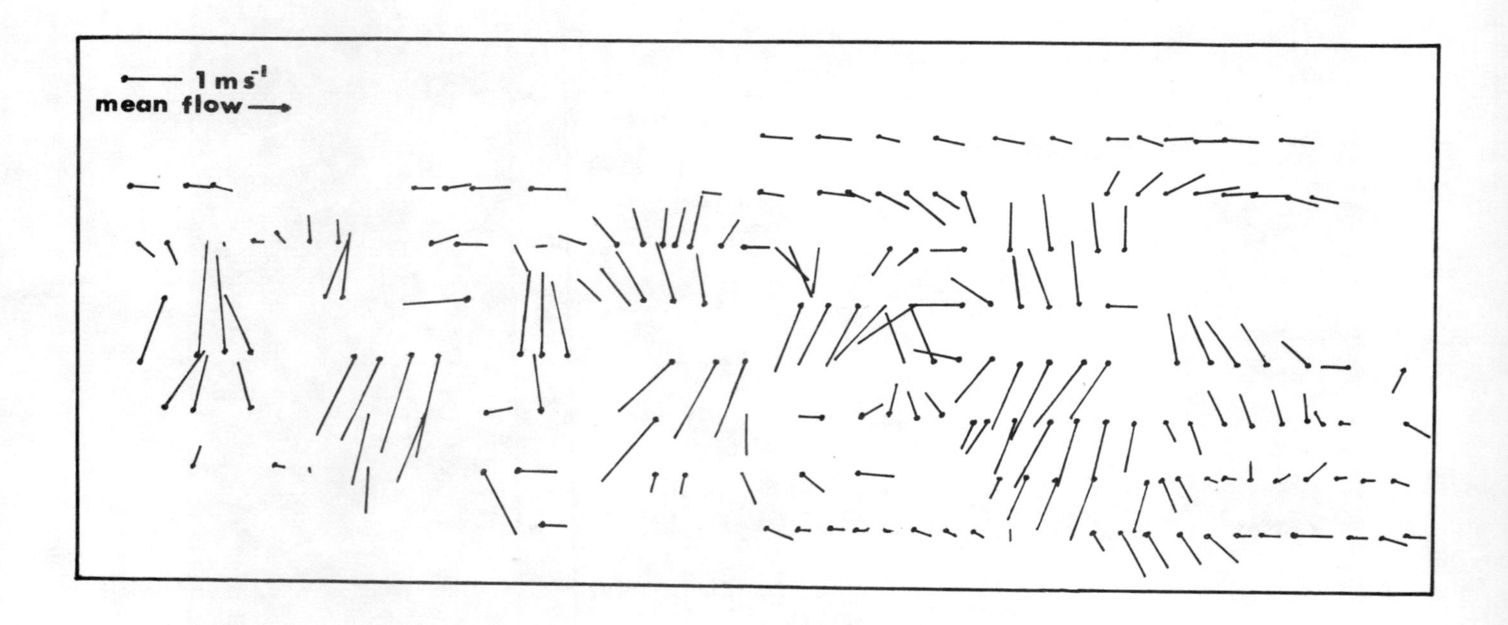

Figure 3: Velocity Vectors Calculated From Flow Shown in Plate 2.

Figure 4: Example of Flow Field With, and Without Removal of Mean Convection Velocity.

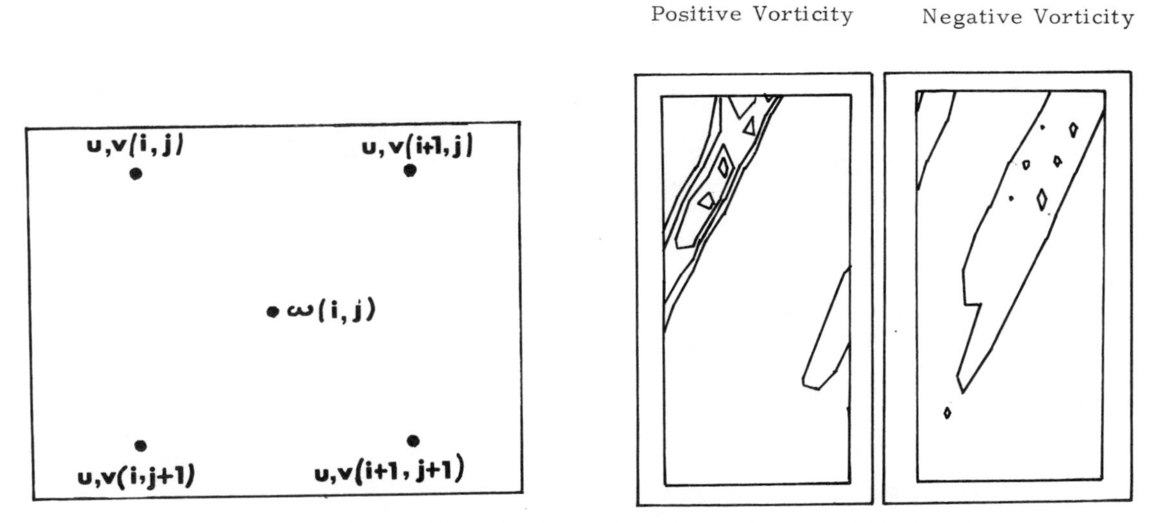

Figure 5: Example of a Vorticity Field Calculated From the Velocity Vectors.

Oblique-optical-axis speckle photography used for measuring 3-D displacements of practical engineering structures

He Yuming, Tan Yushan and Ku Chungshien

Laser and Infrared Technique Lab.
Dept. of Mechanical Engineering, Xi'an Jiaotong University
Xi'an, Shaanxi Province. The People's Republic of China

Abstract

In this paper a new technique for measuring 3-D displacements using oblique-optical-axis speckle photography (OSP) is introduced. The accuracy of seperating the out-of-plane displacement can be improved, and the visual field of the photographic lens can be utilized sufficiently when the OSP method is used. The experimental results of 3-D deformation of both the sample and the machine tool head stock are exemplified. Moreover, the Apple II microcomputer image processing system for fully auto-analysing the Young's fringe patterns is described principally. Lastly, by comparing the theoretical results with the experimental results, the OSP method and the image auto-processing system are satisfactory.

Introduction

The speckle photography technique[1,2,3] can be applied to measuring in-plane and out-of-plane displacements of objects. Asundi and Chiang [4,5] have done some works for measuring 3-D displacements simultaneously. He and Tan [6] have published an article in which the deduction of 3-D displacements equations and the determination of direction of speckle displacement are described. However, the experiments above based on the fact that the optical axis of photographic lens are perpendicular to the surface of the measured object. The applicability of the perpendicular-optical-axis speckle photography (PSP) method is limited by the visual angle of the lens. Because of the limitation, the information of out-of-plane displacement obtained is often unsufficient. Moreover, the visual field can not be utilized sufficiently when the PSP method is used, so that it is difficult to measure accurately the 3-D displacements of larger engineering structure. In order to resolve the problems, the OSP method used for seperating 3-D displacements is put forward. The fully automatic or semi-automatic processes of Young's fringe patterns are discussed in several papers. [7,8] In this paper, the basic principle and method of using an Apple II microcomputer for analysing the Young's fringe pattern are introduced.

Theory of OSP method

1. Focusing principal

As the optical axis is not perpendicular to the measured surface of the object when the OSP method is used, it is necessary to find out the correct way to focus the surface. In Fig.1(a), θ is the angle between the normal of the surface and the optical axis (which is only rotated in the horizontal plane), the point A is the intersection of the surface with the optical axis, and the A' is the corresponding point in the image plane; B is the arbitrary point on the surface. Now, the angle θ_1 can be proved as a constant, i.e. the object surface can be correctly focused after the image plane is rotated to angle θ_1.

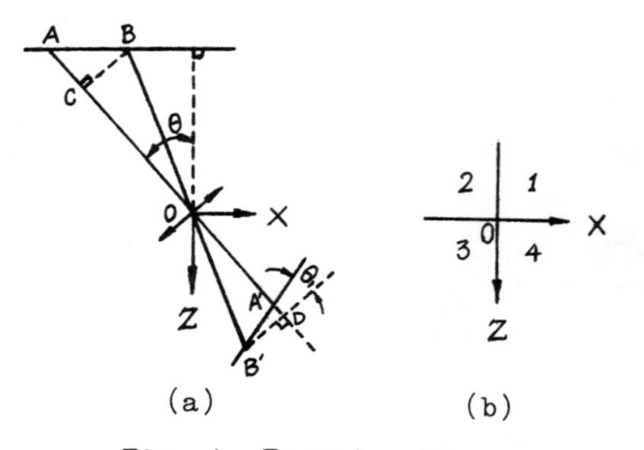

Fig. 1 Focusing diagram

From the definition of magnification

Let $M_1 = A'O/AO$ (1-a)

$M_2 = B'O/BO = DO/CO$ (1-b)

and $CO = AO - AB\cdot\sin\theta$

$DO = A'O + DA'$

$= M_1\cdot AO + M_2\cdot AB\cdot\cos\theta\cdot tg\theta_1$

By substituting above expression into (1-b), it results

$$AO\cdot(M_1 - M_2) = -M_2\cdot AB\cdot(\cos\theta\cdot tg\,\theta_1 + \sin\theta) \tag{2}$$

From the imaging formula of geometric optics, following expressions can be obtained

$$A'O = f\cdot AO/(AO-f) \tag{3-a}$$

$$D'O = f\cdot CO/(CO-f) \tag{3-b}$$

Where f is the focal length.

Substitute (1) and (3) into (2) and simplify

$$tg\,\theta_1 = f\cdot tg\,\theta / (AO-f) \tag{4}$$

From the expression (4), we know that θ_1 is a constant when the position of the camera is determined, so that after the image plane is rotated to θ_1 angle, the object surface can be correctly focused. Moreover, from (4), it can be found that when $\theta=0, \theta_1=0$, this is the situation of optical axis being perpendicular to the object surface.

2. Calculating magnification

In Fig. 1(a), the XOY coordinate plane is parallel to the measured object surface. If the optical axis has a rotation of angle θ in the XOY coordinate plane, the magnification of an arbitrary point on the surface will be the function of x.

From Fig. 1(a), the magnification of the point B on the surface can be written as :

$$M_b = f / (CO-f)$$

and $$CO = AO - (x_b - x_a)\cdot\sin\theta$$

so $$M_b = f/[AO-f-(x_b-x_a)\cdot\sin\theta] \tag{5-a}$$

If the point B has the coordinate (x,y,z), then (5-a) may be expressed as follows :

$$M = f / [\,|z/\cos\theta| - f - (x-x_a)\cdot\sin\theta\,] \tag{5-b}$$

The sign of θ should be determined according to following rule :

If the position of the optical axis is in the 1 and 3 quadrants of ZOX coordinate plane (as shown in Fig.1 (b)), the sign is positive, otherwise it is negative.

3. Correction of in-plane displacement formula

Because there is an angle between the image plane and the object surface, it is necessary to correct the speckle displacements on the image plane for calculating the in-plane displacements on the object surface. The oliquity of the optical axis in horizontal plane makes the magnification be the function of x. So that after a straight line on object surface in horizontal direction (or in x direction) is photographed, it will be an oblique line on the image plane. According to this, if a certain point on the object has a displacement in x direction, the displacement will contribute to the displacement of y direction on the image plane.

In Fig. 2, $\Delta P_2'EP_1'$ is on the image plane. P_1 and P_2 are the points on the object surface, and C is the intersection of the straight line P_1P_1' with the plane which is parallel to the XOY coordinate plane and go through the point P_2'. It is obvious that the line $P_2'C$ is parallel to the line P_1P_2. The point O is the optical center of the photographic lens. Let M_1 and M_2 represent the magnification at P_1 and P_2 respectively, and one may have :

$$CP_1' = M_1\cdot OP_1 - OC$$

and $OC = M_2 \cdot OP_1$

so $CP'_1 = (M_1 - M_2) \cdot OP_1$

and $EC \,//\, OP_1$, $EF \,//\, OZ$

then $CE = P'_2E \cdot tg(\theta + \theta_1)\cos\gamma$

$EP'_1 = CE/tg\beta$

$= y \cdot P'_2E \cdot tg(\theta + \theta_1) / |z|$ (6)

$P'_2C = P'_2E\,[\cos(\theta + \theta_1) + \sin(\theta + \theta_1) \cdot tg\gamma]$

$= P'_2E\,[\cos(\theta + \theta_1) - x\sin(\theta + \theta_1)/|z|]$ (7)

Let $K_x = \cos(\theta+\theta_1) - x \cdot \sin(\theta+\theta_1)/|z|$ (8-a)

$k_y = y \cdot tg(\theta + \theta_1)/|z|$ (8-b)

So that (7) and (8) can be written as :

$EP'_1 = K_y \cdot P'_2E$ (9-a)

$P'_2C = K_x \cdot P_2E$ (9-b)

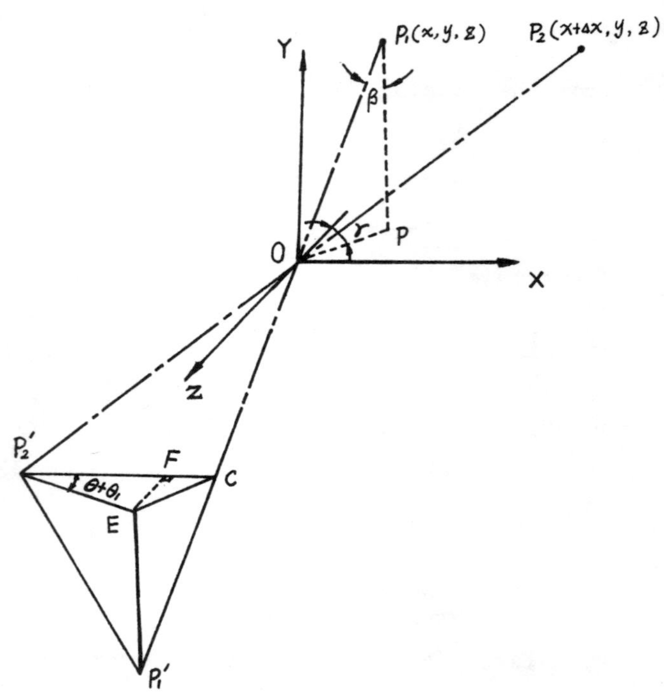

Fig.2 Geometric relation of image plane with object surface

In expression (9-a), EP'_1 is the contribution of P'_2E to y direction. So that, for calculating the displacement in y direction, this term must be subtracted. When the pointwise filtering method is used, the in-plane displacement calculating formula can be corrected as :

$$d_x = K_x \cdot \lambda \cdot L \cos\alpha / S \cdot M \quad (10\text{-a})$$

$$d_y = \lambda \cdot L(\sin\alpha - K_y \cdot \cos\alpha) / S \cdot M \quad (10\text{-b})$$

Where L is the distance from the speckle photogram to the screen, λ is the wavelength of the laser used for analysing the photogram, S is the space of Young's fringes, and α is the angle between the axis along the x direction and the speckle displacement vector which is perpendicular to the Young's fringes. From (8) and (10), it is known that, when $\theta = o$, $K_x = 1$, $K_y = o$, and the formula (10) is same as that of the optical axis being perpendicular to the object surface.

4. Correction of 3-D displacement equations

In order to measure 3-D displacements using speckle photography technique, two speckle photographic cameras are needed and they should be set at different positions and exposured simultaneously. When the optical axis is perpendicular to the object surface, the 3-D displacements of the arbitrary point on the object can be calculated from the following equations :[6]

$$\begin{bmatrix} 1 & 0 & x_I/|z_I| & 0 \\ 0 & 1 & y_I/|z_I| & 0 \\ 1 & 0 & x_{II}/|z_I| & 0 \\ 0 & 1 & y_{II}/|z_I| & 0 \end{bmatrix} \begin{bmatrix} dx \\ dy \\ dz \\ 0 \end{bmatrix} = \begin{bmatrix} dx_{tI} \\ dy_{tI} \\ dx_{tII} \\ dy_{tII} \end{bmatrix} \quad (11\text{-a})$$

Where I and II represent the numbers of the coordinate systems. The origins of these coordinate systems are the optical centers of the photographic lenses. The right terms of the above equations are the speckle displacements of the 3-D displacement vector on the plane which is parallel to the object surface and through the end of the vector. For the OSP method, the speckle displacements can be changed into the ones obtained by using the PSP method through the expression (10). They may be written as follows :

$$\begin{bmatrix} dx_{tI} \\ dy_{tI} \\ dx_{tII} \\ dy_{tII} \end{bmatrix} = \begin{bmatrix} K_{xI} \cdot dx^o_{tI} \\ dy^o_{tI} - Ky_I \cdot dx^o_{tI} \\ K_{xII} \cdot dx^o_{tII} \\ dy^o_{tII} - Ky_{II} \cdot dx^o_{tII} \end{bmatrix} \tag{11-6}$$

where $dx^o_{ti} = L \cdot \lambda \cdot \cos\alpha i / Si \cdot Mi$

$dy^o_{ti} = L \cdot \lambda \cdot \sin\alpha i / Si \cdot Mi$ (i=I,II)

(11-a) and (11-b) are the equations for calculating the 3-D displacements when the OSP method is used, naturally the equations also can be applied to the PSP method.

Method for increasing the sensitivity of measurement

If the displacement of the object is smaller than the speckle size,the information of the displacement can not be obtained. In order to draw away the distance of the speckles before and after the deformation of the object (or to make 'double holes ' speckle pattern), the method of moving the photographic plate is applicable. The recorded speckle displacement is the sum of the plate displacement with the speckle displacement of the deformed object. To pick out the plate displacement, a reference object should be put beside the sample. Following expression may be used for picking out the additional displacement.

$$dx^o_t = (\cos\alpha_t / S_t - \cos\alpha r / S_r) \cdot \lambda \cdot L / M \tag{12-a}$$

$$dy^o_t = (\sin\alpha t / S_t - \sin\alpha r / S_r) \cdot \lambda \cdot L / M \tag{12-b}$$

Where St, Sr ; αt, αr are the spaces of Young's fringes and angles between the normal of Young's fringes and x axis respectively on the deformed object and the reference object.

Smaller displacement may be measured, and the real direction of the speckle displacement can be recognized by applying additional displacement. If the quantity of the additional displacement is suitable, the accuracy of measurement will be satisfactory.

Verification

1. Comparison of the OSP method with the PSP method

In Fig.3, the sample has an out-of-plane displacement. The OSP method and the PSP method are used for measuring the same point A. In the experiment, the magnifications of the two cameras are same. The maximum visual angle of the lens is α_m (which is about 24°). From Fig.3, the absolute value of coordinate x_2 is always bigger than that of x_1. So that, in Fig. 4, the information of the out-of-plane displacement obtained by using the OSP method is greater than that of PSP method.

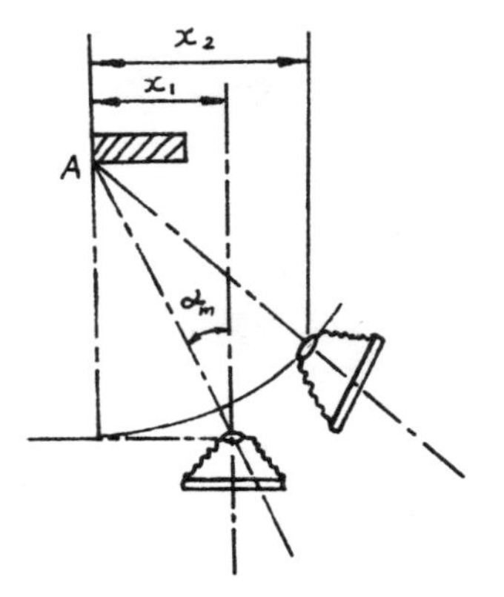

Fig. 3 Comparisòn of OSP with PSP

2. Measurement of 3-D displacement

To verify the validity of the OSP method for measuring 3-D displacement, the deformation of the cantilever having theoretical values along the x axis and z axis is measured. The beam is 215 mm long and fixed in one end. The free end of the beam is acted a load as shown in Fig.5. Two slide projectors were used as the illuminating source. The random speckle pattern is created by brushing the retro-reflective paint on the surface of the beam. The pattern before and after deformation of the beam was recorded on the holographic films using two lenses with focus of 150 mm at 0.27 and 0.25 magnifications respectively in positions I and II. The positions of the cameras are shown in Fig. 5.

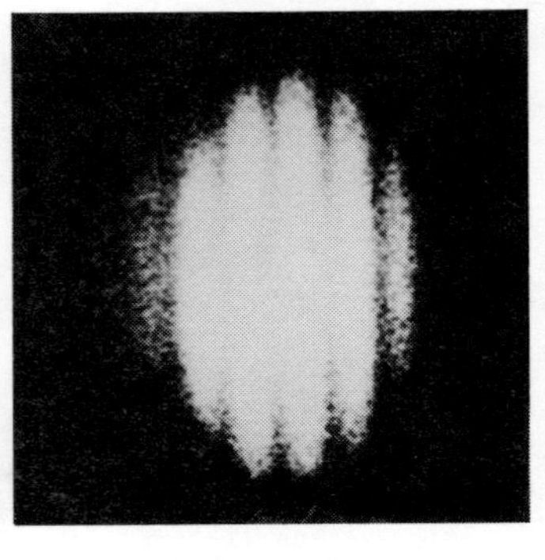

(a) OSP (b) PSP

Fig. 4 Young's fringes

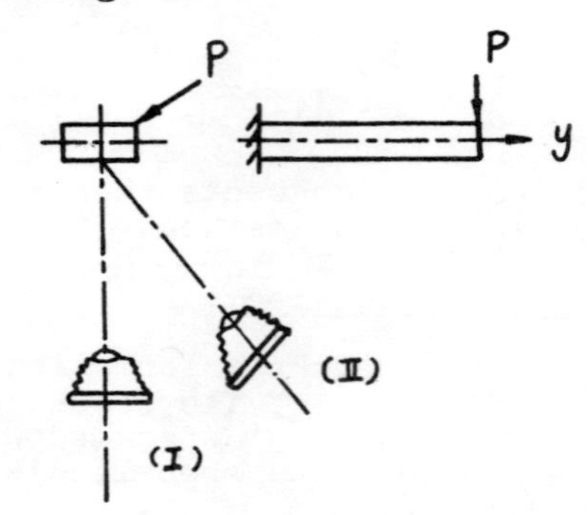

Fig. 5 Measurement of 3-D displacement

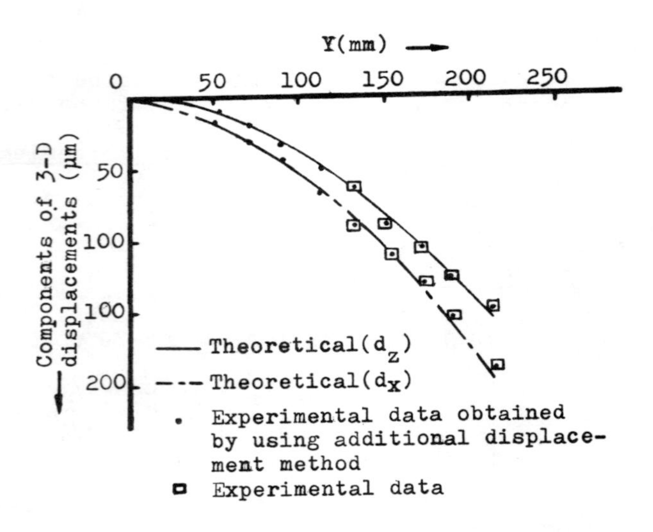

Fig. 6 Experimental values and theoretical curves

The experimental values and the theoretical displacement curves are shown in Fig.6. The experimental values fit well with the theoretical curves. Moreover, it can be seen from the Fig. 6 that smaller displacement can be measured if the additional displacement method is used.

3. Advanced application

According to the theory and measurement techniques mentioned above, we applied the white light OSP method to measure the 3-D thermal deformation of the head stock of a precision lathe. The experiment was done in a workshop and two slide projectors were used as the illuminating source. The double exposure method was used. The first exposure was made before the lathe operated, after it ran for 2 hours at a speed of 1500 r.p.m, we stopped it and made the second exposure. The result of the thermal deformation of the head stock is shown in Fig. 7.

Automatic process of Young's fringes

1. System introduction

Fig. 8 is the diagram of the system for analysing Young's fringes automatically. The analysed specklegram is set on the X-Y work table driven by two step-motors which are controlled by the microcomputer. When a thin laser beam goes through the analysed point on the specklegram, Young's fringes may be observed on the ground glass screen. The light intensities of this pattern are converted to electric signals and sampled to yield a digital picture made up of 128 x 128, each of which is quantitized 256 discrete grey levels. The digital picture is stored directly in the RAM of the Apple II microcomputer. The spaces of Young's fringes in x and y directions can be recognized by using the technique of digital picture process and recognization. Lastly, the calculated speckle displacements are printed by a printer. The process from the input of the image to the output of the speckle displacemtnt is completed automatically. The method for recognizing the spaces of Young's fringes in x and y directions is discussed below.

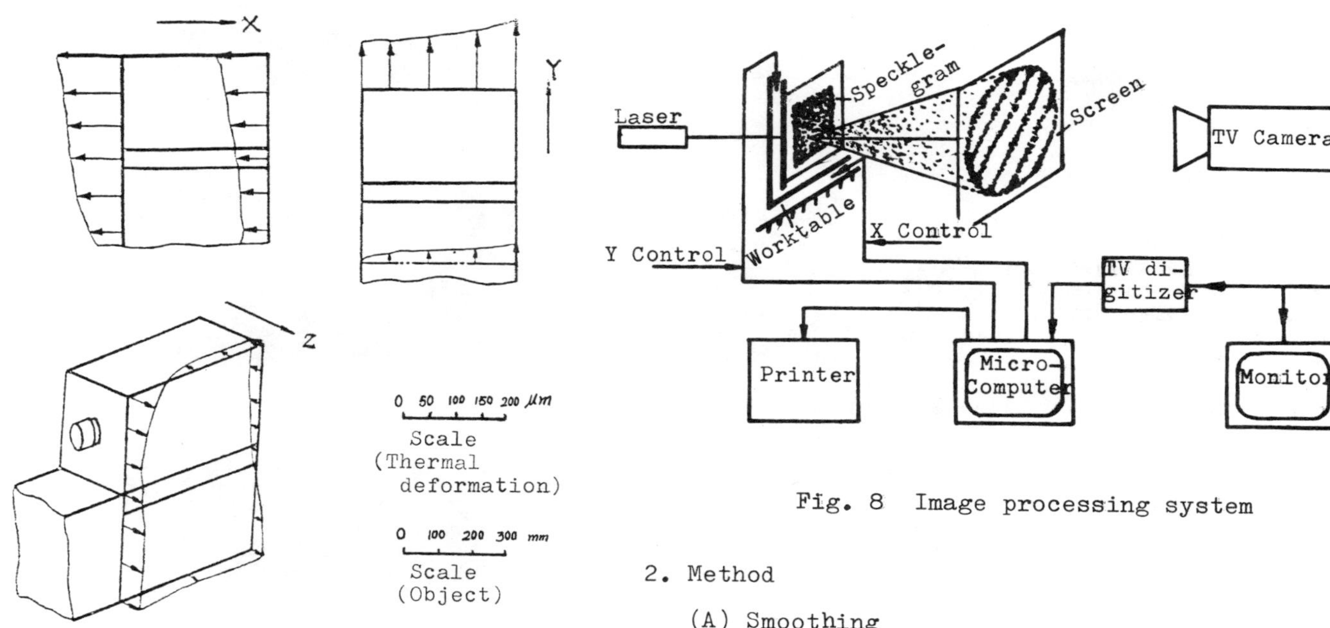

Fig.7 Thermal deformation of the head stock of the lathe

Fig. 8 Image processing system

2. Method

(A) Smoothing

Speckles yield to noise in the speckegram and so deteriorate the Young's fringes quality, so that the fringes digital image should be smoothed before doing recognization works.

(B) Peak value detection

After Young's fringe pattern is filtered suitably, the peak value of the light fringes are detected by the microcomputer. Here, the one dimention scanning method for detecting the peak values is used. First, the scanning line which is closest to the normal of Young's fringes is chosen from the four directions as shown in Fig.9. It is according to the fact that in the same scanning distance, the numbers of fringes along the scanning line is the largest, and the space of the fringes is the smallest. Then along the direction of the scanning line all of the peak values are detected over the image. Fig.10(a) is the Young's fringes. Fig.10(b) and Fig.10(c) are the results of detection.

90° 135° 45° 0°

Fig.9 Four scanning directions

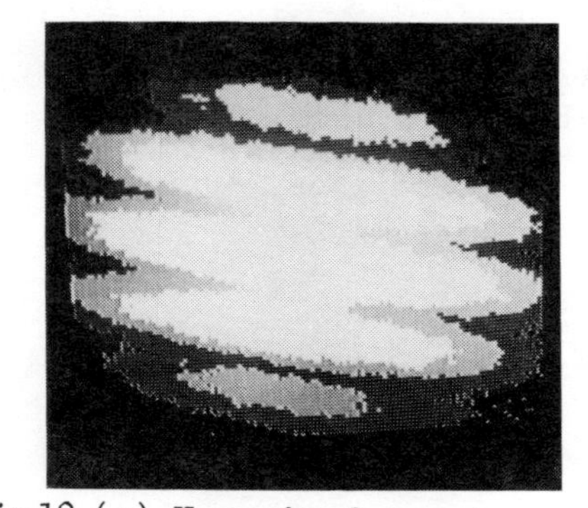

Fig.10 (a) Young's fringes

(C) Recognization of fringe spaces

When the pointwise filtering method is used, the components of speckle displacements in x and y directions can be found out by the following expressions :

$$dx = L \cdot \lambda / S_x \cdot M \qquad (13\text{-a})$$

$$dy = L \cdot \lambda / S_y \cdot M \qquad (13\text{-b})$$

Where S_x and S_y are the spaces of Young's fringes in x and y directions respectively. So that, if S_x and S_y are recognized, the speckle displacement component dx and dy can be calculated from expressions (13-a) and (13-b). Because the center lines of light fringes have been extracted, the spaces S_x and S_y can be identified by scanning the image along x and y directions respectively. The final results of S_x and S_y are the average values of all the real fringe spaces.

After having speckle displacements calculated, the microcomputer sends out the control signals to the step-motors. Then the step-motors drive the X-Y work table. Lastly, the specklegram is moved to a new analysing point and the processes mentioned above are repeated.

(D) Results

Young's fringe patterns of various fringe densities and angles have been processed by using the image processing system. The relative error between the result obtained from the system and the result measured artificially is within five percent, when the numbers of fringes are in the range of five to twenty. Actually, as the result obtained from the system is the averages of whole Young's fringe spaces, so it is more accurate than that measured artificially.

Conclusion

The OSP method and the theory for analysing 3-D displacements discussed in this paper can be applied to either the white light speckle photography or laser speckle photography. The presentment of OSP method makes it possible to measure the 3-D displacements of larger object more accurately, and this has an important sense in measuring displacement field of engineering structures. To process automatically the Young's fringe patterns makes the speckle photography technique tends to be more practical.

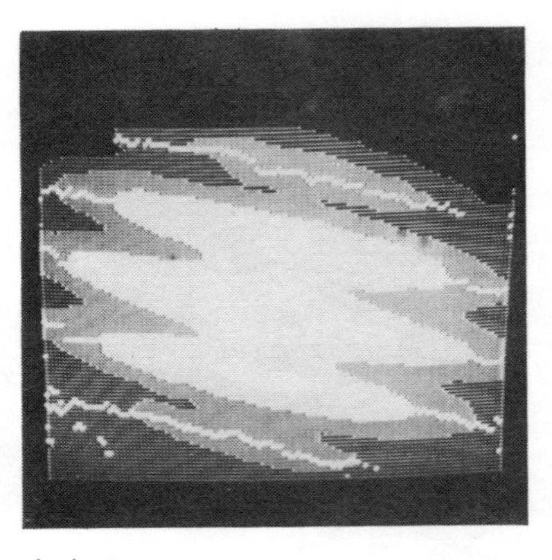

(b) Detecting peak values

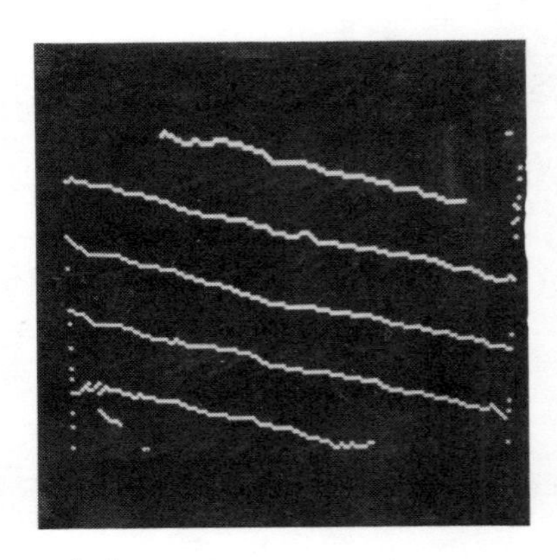

(c) Binary result

Fig. 10 Detection of peak values

Acknowledgement

The authors would like to thank Mrs.H.B.Yien, Mr.X.G.Guo and Prof.Y.L.Cai of Image Process Laboratory, and thank Mr.H.Zhao,Z.W.Wang,Mr.J.M. Wang and Mr.X.Z.Wang of laser and infrared Laboratory in our university,for their help in this paper.

References

1 F.P. Chiang, Applied Optics, Vol.21 No.3, May/June 1982, 379-390.
2 R.P. Khetan, F.P. Chiang, Applied Optics, Vol.15 No.9, Sept. 1978, 2205-2215.
3 A.E.Ennos, M.S.Virdee, Optical Engineering, Vol.21 No.3, May/June 1982,478-482.
4 A.Asundi, F.P.Chiang, Optical Engineering, Vol.21 No.4, July/August 1982,570-580.
5 A.Asundi, F.P. Chiang, Optics and Laser Technology, Feb. 1983,41-45.
6 He Yuming, Tan Yushan, Wan Jienming and Wan Xizou, Laser Journal (published in China), Vol.6 No.5, 1985, 232-235.
7 B.Ineichen,P.Eglin, and R.Dandliker, Applied Optics, Vol.19 No.13, July 1980,2191-2195.
8 David W.Robinson, Applied Optics, Vol.22 No.4, July 1983, 2169-2176.

Measuring rotating component strains using ESPI

Richard W.T. Preater

Department of Mechanical Engineering, The City University,
Northampton Square, London, EClV OHB, U.K.

Abstract

A non-contact method of experimental stress analysis which does not suffer from many of the restrictions of conventional methods and may be used under service environmental conditions is doubly attractive. Pulsed laser holographic techniques display considerable potential for use in this way. One of these methods, Electronic Speckle Pattern Interferometry, is being developed for measuring rotating component strains, at The City University. Results achieved so far display interference fringe information for component tangential velocities over the whole range 0-150 m s^{-1}.

Introduction

Model testing of engineering components may not be truly representative of actual component behaviour in service. Theoretical predictions are often subject to assumptions which although providing a solution may again provide unrealistic results. Prototype testing under service operating conditions alone can provide the actual record of the component performance. Conventional testing methods, however, may require plant shut-down and careful surface preparation which is costly in time and lost production. Holographic techniques, in particular Electronic Speckle Pattern Interferometry (ESPI), using a pulsed laser for illumination offers the possibility of providing a non-contact method requiring the minimum of surface preparation and application under normal operating conditions. This form of approach is very attractive.

ESPI

The ESPI technique uses two beam illumination of the component surface which may be provided by the simple optical system shown in Figure 1. The direction of the resolved displacements is in the plane of the direct and reflected illuminating beams. This method of in-plane displacement measurement was pioneered by Butters and Leendertz [1] at Loughborough University for static components. Development of the technique for application to rotating components is being undertaken at The City University.

Use of a pulsed laser in place of the c/w laser removes the rigorous stability requirements of the system, as it does for conventional holography. The pulse width of 50-70 ns freezes the component motion and provides an interferometric speckle image of the component surface. Providing an initial no-load and live-load speckle images may be recorded both steady state and transient loading conditions may be analysed.

A schematic diagram of the electronic system required is shown in Figure 2. The initial no-load speckle image is recorded digitally and continuously replayed on the monitor. The live-load image is then subtracted from the first and the resulting interference fringe pattern of the component displacement field is recorded on video tape. A single frame search may then be carried out using slow motion replay to display the frame of interest.

In order to achieve satisfactory subtraction of speckle images of the component many revolutions apart precision triggering of the laser Q-switch is required to give image location to within 1 speckle. Timing errors will give superimposed errors distorting the fringe pattern. Camera blanking is required for the high resolution tv-camera used so that complete tv-frames are recorded and subtracted even though firing of the laser may not coincide with the start of a tv-frame. Reliance is made on the persistence of the tv-tube in the absence of electron beam scanning during blanking, to allow recording to await the arival of the next tv-frame pulse.

Early work [2] using this technique has already shown that the suggested limiting component tangential velocity of 2 m s^{-1}, by Cookson et al [3], is unrealistic. Introduction of high resolution tv-equipment has considerably increased the apparent speed range which may be covered. Even then at speeds above 5000 r.p.m. however using the standard plane mirror optical system, some progressive reduction in the useful area of fringe information occurs. Fringe "blurring out" takes place reducing the fringe pattern to a narrow band across the component. Changing the optical system to provide radial sensitivity rather than horizontal alone, removes the restriction and restores full field information for speeds in excess of 12,500 r.p.m. and tangential velocities of 150 m s^{-1}.

In general all the components tested have been coated with matt white paint so as to scatter sufficient light towards the tv-camera, mounted normal to the component surface. However, not all components may be painted in this way and recent tests [4] on an automobile brake disc have shown that using the modified optical system in conjunction with high resolution equipment, then simple machined surfaces still scatter sufficient light to produce satisfactory fringe patterns. Tests on a model aircraft propeller blade show also that the technique is not restricted to plane surface components and that use of a macro-zoom lens gives a reasonable "depth of focus".

Experimental results

The displacement interferograms presented here are chosen to show the considerable improvement that has taken place in the development of the technique.

Figure 3 shows an interferogram of horizontal displacements in a tension strip containing three holes of 1, 2 and 5 mm diameter. This tension strip was mounted radially on a rotating component to display an anticipated form of fringe pattern in some of the earliest tests carried out at 500 r.p.m. and tangential velocity of 5 m s^{-1}.

The results shown in Figures 5 and 6 show the blurring out of useful fringe information at speeds above 5000 r.p.m. with the plane mirror optics. Figures 7 and 8 show the effect of optics modification to produce radial sensitivity rather than horizontal alone, and the restoration of full field information.

The interferogram for a model aircraft propeller blade is shown in Figure 4. Clear fringes may be seen on the disc some 50 mm behind, seen here above and below the blade. This shows the apparent "depth of focus" of the system.

Results for an unpainted component, an advanced design vented brake disc, are shown in Figures 9 and 10. Using modified optics to give radial sensitivity removes troublesome specular reflections and provides more even illumination. Sufficient light is scattered from the machined surface for the static disc in Figure 9 and when rotating at 1000 r.p.m. in Figure 10.

Conclusions

Use of high resolution tv-equipment in conjunction with pulsed laser ESPI now provides clear high contrast fringe patterns over a wide range of component speeds. Full field fringe information has been demonstrated for component tangential velocities up to 150 m s^{-1} using a novel optical system. The technique is not restricted to plane components, use of a macro-zoom lens gives a reasonable "depth of focus". With the modified optics sufficient light may also be scattered from simple machined surfaces to give satisfactory fringe patterns in cases where painting the surface is not feasibile. Tests on particular engineering components are now in progress.

Acknowlegements

The development of ESPI for the analysis of rotating component strains is supported by the Department of Trade & Industry and British Rail. Automobile brake parts have been provided by B.L.Technology Ltd.

References

1. Butters, J.N. & Leendertz, J.A. A Double Exposure Technique for Speckle Pattern Interferometry. J.Phy.E.,Sci.,Instrum., Vol.5., pp.272-279, 1971.
2. Preater, R.W.T. In-plane Strain Measurement on Rotating Components using Pulsed Laser Electronic Speckle Pattern Interferometry. Proc. ICO13 Optics in Modern Science & Technology, Sapporo, Japan, A8-10, pp.674-675, 1984.
3. Cookson, T.J., Butters, J.N., & Pollard, H.C. Pulsed Lasers in Electronic Speckle Pattern Interferometry. Optics and Laser Technology, June, pp.119-124, 1978.
4. Preater, R.W.T. Pulsed Laser ESPI applied to Particular Rotating Component Problems. 2nd ITS on Optics and Electro-Optical Applied Science and Engineering, Cannes, 1985, Proc.to be publ. by SPIE, Vol.599-027.

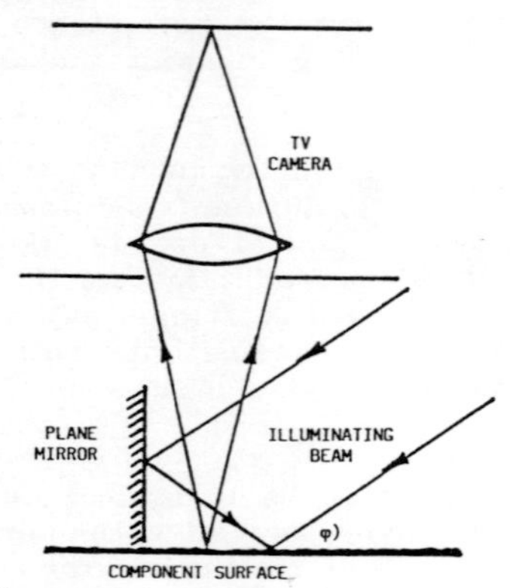

Figure 1. Plane mirror optical component layout.

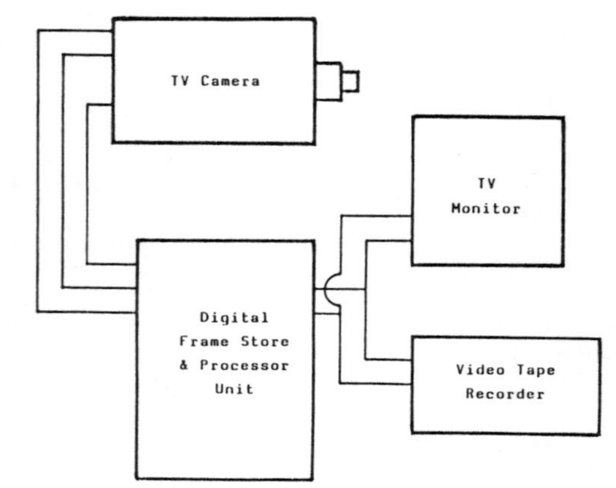

Figure 2. Schematic diagram of the electronic equipment.

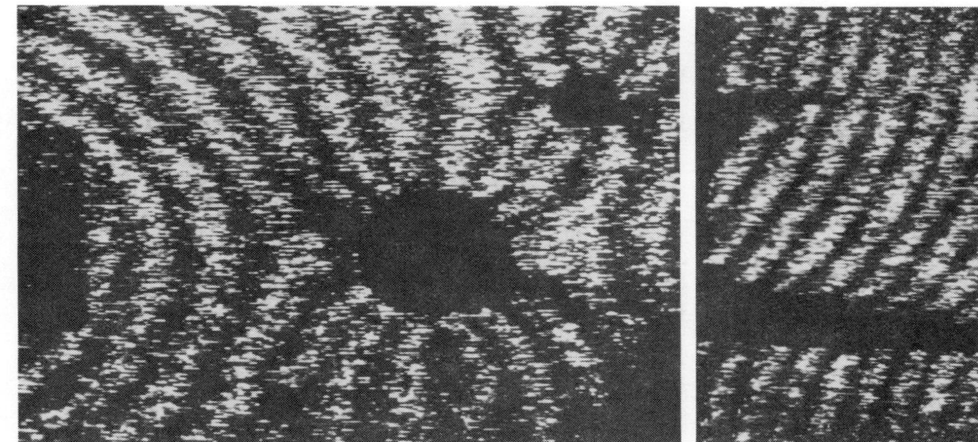

Figure 3. Interferogram of horizontal displacement of a tension strip with holes of 1, 2 and 5 mm diameter rotating at 500 r.p.m. - 5 m s^{-1}.

Figure 4. Interferogram of displacement fringes on a model aircraft propeller blade rotating at at 2750 r.p.m. - 27.5 m s^{-1}.

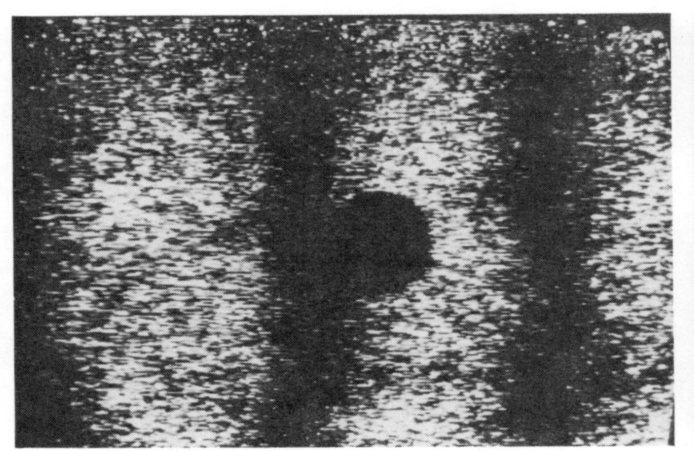

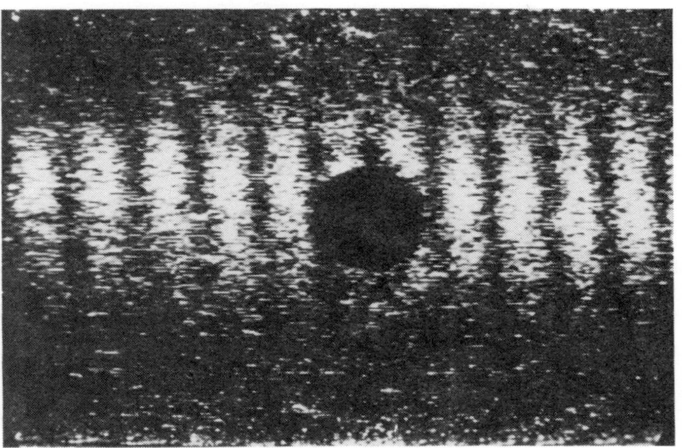

Figure 5. 1000 r.p.m. - 10 m s^{-1}. Figure 6. 10,000 r.p.m. - 100 m s^{-1}.
Interferograms of simulated displacement fringes at different speeds showing the effect of fringe "blurring out" at high speeds using the optical system in Figure 1.

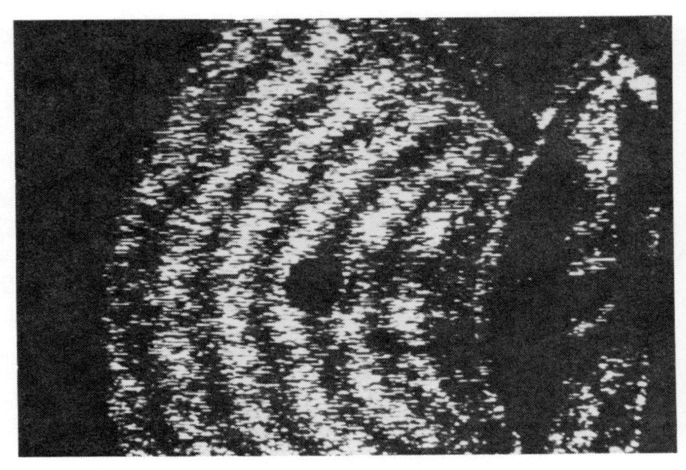

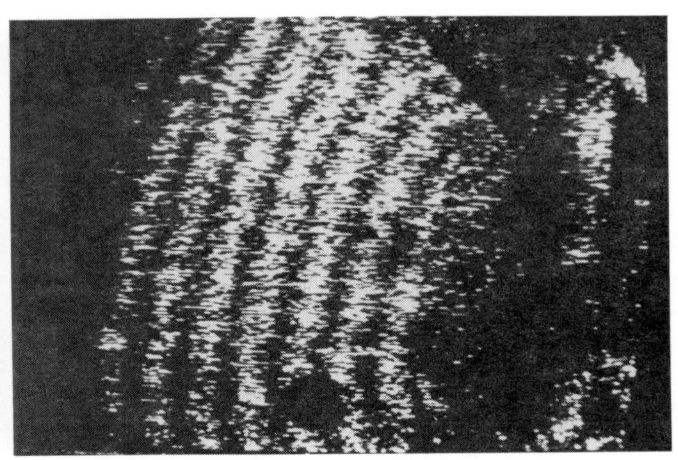

Figure 7. 4200 r.p.m. - 42 m s^{-1}. Figure 8. 11,200 r.p.m. - 112 m s^{-1}.
Full field fringe information produced over a wide range of component speeds using the modified optics.

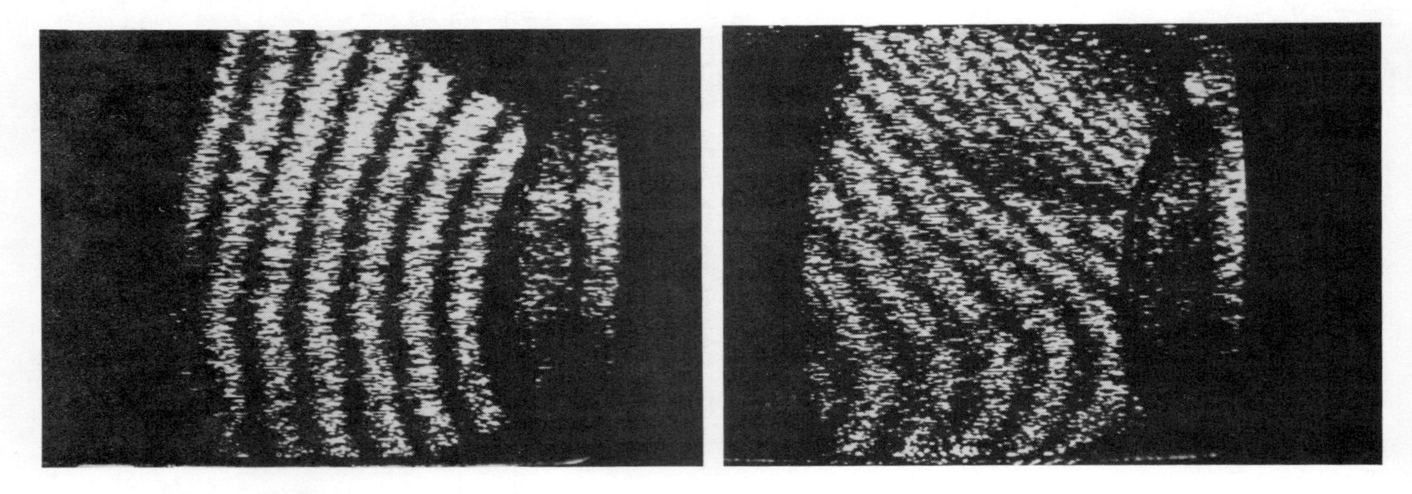

Figure 9. Static Test Figure 10. Rotation at 1000 r.p.m.
Interferograms of simulated displacements on an unpainted brake disc. Sufficient light is scattered from the machined surface to give full field fringe distributions.

A new method of surface roughness measurement using far field speckle

Liu Rong-ping   Liu Pei-sen

Department of Optical Engineering, Beijing Institute of Technology
Department of Optical Engineering, Beijing Institute of Technology, Beijing, P. R. China

Abstract

In this paper, an expression which relates the normalized ensemble average of central peak intensity in the far field speckle pattern to the variance and correlation length of a rough surface under measurement is derived theoretically. On this basis, a new method of measuring surface roughness is suggested, and computer simulation experiment is conducted. The results obtained are in good agreement with the theoretical analysis.

Introduction

Since the applications of laser speckle came into attention, speckle techniques of measuring surface roughness have been developed. W.T.Welford [1] and T.Asakura [2] have made reviews in this field and supplied a detailed collection of references for readers. In recent years, the achievements in this aspect also have been made in China.[3-6]

This paper, based on researching the theoretical dependence of the central peak intensity of the far-field speckle formed by a rough surface whose height fluctuation is far smaller than the wavelength of illumination light upon the surface roughness, suggests a new method of measuring statistical parameters of surface roughness (standard deviation and correlation length), and conducts successfully a computer simulation experiment to prove it.

Principle

The principle scheme of the optical setup for measuring surface roughness is shown in figure 1. A polarized He-Ne laser beam which is expanded by lens L and collimated by lens L is normally incident upon a transparent scattering plate R (e.g. ground glass). The observation plane of the far-field speckle is located at the focal plane of the transform lens L (spectral plane).

Consider an average plane of the height fluctuation on the right side of the plate R, which is perpendicular to the optical axis. (In the following, it is briefly called the exit plane.) Take coordinates (x,y) in the exit plane and (ξ,η) in the observation plane. For mathematical convenience only a one-dimensional case is considered here. When a plane wave of unit intensity is normally incident upon a completely smooth and non-absorbing surface the complex amplitude distribution in the exit plane is given by

$$A_o(x) = \text{rect}\left(\frac{x}{D}\right) \tag{1}$$

where D represents the width of illumination. Then the corresponding intensity distribution in the spectral plane is given by

$$I(\xi) = A(\xi)A^*(\xi) = D^2\text{sinc}^2(D\xi) \tag{2}$$

The intensity at the center is

$$I(0) = I_o = D^2 \tag{3}$$

Suppose the height fluctuation h(x) of the weak scattering surface to be ergodic random process, whose mean and variance are zero and respectively. Then the phase function of the light field in the exit plane, which can be represented as

$$\varphi(x) = \frac{2\pi}{\lambda}(n-1)h(x) \tag{4}$$

where n is the refractive index of the plate, is also an ergodic random process with zero mean and variance σ_φ^2, and there is the relation

$$\sigma_\varphi^2 = \left(\frac{2\pi}{\lambda}\right)^2(n-1)^2\ \sigma_h^2 \tag{5}$$

Neglecting the effect of Fresnel transmission coefficients, the complex amplitude distribution of light field in the exit plane can be written as

$$A_o'(x) = e^{j\varphi(x)} \text{ rect } (\frac{x}{D}) \quad (6)$$

Here it is generally assumed that the polarization state is unchanged. Consequently, the complex amplitude in the spectral plane is given by

$$A'(\xi) = \int_{-\infty}^{+\infty} [e^{j\varphi(x)} \text{ rect}(\frac{x}{D})] e^{-j\frac{2\pi}{\lambda f}\xi x} dx \quad (7)$$

where f is the forcal length and the ensemble average of the peak intensity at the spectral center is given by

$$\langle I'(0)\rangle = \langle A'(0)A'^*(0) \rangle =$$
$$= \iint_{-\infty}^{+\infty} \langle e^{j[\varphi(x)-\varphi(x')]} \rangle \text{ rect}(\frac{x}{D})\text{rect}(\frac{x'}{D})dxdx' \quad (8)$$

For a weak scattering surface there exists

$$\varphi(x) - \varphi(x') = \frac{2\pi}{\lambda}(n-1)[h(x)-h(x')] \ll 1 \quad (9)$$

Expanding the exponential function in (8) into a series and neglecting the terms over the second order, the ensemble average of the peak intensity can be written as

$$\langle I'(0)\rangle = (1-\sigma_\varphi^2)I_o + \sigma_\varphi^2 D \int_{-\infty}^{+\infty} \Lambda(\frac{\Delta x}{D}) r_\varphi(\Delta x) d\Delta x \quad (10)$$

where $\Lambda(\frac{\Delta x}{D})$ is a normalized triangular function and $r_\varphi(\Delta x)$ is a normalized statistical autocorrelation function of the phase fluctuation in the exit plane. When the illumination width is far larger than the central peak width of the autocorrelation function $r_\varphi(\Delta x)$(the region $|\Delta x| < L$, within which $r_\varphi(\Delta x)$ reaches its first zero at $x = \pm L$, is called central peak area.) so that $\Lambda(\frac{\Delta x}{D})$ can be regarded as constant, the ensemble average of the peak intensity can be approximately written as

$$\langle I'(0)\rangle = (1-\sigma_\varphi^2)I_o + \sigma_\varphi^2 DC \int_{-\infty}^{+\infty} r_\varphi(\Delta x) d\Delta x \quad (11)$$

On the supposition that the random process is ergodic the statistical mean and autocorrelation function is equal to its spatial mean and autocorrelation function respectively. Thus, due to the zero mean, it can be proved that

$$\int_{-\infty}^{+\infty} r_\varphi(\Delta x)\, d\Delta x = 0 \quad (12)$$

and we find

$$\langle I'(0)\rangle = (1-\sigma_\varphi^2)I_o \quad (13)$$

After being normalized it becomes

$$\bar{I}(0) = \frac{\langle I'(0)\rangle}{I_o} = 1 - (\frac{2\pi}{\lambda})^2(n-1)^2 \sigma_h^2$$
$$= 1 - (\frac{2\pi}{\lambda})^2 \sigma_{h_\Delta}^2 \quad (14)$$

Then, it comes to the conclusion that under the condition that illumination field is sufficiently large the normalized ensemble average of the peak intensity at the spectral center is parabolically related to the standard deviation σ_h of the height fluctuation h(x). According to this relation, we can get the standard deviation σ_h of a rough surface by measuring its normalized ensemble average of the peak intensity at the spectral center.

In order to evaluate the correlation length of the sample, consider the integral in (10). When the illumination width is small so that the effect of triangular function $\Lambda(\frac{\Delta x}{D})$ on the autocorrelation $r_\varphi(\Delta X)$ can not be regarded as constant, the value of the integral in (10) is no longer equal to zero and would have a contribution to the measured intensity. Especially, if the contribution is mainly from the central peak area of autocorrelation function $r_\varphi(\Delta x)$, then, according to customary definition that the half width of central area of autocorrelation function is the correlation length of the function $\varphi(x)$, we can evaluate the correlation length by the value of the integral.

Suppose that the illumination width reduces to D_1, which makes the contribution of the integral mainly from the central peak area of $r_\varphi(\Delta x)$. Substituting the expression of triangular fanction into (10) we obtain

$$\langle I'(0)\rangle = (1-\sigma_\varphi^2)I_o + 2\sigma_\varphi^2 D_1 \int_0^{D_1} (1-\frac{\Delta X}{D_1})\, r_\varphi(\Delta X) d\Delta X \quad (15)$$

Generally, the concrete form of the autocorrelation function is much more complicated but for estimating the average size of the correlation area we may choose triangular function to approximately represent the actual autocorrelation function $\gamma_\varphi(\Delta x)$ within the central peak area, i.e.

$$\gamma_\varphi(\Delta x)=\begin{cases}1-\dfrac{|\Delta x|}{L} & |\Delta x|\leq L\\ 0 & |\Delta x|>L\end{cases} \tag{16}$$

where L is the correlation length as defined above. Substituting the expression into (15) we get

$$\begin{aligned}\langle I'(0)\rangle &= (1-\sigma_\varphi^2)I_0+2\sigma_\varphi^2 D_1\int_0^L \left(1-\frac{\Delta x}{D_1}\right)\left(1-\frac{\Delta x}{L}\right)d\Delta x\\ &=(1-\sigma_\varphi)I_0+D_1\sigma_\varphi^2\left(L-\frac{L^2}{3D_1}\right)\end{aligned} \tag{17}$$

Because of $D_1 > L$, the value of the second term in (17) is positive. That is to say, for the same variance σ_φ^2 the intensity at the spectral center will increase when the illumination width reduces. Normalizing (17)

$$\begin{aligned}\bar{I}_L &= \frac{\langle I'(0)\rangle}{I_0} = (1-\sigma_\varphi^2)+\frac{\sigma_\varphi^2}{D_1}\left(L-\frac{L^2}{3D_1}\right)\\ &=\bar{I}+\frac{\sigma_\varphi^2}{D_1}\left(L-\frac{L^2}{3D_1}\right)\end{aligned} \tag{18}$$

and regularizing (18) we get

$$\Delta\bar{I}=\bar{I}_L-\bar{I}=\frac{\sigma_\varphi^2}{D_1}\left(L-\frac{L^2}{3D_1}\right) \tag{19}$$

After measuring the difference $\Delta\bar{I}$ of the peak intensity at the spectral center we can find the correlation length of the sample from (19).

Computer simulation experiment and result analysis

We conducted a simulation experiment with an IBM/PC computer, and got the I - $\sigma_{h\Delta}$ curves as shown in figure 2. The curve formed by dots "." represents the parabolic curve calculated from the theoretical formula (14) and the curve formed by "+" represents the experimental values obtained by measuring the normalized ensemble average of intensity at the spectral center. It shows that experimental values coincide well with the theoretical curve. The maximum error is about 0.0004μ and the relative error is less than 1% as $\sigma_h \leq 0.048\mu$. The curves in figure 3 are obtained by measuring the three samples with different correlation lengths. The figure shows that under the wide illumination field the measurement curves of I - $\sigma_{h\Delta}$ are identical with each other. Thus, it is experimentally proved that the approximative process in theoretical analysis is reliable.

In order to measure the correlation length of the rough surface, we reduce the width of the illumination under the condition that we have got the standard deviation of the surface height fluctuation. To the measurement of correlation length it is important to properly choose the small illumination width. In the experiment, we measured three samples whose correlation lengths are $L_1 = 70\mu$, $L_2 = 85\mu$ and $L_3 = 145\mu$ respectively under a mumber of different illumination widths, and obtained the curves shown in figure 4. Some features can be obviously seen in this figure. 1) After the illumination width increases up to a certain value the measured values of the correlation length reduce rapidly and these measured values of samples with different correlation length all tend to zero. This indicates again that the larger the width of the illumination, the smaller the affection of the normalized ensemble average of intensity. 2) There is a maximum of the measured correlation length, and 3) The value of the measured correlation length keeps stationary within a certain extent of the illumination width. The stationary area is nearby the maximum of the measured correlation length. And these measured values in the stationary area are approximately equal to the actual correlation length of the samples.

Thus, for different types of samples to be measured we can reduce the illumination field so that the measured values L jump into the stationary area. With such small illumination field we can measure the correlation length of the rough surfaces. In our experiment, we chose the small illumination width to be 1400μ, and the measured values of the correlation length were about $L_1 = 71\mu$, $L_2 = 81\mu$ and $L_3 = 141\mu$ respectively, as the dotted line shown in figure 4. It shows that the measured values agree with the actual ones of the samples. The values of the correlation length measured in this way are almost unchanged for the samples with the same correlation length but different standard deviation.

Conclusion

The method of measuring surface roughness, suggested in this paper, is an uncontact and two-dimensional method. Its theory is concise and it is desirable to use the method in on-line examination of product quality by a simple arrangement. The roughness measurement extent depends on the wavelength of the light under use and the demanded precision. In the computer simulation experiment we used a He-Ne laser, whose wavelength is 0.6328μ, as an illumination light. The maximum relative error is about 3.5% as $\sigma_h \leq 0.08\mu$. The measurement extent can be further expanded if infrared radiation, which has a longer wavelength, is used.

References

1. W.T.Welford, Opt. Quantum Electron, 9, pp.269, (1977).
2. R.K. Erf "Speckle Metrology" (Academic Press Inc., 1978), pp.11-49.
3. Cheng Lu, and Zhang Bing-quan, Acta Physica Sinica 29, pp.1570, (1980).
4. Mao Wen-yi, Bao Xue-Cheng and others, Acta Optica Sinica, Vol. 5, No.8, pp.724 (1985).
5. Cheng Lu "Optics —— principles and developments" (in chinese), Academic Press, Beijing, China (to be published).
6. Zheng Yue-ming, Wang Ce and Ling De-hong, Acta Optica Sinica, Vol.5, No.3, pp.248, (1985).

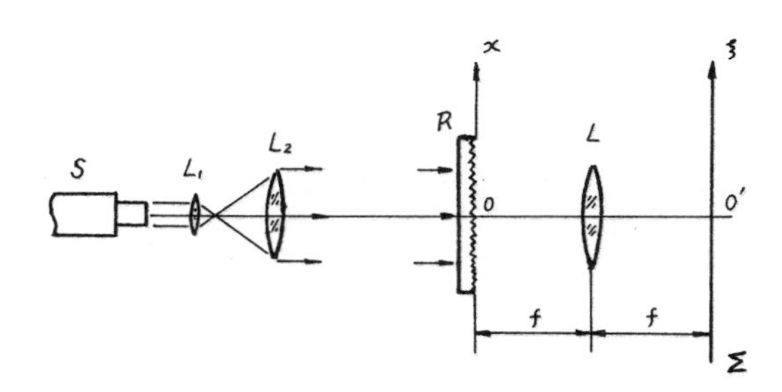

Figure 1. Principle scheme of the experiment

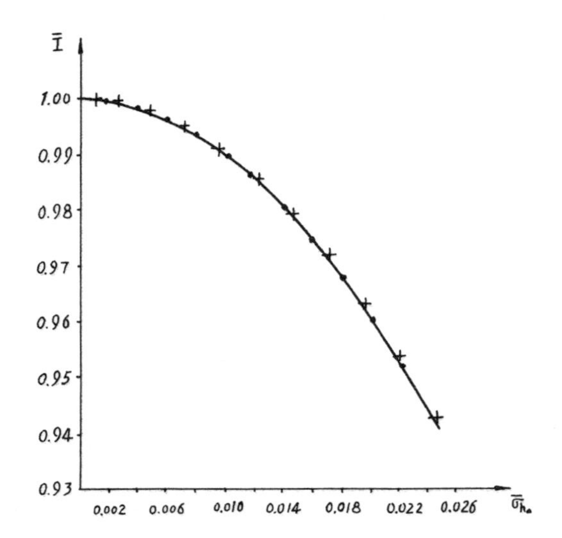

Figure 2. The curves of relation between the normalized ensemble average of the peak intensity $\bar{I}$ and the standard deviation $\bar{\sigma}_{h\Delta}$.

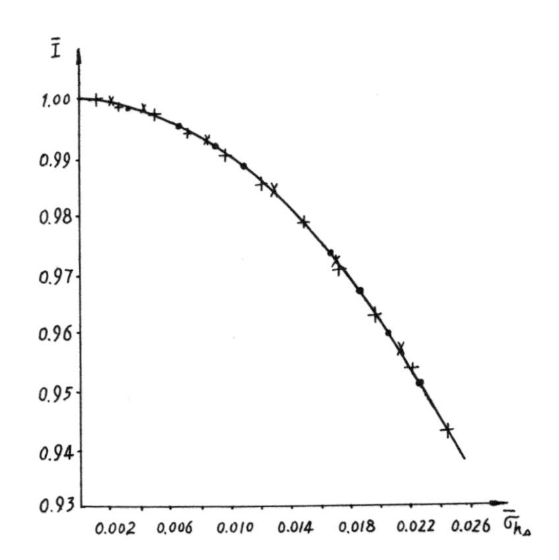

Figure 3. The I-$\sigma_{h\Delta}$ curves of the samples with different correlation length. "+", "." and "x" represent respectively L=70μ , L=85μ and L=145μ .

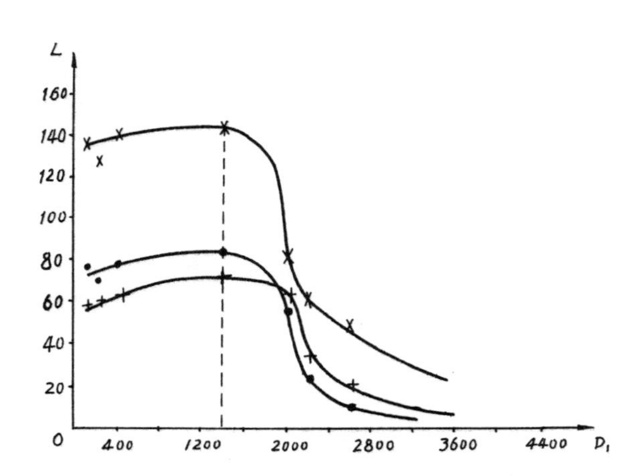

Figure 4. The curves of measured correlation length under different illumination widths.

Forming phase diffusers using speckle

M. Kowalczyk

Institute of Geophysics, University of Warsaw, Pasteura 7, 02-093 Warsaw, Poland

Abstract

Formation of phase diffusers by means of nonlinear record of speckle pattern in the phase photosensitive material is proposed. The essential parameters of diffusers for Fourier transform hologram recording are calculated for analytically approximated tipical nonlinearities.

Introduction

Formation of random phase diffusers (RPDs) by means of a speckle pattern recording in phase photosensitive material was proposed by several authors.[1,2,3] In the References 1, 2, and 3 a linear conversion of exposure into phase shift was assumed. By phase shift or, shortly, phase we mean the argument of the complex amplitude transmittance of diffuser. In this paper a nonlinear conversion is employed to obtain a desired form of the probability density function of phase. This density is of importance when one is seeking to fabricate a RPD which does not transmit a specular component of incident radiation and whose surface relief is quite smooth. Such a diffuser provides a linear record of Fourier transform hologram and low level of coherent noise in the reconstructed image.

Analysis

We consider the following nonlinear characteristic curves of phase photosensitive material:

$$\phi_1(E_1) = \pi\left[1-2\exp(-E_1/\langle E_1\rangle)\right] \tag{1}$$

where ϕ denotes phase, E is for exposure and $\langle\rangle$ denotes statistical average,

$$\phi_2(E_2) = \pi \operatorname{erf}\left[0.5\sqrt{2}(E_2-\langle E_2\rangle)/\sigma_{E_2}\right] , \tag{2}$$

where

$$\operatorname{erf} s = 2\pi^{-1/2}\int_0^s \exp(-t^2)\,dt , \tag{3}$$

and σ_E^2 denotes the variance of exposure, and

$$\phi_3(E_3) = j_{0,1}\sin\{0.5\pi \operatorname{erf}\left[0.5\sqrt{2}(E_3-\langle E_3\rangle)/\sigma_{E_3}\right]\} , \tag{4}$$

where $j_{0,1}$ denotes first zero of Bessel Function $J_0(\cdot)$. Equation (1) approximates the characteristic curve of a tipical photoresist[4] (Figure 1), Equations (2) and (4) approximate the characteristics of a bleached silver-halide emulsion. The curve given by Equation (4) corresponds to the emulsion of higher contrast than that of Equation (2). Diffusers can be recorded in free propagation geometry as it is shown in Figure 2.

If E_1 obeyes the negative exponential density (this is the case when a normal speckle pattern[1] is recorded[5]) and E_2 is Gaussian (the case of incoherent sum of not correlated normal speckle patterns[6]) then ϕ_1 and ϕ_2 are uniformly distributed in the $(-\pi,\pi)$ interval:[7]

$$W(\phi_1) = W(\phi_2) = \begin{cases} 1/2\pi , & |\phi| \le \pi \\ 0 , & |\phi| > \pi \end{cases} , \tag{5}$$

where $W(\phi)$ denotes the probability density function of phase. If E_3 is Gaussian then ϕ_3

obeyes the following density function[8]

$$W(\phi_3) = \begin{cases} [\pi^2 (j_{0,1}^2 - \phi_3^2)]^{-1/2}, & |\phi_3| < j_{0,1} \\ 0, & |\phi_3| \geq j_{0,1} \end{cases} \tag{6}$$

It was shown that exposition of about 15 times to a normal speckle pattern is sufficient to obtain gaussian distribution of exposure, provided succesively exposed patterns are not correlated to each other[3].

The density functions $W(\phi_1)$, $W(\phi_2)$, and $W(\phi_3)$ fulfill the strong scattering condition:[9]

$$\left| \int_{-\infty}^{\infty} W(\phi) \exp(i\phi) d\phi \right|^2 = 0 \quad , \tag{7}$$

so that corresponding diffusers are strong. These diffusers are quite smooth as theirms rms phase is of order 1.8 rad. Among diffusers that we consider one whose density function of phase is given by Equation (6) is of smallest rms phase. Under the assumption that refraction index of diffuser material equals 1.5 its rms roughness is equal to only 0.54 of a wavelength.

Since the average intensity in the recording plane of a Fourier transform hologram is given by[7] (without the reference beam):

$$\langle I(\xi,\eta) \rangle \propto |T_a(\omega_x,\omega_y)|^2 \ast S_t(\omega_x,\omega_y) \tag{8}$$

where $T_a(\omega_x,\omega_y)$ is the spatial frequency spectrum of recorded object, $S_t(\omega_x,\omega_y)$ denotes the power spectrum of the diffuser transmittance, $\omega_x=\xi/\lambda f$, $\omega_y=\eta/\lambda f$, and f is the focal length of the Fourier transformer, we shall calculate the power spectrum for the diffusers under investigation. Then we calculate some diffuser parameters that influence the level of the coherent noise in the image plane. In order to demonstrate the method of calculations we perform them for diffuser of the smallest roughness. For uniform diffusers only final results are included. Some of them are recalled from the Reference 7, for the benefit of the reader.

In order to calculate the autocorrelation function $R_{t_3}(\tau)$ of the diffuser transmittance $t_{d3}(x,y)$ we assume that $E_{31} \equiv E_3(x_1,y_1)$ and $E_{32} \equiv E_3(x_2,y_2)$ are Jointly Gaussian with the correlation coefficient $r_{E_3}(\tau)$[3], so that their joint density is as follows: (also the isotropy of the random field $E_3(x,y)$ is assumed)

$$W(E_{31},E_{32};\tau) \cong \frac{[1-r_{E_3}^2(\tau)]^{-1/2}}{2\pi\sigma_{E_3}^2} \exp\left[- \frac{(E_{31}-\langle E_3\rangle)^2 + (E_{32}-\langle E_3\rangle)^2 - 2r_{E_3}(\tau)(E_{31}-\langle E_3\rangle)(E_{32}-\langle E_3\rangle)}{\sigma_{E_3}^2 [1-r_{E_3}^2(\tau)]} \right] \tag{9}$$

This is an assumption indeed since this does not follow from the fact that first order density is Gaussian. Also, the central limit theorem may not be applied here since we assume that successively recorded patterns are not correlated. Thus they may be statistically dependent. Under this assumption autocorrelation $R_{t_3}(\tau)$ is given by:

$$R_{t_3}(\tau) = \int_0^{\infty}\int_0^{\infty} \exp\{i[\phi_3(E_{31}) - \phi_3(E_{32})]\} W(E_{31},E_{32};\tau) dE_{31} dE_{32} \quad , \tag{10}$$

where $\phi_3(E_3)$ is given by Equation (3) and $W(E_{31},E_{32};\tau)$ by Equation (9). Expanding $W(E_{31},E_{32};\tau)$ into a sreies of Hermite polynomials $H_l(\cdot)$ and substituting Equation (4) to Equation (10) yields

$$R_{t_3}(\tau) = \sum_{l=0}^{\infty} \frac{a_l}{l!2^l} r_{E_3}^l(\tau) \quad , \tag{11}$$

where

$$a_l = \frac{1}{\sqrt{\pi}} \int_{-\infty}^{\infty} H_l(t) \exp\left[-t^2 + i j_{01} \sin(0.5\pi \operatorname{erf} t)\right] dt \tag{12}$$

and the following definition of Hermite polynomial is used:

$$H_l(t) = (-1)^l \exp(t^2) \frac{d^l}{dt^l}\left[\exp(-t^2)\right] \quad , \tag{13}$$

For the uniform diffuser that is recorded by means of single exposition to a normal speckle pattern we have

$$R_{t_1}(\tau) = \sum_{n=0}^{\infty} |c_n|^2 r_{E_1}^{n}(\tau) \quad , \tag{14}$$

where

$$c_n = \int_0^1 L_n(-\ln t) \exp(-2\pi i t) dt \quad . \tag{15}$$

In this case the second-order density function of exposure is expanded into series of Laguerre polynomials that are defined as follows:

$$L_n(t) = \frac{\exp t}{n!} \frac{d^n}{dt^n} t^n \exp(-t) \quad . \tag{16}$$

For the uniform diffusers that is recorded by means of multiple exposition to a normal speckle pattern we have:

$$R_{t_2}(\tau) = \sum_{m=0}^{\infty} \frac{|d_m|^2}{m!} \frac{r_{E_2}^{m}(\tau)}{2^m} \tag{17}$$

where

$$d_m = \frac{1}{\sqrt{\pi}} \int_{-\infty}^{\infty} H_m(t) \exp(-t^2 - i\pi \operatorname{erf} t) dt \quad . \tag{18}$$

If diffusers that we consider are recorded in the same system that is shown in Figure 1 then

$$r_{E_1}(\tau) = r_{E_2}(\tau) = r_{E_3}(\tau) = r_I(\tau) \quad , \tag{19}$$

where $r_I(\tau)$ is the correlation coefficient of intensity in the normal speckle pattern to which the phase recording material is exposed.

Since $r_I(0)=1$ (by definition) and $R_t(0)=1$ (t_d represent sample functions of unity-modulus complex random process) the coefficients of series (11), (14), and (17) fulfill the following relation:

$$\sum_{m=0}^{\infty} \frac{|d_m|^2}{m!\,2^m} = \sum_{n=0}^{\infty} |c_n|^2 = \sum_{l=0}^{\infty} \frac{|a_l|^2}{l!\,2^l} = 1 \quad . \tag{20}$$

Equation (20) can be used to estimate accuracy when infinite series are approximated by finite number of terms.

In accordance with the Wiener-Khintchine theorem the power spectrum of the amplitude transmittance can be calculated as follows:

$$S_t(\rho) = 2\P\int_0^\infty \tau J_0(2\P\rho\tau) R_t(\tau)\, d\tau \quad . \tag{21}$$

To proceed further we assume special case of a uniform and circular scattering spot (Figure 2). In this case[6]

$$r_I(\tau) = \left[2J_1\left(\frac{\P L\tau}{\lambda z}\right) \Big/ \frac{\P L\tau}{\lambda z}\right]^2 \quad . \tag{22}$$

Combining Equations (11), (21), and (22) yields:

$$S_{t_3}(\rho) = 2\P\int_0^\infty \tau J_0(2\P\rho\tau) \sum_{l=0}^{\infty} \frac{|a_l|^2}{l!2^l}\left(\frac{2\lambda z}{\P L\tau}\right)^{2l} J_1^{2l}\left(\frac{\P L\tau}{\lambda z}\right) d\tau = \sum_{l=0}^{\infty} \frac{|a_l|^2}{l!2^l}\left[\frac{1}{\P}\left(\frac{2\lambda z}{L}\right)^2 \mathrm{circ}\frac{2\rho\lambda z}{L}\right]^{\ast 2l} , \tag{23}$$

where $(\cdot)^{\ast 2n}$ denotes self-convolution of the function in parenthesis, iterated 2n-1 times. For uniform diffusers we have:

$$S_{t_1}(\rho) = \sum_{n=0}^{\infty} |c_n|^2 \left[\frac{1}{\P}\left(\frac{2\lambda z}{L}\right)^2 \mathrm{circ}\frac{2\rho\lambda z}{L}\right]^{\ast 2n} , \tag{24}$$

$$S_{t_2}(\rho) = \sum_{m=0}^{\infty} \frac{|d_m|^2}{m!2^m} \left[\frac{1}{\P}\left(\frac{2\lambda z}{L}\right)^2 \mathrm{circ}\frac{2\rho\lambda z}{L}\right]^{\ast 2m} , \tag{25}$$

The curves $S_{t_1}(\rho)$ and $S_{t_2}(\rho)$ are drawn in Figure 3. They are evaluated when 16 terms of series (24) and (25) is taken into account.

Iwamoto shown that intensity fluctuations in the image plane, caused by finite bandwidth of hologram and not perfect image focusing, depend on spatial derivatives of phase.[10] Therefore for determination of statistical parameters of intensity fluctuation we must at first determine density function of spatial derivatives of phase. To do this we have to start with calculation of the second-order density of phase. If we make use of the usual transformation technique of random variables[11] then Equations (6) and (9) yield

$$W(\phi_{31}, \phi_{32}; \tau) = \frac{[1-r_I(\tau)]^{-1/2}}{\P^2 j_{01}^2} \frac{\exp\left[\dfrac{2r_I(\tau)\Theta(\phi_{31})\Theta(\phi_{32}) - r_I^2(\tau)\left[\Theta^2(\phi_{31}) + \Theta^2(\phi_{32})\right]}{1-r_I^2(\tau)}\right]}{\prod_{l=1}^{2} \cos\left(\arcsin\dfrac{\phi_{3l}}{j_{0,1}}\right)} , \tag{26}$$

where

$$\Theta(\phi_3) = \mathrm{inverf}\left(\frac{2}{\P}\arcsin\frac{\phi_3}{j_{01}}\right)$$

and inverf$(\cdot)$ is the inverse of the error function.[12] To proceed further we introduce two auxiliary random variables ζ_3 and ψ_3:

$$\zeta_3 = \frac{\phi_{32}(x+\Delta x, y) + \phi_{31}(x,y)}{2} , \tag{27a}$$

$$\psi_3 = \frac{\phi_{32}(x+\Delta x, y) - \phi_{31}(x,y)}{\Delta x} . \tag{27b}$$

It may be shown that[13]

$$W\left(\phi_3,\frac{\partial\phi_3}{\partial x}\right)=\lim_{\Delta x\to 0} W(\zeta,\psi;\Delta x). \tag{28}$$

Making use of the Equations (26) and (27) to calculate the density $W(\zeta,\psi;\Delta x)$, and then using relation (28) we obtain

$$W\left(\phi_3,\frac{\partial\phi_3}{\partial x}\right)=\frac{\exp\left[\Theta^2(\phi_3)\right]}{\pi^2\sqrt{-\ddot{r}_I(0)}\,(j^2_{0,1}-\phi^2_3)}\exp\left[\frac{\left(\frac{\partial\phi_3}{\partial x}\right)^2\exp\left[2\Theta^2(\phi_3)\right]}{\pi\ddot{r}_I(0)\,(j^2_{0,1}-\phi^2_3)}\right], \tag{29}$$

where $\ddot{r}_I(0)=\partial r_I^2(0)/\partial\tau^2$. For the uniform diffusers we have

$$W\left(\phi_1,\frac{\partial\phi_1}{\partial x}\right)=\frac{\exp\left[\dfrac{-\left(\frac{\partial\phi_1}{\partial x}\right)^2}{2(\pi-\phi_1)^2\ddot{r}_I(0)\ln\left[(\pi-\phi_1)/2\pi\right]}\right]}{2\pi\sqrt{2\pi}(\pi-\phi_1)\{\ddot{r}_I(0)\ln\left[(\pi-\phi_1)/2\pi\right]\}^{1/2}}, \tag{30}$$

$$W\left(\phi_2,\frac{\partial\phi_2}{\partial x}\right)=\frac{\exp\left[\mathrm{inverf}^2(\phi_2/\pi)\right]}{4\pi^2\sqrt{-\ddot{r}_I(0)}}\exp\left[\frac{\left(\frac{\partial\phi_2}{\partial x}\right)^2\exp\left[2\,\mathrm{inverf}^2(\phi_2/\pi)\right]}{4\pi\ddot{r}_I(0)}\right]. \tag{31}$$

It is rather difficult to calculate marginal densities $W(\partial\phi/\partial x)$ integrating Equations (29), (30), and (31) over ϕ in a closed form. Nevertheless in particular cases we can evaluate some statistical parameters of image-plane intensity fluctuations without such an integration. For example, for uniform diffusers Equations (30) and (31), and assumption on isotropy of a random field $I(x,y)$ are sufficient to calculate average contrast of intensity fluctuations.[7]

Conclusions

The main purpose of this theoretical study has been to show that by means of nonlinear record of speckle pattern in phase materials one can form random diffusers of quite interesting properties. From our analysis it follows that key parameters of the method that we propose are the correlation coefficient of the speckle intensity and the characteristic curve $\phi(E)$. They influence both spectral and imaging properties of diffusers. We suggest that experimental realizations of diffusers that we described should be performed using photoresist materials as their characteristic $\phi(E)$ does not on spatial frequency in a wide range of frequencies.[4]

Acknowledgments

The research work leading to this paper was performed under research projects MR I/5 and CPBP 01.06.

References

1. R.C.Waag and K.T.Knox, "Power spectrum analysis of exponential diffusers," J.Opt.Soc.Am. 62, 877-881(1972).
2. P.F.Gray, "A method of forming optical diffusers of simple known statistical properties," Opt.Acta 25, 765-775(1978).
3. B.M.Levine and J.C.Dainty, "Non-Gaussian image plane speckle: measurements from diffusers of known statistics," Optics Comm. 45, 252-256(1983).
4. R.A.Bartolini, "Photoresists," in Holographic Recording Materials, H.M.Smith, ed. (Springer-Verlag, New York, 1977).
5. J.C.Dainty, "The statistics of speckle patterns," in Progress in Optics, Vol.14, E.Wolf, ed. (North-Holland, New York, 1975).
6. J.W.Goodman, "Statistical properties of laser speckle patterns," in Laser Speckle and Related Phenomena, (Springer-Verlag, New York, 1975).
7. M.Kowalczyk, "Spectral and imaging properties of uniform diffusers," J.Opt.Soc.Am. A 1, 192-200(1984).
8. C.N.Kurtz, H.O.Hoadley, and J.J.DePalma,"Design and synthesis of random phase

diffusers," J.Opt.Soc.Am. 63, 1080-1092(1973).
9. M.Kowalczyk,"Strong phase diffusers with the minimized standard deviation of roughness," J.Opt.Soc.Am.A 3, xxx(1986) to be published in August issue.
10. A.Iwamoto,"Artificial diffuser for Fourier transform hologram recording," Appl.Opt. 19, 215-221(1980).
11. A. Papoulis, Probability, Random Variables, and Stochastic Processes, (McGraw-Hill, New York, 1965).
12. A.J.Strecok, "On the calculation of the inverse of the error function," Math.Comp. 22, 144-158(1968).
13. B.R.Levin, Teoretitcheskie Osnovy Statistitcheskoj Radiotekhniki, (Sovetskoe Radio, Moscow, 1974), Vol.1.

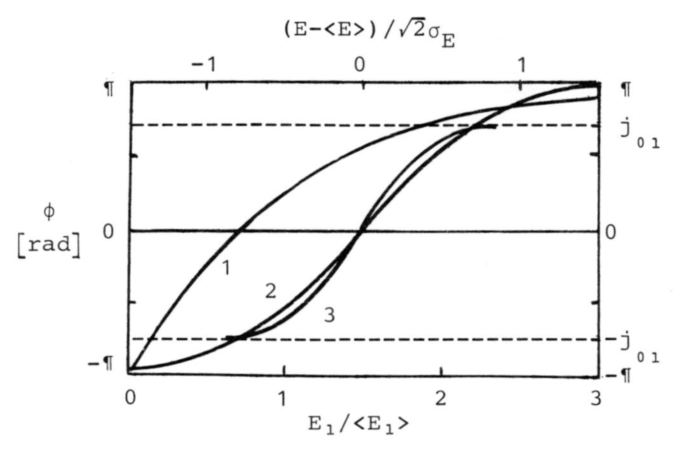

Figure 1. Characteristics of phase materials for recording strong and smooth RPDs. Curve 1 and lower horizontal axis correspond to Equation (1), upper axis and curves 2 and 3 correspond to Equations (2) and (4) respectively.

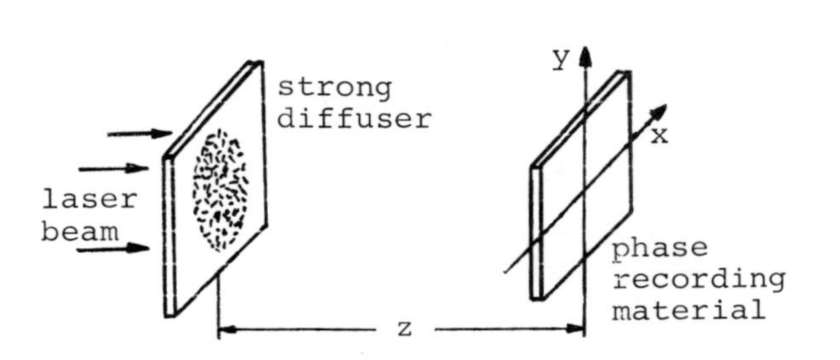

Figure 2. System for forming RPDs. In order to ensure stationarity and isotropy of diffuser transmittance illumination of scattering spot should be circularly symmetric and central part of speckle pattern should be utilized only.

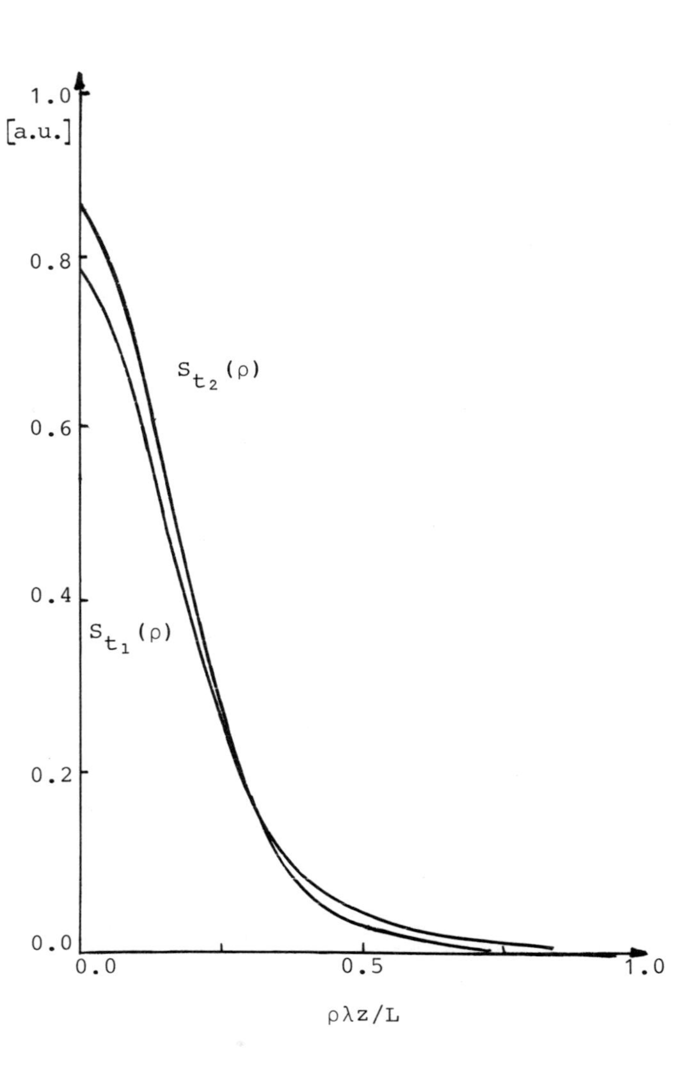

Figure 3. Power spectra of uniform diffusers.

THE STUDY OF SPECKLE PATTERN AND SURFACE ROUGHNESS MEASUREMENT WITH IMAGE PROCESSING TECHNIQUE

Zhang Qing-chuan Xu Bo-qin Wu Xiao-ping
Department of Modern Mechanics
University of Science and Technology of China

ABSTRACT

Digital image processing technique for investigating the statistical properties of speckle pattern and measuring surface roughness is described firstly. With the aid of Model-75 Image Processing System , a speckle pattern can be processed within a minute.

SUMMARY

The special spectrum of objective speckle pattern depends on the surface microstructure or roughness of diffuse object. The reciprocal of cut-off frequency is a reasonable measurement of the mean size of speckles. Using image processing technique, we can get the special spectrum of speckle pattern. According to the spectrum, the roughness of diffuse surface can be measured or the statistical properties of speckle pattern can be studied if the surface roughness has been known.

Combining digital image processing technique with the analysis of speckle pattern, we have obtained a series of measuring results of statistical properties of speckle patterns resulted from various different rough surfaces. During the processing procedure, by Model-75 Image Processing System the speckle pattern is converted into digital picture made up of 512 512 sample points with 256 gray levels . Then VAX-11 minicomputer calculates the two-dimensional Fourier Transform of speckle pattern.

Experimental results have shown that:(1) If $\lambda \ll \sigma_z$ (here σ_z is the root-mean-square deviation of a rough surface, λ is the laser wave length), the relation between the special spectrum of the speckle pattern and the change of roughness is not quite obvious. (2) If σ_z approach λ or is less than λ , the relation will become quite obvious. In the second case, with the descreasing of surface roughness, the high frequency part of the spectrum of speckle pattern is cut down, the cut-off frequency reduses , the mean size of speckles becomes larger. According to the cut-off frequency, the surface roughness of diffuse object can be measured.

Combined with digital image processing technique,the method becomes convenient, realiable and quick.The average time of processing a speckle pattern picture is less than a minute. Possessing above advantages, the method is quite practical.

INTRODUCTION

If a speckle pattern is produced by coherent light incident on a rough surface then surely its statistics will depend on the detailed surface properties. Whilst this is undoubtedly true, it is generally very difficult to extract meaningful surface parameters from speckle pattern,especially for very rough surfaces in monochromatics light where the dependence on roughness is almost negligible. This difficulty is common to many optical methods of evaluating surface structure. In recent years, the methods of measuring roughness with speckle pattern contrast, with ratio of intensity between reflective beam and scattering beam, and with optical Fourier transformation, have been reported. [2] [3]

The width of the autocorrelation function of laser speckle pattern provides a reasonable measure of "the average width" of speckles. But practically it is difficult and requires much work to make numerical computation of autocorrelation function, and till now there hasn't been any computational mothed which is fast enough. However, since the Wiener spectrum and the autocorrelation function are related by a two-dimensional Fourier Transform, knowledge of one will always imply knowledge of the other, and so it will only be necessary to speak in terms of one or the other. By using digital image processing mothed and two-dimensional Fast Fourier Transform (FFT) of the speckle pattern image, The speckle pattern Wiener spectrum, and the zero moment, the first moment and the second moment of spectrum can be obtained conveniently. So the width of the autocorrelation function can be deduced reasonably from the reciprocal of the width of Wiener spectrum.

In this paper the procession of a series of speckle pattern images which have various roughnesses of surfaces with digital image processing technique is reported. The pattern image is converted into a digital picture made up of 512×512×8 bit. According to the polishing direction of diffuse surface, and to make full use of the computer internal memory, we choose a square window and a striped window to get two-dimensional FFT of a digital picture, then normalize the computed results with reasonable mothed, and use a kind of function curve to approach the computed results. We get a series of curves of various rough surfaces. With variation of the parameter of rough surface, the approaching curves of Wiener Spectrum of a speckle pattern pictures also have a good tendency.

BASIC THEORY

Consider the free space propagation geometry shown in Fig.1 . monochromatic polarized coherent light is incident on a rough surface ,and the scattered light is observed at some distance z, without any intervening optical elements. The complex field $A(x,y)$ is observed across Fraunhofer plane paralleling with the (ξ,η) plane . The autocorrelation function of the intensity distribution $I(x,y)=|A(x,y)|^2$ in the (x,y) plane is

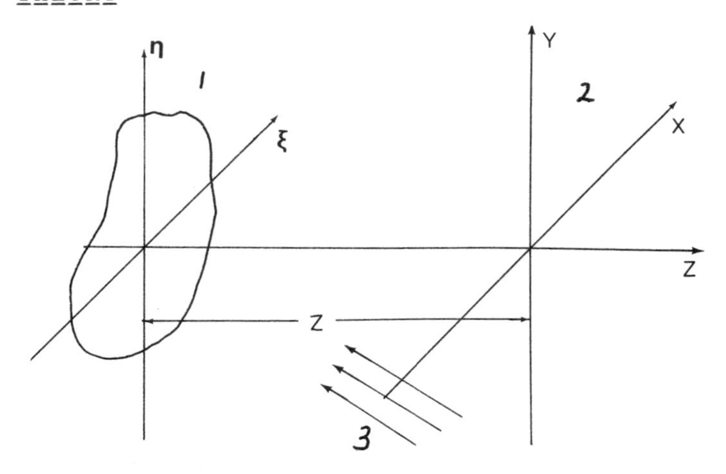

Fig. 1
1:Rough Surface
2:Observation Plane
3:Incident Beam

$$Ri(x1,y1,x2,y2)=\langle I(x1,y1)I(x2,y2)\rangle \quad \text{------------------(1)}$$

Where the average is over an ensemble of rough surface. The autocorrelation function of the field or the mutual intensity of field:

$$Ja(x1,y1,x2,y2)=\langle A(x1,y1)A^*(x2,y2)\rangle \quad \text{------------------(2)}$$

Let the diffuse field obey circular complex Gaussian statistic. The required relation between Ri and Ja is

$$Ri(x1,y1,x2,y2)=\langle I(x1,y1)\rangle\langle I(x2,y2)\rangle+|Ja(x1,y1,x2,y2)|^2 \quad \text{--------(3)}$$

Let J_α be the mutual intensity in the (ξ,η) scattering plane. Following Van Cittert-Zernike theorem:

$$Ja(x1,y1,x2,y2)=\frac{1}{\lambda^2 z^2}\iiiint_{-\infty}^{\infty} J_\alpha(\xi_1,\eta_1,\xi_2,\eta_2)\,EXP\left[i\frac{2\pi}{\lambda z}(x1\xi_1+y1\eta_1-x2\xi_2-y2\eta_2)\right] d\xi_1 d\eta_1 d\xi_2 d\eta_2 \quad \text{------------------------(4)}$$

here $i=\sqrt{-1}$

In our privious discussion, we have made two simplifications in that general relationship:(1) Since we are concerned here only with the modulus of Ja, we drop the initial exponential factor in (x1,y1) and (x2,y2); (2) the observing plane is in the region of Fraunhofer diffraction, the exponential factor in (ξ_1^2,η_1^2), and (ξ_2^2,η_2^2) in the integration is dropped.

Let the function $P(\xi,\eta)$ represents the amplitude of the incident field and consider the structure of the incident fields, $P(\xi,\eta)$ is extremely coarse compared width of the complex coherence factor $u_\alpha(\Delta\xi,\Delta\eta)$ and hence to an excellent approximation.

$$J_\alpha(\xi_1,\eta_1,\xi_2,\eta_2)=k|P(\xi_1,\eta_1)|^2 u_\alpha(\Delta\xi,\Delta\eta) \quad \text{----------------------(5)}$$

here $\Delta\xi=\xi_1-\xi_2$, $\Delta\eta=\eta_1-\eta_2$, under the above assumption, equation (4) becomes of this form:

$$J_A(x1,y1,x2,y2)=\frac{1}{\lambda^2 z^2}\iiiint_{-\infty}^{\infty} k|P(\xi_1,\eta_1)|^2 u_\alpha(\Delta\xi,\Delta\eta)\,EXP\left[-i\frac{2\pi}{\lambda z}(x1\xi_1+y1\eta_1-x2\xi_2-y2\eta_2)\right] d\xi_1 d\eta_1 d\xi_2 d\eta_2 \quad \text{-----------------(6)}$$

The complex coherent factor takes the form:

$$u_A(x1,y1,x2,y2)=\frac{Ja(x1,y1,x2.y2)}{[Ja(x1,y1,x1,y1)Ja(x2,y2,x2,y2)]^{1/2}}$$

$$=\frac{\iiiint_{-\infty}^{\infty}|P(\xi_1,\eta_1)|^2 u_\alpha(\Delta\xi,\Delta\eta)\,EXP(-i\frac{2\pi}{\lambda z}(x1\xi_1+y1\eta_1-x2\xi_2-y2\eta_2)]d\xi_1 d\eta_1 d\xi_2 d\eta_2}{\iiiint_{-\infty}^{\infty}|P(\xi_1,\eta_1)|^2 u_\alpha(\Delta\xi,\Delta\eta)\,d\xi_1 d\eta_1 d\xi_2 d\eta_2} \quad \text{--(7)}$$

From the eq.(3), eq.(4) and eq.(7), the intensity autocorrelation function is

$$Ri(\Delta x,\Delta y)=\langle I\rangle\ [1+|u_A(x1,y1,x2,y2)|^2] \quad \text{---------------------(8)}$$

here $\Delta x=x1-x2$, $\Delta y=y1-y2$.

Since the power spectral density G(vx,vy) of I(x,y) is given by the Fourier transform of the autocorrelation function $Ri(\Delta x,\Delta y)$

$$G(vx,vy)=\mathcal{F}\{Ri(\Delta x,\Delta y)\}$$
$$=\mathcal{F}\{I(x1,y1)\times I(x2,y2)\}=|\mathcal{F}\{I(x1,y1)\}|^2$$
$$=\langle I\rangle\ \{\delta(vx,vy)+\mathcal{F}\{|u_A(\Delta x,\Delta y)|\ \} \quad \text{------------------------(9)}$$

State in words for the eq.(9),under those previous assumption, the power spectral density of speckle pattern consists of a δ function component at zero frequency (vx=vy=0) plus a component extented over frequency and having the shape of the normalized autocorrelation function of the intensity distribution incident on the scattering spot. Knowing from eq. (7) and eq.(9), the component extended over frequency factor is related with the integral of the complex coherent factor of scattering wavefront at rough surface. In other words,it implies the properties of the surface roughness.

In general case,the power spectrum should be calculated according to eq.(9). In this paper,with the digital image processing technique,we shall measure the power spectrum curve represented by eq.(9).

Experiment Research

The experimental arrangement is illustrated in fig.2. The record plane is located in Fraunhofer diffraction region. The x,y axes parallel the ξ, η axes respectively, the z axis direction is the mirror reflection of the rough surface. the other assumed conditions are the same as in Fig.1. Let the polish direction of the rough surface be the η direction. In the observing plane we will get a diffuse bright speckle strip paralleling x axis, the center of which coincides with x axis. As the roughness is cut down , or σ_z decreases. the diffuse bright speckle strip gets more bright, and the "mean width" of speckle pattern gets bigger gradually. That is, the diffraction spectrum of objective speckle pattern will vary with the variation of the roughness of diffuse surface .

The speckle pattern pictures got in observing plane are dealed with the I^2S digital image processing system, after being sampling and quantizing to 512×512×8 bit. This system consists of three parts: VAX-11/750 minicomputer, Model-75 Image Processing machine, and the Image Input and Output system (I/O). VAX-11/750 minicomputer possesses two million internal memory. The I/O system has a DAGE-68 video camera which deals with the recorded speckle pattern picture by video scanning and high fast A/D alternating and turns the picture into a digital picture with 512 lines and 512 collumns and with 256 levels, then sends digital picture into the image processing machine.

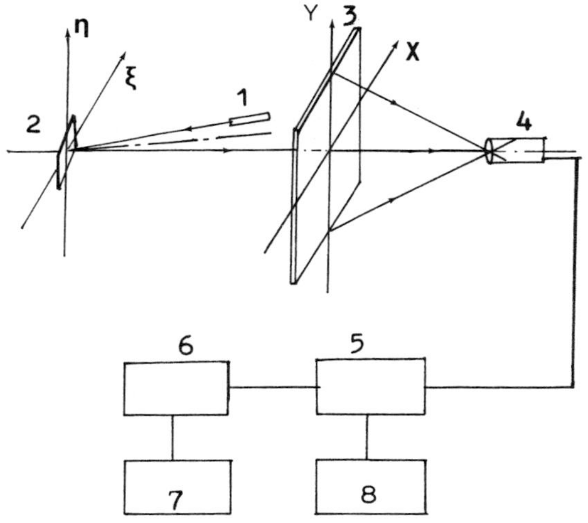

Fig.2
1:Laser 2:Rough Surface 3:Observation Plane
4:Video Camera 5:Model 75 6:VAX-11/750
7:Printer 8:Monitor

Because picture Fourier transform is concerned with the complex numerical computing, to make full use of the internal memory of minicomputer, to collect as many as possible meaningful picture samples, and to make the computation convenient and fast, we use a 128×128×8 bit sample square window and 32×512 sample stripe window to get two-dimensional Fast Fourier Transform (FFT) of the 512×512×8 bit digital picture. The stripe of 32×512 samples is parallel with vx direction. Before FFT, the spectrum center should be moved to the center of the transformation window. After transform, we get the speckle intensity Fourier system. Then we calculate the square of spectrum and get the orignal speckle picture Wiener spectrum represented by eq.(9). The function at zero frequency must be avoided and suitable number to normalize the two-dimension image power spectrum should be chosen. Let maxium of the power spectrum be at zero frequency and the second maxium number be the normalizing number. Then we get the power spectrum represented by eq.(9). We take one-dimension power spectrum data to analyse, for example, choose $G(vx,vy)|_{vy=0}$, $G(vx,vy)|_{vx=0}$, and the vx is the diffuse bright stripe direction of polish rough surface. The vx and vy scales are the sample point numbers.

Dealing with the digital picture by FFT after a continious picture is sampled and quantized, the results must be dispersed to some extent, so these dispersed power spectrum should be approached with some apprapriate curves with the least square method.

In this paper, we take flat polish surface as example, and measure the spectrum of four kinds rough surfaces which the σ_z=1.1, 0.8, 0.6, 0.4μm respectively. Let the symbol 1,2,3,4, represent that cases which σ_z equals to 1.1, 0.8, 0.6, 0.4μm respectively. The symbol Pbi|i=1,2,3,4 represent the cases of FFT window of 128×128 samples, the symbol Pci|i=1,2,3,4 represent the cases of window of 32×512 samples , the G(vx) is the dispersed spectral data, Ui|i=0,1,2, are the zero, first and second moments which are shown as follow:

$$U0=\left(\sum_{i=1}^{N} G(vx_i)\right)/N \qquad U2=\left(\sum_{i=1}^{N} G(vx_i)\times vx_i^2\right)/N$$

$$U1=\left(\sum_{i=1}^{N} G(vx_i)\times vx_i\right)/N \qquad U3=\left(\sum_{i=1}^{N} G(vx_i)\times vx_i^3\right)/N$$

In above eq. the N=128 for the square window, and N=512 for the striped window . the computation results are shown in table one, and the symbols of PBX and PBY in table 1 represent the data of G(vx,vy)|vy=0 and G(vx,vy)|vx=0 respectively by the FFT window of 128×128 samples, and PCX, the data of G(vx,vy)|vy=0 by the FFT window of 32×512 samples. As the σ_z reduces, the moments also reduce. The reduction of zero moment shows that the scattering spectral energy reduces and the reflecting spectral energy increase. The first moment is the measure of mean scale of scattering spectrum and its reciprocal is a reasonable measure of speckle pattern mean size.

We approach the separated power spectrum with the function curve of $EXP\{-a*x-b*x^2-c*x^3\}$. The parameters of approach curve, a,b,c,and the correlation coefficient R between the seperated data and the approach curve have been computed and shown in table 1. We drafted the curves represented by the function in Fig.3 , Fig.4 and Fig.5. Just as shown in Fig.3, Fig.4 and Fig.5, as the surfaces get rougher,and σ_z gets bigger,the part of high frequency of diffuse power spectrum increases, and the power spectral curves which are symmetric about y axis become wider.

Table 1

	no	a	b	c	R	U0	U1	U2	U3
PBX	1	.158E0	-.303E-2	.293E-4	.802	.273E2	.362E3	.891E4	.285E6
	2	.190E0	-.314E-2	.237E-4	.815	.196E2	.216E3	.484E4	.145E6
	3	.208E0	-.265E-2	.122E-4	.810	.138E2	.124E3	.236E4	.568E5
	4	.230E0	-.330E-2	.170E-4	.668	.102E2	.942E2	.183E4	.426E5
PBY	1	.183E0	-.385E-2	.273E-4	.904	.304E2	.465E3	.155E5	.685E6
	2	.192E0	-.411E-2	.331E-4	.882	.253E2	.369E3	.116E5	.486E6
	3	.199E0	-.275E-2	.161E-4	.810	.165E2	.142E3	.289E4	.730E5
	4	.223E0	-.290E-2	.123E-4	.841	.151E2	.101E3	.170E3	.366E4
PCX	1	.266E-1	-.929E-4	.249E-6	.768	.645E2	.317E4	.251E6	.249E8
	2	.429E-1	-.389E-3	.158E-5	.781	.585E2	.302E4	.242E6	.224E8
	3	.578E-1	-.474E-3	.165E-5	.770	.336E2	.173E4	.140E6	.142E8
	4	.714E-1	-.646E-3	.218E-5	.843	.273E2	.154E4	.114E6	.120E8

PBX

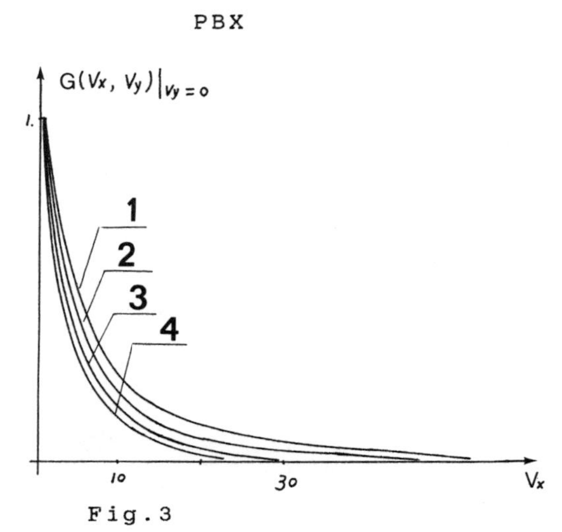

Fig.3

PBY

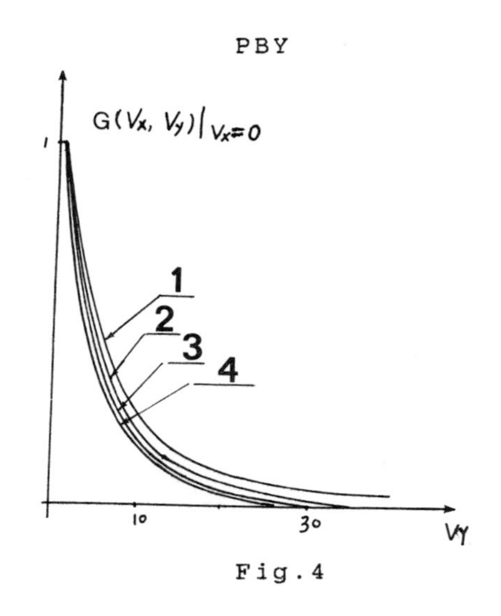

Fig.4

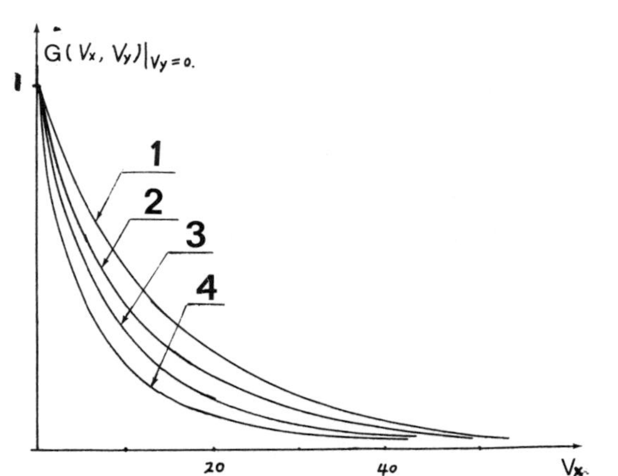

Fig.5

Refrences

[1] J.C.Dainty, Laser Speckle And Related Phenomena,Springer-verlag,1975

[2] H.Fujii, T.Asakura, OPt. Commun. 11, 35, 1974

[3] H.M.Pederson, Opt. Commun. 12, 156, 1974

[4] Joseph W. Goodman, Introduction to Fourier Optics, Mcgraw-hill,Inc.,1968

Session 11

Holographic Applications in Medicine and Speckle Techniques

Chairmen
R. J. Pryputniewicz
Worcester Polytechnic Institute, USA
Yu-Shan Tan
Xian Jiaotong University, China

A method of determining minute deformation and displacement of skull by speckle photography

Liu Fuxiang

West China University of Medical Science, Chengdu, China

Wang Shifan

Chengdu Institute of Radio Engineering, Chengdu, China

Abstract

This paper presents a method of measuring 3D minute displacement field of craniofacial bone by speckle photography. Multiple specklegrams are used in this method. Real spatial displacement is computed by the measurement of in-plane translations perpendicular to some viewing directions and the use of the projective transformation relation of the displacement vector. As a calculating example, the initial reaction of craniofacial bone produced by orthodontic forces is measured. Experimental results are in agreement with the results obtained by holographic interferometry.

Introduction

It is an important factor that orthodontic forces, which act on dentition and produce craniofacial skeleton deformation and displacement, finally correct the malocclusion. The tendency of secondary biologic remodeling of the jaw depends on the transmission and distribution of orthodontic forces acting on the craniofacial bone. Difference in the distribution of orthodontic forces make the deformation of craniofacial bone in the mechanical sense different, and so, form a complex 3D minute displacement field. The extent of the change relates directly to the characteristics of the acting forces and the structure of the skeleton itself. Undoubtedly, it is significant to study the initial changes of the craniofacial skeleton under orthodontic treatment.

This paper presents a method of how to determine 3D minute displacement field of craniofacial bone by speckle photography. The special feature of this method consists in using multiple double-exposure specklegrams. Spatial displacement is computed by the measurement of in-plane translations and the use of the projective transformation relation of the displacement vector.

Compared with holographic interferometry, the primary advantages of speckle photography are its simple optical setup, decreased mechanical stability, readily adjustable measuring sensitivity and ease of measuring in-plane displacement. So it is used in the research of biomechanics and some branches of engineering.

Theory

Recording and handling of double-exposure specklegrams

Fig. 1 shows the optical setup for recording a specklegram. Double-exposure specklegram photos are taken to the measured object before and after displacement.

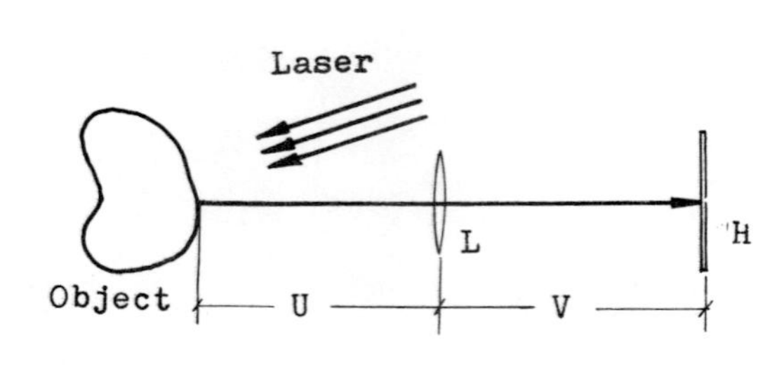

Figure 1. Optical setup for recording a specklegram.

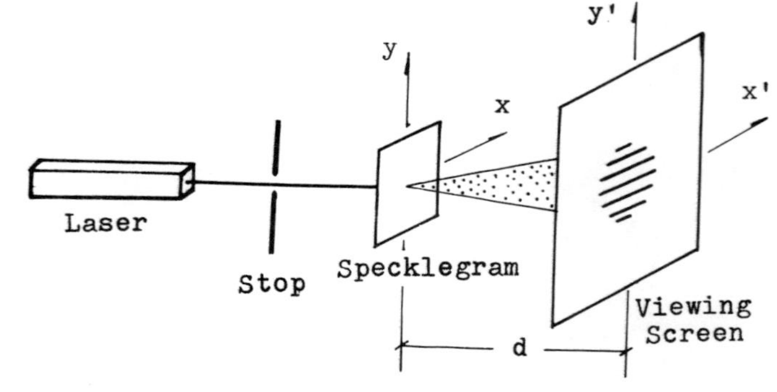

Figure 2. Handling of specklegram—point by point analytical method.

Figure 2 shows the optical setup for handling specklegrams using the point by point analytical method. At this time, Young's fringes can be viewed on the viewing screen at a distance d from the speckle photograph. Young's fringes are generated by more than one million speckle pairs with nearly identical displacement in the small illuminated region of speckle photograph. They are a set of parallel lines. Their spacing indicates the magnitude of displacement, and the line joining the double-aperture of speckle pairs shows the direction of displacement. The spacing Δt between adjacent bright (or dark) fringes is determined by following equation:

$$l = \frac{\lambda d}{\Delta t} \quad (1)$$

where l is the spacing of double-aperture of speckle pairs; d is the distance between the speckle photograph and the viewing screen; λ is the wavelength of laser light. On the assumption that M is lateral magnification of image speckle, the in-plane translation is

$$L = \frac{l}{M} = \frac{\lambda d}{M \Delta t}, \qquad M = \frac{v}{u} \quad (2)$$

where u is object distance and v is image distance. Their values are determined by the following equations:

$$\frac{1}{u} + \frac{1}{v} = \frac{1}{f} \quad (3)$$

$$u + v = |\mathbf{R}_2| \quad (4)$$

where f is focal length of lens, $\mathbf{R}_2$ is space vector of the observed point.

The formula used for calculating spatial displacement

Speckle photography is sensitive to measurements of in-plane translation, but is not sensitive to measurements of out-of-plane translation. For the sake of measuring spatial displacement, consider in-plane translation to be the projection of spatial displacement on the given plane, as shown in Figure 3. $\mathbf{L}$ represents spatial displacement, $\mathbf{L}_p$ represents its projection on the plane P, $\mathbf{K}_2$ represents the P's normal.

Their relationship is as follows:

$$\mathbf{L}_p = \mathbf{L} - \mathbf{K}_2(\mathbf{K}_2 \cdot \mathbf{L}) \quad (5)$$

or in matrix form:

$$\begin{pmatrix} L_{px} \\ L_{py} \\ L_{pz} \end{pmatrix} = \begin{pmatrix} L_x \\ L_y \\ L_z \end{pmatrix} - \begin{pmatrix} K_{2x} \\ K_{2y} \\ K_{2z} \end{pmatrix} (K_{2x} \; K_{2y} \; K_{2z}) \begin{pmatrix} L_x \\ L_y \\ L_z \end{pmatrix}$$

$$= \left[\underset{\sim}{I} - \begin{pmatrix} K_{2x} \\ K_{2y} \\ K_{2z} \end{pmatrix} (K_{2x} \; K_{2y} \; K_{2z}) \right] \begin{pmatrix} L_x \\ L_y \\ L_z \end{pmatrix} \quad (6)$$

$\underset{\sim}{I}$—identity matrix.

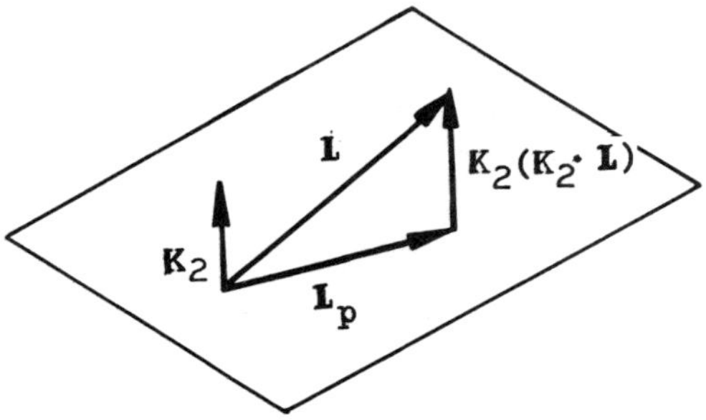

or in simplified form:

$$\mathbf{L}_p = \underset{\sim}{P}\mathbf{L} \quad (7)$$

where

Figure 3. Vector projection.

$$\underset{\sim}{P} = \underset{\sim}{I} - \begin{pmatrix} K_{2x} \\ K_{2y} \\ K_{2z} \end{pmatrix} (K_{2x}\ K_{2y}\ K_{2z}) \tag{8}$$

$\underset{\sim}{P}$ is known as the projection matrix.[1] It is only related to space vector $\mathbf{K}_2$. If the measured object's point is taken as the origin of the rectangular coordinate system, then

$$\mathbf{K}_2 = \frac{\mathbf{R}_2}{|\mathbf{R}_2|} \tag{9}$$

It is easily proven that

1) $\underset{\sim}{P}$ is a symmetrical matrix and $\underset{\sim}{P}\underset{\sim}{P} = \underset{\sim}{P}$

2) $|\underset{\sim}{P}| = 0$, $\underset{\sim}{P}$ is singular and, therefore, it does not have an inverse. So it is not possible to obtain $\mathbf{L}$ directly from Equation (7). It is necessary to form an overdetermined set of simultaneous equations. Hence, we need multiple specklegrams photographed simultaneously from different viewing directions. To every specklegram there corresponds one equation similar to Equation (7), so n specklegrams correspond n equations. That is

$$\begin{pmatrix} \mathbf{L}_{p1} \\ \mathbf{L}_{p2} \\ \vdots \\ \mathbf{L}_{pn} \end{pmatrix} = \begin{pmatrix} \underset{\sim}{P}_1 \\ \underset{\sim}{P}_2 \\ \vdots \\ \underset{\sim}{P}_n \end{pmatrix} \mathbf{L} \tag{10}$$

Multiplying both sides of Equation (10) by the matrix $(\underset{\sim}{P}_1^T\ \underset{\sim}{P}_2^T \cdots \underset{\sim}{P}_n^T)$ and noting that $\underset{\sim}{P}$ is symmetric matrix, we can obtain:

$$\mathbf{L} = \left(\sum_{i=1}^{n} \underset{\sim}{P}_i \right)^{-1} \left(\sum_{i=1}^{n} \mathbf{L}_{p_i} \right) \tag{11}$$

The above equation is the fundamental formula for calculating spatial displacement. Quantities to be measured are:

—space vector: $R_2(i)$

—distance between the speckle photograph and the viewing screen: d

—focal length of the lens: f

—fringes spacing: Δt

—fringe's azimuth angle: θ

and then substituting these values into expressions (2), (7), (11) to determine $\mathbf{L}$.

We compiled a computer program of the above calculating process with FORTRAN 77 language and used this technique to research initial change of some regions of craniofacial bone produced by orthodontic forces.

When the position of the measured object point needs changing, all we need to do is change the input data, according to the coordinate difference calculated for the theoretically corresponding point.

The method of the experiment

The object of study was a corpse skull with initial mixed dentition and complete periodontium and bone suture fibre. The removable appliance was designed and make according to the treatment of the mexilla complex protrusion and was fixed on the upper dentition of the skull. Backward traction of about 400g and 800g was loaded prelaterally along the occlusion plane on the upper dentition between the two exposures. In order to study the movement tendency of the craniofacial bone while the upper dentition was loaded, we measured the 3D displacements of the 14 points (Figure 4 consisting of the alveolus area 4 points, the zygomatic area 3 points, the temple area 7 points) on the surface of the skull. Figure 5 shows the experimental specimen.

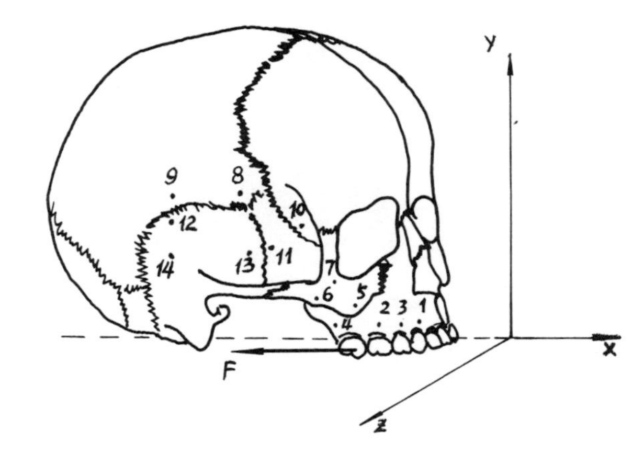

Figure 4. Measured regions.

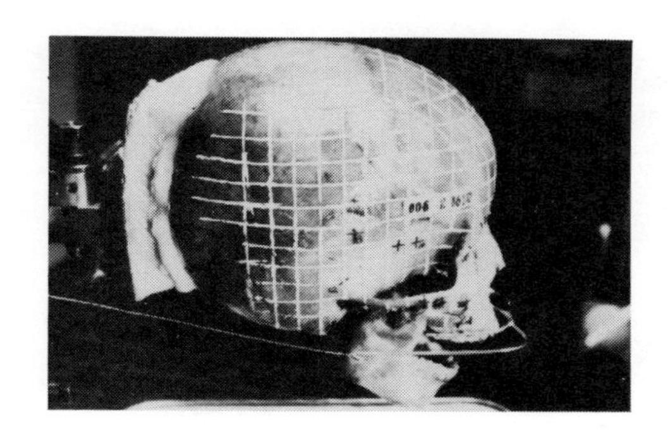

Figure 5. The experimental specimen.

The optical setup is indicated in Figure 6. Here we use two speckle photographs for simplicity. The surface of the skull is given a coat of white paint, and then the measured points are clearly indicated. The object is illuminated by a collimated He-Ne laser beam. Clear images of the object are obtained on the recorder plane H_1, H_2 with the help of the imaging lens. The determination of the distance between the object and the image depended on the preliminary estimate for the displacement range measured. After the photographing of the skull was finished, the photo's plane was developed normally. Then the spacing and orientation of Young's fringes was measured point by point and the optical setup geometry parameters were measured. At last, all of the collected data was used to form the data file and was fed into the computer to find the value of the 3D displacement.

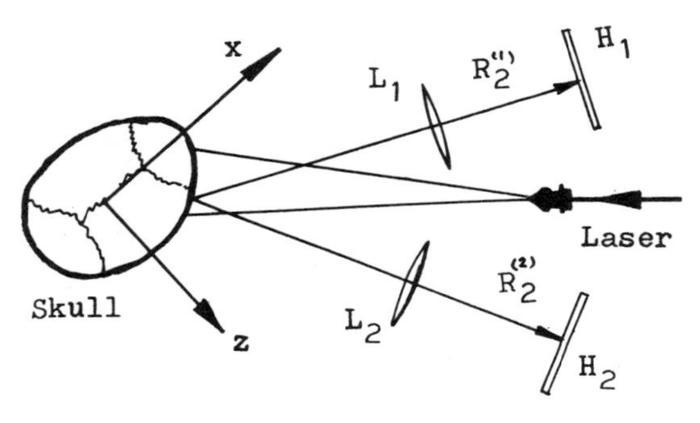

Figure 6. Optical setup for experiment.

Experimental results and calculations

The fringe pattern photographed on the viewing screen is given example in Figure 7. The example of the original experimental data is cited in Table 1.

Table 1. The example of the original experimental data

Measured point	Specklegram	u(cm)	v(cm)	Δt(cm)	θ(rad)	R_{2x}(cm)	R_{2y}(cm)	R_{2z}(cm)
1	1	21.51	36.19	1.70	1.347	25.04	8.40	51.32
	2	19.36	44.59	1.10	1.373	-22.55	9.50	59.08
2	1	21.85	35.32	1.90	1.548	27.86	8.00	49.28
	2	19.63	43.25	1.10	1.485	-20.03	9.70	58.81

When the acting force is 400g, the formed fringes show us that there are moderate fringes spacing on the image point of the upper alveolus area (Figure 7a). It is sparse on the zygomatic area but still can be measured (Figure 7b) and it is too sparse to measure on the temple area (Figure 7c).

When the force is increased to 800g, the fringes which can be precisely measured present on the image point of the temple area and the density of its Young's fringes increase on the zygomatic and alveolus area very much (Figure 8). It is indicated that the displacement of the upper alveolus, zygomatic and temple areas decreases in direct proportion to the distance from the upper dentition when experiencing load force. It has been proven by following the calculation result. Table 2 presents the results of spatial displacment of skull.

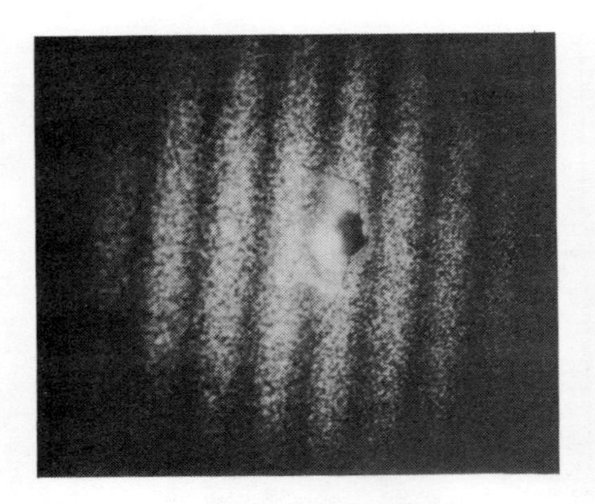

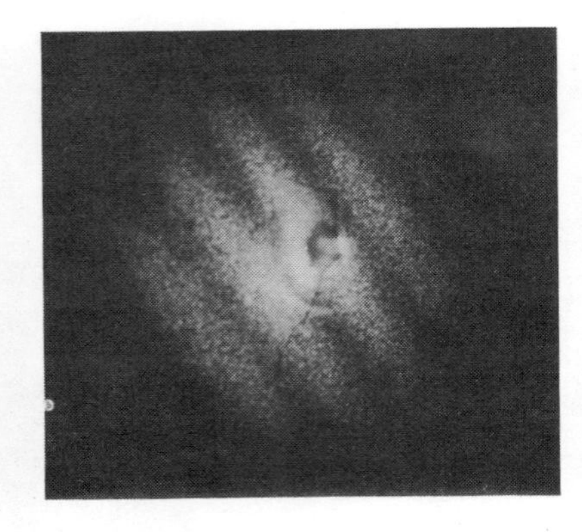

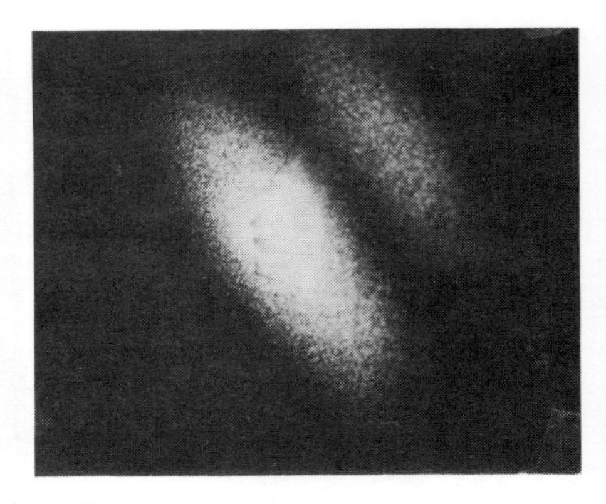

a. The upper alveolus area b. The zygomatic area c. The temple area

Figure 7. Fringe pattern photographed when the acting force is 400g.

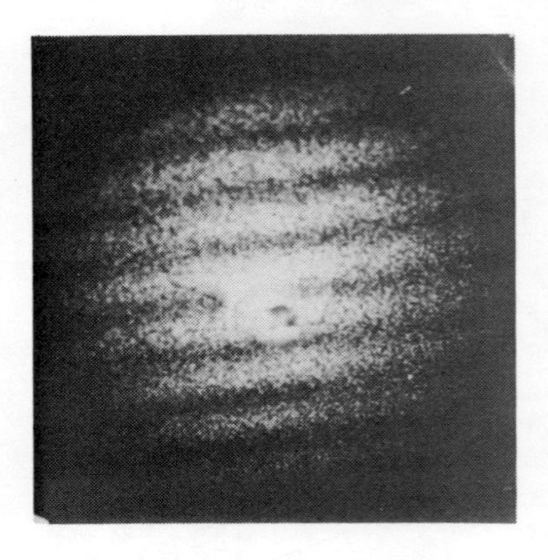

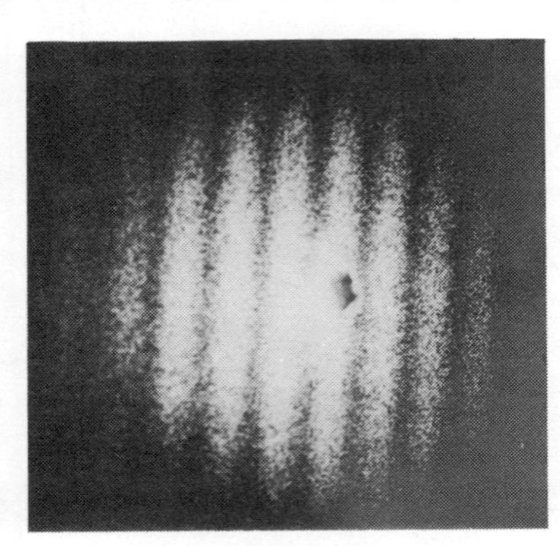

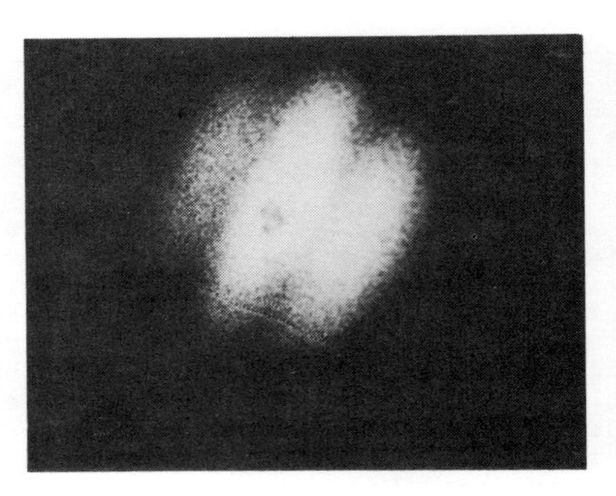

a. The upper alveolus area b. The zygomatic area c. The temple area

Figure 8. Fringe pattern photographed when the acting force is 800g.

Table 2. The results of spatial displacement of skull

acting force	measured point		$L_x(\mu m)$	$L_y(\mu m)$	$L_z(\mu m)$	$L(\mu m)$
400g	alveolus area	1	-17.777	-4.271	-4.752	18.890
		2	-18.274	-1.947	-3.999	18.808
		3	-17.878	-2.977	-4.793	18.747
		4	-17.750	-3.422	-4.743	18.689
400g	zygomatic area	5	-6.059	-4.062	-3.523	8.101
		6	-6.620	-3.025	-5.129	8.904
		7	-4.284	-3.476	-4.680	7.236
800g	temple area	8	-5.260	-4.747	5.332	8.868
		9	-5.171	-3.595	4.454	7.714
		10	-6.601	-3.747	4.022	8.590
		11	-5.857	-3.143	3.633	7.575
		12	-7.968	-3.285	2.536	8.984
		13	-6.966	-3.337	2.850	8.233
		14	-6.568	-3.568	3.055	8.075

Discussion

1. Measuring the minute displacement of diffused reflection surface by speckle photography has a simple optical setup, lower requirements for mechanical stability and adjustable measuring sensitivity at the certain range. Therefore this technique has some special advantages in solving problems in stomatology and biomechanics.

2. The information obtained in the measurement is the alternate bright and dark, equal wide fringes. They are recognized and handled easily by computer. So it is possible to construct a highly precise real time measuring system of the minute 3D displacement.

3. The sensitivity of measurement by speckle photography is limited by the characteristic speckle size and the correlation of the speckle field. The magnification of the speckle pattern can be changed because it depends on the imaging system. Therefore, the sensitivity of measurement using speckle photography is greater than that using holographic interferometry.

4. The experimental results show that the maxilla, zygomatic and temple bones moved backward and downward when the maxilla was pulled back along the occlusion plane by the removable appliance. These results are the same as those measured by holographic interferometry[3] and identical with the effect of the bone movement caused by analogous forces in clinical practice[2]. The calculated results indicate that the craniofacial bone, when acted on by force, produces an unhomogeneous 3D displacement field. An obvious tendency is that the more distant from the load force area, the less displacement. That's obviously related to structure of the skull itself.

The human skull consists of several thin, shell-like bones which are connected together by very strong fibre connective tissue. It is an optimal mechanical structure to buffer the occlusion and impulsive force during chewing. There is proof that the cerebrum is not damaged by strong impacts or impulsive force even after long chewing. This system has two effects: One, it presents the impact from damaging the cerebrum during mastication and the second, it decreases evidently the therapeutic effect of orthodontics force in clinical practice. It is a result of this buffer effect that displacement is attenuated step by step on the craniofacial skeleton surface. This information indicates the requirement for nearly designed appliances for use in clinical treatment.

When the skull accepts a dead load (orthodontics force) for a long time, the fibre connective tissue between two bones is compressed or stretched. During this process, elastical potential energy is accumulated in the fibre. This stretching and compression may make the osteoblast and osteoclast active and cause the accumulation or resolution of osteal tissue, thus giving rise to the histologic remodeling of the bone. Therefore the initial movements of the bone may indicate the tendency of the histological remodeling. Table 2 suggests that the amount of movement of the upper alveolus is larger than the one of zygomatic bone and that the movement of the temporal area is very small during the treatment.

To sum up, the complex 3D displacement and deformation field of a diffuse reflected object can be determined using speckle photography. Once this data is obtained, we can study the initial reaction of craniofacial bone due to force using this technique. The information obtained from this study is very important in the teaching and clinical practice of orthodontics. We hope that the method will prove to be a useful tool in the research of biomechanics and various branches of engineering.

References

1. Stetson, K. A., "Use of Projection Matrices in Holographic Interferometry," J. Opt. Soc. Am., Vol. 69, No. 12, December 1979, pp.1705-1710.
2. Badell, M. C., "An Evaluation of Extraoral Combined High-pull Traction and Cervical Traction to the Maxilla," Am. J. Orthod., 69:431, 1976.
3. Liu Fuxiang, The Study of Holographic Interferometry for the Initial Reaction of Maxilla Pulled Backward by Heavy Force, M.Sc. thesis, West China University of Medical Science, Chengdu, China, February 1986.

Manifestation of Gabor's holographic principle in various evolutionary stages of the living material

Pal Greguss

Applied Biophysics Laboratory, Technical University Budapest
P.O. Box 91, Budapest, Hungary H-1502

Abstract

Based on Gabor's holographic principle a generalized, qualitative model of the interaction of any signal carrier with the living material is presented. The model is applied to describe various physiological and psychophysical events, including brain functioning.

Historical introduction

The fourteenth anniversary of a tennis play should be celebrated the coming Easter, the result of which presented mankind with a remarkable idea, and of which a new trend of physics has been evoked and a new industry is rapidly developing: holography.

At Easter 1947, a Hungarian born engineer working on the problem of improving electron microscopy was awaiting his turn to play a game with his wife Marjorie in Rugby, when all of a sudden the solution presented itself to him in a flash, "why not to take an electron picture, one which contains the whole information, and correct it by optical means. To capture the whole information including the phase, the coherent background must be supplied by the same electron beam, which will therefore produce interference fringes: photograph these and then illuminate this photograph with light and focus it onto a photographic plate." He called the electron diffraction pattern a "hologram" because it contained the whole information, amplitude and phase, at a time.

The first time I met the word hologram and the name of Dennis Gabor was in the first half of the fifties when I started to investigate how ultrasonic waves could be used in NDT as signal carriers. I wish to emphasize the word 'signal carrier' - and not 'information carrier' as notation - since at that time we already pointed out that the information content of a signal pattern strongly depends upon the philosophy and strategy how it is processed. I was really fascinated with the elegant way Gabor solved the problem of how to record, at a time, amplitude and phase bound signals by using coherent radiation as signal carrier, since working with ultrasonic waves I was faced in my everyday practice with consequences of coherence, e.g., in the form of speckles. Nevertheless, it took me several years until I realized during a talk given at the ISFA Research Film meeting at Teddington 1964, that the plane in which the ultrasonic beam of an ophthalmic transducer and the beam reflected from the retina intersect can be regarded as a Gabor type hologram. Fortunately, at that time we had already developed various types of sonosensitive plates, thus, returned to Budapest we could experimentally confirm this recognition by placing the sonosensitive plate into the beam during ultrasonic ophthalmic investigation, as shown in Fig. 1.[1, 2] Acoustic holography was born.

After having verified that holography can be performed with both electromagnetic and acoustic signal carriers, we got the impression that the concept of holography is not only a tool for solving technical problems. It is also a fundamental idea for the better understanding of nature, the living matter, since Szent-Györgyi[3] pointed out several times that life is based on order and pattern, which conforms well with the fundamental pillars of holography, namely, coherence and interference.

Working in 1966 in India together with researchers studying the life of bats and their echolocating systems, I was faced with the problem of how to interpret the estonishing capabilities of these creatures, - i.e., that they can distinguish between targets of various shapes, they differentiate according to the quality of the target, e.g., if it is food or it should be avoided, if the obstacles are horizontal or vertical, etc. - since these capabilities far exceed what conventional echolocation may offer. Drawing a parallel between acoustical holography and the behavior of echolocating bats, we developed a "bioholographic" model capable to describe these performances and predicted that the bat's brain has a structure which permits a signal processing according to the principles of holography, i.e., it creates a reference pattern coherent with the emitted and from the target reflected signal pattern.[4] Physiological and histological evidences of these assumptions were found by Ayrapetyans[5] at Leningrad.

About the same time Pribram,[6] fascinated by the unique property of the hologram, i.e., that any small part of it with an area above a certain minimum is capable of producing reconstruction of the entire original signal pattern, suggested - independently from us - possible parallels between the coding of neural processes and the holographic process, and tried to set up a model capable of furnishing a mathematically precise analogy to the well-known neurophysiological property of distributedness. This theory laid down conditions which are necessary but not sufficient for establishing a holographic model of the living matter, because it does not really distinguish between relevant and irrelevant kinds of similarity. Nevertheless, the fundamentals of bioholography have been laid in 1966, and since then, several hundreds of papers - some rather controversial - have been published on this subject. A good, although not complete, literature survey has been compiled by Ibn Ravn.[7]

In summary, at present, applying the principle of holography to the field of signal processing opens three new areas:

1. where the signal carrier is of electromagnetic nature, referred to as light holography

2. where the signal carrier is of mechanical nature, referred to as acoustical holography

3. where the signal carrier is of biological-physiological nature, referred to as bioholography.

This paper will deal only with the basic problems of bioholography.

The living matter and its environment

Before applying the principles of holography to describe the signal processing mechanism and the following information in living matter, one has to make clear what can be considered as living matter, and what cannot. The living matter, namely, has not only a physical and chemical character but also a biological nature, i.e., it is a regulatory, homeostatic system. The degree of homeostasis, however, depends upon the evolutionary stage of the living matter in question; it is lowest at biomolecular, and highest at the vertebrate level. Further, there is a hierarchy among the various interacting regulatory schemes, which all are of oscillatory nature.[8]

The living matter may perceive the changes in the parameters of its environment as signals (in biological language: stimuli), to be processed as signals, however, there is a fundamental difference between the signal processing of living matter and signal processing in a cybernetic system. As long as computer signals propagate in an essentially constant substrate, the stimulus propagation and transmission depends upon the functional mobility of the living matter, which, however, permanently changes during its activity, as demonstrated by Uhtomskiy and others,[9] the consequence of which may be that the transfer function τ of the living matter may vary too. With other words this means that the information content of a signal pattern affecting the living matter depends upon the location and the conditions at the time the signal pattern is acting upon the living matter. At the highest evolutionary level, in the case of human organism, the social environment is also a part of this functional mobility. Therefore, the signal processing of the living matter seems to be much less precise as compared to the functioning of a computer. This scantiness we think is fundamental to "intelligence" because it makes it possible to establish similarities and analogies. As we shall see, by thinking along holographic principle line, a better interpretation of these considerations could perhaps be achieved. Approaching this problem from the side of social development (evolution) Ibn Ravn[7] uses the term holonomy (holos = whole, nomous = law) to denote dynamic orders (coherence), i.e., to indicate that the social environment is also a part of functional mobility.

Before setting up any signal processing model of the living matter, first one has to clear the criteria that determine whether a signal carrier is treated by the living matter as signal carrier, i.e., it is processed to become information for the living matter, which has to be "answered" and which manifest itself in "functional characteristics".

Assuming that signal processing in living matter is based on holographic principles, the first answer one gets matches well the observation of the biologists if one starts from the basic equation of holography. This states that the resulting intensity after signal processing is

$$I = I_S + I_R + (I_S I_R)^{1/2} \cos \varphi \qquad (1)$$

where I_S is the intensity of the stimulus proportional to the intensity of the signal carrier, and I_R is the intensity of the stimulus involved in the signal processing.

Transcribing Equ. 1 by using energy terms:

$$E = E_S + E_R + (E_S E_R)^{1/2} \cos \varphi \qquad (2)$$

we can demonstrate that if the energy of the stimulus E_S is smaller than a certain minimum E_o, no holographic type signal processing will take place. This is so because if $E_S < E_o$, the third term in the right side of Equ. 2, the one that carries the signal pattern, becomes smaller than the square of the random deviation of the reference background

$$E_S E_R < \overline{(\delta E_R)}^2 \tag{3}$$

According to Gabor,[10] if $E_S = E_o$, the formation of an interference pattern equal to a hologram is just possible, thus, until the threshold level is not reached, the "biological" part of the living matter - which is responsible for the signal processing - is not switched on, the stimulus threshold is not reached. In this case

$$\delta \frac{\overline{E_R}}{E_o}^{2} = \frac{\overline{E_R}}{E_o} \tag{4}$$

Since both sides are dimensionless, we can write:

$$\overline{\delta N^2} = \overline{N} \tag{5}$$

which is the Poisson formula for the alternation of "seldom" events, indicating that the monochromatic signal carrier becomes stimulus only if its energy exceeds a given value, otherwise the living material behaves itself only as a "nonliving" system, none of the homeostatic schemes have been activated. Therefore, when speaking of biological signal processing, not the intensity of the signal carrier but its spectral irradiance ($mW.cm^{-2}.nm^{-1}$) should be considered.

As shown in Fig. 1 - which is a qualitative but generalized description of the behavior (functional characteristics) of the living matter when a signal pattern is acting upon it - the same signal pattern provokes different behavioral patterns, depending not only upon the spectral irradiance S_i of the carrier λ_1, λ_2, ... but also upon the exposure time t. So, e.g., signal patterns of monochromatic carrier λ_1 or λ_2 described by $f(S_i, t)$ will never be considered by the living matter as signal pattern because their energy content E_S never reaches the threshold needed to activate the "biological part" of the living matter which then processes the signal pattern in such a way that it becomes information which influences the functional characteristics, the reactions, the behavior of the living matter. With other words, as long as E_S does not reach or exceed E_o, the response of the living matter is of pure physical and/or chemical nature such as rise in local temperature, change in pH, etc. However, in some cases - depending upon the character of the signal carriers acting at a time - these pure physical and/or chemical changes (especially the latter ones) may pile up in such a way that a signal carrier of spectral irradiance $E_S < E_o$ may trigger one or more local homeostatic schemes, the result of which, however, is not information, only a computer like feedback. This is one reason why it is sometimes hard to tell whether a given signal carrier is biospecific or not. This situation is represented by signal carrier λ_3 in Fig. 1.

The situation is quite different when $E_S \geqq E_o$, even if the signal carrier is the same. (Fig. 1, λ_1', λ_2') In this case, the signal pattern is not only to activate feedback mechanisms, but it is also processed to become information needed by the living matter to survive or to change. If information is not obtained from the signal pattern at the level demanded by the living matter as a consequence of its evolution, then its homeostatic system does not function adequately at that level, and this condition is also signaled to all homeostatic schemes, and this is what we generally call, at human level, "bad feelings." This processing procedure we believe is based on holographic principles, although at present it is rather difficult to prove this assumption experimentally, especially, at the lower levels of evolutionary stage. Nevertheless, observations, as reported by Bishop and others[11, 12] that only those incoming stimuli are recorded that are synchronous with the positive phase of the quiescent period of activity of the brain, strongly suggest the existence of a holography-based processing and storage

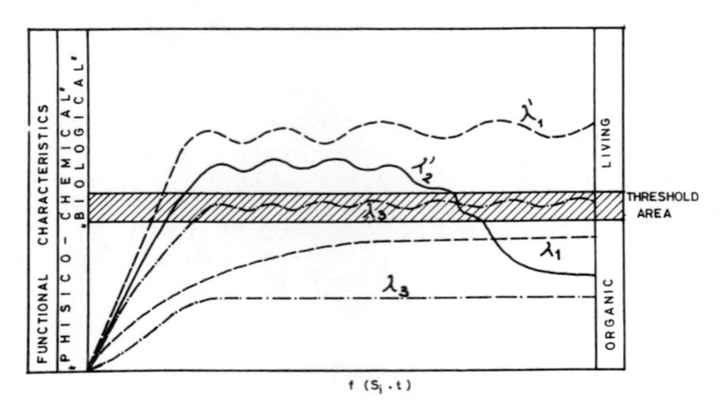

Fig. 1 Changes in functional characteristics of the living matter when a signal pattern is acting upon it.

strategy since only the component of the complex amplitude is recorded that is in phase with the reference background.

Coherence requirements

Assuming biological signal processing mechanisms based on holographic principles, the first question is how the living matter provides the coherent background necessary to this type of signal processing, and what the coherence requirements are at all. The answer to the first part of the question sounds rather simple: always one internal signal (stimulus pattern) which is part of the homeostatic schemes not affected by the signal pattern to be processed serves as reference background, but to prove this statement experimentally, especially at evolution level such as of a cell, is at present rather difficult. This is the reason why most bioholographic models are dealing with signal processing schemes at the level of the central nervous system.

The coherence requirements for a holographic type of biological signal processing may be estimated as follows:

A signal pattern that provokes a stimulus pattern, the carrier of which can be characterized by a travelling speed c and an angular frequency $\omega = 2\pi n$, and one can assume that at the moment of the information processing ($t = 0$) it is in phase with the internal signal carrier characterized by $(N - \Delta N)$. It can be shown that after a time elapse of t the phase difference will be

$$\Delta\varphi = 2\pi \Delta Nt. \quad (6)$$

After differentiation, Equ. 6 yields

$$\Delta\varphi = -2\pi[c \Delta\lambda t/\lambda^2]. \quad (7)$$

The time necessary for the difference of frequency of the two carriers to become $\pi/2$, i.e., the timeone needs for a holographic signal processing can be calculated using Stroke's approach[13]: $t_{\pi/2} = 0.25\ \lambda^2/c\Delta\lambda$.

Substituting physiological parameters into Equ. 7 one obtains values that are well in the range of the known reaction times, indicating that in spite of the functional mobility of the living matter the possibility of a holographic signal processing mechanism exists.

Difference between living and organic material

The idea that in bioholography the reference pattern used for signal processing issues from a part of the whole homeostatic scheme system of the living material raises the question: When does the always present signal-to-noise ratio (noise = homeostatic schemes activity not involved in the particular signal processing) prevent a meaningful holographic signal processing?

If the complex amplitude of the signal carrier is A_S, that of the reference background A_R, and that of the noise A_N, then the entire amplitude to be handled is $A_S + A_R + A_N$, but the resulting interference pattern, the "biohologram", is formed only by A_S and A_R, and not by A_N, since the carrier of the latter one is not coherent either with that of A_S or A_R.

Let us now introduce the product ($\mathcal{E}$) of the average stimulus intensity, and the duration of the stimulus, t, which then is related to the transfer function τ, as $\tau = f(\mathcal{E})$. During signal processing the noise level is given similarly to the product of the intensity of the reconstructing stimulus and of the noise level of the holographically processed signal pattern, thus, it can be characterized as

$$N = g(\tau_o^2)\ A_S^2 \quad (8)$$

According to definition, the square of the absolute value of A_S, A_R, and A_N, respectively, is equal to the intensity of the stimulus, and so one can write:

$$\mathcal{E} = (A_S^2 + A_R^2 + A_N^2 + A_S A_R^* + A_S^* A_R)t \quad (9)$$

The quantities in parentheses are mean values during time t, provided that $A_S^2 \ll A_N^2$, which is in general the case with [illegible]ving matter. Thus,

$$\mathcal{E} \sim \mathcal{E}_o + \mathcal{E}_1 + \mathcal{E}_1' \quad (10)$$

where

$$\mathcal{E}_o = (A_R^2 + A_N^2)t; \quad \mathcal{E}_1 = A_S A_R^* t; \quad \mathcal{E}_1' = A_S^* A_R t$$

where * denotes the complex conjugates.

Since ε_1 and ε_1' are smaller than the mean value of ε_o, and the activity of the living material before signal processing can be described by $\tau \cong f(\varepsilon)$, we can write

$$t = \tau_o + \tau_1 + \tau_1' \tag{11}$$

where

$$\tau_o = f(\varepsilon_o); \quad \tau_1 = f'(\varepsilon_o)\,\varepsilon_1; \quad \tau_1' = f'(\varepsilon_o)\,\varepsilon_1'$$

where $f'(\varepsilon_o)$ is the meanvalue of the derivate of $f(\varepsilon)$ at the point $\varepsilon = \varepsilon_o$ if $\tau_o = f(\varepsilon_o)$.

Since according to the hologram principle the amplitude of the reconstructed signal carrier is A_R times a constant, and since both the intensity of the signal carrier and that of the noise are proportional to the intensity of the reconstructing signal carrier, when calculating the signal-to-noise ratio, this constant can be taken as 1. As a consequence, one obtains for the amplitude of the reconstructed signal

$$A_R\tau_1 = f'(\varepsilon_o)A_S A_R^{\,2} t. \tag{12}$$

Thus, the intensity of the stimulus carrying the processed signal pattern as an information pattern will be

$$I = [f'(\varepsilon_o)]^2 A_S^{\,2} A_R^{\,4} t^2 = [A_S^{\,2} A_R^{\,2}/(A_R^{\,2} + A_N^{\,2})][f'(\varepsilon_o)]^2 \varepsilon_o^{\,2} \tag{13}$$

The second form of Equ. 13 follows from Equ. 10.

Dividing Equ. 13 by Equ. 8 one obtains for the signal-to-noise ratio

$$I/N = [4A_S^{\,2} A_R^{\,2}/(A_R^{\,2} + A_N^{\,2})^2]\{[f'(\varepsilon_o)]^2\,\varepsilon_o^{\,2}/4g(\tau_o^{\,2})\}. \tag{14}$$

These associations of ideas are along the same line as those of Hamasaki[14] for recording holograms in strong incoherent light, the question, however, is whether it can possibly be used to explain some behavior of the living material if a signal pattern acting upon it becomes a stimulus pattern to be processed.

Equ. 14 states that the quality of the signal processing is given by a product of two factors:

1. A simple function of the intensity of the signal processing stimulus represented by $4A_S^{\,2}A_R^{\,2}/(A_R^{\,2} + A_N^{\,2})^2$ indicates that if the signal carrier remains constant, and if the intensity of the processing stimulus approaches the intensity of that of the general homeostatic schemes, then, the first factor of I/N increases. In case of $A_R^{\,2} = A_N^{\,2}$, the situation is optimal. Thus, if the living system is capable to increase the intensity of the signal processing stimulus, it can optimize the signal-to-noise ratio. This conclusion issuing from the bioholographic signal processing model is in good agreement with the experimental findings of Ayrapetyans,[13] who reported that in the case of echolocating bats the intensity of the signal pattern received may be lower than the average noise level by a factor of about 2000 without causing any trouble, because the bat when background noise gets higher increases only the intensity of the signal processing stimulus, and not the intensity of the emitted ultrasonic pulse, as one would expect.

2. In order to consider the real meaning of the second factor of Equ. 14, analytic expressions would be needed for the transmission characteristic $\tau = f(\varepsilon)$ and for the noise function $g(\tau^2)$. Although analytic expressions which correspond well to biological parameters have not yet been really found, nevertheless, there are strong indications that this factor describes the permanently <u>changing</u> functional <u>mobility</u> of the living matter, and, as a consequence, it describes that a material is living and stays alive as long as these factors function, otherwise it is only "organic", a material which has only physical and/or chemical character, as shown in Fig. 1.

Some consequences of Equ. 14

Although at present there are not enough experimental data for an unambiguous interpretation of Equ. 14, nevertheless, it teaches us quite a lot. So, among others, it gives us a hint as to how to interpret, e.g., "pain", and even "illness".

Pain

In general, pain is considered as a warning signal, however, this is true only in some situations, since, e.g., in the case of the pain of cancer, far long the time pain is perceived, the disease is well advanced. On the other hand, there are diseases that are not accompanied by pain, like diabetes. Closer to reality is perhaps the definition that pain is a total response to a signal pattern which, when processed, is modified by all manner of internal and external conditions.

In the light of Equ. 14 this means that pain is the result of a nonadequate signal-to-noise ratio that is needed for proper signal processing, and the living system is unable to increase - for one reason or another - the intensity of the signal processing stimulus. hether this situation is recognized (perceived) as "pain" by the living system, or not, depends on the one hand on its evolutionary stage, on the other hand on how the various homeostatic schemes are interrelated. Thus it follows from Equ. 14 that there are two ways to eliminate pain: one is to prevent the signal pattern causing "pain" to reach the signal processing site where the signal-to-noise ratio is not adequate, the other, to improve the signal-to-noise ratio needed to process the signal pattern in question.

In pain treatment generally accepted today the first approach is followed, which, however, is only a symptomatic treatment, while the endeavors to use, e.g., electric impulses for relieving pain[16] are trying to provide appropriate coherence to process the signal pattern to information needed for restoring the adequate relation among the various homeostatic schemes. Most probably this happens during acupuncture too - when it works.[17] The same holds for the status "hunger" or even "illness".

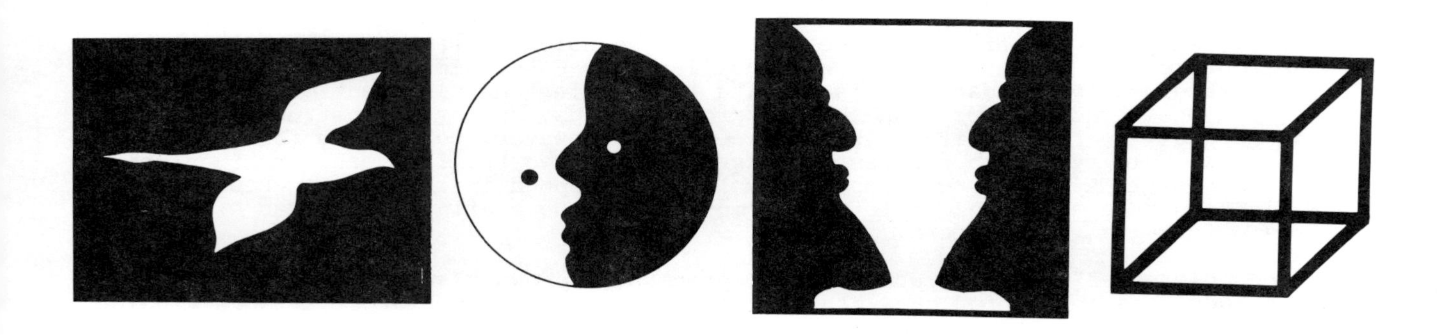

Fig. 2 Ambiguous figures

Ambiguous figures are perhaps the best examples that information depends upon how a signal pattern has been processed, as it is demonstrated in Fig. 2, which shows a set of well known ambiguous figures. Every time the same signal pattern reaches the retina, nevertheless, the information derived from it is different and valid only at a time, i.e., the meaning (information) of the signal pattern fluctuates. Although many proposals have been made to explain this phenomenon - which is not restricted to human vision or vision at all, - none of them could present a biological information processing model which at least renders a credible description of the happenings.[18, 19] The bioholographic model, however, does this.

Fig. 3 Is it a man or a girl?

Let us suppose that the interesting pattern of Fig. 3 consists of two parts: a "man" (m) and a "girl" (g) part, respectively, and the designated stimulus of m is M, and that of g is G. In accordance with the bioholographic principle, the transfer function τ of the stimulus pattern resulting from the holographic signal processing - one may call it holographic memory, HM - can be described as

$$\tau = K(MM^* + GG^*) + KMG^* + KM^*G \quad (15)$$

where K is a constant containing the second factor of Equ. 14, and * is the complex conjugate.

When the signal pattern of Fig. 3 is presented, first an arbitrary stimulus S_A is acting upon it, thus, the resulting "information" θ_o will be

$$\theta_o = S_A \tau_{MH} \quad (16)$$

Performing this operation one obtains four terms, two of which refer to the virtual part of the stored signal pattern composed of m and g, which we designate by S_{mg}. Then

$$\theta_{mg} = CM(G^*S_A) + CG(M^*S_A) \tag{17}$$

where C is a constant. To get the information "man", the following equation has to be fulfilled:

$$M(G^*S_A) = M, \quad \text{and for the information "girl"} \quad G(M^*S_A) = G. \tag{18}$$

It is obvious that Equ. 18 is satisfied if $S_A = G$, or if $S_A = M$, respectively, i.e., the information at a time may be only "man" or "girl", but never both.

Somewhat similar signal processing mechanism may be the cause of the success of "mnemonics" when well-known words or notions help to memorize a new signal pattern. Presumably the stimulus pattern attached to the word or notion plays the role of the reference background in the particular signal processing, losing thereby its information content, which is in good agreement with the properties of holograms recorded without external source of reference beam.[20]

How "complete" is holographic signal processing?

Gabor named his signal processing method HOLOGRAPHY because he wanted to emphasize that the resulting interference pattern, the hologram, records all information, and this is also one reason why it has always been assumed that holographic reconstruction yields an image exactly at the same place where the object was when the hologram was recorded. This, however, is not quite true because - as pointed out later by Gabor himself[21] - "the holographic plate can record only one datum; this results in dropping one component of the imaging amplitude, the one in quadrature with the background", thus, during reconstruction some information, generally not recognized, is lost. One can, however, compensate for this loss by recording two holograms with reference backgrounds delayed $\lambda/4$ from each other, since then the second will have just the component which was dropped in the first. Reconstructing the two holograms at a time the complete information will be available, one can exactly reproduce the original signal pattern.

This "incompleteness" of holographic signal processing shows also up in the presented holographic model, as we pointed out when we tried to interpret the rather strange finding that some cetaceans use double pulses during echolocating, especially when they are in "trouble", i.e., when they really need the complete information carried by the received signal pattern. Our calculations showed[22] that the delay between the impulses corresponds to $\lambda/4$, i.e., in this case, the cetacean holographic type signal processing yields the complete original information pattern.

Recently, A.H. Fry and R.D. Fry[23] demonstrated in a very elegant psychophysical experiment that the perceived distances of holographically reconstructed objects are not equivalent to the perceived distances of the actual objects, and they concluded that "a fundamental assumption in holography has not been supported by experimental data." But just in contrary, these wonderful experiments underline what Gabor already emphasized, i.e., that when recording a single hologram one component of the imaging amplitude, the one in quadrature with the background, is always lost, i.e., the reconstructed image cannot be exactly identical with the original object. What, however, is really estonishing is that our visual system detects this minor "incompleteness" of the holographic signal processing technique, thus, one can completely agree with the conclusions of the authors that holography may provide us with a new means to advance our limited understanding of distance perception, which, however, is strongly related to the problem of how living systems handle spatial signal patterns.

Life in space

Speaking of space there is a lot of misunderstanding about its real meaning. In general, "visual space" is regarded as prototypical space from which all the kind space invented by physicists and mathematicians were developed, and is based on a concept evolved by "touch" experiences. In contrary, real "physical space" has properties which visual space do not fit, and it cannot be described by terms height, width, depth, especially not by up-down, right-left, near-far, etc.

Since the living material - independently of its evolutionary level - receives signal patterns from a physical, and not a visual, space, it has to use a two-stage signal processing: first encoding not only the amplitude and phase parameters, but also the arriving direction of the signal carrier in respect to its own coordinates, and then, when decoding, (information retrieval) all of the signal parameters have to be available for further processing, i.e., the signal processing mechanism has to show selfreciprocity. According to our present knowledge, only Gabor's holographic principle fulfills these requirements, independently of the fact whether the result of the operation is the sum of the parameters, as in the case of Fresnel transform description, or of their product, as in the case of Fourier transform description.

Since the living matter receives the signal pattern in a physical space, for the sake of not to lose any of its information carrying parameters, the living matter needs a processing method of analyzing signal in which time and frequency play a symmetrical part. This recognition, however, is the main point in holographic signal processing and is the consequence of Gabor's theory of communication.[24] According to Gabor, the elementary signal can be described by the incertainty relation $\Delta t \Delta f \geq 1/2$, where Δt is the "effective duration" and Δf is the "effective frequency width", i.e., the elementary signal may be represented in the signal pattern by a rectangle with sides Δt and Δf, and is one half centering at the point (t_o, f_o) representing the elementary quantum of information attached to the signal with two orthogonal carriers 90° out of phase. The elementary signal was named by Gabor as logon, and we proposed to call its equivalent in the stimulus domain "biologon".

As already mentioned, it is rather difficult to verify experimentally the presence of holographic signal processing at lower levels of evolution, nevertheless, some indications have already been found that the structure of cell membrane may be regarded as a (volume) hologram, and may serve in one of the homeostatic schemes as memory or filter. There are proposals for analog biochips for performing tasks such as pattern recognition based on bioholographic principle. In contrary to conventional digital computers using Neumann architectures, systems based on holographically structured biochips would be superior at context-dependent data processing, in tasks that require complex weighting and recognition functions and integration of signals of information carrying signals, and computations from different sources. These systems, as the living material, would be capable of saying not only yes and no, but also may be.

The meaning of "perspective"

Vision is based on the fact that the signal carrier (visible spectrum of electromagnetic radiation) projects signal patterns from a part of the physical space onto a 2-D surface (retina) assumed to be euclidian. This, however, means that that part of the signal carrier that holds "time" related information (in visual space: depth) will be never parallel to any dimensions of the retina, nevertheless, the corresponding stimulus has to be processed in such a way that depth perception is not lost. In general it is believed that this is accomplished by the "vanishing" phenomenon, i.e., depth appears to vanish with distance, creating the sensation of convergence, called "perspective". However, an object goes out of sight by becoming too far to see, not by vanishing into a point. The experienced vanishing sensation is a "learned", an "acquired" feeling, and has nothing to do with the properties of the physical space, and even not with the primary biological processing event. It is based on secondary, culturally determined schemes.

Accepting that the stimulus pattern equivalent to the signal pattern originating from the physical space is processed according to the holographic principles, during reconstruction two stimulus patterns have to emerge at a time: one showing the vanishing phenomenon and another just the opposite, an "inverted" perspective. This phenomenon is well observed in young children's drawing as long as they have not been affected by their surrounding saying that "objects nearer to you 'have to be' drawn larger than those farther away to be 'realistic'" (geometric perspective). In cultures where geometric perspective is not obligatory, e.g., in oriental and icon paintings or in the works of some artists as in the altar painting "Madonna enthroned" of Giotto, and in the painting of Picasso "Bird cage and playing cards," both perspectives are present.

The living matter as diffractive medium

Applying Gabor's holographic principle to describe and interpret phenomena associated with life means that living matter can be considered as a diffracting medium, i.e., in the interpretation of its functioning Huygens' principle may be applicable. In this case, however, diffraction pattern cannot be regarded - as by Abbe - as consisting of a number of discrete "beams", but Gabor's "expansion theorem"[25] has to be applied, according to which information attached to the signal pattern is not carried by "rays", but by a certain "tube of rays" the cross section of which is proportional to the square of the signal carrier's wavelength. This, however, means that the central nervous system (CNS) (including brain functioning) may be treated as a source of waveguide system, however, to describe its functioning only by simple interference phenomenon falls short at explaining signal pattern storage in CNS. The notion of a signal interacting with the delayed part of itself may explain association recall, but it does not explain recognition of the signal pattern itself. Furthermore, it is not applicable to the recognition of new signal patterns. The various bioholographic models, however, throw light upon these relationships difficult to understand. This is not the place to go into details in discussing holographic brain functioning descriptions,[26, 27, 28] thus, we mention only a rather controversial problem, the relation between the intellectual capacity and degree of convolutions in the brain to demonstrate the ideas behind bioholographic models. It is without question that a larger volume of higher degree of convolution does not necessarily mean that they match a higher intellectual faculty, but rather a better "resolving power" of the signal processing mechanism is responsible for a higher intellectual performance. According to the bioholographic model, however, this does not mean a necessarily higher neural density/unit volume. The neuron, e.g., may be regarded as a combination of various colloidal solutions, the ratio of which may influence the performance (the resolving power) of the CNS in a somewhat similar way as the gelatine structure of the photographic emulsions determines the resolving power.

Cell division and holography

To understand cell division it is not enough to be able to describe how a cell divides itself, but one has also to tell how the cell "knows" that it has to divide itself into a given number of identical cells, and how these cells should be distributed in the physical space in reference to the "mother" cell. Since one can assume that a cell as living matter is a diffracting medium, the signal processing mechanisms involved in cell division may be interpreted on holographic principle. If the cell can be described by a signal pattern $f(x', y')$, and the "message" how many identical cells have to be made and how they have to be spatially distributed is stored in a holographic form characterized by $S(u, v)$, and due to Fourier operations, one gets in the x,y plane a signal pattern

$$h(x, y) = f(kx, ky) * S(x, y) = \iint_{-\infty}^{\infty} f(kx'', ky'')\, s(x - x'', y-y'')\, dx''dy'' \tag{19}$$

where k is a constant, and $s(x, y)$ the inverse Fourier transform of $S(u, v)$. With other words, the cell forms in the x,y plane the convolution integration between the functions $f(kx, ky)$ and $s(x, y)$. If now $s(x, y)$ is an array of δ functions,

$$s(x, y) = \sum_{p,q} \delta(x-a_p)\, \delta(y-b_q) \tag{20}$$

and we get

$$h(x, y) = \sum_{p,q} f[k(x-a_p), k(y-b_q)] \tag{21}$$

what is nothing more or less than a series of functions describing the signal pattern of cells identical to the "mother" cell, i.e., the message needed is transmitted. This signal processing mechanism, however, works only if $s(x, y)$ is an array of δ functions; if not, then errors may be observed in cell multiplication. Further developing this idea, e.g., by combining it with those inherent in Equ. 14, a new description of "malignancy" may be obtained, which could lead us to a better understanding of what is really going on in one or another form of cancer. This may sound even more plausible if we consider that the cell membrane is not only a passive barrier of diffusion but also plays an active role in signal processing through material and energy transport, since in the structure of the membrane pattern Equ. 19 - which represents the multiplication "message" in a Fourier transformed holographic form - may be encoded, fixing in space cooperatively controlled catalytic sites which interact via diffusable chemicals. However, one has to emphasize that this is not the only possible bioholographic signal processing description at cellular level.

In conclusion

We hope to have demonstrated that signal processing based on holographic principles is most probably current in living matter. Its form and appearance is strongly determined by philogenesis of the living matter in question, its ontogenesis and ecological specialization of analysatory systems. Just two examples to back up this statement:

It has been demonstrated that the lateral line of fish and amphibians processes spatial signal patterns similarly to side looking radar, i.e., on holographic principles. However, the lateral line of salamanders (Triturus vulgaris) is totally degenerated during terrestrial life, but it redevelops together with the fin during breeding season when the animals return into water, where more phase processing is needed than on the 2-D solid surface.

In the case of "nonflying" birds such as kiwi, ostrich, etc., the manifolded, comb like annex in the eye, the pecten, is poorly or not at all developed in contrast to the eyes of flying birds, indicating that the pecten has something to do with the need of space processing. Indeed, the pecten most probably acts as a coded aperture in the visual signal processing scheme of these species. This assumption is backed up by the fact that fish, which philogenetically proceed birds and need more phase information processing than do terrestrial species, have a structure analogous to the pecten: the falciformis.

At the end, I would like to express my belief that the holographic concept of Gabor is as fundamental as the general relativity theorem of Einstein, and it has to be explored further for a better understanding of nature in which we live. (1920)

References

1. Greguss, P., "Ultraschall-Hologramme," Research Film, Vol. 5, pp. 330-337. 1965.
2. Greguss, P., Bertenyi, A., "The possibility of using holography in ultrasonic diagnostics applied to ophthalmology," Proc. of SIDUO II, Brno 1967, pp. 133-136.
3. Szent-Györgyi, A., "Cell division and cancer," Gordon H. Scott Lecture, Detroit, Michigan, May 1972.
4. Greguss, P., "Bioholography - a new model of information processing," Nature, Vol. 219, pp. 482-483. 1968.
5. Ayrapetyans, E. S., Konstantinov, A. T., Voprosii bioniki, Nauka USSR 1967
6. Pribram, K., "Holographic memory," Physiology Today pp. 71-84, 1979.
7. Ibn Ravn, Holonomy: a bibliography, Department of Social Systems Sciences, University of Pennsylvania, Philadelphia PA-19104, 1986.
8. Greguss, P., "Interaction of optical radiation with living matter," Optics and Laser Technology, Vol. 17, pp. 151-158, 1985.
9. Uhtomskij, A. A., "Concerning the condition of excitation in dominance," Prihod Abstr., No. 2388, 1927.
10. Gabor, D., "Light and Information," in: Progress in Optics, Ed. E. Wolf, Vol. 1, pp. 111-156, 1961.
11. Bishop, G. H., "The interpretation of cortical potentials," Quant. Biol., Vol. 4, pp. 305-319, 1936.
12. Clare, M. H., Bishop, G. H., "Potential wave mechanism in cat cortex," Electroenceph. Clin. Neurophysiol., Vol. 8, pp. 583-602, 1956.
13. Stroke, G. W., An Introduction to Coherent Optics and Holography, Acad. Press, New York 1966.
14. Hamasaki, J., "Signal-to-noise ratios for hologram images of subjects in strong incoherent light," Appl. Opt., Vol. 7, pp. 1613-1620, 1968.
15. Ayrapetyans, E. S., Konstantinov, A. T., Echolocation in Nature, Nauka, Leningrad 1974.
16. Indeck, W., Printy, A., "Skin application of electrical impulses for relief of pain," Minnesota Medicine, Vol. pp. 305-309, 1975.
17. Greguss, P., "A model for making acupuncture consistent with Western concepts of biological information processing," Proc. NIH Acupuncture Research Conference, Bethesda, MD 1973, pp. 95-99.
18. Bach-y-Rita, P., Brain Mechanisms in Sensory Substitution, Academic Press, New York 1972.
19. Euler, M., "Nonlinear systems, self-organisation and acoustical perception", Microscience, (Ed. Marx, G., Szücs, P.) International Center for Educational Technology, Veszprem 1985.
20. Rosen, L., Clark, W., "Film plane holograms without external source reference beam," Appl. Phys. Lett., Vol. 10, pp. 140-142, 1967.
21. Gabor, D., "Light and information," in: Astronomical Optics and Related Subjects, Ed. Kopal, Z., North Holland Publ. Co., Amsterdam 1956.
22. Greguss, P., "Bioholography," Spie Seminar Proc., Vol. 25, pp. 55-83, 1971.
23. Fry, A. H., Fry, R. D., "Distance perception with holographically reconstructed objects," J. Opt. Soc. Am. A, Vol. 2, pp. 1217-1219, 1985.
24. Gabor, D., "Theory of communication," J. IEE, Vol. 93, III, pp. 429-257, 1946.
25. Gabor, D., "Imaging with coherent light", Seminar Lectures, CBS Laboratories, 1965.
26. Westlake, P. R., Towards a Theory of Brain Functioning: A Detailed Investigation of the Possibilities of Neural Holographic Processes, University Microfilms, Ltd., High Wycomb, England 1970.
27. Cavanagh, J. P., Holographic Processes Realizable in the Neural Realm: Prediction of Short-term Memory Performance, Ph.D. Dissertation, Université de Montreal 1972.
28. Ladik, J., Greguss, P., "Possible molecular mechanism of information storage in the long-term memory," in: Proc. Symposium on Biology of Memory, Tihany 1969, pp. 343-355.

Holographic testing of human vision.
Part 1. Shapes differentiation

B.Smolińska

Institute of Physics,Warsaw Technical University,Poland, 00662 Warsaw

Abstract

Visual perception has been investigated by holographic methods. The comparison between human sense of similarity and measurements made in optical correlator system leads to the conclusion of delineating procedure in human vision tract.

Introduction

The question is : what is the principal mechanism of pattern recognition in human perception, whether the decomposition of observed images in spatial frequency channels or in bar detection units [1] takes place in the vision tract.Hre we have tried to exploit holography for solving the problem quantitatively and independently on the subject.

Similarity measurement

Holography enables measurements of similarity ratio between optical signals. Generally the value of vectors product in vector signal space can serve as similarity gauge I.[2] The value I defined as

$$I(\vec{x}_0) = \frac{\vec{S}_1(\vec{x}_0)\,\vec{S}_2(\vec{x}_0)}{|S_1(\vec{x}_0)||S_2(\vec{x}_0)|},$$

where S_1 and S_2 are signals compared , could be easily measured in optical correlator systems. Schematic diagram of generalized correlator system is shown in Fig 1. There the signals compared S_1 , S_2 could be real patterns or its Fourier spectra as well. The intensity distribution in the output plane mapped correlation function. The light intensity measured exactly in the center of the correlation image , normalized energetically, is the searched similarity value I. It obtaines maximal value = 1 in the case of signals identity.

$$I(x=0) = \left| \frac{S_1 \star S_2}{\int S_1(x_0)\,dx_0 \int S_2(x_0)\,dx_0} \right|^2$$

Experimental

We tested a set of intentionally chosen patterns in optical correlator system. The patterns are polygons with deminishing rotational symmetry ,from a circle to triangle. We expected that the number of vertexes n in the polygons is connected with the similari degree between the figures .Namely the fixation points during eye movements concetrate inside the angles. One set of the patterns consists of full polygons, and the other of linear ones.Three types of correlator systems are used, image to image , quasi Fourier and van der Lugt . The results do not depend on correlator system used.

The results are shown in Fig 2 and 3. The similarity ratio I,as intensity measured exactly in the center of correlation image,at the output of van der Lugt correlator system,versus n,the number of vertexes in every one polygon, is there plotted.

Conclusions

For human being the polygons in both sets are ditinctly differentiated - the similarity value between the polygons decreases rapidly with decreasing number of vertexes n.

For correlator system all the full polygons are nearly identic.

For correlator system the linear polygons are well discriminated , the similarity value I decreases rapidly with decreasing n .

From mathematical point of vieux results obtained in correlator system are correct ,in the contrary a man differentiates shapes of full figures and linear figures as well.

The results point to the decisive role of delineating procedure which must take place in human vision tract just at beginning. The results confirm previous biological investigations.[3]

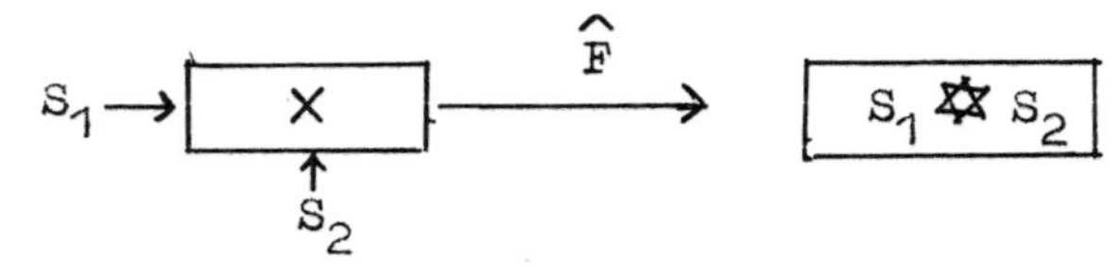

Fig.1

Schematic diagram of optical correlator system

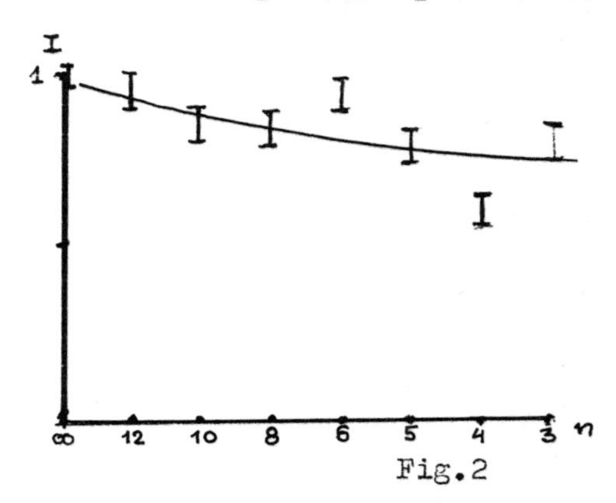

Fig.2

Measured similarity ratio versus the number of vertexes n in full polygons

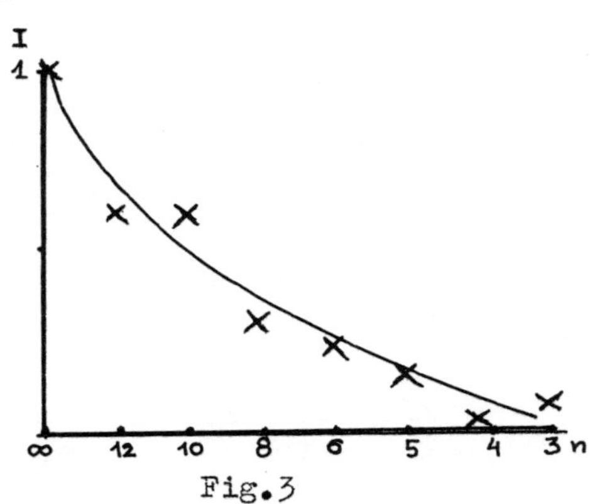

Fig.3

Measured similarity ratio versus the number of vertexes n in linear polygons

References

1. Mac Leod I., Rosenfeld A." The visibility of gratings :spatial frequency channels or bar detection units",Vis.Res. ,14 , 909 ,(1974).
2. Tow J., Gonzales R. " Pattern Recognition Principles " Addison Wsley Co ,(1974)
3. Hartline H.K. " The response of single optic nerve fibers of the vertebrate eye to illumination of the retina " ,Am.J.Physiol.,121,400,(1938) .

Holographic testing of human vision. Part 2. Optical correlator matched with human eye.

M.Komarnicki, B.Smolińska
Institute of Physics,Warsaw Technical University ,Poland, 00 662.

Abstract

A spatial filter simulating human eyes transfer function is constructed and used in optical correlator system in order to equalize the recognition results obtained in the optical system with the human valuation of patterns identity.

Introduction

Optical transfer function of a lens differs in some aspects from transfer function of human vision tract. The former can be approximated by a diffraction limited ideal system.The latter is tied with the data processing which proceeds in the vision nerve just at the beginning[1]. Moreover the retina is a discrete light intensity detector with limited number of receptors .The eyes pupil plays here a minor role.The grave differences between optical system and the vision system lead to misunderstandings in testing the results of recognizing in optical correlators.The aim of our work is matching correlator system to human vision.

Spatial filter " eye " construction

Van der Lugt correlator system is tested with a set of patterns . These are:binary master pattern and a set of replicas,smeared increasingly during copying.

The filter which simulates eye transfer function obeys two conditions 1) it must trunkate the Fourier spectrum in correlator system at 100 1 / mm ; the transmittance of the filter decreases slowly accordingly the ln function.2) the zeroth order spatial frequency must be suppressed;this leads to deminishing of steady illumination and enhancement of edges.The width of the central masking part in the filter is calculated from inspected pattern maximal dimensions. The transmittanc plot of the filter " eye " is shown in Fig 1.

Results

The results of measurements are given in the Table 1.

Table 1.
Normalised intensity values in center of correlation

Distance d in the copying process from master pattern [mm] .	Intensity	
	without filter	with filter
0	1,0	1,0
0,01	0,3	0,6
0,01 ORWO LP1	0,2	0,5
0,04	0,15	0,3
0,1o	0,1	0,3
completely different pattern	0,17	0,2

Conclusions

The filter constructed by us reduces the level of senstivity of the correlator system. The smeared pattern in a degree nonvisible for a man are classificated in the corelator without the filter as completely different signals , and with the use of the filter " eye " as highly similar ones. At the same time the **discrimination** power between completely different pattern is saved. Newertheless the filter " eye " we constructed is jet insufficient.

References

1. Marr D. " Vision " , W.H.Freeman and Co (1982)

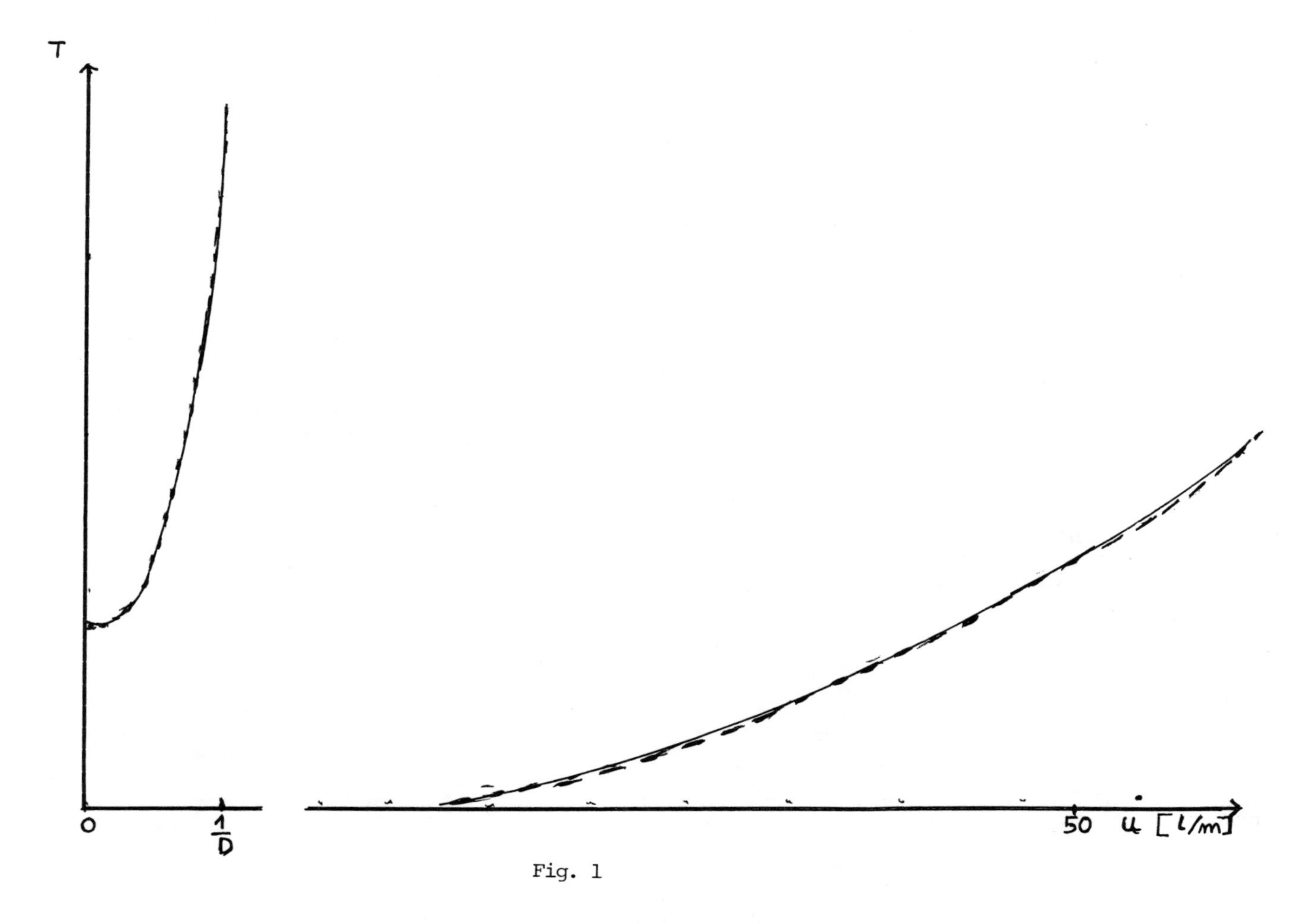

Fig. 1

The transmittance plot of spatial filter "eye".

Study of speckle multiple-shearing interferometry

Ma Yaowu , Ke Jingtang

Holographic Interferometry Lab., Zhengzhou Institute of Technology
10 Wenhua Road, Zhengzhou, Henan, People s Republic of China

Abstract

Based on speckle multiple-shearing interferometry advanced by Prof. Y.Y.Hung [1] , the paper deals with further study of its principles and proposes several kinds of new wedges, as well as analyzes the frequency spectra of double-exposure specklegrams. Discussions are also made on second order statistical characteristics of speckles when non-coherent superposition is effected for two speckle fields with no correlation at all. In order to raise the cut-off frequency of Fourier's filtering system, it is recommended in the paper that non-coaxial system be employed in place of coaxial system commonly used. Formulas for computation of changes in phase difference in speckle shearing interferometry have also been derived, and results of computations are compared with those of Prof.Hung.

Introduction

Of all methods of measurement in experimental mechanics, holographic interferometry and speckle shearing photography are particularly characterized by high sensitivity and non-contacting measurements, hence fit for the study of displacements and strain fields at the surface of an object. Only direct measurements of displacement field, however, are mostly effected. In order to determine the stresses, the study of a method of directly measuring the derivatives of displacements will be of significance. Such is accomplished by using laser speckle shearing interferometry, while at the same time obtaining fringe patterns of high quality. Compared with speckle shearing interferometry in one direction, multiple-shearing interferometry is instrumental to acquiring multiple (3 or more) fringe patterns from single double-exposure specklegram. In measuring derivatives of three-dimensional displacements, the number of records of double-exposure specklegram may be reduced, simplifying thus the optical scheme as well as test equipment. On the basis of work previously done by Hung [1,2] , and reference made on former work of the authors [3] , further study has been made on multiple-shearing interferometry, with results which may be found useful.

Multiple-shearing camera and principles of multiple shearing

Multiple-shearing camera is a combination of ordinary camera and multiple-shearing wedges placed in front of, in-between or behind the lenses.

Some multiple-shearing wedges are shown in Fig.1. The wedges presented by Hung enable shearing to realize in three directions, along the x-axis, y-axis and and at an angle of 45°. The use of wedges as such facilitates only gathering of two sets of fringes out of the total of three, the remaining one being difficult to obtain as a result of interference of the two aforesaid sets of fringes and a bright background. The third fringe pattern is readily obtainable through the use of bisecting and non-bisecting right-angled shearing wedge, as is presented here in the paper, also proposing another kind of multiple-shearing wedge --- bisecting 60° shearing wedge --- enabling shearing in the direction of 0°, 60° and 120°, thus most convenient for use in solving axial symmetric problems.

The principle of shearing may be expressed thus: Take the wedges shown in Fig.1(b) as example. Part 1 is a piece of flat crystal, and part 2 and 3 are wedges, the angle of inclination of part 2 being in the plane XOZ and that of part 3 in plane YOZ. The wedges as a whole are functional in resolving light waves from the surface of the object passing the wedges into three components, each focussed on a point on the image plane. In other words, except for points on the edges of the image in its plane, light wave at any point on the image is the result of superposition of light waves from 3 points on the surface of the object, the relative position of the points being shown in Fig.2(a). Mutual interference of any two points effects shearing interference in the direction along x-axis, y-axis and at an angle of 45°, the amount of shearing being δx, δy and δl_{45^o}. Based on the same principle, when the wedge shown in Fig.1(b) is used, any point on the image receives light waves from 3 points on the surface of object, as shown in Fig.2(b), superimposed on the image plane to realize shearing interference in the direction of 0°, 60° and 120°, with amount of shearing of δl_0, δl_{60} and δl_{120}.

When an object is illuminated by coherent light sources, photos may be taken by using multiple-shearing camera, as shown in Fig.3. Due to the function of multiple-shearing wedge, a number of sheared images modulated by speckles are obtained on the image plane. The images are sheared along the x-axis, y-axis and at an angle of 45°, or in the direction of 0°, 60° and 120°.

Prior and subsequent to deformation of an object, double exposures are made with multiple-shearing camera. After processing, the negatives are placed into the Fourier filtering system to obtain isopleths of changes in phase difference Δ_i, related to derivatives of displacements in the 3 directions of shearing subsequent to high-pass filtering in the 3 directions. When

$$\Delta_i = 2\pi N_i \qquad (i = 1, 2, 3; \quad N_i = 0, \pm 1, \pm 2, \ldots) \tag{1}$$

bright fringes will appear. Here N_i is the order of the fringe.

Formation of fringes

When the first exposure is being made, the light wave at any point on the image plane is formed through superposition of light waves from 3 points on the surface of the object, so that we may write:

$$U_1 = U(x,y) + U(x+\delta x,y) + U(x,y+\delta y) \tag{2}$$

Now let

$$U(x,y) = a(x,y)\exp\left[\theta(x,y)\right] \tag{3}$$

and the corresponding light intensity is

$$\begin{aligned} I_1 &= U_1 U_1^* \\ &= a^2(x,y) + a^2(x+\delta x,y) + a^2(x,y+\delta y) + 2a(x,y)a(x+\delta x,y)\cos\phi_x(x,y) + 2a(x,y)\cdot \\ &\quad a(x,y+\delta y)\cos\phi_y(x,y) + 2a(x+\delta x,y)a(x,y+\delta y)\cos\phi_{xy}(x,y) \end{aligned} \tag{4}$$

Here,

$$\begin{aligned} \phi_x(x,y) &= \theta(x+\delta x,y) - \theta(x,y) \\ \phi_y(x,y) &= \theta(x,y+\delta y) - \theta(x,y) \\ \phi_{xy}(x,y) &= \theta(x+\delta x,y) - \theta(x,y+\delta y) \end{aligned} \tag{5}$$

The second exposure is made after deformation of the object. If the deformation is small, displacement of speckles of points due to displacement of the object may be neglected and only phase difference is introduced, For light waves received on the negative we may write:

$$U_2 = U(x,y)\exp\left[\Delta(x,y)\right] + U(x+\delta x,y)\exp\left[\Delta(x+\delta x,y)\right] + U(x,y+\delta y)\exp\left[\Delta(x,y+\delta y)\right] \tag{6}$$

and the corresponding light intensity is:

$$\begin{aligned} I_2 &= a^2(x,y) + a^2(x+\delta x,y) + a^2(x,y+\delta y) + 2a(x,y)a(x+\delta x,y)\cos\left[\phi_x(x,y) + \Delta_x\right] \\ &\quad + 2a(x,y)a(x,y+\delta y)\cos\left[\phi_y(x,y) + \Delta_y\right] + 2a(x+\delta x,y)a(x,y+\delta y)\cos\left[\phi_{xy}(x,y) + \Delta_{xy}\right] \end{aligned} \tag{7}$$

in which Δ_x, Δ_y and Δ_{xy} are respectively changes in phase difference of shearing interference in the 3 corresponding directions of shearing:

$$\begin{aligned} \Delta_x &= \Delta(x+\delta x,y) - \Delta(x,y) \\ \Delta_y &= \Delta(x,y+\delta y) - \Delta(x,y) \\ \Delta_{xy} &= \Delta(x+\delta x,y) - \Delta(x,y+\delta y) \end{aligned} \tag{8}$$

Add Eq.(4) and (7) to get the total light intensity on the negative:

$$\begin{aligned} I &= I_1 + I_2 \\ &= 2a^2(x,y) + 2a^2(x+\delta x,y) + 2a^2(x,y+\delta y) + 4a(x,y)a(x+\delta x,y)\cos\left[\phi_x(x,y) + \Delta_x/2\right]\cos\Delta_x/2 \\ &\quad + 4a(x,y)a(x,y+\delta y)\cos\left[\phi_y(x,y) + \Delta_y/2\right]\cos\Delta_y/2 \\ &\quad + 4a(x+\delta x,y)a(x,y+\delta y)\cos\left[\phi_{xy}(x,y) + \Delta_{xy}/2\right]\cos\Delta_{xy}/2 \end{aligned} \tag{9}$$

Information of fringes are obtained through optical information processing subsequent to processing of the negative. Changes in frequency spectra due to Δ_x, Δ_y and Δ_{xy} are respectively in the directions of 3 frequency spectra. Filtering in 3 directions will be instrumental to obtaining respective isopleths of Δ_x, Δ_y and Δ_{xy} on the image plane.

When

$$\Delta_i = (2N_i + 1)\pi \qquad (i = x, y, xy; \quad N_i = 0, \pm1, \pm2, \ldots) \tag{10}$$

dark fringes appear. Here, N_i is the order of the fringe.

Characteristics of speckle frequency spectra

Study on characteristics of speckle frequency spectra on negatives from single-direction shearing interferometry was made in reference [4]. Based on previous work, in-depth discussions are also presented in this paper. According to reference [4], the wedge used in single direction shearing interferometry is shown in Fig.4. Where changes in phase difference $\Delta = 2\pi N$ ($N = 0, \pm1, \pm2, \ldots$), the light waves recorded on the negative through double-exposure are due to superposition of coherent light waves passing through two semi-circular image-forming openings, so that the speckle frequency spectra on the negative are dependent of the diameter of the entire circular hole. The halo of frequency spectrum is a circle. Where changes in phase difference $\Delta = (2N + 1)\pi$, the light waves recorded on the negative through double-exposure are due to non-coherent superposition of light waves respectively from two semi-circular image-forming holos, so that the speckle frequency spectra on the negative are consequent on combined effect of the two semi-circular apertures.

As peripheries of halos of speckle frequency spectra formed by using many other sizes of image-forming apertures are employed in discussing multiple-shearing interferometry, the general methods of determining the peripheries are to be discussed first. As the speckle frequency spectrum is a self-correlating function of the size of image-forming aperture, a "graphical method" is presented here for determination of the periphery of the halo of the spectrum.

When making the diagram, the image-forming aperture through which light passes is sheared, to form a diagram of two apertures, with no overlap. When one of the diagrams is held in position and the other is sheared in different directions relative to the first one, the closing locus of any point on the diagram is the curve of the periphery of the halo of speckle spectrum, some of these being shown in Fig.5 which correspond to different image-forming aperture sizes. Fig.6 shows the corresponding experimental results.

As regards second-order statistical characteristics of speckles subsequent to non-coherent superposition of two speckle fields not at all correlated, we may assume that speckle field I is obtained through superposition of two entirely non-correlated speckle fields A and B. As one of the speckle fields may be taken as a random process, the self-correlating function of speckle field I may be written thus:

$$R_{II}(\Delta_x, \Delta_y) = R_{AA}(\Delta_x, \Delta_y) + R_{AB}(\Delta_x, \Delta_y) + R_{BA}(\Delta_x, \Delta_y) + R_{BB}(\Delta_x, \Delta_y) \tag{11}$$

Now assume that the mean values of speckle field A and B are all $\langle I \rangle$, these being non-correlated, the function of mutual correlation in the above equation reads thus:

$$R_{AB}(\Delta_x, \Delta_y) = R_{BA}(\Delta_x, \Delta_y) = \langle I \rangle \langle I \rangle = \langle I \rangle^2 \tag{12}$$

and Eq.(11) may be rewritten as follows by using the expression for self-correlating function of speckles presented by J.W.Goodman [5]:

$$R_{II}(\Delta_x, \Delta_y) = \langle I \rangle^2 \left\{ 4 + \left| \frac{\iint_{-\infty}^{+\infty} |P_A(\xi,\eta)|^2 \exp\left[j\frac{2\pi}{\lambda z}(\xi\Delta x + \eta\Delta y)\right] d\xi d\eta}{\iint_{-\infty}^{+\infty} |P_A(\xi,\eta)|^2 d\xi d\eta} \right|^2 + \left| \frac{\iint_{-\infty}^{+\infty} |P_B(\xi,\eta)|^2 \exp\left[j\frac{2\pi}{\lambda z}(\xi\Delta x + \eta\Delta y)\right] d\xi d\eta}{\iint_{-\infty}^{+\infty} |P_B(\xi,\eta)|^2 d\xi d\eta} \right|^2 \right\} \tag{13}$$

From the foregoing equation it may be seen that as the contrast of speckle field I diminishes, the directional characteristic diameter of speckles in the field is the largest value of those of two fields subject to non-coherent superposition.

Through Fourier transformation of Eq.(13), we get the function of density of power spectrum. It may be seen that the density of power spectrum of speckles concentrates toward zero frequency. The directional diameter of the periphery of halo of the spectrum is thus governed by the largest value of directional diameter of the peripheries of frequency spectra of two speckle fields.

From the above discussions, it may be seen that the halo of frequency spectrum of speckles coherently superimposed in single-direction shearing interferometry is represented by the area of a large circle shown in Fig.6, whereas that of non-coherent superposition is shown by the area surrounded by the shaded zone. When subject to Fourier filtering, the filter on the spectral plane only let the light from the shaded area pass, and isopleths of Δ may be obtained at the output plane.

Similar discussions may be carried out for multiple-shearing interferometry. In Fig.7 (b), speckle frequency-spectrum halos are shown on the negative subject to double exposures when multiple-shearing wedges as shown in Fig.1(b) are used. The full line, dotted line and dot-and-dash line respectively represent the curves of peripheries of spectrum halos where changes in phase difference $\Delta_{jk} = 2N_{jk}\pi$ where $N_{jk} = 0, \pm1, \pm2, \ldots, j, k = 1, 2, 3$. And when $\Delta_{jk} = (2N_{jk} + 1)\pi$, the curve of periphery of frequency-spectrum halo will not exceed the area enveloped by the periphery when coherent superposition is effected. When filtering, therefore, light passing through areas T_{12}, T_{13} and T_{23} are to pass through filters, to obtain respective isopleths of Δ_{12}, Δ_{13} and Δ_{23} on the output plane. For the case of wedge presented by Hung (Fig.1(a), the frequency-spectrum halo of speckles on double-exposure specklegram is shown in Fig.7(a). It is seen that the area of T_{13} and T_{23} are rather large, so that information about Δ_{13} and Δ_{23} will be readily obtained, whereas T_{12} is shaded by the area of frequency-spectrum halo of speckles formed by light waves passing through wedge 3, and when the light waves passing through wedges 1 and 3 are coherently superimposed to those passing through wedges 2 and 3, T_{12} is also totally shaded off by the area of frequency-spectrum halo of speckles thus formed. Hence, there is no independent and separable zone for T_{12}, and if the filtering aperture is opened at T_{12}, the fringes thus obtained will be interfered by a bright background and the two other sets of fringes, so that the quality of fringes is very poor.

With the two kinds of wedges shown in Fig.1(c) and (d), the speckle frequency-spectrum halos on the double-exposure specklegram are those shown in Fig.7(c). It may be seen that the areas of T_{12}, T_{13} and T_{23} are equal, all of them being considerably large, so that information of fringes is readily obtained through filtering, mutual interference between fringes being slight.

In speckle shearing interferometry, deformation of the object brings about displacement of speckles of the image points, thus exhibiting speckle photographic effect, which is very slight, or even vanishes in case the displacements of speckles are not appreciable. It is therefore known that the sensitivity of speckle shearing interferometry is not related to the radial position of filtering holes. Also, we know that the sensitivity of speckle photographic full-field analysis is the sensitivity of wave filtering; in other words, the sensitivity of speckle photography varies with the radial position of filtering apertures. To eliminate the speckle photographic effect, therefore, directional high-pass filtering apertures or fan-shaped ones are used instead of apertures of small sizes. In doing so, the information of speckle photography and speckle shearing interference are to pass directional or fan-shaped filtering apertures. On the output plane, however, speckle photographic information of different sensitivities or overlapping isopleths of different spacings become blurred, whereas the isopleths of speckle shearing interference become more outstanding and are readily separable.

Fourier filtering system

In speckle shearing interferometry, speckles are random carrier waves, maximum frequency of which to reach [5]

$$f_{max} = \frac{D}{\lambda z} = D/\lambda f(1+M) = 1/(1+M)\lambda F' \qquad (14)$$

With magnification of $M = 0.5$, relative diameter of aperture $F' = 2$, length of light wave $\lambda = 0.6 \times 10^{-3}$ mm, then $f_{max} = 556$ lines/mm. In the case of Fourier filtering system, if two identical lenses are used for Fourier transformation, the cut-ff frequency of co-axial 4F system can hardly exceed $1/2\lambda F$ [6]. In fact, the relative diameter of apertures for lenses for Fourier transformation cannot be very large, being 1:3 to 1:10 in general, so that with length of light wave $\lambda = 0.6 \times 10^{-3}$ mm, the highest value of cut-off frequency of the system is

$$f = 1/2\lambda F = 1/2\times0.6\times10^{-3}\times3 = 275 \text{ lines/mm} \qquad (15)$$

It is therefore seen that the cut-off frequency of Fourier filtering system is much lower than the maximum frequency of speckles, so that the former is inadequate for gathering information of fringes, which become actually invisible when filtering system as such is used under certain conditions.

Several kinds of filtering system are presented in reference [6] , all with low cut-off frequency, attributed mainly to small diameter of relative aperture of the lenses for transformation. In practice, it is difficult to make transformation lenses of large relative diameters of apertures ($F > 3$), and the cost may be very high. In the paper, therefore, non-coaxial system is adopted instead of coaxial system in Fourier filtering, to raise the cut-off frequency. By using non-coaxial system, satisfactory results may be achieved in speckle shearing interferometry. Two kinds of non-coaxial 4F system are shown in Fig.8.

Formula for computation of changes in phase difference in speckle shearing interferometry

Referring to Fig.9, light source for illumination being at S (x_s, y_s, z_s) and the shearing camera at O (x_o, y_o, z_o), deformation of the object will cause a point on the object $P_1(x, y, z)$ to shift to P_1' ($x+u$, $y+v$, $z+w$), leading up to difference in light travel

$$\delta l_1 = (SP_1' + P_1'O) - (SP_1 - P_1O)$$

$$\approx (\vec{SP_1} - \vec{P_1O}) \cdot \vec{L}$$

$$= \left(\frac{x - x_o}{R_o} + \frac{x - x_s}{R_s}\right)u + \left(\frac{y - y_o}{R_o} + \frac{y - y_s}{R_s}\right)v + \left(\frac{z - z_o}{R_o} + \frac{z - z_s}{R_s}\right)w \tag{16}$$

in which

$$R_s = \sqrt{(x-x_s)^2 + (y-y_s)^2 + (z-z_s)^2}$$

$$R_o = \sqrt{(x-x_o)^2 + (y-y_o)^2 + (z-z_o)^2} \tag{17}$$

Similarly, when point $P_2(x+\delta x, y+\delta y, z)$ moves to P_2' ($x+\delta x+u+\delta u$, $y+\delta y+v+\delta v$, $z+w+\delta w$) subsequent to deformation of the object, the difference in light travel will be:

$$\delta l_2 = \left(\frac{x - x_o}{R_o'} + \frac{x - x_s}{R_s'}\right)(u + \delta u) + \left(\frac{y - y_o}{R_o'} + \frac{y - y_s}{R_s'}\right)(v + \delta v)$$

$$+ \left(\frac{z - z_o}{R_o'} + \frac{z - z_s}{R_s'}\right)(w + \delta w) \tag{18}$$

As $\delta x, \delta y$ are negligibly small compared with $|z - z_s|$, $|z - z_o|$, R_s' and R_o' in the above expression may be taken as R_s and R_o respectively, or

$$R_s' = R_s$$

$$R_o' = R_o \tag{19}$$

As a result of shearing, the points P_1 and P_2 on the object will coincide on the image plane, so that the change in phase difference due to deformation of the object may be expressed thus:

$$\Delta = \frac{2\pi}{\lambda}(\delta l_2 - \delta l_1)$$

$$= \frac{2\pi}{\lambda}(A\delta u + B\delta v + C\delta w) \tag{20}$$

in which

$$A = \frac{x-x_o}{R_o} + \frac{x-x_s}{R_s}$$

$$B = \frac{y-y_o}{R_o} + \frac{y-y_s}{R_s} \tag{21}$$

$$C = \frac{z-z_o}{R_o} + \frac{z-z_s}{R_s}$$

Expression similar to Eq.(20) appeared in the formula derived by Y.Y.Hung [1, 2] , but it was taken that $x = y = z = 0$ in Eq.(17), i.e., SP_1 and OP_1 are respectively equal to the distance of point S and O from the origin.

$$R_s = \sqrt{x_s^2 + y_s^2 + z_s^2}$$
$$R_o = \sqrt{x_o^2 + y_o^2 + z_o^2} \qquad (22)$$

This is true only when the size of object is very small, appreciably large errors being brought about in case of large field of view.

Relation between different fringe patterns in multiple-shearing interferometry

If shearing is to take place in the direction of x-axis, y-axis and at an angle of β in multiple-shearing interferometry, and the amount of shearing is small enough, the 3 fringe patterns will respectively correspond to isopleths expressed thus:

$$\Delta_x = \frac{2\pi}{\lambda}\left(A\frac{\partial u}{\partial x} + B\frac{\partial v}{\partial x} + C\frac{\partial w}{\partial x}\right)\delta x$$
$$\Delta_y = \frac{2\pi}{\lambda}\left(A\frac{\partial u}{\partial y} + B\frac{\partial v}{\partial y} + C\frac{\partial w}{\partial y}\right)\delta y \qquad (23)$$
$$\Delta_\beta = \frac{2\pi}{\lambda}\left(A\frac{\partial u}{\partial l_\beta} + B\frac{\partial v}{\partial l_\beta} + C\frac{\partial w}{\partial l_\beta}\right)\delta l_\beta$$

In accordance with the concept of directional derivatives, we have:

$$\frac{\partial u}{\partial l_\beta} = \frac{\partial u}{\partial x}\cos\beta + \frac{\partial u}{\partial y}\sin\beta$$
$$\frac{\partial v}{\partial l_\beta} = \frac{\partial v}{\partial x}\cos\beta + \frac{\partial v}{\partial y}\sin\beta \qquad (24)$$
$$\frac{\partial w}{\partial l_\beta} = \frac{\partial w}{\partial x}\cos\beta + \frac{\partial w}{\partial y}\sin\beta$$

and

$$\delta l_\beta = (\delta x)_\beta \cos\beta + (\delta y)_\beta \sin\beta \qquad (25)$$

in which $(\delta x)_\beta$ and $(\delta y)_\beta$ are respectively projection of δl_β on the x-axis and y-axis, so that the third expression in Eq.(23) may be written thus:

$$\Delta_\beta = \frac{2\pi}{\lambda}\Big[A\left(\frac{\partial u}{\partial x}\cos\beta + \frac{\partial u}{\partial y}\sin\beta\right) + B\left(\frac{\partial v}{\partial x}\cos\beta + \frac{\partial v}{\partial y}\sin\beta\right) + C\left(\frac{\partial w}{\partial x}\cos\beta + \frac{\partial w}{\partial y}\sin\beta\right) \cdot$$
$$\left((\delta x)_\beta \cos\beta + (\delta y)_\beta \sin\beta\right)\Big] \qquad (26)$$

The first terms of Eq.(23) are respectively multiplied by $\frac{(\delta x)_\beta}{\delta x}\cos\beta$ and $\frac{(\delta y)_\beta}{\delta y}\sin\beta$ and then added up to get

$$\Delta_x \frac{(\delta x)_\beta}{\delta x}\cos\beta + \Delta_y \frac{(\delta y)_\beta}{\delta y}\sin\beta = \Delta_\beta \qquad (27)$$

A useful conclusion is thus reached: in case of right-angled multiple-shearing interferometry, only two of the 3 fringe patterns are independent.

The above conclusion is general in character, and similar conclusions may be drawn for multiple shearing in other directions, using for instance 60° shearing wedges. Even if shearing is to realize in more than 3 directions, or infinitely large number of directions, only two independent fringe patterns are obtainable. The conclusion is also applicable to multiple-aperture, grating and other shearing interferometry.

Experiments

As typical test, measurements were carried out by applying multiple-shearing interferometry to thin rectangular plate fixed along 4 edges and subject to concentrated load at the center. Hard aluminium plate of 120 x 90 x 3 mm was used. Right-angled shearing wedges as shown in Fig. 1(b) were used for multiple shearing. The wedge angle was $\alpha = 0.4^\circ$, and shearing was effected along the x-axis, y-axis and at an angle of 45°.

Pictures were taken with multiple-shearing camera normal to the rectangular plate, illumination being effected by using He - Ne laser. Holographic plates of Tianjin-I were used for double exposures prior and subsequent to deforming the plate. The loading was 40.0 μm. After processing, the plates were placed into non-coaxial Fourier filtering system (Fig.8b) to gather information of fringes. The apertures of speckle spectra are shown in Fig.7(a). The surface of the object was turned to effect filtering in 3 directions, to get fringe patterns as shown in Fig.10. The 3 directions of filtering were respectively in the direction of T_{12}, T_{13} and T_{23}, as shown in Fig.7(a).

Conclusions

As a kind of speckle measuring technique, laser speckle multiple-shearing interferometry may be used for direct measurement of displacement derivatives, producing full-field fringes of rather high quality. As compared with single-direction shearing interferometry, the method of multiple shearing enables us to get a number of fringe patterns (3 or more) from a single double-exposure specklegram, thus facilitating measurements for solving problems related to fields of 3 groups of displacement derivatives, especially in case of applying load but once, owing to practical limitations. The optical scheme is rather simple. When carrying out quantitative computation, the use of surplus fringe pattern may bring about increased precision in case only a small amount of experimental work can be done.

Shearing interferometry is most effective in measuring the field of off-plane displacement gradients, simultaneous solutions for successive points in 2 or 3 fringe patterns being needed, however, for the case of field of in-plane displacement gradients.

The authors have applied multiple-shearing interferometry to the measurement of residual stresses, see reference 3 .

References

1. Hung,Y.Y. and Liang C.Y., Appl. Opt., 18, 1046, 1979.
2. Hung,Y.Y. and Durelli,A.J., J.Strain Analysis, 14, 81, 1979.
3. Ke, Jingtang, Ma, Yaowu and Zhao, Caifu, Proc. of SPIE, vol. 599, 216, 1985.
4. He, Yuening, J. of Optics, 5, 241, 1985.
5. Dainty, J.C., Laser Speckle and Related Phenomena, Springer-Verlag, 1975, Chapter 2.
6. Yu Meiwen, Optical holography and information processing, National Defence Ind.Press, 290, 1984.

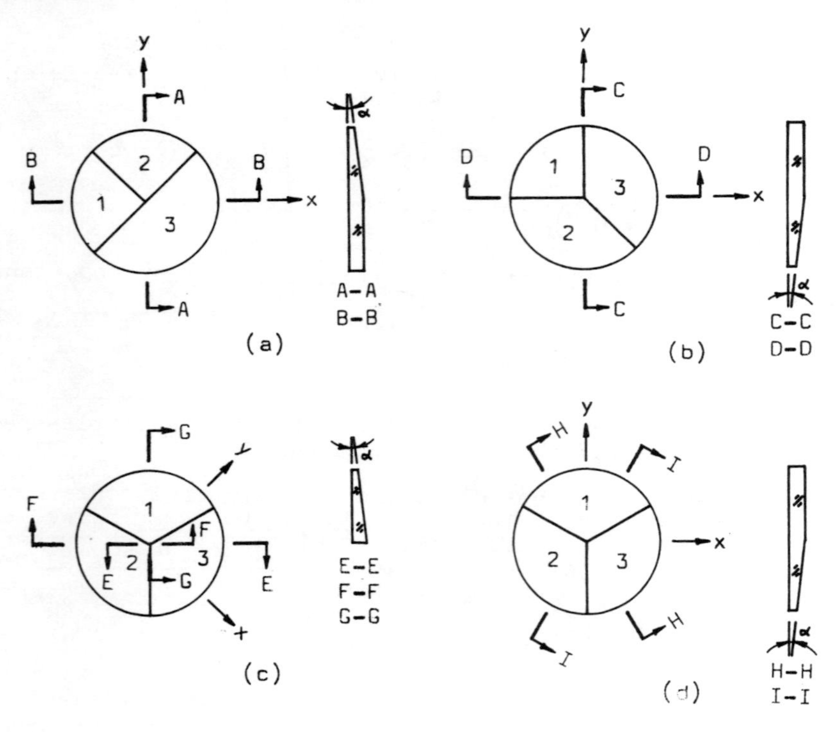

Fig.1 Different kind of multiple shearing glass wedges

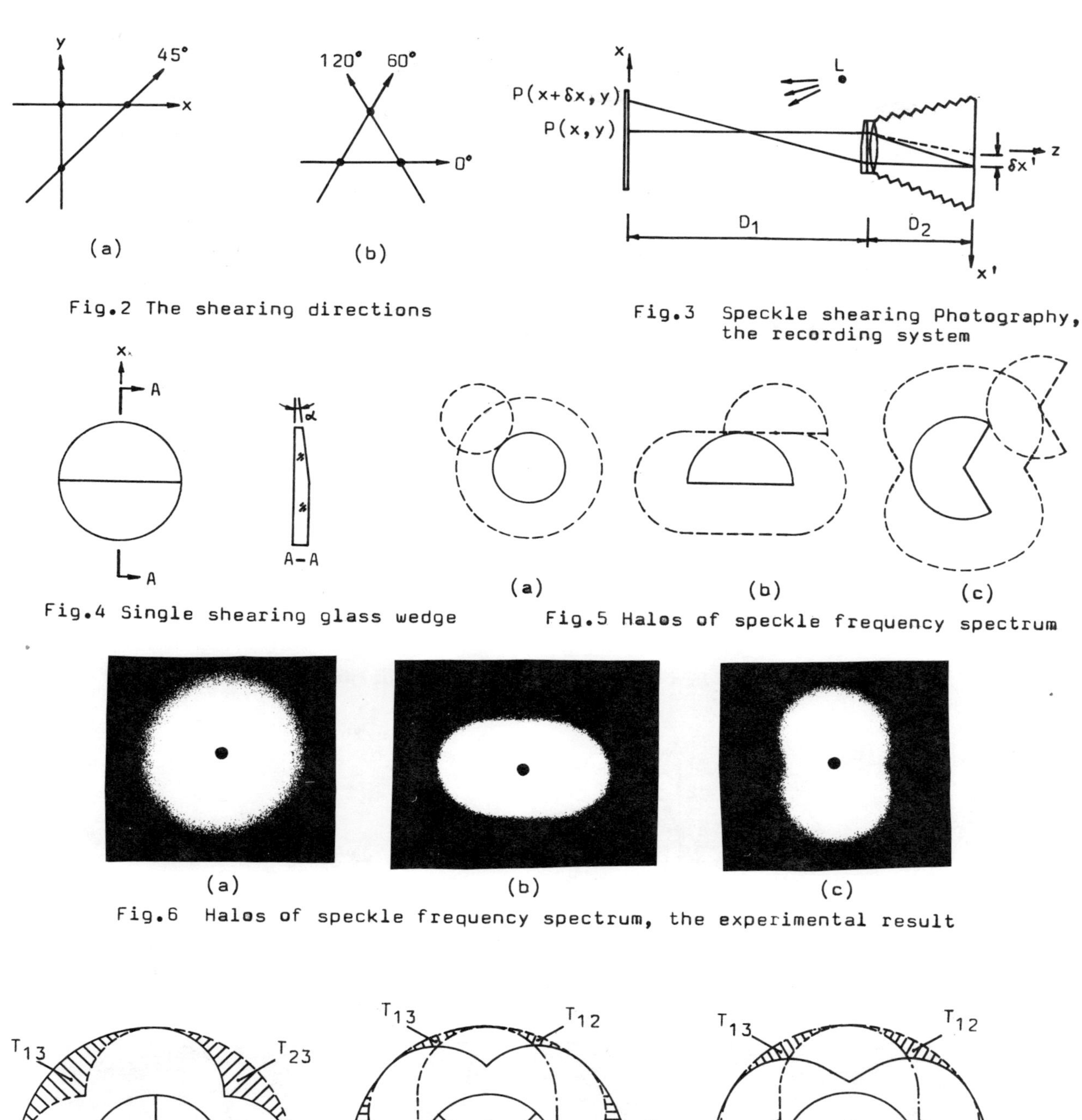

Fig.2 The shearing directions

Fig.3 Speckle shearing Photography, the recording system

Fig.4 Single shearing glass wedge

Fig.5 Halos of speckle frequency spectrum

Fig.6 Halos of speckle frequency spectrum, the experimental result

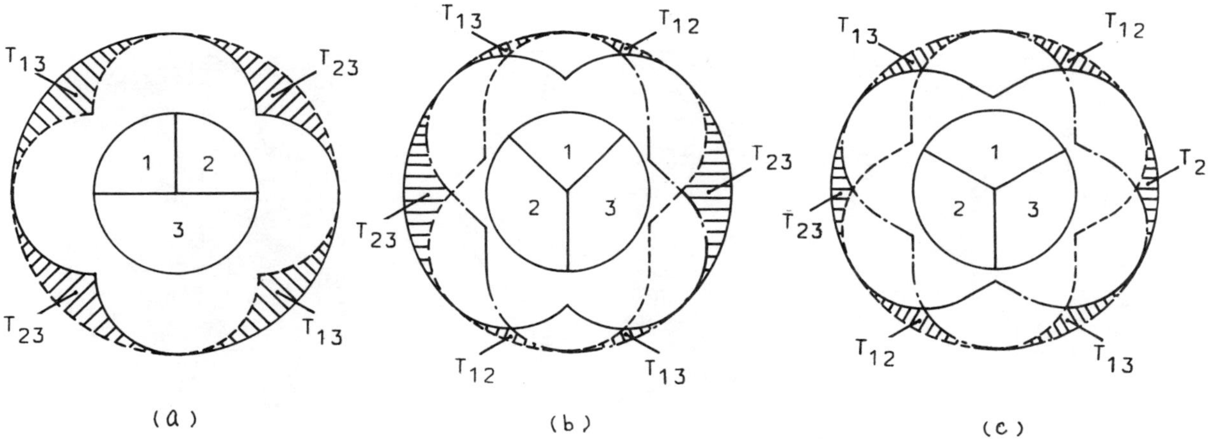

Fig.7 Halos of speckle frequency spectrum from different multiple shearing wedge

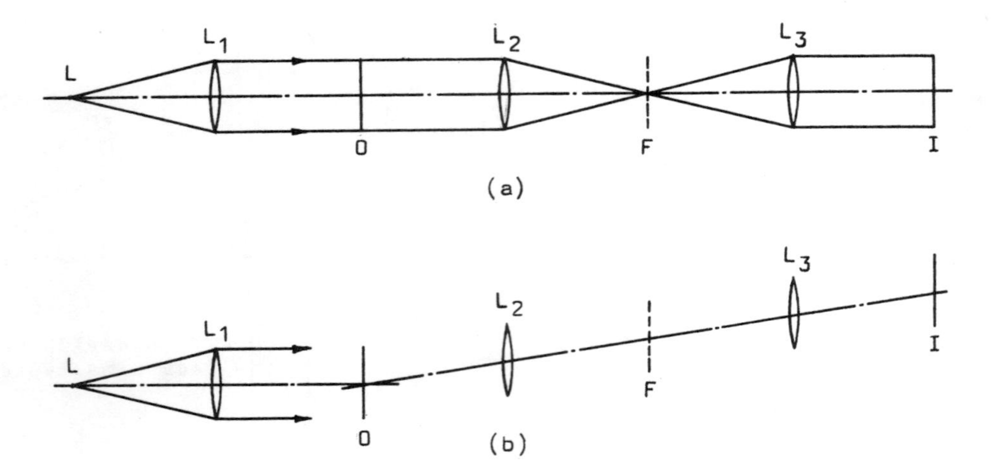

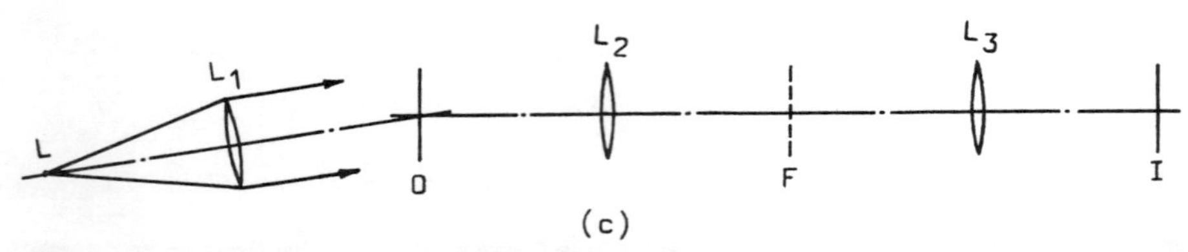

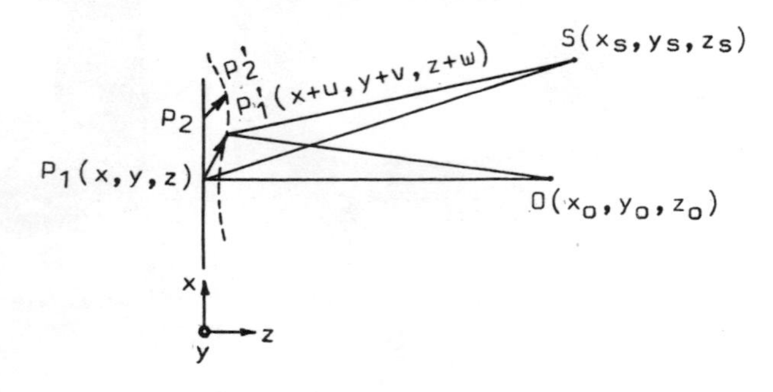

Fig.8 Non-coaxial system for Fourier filtering

Fig.9 Optical path diagram

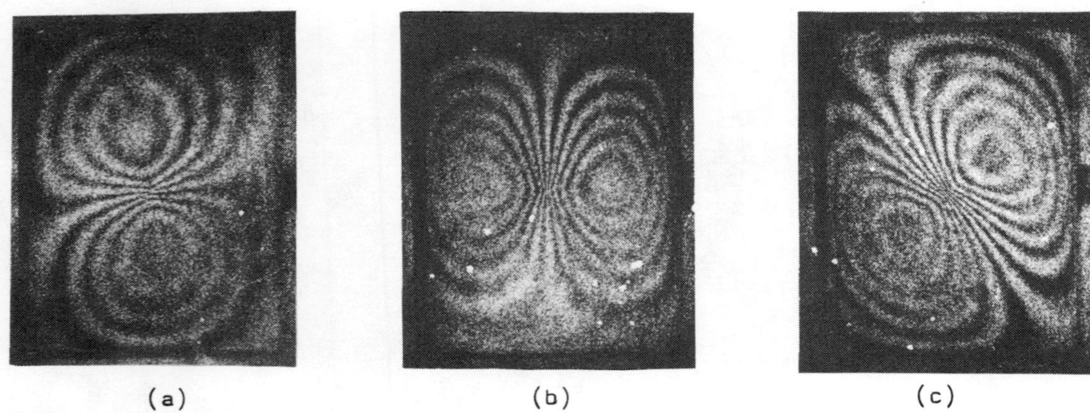

Fig.10 Measurement of speckle multiple image-shearing interferometry
(a) Shearing in direction y
(b) Shearing in direction x
(c) Shearing at an angle of 45°

Session 12

Holographic Recording Materials and Devices I

Chairmen
Tung H. Jeong
Lake Forest College, USA
Mei-Wen Yu
Beijing Institute of Technology, China

Optimum holographic disk scanners with bow-locus corrections

Yukihiro Ishii and Kazumi Murata

Department of Applied Physics, Faculty of Engineering, Hokkaido University
N-13, W-8, Sapporo 060, Hokkaido, Japan

Abstract

A design of disk-type holographic diode laser beam scanner exhibiting the bow-free and aberration-reduced properties is described, which is based on ray tracing and least-squares optimization technique. Scanning with a small bow deviation of less than ±130 μm on a 30-cm scan length has experimentally demonstrated for a holographic disk scanner recorded with the aid of a computer-generated hologram.

Introduction

Laser beam scanning has increased in importance as a recording device. The scanning line of disk-type rotating holographic grating has an unwanted curved scan locus. Such applications require flat-field straight line scanning. Recently diode lasers are compact and potentially reliable for wide uses of key devices in optical electronics. It is desirable to make a holographic grating scanner recorded on silver halide material at a wavelength that is different from the readout wavelength of diode laser, because it is relatively insensitive to exposures at near infrared wavelengths. The major problems associated with making this scanner are that, the one is, the aberration due to a recording-readout wavelength shift is seriously existed for a fixed geometry and that, the other is, the scan spots diffracted from a hologram suffer from aberrations since the recording geometry differs from that of the reconstructed hologram.

This paper provides a design of a holographic scanning method with a diode laser to correct for bow-scanning locus and to reduce the aberration due to a recording-readout wavelength shift. The design of the scanner based on a ray tracing and iterative optimization techniques[1,2] such as the damped least-squares (DLS) method allows for a bow-free scan at wide scan angles. The desired holographic scanner is interferometrically fabricated with a diffracted optimum aspherical wavefront from a computer-generated hologram (CGH)[3] and a spherical wavefront as the reference wave. Experiments demonstrating the feasibility of this scanner are demonstrated.

Optimum phase transfer function for holographic disk scanners

A discussion of the phase transfer function to be optimized is appropriate in this section. An x-y-Z coordinate system is taken in the plane of the flat holographic optical element (HOE) whose center of rotation is o, and an X-Y-Z coordinate is chosen in a fixed coordinate system located on the center O as shown in Figure 1. The corresponding coordinates denoted by the prime are taken in the scanning plane. The formation and reconstruction distances stand for the subscripted variable F. The radius of a disk is R, and the distance from point sources to aperture center is a. Thus, we make a holographic zone plate on the center $H(x_j, y_j)$ by interfering a divergent spherical wave with a convergent spherical wave by using a He-Ne laser line. If the HOE rotates about the axis Z by angle θ_j, the focused laser beam may be scanned along a curved line as shown by the dashed line of Figure 1. The bow-locus correction can be incorporated into the object wave together with a aspherical wave produced by a CGH. The readout diode laser light source locates on the distance c from the axis oo'.

The phase transfer function can be expressed in terms of the point coordinate (x_j, y_j) instead of the scan angle θ_j as follows. The x-y rotating coordinate as a function of the X-Y fixed coordinate is given by

$$\begin{pmatrix} x + R\cos\theta_j \\ y - R\sin\theta_j \end{pmatrix} = \begin{pmatrix} \cos\theta_j & \sin\theta_j \\ -\sin\theta_j & \cos\theta_j \end{pmatrix} \begin{pmatrix} X \\ Y \end{pmatrix} , \tag{1}$$

with the aid of the relationship of coordinate transformation as depicted in Figure 2. The point $H(x_j, y_j)$ of HOE center and the point $S(x_j^s, y_j^s)$ of the readout light source is written as

$$\begin{pmatrix} x_j \\ y_j \end{pmatrix} = \begin{pmatrix} R(1 - \cos\theta_j) + a \\ R\sin\theta_j \end{pmatrix} \quad \text{and} \quad \begin{pmatrix} x_j^s \\ y_j^s \end{pmatrix} = \begin{pmatrix} c\cos\theta_j \\ -c\sin\theta_j \end{pmatrix} , \tag{2}$$

respectively, by substituting the relations of $H[(R + a)\cos\theta_j, (R + a)\sin\theta_j]$ and $S(R + c,0)$ expressed by the fixed coordinate system into Equation (1).

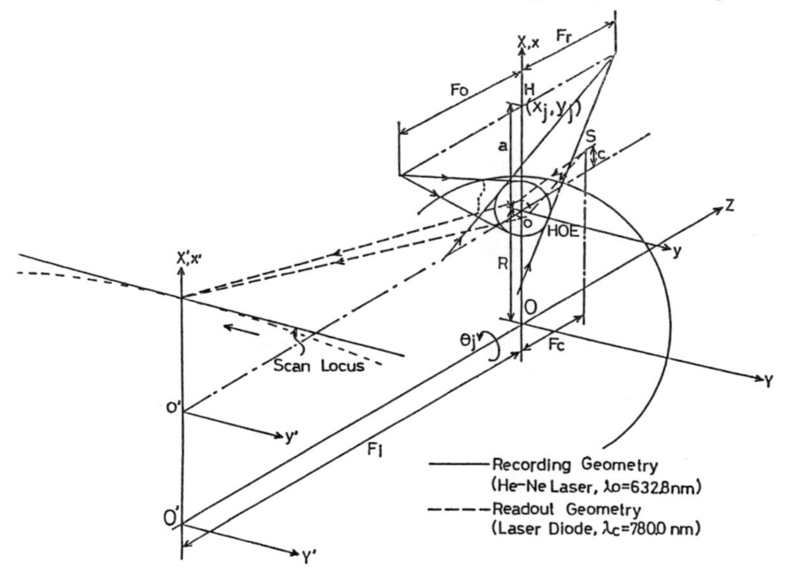

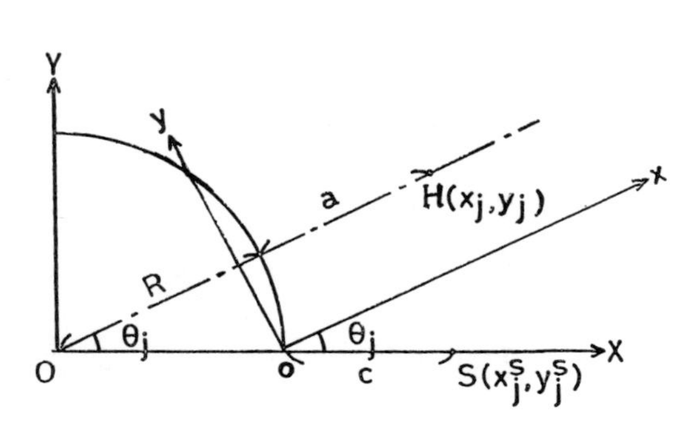

Figure 2. Schematic diagram for computing the HOE center H and the point S of readout light source with the coordinate transformation.

Figure 1. Hologram recording geometry is shown by solid lines and readout geometry by dashed lines for a disk scanner at scan angle $\theta_j=0°$.

If the HOE is recorded at wavelength λ_o by interfering between the object wavefront $\phi_o(x,y)$ and the reference wavefront $\phi_r(x,y)$, the generalized phase transfer function $\phi_H(x,y)$ with the correction phase $\phi_G(x,y)$ is described as

$$\phi_H(x,y) = \phi_o(x,y) + \phi_G(x,y) - \phi_r(x,y). \tag{3}$$

The phase ϕ_G having an even power of x and y,[1] is given by

$$\phi_G(x,y) = \frac{2\pi}{\lambda_o}W_G(x,y) = \frac{2\pi}{\lambda_o}(C_{20}x^2 + C_{40}x^4 + C_{60}x^6 + C_{80}x^8 + C_{02}y^2 + C_{04}y^4 + C_{06}y^6 + C_{08}y^8 + C_{22}x^2y^2 + C_{44}x^4y^4), \tag{4}$$

where C values are the unknown coefficients used as parameters to be optimized, and W_G is the optical path difference (OPD) of correction phase. The phase $\phi_i(x,y)$ of the diffracted wave from HOE when illuminated by the phase $\phi_c(x,y)$ of the reconstruction wave at the wavelength λ_c is written as

$$\phi_i(x,y) = \phi_H(x,y) + \phi_c(x,y). \tag{5}$$

According to Figure 1, all phases have the form

$$\phi_o = \frac{2\pi}{\lambda o}[(x^2 + y^2 + F_o^2)^{1/2} - (F_o^2 + a^2)^{1/2}], \tag{6}$$

expressing the divergent spherical object wavefront,

$$\phi_r = -\frac{2\pi}{\lambda_o}[(x^2 + y^2 + F_r^2)^{1/2} - (F_r^2 + a^2)^{1/2}], \tag{7}$$

expressing the convergent spherical reference wavefront, and

$$\phi_c = -\frac{2\pi}{\lambda_c}[(x^2 + y^2 + F_c^2)^{1/2} - (F_c^2 + c^2)^{1/2}], \tag{8}$$

expressing the divergent spherical reconstruction wavefront. If the HOE moves along the x direction by x_j and the y direction by y_j as shown in Figure 2, the phase transfer function in such a case given by Equation (3) is rewritten as

$$\phi_H(x - x_j, y - y_j) = \phi_o(x - x_j, y - y_j) + \phi_G(x - x_j, y - y_j) - \phi_r(x - x_j, y - y_j). \quad (9)$$

The corresponding phase of image ray ϕ_i in Equation (5) is replaced by the new one ϕ_{ij}, giving

$$\phi_{ij}(x,y) = \phi_c(x - x_j^s, y - y_j^s) + \phi_H(x - x_j, y - y_j), \quad (10)$$

which can be specified by using Equation (2). The problem then is determine the OPD W_G, so that the mean-squared spot radius and the mean-squared centroid of spot diagram are minimum as stated below.

Prior to the optimization of phase transfer function in Equation (9), it is useful to make the experimental parameters of scanner be optimum, because the optimized experimental parameters may be helpful to find the reasonable local minima of the merit function that describes the state of correction. We beforehand need to derive the positions of scanned spot. The direction cosines l_{ij}^k, m_{ij}^k, and n_{ij}^k of the image ray propagating in air with respect to the x, y, and Z axis, respectively, can be computed from the partial derivatives of the phase function in Equation (10); superscript k represents the k-th ray $(x=x^k, y=y^k)$ within the input aperture, i.e.,

$$l_{ij}^k = \frac{\lambda_c}{2\pi}\left(\frac{\partial\phi_{ij}}{\partial x}\right)_{Z=0}, \quad m_{ij}^k = \frac{\lambda_c}{2\pi}\left(\frac{\partial\phi_{ij}}{\partial x}\right)_{Z=0} \quad \text{and} \quad n_{ij}^k = \pm[1 - (l_{ij}^k)^2 - (m_{ij}^k)^2]^{1/2}, \quad (11)$$

where the sign choice in Equation (11) is used to select either the transmission (-) or the reflection (+) from Figure 1 in which we have chosen - sign. Therefore, the ray-traced image points x'^k_j, y'^k_j are given by the equations

$$x'^k_j = (x^k - x_j) + \frac{l_{ij}^k}{n_{ij}^k} F_i \quad \text{and} \quad y'^k_j = (y^k - y_j) + \frac{m_{ij}^k}{n_{ij}^k} F_i \quad (12)$$

using Equation (11). The ray-traced position in Equation (12) can be transformed into the fixed coordinate system (X'^k_j, Y'^k_j) in the scanning plane following the same manner described by Equation (1) as

$$\begin{pmatrix} X'^k_j \\ Y'^k_j \end{pmatrix} = \begin{pmatrix} \cos\theta_j & -\sin\theta_j \\ \sin\theta_j & \cos\theta_j \end{pmatrix} \begin{pmatrix} x'^k_j + R\cos\theta_j \\ y'^k_j - R\sin\theta_j \end{pmatrix}. \quad (13)$$

The mean-squared spot radius σ_j^2 at each j-th scanned beam is expressed with the ray-traced spot positions given by Equation (13) as

$$\sigma_j^2 = \frac{1}{N}\sum_{k=1}^{N}[(X'^k_j - \langle X'^k_j\rangle)^2 + (Y'^k_j - \langle Y'^k_j\rangle)^2], \quad (14)$$

for total number of rays N = 37, forming a hexapolar array, where $\langle\ldots\rangle$ denotes the average over spot diagram and then a spot center. The other performance function f for bow deviation consists of the average of the difference between the j-th spot center and the average of j-th spot center over thirteen scanned beam (M=13) at half of the scanning plane inside the scan angle 12°, giving

$$f = \frac{1}{M}\sum_{j=1}^{M}[\langle X'^k_j\rangle - \frac{1}{M}\sum_{j=1}^{M}\langle X'^k_j\rangle]^2. \quad (15)$$

Thus, the optimum experimental parameters a, c, F_c and F_i for the fixed parameters $F_o=F_r$=100 mm and $\phi_G=0$ are determined so as to set a scanned spot size σ_j^2 in Equation (14) of less than 0.5 mm^2, a bow deviation $(f)^{1/2}$ in Equation (15) of less than 0.5 mm, and a scan length of more than 400 mm, by using a nonlinear least-squares method. The results are as follows: a=16.9 mm, c=1.2 mm, F_c=48.3 mm and F_i=265.3 mm.

Using the determined geometry as a starting initial guess, we further optimize a scanner with the DLS method, introducing the merit function mentioned below. The merit function Γ is given as the sum of a weighted (W_1) average of mean-squared spot radius shown in Equation (14) and a weighted (W_2) average of mean-squared centroid shown in Equation (15) expressing a bow deviation over thirteen scanned beams, i.e.,

$$\Gamma(\mathbf{C}) = [W_1(\frac{1}{M}\sum_{j=1}^{M}\sigma_j^{\,2})]^2 + (W_2 f)^2, \tag{16}$$

where the vector $\mathbf{C} = (C_{20},\ldots,C_{44})$ having in mind equation (4) and W's are the important parameters which are taken as adaptive numbers to arrive at the reasonal solutions. Our problem is to solve $\mathbf{C}$ values that minimize Γ by DLS method with an appropriate weight ratio W_1/W_2. Starting with $\mathbf{C}$=O where the spherical wavefront in Equation (9) is noted, all ten coefficients of $\mathbf{C}$ vector are allowed to vary during DLS optimization. The DLS procedures are repeated while taking succesively the step length in $\mathbf{C}$ vector until the convergence is confirmed within the required accuracy.

Numerical results

This section presents the numerical results for each step solution in the HOE design. The wavelengths λ_o=632.8 nm and λ_c=780.0 nm are used for the calculation. Figure 3 shows the plots of the square of rms spot radius in Equation (14) as a function of scan angle which compare an optimized scanner with the correction phase (denoted by the asterisk) with a scanner made with the optimum experimental geometry and recorded only with the spherical wave (denoted by the circle). Many times during optimization routines when varying the weighted parameters designated in Equation (16), we see that, as a ratio W_1/W_2 increases, the spot size decreases for large scan angles over 6°. For $W_1/W_2=10^3$, the rms spot size becomes minimum, such that on the other hand, the bow is overcompensated for a range of scan angles over 10°, because W_2 is underweighted as depicted in Figure 4. This figure indicates the bow deviation expressed by $[\langle X'^{k}_{j}\rangle - \langle X'^{k}_{0}\rangle$ (at $\theta_j=0°$)] as a function of scan angle. The same sign as that of Figure 3 is marked in Figure 4. But, it is noticeable that inside the scan angle 10°, the minimum bow deviation can be obtained as ±130 µm. Bow is well compensated. Thus, we adopt as this weight ratio to get the optimum solution. The merit function can rapidly converge with only a few repetitions with the help of the predetermined, optimum experimental geometry. This design method is demonstrated for a 32-cm scan length with about 350-µm aberration-reduced spot diameter.

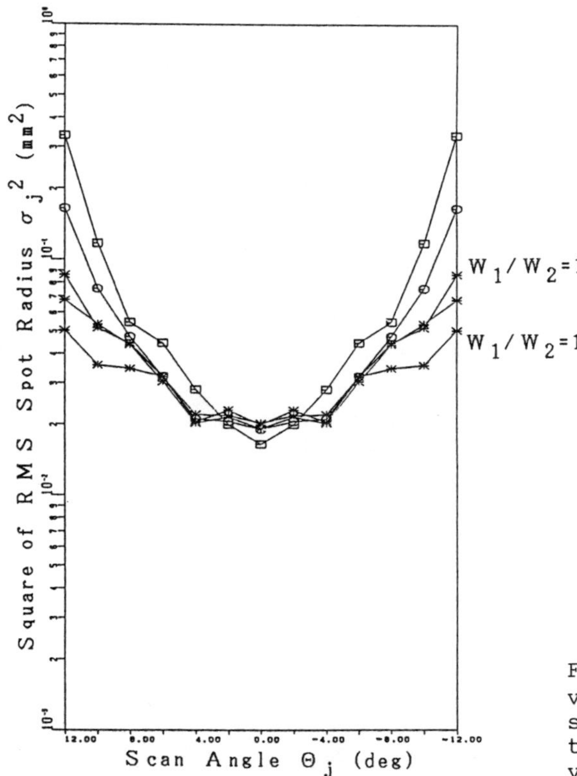

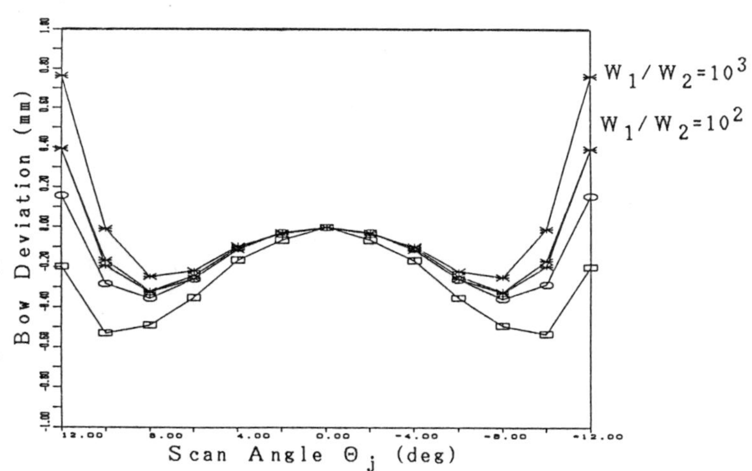

Figure 4. Plots of the bow deviation versus scan angle θ_j. The sign choice is the same manner as that of Figure 3.

Figure 3. Plots of the square of rms spot radius σ_j^2 versus scan angle θ_j. The circle sign corresponds to a scanner with the optimum experimental parameters and the asterisk sign to an optimized scanner with the variation of a weight ratio W_1/W_2.

The resultant optimized coefficients at the margin of the HOE (x=22.1 mm,y=23.8 mm) are shown in Table 1. The coefficients of an aspherical wavefront are normarized to have units of wavelength (λ_o). The maximum amount of phase function W_G at the margin is 20.1 λ_o. The dominant term in Equation (4) is C_{44}. A three-dimensional plot of the optimized wavefront used for correction is shown in Figure 5.

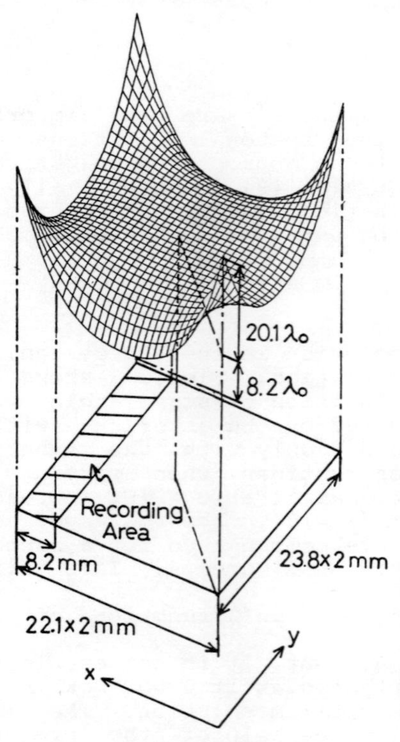

Table 1. Optimized Coefficients of Aspherical Wavefront at x = 22.1 mm and y = 23.8 mm for λ_o = 632.8 nm

$C_{20}x^2$	= 1.44	$C_{40}x^4$	= -2.40	$C_{60}x^6$	= -2.46
$C_{80}x^8$	= 1.85	$C_{02}y^2$	= -0.43	$C_{04}y^4$	= -3.53
$C_{06}y^6$	= -5.14	$C_{08}y^8$	= 0.87	$C_{22}x^2y^2$	= 11.68
$C_{44}x^4y^4$	= 18.20				

Figure 5. Three-dimensional plot of the optimized correction wavefront. The recording area of hologram is shown by hatched lines.

Figure 6 illustrates the numerical results of a conventional uncorrected scanner whose optimization is not made. Figure 6a shows the plots of rms spot radius as a function of scan angle and Figure 6b is the plots of the bow deviation as a function of scan angle. The experimental parameters used for the calculation are as follows: a=22.0 mm, c=12.0 mm, F_c=47.0 mm and F_i=277.2 mm. In such a case, the diameter of scan spots is about 1.5 mm and the maximum bow deviation is 8 mm.

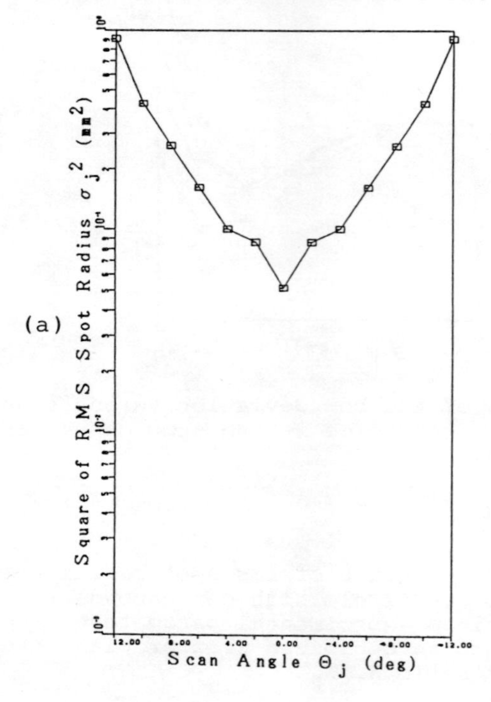

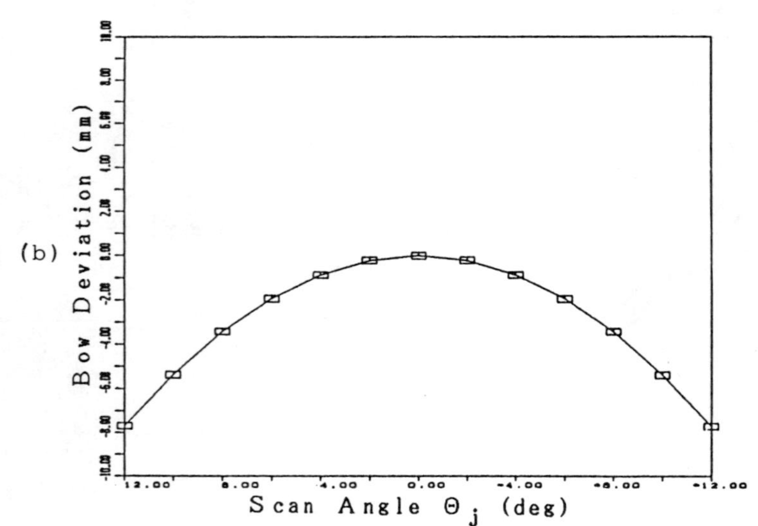

Figure 6. (a) Plots of the square of rms spot radius σ_j^2 versus scan angle θ_j and (b) plots of the bow deviation versus scan angle θ_j for a conventional uncorrected scanner.

Experiments with ray-traced spot diagrams

The optimum holographic disk scanner with bow-locus correction as will be described in this section is recorded using the optical system diagrammed in Figure 7. The desired holographic scanner is interferometrically fabricated with a diffracted aspherical wavefront from the CGH and a convergent spherical wavefront at 19.2° offset angle on an Agfa 8E75 HD plate. The CGH is tilted so that it and the HOE plane are in conjugate image planes, and subsequently its image is distorted by less than 1 %. The CGH that recorded the phase variation ϕ_G of Equation (4) is an interference-type CGH. This correction phase has large, but the recording area shown by hatched lines in Figure 5 is truncated such that the phase variation to be recorded with a CGH can be considerably reduced. The required number of carrier fringes to separate first and second orders along the x direction is 110 fringes. Figure 8a is an interferogram of the correction phase generated by a CGH, which can be obtained by simultaneously illuminating the developed hologram with both divergent spherical and convergent reference waves, as compared with the drawing of a computational interferogram shown in Figure 8b. Good agreement results. Here, the y direction is scanning direction. A holographic lens with large aperture is required to generate the reference convergent wave because of a small F/No. optical system. The holographic lens, produced by the interference between a divergent spherical wave and a plane wave, is used with backillumination to generate a convergent spherical wave. After bleaching the plate with Brom gas, the plate is again placed on the HOE plane, and a divergent spherical wave from a temperature-stabilized (AlGaAs) diode laser emitting the 780-nm single line impinges on the scanner.

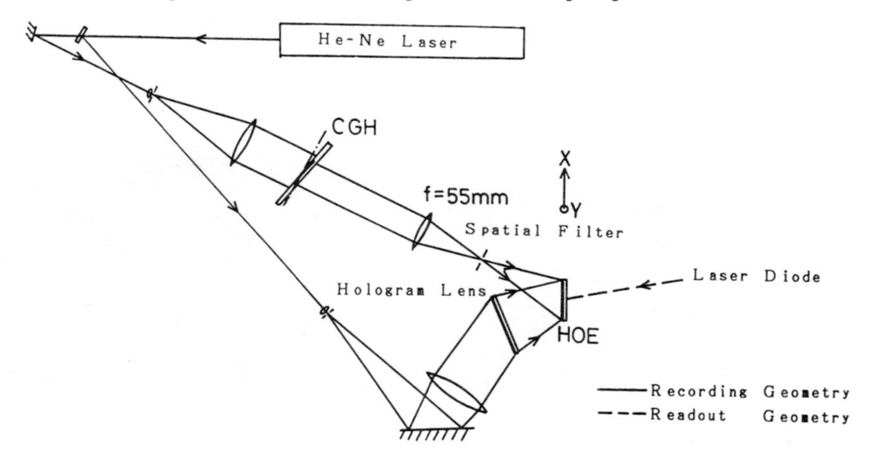

Figure 7. Experimental setup for making a corrected scanner using a CGH. The readout diode laser light impinges on HOE plane shown by dashed lines.

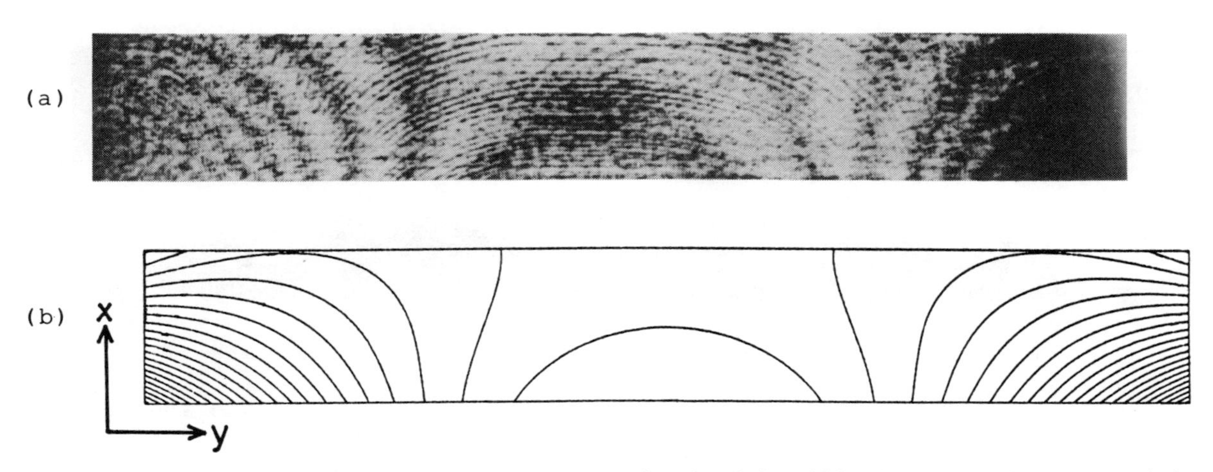

Figure 8. (a) Holographic interferogram obtained by illuminating the HOE with both the divergent and convergent spherical waves. (b) Computational interferogram of aspherical wavefront defined by Table 1.

A disk-rotating stage attaching the scanner is positioned sequentially at desired scan angles and the resulting scanned spots are recorded on film. The comparison of the computa-

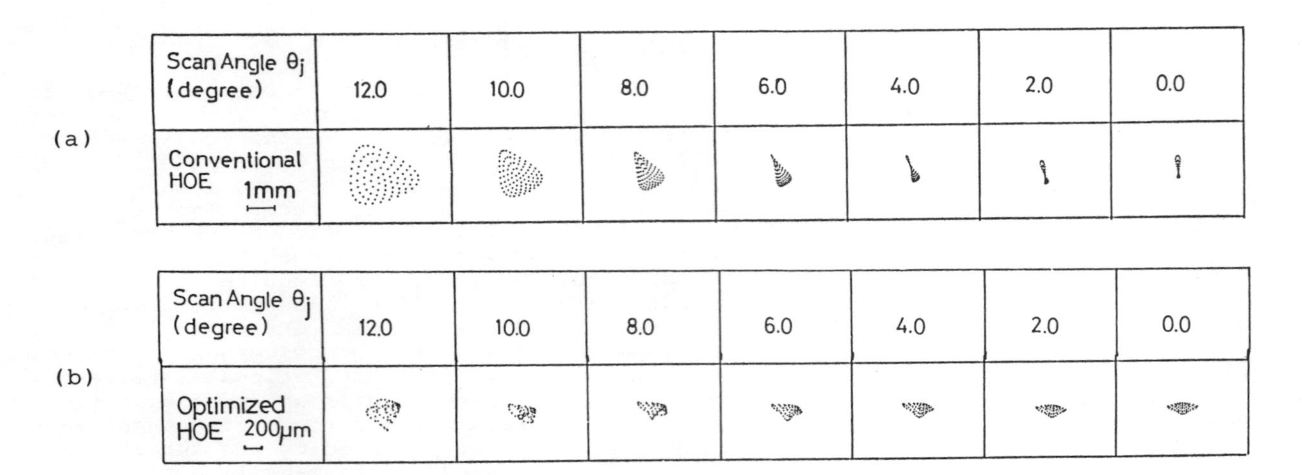

Figure 9. Computational spot diagrams of (a) a conventional uncorrected scanner and (b) an optimized corrected scanner.

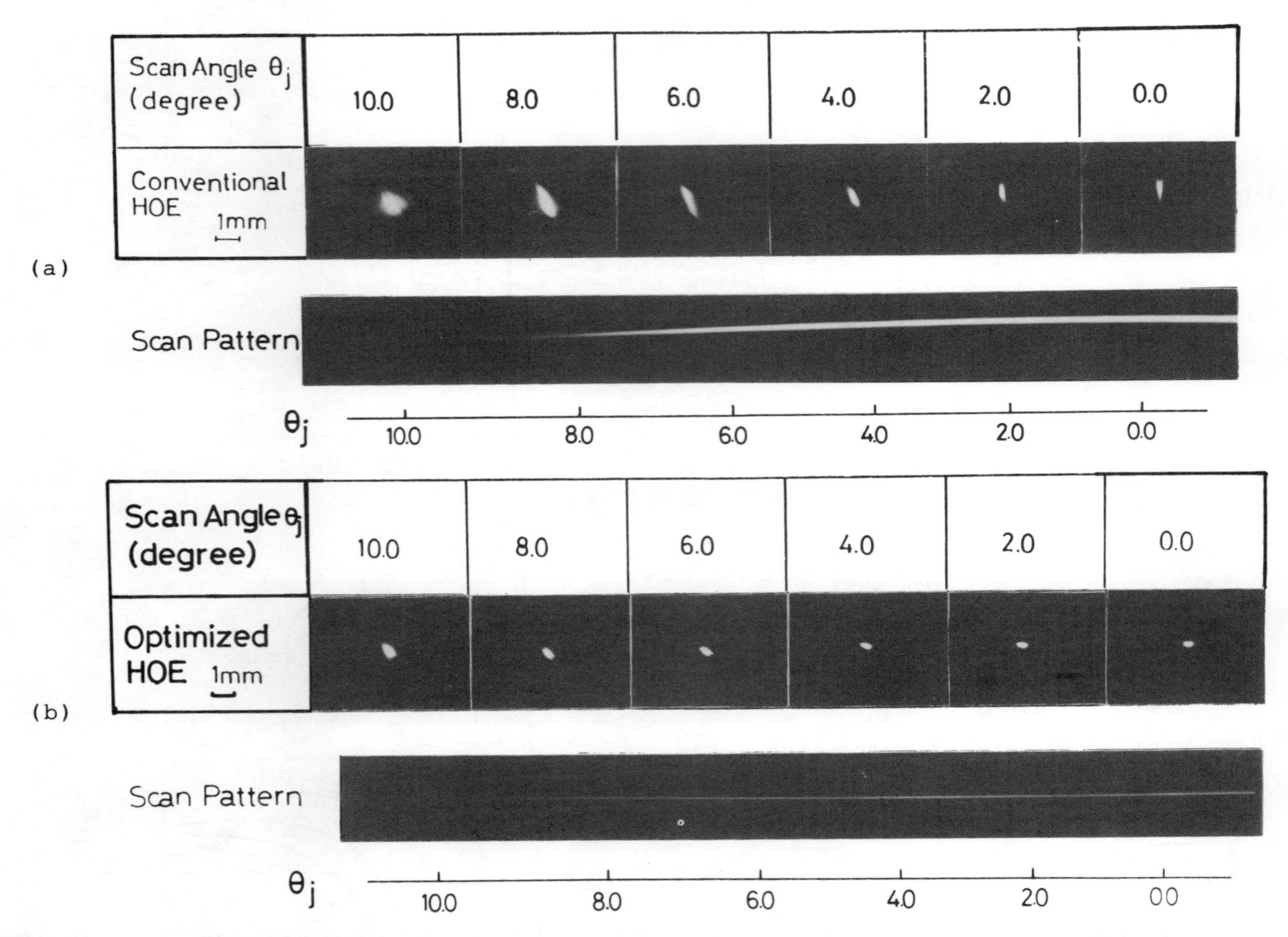

Figure 10. The magnified scan spots and corresponding scan patterns from (b) the corrected scanner with those from (a) the conventional scanner. A significant improvement in both scan spots and scan loci can be seen.

tional scanned spot diagrams of Figure 9 with the corresponding magnified scan spots at different scan angles of Figure 10 indicates good agreement. The scan spots are displayed on half of the flat scanning plane, for simplicity, because it is symmetric with respect to the center. These spots are made with an input aperture diameter of 6 mm. The spot diagrams are generated by raytracing thirty-seven rays (N=37) forming a hexapolar array through a flat HOE. Note that the different horizontal scales in the Figures give for the different cases. The numerical and experimental results of Figures 9a and 10a for a conventional scanner, and Figures 9b and 10b for a corrected scanner are well supported with the theoretical results shown in Figure 6, and Figures 3 and 4, respectively. The results shown in Figures 9a and 10a are demonstrated for large aberrations and curved scannings. The optimized corrected scanner shown in figures 9b and 10b exhibits the straight line scanning and the aberration-reduced spot size is about 350 µm in diameter. A small bow-deviation capability in the scan spots, ±130 µm inside the scan angle 10° is perceptible for a measured 30-cm scan length in the scanning plane.

Additional remarks

The other methods for bow-scanning corrections have been proposed by several authors,[4,5] in which a straight line scanning can be realized by reconstructing from a geometrical zone plate (GZP). The GZP can be synthesized with the plural spherical waves and/or the hologram illuminated at longer wavelengths differing from the forming wavelength.

This section briefly shows that our designed HOE without aspherical wave approaches to the GZP. When $\phi_G=0$ and $F_o=F_r=F$, and if a unit plane wave is assumed to be used as a readout light, the phase of image ray is written using Equation (10) as

$$\phi_{ij}(x,y) = \frac{2\pi}{\lambda_o}\cdot 2\{[(x-x_j)^2 + (y-y_j)^2 + F^2]^{1/2} - F\}. \tag{17}$$

The diffraction angle ψ from HOE at a chief ray (x=0,y=0) can be put in the form

$$\sin\psi = [(l_{ij})^2 + (m_{ij})^2]^{1/2}_{x=0,y=0} = \frac{(x_j^2 + y_j^2)^{1/2}}{[(x_j^2 + y_j^2)/\alpha^2 + (F/\alpha)^2]^{1/2}} \tag{18}$$

by using Equations (11) and (17), where $\alpha=2\lambda_c/\lambda_o$. Since the relation, $F^2 \gg (x_j^2+y_j^2)$ is established and if $\alpha\to\infty$ under the condition $\lambda_c/\lambda_o > 1$, Equation (18) is reduced to $\sin\psi \simeq (x_j^2+y_j^2)^{1/2}/(F/\alpha)$ which is equivalent to the diffraction angle of GZP. Thus, the original HOE is close to the GZP with the factor of division by α, which becomes effective in employing the property of bow-free scanning for the design of HOE.

Summary

The conditions of the straight line scanning and reduction of aberrated scan spots have been derived for a disk-type holographic diode laser beam scanner employing the correction phase produced by a CGH. Scanning with a small bow deviation has experimentally demonstrated by using the overall DLS optimization techniques of both experimental parameters and aspherical wavefront without auxiliary optics. Experimental results demonstrate the feasibility of this method.

References

1. Fairchild, R. C. and Fienup, J. R., "Computer-Originated Aspheric Holographic Optical Elements," Opt. Eng., Vol. 21, pp. 133-140 1982.
2. Ishii, Y. and Murata, K., "Flat-Field Linearized Scans with Reflection Dichromated Gelatin Holographic Gratings," Appl. Opt., Vol. 23, pp. 1999-2006 1984.
3. Lee, W-H., "Computer-Generated Holograms: Techniques and Applications," in Wolf, E., ed., Progress in Optics, Vol. 16, North-Holland 1978.
4. Ono, Y. and Nishida, N., "Holographic Disk Scanners for Bow-Free Scanning," Appl. Opt., Vol. 22, pp. 2132-2136 1983.
5. Hasegawa, S., Yamagishi, F., Ikeda, H. and Inagaki, T., "Holographic Line Scanner using Different Wavelength Method," Conference Digest of ICO-13, Sapporo, pp. 692-693 1984.

Hologram filters for optical image processing

Kazumi Murata

Department of Applied Physics, Faculty of Engineering, Hokkaido University
N13, W8, Kitaku, Sapporo 060, Japan

Abstract

Recent development of hologram filters for optical image processing is reviewed. Designing and fabricating techniques of the hologram filters are discussed, and the applications of the filters to optical image processings are shown, which include image operation, image restoration, matched filtering, image transformation and wavefront conversion.

Introduction

In optical image processing[1,2], a hologram filter (HF) often plays an important role as a spatial filtering element which spatially modulates or transforms the input object into the output image

Hologram filters (HFs) have advantages over conventional optical filters as follows: 1) capability of complex (amplitude and phase) spatial filtering, 2) easiness for fabrication and duplication, and 3) conveniency in thin form and with light weight. On the contrary the drawback of HFs is low diffraction efficiency.

Depending on the fabricating procedure, there are two kinds of HF. The one is a usual optical hologram (OH), which is the record of interference fringes of an object-wave and a reference-wave, and the other is a computer-generated hologram (CGH)[3], which is synthesized by a digital computer and a plotting device.

OHs are better than CGHs in diffraction efficiency and space-bandwidth product. On the other hand, CGHs are superior to OHs in flexibility for encoding a desired signal wavefront. In optical image processing, CGHs are frequently used as HFs when the filter function is too complicated to be realized by optical holography. To increase the diffraction efficiency and the space-bandwidth product of the HF, CGHs are optically copied on volume phase materials such as dichromated gelatin[4]. This hybrid construction of a HF is one of the tendencies in these ten years.

HFs have been extensively applied to wide variety or optical image processing, such as image operation, image restoration, matched filtering, image transformation, wavefront conversion and so on. In the following sections, recent developments of the above applications are individually described.

Image operation

Various mathematical operations to images[5]; for example, subtraction[6]; differentiation[7] or correlation[8] can be performed by optical image processing with HFs.

The holographic subtraction of two images is applied to submicron defect detection of integrated circuits[9]. A differentiation filter has an amplitude transmittance linear to spatial frequency along the differentiating direction. One dimensional differentiating HF has been optically constructed by recording a moire fringe of two interference fringes having slight different spatial frequencies[10], and applied to the measurement of a phase object[11]. Bidirectional spatial differentiation has been realized by means of CGHs[3] and applied to extract informations from radar and aerial images [12]. Cross- or auto-correlating operation is also very useful in optical image processing. Optical transfer function of a lens can be measured two-dimensionally by a HF, which performs the auto-correlation of the pupil function of the lens.[13,14]

Image restoration

Image restoration from a blurred image can be accomplished with use of the inverse filter. The filter has the complex transmittance $1/H(u,v)$, which is an inverse of the blurring transfer function $H(u,v)$, and it is holographically realized by combining two filters H^* and $1/|H|^2$ as a sandwich[15]. The real amplitude component can be recorded photographically, and the complex component holographically. Holographic inverse filters are also useful, for example, to restore the radiologic images degraded by the finite-size focal spot of the X-ray tube.[16]

An inverse HF not as a sandwich-type but as a single filter film can be realized [1,7] In the first step, an amplitude filter $1/|H|$ is made photographically. In the second step, H is holographically recorded with a reference wave R on a film, in front of which the first filter $1/|H|$ is directly inserted. After developing the film with $\gamma = -2$, we obtain a HF which contains the desired term $1/H$. In comparison with a sandwich-type inverse filter, this filter decreases the number of components, requirements for adjustment, and spatial extent of the setup. Some provisions are effective for noise reduction and for scale change[18].

A volume hologram of a single grating with a modulation varying in depth shows the narrow angular selectivity and acts as a spatial frequency filter. A linearly smeared picture can be restored by such a volume hologram.[19] The filter is merely placed in contact with the smeared photograph and the resulting diffracted wave represents the deblurred picture. A simple way for producing Wiener-filters has been proposed[20], in which the amplitude of the filter function is exposed on a film through the master grating and the phase reversal is given by a translation of half a period of the grating. The HF is applied to deblurring high resolution electron micrographs of weak phase objects. Deblurring filters are also available by CGHs[3,21], which are used in the coherent filtering systems.

Matched filtering

Since the first matched spatial filter (MSF) was holographically realized by Vander Lugt[4] in 1964, much interest has been paid to MSFs because of their application to optical pattern recognition. Various improvements of modifications of the Vander Lugt filter have been reported.

To increase the discrimination of MSF, supperssion of zero-frequency component or high pass filtering is effective[22]. Production of a MSF for detecting gradient correlation has been presented, where a multiple-exposure technique is used[23]. As MSFs are inconveniently sensitive to scale and orientations of the input pattern, much efforts have been done to overcome these disadvantages. For example, an averaged multiplexed MSF for recognition of some diatoms having various size and depth of focus is constructed[24] and rotated at the coherent filtering plane[25]. On the other hand, for utilization of high sensitivity to rotation, a multiplexed MSF for the objects oriented variously can extract the rotation of the object.[26]

An optical correlator has been proposed for the real-time recognition and control of mass-produced pieces in automatic machining and assembling lines[27]. The information of a master piece is stored in a computer generated MSF, which is used to inspect and control the shape of mass-produced mechanical elements.

Phase-only MSF gives us autocorrelation spots narrower and brighter than those obtained by Vander Lugt filter. It has been shown that the ability of a phase-only MSF to discriminate the characters between closely similar ones in considerably high[28]. The phase-only MSF technique is also used in a broad spectral band or white-light optical correlator.[29]

Reflection-type volume holograms have been utilized in matched filtering with incoherent light[30]. The blur of correlation output when the MSF is illuminated by a broad-spectrum light source is evaluated. The optical filtering system is shortend by means of the reflection-type MSF.

The use of the color-sensitive MSF with a volume hologram has been investigated. A transmission-type[31] and a reflection-type[32] MSF are represented for color-coded character recognitions.

Image transformation

Another interesting optical processing with HFs is geometrical image transformation, in which the shape or intensity distribution of an input object can be changed in the output image[33]. CGHs are suitable for the image transformation because the phase variation over a HF is somewhat complicated and can not be encoded by an optical system. An image transforming HF gives the correct output image from a distorted input image, or vice versa[34]. The shape of the input can be transformed into the quite different one, for example, from a ring to a point, or from a uniform circle to a line, which may be applied to a line scanner[35]. Likewise, a Gaussian beam can be redistributed on the squared frame of a hollow box[36].

Wavefront conversion

One of the important features of HFs is the wavefront converting ability. Reconstructing process from the hologram of an object wavefront is equivalent to a wavefront converting process. A simple holographic grating serves as a beam splitter or a beam combiner like as a half-mirror, and, if necessary, it can bear some focal power for imaging the input object at desired distance. Hologram beam combiners made by dichromated gelatin[37] have been used for various visual display systems,[38] such as head-up display, helmet-mounted display[39] or visual simulation systems.

A wavefront aberration or an aspheric wavefront can be encoded in a CGH and optically reproduced. If a particular aberration or a combination of different aberrations are encoded in a CGH, the diffraction patterns suffering from these aberrations can be displayed, and the effects of defocusing and stopping down can also be observed.[40,41]

In interferometric testing of an aspheric mirror surface, a CGH is often used as a HF which reconstructs an aspheric wavefront to be compared with the one coming from the aspheric surface under test[42,43,44]. On the other side, some non-interferometric testing have been presented, for example, optically recorded holograms are used to detect the surface error of a precise gear[45] or to detect the submicron defect of an integrated circuit photomask[9].

Holographic zone plates have a focusing property and are used as laser beam scanning elements for facsimile-reading and -recording, pattern generation, computer output printing and so on. CGH is suitable for these scanning elements because these should encode a complicated aspheric wavefront which compensate the aberration of the zone plate to retain the flat and linear scanning in the adequate field[3]. There have been reported many variations of CGH scanner. For instance, a reflection volume hologram is realized as a scanning element with field-curvature corrections[46], and a transmission-type hologram scanner is developed for the scanner with a diode laser[47].

In many laser applications, the conversion of the Gaussian distribution into a uniform one is effective. Interlaced diffraction gratings are used to obtain a uniform intensity at the focal plane[48]. To have a uniform and well-collimated laser beam, a set of two CGH filters are fabricated and used for intensity reshaping of an incident Gaussian beam[49,50]. The other CGH can correct an aberrated wavefront reflected by an off-axis spherical mirror in case of a large aperture mirror collimator[51].

Conclusions

HFs have been extensively used in optical image processing. We have reviewed recent developments of the HF, especially being applied to coherent optical processing namely, image operation, image restoration, matched filtering, image transformation and wavefront conversion. To incoherent or polychromatic image processing, HFs can also be applied[52], however, we have to give up this interesting field for want of space.

The techniques and applications of HFs will be still more developed in future togather with the developments of recording materials and with the advances of digital methods by the help of computers. For instance, one of the new techniques is the fabrication of CGHs by electron-beam lithography.[53,54]

References

1. Casasent, D., Optical data processing, Springer-Verlag, 1978
2. Lee, S. H., Optical information processing, Springer-Verlag, 1981
3. Lee. W. H., "Computer-generated holograms: Techniques and applications," Progress in Optics, Vol. 16, pp. 119-232,
4. Lowenthal, S., and Chavel, P.. "Reduction of the number of samples in computer holograms for image processing," Appl.Opt., Vol. 13, pp.718-720, 1974
5. Lee, S. H., "Mathematical operations by optical processing," Opt. Eng., Vol. 13, pp. 208-218, 1974
6. Ebersole, J. F., "Optical image subtraction," Opt. Eng., Vol. 14, pp.436-447,1975
7. Yao, S. K., and Lee, S. H., "Spatial differentiation and integration by coherent optical-correlation method," J. Opt. Soc. Am., Vol. 61, pp.474-477 ,1971
8. Vander Lugt, A., "Signal detection by complex spatial filtering," IEEE Trans. Inform. Theory, Vol.10, pp. 139-145, 1964
9. Fusek, R. L., et al, "Holographic optical processing for submicron defect detection," S.P.I.E., Vol. 523, Applications of Holography, pp. 54-59, 1985
10. Ishii, Y., and Murata, K., "Some simple methods for making a holographic differentiation filter and its related applications," Japn. J. Appl. Phys., Vol. 14, Suppl.I, pp.229-233, 1975
11. Kamemaru, S., and Murata, K., "Hybrid processing for phase determination using a holographic filter," Optik, Vol. 69, pp. 80-84, 1985
12. Soubri, E. H., Grosmann, M., and Meyrueis, P., " A method of image processing through computer made hologram and some applications," S.P.I.E., Vol. 437, pp.182-190, 1983
13. Murata, K., and Fujiwara, H., "Optical transfer function measurement by means of hologram filter," Opt. Laser Tech., Vol. 2, pp. 182-184, 1970
14. Murata, K., Fujiwara, H., and Sato, R., "Two-dimentional measurement of optical transfer function by holographic techniques," Proc. S.P.I.E., Vol. 46, pp. 120-123, 1974

15. Zhuang, S. L., Chao, T. H., and Yu, F. T. S., "Smeared-photographic-image deblurring utilizing white-light-processing technique," Opt. Lett., Vol. 6, pp. 102-104, 1981
16. Krusos, G. A., "Restoration of radiologic images by optical spatial filtering," Opt. Eng., Vol. 13, pp. 208-218, 1974
17. Zetzsche, C., "Simplified realization of the holographic inverse filter: a new method" Appl. Opt., Vol. 21, pp. 1077-1079, 1982
18. Jiang, Y. G., and Xu, Y. R., "Simple method for image deblurring", Appl. Opt., Vol. 22, pp. 784-786, 1983
19. Peri, D., and Friesem, A. A., "Volume holograms for image restoration," J. Opt. Soc. Am., Vol. 70, pp. 515-522, 1980
20. Herrmann,K.H., Reuber, E., and Schiske, P., "A simple way for producing holographic filters suitable for image improvement," Proc. 9th Int. Congr. Electron Microscopy, Vol. 1, pp. 226-227, 1978
21. Campbell,K., Wecksung, G. W., and Mansfield, C. R., "Spatial filtering by digital holography," Opt. Eng., Vol. 13, pp. 175-188, 1974
22. Lowenthal, S., and Belvaux, Y., "Reconnaissance des formes par filtrage des fréquençes spatiales," Opt. Acta, Vol. 14, pp. 245-258, 1967
23. Petrosky, K. J., and Lee, S. H., "New method of producing gradient correlation filters for signal detection," Appl. Opt., Vol. 10, pp. 1968-1969, 1971
24. Almeida, S. P., and Eu, J. K. T., "Water pollution monitoring using matched spatial filters," Appl. Opt., Vol. 15, pp. 510-515, 1976
25. Fujii, H., Almeida, S. P., and Dowling, J. E., "Rotational matched spatial filter for biological pattern recognition," Appl. Opt., Vol. 19, pp. 1190-1195, 1980
26. Horner, J. L., and Caulfield, H. J., "Parameter extraction by holographic filtering," Appl.Opt., Vol . 21, pp. 1599-1601, 1982
27. Tschudi, T., "Computer-generated holograms for the quality control in micromechanics," S.P.I.E., Vol. 437, pp.176-181, 1983
28. Horner, J. L., and Gianino, P. D., "Phase-only matched filtering," Appl.Opt., Vol. 23, pp. 812- 816, 1984
29. Javidi, B., and Yu, F. T. S., "Performance of noisy phase-only matched filter in a broad spectral band optical correlator," Appl. Opt., Vol. 25, pp. 1354-1358, 1986
30. Zhang, E. Y., Ishii, Y., and Murata, K., "Matched spatial filtering using a reflection-type volume hologram." Optica Acta, Vol. 29, pp. 1049-1060, 1982
31. Case, S. K., "Pattern recognition with wavelength-multiplexed filters," Appl. Opt., Vol. 18, pp. 1890-1894, 1979
32. Ishii, Y., and Murata, K., "Color-coded character-recognition experiment with wavelength-triplexed, reflection-type holographic filters," Opt. Lett., Vol. 7, pp.230-232, 1982
33. Bryngdahl, O., "Geometrical transformation in optics," J. Opt. Soc. Am., Vol. 64, pp. 1092-1099, 1974
34. Bryngdahl, O., "Computer-generated holograms as generalized optical components," Opt. Eng., Vol. 14, pp. 426-435 1975
35. Cederquist, J., and Tai, A. M., " Computer-generated holograms for geometric transformations," Appl. Opt., Vol. 23, pp. 3099-3104, 1984
36. Case, S. K., Haugen, P. R., and Lokberg, O. J., "Multifacet holographic optical elements for wave front transformations," Appl. Opt., Vol. 20, pp. 2670-2675, 1981
37. Chang, B. J., "Dichromated gelatin holograms and their applications," Opt. Eng., Vol. 19, pp. 642-648, 1980
38. McCauley, D. G., Simpson, C.E., and Murbach, W. J., "Holographic optical element for visual display applications," Appl.Opt., Vol. 12. pp.232-242, 1973
39. Withrington, R. J., "Optical design of a holographic visor helmet-mounted display," S.P.I.E., Computer-Aided Optical Desogn, Vol. 147,,pp. 161-170, 1978
40. Ishii, Y., Maeda, J., and Murata, K., " Holographic display of diffraction patterns suffering from third- and fifth-order aberrations," Optica Acta, Vol. 26, pp. 969-983, 1979
41. Ishii, Y., and Murata, K., "Simulation of aberrations using binary computer-generated holograms," Optik , Vol. 69, pp. 65-72, 1985
42. Wyant, J. C., and Bennet, V.P., "Using computer generated holograms to test aspheric wavefronts," Appl. Opt., Vol. 11, pp. 2833-2839, 1972
43. Dörband, B., and Tiziani, H. J., "Testing aspheric surfaces with computer-generated holograms: analysis of adjustment and shape errors", Appl. Opt., Vol. 24, pp. 2604-2611 1985
44. Ono, A., and Wyant, J. C., "Aspherical mirror testing using a CGH with small errors" Appl. Opt., Vol. 24, pp. 560-563, 1985
45. Honda, T., Ichimura, I., Jin, G. X., and Tsujiuchi, J., "Holographic rapid error detection of precise gear surfaces," S.P.I.E., Vol. 523, Applications of Holography, pp.150-154, 1985
46. Ishii, Y., "Reflection volume holographic scanners with field-curvature corrections," Appl. Opt., Vol. 22, pp. 3491-3499, 1983
47. Ishii, Y., and Murata, K., "Optimized holographic scanner with diode lasers," Digest Topical Meeting on Holography, Opt. Soc. Am., p. 86, 1986
48. Veldkamp, W. B., "Laser beam profile shaping with interlaced binary diffraction

gratings," Appl. Opt., Vol. 21, pp.3209-3212, 1982
49. Han, C. Y., Ishii, Y., and Murata, K., "Reshaping collimated laser beams with Gaussian profile to uniform profiles," Appl. Opt., Vol. 23, pp. 3644-3647, 1983
50. Ishii, Y., and Murata, K., "Flat-filed linearized scan with reflection dichromated gelatin holographic gratings," Appl. Opt., Vol. 23, pp. 1999-2006, 1984
51. Murata, K., Ishii, Y., "Laser beam wavefront converters using hologram filters," Digest, 5th Intern. Conf. on Lasers and their Applications, p. 77, 1985
52. Yu, F. T. S., Optical information processing, John Wiley & Sons, 1983
53. Freyer, J. L., Perlmutter, R. J., and Goodman, J. W., "Digital holography: Algorithms, e-beam lithography and 3-D display," S.P.I.E., Vol. 437, pp. 38-47, 1983
54. Arnold, A. M., "E-beam fabrication of computer-generated holograms (CGH)", S.P.I.E., Vol. 523, pp. 285-291, 1985

Dichromated Gelatin Holographic Scanner

Christian LIEGEOIS - Roma PIEL

X-IAL, 7, rue de l'Université, 67084 STRASBOURG CEDEX, FRANCE

Patrick MEYRUEIS

Ecole Nationale Supérieure de Physique de Strasbourg (ENSPS)
7, rue de l'Université, 67084 STRASBOURG CEDEX, FRANCE

Abstract

This paper after recalling the advantages and shortcommings of dichromated gelatin explains the process of holographic image formation in the medium with the quantitative results we obtained. As an example we apply the technique of Holographic Optical Elements to the realisation of laser scanners.

Introduction

The development of a new type of optical elements, called Holographic Optical Elements (HOE's), working not as conventional elements (via Reflexion or Refraction) but instead by diffracting the incoming light beams, allows for some major developments in Optics. The interest, however for such elements remained weak, and the few applications they found were very specialized, mainly because of the low efficiency allowed by the existing photosensitive materials. But considering the inherent advantages of such elements coupled to the growing needs, especially in the airplane industry for new HUDs, pushed the scientific community to investigate for a new efficient, nearly ideal photosensitive material. These investigations ended up with the development of HUD's on dichromated gelatin (DCG). This material has, for the realisation of HOE's, some definite advantages, and is considered as being the best material available today.

After the development of HUD's, DCG found some other applications as for example: Scanners (for Bar Code readers, laser printers), lenses, beamsplitters, coupling elements, warelength mux and demux, display,...

I Dichromated gelatin general features

Shortcommings :

- low sensitivity
- sensitive to H2O
- requires precise environment conditioning

Advantages :

- low diffusivity
- high diffraction efficiency
- high resolution $\gg$ 5 000 l/mm
- tuning of diffracted peak wavelength
- tuning of wavelength range
- tuning of angular selectivity
- can be reprocessed
- controlable thickness

The main parameters affecting the DCG's performances are :

- the initial thickness of the layer
- the initial hardness of the layer
- the sensitizer concentration
- the layer's drying conditions (humidity, temperature and duration for a good nucleation, gelation balance)
- the recording wavelength
- the exposure-development delay
- the development bathes (concentration, temperature, duration)
- the recording energy

All these parameters will affect the resulting HOE's parameters, namely :

- diffraction efficiency
- noise
- reconstructing wavelength
- angular and spectral selectivity.

II Formation of a hologram in a DCG layer (fig. 1) /1/

Consider a pure layer of gelatin coated on a substrate (glass, PVC, ...). By dipping it in a solution, the layer will swell by an amount modulated by his solubility in the solution. In the case of a dichromate solution, the solubility of the gelatin layer is determined by the initial hardness of the layer, as well as the solutions pH and temperature.

The gelatine swells by absorbing water which is the dichromates transporter. The quantity of absorbed dichromate is a function of both dichromate concentration and immersion time.

When drying, dichromate remains and water is eliminated thus shrinking the layer. After drying this sensitized layer is exposed to light, the introduced Cr^{6+} undergoes a photoreduction to Cr^{3+} (via intermediate states Cr^{5+} and Cr^{4+}) which creates through carboxyle groups, links (called crosslinks) between the gelatin's molecules thus creating a hardening of these areas.

The development than converts this latent image in an index modulation corresponding to the formerly created crosslinks.

After exposure, the remaining unreacted dichromate is extracted by soaking the layer in water.

The gelatin, swells again, by an amount which is now a function of initial hardness, black reaction hardness, exposure hardness, temperature and pH.

Since the lighting was not homogeneous over the whole layer, the crosslink creation was also unhomogeneous, thus creating a differential hardness throughout the layer which will induce through the layer a swelling gradient.

After rinsing the plate is imersed in various water-alcool bathes, eliminating the water from the layer, and leaving a strong index modulation. Concentration and temperature of these bathes depend mainly on the layers thickness and hardness, and will inturn mainly determine the HOE's noise, reconstructing wavelength and wavelength selectivity.

The most delicate operation is probably the wavelength selectivity adjustment.

Remember that the wavelength selectivity is mainly determined by the layers thickness and the Δn amplitude.

III Obtained results

Example 1 : $450\ nm \leqslant \lambda \leqslant 680\ nm$
$\eta \geqslant 99\ \%$
$5\ nm \leqslant \Delta\lambda_{50\%} \leqslant 70\ nm$
$0{,}8° \leqslant \Delta\Theta_{50\%} \leqslant 20°$

Example 2 $\eta \geqslant 94\ \%$ from $\lambda = 450$ nm to $\lambda = 590$ nm

Example 3 $\eta > 60\ \%$ at $\lambda = 850$ nm

IV Holographic Scanners /2,3/

Main problems :

- curvature of scanning line
- disc vibrations
- chromatic aberration (when recording and reconstructing wavelength are different)
- diffraction efficiency
- exposure inhomogeneous

The realised scanner is nearly linear, insensitive to centering and alignment errors, as well as insensitive to vibrations.

A. Deflection and scanning

The physical law ruling a transmission grating is : $\sin\Theta i + \sin\Theta d = m\lambda/d$

The choice of Θi and Θd depends mainly on two parameters : scanning line curvature and vibrations of the disk.

1) Curvature

When rotating a grating in his plane, the diffracted ray describes a circle around the incident ray.

In the plane defined by the incident ray and the angle to the grating's vector, the incidence angle depends on the rotation Θr of the grating, following :

$$\sin \Theta i\ (\Theta r) = \sin (\Theta i) \cos \Theta r$$

when considering the first order.

Also Θd depends on Θi, following :

$$\sin \Theta d\ (\Theta i) = \lambda / d - \sin [\Theta i\ (\Theta r)]$$

when Θi varies, Θd varies also.

It can be demonstrated that, for $\Theta i + \Theta d = 90°$, the scanning lines curvature is reduced.

Also the variation of $\Theta i + \Theta d$ vs. Θr has to be minimised. By computing these variations as a function of Θr, it appears that they are minimum for $\Theta i = 45{,}55°$ and $\Theta d = 44{,}45°$.

These values correspond to a curvature variation of less than 20 seconds of arc. (fig. 2)

2) Vibrations (fig. 3)

The vibrations correspond to a tilt variation of the grating which comes out as a variation of Θd vs. $d\,\emptyset$ (tilt angle of the grating from the plane normal to his rotation axis).

In the case of a transmission grating this variation is given by :

$$d\,\Theta d = [1 - \cos (\Theta i + \emptyset) / \cos (\Theta d - \emptyset)]\, d\,\emptyset$$

It appears that this effect can be compensated and is minimal if Θi and Θd keep close.

B. Wavelenght change between recording and reconstruction

It can be compensated for via one or a conbination of the following methods :

- create recording "errors" (to compensate for wavelength shift and associated aberrations)
- swell or shrink the layer
- if possible change reconstruction angle
- make a computer generated hologram corresponding to the reconstruction wavelength

C. Diffraction Efficiency (DE)

Once the reconstruction conditions are respected, the diffraction efficiency is only determined by the photosensitive material used.

Usually with conventional materials (silver halide) the DE is around 30 %. This value can be increased by bleaching the hologram, but at the same time noise and optical qualities are degraded.

The mean values we obtained with DCG, at a given wavelength, are of the order of 95 %.

The diffraction efficiency of a sinusoidal thick phase hologram is given by /4/ :

$$\eta = \sin^2 (\xi^2 + \nu^2)^{1/2} / (1 + \xi^2/\nu^2)$$

$$\text{where } \nu = \Pi\, n1\, T / [\cos \Theta_0 - \cos (\Theta_0 - 2\,\emptyset)]^{1/2}$$

$$\xi = \Theta T / 2 \cos (\Theta_0 - 2\,\emptyset)$$

with $n1$ = index modulation
λ = wavelenght in air
T = medium thickness
Θ_0 = Braggangle in the medium
$\emptyset$ = fringe slant
Θ = dephasing vs. incidence to Braggangle

When the hologram rotates Θi varies (for $\Theta r = 15°$, $\Delta \Theta i \simeq 2°$) that means that also η will change.

Also η will change with the variation of the holograms inclination, so we have one more reason to avoid vibrations.

In order to avoid important DE variations, so that it remains constant for $\Delta \Theta i \simeq 2\text{-}3°$ we can act chemically on the medium, since the DCG's diffraction peak can be tuned to be narrow or large.

D. Exposure inhomogeneities from facet to facet

The exposure inhomegeneities are due to the laser's power fluctuation (at recording) and/or differential phase fluctuations between the interfering waves.

The power fluctuations can be easily avoided by monitoring the lasers output and by taking care to expose each facet of the scanner at the same energy.

The differential phase fluctuations can be compensated by using the fringe pattern formed by the two interfering waves to adjust the phase of one of the beam.

E. Diffraction spot

Suppose a plane linear grating, diffracting a plane wave, and that the focalisation is obtained by a lens situated between a screen and the disc.

The beam has a Gaussian distribution and the spot diameter will be :

$$D = 1{,}27\,\lambda f/d_0 \text{ (at } 1/e^2\text{)}$$

where d_0 is the diameter of incident beam and f focal of the lens. In fact due to lenticular and holographic aberrations D will be larger.

The spot diameter will determine the number of obtained points ruled by :

$$D = 2\, f\, \Theta s/N$$

where N is the number of points/line that we want (Rayleigh criteria, distance between center = d/2 and Θs scanning angle.

Example : for f = 500 mm, Θs = 0,6 rad and N = 5 000, we have :
D = 120 μm
so d_0 = 3,3 mm (λ = 632,8 mm)
d_0 = 2,7 mm (λ = 514,5 mm)

F. Number of facets (F)

The relation between rotation angle and scanning angle is :

$$\text{tg}\, \Theta s = \text{tg}\, \Theta r\, \lambda/d$$

This relation gives us the usefull rotation angle for each grating. F is equal to the even part of $2\Pi/\Theta r$. But the real angle of rotation is $2\Pi/N$. The difference between the two values corresponds to the parts where the beam moves from one facet to the other.

V Obtained results

1) One line scanner
Disc diameter 8,5 cm
Number of facets : 10
$\Theta s = 46°$
$\Theta i = \Theta d = 45°$
η = 85 %
$\Delta\eta$ = 10 %
Scanning line 350 mm
λ recording 488 nm
λ reconstruction 633 nm

2) Bar code reader
Disc diameter 20 cm
Number of facets 20
$\Theta i = 22°$
$\Theta d = 30°$
Number of lines 20
f = 250 mm
η = 85 %
$\Delta\eta$ = 10 %
λ recording 488 nm
λ reconstruction 633 nm

Conclusion

We showed the main fundamental process allowing for the required index modulation in a DCG hologram. Once a chemical DCG procedure is developed, one can obtain almost everything with this medium. But in order to obtain wat is wanted, a fine tuning procedure is needed which is perhaps more difficult to obtain.

The given HOE examples, show how it is possible with no investment to obtain quickly some already usable scanners. The quality of these scanners can of course be far more improved by rising the quality of the environment as well as the quality of the optical and mechanical equipment.

References

1/ C. LIEGEOIS, R. PIEL, Annales des Télécommunications, 41, n°1-2, 1986, 66
2/ C.J. KRAMER, SPIE, 390, 1983
3/ L.D. DICKSON etal, IBM J. Res. Develop., 26, n°2, 82
4/ H. KOGELNIK, Bell Syst. Techn. J., 48, n°9, 1969, 2909

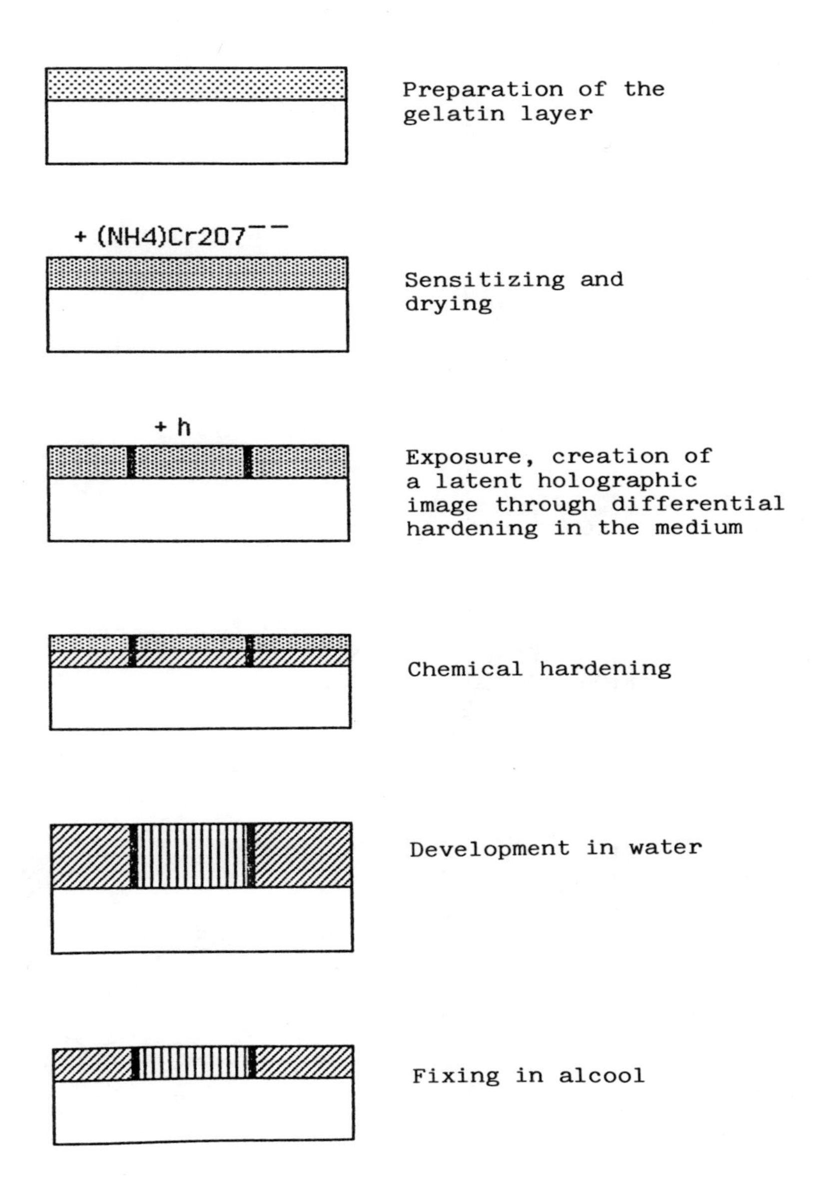

Fig. 1 : Hologram Formation in a DCG layer

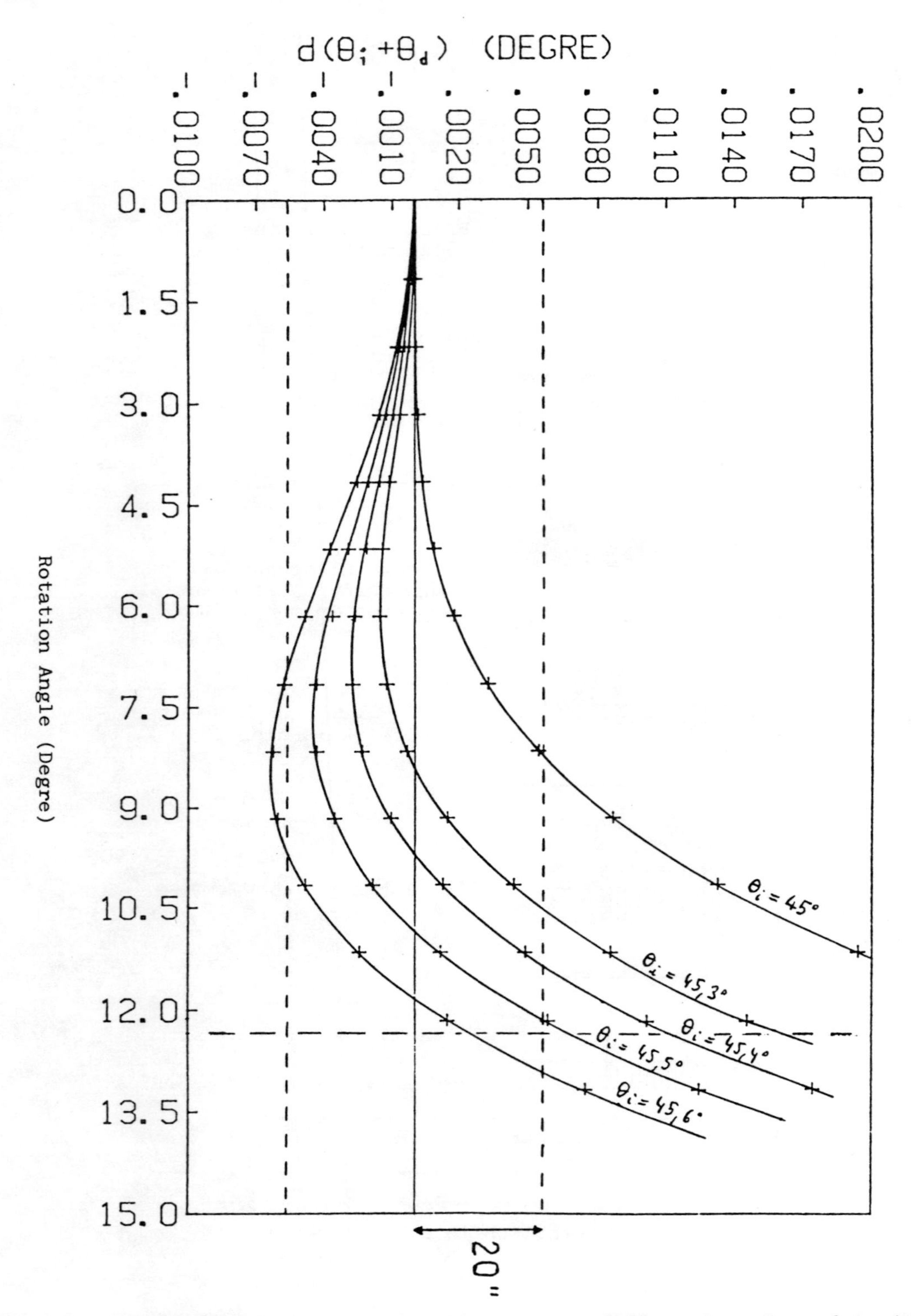

Fig. 2 : Curvature of the scanning line versus different angles of incidence

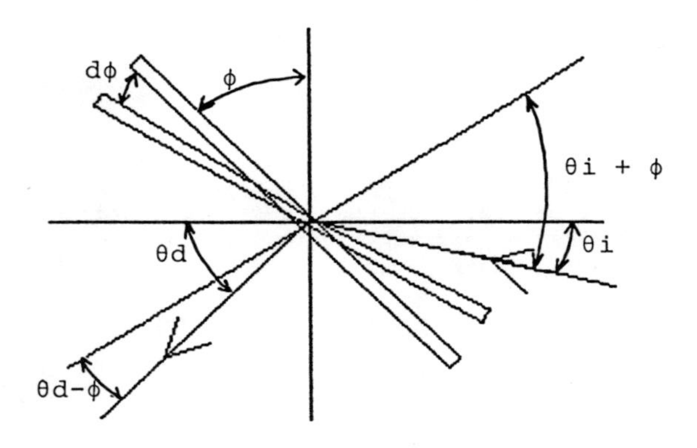

Fig. 3 : Variation of θ d vs. d $\emptyset$

Oblique incidence interferometer using holographic optical elements

Masane Suzuki, Motonori Kanaya, Takayuki Saito, Kenji Yasuda

Optical laboratory, Fuji Photo Optical Co., Ltd.
1-324, Uetake-cho, Omiya city, Saitama pref. Japan

Abstruct

Optical elements made by holographic technology are called Holographic Optical Elements (HOEs). HOEs have some advantages as conventional optical elements are not equipped and have been applied in various optical devices. We have applied HOEs, a pair of holographic gratings and a holographic lens, in an oblique incidence interferometer. In following lines we call it OII as Oblique Incidence Interferometer.

Introduction

Fizeau interferometer is popular as a measuring device for the flatness of optical elements and the wavefront aberration of lenses. And it is suitable to measure the unevenness of surfaces within 10 times of wavelength of used light source of an interferometer. For example, it is used to measure the flatness of precise glass optics, polygonal mirror and the wavefront aberration of D.A.D. pick-up lens.

There is a great demand to measure a little more uneven surface which exceeds the measurable extent of Fizeau interferometer. For that purpose, an OII has been planned by K.G.Birch. (1) The measuring sensitivity of an OII can be lowered than Fizeau interferometer. So OII is suitable to measure an IC wafer or an optical memory disc.

In the past, we have applied a holographic lens in Fizeau interferometer as a collimating lens and it showed its good performance. (2) This is a report on the development of our new interferometer, an OII with HOEs.

Principle

An optical system of Birch's OII is shown in Fig.1. Collimated laser light strikes a diffraction grating G1 and is divided into two lights, one goes straight forward and the other is diffracted. The former directly strikes a grating G2. The latter also strikes G2 but after being reflected from on the surface of an object.

In this case, the incident angle is $(\pi/2-\theta)$. Diffracted light by G2 and the light from an object interfere mutually and show the unevenness of an object's surface as the interference fringes pattern. Assuming that "d" is the pitch of grating G1 and G2, "θ" is the diffraction angle and "λ" is the wave length of laser to be used, we can get an equation as

$$d = \lambda / \sin\theta$$

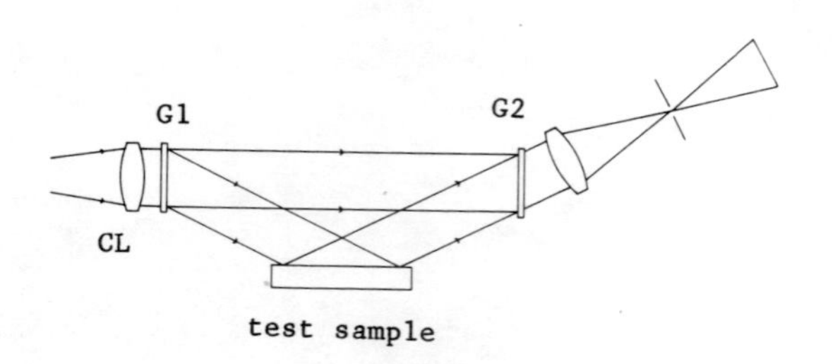

Fig. 1 Birch's OII

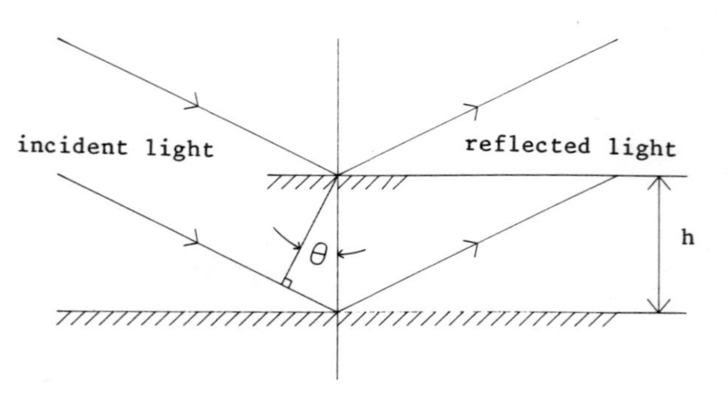

Fig. 2 Sensitivity of OII

And referring to Fig.2, the optical path difference "Δ", which is according to unevenness "h" of the surface is

$$\Delta = 2h\sin\theta$$

so the interference finges are given by "h"

$$h = \lambda/2\sin\theta$$
$$= d/2$$

In other words, sensitivity of an oblique incidence interferometer equals one twice of the grating's pitch. When 4μm pitched grating is used for this interferometer, its sensitivity becomes 2μm. It is possible to measure the surfaces with a little larger unevenness.

An oblique incidence interferometer with HOEs

Our OII uses holographic gratings HG1 and HG2, besides, uses a holographic lens for observing in the optical system.

In Birch's original OII, the plane wave diffracted by G2 and the light reflected from sample interfere and measure the uneveness of a surface of a sample. But in this way, measuring accuracy is limited by the accuracy of a collimating lens CL and gratings G1 and G2. So, if the accuracy of CL, G1 or G2 is low, the reference plane wave diffracted by G2 is distorted by them and makes measuring accuracy low.

But in the case of our OII with HOEs, the reference plane wave is reconstructed by HG2 (this wave is same as the wave from the reference mirror.) and its measuring accuracy doesn't depend on the accuracy of a collimating lens and the gratings so seriously. And it becomes possible to observe the full view of the test sample by using holographic lens. It can convert an oblique image into an image which is observed in perpendicular. Its optical system is shown in Fig.3.

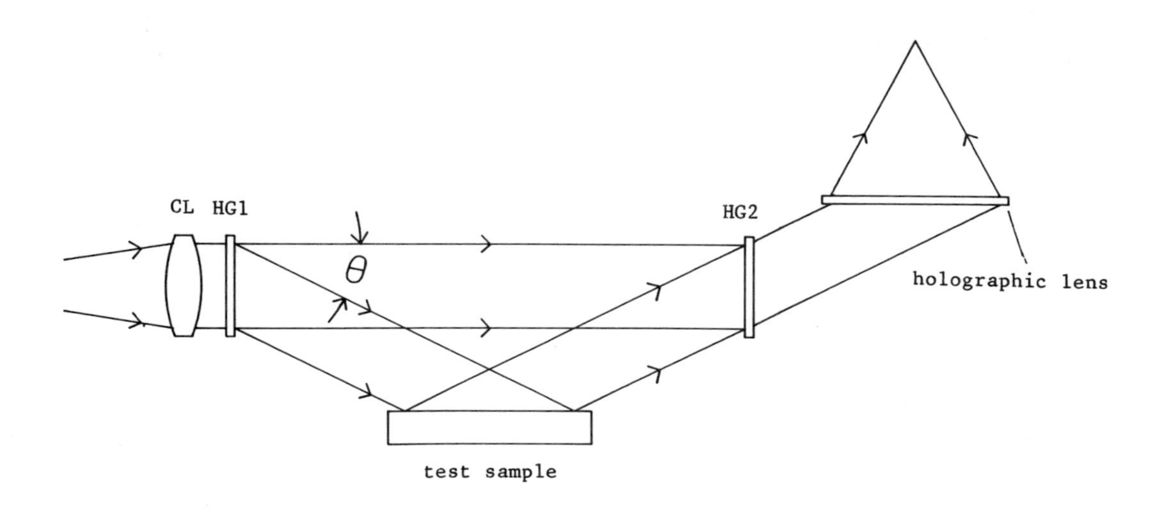

Fig. 3 Optical System of OII with HOEs

Production of a holographic grating HG1

Holographic grating HG1 is made by two bundles of collimated light, of which intersection angle is 9.1° using He-Ne laser, so that the grating's pitch becomes 4μm. (One collimated light strikes a hologram perpendicularly.) Its optical system is shown in Fig.4.

Production of a holographic grating HG2

Holographic grating HG2 is made by using holographic grating HG1 and the reference flat mirror. Its optical system is shown in Fig.5. Collimated light strikes HG1 and is divided into transmitted light and diffracted light. HG2 is made by these two bundles of light, one is the light reflected from on the referene mirror and the other is direct transmitted light through HG1. The flatness of a refrence mirror is about $\lambda/20$.

When we measure a sample actually, we put a sample in place of a reference flat mirror. The reflected light from the surface of a sample and the reconstructed light by HG2 interfere each other. The latter has the reference wavefront which is given by a reference mirror.

Production of a holographic lens

As shown in Fig.6, a holographic lens is made by two bundles of light, one is collimating light, of which incidence angle is 80.9°, and the other is diverging light. When a holographic lens is installed into the interferometer, the laser light with the information of uneveness of a sample strikes it onto the back face, and the light is diffracted and focused at the specified location.

So, we can observe the full image of the surface from front side by placing eyes or a TV camera at the specified location.

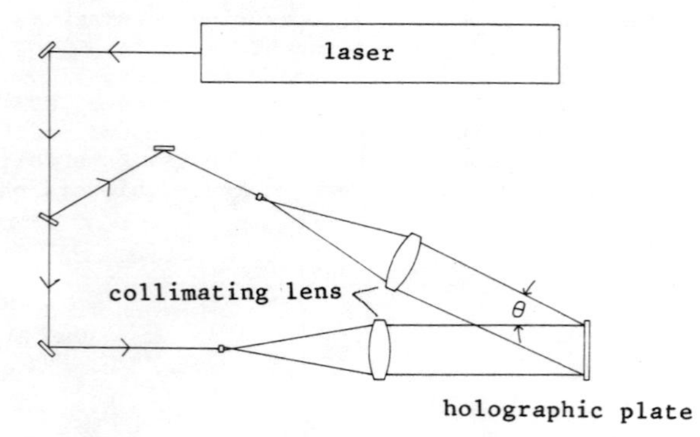

Fig. 4 Production of a holographic grating HG1

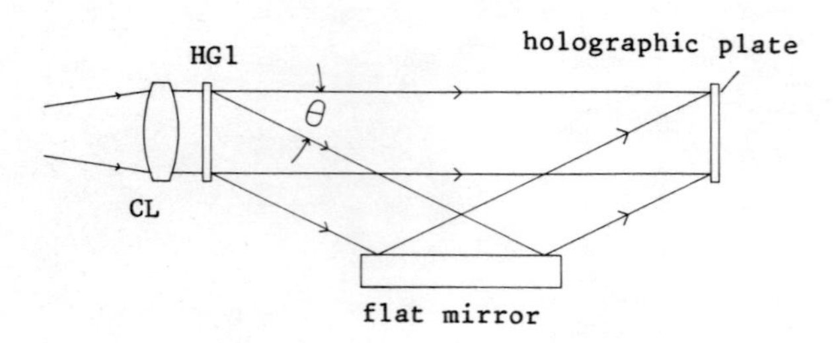

Fig. 5 Production of a holographic grating HG2

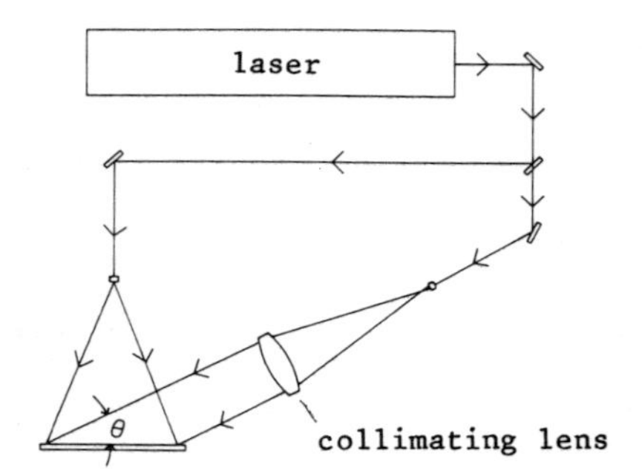

Fig. 6 Production of a holographic lens

Experimental results

We measured the flatness of I.C.wafers, optical discs, metal and glass optics by our OII with HOEs. As the result, we found it is possible to measure the a little rough surfaces, for example, a machined metal element and ground glass.

Fig.7 shows our OII with HOEs, model LSI 1501. This interferometer have a 15mW He-Ne laser as a light source. Diameter of the collimating lens is 150mm and size of holographic grating is 120x165mm.

This is suitable to measure a pre-completed I.C.wafer, a patterned wafer, and an optical memory disc. The measuring sensitivity of our OII is 2um and it can measure the object with diameter up to 6 inches.

The measured results are shown in Fig.8 to Fig.11. Fig.8 shows a machined metal component. Four interference fringes express that the center portion is higher about 8um than its edge.

Fig.9 shows a pre-completed I.C.wafer with 3 inches diameter. The pre-completed surface is rough. In the factories, an interferometer which can measure such a rough surface is demanded.

Fig.10 shows a 3 inches diameter patterned I.C.wafer. It is thought to be very useful to measure the flatness of I.C.wafer through its machining process.

Fig.11 shows a measured result of optical disc surface. Our OII with HOEs makes it possible to measure the samples which cannot be measured by a current Fizeau interferometer.

Fig. 7 OII with HOEs

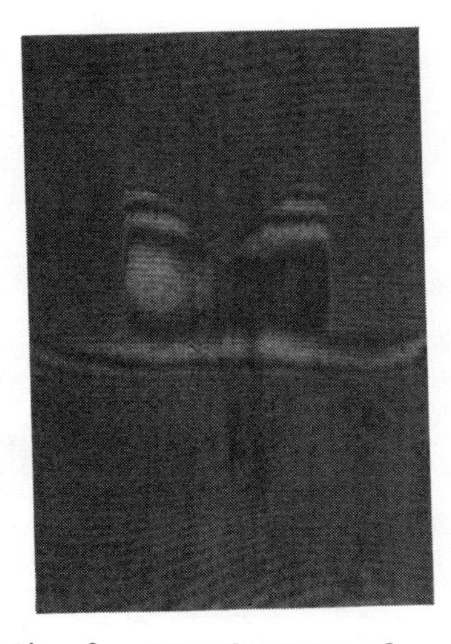

Fig. 8 Interferogram of machined metal component

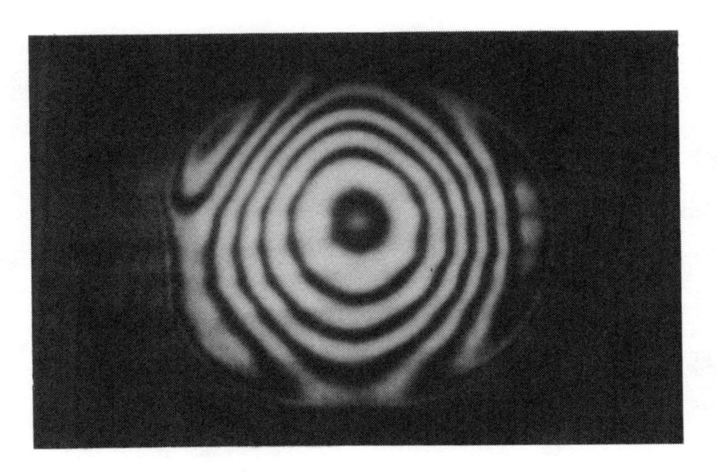

Fig. 9 Interferogram of pre-completed I.C. wafer

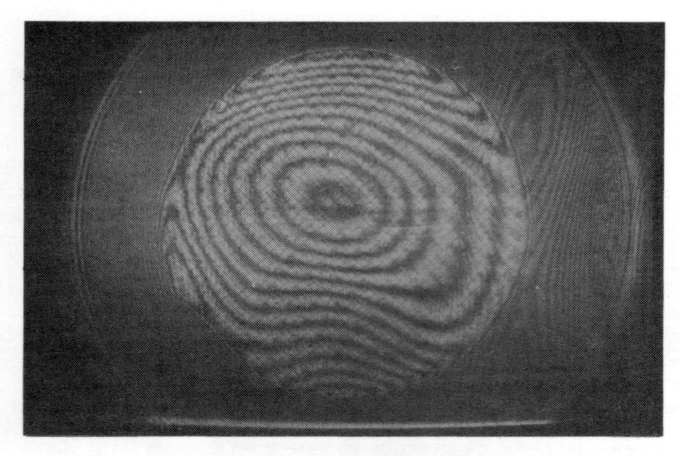

Fig. 10 Interferogram of patterned I.C. wafer

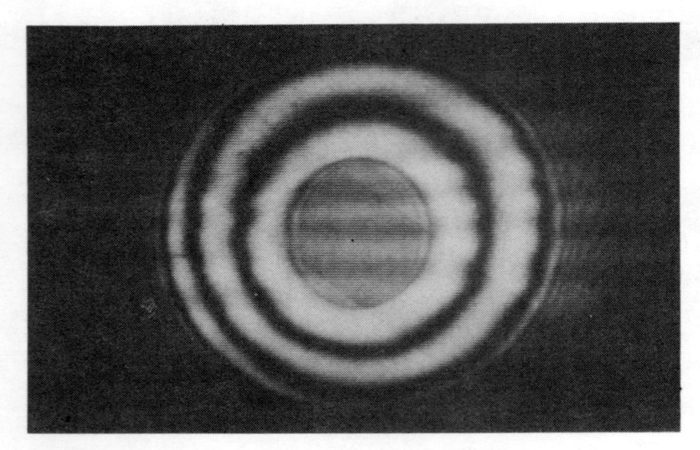

Fig. 11 Interferogram of optical disc surface

Summary

As we said before, our OII with HOEs can measure the surfaces in good measuring accuracy even if the optics in the interferometer have some aberration. This merit depends on making HG2 (in fig. 5) holographically.

In Birch's interferometer, there appears some fringes on interferogram if CL or G1 have some aberration.

In the case of our OII with HOEs, the aberration on CL or ref. mirror are canceled by making HG2 holographically and the fringes by the aberration don't appear. When we exchange ref. mirror for an object, appearing fringes depend on only the difference between ref. mirror and an object. So, its measuring accuracy depends on only the accuracy of the ref. mirror.

And by a holographic lens, it is possible to convert an oblique viewing image into a image which is observed in perpendicular.

Reference

(1) K.G.Birch;J.Phuy.E:S 6(1973)1045

(2) M. Suzuki et al;Proc.Int.Conf.on Lasers'83 (1983)562

Formation of amplitude grating in real-time holographic recording medium BSO crystal

Yan Yuan, Wu Wei-shu, Chen Ying-li

Department of Applied Physics, Shanghai Jiao Tong University, Shanghai, PRC.

Abstract

The intensity-dependent absorption was discovered through experiments in photorefractive crystal BSO. The nonlinear coupled-wave equation describing the dynamic mixed volume gratings was derived, in which self-diffraction and the intensity-dependent absorption was considered. The calculated results are in agreement with the experiment.

Introduction

About twenty years ago photorefractive effect was first discovered by researchers at Bell Laboratory. Since then, this effect has been found in more and more crystals such as $LiNbO_3$,[1] $KNbO_3$,[2] $BaTiO_3$,[3] $Bi_{12}SiO_{20}$ and $Bi_{12}GeO_{20}$.[4] The possibility of using photorefractive crystals as holographic recording media was proposed in 1968. So far, many theoretical investigations, as well as experimental studies, have been making of hologram application, especially for $LiNbO_3$ crystals. As we all know, the holographic recording, in principle, is fulfilled through the formation of volume grating which deals with the exciting, migration and trapping of ligh-induced carriers.

With the development in the research on photorefractive effects, the photorefractive crystals mentioned above have been widely used in information storage, real-time holographic technique, optical signal processing and many other fields in optics.

Recently, much attention was paid to photorefractive crystal bismuth silicon oxide (BSO). Because of many advantages such as fine optical quality, high photorefractive sensibility and easy preperation, BSO crystals are widely used in two-wave interaction and four-wave mixing. In addition, the research on application in holographic recording using BSO crystals was made and many theories were set up to describe the recording process.

Discovery of intensity-dependent absorption

As we know, the holographic recording in photorefractive crystals is usually realized through the formation of phase gratings. The phase gratings result from index modulation via electro-optical effect. In the previous theories, the effects of absorption on the gratings are ignored or simply taken as $\exp(-\alpha d)$. As a matter of fact, the effect of absorption of light in BSO crystals is not so simple.

In our cases, the absorption coeffient of BSO crystal is about 2.1 cm^{-1}, which is much larger than that of $LiNbO_3$. Therefore we must take the absorption into account. Furthermore, the absorption coefficient of BSO will vary with the incident light intensity. Letting a beam of He-Ne laser penetrate the BSO crystal we measured its intensity with a powermeter. Then a beam of Ar^+ laser was allowed to pass the crystal. There was an angle (about 30°) arranged between two beams so that the Ar^+ beam could not enter the detector. Again, we measured the transmitted intensity of He-Ne beam. The intensity decreased noticeably. When we used a piece of normal glass instead of BSO crystal and repeated the measurement, this phenomenon didn't occur at all. This experiment tells us that the absorption coefficient does change with the incident light intensity. We call it intensity-dependent absorption.

In another experiment we measured the output light intensity versus the input intensity. The experimental arrangement is showed in Fig.1. Using two well-corrected powermeters, we measured the input power and output power simuteneously. If the crystal had no intensity-dependent absorption, the output power would increase linearly with the input power just as the case of normal glass. However, we found the behavior of BSO crystal not like that. When the input light intensity increased, the output light intensity did not increase linearly. In other words, the transmittance became smaller and smaller. Since all the conditions (such as reflection, scattering etc.) remained unchanged we concluded that the absorption coefficient of BSO crystal increased with the intensity.

Theoretical description of the formation of grating

When two coherent plane waves propagate in BSO crystal the interference fringes are formed throughout the crystal. Since the absorption of BSO crystal is nonuniform and gives the same modulation frequency as that of interference fringes, i.e. an amplitude volume grating is formed in BSO crystal, which is added to the phase grating to form a mixing grating.

Compared with the phase gratings in BSO crystal the amplitude grating have these properties:
First, as the grating formation is unrelated to the trasport process of light-induced carriers, the recording process is a fast process which differs from that of phase grating, and the recorded grating will not remain when recording light beams are removed. In other words, only the recording process and real-time grating will be influenced by the amplitude gratings.

Second, the responce of crystal absorption to intensity is localized. So there is no phase shift between the amplitude grating and interference fringes, which usually exists in phase grating.

Third, it is clear that the modulation (hence the diffraction efficiency) of the amplitude grating depends on both intensity and intensity ratio, whereas that of the phase grating depends on the intensity ratio only.

Of course, in describing the mixing grating we must take the effect of self-diffraction into account, which makes the intensity distribution in steady state different from that of the initial state.

Based on these considerations, we began to set up the nonlinear coupled-wave equations describing the formation of holographic volume mixing grating.

In isotropic medium BSO crystal, the material equations can be written as

$$\begin{cases} \vec{J} = \sigma \vec{E}(\vec{r}, t) \\ \vec{D} = \varepsilon_0 \varepsilon \vec{E}(\vec{r}, t) \\ \vec{B} = \mu_0 \vec{H}(\vec{r}, t) \end{cases} \quad (1)$$

where $\vec{E}(\vec{r}, t)$, $\vec{H}(\vec{r}, t)$ are electric and magnetic vector of the light field, respectively. σ is used for describing the loss in the crystal. ε is dielectric constant. ε_0 and μ_0 are permittivity and permeability in vacuum respectively.

In the case of TE waves, we can use scalar quantity Maxwell's equations. Substitution of Eq. (1) into Maxwell's equations gives

$$\nabla^2 E(\vec{r}, t) - \frac{\varepsilon}{c^2} \ddot{E}(\vec{r}, t) = \sigma\mu_0 \dot{E}(\vec{r}, t) \quad (2)$$

For a monochromatic light, let $E(\vec{r}, t) = E(\vec{r}) e^{j\omega t}$, then we have

$$\nabla^2 E(\vec{r}) + k^2 E(\vec{r}) = 0 \quad (3)$$

where

$$k^2 = \frac{\omega^2}{c^2}\varepsilon - j\omega\mu_0\sigma = \beta^2 - 2j\alpha\beta \quad (4)$$

$$\beta = \frac{2\pi}{\lambda}\sqrt{\varepsilon}, \qquad \alpha = \frac{\mu_0 c \sigma}{2\sqrt{\varepsilon}}$$

As shown in Fig.2, two monochromatic plane waves fall on the crystal surface at the same angle θ, and with the same polarization perpedicular to the plane of incidence. The electric field intensity can be expressed as

$$E = Re^{-j\vec{\rho}\cdot\vec{r}} + Se^{-j\vec{\sigma}\cdot\vec{r}} \quad (5)$$

where $\vec{\rho}, \vec{\sigma}$ represent the wave vectors of reference beam and signal beam respectively, and R,S the amplitude of the two beams. The grating vector formed in crystal is $\vec{K} = \vec{\rho} - \vec{\sigma}$ with magnitude

$$K = \frac{4\pi}{\lambda}\sin\theta \quad (6)$$

According to the experimental results, the change of absorption coefficient is approximately proportional to the incident intensity, i.e. $\Delta\alpha = bI$ (7)

In the common approach of SVA along z-axis, we can obtain following coupled-wave equation by substituting Eq. (5) into Eq.(3) and ignoring the higher order Bragg diffraction term.

$$\cos\theta\frac{\partial R}{\partial Z}=-[\alpha_0+b(RR^*+SS^*)]R-\left(j\frac{\pi n_0^3 r_{41}}{2\lambda}|E_1|e^{-j\phi}+bRS^*\right)S$$

$$\cos\theta\frac{\partial S}{\partial Z}=-[\alpha_0+b(RR^*+SS^*)]S+\left(j\frac{\pi n_0^3 r_{41}}{2\lambda}|E_1|e^{j\phi}-bR^*S\right)R \qquad (8)$$

where, E_1 is the complex amplitude of the space-charge field in crystal. It is described by the following equation derived from Ref.5 by changing the normalized factor:

$$\frac{\partial}{\partial t_N}\left(\frac{E_1}{E_q}\right)+A\left(\frac{E_1}{E_q}\right)=B\frac{R^*S}{RR^*+SS^*} \qquad (9)$$

where

$$A=(1+E_T/E_q+jE_0/E_q)/D$$
$$B=(-E_0/E_q+jE_T/E_q)/D$$
$$D=1+E_T/E_M+jE_0/E_M$$

and E_T is the effective diffusion field, E_q is the maximum space-charge field, E_0 is external field. $E_M=(K\mu\tau_r)^{-1}$, μ is electron mobility and τ_r the carrier recombination time, $t_N=t/\tau_d$, τ_d is the dielectric relaxation time.

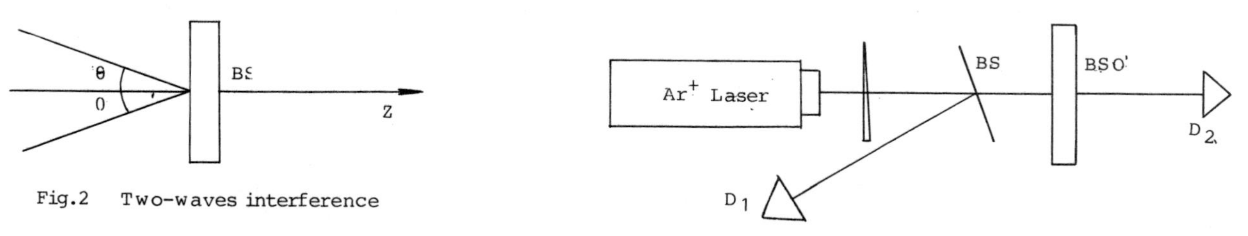

Fig.2 Two-waves interference

Fig.1 Measurement of Absorption

Experimental results

As mentioned above, the effect of the absorption has been considered in the coupled-wave equations. The last term in the right hand side of the equation shows the existance of the amplitude grating. The following example will show the effect of the grating on the light coupling in crystal.

We define a parameter γ_0 as effective signal light gain:

$$\gamma_0=\frac{I_S(I_R\neq 0)}{I_S(I_R=0)}$$

Usually, to find an analytical solution of Eq.(8) is very complicated and difficult. Only with some approximations can it be obtained. On the other hand, we can obtain the numerical solution of coupled-wave equation by means of computers. Therefore, the characteristics of γ_0 versus other experimental parameters are obtained. For example, we choose the experimental condition used by Gunter et al.[6] into the coupled-wave equations. The results show that our theoretical value $\gamma_0=1.3$ is much closer to Gunter's experimental value 1.12 than his theoretical value 3.5. Another caculated result is given in Fig.3.

The experimental measurement of γ_0 is also performed. The experimental arrangements are shown in Fig.4. The intensity of Ar^+ laser beam was stablized by an external unit which could decrease the beam intensity fluctuation. A comparison of experimental results to theoretical results is given in Fig.5.

Effective signal gain (Fig.5) in agreement with the theoretical caculation reveals grouth with incident angle θ, When θ is less than 10°. However when θ is more than 10°, the experimental results become much lower than the theoretical values. The reason for this is that in the theoretical caculation, we ignore the finite dimension of the crystal, but in fact, when the angle θ increases, the effective coherent area decreases, especially in the case of large θ.

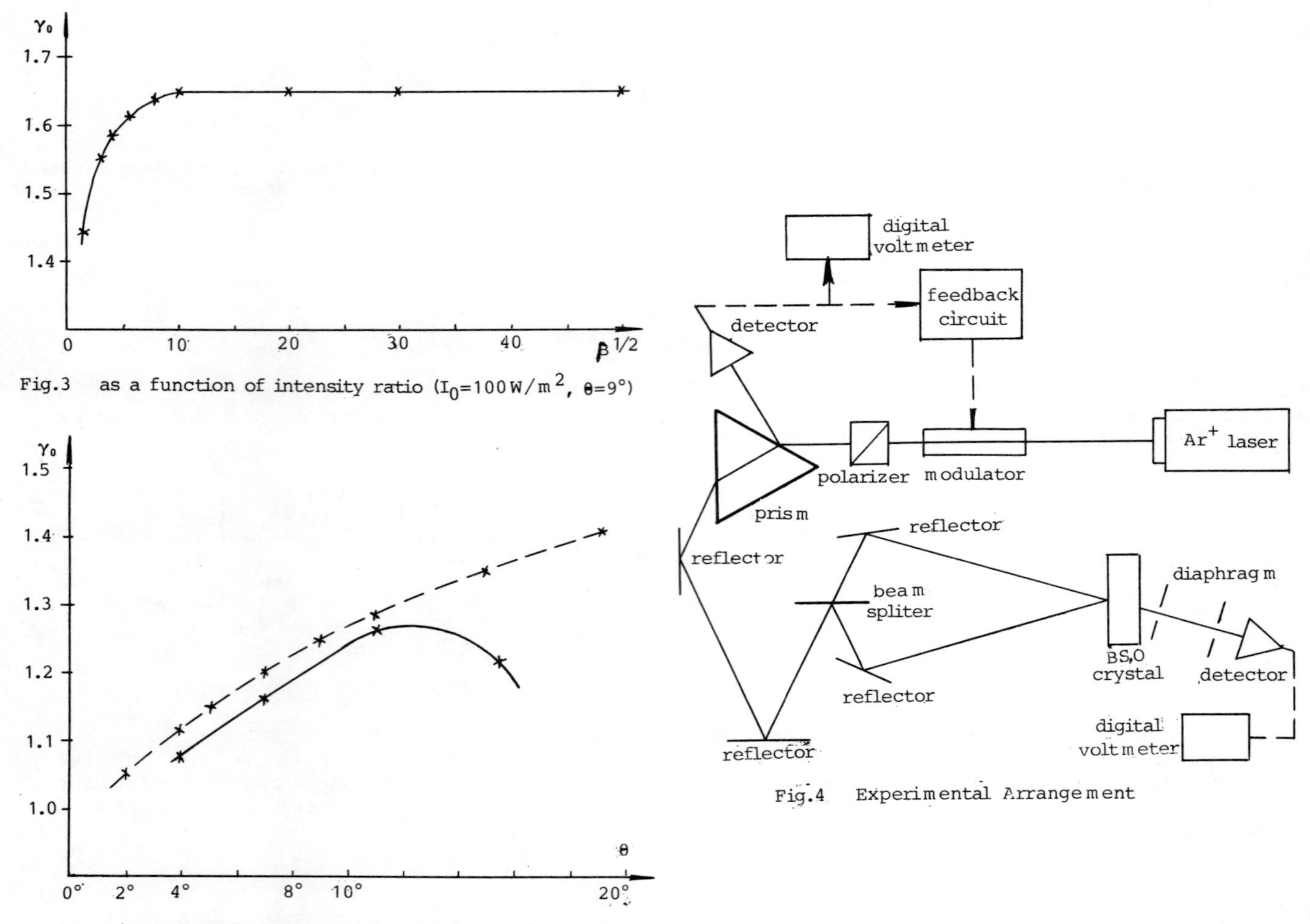

Fig.3 as a function of intensity ratio (I_0=100 W/m^2, θ=9°)

Fig.4 Experimental Arrangement

Fig.5 As a function of angle θ, solid line: experimental, dashed line: theoretical

References

1. Ashkin A. et al. "Optically-induced Refrective Index inhomogeneities in $LiNbO_3$ and $LiTaO_3$," Appl. Phys. Lett. Vol.9, No.1, pp. 72-74. 1966.

2. Gunter P. et al. "Photorefractive Effects and Photocurrents in $KNbO_3$:Fe," Ferroelectrics Vol. 18, pp. 27-38, 1978.

3. Jack Feinberg et al. "Photorefractive Effects and Light-induced Charge Migration in Barium Titanate," J. Appl. Phys. Vol. 51, p. 1291. 1980.

4. Huignard et al. "High-sensitivity Read-Write Volume Holographic Storage in $Bi_{12}SiO_{20}$ and $Bi_{12}GeO_{20}$ Crystals," Appl. Phys. Lett., Vol. 29, No. 9, pp. 591-593. 1976.

5. Heaton J. M. and Solymar L. "Transient Energy Transfer During Hologram Formation in Photorefractive Crystals," Optica Acta, Vol. 32, No. 4, pp. 397-408. 1985.

6. Gunter P. "Holography, Coherent Light Amplification and Optical Phase Conjugation with Photorefractive Materials," Physics Reports, Vol. 93, pp. 199-299. 1982.

NOTEWORTHY QUALITIES OF HOLOGRAPHIC GRATING IN THE DESIGN OF ANALYTICAL SPECTROMETER

JP. LAUDE

OPTO ELECTRONIC Department INSTRUMENTS SA Division JOBIN YVON
16 rue du canal BP 118 91160 LONGJUMEAU FRANCE

ABSTRACT

High frequency blazed holographic grating (1800 and 3600 lines/mm) were used to design low detection limit emission spectrometer (sub ppb for several elements) and high resolution Raman spectrometer (oxygen line at 2 cm^{-1} well resolved on the air spectrum).

INTRODUCTION

The 3 main qualities of holographic gratings over those made by conventional methods are : small amount of stray light ; possibility of intrinsic aberration corrections ; higher resolution with an easier manufacturing of large gratings with fine spacing.

During several years these advantages were, to a certain amount, counter balanced by a lower efficiency, at least for gratings with a groove spacing much larger than the wavelength.

With the development of several new methods for blazing the holographic gratings (mainly adjustment of the grooves depth recording of standing waves on a tilted substrate (figure 1), oriented ion etching (figure 2) and oriented chemical etching), the holographic gratings are now quite efficient (figure 3).

The research and development teams of JOBIN YVON have used these unique properties to design several interesting new spectrometers.

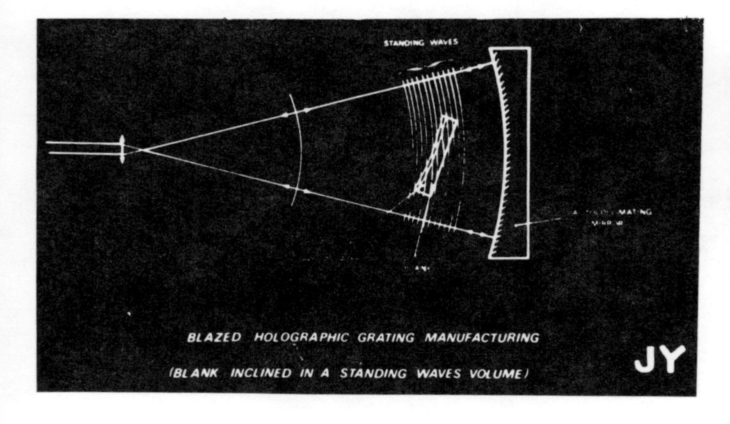

Figure 1

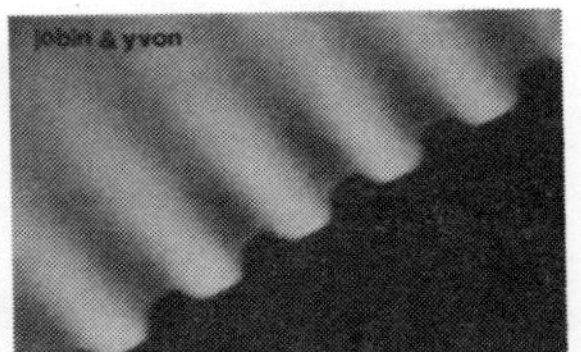

Groove shape of the holographic mask (1200 gr / mm)

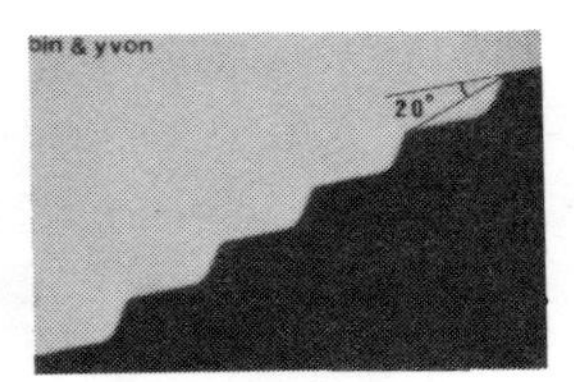

Triangular structure brought into the glass substrate after oriented ion etching

RHL / 62

Figure 2

RAMAN SPECTROMETER

On our Raman spectrometer U 1000 (figure 4) we use a double monochromator. The focus is 1 meter. The 2 monochromators are classical, CSERNY TURNER mounts using plane holographic master gratings.

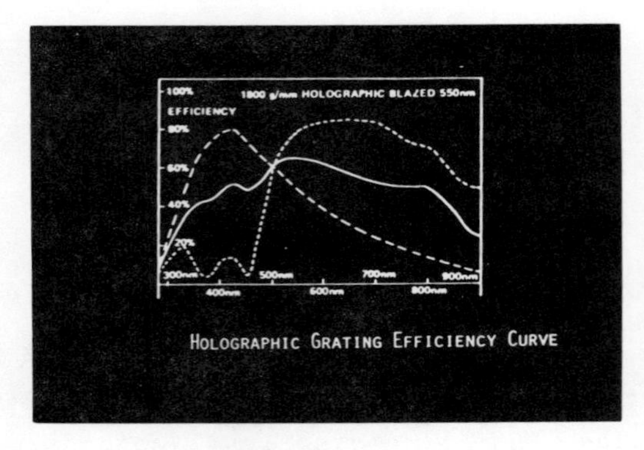

Figure 3

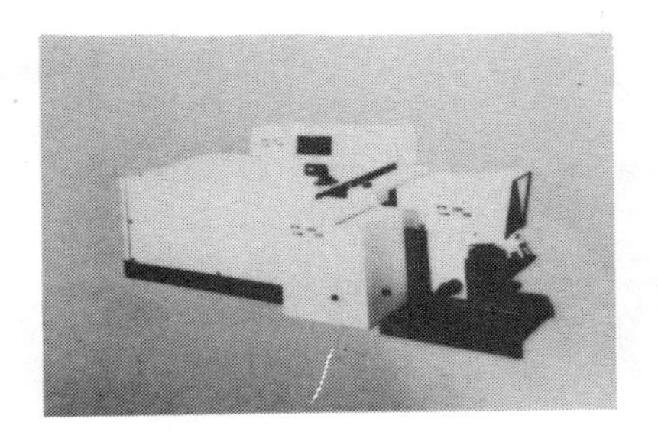

Figure 4

The number of lines per mm is 1800. The gratings are "blazed" between 500 and 950 nm. Master gratings are prefered in order to avoid the small addition of stray light always brought by the copying even if the best process is used. This have permitted to get a stray light level as low as that of the high quality triple monochromators with in addition a better throughput. This is illustrated on the Raman spectrum of the air (figure 5). Onecan see that the rotation line of the oxygen at 2 cm^{-1} can be easily resolved. To our knowledge the U 1000 is the sole instrument able to give this result up to now.

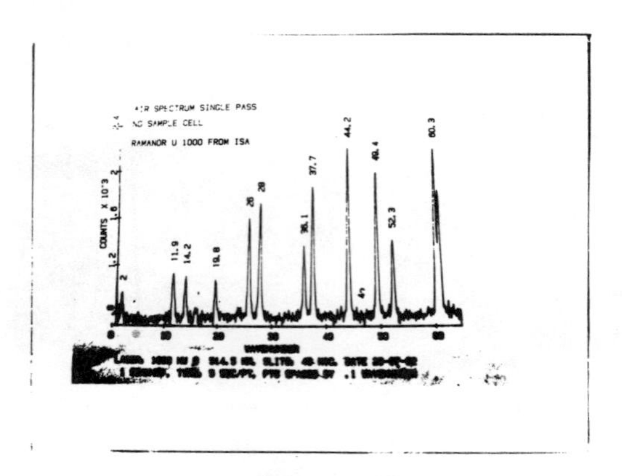

Figure 5

EMISSION SPECTROMETER

The holographic gratings have also greatly raised the performances of the emission spectrometers. Let us take the example of ICP. In ICP, the sample is nebulized and introduced in an argon flow (figure 6) in the plasma torch. The main lines are found in the near vacuum UV and blue, but some elements like Na K Cs have their lines in the visible and near IR. The configuration is presented on figure 7 (ref 1) : an holographic grating, concave and highly dispersive (3600 lines/mm) is used on our JY 32 instrument to disperse the lines from 1700 to 4100 Å in a Pashen Runge configuration.

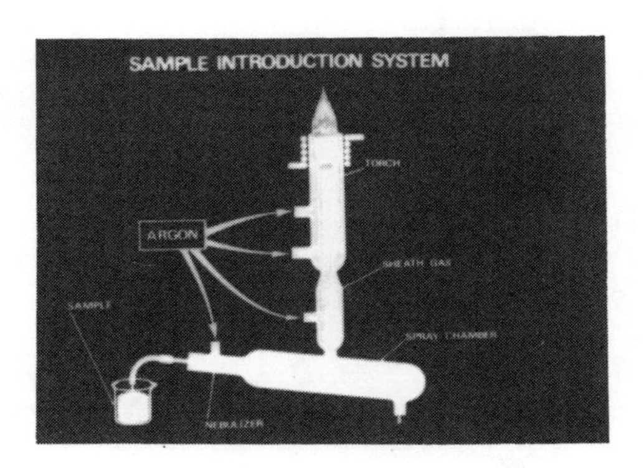

Figure 6

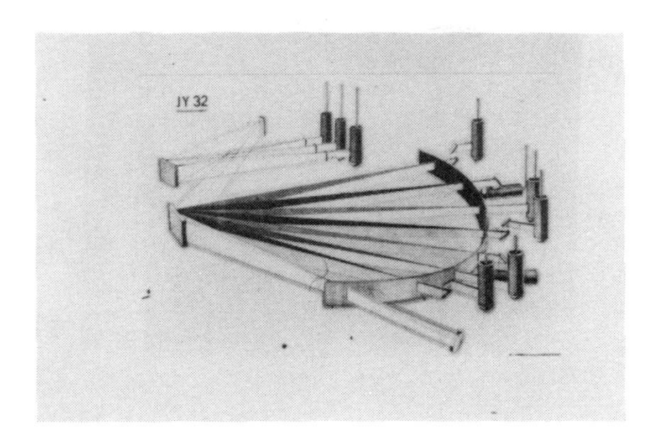

Figure 7

The linear dispersion is 5,5 Å/mm in the first order and 2,75 Å/mm in the second order. The longer wavelengths are deflected in the zero order by this holographic grating in the flat field configuration (ref 2). This instrument is shown on figure 8. Some detectors limits obtained in lubrificant oil without oil dilution are given in the table 1. One can see very low standart deviation on a set of measurements and low detection limits in such a complex matrix. On table 2 are presented results on Zn Fe Mg Cu in serum.

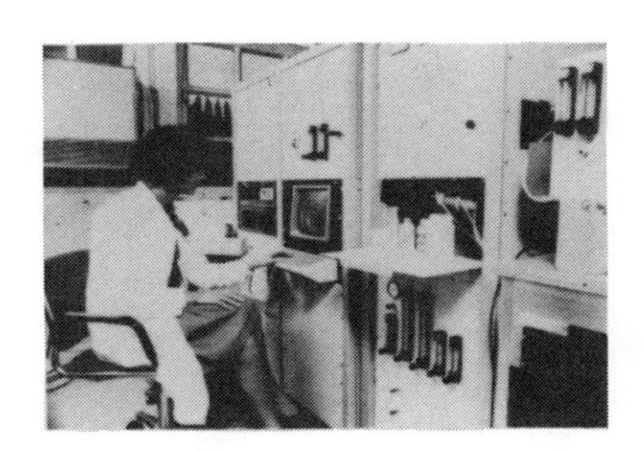

Figure 8

ELEMENT	Å	0 PPM Average 9 Mesures	0 PPM Std Dev	0 PPM Detection Limit PPB(3σ)	1 PPM Average 9 Mesures	1 PPM Std Dev	3 PPM Average 9 Mesures	3 PPM Std Dev
Fe	2599	0.00080	0.0010	3	1.023	0.004	3.099	0.010
Cu	3247	0.0016	0.00056	1.6	1.005	0.004	3.037	0.004
Ag	3280	0.0046	0.0010	3	1.035	0.003	3.115	0.005
Ti	3349	0.0015	0.00047	1.5	1.009	0.002	3.068	0.003
Al	3961	0.0129	0.0035	10	1.023	0.006	3.069	0.005
Mg	2851	0.00011	0.00067	2	1.031	0.003	3.036	0.005
Cr	4254	0.00040	0.0049	15	1.040	0.009	2.975	0.005
Ni	2316	0.005	0.0010	3	0.998	0.006	3.026	0.007
Pb	2203	0.023	0.020	60	1.088	0.014	3.182	0.020
Sn	1899	0.042	0.080	240	1.154	0.050	3.313	0.054
Na	5895	0.049	0.055	165	1.112	0.052	3.073	0.071

LUBRICANT OILS FOR AIRCRAFT (NYCO 13B)

Tableau 1

	SERONORM			PATHONORM H			PATHONORM L		
	EXPECTED CONCENT.	ICP RESULT	RSD %	EXPECTED CONCENT	ICP RESULT	RSD %	EXPECTED CONCENT	ICP RESULT	RSD %
Zn	0.93	0.77	2.90	1.21	1.10	2.00	0.78	0.76	2.23
Fe	1.60	1.67	1.40	1.96	2.03	2.76	0.87	0.99	1.22
Mg	25.9	2.33	1.10	34.8	31.1	0.83	13.7	13.2	0.72
Cu	1.08	0.97	1.46	1.01	1.02	1.16	0.60	0.61	1.70

Tableau 2

CONCLUSION

The holographic gratings have allowed the design of efficient new configurations in analytical spectrometry. They are used routinely on all kind of spectrometers. Particularly their qualities of high resolution and low stray light level find application in Raman and emission spectrometer.

Acknowledgment

We would like to thank : Dr Nguyen Anh Tuan for his recordings of the Raman spectrum and the applications Laboratory of JOBIN YVON, Dr ALAIN of CHU in Angers and the SNECMA for their results in emission spectrometry.

REFERENCES

1 - Dispositif optique pour spectromètre d'émission - French Patent 5/2/82 A. THEVENON
2 - Spectrographe à champ plan pour un domaine spectral étendu utilisant un réseau holographique concave - French Patent 10/12/75 - G. PASSEREAU.

Development of high resolution zone plate by holography and ion-etching techniques

Yong-Gang SU, Shao-Jun FU, Xiao-Min TAO, Yi-Lin HONG and Yun-Wu ZHANG

National Synchrotron Radiation Laboratory, University of Science and Technology of China
24 Jin-zhai Road, Hefei, Anhui, People's Republic of China

Abstract

The purpose of development and the processing steps to fabricate high resolution zone plates, especially holographic exposure and ion-etching, are described. Also the experimental results and some prospects are presented.

Introduction

In the last few years, with the rapid progress of space and astronomic optics, plasma physics, especially synchrotron radiation - a novel light source available in laboratories and having very wide wavelength range, the requirements to develop excellent soft x-ray optical components become more and more urgent. The conventional optical elements can't image to x-ray because the real part of the refractive index of all optical material to soft x-ray is less than unity by only a very small amount and so the refraction is negligible.[1] But many researches have showed that the zone plate has lots of advantages and vast vistas as a dispersing component in synchrotron radiation linear monochromator and as an imaging element in a soft x-ray microscope.[2]

Zoneplate-diaphragm monochromator

To a soft x-ray microscopy beam line being constructed at Hefei National Radiation Laboratory, we did a design of a monochromator utilizing a condenser zone plate (CZP) and a diaphragm with a diameter d.[3] Denoting outermost zone diameter, total zone number and width of the CZP by D, n and ΔR_n respectively, the spectral resolution of this monochromator is $\lambda/\Delta\lambda \approx D/2d \approx 200$ which satisfies the temporal conherence requirement of the soft x-ray microscope downstream. Meanwhile the spatial coherence degree $|M| = |2J_1(v)/v| > 80\%$ is enough for the illumination of the micro zone plate (MZP) in the microscope to form a diffraction-limited spot with a diameter of about 0.2 μm at the sample, where $J_1(v)$ is a Bessel function of first kind and first order, the variable $v = \pi dD'/2\lambda g'$, D' is the outermost zone diameter of the MZP and g' is the distance between diaphragm and MZP in the microscope. Fig. 1 presents the schematic of the optical arrangement of a zoneplate-diaphragm monochromator in a soft x-ray beam line and a microscope downstream at Hefei National Synchrotron Radiation Laboratory.

The parameters of three CZPs in a new design for our microscopy beam line are as following: D = 2.8 mm, n = 1506, 1082 and 770, ΔR_n = 0.465 μm, 0.647 μm and 0.909 μm for the wavelength ranges 2.0-2.8 nm, 2.7-3.9 nm and 3.8-5.4 nm respectively. We have been trying to use the holography-ionetching techniques to fabricate so fine zone plates which finally have only the polyimide as a transparent layer to soft x-ray, the gold with Fresnel zone plate pattern as an opaque layer and some support structure.

Processing steps for CZP fabrication

1) Application and curing of polyimide (~2.0 μm thick) on a polished glass.
2) Coating of about 0.3 μm gold layer on the polyimide.
3) Spin 0.2 μm AZ-1350J photoresist on the gold.
4) Holographic exposure with an A_r+ laser.
5) Development and fixing of the exposed photoresist.
6) Ion-etching.
7) Etching the glass with hydrofluoric acid from the back face and making the support structure.

We will describe some details of the two key steps in the following sections.

Holographic exposure

Fig. 2 shows an optical arrangement for holographic exposure to produce the Fresnel zone plate pattern in the photoresist spun on the gold layer. The beam, generated from an A_r^+ laser whose 457.9 nm wavelength output is used now, is split into two branches. After passing through some optical components, one become collimated and another convergent in the same direction. Then the photoresist layer is exposed in the interference field formed by the superposition of these two branches. The zone plate pattern is thus obtained in photoresist layer as a mask for the following ion-etching step after such holographic exposure, development and fixing.

In our experiment, the geometry of the zone plate relief pattern profile in the photoresist is very important to the qulity of holography-ionetched zone plate. Fig. 3 indicates the ideal geometry of photoresist layer profile.

So we should properly control the thickness of photoresist, exposure quality,development condition, and so forth in order to obtain a better geometry of photoresist profile. For example, if the photoresist is thick enough to resist the ion-etching for shading the gold, we might reduce the thickness of photoresist as much as possible to shorten the exposure time and to lower the affection of the circumstance such as temperature and vibration. We have the empirical formula for the photoresist thickness required after spinning and before exposure

$$\text{Thickness(photoresist)} = \frac{1}{4}\text{Thickness(gold)} + vt \qquad (1)$$

after considering the following two factors: (1) the etching rates to photoresist and to gold are different. Under the ordinary etching condition such as 500 eV ion energy and 1mA ion current, the etching rate to AZ1350J photoresist is about one quarter of that to gold. (2) while developing the exposed photoresist layer with alkaline developer, even in the unexposed area there still is an etching rate v which reduce the thickness of unexposed layer vt, where t is the exposure time.

On the other hand, in view of the big aberration caused by the fact that the zone plate pattern is produced with ultra violet or visible light such as 457.9 nm here and the zone plate will be used in the soft x-ray region such as 2.0-5.0 nm which is a factor of about one hundred shorter, we are designing a special optical system instead of the seperated optical components used now and showed in Fig. 2 to solve this problem for the fabrication of very fine zone plates.

Ion beam etching

Ion beam etching technique can be used to fabricate a variety of devices with micron or submicron structure including fine gratings, zone plates, large scale integrated circuits, microwave and acoustic surface wave devices, etc. We are using a LKJ-1B type ion beam etching machine manufactured by the Ministry of Space Industry of China for us.

The rate and the qulity of ion beam etching depend on the kind of etched material, ion energy, ion beam density, incident angle, cleaning status and cooling condition and so on. After a large number of experiments, we select the ion beam etching conditions as below. Ion energy: 600-700 eV (while good water-cooling). Ion beam density: 0.3-0.6 mA/cm^2. Ion beam incident angle: 0°. Then the ratio of the etching rate of gold to that of AZ1350J photoresist reaches 4.0:1 - 4.5:1.

The parameters of a holography-ionetched zone plate made at our laboratory are as follows. D = 3.8 mm, n = 1350 and $\Delta R_n = 0.7\ \mu m$. Fig. 4 and 5 are the scanning electron microscope (SEM) micrographs of another holography-ionetched zone plate pattern and the profile of its photoresist and gold layers on the glass substrate.

Prospects

In the near future, after the completion of the test of all the processing steps for zone plate fabrication we will begin to fabricate holography-ionetched gratings for the synchrotron radiation beam lines and spectrometers. Next year we hope that the fabrication of CZP using the new special optical system would be started.

Acknowledgments

The authors wish to acknowledge Dr. R. L. Johnson, Prof. G. Schmahl and his group, Prof. I. Lindau and his group, Prof. E. Spiller for their suggestions, encouragements and much help. Thanks are also due to Prof. Guo-guang Yang and other chinese colleagues for the beneficial discussions.

References

1. Underwood, J. H. and Attwood, D. T., "The Renaissance of X-ray Optics", Physics Today, pp. 44-52, April 1984.
2. Schmahl, G. et al, "Zone Plates for X-ray Microscopy", in: X-Ray Microscopy, Edited by G. Schmahl and D. Rudolph, Springer-Verlag, Berlin, pp. 63-74, 1984.
3. Su, Yong-gang and Fang, Lu, "A Preliminary Optical Design for a Microscopy Beam Line at Hefei", Nucl. Instr. and Methods, Vol. 222, pp. 9-10, 1984.

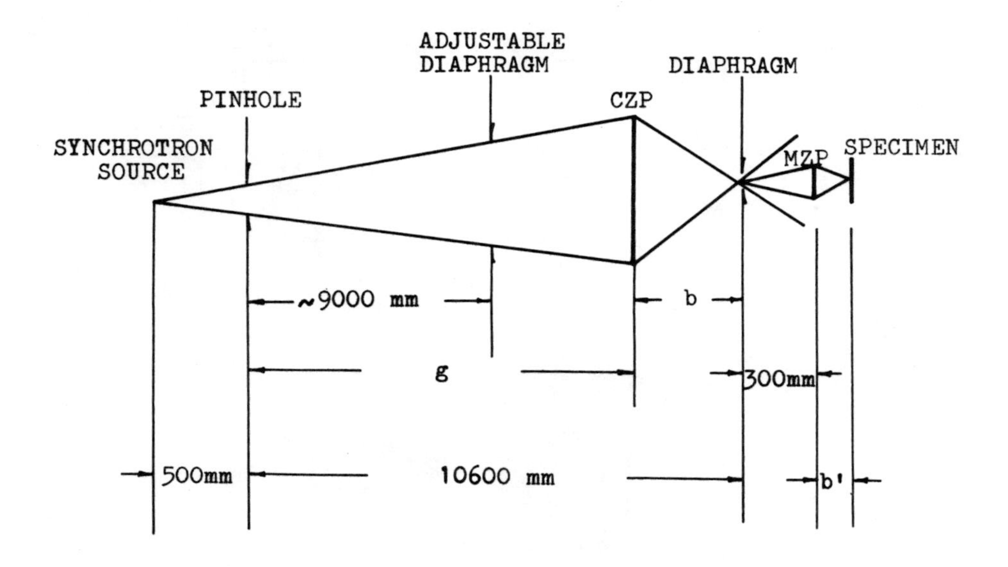

Fig. 1. Schematic of a soft x-ray beam line, which mainly consists of a zoneplate-diaphragm monochromator, and a microscope optics at Hefei National Synchrotron Radiation Lab.

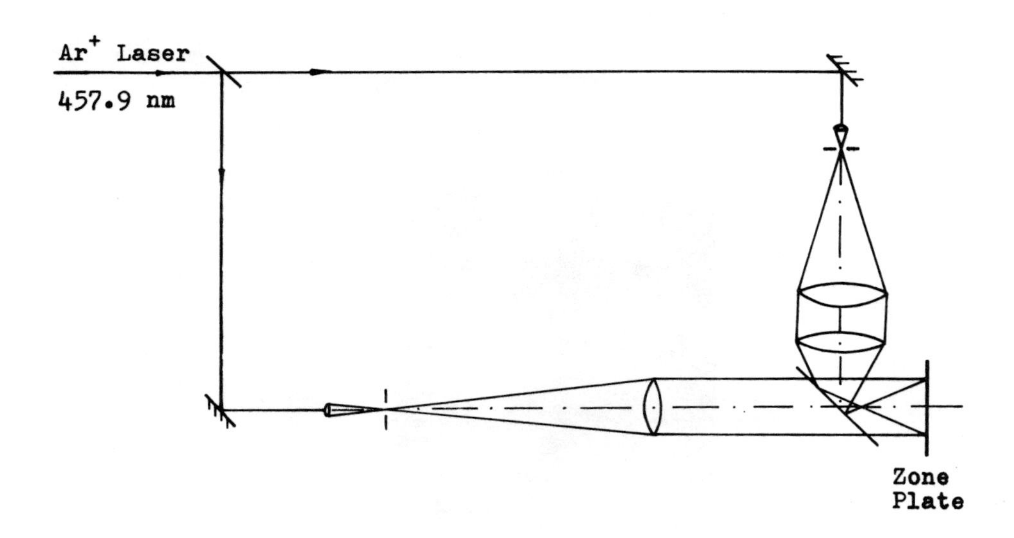

Fig. 2. Schematic of an optical arrangement for holographic exposure to the AZ-1350J photoresist layer on the wafer.

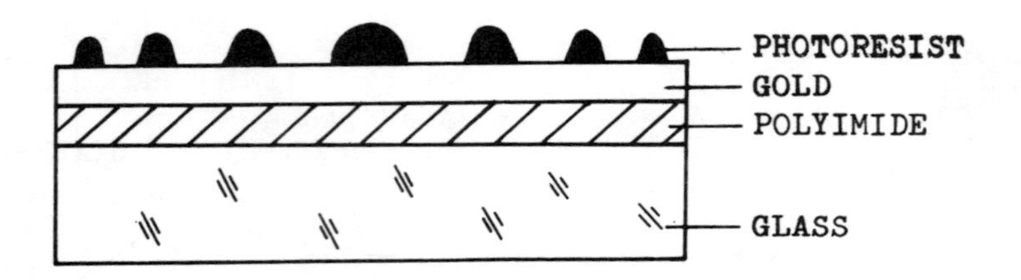

Fig. 3. The ideal geometry of photoresist layer profile after holographic exposure and development.

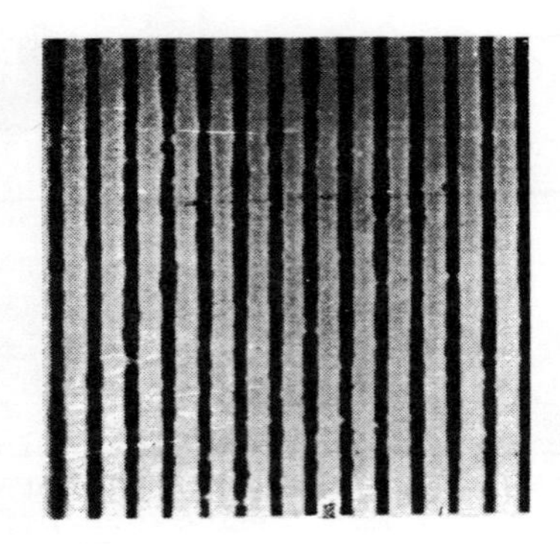

Fig. 4. The SEM micrograph (×2000) of a zone plate pattern with the following parameters: outermost zone diameter D = 20 mm, total zone number n = 3500, width of outermost zone $\Delta R_n \approx 1.4 \mu m$.

Fig. 5. The SEM micrograph (×5000) of the relief profile of the holography-ionetched zone plate in Fig. 4.

THE STUDY ON THE NARROW BAND HOLOGRAPHIC REFLECTION FILTERS

Liu Dahe, Tang Weiguo, Chou Jing, Huang Wanyun

Department of Physics, Beijing Normal University, Beijing, China

Abstract

This paper deals with the research works in constructing a kind of narrow band holographic reflection filters based on the coupled wave theory. The methods of construction were discribed, and properties of the filters were analyzed. A few points relevant to the coupled wave theory were discussed, and suggestions for future practical use of the filters were given. Data proved that the holographic filters have narrow bandwidth, high efficiency, tunability, and large area filters can readily be constructed with very low cost.

Introduction

The narrow band filters commenly used in optical systems are of multilayer dielectric interference type. This kind of filters are not satisfactory enough in that they usually have more than one peak wavelengthes and their efficiency is low. Rapid development of holographic technique in recent years introduced an entirely new principle for the construction of filters with single peak and high efficiency. The coupled wave theory points out that volume holograms have high diffraction efficiency and narrow bandwidth. according to this theory, it is possible to construct a kind of narrow band reflection filters with good optical properties. S.S.Duncan had reported such kind of filters working in the near infrared region[1].

Theoretical consideration

For a volume hologram, its maximum intensity of diffraction follows the Bragg condition

$$2nd\sin\theta=\lambda \tag{1}$$

where, n is refractive index of the medium, d is the spacing of fringes, θ is Bragg angle, λ is wavelength of the diffraction beam. Thus, when a beam of white light stricks on a volume hologram, a beam of diffraction with narrow band can be obtained with high efficiency. If the spacing of fringes of the volume hologram is controlled precisely by using a Laser of suitable wavelength with suitable incident angle during its construction, the wavelength of the beam of diffraction could be controlled as desired. In fact, this kind of volume hologram works well as a kind of reflection filter with narrow bandwidth and high diffraction efficiency.

For a volume hologram having certain value of d, when Bragg angle θ is changed, diffraction beam with different wavelength can be obtained from corresponding directions. It means the reflection holographic filterrs are tunable within a certain range.

When absorption by medium is neglected, the coupled wave theory gives out the formula of diffraction efficiency at peak wavelength as

$$\eta = th^2\left(\frac{\pi \Delta nT}{2nd\sin^2\theta}\right) \tag{2}$$

and the formula of band width at full width half maximum (FWHM) as

$$\Delta\lambda = C\,\frac{2nd^2\sin\theta}{\pi T} \tag{3}$$

where, T is the thickness of medium, Δn is the amplitude of index modulation, C is a constant concerned with diffraction efficiency.

Formula (3) shows that , by increasing the thickness of medium, the bandwidth can be narrowed down. This provides a way to get filters with narrow band.

Theoreetical analysis illustrates, when absorption by medium is considered, due to the effect of absorption as the light beam penetrates the medium, the intensity of the light at different depth Z are different. Thus during construction of filter, the visibility of fringes at different depth within the medium are also different. They are calculated as

$$V=\cosh\left[\alpha\left(\frac{T}{2} - Z\right)\right] \tag{4}$$

where, α is the absorption coefficient. The non-uniform distribution within the medium makes the diffraction efficiency low, and the relationship between diffraction efficiency and the thickness of medium become markdly different from formula (2). This point will be further discussed in another paper.

Experiments

Methods

The recording medium used in the following experiments is dichromated gelatin (DCG). Because of the existence of absorption, the intensities of light at different depth within the medium are different. If the exposed fringes are slanted with respect to the surface of substrate, different intensity of light of exposure will result in different degrees of sweling in the medium, it will cause the developed fringes within the medium curved, and this will affect diffraction efficiency and bandwidth of the filter[3]. In order to overcome this effect and to reduce the non-uniform distribution of visibility of the fringes, the symmetric exposure geometry as showed in Figure (1) was used.

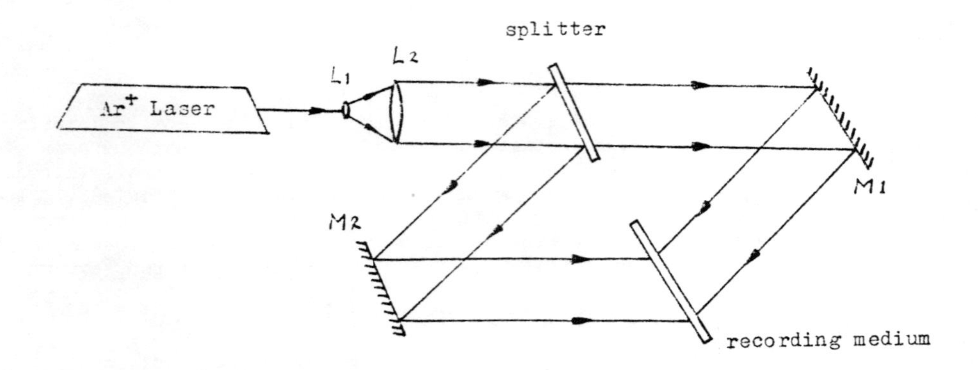

Figure 1. Experimental set-up for exposure of holographic filters

where, i_0 is the incident angle of light beam, S is a splitter with split ratio of 1:1, M is the mirror, H is the recording medium.

For findind out the real relationship between diffraction efficiency, bandwidth and thickness of media, DCG plates with seven different thickness were used. The absorption coefficient α of these DCG plates were measured and given in Table (1).

Table (1). Absorption coefficient of DCG plates

T (μm)	12	24	36	48	60	72	84
t	0.58	0.42	0.33	0.25	0.17	0.12	0.08
α(1/μm)	0.045	0.037	0.031	0.029	0.030	0.029	0.030

here, t is the transmition rate of DCG plate.

The measuring set-up is shown in Figure (2).

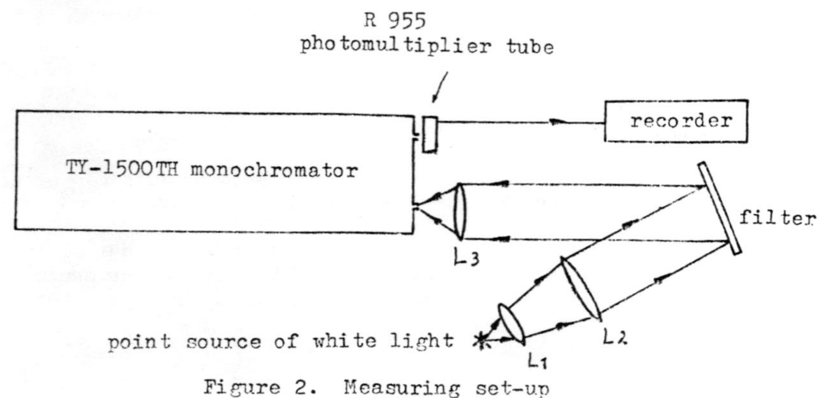

Figure 2. Measuring set-up

where, i is the incident angle of beam during measuring. As could be seen, all the measurements are made using reflected but not transmitted beam, because the filters are going to be used as reflection filters.

In the measuring set-up, the monochromator used was the model TY-1500TH, the photomultiplier tube was the model R 955. The incident angle i_0 was arranged at 0°, 20°, 40°, respectively, during construction of the filters. For each i_0 a group of filter was constructed with DCG plates of seven different thickness. About 300 plates were exposed altogether, the following data were obtained from typical examples.

Results

1. The relationship between peak wavelength and angle

Table (2) gives out the calculated and measured peak wavelength at different angles under the condition: i_0=0°, T=24μm. Figure (3) is the graph corrisponding to Table (1), where the curved line shows the theoretical value given by Bragg formula (1), the dots show the measured results.

Table (2). Peak wavelength at different angles

Incident Angle (°)	5	10	15	20	25	30	35	40	45	50	55	60
Theoretical Value (nm)	547.7	545.0	540.6	534.3	526.9	518.1	508.0	497.1	485.6	473.8	462.1	450.8
Measured Value (nm)	547.7	547.2	542.4	531.6	528.9	518.0	507.0	496.4	484.3	472.0	463.9	453.1

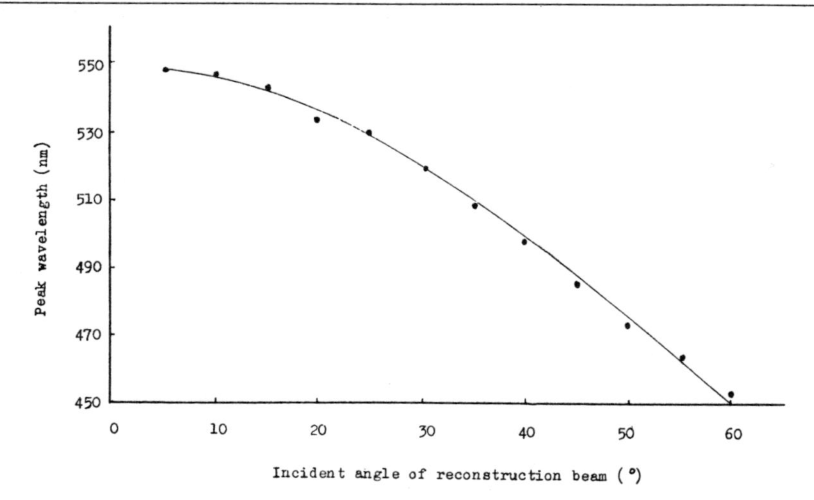

Figure 3. Peak wavelength at different angles

(i_0= 0 , T=24μ)

The tunable range of a filter with different i_0 at T=48μm were shown in Table (3).

Table (3). Tunable range

Incident angle of Construction (°)	0	20	40
Tunable Range (nm)	423-530	449-552	503-624

2. The propties of filters with different thickness

Table (4) gives out the relationship between bandwidth $\Delta\lambda$ and media thickness T. Figure (4) is the graph corresponding to Table (4).

Table (4). Relationship between $\Delta\lambda$ and T

T (μm)	12	24	36	48	60	72
$\Delta\lambda$(nm)	20.8	16.2	13.4	9.4	7.5	4.0
$\Delta\lambda/\lambda$	0.039	0.032	0.026	0.019	0.013	0.0076

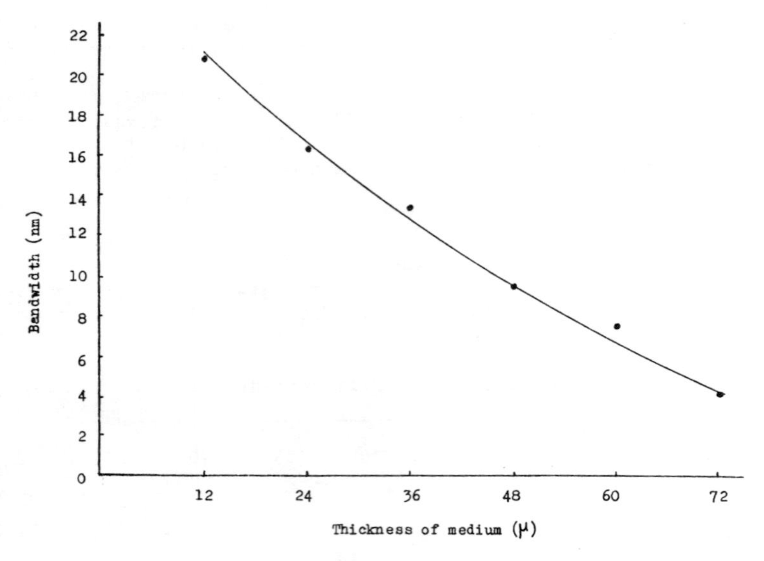

Figure 4. Relationship between $\Delta\lambda$ and T

Figure (5) shows the filtering property when i_0=0°, Figure (5)-A is the curve of T=24μm, Figure (5)-B is the curve of T=72μm.

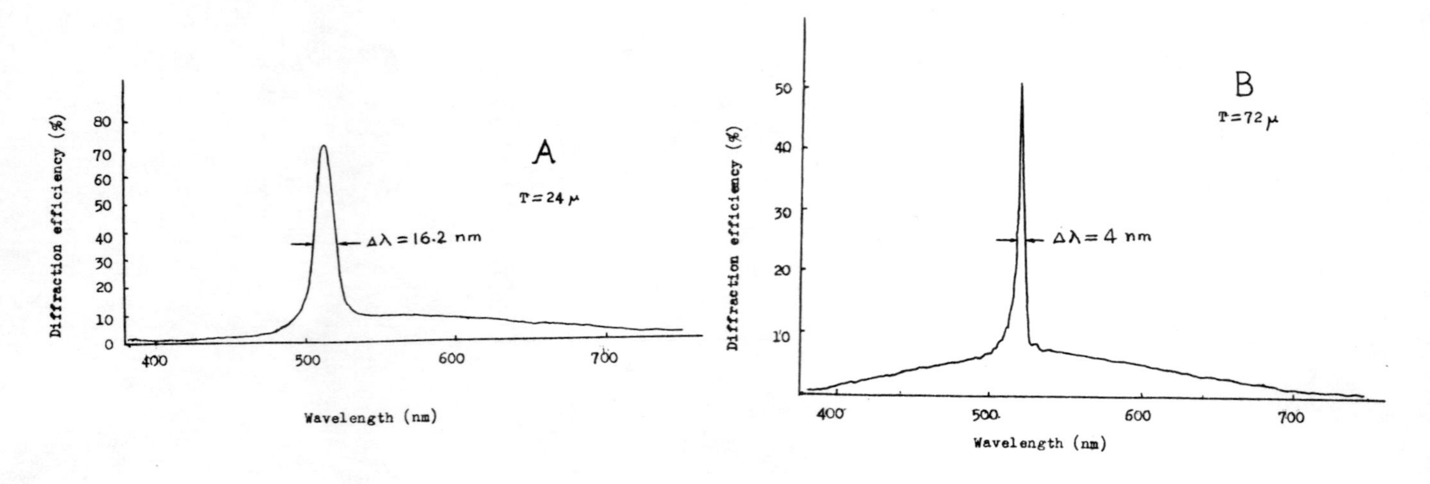

Figure (5). Filtering characteristic
(i_0=0°)

Figure (6) draws out the tendency in variation of bandwidth with the chage of angle. It can be seen that the bandwidth reached its minimum at a certain incident angle, but the variation of bandwidth is only slight.

3. Variation of diffraction efficiency with different thickness and angles

The relationship between η and T is given in Table (5), and the graph corresponding to Table (5) is given by Figure (7).

Table (5). Relationship between η and T

T (μm)	24	36	48	60	72	84
η	72%	76%	86%	77%	54%	34%

The variation of diffraction efficiency with the change of incident angle is shown in Figure (8). It shows that the diffraction efficiency reached its maximum at a certain incident angle.

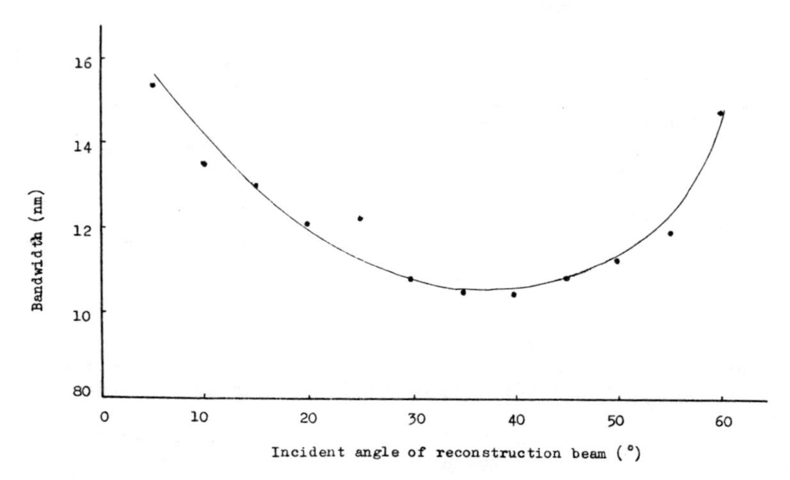

Figure 6. Relationship between $\Delta\lambda$ and i

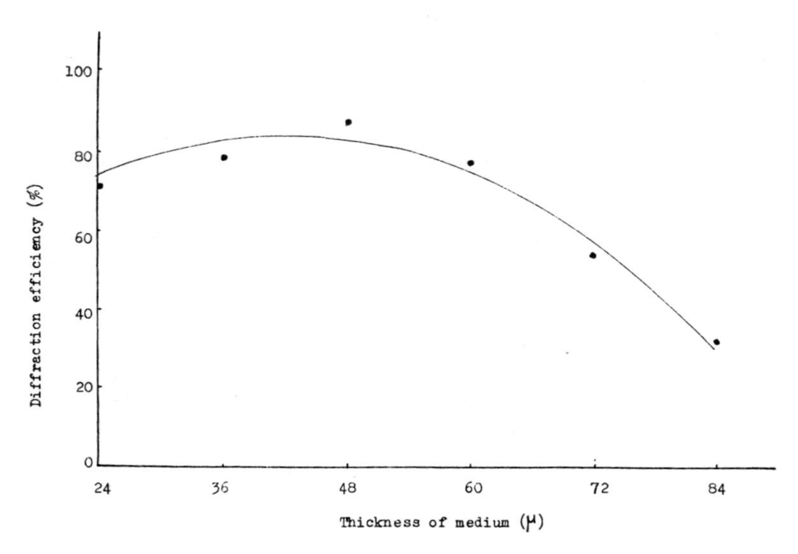

Figure 7. Relationship between η and T

Figure 8. Relationship between η and i
($i_0 = 40°$, $T=24\mu$)

Discussion

As could be seen from the above experimental results, the holographic reflection filters possess many outstanding features.

1. Excellent single peak character

From Figure (4) it can be seen that the holographic filters have excellent single peak character. In the measured region from 350nm to 750nm, only a single peak was seen, leaving the basal portion of the curve very smooth. As we know, the conventional multilayer dielectric interference filters usually have several peaks, when measures were taken to reduce the number of peaks, their efficiency would become lower.

2. Relationship between bandwidth and thickness of media

The curve in Figure (4) showed that the bandwidth narrowed down with the increase of media thickness. This is essentially in consistency with the formula

$$\Delta\lambda = C\,\frac{2nd^2\sin\theta}{\pi T}$$

which did not take into consideration the effect of absorption during construction. This means that absorption by medium does not show an important influence on bandwidth during construction of holographic filters.

In the present experiments, when the thickness of the medium is 72μm, the bandwidth at 521.8nm has reached 4nm, the relative bandwidth $\Delta\lambda/\lambda$ is only 0.0076, and the diffraction efficiency still reached 50%. These mean that the holographic filters can meet most of the requirments in many respects. It is postulated that the bandwidth could be narrowed down further to a certain extent by increasing the thickness of media if desired.

3. Relationship between diffraction efficiency and thickness of media

According to the formula

$$\eta = th^2\left(\frac{\pi\Delta nT}{2nd\sin^2\theta}\right)$$

where, the effect of absorption was not taken into consideration during construction, the diffraction efficiency will increase with the increase of thickness of media. But the results of the present experiments, as illustrated in Figure (7), did not follow this rule. When medium is considerably thin, the diffraction efficiency increased only slightly with the increase of media thickness, but when thickness increased beyond a certain value, the efficiency decreased rather than increased with further increase of media thickness. This means that the absorption by media during construction is a factor of practical importance in the construction of holographic filters, especially when the medium is considerably thick. Never the less, this point has seldom been discussed in discriptions relating the coupled wave theory.[2,4]

Now we are confronted with a problem in practice of construction of holographic filters that the bandwidth could be narrowed down only at the expense of sacrificing the efficiency. It is highly desirable that the problem could be solved, the further study on this topic are now undergoing in our laborotory.

4. Tunability

From Table (2) and (3), it is seen that the peak wavelength of diffraction beam can be perfectly controlled by incident angle of reconstruction beam in quite wide range of wavelength. Figure (3) shows that the variation of peak wavelength in accordance with different angles coincides very well with Bragg formula. In experiments, the maximum relative deviation of experimental data from theoretical value was less than 3%. When a beam of white light illuminates on the filter, a series of saperate peaks of wavelength can be obtained from different directions, if some special methods are used. This quality of tunability is very important characteristic of the holographic filters, and this quality the interference filters do not possess. It enables the holographic reflection filters to be applied in optical communication.

5. Large area and low cost

Large area narrow band filters can readily be constructed with holographic technique, in our experiments, the filters of 100mm diameter had been constructed with very low cost. It is extremely difficult to achieve this with interference filters.

Acknowledgements

The authors wish to thank Zhao Lin, of The Chinese Academy of Science, for the provision of the recording media. The authors also want to show the gratefulness to Prof. Sun Yinguan and Mr. Wang Ning for their helps.

References

1. S.S. Duncan, J.A. McQuoid, D.J. McCartney, "Tunable holographic filters in dichromated gelatin operating in the near infrared region", Vol.523, Proc SPIE Conf, pp. 196-202, 1985.
2. R.J. Collier, C.B. Burckhardt, L.H. Lin, ch. 9, "Optical Holography", ACADEMIC PRESS INC (USA), 1971.
3. Seven. Sjolider, "Bandwidth in dichromated gelatin holographic filters", OPTICA ACTA.(USA), Vol. 31, No. 9, pp. 1001-1012, 1984.
4. H.Kogelnik, "Coupled Wave Theory for Thick Hologram Gratings", Bell. Syst. Tech. J., 48, 2909, 1969.

Session 13

Holographic Recording Materials and Devices II

Chairmen
Charles S. Ih
University of Delaware, USA
Y. Ishii
Hokkaido University, Japan

A THERMOPLASTIC HOLOGRAPHIC CAMERA FOR INDUSTRIAL NON DESTRUCTIVE TESTING : DATA ACQUISITION AND IMAGE ANALYSIS

C. LIEGEOIS, P. MEYRUEIS
Ecole Nationale Supérieure de Physique de Strasbourg
7, rue de l'Université - 67000 Strasbourg - France

J. FONTAINE
Ecole Nationale Supérieure des Arts et Industries de Strasbourg
24, boulevard de la Victoire - 67000 Strasbourg - France

Abstract

We present a 3-D holographic inspection device for laboratory and industrial use. The photosensitive recording medium is a thermoplastic film with an effective aperture size of 70 x 50 mm. The physics of holographic recording on a thermoplastic film is well-known. This medium has numerous advantages over traditionnaly used holographic recording media : it is efficient, cheap and allows for dry, in-situ processing. However many technical problems had to be solved in order to develop an easy-to-use thermoplastic holographic camera. We will describe this camera and give some examples of its use for non destructive testing in laboratory and industrial environment. We will also point out the future possible developments of this device and the specific improvements we are expecting, in particular in the processing of the interference fringe pattern.

Introduction

Usually an optical medium requires chemical wet processing after recording in order to transform the information into a suitable form for readout. For some applications (holographic industrial interferometry, optical memories, optical information processing) this processing should be in-situ and quick.

The usual photosensitive materials used in holographic industrial interferometry are silver halide emulsions. These emulsions are very sensitive and have a high resolution but require after exposure a conventional photographic development procedure. This procedure implies :
- removing the plate from the optical set-up,
- development and fixing in chemical bathes,
- drying of the layer,
- repositioning in the initial set-up.

Such a procedure has following main drawbacks :
- swelling and shrinking of the photosensitive medium, which generates, at the reconstruction state, interferometric fringes due to layers thickness differential between recording and reconstruction,
- it is time consuming,
- needs a chemical laboratory,
- needs exact repositioning, in case of real time holography otherwise the object will be covered with interference fringes due to repositioning errors,
- it is expensive (silver halide plates or films + chemicals + chemical facility).

We propose here an alternative recording medium which allows for performances comparable with silver emulsions less all the above mentionned drawbacks.

I THE THERMOPLASTIC FILM

The photosensitive support of our camera is a thermoplast (PT 1000 S by KALLE-HOECHST Comp.) and concists in three layers mamely, the thermoplastic material coated on a photoconductive layer, both of them supported by a strong, transparent plastic carrier.

The recording of a hologram on such a material consists in four steps.
First, the film is sensitized by a uniform charge deposition, on both thermoplastic and photoconductive layer. This operation is performed by Corona effect.
Then the film is ready to be exposed to light. The light has no incidence on the thermoplastic's charge distribution but redistributes the charges in the photoconductive layer. So we have now two different charge distributions in the themoplastic and photoconductive layers, thus creating an electrostatic force pattern which is the exact replica of the incident light distribution.

Now that the hologram is stored in terms of an electrostatic force pattern, we have to transfer this information in a more practical form. This operation, consisting in fact in the development procedure, is done by heating the film. The film is heated in such a way that the thermoplast becomes soft and is deformed under the electrostatic force pattern.

The fourth and last step is the fixing. Once the film was heated beyond his softening point, and each point in the film has reached an equilibrium between the surface tension and the electrostatic forces, we have a surface modulated film which is again the exact replica of the recorded interference pattern. All what is needed now to freeze this surface modulation is to cool the film. This fixing operation is instantaneous, since the deformed film hardens when returned to ambiant temperature.

The most important and delicate operations in this process are the charge deposition on the film as well as the heating of the film. To obtain a homogeneous film sensitivity across the whole active area, the surface charge has to be perfectly uniform, because the sensitivity of the film depends at least quadratically on the electric field or the charge density. Usually, the surface has to be charged to several hundreds of volts, which is easiest obtained by CORONA discharge in air.

To perform the phase transition of softening, a certain amount of heat has to be delivered to the thermoplastic layer. This operation has to be done rather quickly in order to avoid surface charge losses due to conductivity of the thermoplastic film which rises drastically near the melting point. To avoid these losses, we heat for a short time the material far above the real melting temperature in order to pump the needed development heat in the material in a very short time.

The obtained hologram, is a phase hologram stored as thickness variations in a transparent media. The average diffraction efficiencies obtained are around 30 % (34 % theoretical value of a one-point hologram) and thus are the same as those obtained with classical silver halide emulsions. The time needed for the entire process is very short, since the sensitizing is less than 5 sec., and the heating 1,5 sec. The recording time depends on the available laser power, but is fairly short since the film sensitivity is high (0,8 uJ/cm^2).

II THE CAMERA

The camera concists of two units, the power supply and the sensitive or recording head with an aperture of 50 x 70 mm (Figs. 7 and 8). In the recording head, the thermoplastic film is stretched accross a carrier plate made of insulating material.

A glass plate with a vaccuum-deposited, transparent resistive layer on the surface is embedded in the carrier plate in such a way that the film is guided exactly along its coated surface. The resistive layer on the plate is equipped with contacts in the form of thicker, vaccuum-deposited conductive strips along two parallel edges of the plate. Two thick copper mesh strips are pressed by rubber cushions against these plate contacts. One of the plate contacts is grounded.

At sensitizing phase, a carriage for CORONA charging is slowly moved at a constant distance along the film surface. The films adhesion to the glass is provided by the static charges on the film.

For film transport, the carriage lifts the adhering film from the surface of the glass plate, and a motor winder transports the film whose total length is 27 m.

To ensure a perfectly uniform charge distribution on the film and thus a uniform sensitivity across the whole area, we use a conductive wire which is somewhat longer than the film width in order to avoid edge effects. The distance of the wire to the thermoplastic surface is optimized to allow for homogeneous charge distribution without any charge concentration in the central part of the film area and to prevent local breakdowns which would damage the transparent resistive layer on the glass.

For development, a current pulse is fed to the resistive layer on the glass substrate, which acts as a resistive heater and simultaneously as a resistive temperature sensor, allowing the temperature of the heater to be measured and the heating to be turned off at the appropriate moment.

III EXPERIMENTAL RESULTS

The results shown, were obtained with the thermoplastic camera installed in a holographic set-up. The tests chosen are : the transfer characteristics, the diffraction efficiency characteristics and the spatial frequency response. The recording and reconstruction wavelength, for all results presented, was the 514,5 nm line of an Argon laser.

Figure 1 shows the curve of brightness in the reconstructed image as a function of exposure for holograms of a diffuse object. We see that the exposure sensitivity of the film is high and is comparable with that of high resolution photographic emulsions with a maximum at 0,8 uJ/cm^2.

For the SNR measure, we used a white screen with a central opaque strip as an object. After recording, we scanned the reconstructed real image with a pinhole PM and computed the ratio of averaged intensities across the illuminated and the opaque part of the image. The result showed is obtained at a spatial frequency of 650 l/mm with a 14,5 kV corona voltage and a reference to signal beam ration of 1.

Figures 2 and 3 show the dependence of the brightness of the reconstructed image on the development temperature. We see, especially in Fig. 2, that the reproducibility in image brightness, depends strongly on the development heat. But Fig. 3 shows us clearly that the duration of the heating pulse has no significant influence on the brightness. A deviation of $\pm$ 0,5°C from the optimum development temperature results tipically in a decrease of 25 % brightness.

In Fig. 4, the brightness in the reconstructed image at a spatial frequency of 650 l/mm is shown as a function of the beam ration. The maximum brightness is not coincident with the highest SNR.

Fig. 5 shows the maximum diffraction efficiency obtained at 600 l/mm. The diffraction efficiency is the percentage of the incident light diffracted into the first order by the hologram.

As shown in Fig. 6, we obtained for a diffuse object a bandpass response for spatial frequencies with a maximum at 550 l/mm and a useful range from 350 l/mm to 900 l/mm. I_o^* (NORM) is the distribution of the brightness in the reconstructed image divided by the light distribution of the illuminated object. Short exposure times lead to additional suppression of higher frequencies.

IV INDUSTRIAL APPLICATIONS

Holographic interferometry is an optical non destructive testing method which becomes more and more widely used in different industrial fields. The main advantages of holographic techniques are to be global, requiring no contact with the object, they are applicable to any subject of any shape and any surface structure, the object beeing static or undergoing a dynamic process. The applications in industry are numerous and very diversified, we give hereafter some examples of applications where thermoplastic cameras are already in use :

- homogeneity defect detection and analysis on composite or honeycomb structures,
- fatigue and resistance control,
- Eigenmode detection and analysis of compressor and turbine blades, motor parts, transmission, reservoirs, tires...
- homogeneity control of assembled structures, pressure fits, rivetting, welding...
- fluid flow or thermal gradient inspection,
- detection of microcracks,
- cristal growth studies.

The most advantages of the thermoplastic camera are :

- very low cost/hologram,
- no need for chemicals for the development,
- no need for a separate development laboratory,
- in-situ development,
- high sensitivity,
- versatility, since the holographic camera is applicable as well for real time, time average as double exposure holography.

V AUTOMATIC FRINGE ANALYSIS METHOD

The efficiency of the thermoplastic camera can be very improved when coupled to an automatic fringe analysis system. Such a system is currently under development and will be on the market in the future together with the Micraudel Holodata Camera.

METHOD

The information stored in the thermoplastic camera is picked-up by a vidicon or CCD camera, then digitized and processed in a computer.

The first operation of such a system is to process the noise carried by the signal. This noise has different origins and thus each time requires the adapted correction procedure. The different types of noise have different sources and can be described as follow :

- Imaging noise :
  This noise is due to the data acquisition camera and has two origins :
  . imaging optic noise
  . electronic noise (vidicon tube or CCD)
  The reponse of both systems to an impulse is very different. For the first, the respon-curve is Gaussian (response is a function of the optical axis), as for the second we have a saturation phenomenon in amplitude essentially due to the electro-optical characteristics of the hardware.

- Digitization noise :
  Its a randomly varying noise as well versus time as versus space. It is a weak noise which can be a problem in the case of very low amplitude signals.

- Speckle noise :
  This noise is due to the coherence of the laser light which creates interferences with the light reflected from the objects surface. This noise is strongly decorelated, it varies a lot over small distances, and is mainly a function of the optical system's aperture.

- Emulsion noise :
  This type of noise depends on the type of photosensitive material used (ie : photographies of interferograms, conventional high resolution photographic plates, thermoplastic...). This noise can be very important in the case of low contrast interferograms.

- Desired shape of the extracted signal :
  There are of course several techniques which allow a SNR improvement on a signal extracted from a photographic emulsion. But in our case we have to develop a very specific technique, because here the only important signal is carried by the shape of the fringes. A binary image of the fringe would be enough to give us the desired information.
  The technique that we elaborate has two main goals :
  . precise shape extraction,
  . suppression of the binary noise.

- Instrumental decorelation :
  The problem of this filtering, is in our case quickly solved, because the instrumental characteristics are strictly positive. The modulation of a cosinusoïdal function by such a function does not modify the minimas of the curve and since our aim is not to extract the exact sinusoïdal signal but the minimas, we can ignore the instrumental characteristics. The treatment of the complete problem would require a large amount of machine computing time (FFT, SVD, etc...).

- Electronic noise :
  This noise as well in time as in space varies randomly. A weight average of this noise is weak, and spatially it can be considered as a centered Gaussian distribution (with an almost zero mean value).
  The filtering method consists in several readings of the signal. The signals amplitude will this way be risen and the noise, since it is decorelated, will be canceled by himself. After the readings, the values are divided by the number of readings. Fifteen readings are a good compromise between signal quality and computing time.

- Speckle noise :
  Speckle is a stationary noise, unless the laser cavity is unstable, its spatial variation is random, it varies a lot over small distances and it's importance depends mainly on the optical system's aperture.
  For the suppression of the speckle we use the fact that it is decorrelated : over small distances, in regard to the holographic support, the intensity of the speckle points can vary a lot, we can also say that the speckle noise is located in the high frequencies of the image ; so the idea is to make a low-pass filtering of the image spatial frequencies, or equivalently to integrate the image over 3 x 3 point blocks by convoluting the image matrix with masks, as for example :

$$M = \frac{8}{9} \begin{bmatrix} 1 & 1 & 1 \\ 1 & 1 & 1 \\ 1 & 1 & 1 \end{bmatrix}$$

- Elimination of the emulsion (or other support) noise :
  This noise is a problem in the case of a low contrast image. The way we treat this

problem can be resumed as follows :

1) Digitization of the picture (picture I1)
2) Spatial non recovering integration (blocks of 4 points) (picture I2)
3) I1 has better defined contours as I2, but the variance of the noise is stronger. The variance of the noise by point in I2 is the fourth of that in I1
4) I2 is redigitized over a uniform scale, so that the grey level quantification are spaced by 4 times the standard deviation of the noise. For a Gaussian noise, this leads to a quantization correct to 95 %
5) Scanning of I2 by blocks of 3 x 3 points. If the 8 points external to the block, have the same grey level value but that the center is different, the center is supposed to be wrong and his value is replaced by that of the eight meighbouring points.

The general procedure is in fact a comparison procedure, the image is recorded and compared to a reference image.

- Recalibration of the fringe pattern :
The two fringe levels (test and reference) are not necessarily idealy identical, they can be different. Since the reference pattern is stored, we can measure the mean distance between the fringes and then adjust by grey level calibration on the test image, the fringe distance of the second image to that of the first, so that the fringes shape comparison is now possible.

- Shape comparison :
At this point, we have to find a description of the shape, by a number or a syntax, so that it is independent of size, scale and shape orientation.
We have developed a concept which allows shape comparison (at a first step for closed curves) working with the theory of Freeman's chains.

- Number or chain shape :
Let C be a closed curve in a rectangle (a,b)
a and b will be respectively called minor and major axes of the shape.
For a given curve, the Freeman chain is obtained by going along the curve clockwise, in the constraints fixed by the discretization mesh. For an implementation of the Freeman number, independent of size and orientation we will code the circuit differently, by a chain that we call Freeman derivative, this number is obtained the same way as formerly, except that we do not code distances but the angles, so that any shape is characterized by a number in base 4. This number depends of course on the point at which we started coding ; from all the numbers obtained that way we take the smallest to characterize the shape.

- Comparison of two shapes :
The number we obtained with our technique varies in his width (we will say order) depending on the defined circuit mesh (net). So the more the order is large the more precisely the shape will be described. Also two shapes can be compared by observation of the order at which the derivatives of Freeman differ.

- Defect classification :
If our aim is to measure defects, we have also the possibility to determine if these defects are internal or external.
External defects are characterized by a fringe pattern disturbed at the first degree (1st degree derivative discontinuity, angular points...) and internal defects are characterized by a discontinuity of the fringes curvature. This can be determined by a specialized routine included in the last part of the analysis.

CONCLUSION

We exposed a complete automatic non-destructive testing system, with in the first part a sensitive and efficient industrial thermoplastic camera, already on the market. In the second part, we described a analysis software of the interference pattern obtained with the camera. This software is currently under development and will be proposed on the market with the Holodata Camera.

REFERENCES

1) C. LIEGEOIS et al, Thermoplastic film camera for holographic recording of extended objects, Applied Optics, Vol. 21, 2 209, 1982
2) R.K. ERF, Holographic non destructive testing, Academic Press, 1974
3) R.K. ERF, Speckle Metrology, Academic Press, 1978

FIGURES

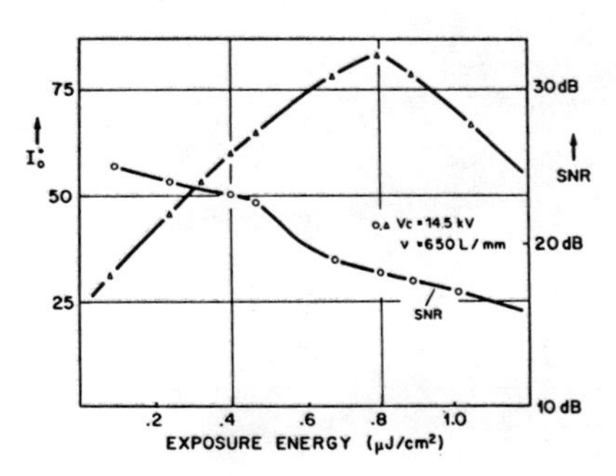

Fig. 1 : Curve of the brightness and SNR in the reconstructed image of thermoplastic holograms as a function of exposure energy

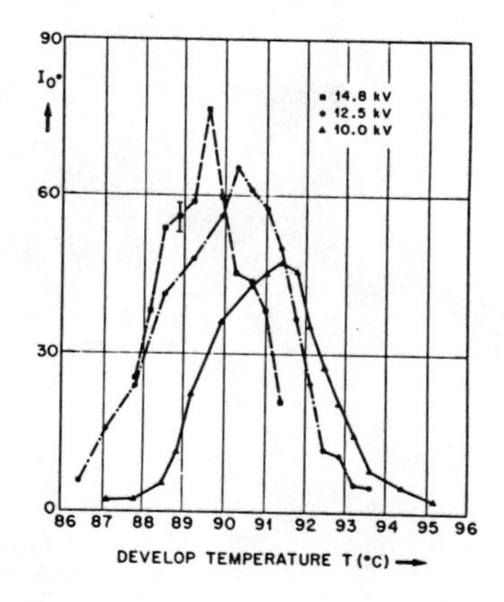

Fig. 2 : Experimental results showing the brightness in the reconstructed image vs development temperature for different corona voltages

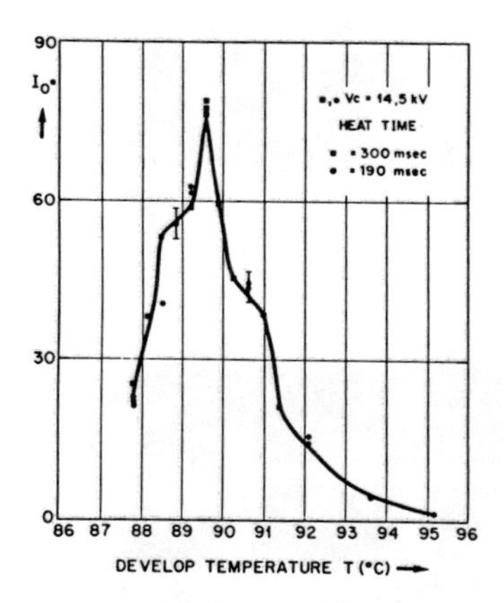

Fig. 3 : Representative curve for development temperature and different heating times

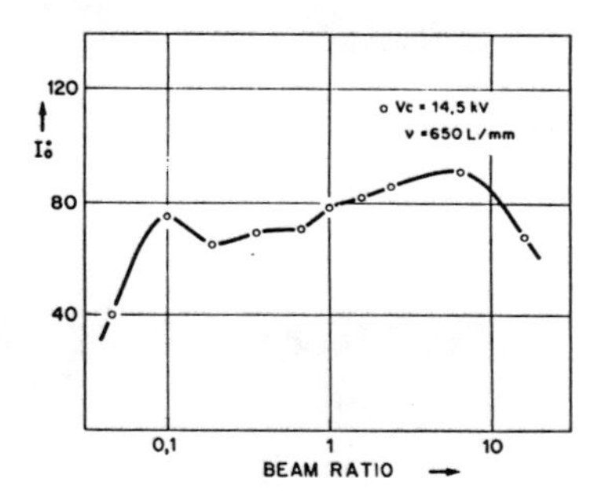

Fig. 4 : Brightness in the reconstructed image vs reference-to-signal beam ration

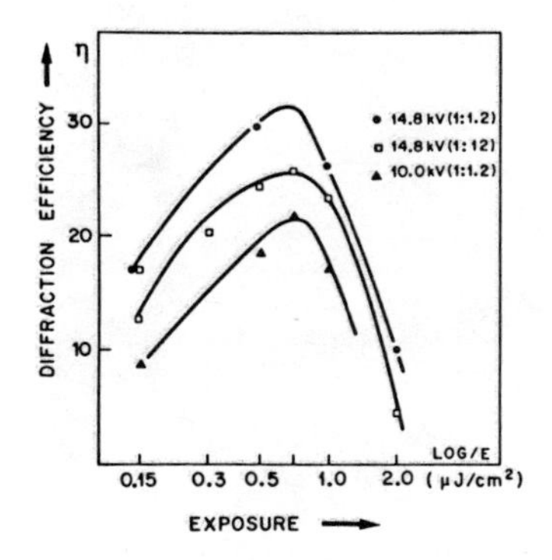

Fig. 5 : Dependence of diffraction efficiency on exposure

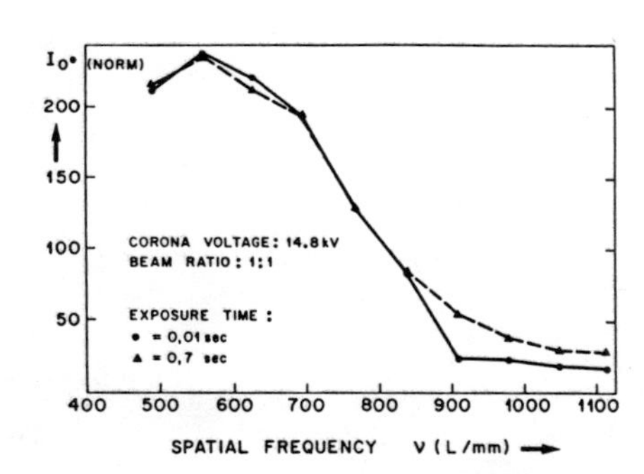

Fig. 6 : Dependence of brightness in the reconstructed image on spatial frequency with different exposures times

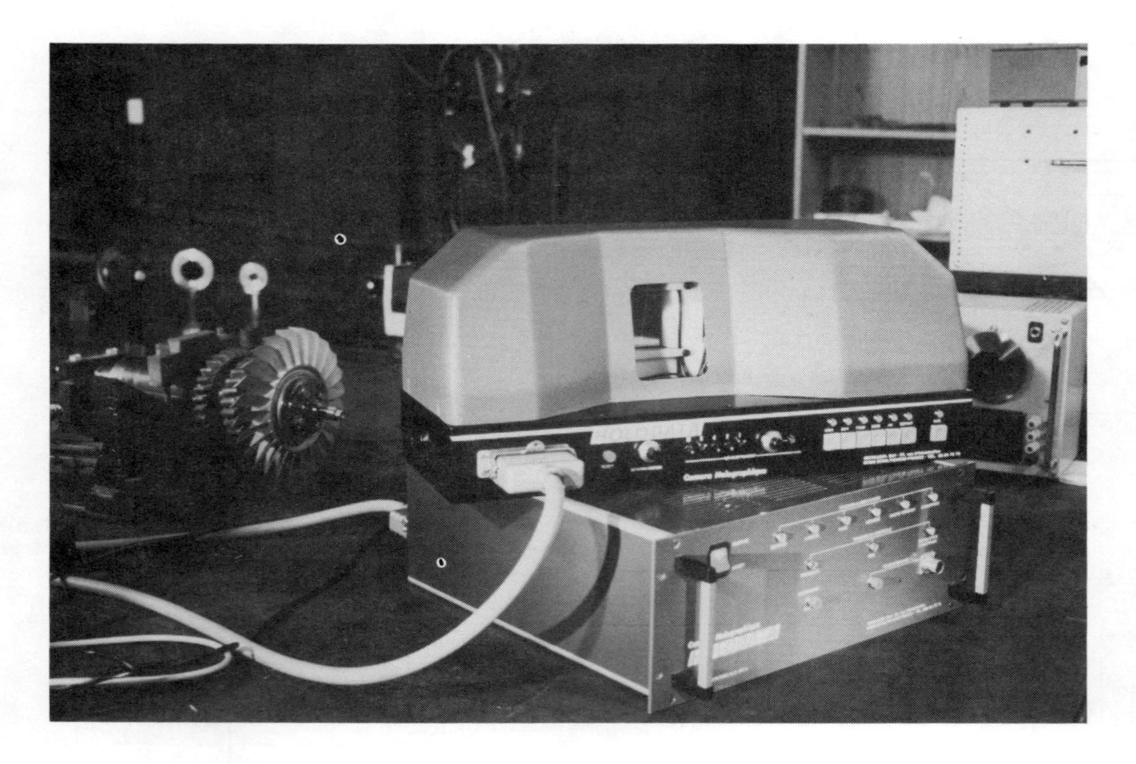

Fig. 7 : Holodata, thermoplastic camera : recording head and power supply

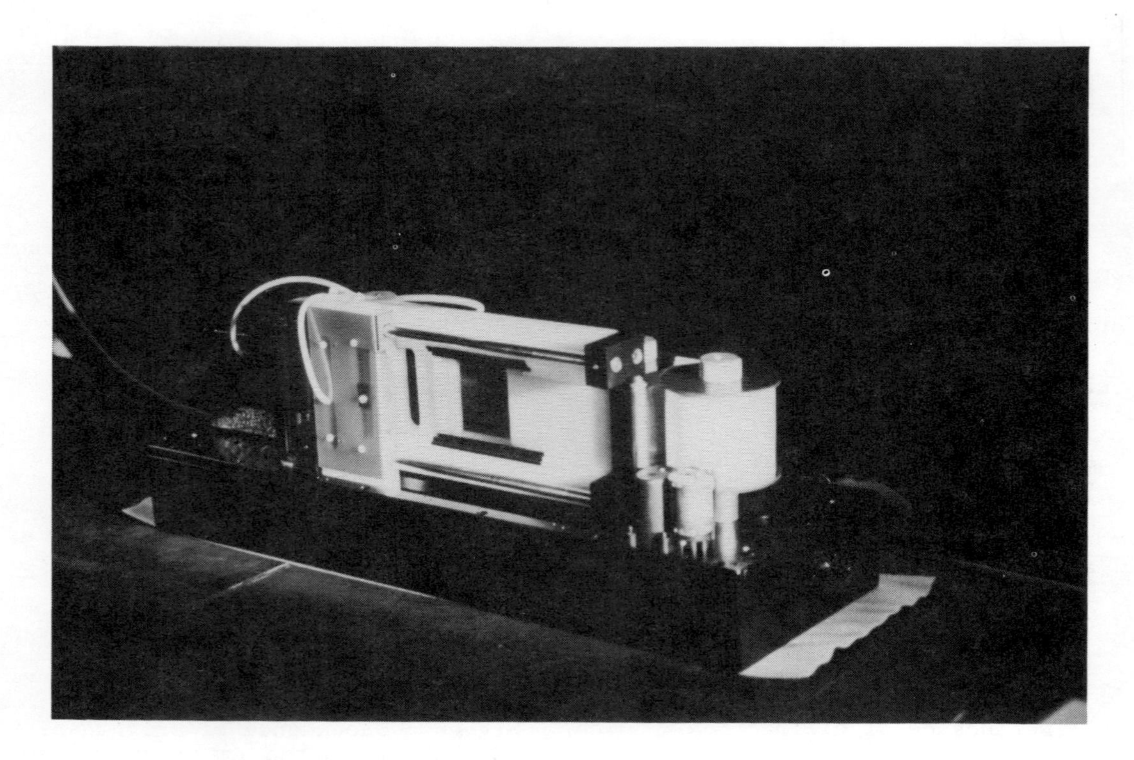

Fig. 8 : Holodata thermoplastic camera : uncovered recording head

HOLOGRAPHIC AQUACULTURE

Richard Ian
Elisabeth King

Advanced Environmental Research Group
100 Memorial Drive, Suite 8-2A
Cambridge, Massachusetts 02142 USA
(617) 864-4982

Abstract

Proposed is an exploratory study to verify the feasibility of an inexpensive microclimatecontrol system for both marine and freshwater pond and tank aquaculture, offering good control over water temperature, incident light flux, and bandwidth, combined with good energy efficiency.

The proposed control system utilizes some familiar components of passive solar design, together with a new holographic glazing system which is currently being developed by, and proprietary to Advanced Environmental Research Group (AERG).

The use of solar algae ponds and tanks to warm and purify water for fish and attached macroscopic marine algae culture is an ancient and effective technique, but limited seasonally and geographically by the availability of sunlight.

Holographic Diffracting Structures (HDSs) can be made which passively track, accept and/or reject sunlight from a wide range of altitude and azimuth angles, and redirect and distribute light energy as desired (either directly or indirectly over water surface in an enclosed, insulated structure), effectively increasing insolation values by accepting sunlight which would not otherwise enter the structure.

Introduction

Attached marine algae (seaweeds) represent a major and growing sector of food and commercial raw material production potential world-wide. In particular, new uses are being discovered for phycocolloids, and phycocolloid-producing species are assuming increasing economic importance. Future growth potential of the industry will depend on optimized microclimate control.

Well-understood passive solar techniques exist which are ideally suited to large, low-temperature applications such as warming water for fish or seaweed culture. The introduction of sunlight into an insulated space, its absorption by a dark surface, and the storage of the extra heat for future use in a thermal mass of some sort (often water) are familiar features of passive solar buildings.

Marine life has been cultivated for centuries in tanks or ponds where algae growth on the walls serves to absorb heat from sunlight, to oxygenate the water, to provide nutrients and to remove the metabolic wastes produced by the fish. Water from such life-intensive tanks is suitable for growing plants hydroponically, and the plants also benefit from the warmth produced by the solar-heated water.

An opportunity exists to combine and enhance these well-understood technologies with a new technique, proprietary to AERG, for passively tracking sunlight holographically, thereby making sunlight available for more hours of the day and at more northerly latitudes, effectively extending the geographical and seasonal range for practical aquaculture.

Technical objectives

The objective is to develop a simple and reliable non-mechanical, low cost holographic optics system which will gather sunlight from wide angles of altitude and azimuth to provide adequate warmth and controlled illumination for both fresh and salt water aquaculture, with good temperature control and sufficient carryover through nighttime and cloudy weather conditions.

Operation of the system

Sunlight enters through a conventional glazing which carries a film incorporating holographic diffractive structures which are capable of collecting sunlight from varying angles of altitude and azimuth and redirecting it away from its normal path to another desired location, in this case, the surface of a culture pond or tank. The sunlight penetrating the water produces a warming effect due to the energy which is absorbed by the (dark-colored and hence heat-absorbent) surfaces of the tank and/or by the mass of algae itself.

After a certain period of exposure to sunlight, the tank water will rise above the design temperature, and at this point a thermally-switched pump will draw cool water from a storage reservoir which is cooled by contact with groundwater temperatures (see diagram, Fig. 1). The cooling water will be diffused into the culture tank. The culture tank will overfill and warm water will be displaced and overflow (through a UV filter) into the thermal storage reservoir, where it will stratify at the top.

If the sun clouds over or when night falls, the culture tank is kept warm by thermal conduction from the large mass of warm water stratified below. This mass might have to be large enough to carry the system over many cloudy days, in some locations, if backup heat is to be avoided. The carryover capacity will be determined by the size and geometry of the thermal storage reservoir.

The UV filter is necessary to prevent the escape of living organisms from the culture tank into the reservoir where they could create problems. For example, if the water in the reservoir is not sterile, when the crop is harvested and the bottom of the culture tank is exposed to full sunlight, the water under the culture tank floor will rise in temperature. This could cause a sudden uncontrolled growth of organisms which, after using up all of the available oxygen will die, creating a severe maintenance problem.

The obvious advantage of using recycled water (instead of groundwater) for cooling is that the PH and mineral content of the water will be more easily controlled, and there is no hazard from pollutants, such as dioxin, benzene, etc. Toxic waste pollution in groundwater supplies is an increasingly important problem all over the world.

When the sun shines again, the temperature of the culture tank will again rise to the point at which it will trigger the sensor which controls the cooling-water pump, and the whole cycle will repeat.

HDSs can be engineered to control intensity and color of light as well as light distribution. Investigations in the study of the effects of light and of photoperiodicity show that growth can be stimulated by high light intensities, whether natural or artificial, incandescent or flourescent, periodic or continuous.

Color of light can also affect growth. Light requirements are intimately linked to water temperatures and vary from one species to the next, for example, Undaria gametophytes are more tolerant of high light intensities than gametophytes of Laminaria and can withstand higher intensity in the culture shed than is suitable for Porphyra. Flourescent light tubes used in aquaculture have a spectrum matching the light-absorbing capacity of chlorophyll and giving a faintly violet or pinkish color.

Construction of the HDSs

The glazing material itself does its sunlight-gathering and redirection with a complex structure of diffracting elements working together, intercepting light from different sun angles at different times of day throughout the seasons of the year. The diffracting structures are created by the interference of two beams of laser light recorded photographically in a photsensitve material. The result is a coded convoluted record of solar angles and paths. In order to translate this coded pattern into an affordable material for practical use, it must first be translated into a relief structure which can then be embossed in a roller-to-roller process at a rate of thousands of feet per second. The anticipated cost will be not much greater than that of treated window films which are presently on the market, although marketed for a different purpose, i.e., heat rejection.

The engineering of the HDS glazing will be site-specific in that sun angles and intensities are specific to certain latitudes, and also in that the term "aquaculture" can encompass the culture of a wide variety of organisms, both animal and vegetable in nature, and each of these organisms has its own particular growth requirements. It is anticipated, however, that this technology allows some fairly broad tolerances, so that the glazing can be somewhat generic in nature, a relatively small number of designs being able to cover a relatively large number of applications.

Conclusion

Attached marine algae represent a major and growing sector of food and commercial raw material production potential world-wide. In particular, new uses are being discovered for phycocolloids, and phycocolloid-producing species are assuming increasing economic importance. Closed aquaculture systems are a potential means of restoring wasteland to productivity and would be viable as a cottage industry in any climate with enough sun, strengthening the local source of high-quality food.

Future growth potential of the industry will depend on optimized microclimate control, including thermal control of the growing medium. Heat storage which is optimized using holographic techniques will have many other commercial applications.

The system we envision could provide a high-quality source of valuable nutritional and industrial raw materials locally, wherever they may be required, and create economic benefits to the economy as a whole.

Acknowledgement

The information presented in this paper has evolved out of work done by, among others, Sally Weber and Don Thornton at AERG together with Professor Hendrik Gerritsen and his students at Brown University, and Professor Stephen Benton and his students at MIT. Funding for the initial study of the use of HDSs for sunlight control was provided by the U.S. Department of Energy. Information on the marine algae was contributed particularly by Jean Higgins, U. of California at Santa Cruz.

References

1. Advanced Environmental Research Group, Holographic Diffractive Structures For Daylighting, Final Report FY85, DE-AC03-84SF15292, U.S. Department of Energy, (October 1985)

2. Ffrench, R.A., Rhodymenia palmata: An Appraisal of the Dulse Industry, Laboratory Technical Report, Atlantic Regional Laboratory, National Research Council Canada, (1974)

3. Ian, R.; King, E.; Weber, S., Holographic Diffractive Structures for Daylighting: Direct Illumination, American Solar Energy Society, Proceedings of the 10th National Passive Solar Conference, Raleigh, North Carolina, (1985)

4. MacFarlane, C.I., The Cultivation of Seaweeds in Japan, and its Possible Application in the Atlantic Provinces of Canada, Project Report No. 20, Nova Scotia Research Foundation, Halifax, Canada, (1968)

5. Mathieson, A.C., Seaweed Aquaculture, MFR paper 1111, Marine Fisheries Review, Vol. 37, No. 1, January (1975)

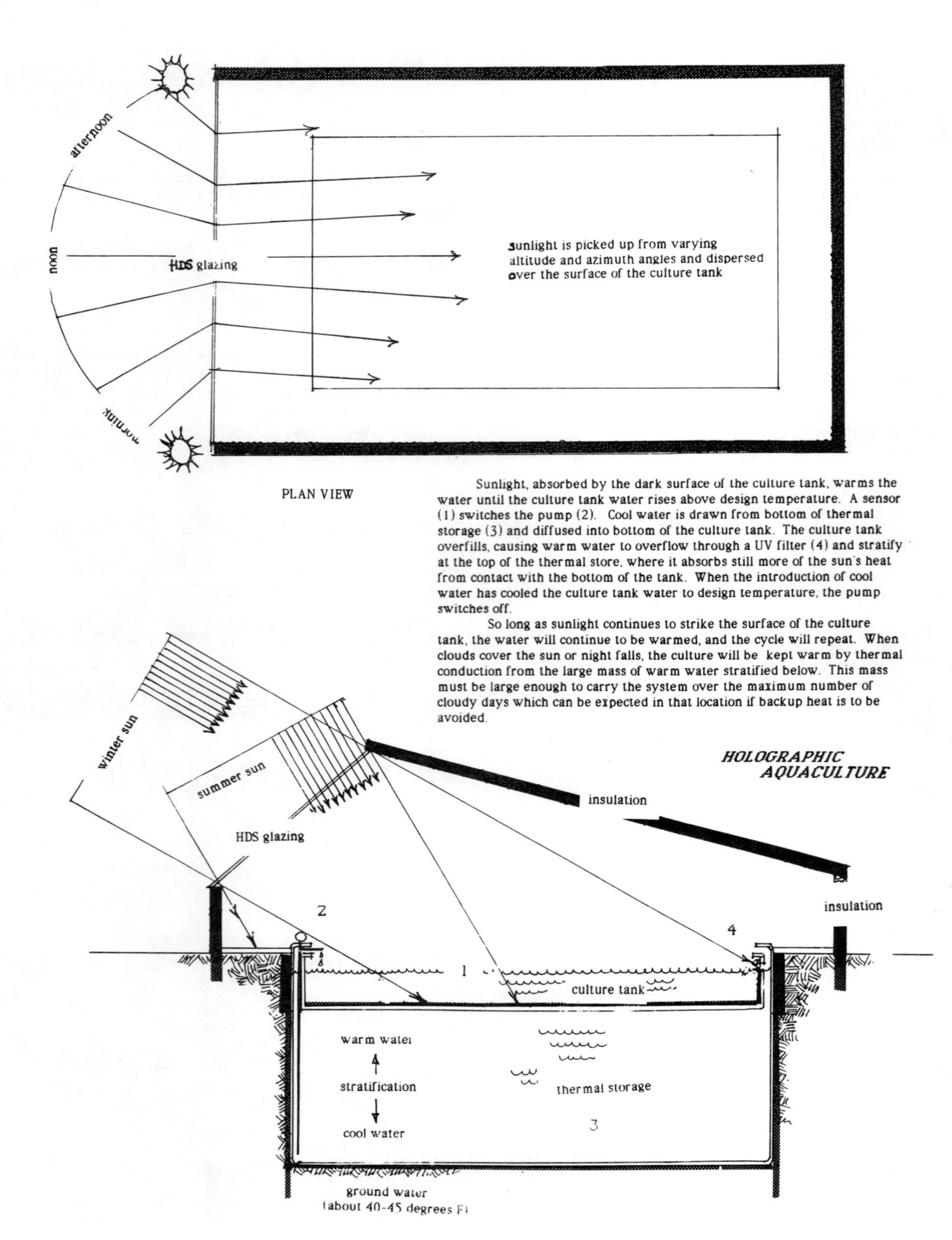
afternoon
noon
morning
HDS glazing
sunlight is picked up from varying altitude and azimuth angles and dispersed over the surface of the culture tank
PLAN VIEW
Sunlight, absorbed by the dark surface of the culture tank, warms the water until the culture tank water rises above design temperature. A sensor (1) switches the pump (2). Cool water is drawn from bottom of thermal storage (3) and diffused into bottom of the culture tank. The culture tank overfills, causing warm water to overflow through a UV filter (4) and stratify at the top of the thermal store, where it absorbs still more of the sun's heat from contact with the bottom of the tank. When the introduction of cool water has cooled the culture tank water to design temperature, the pump switches off.
So long as sunlight continues to strike the surface of the culture tank, the water will continue to be warmed, and the cycle will repeat. When clouds cover the sun or night falls, the culture will be kept warm by thermal conduction from the large mass of warm water stratified below. This mass must be large enough to carry the system over the maximum number of cloudy days which can be expected in that location if backup heat is to be avoided.
HOLOGRAPHIC AQUACULTURE
winter sun
summer sun
HDS glazing
insulation
insulation
2
4
1
culture tank
warm water
stratification
cool water
thermal storage
3
ground water (about 40-45 degrees F)
CROSS SECTION

Holographic Storage Properties of Electrooptic Crystals

E. Krätzig and R.A. Rupp

Fachbereich Physik der Universität Osnabrück, D-4500 Osnabrück, FRG

Abstract

The storage of volume phase holograms yields extremely large recording densities. The physical processes involved and the most important storage properties (sensitivity, density, storage time and hologram stabilization) are discussed for various electrooptic crystals.

Introduction

Holographic storage demonstrates the enormous capabilities of optical methods. Particularly the whole volume of a light-sensitive material - not only the surface - may be utilized. This possibility yields extremely large recording densities: Experimentally, values up to 10^{10} bit/cm^3 have been obtained in electrooptic crystals.

During the recording process laser light passes a transparent pattern containing the information to be stored. This object beam is superimposed by a coherent reference beam in the storage material. By interference an intensity distribution is produced which is recorded as a hologram. During the readout process the storage material is illuminated by the reference beam only. Part of the light is diffracted thus permitting the reconstruction of the original pattern.

If the hologram is recorded in the whole volume of a storage material, reconstruction is possible under a certain angle only. If this Bragg-angle is changed by a small amount, e.g. a tenth of a degree, the reconstructed pattern vanishes and a further hologram may be recorded. By these means several hundred holograms have been superimposed in the same volume under different angles.

This method is of practical interest in the case of phase holograms only yielding large readout efficiencies. Materials are required which change the index of refraction under the influence of light.

Large effects of this kind have been found[1] for various electrooptic crystals. The principle of this photorefractive process is schematically illustrated in Fig. 1. Interfering light beams generate dark and bright regions in an electrooptic crystal. When light of

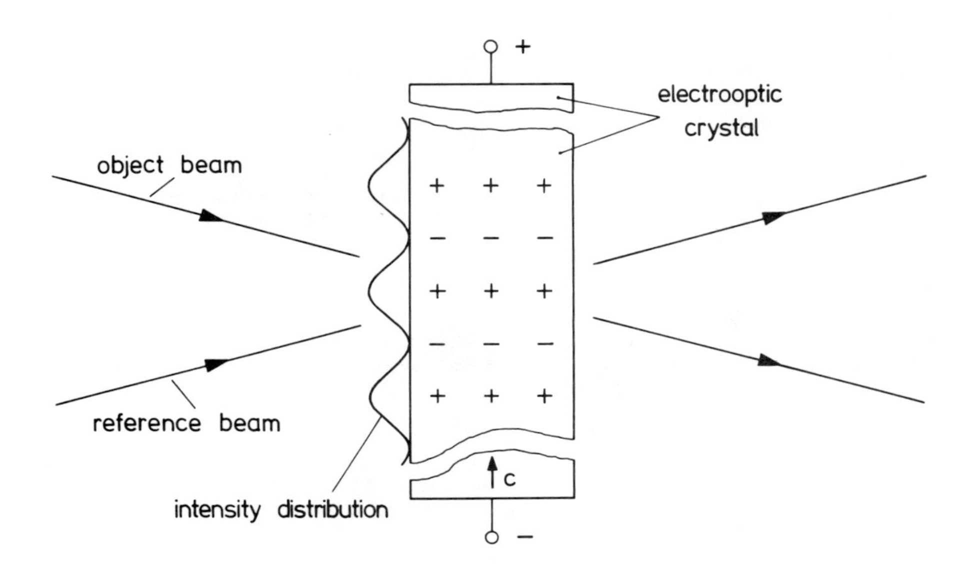

Figure 1: Light-induced refractive index changes in electrooptic crystals

suitable wavelength is chosen, charge carriers - electrons or holes - are excited in the bright regions and become mobile. The charge carriers migrate in the lattice and are subsequently trapped at new sites. By these means electrical space charge fields are set up which give rise to a modulation of the refractive index via the electrooptic effect. Light-induced index changes up to 10^{-3} are obtained. The trapped charge can be released and the field erased by uniform illumination or by heating.

In this contribution we investigate photorefractive processes in various electrooptic crystals. Because an optimization of the material properties for holographic storage requires the knowledge of the microscopic processes involved we also describe the light-induced charge transport which has been intensively studied for $LiNbO_3$ and $LiTaO_3$. Finally we discuss in detail the results obtained so far for the holographic storage properties.

Light-induced charge transport

Many photorefractive studies have been performed with $LiNbO_3$ and $LiTaO_3$ crystals. In these cases transition metal dopants have crucial influences on the photorefractive properties. Of special interest are Fe impurities. It is known from Mössbauer, electron spin resonance and optical absorption measurements that only Fe^{2+} and Fe^{3+} centers exist in these crystals.[2] Suitable thermal treatments - annealing in oxygen atmosphere or in an atmosphere with low oxygen partical pressure, respectively - allow a control of the Fe^{2+}/Fe^{3+} ratio in a wide range.

The migration of the electrical charge under the influence of light is determined by three contributions: By a new photovoltaic effect, by the drift in electrical fields and by diffusion.

The photovoltaic effect[3] is based on asymmetric lattice sites of the impurity centers supplying the charge carriers and cannot be explained by conventional photovoltaic processes in semiconductor crystals containing macroscopic inhomogeneities or by Dember voltages. Homogeneous illumination of single domain $LiNbO_3$ or $LiTaO_3$ crystals yields in a short-circuited configuration steady-state currents and in an open circuit steady-state voltages and fields up to 10^5 V/cm. In first investigations a photovoltaic current has been observed along the polar c-axis only. Further experiments have revealed the tensor properties of the photovoltaic effect.[4]

The drift in electrical fields is mainly determined by the photoconductivity σ_p which is large compared to the darkconductivity σ_d in most cases. Diffusion is of minor importance compared to photovoltaic and drift effects.

Taking into account these contributions the current density j along the c-axis can be approximated:

$$j(z,t) = \kappa_o \alpha I(z,t) + (\sigma_p(z,t)+\sigma_d)E(z,t) + eD\,\frac{\partial N(z,t)}{\partial z}\,.$$

Here κ_o is the photovoltaic constant, α the absorption coefficient, I the light intensity, e the electron charge, D the diffusion coefficient and N the concentration of optically excited charge carriers. The electrical field E may contain contributions of an externally applied field E_{ext} and of an internal space charge field E_i generated by light-induced charge transport.

For Fe-doped $LiNbO_3$ and $LiTaO_3$ crystals the electronic transitions determining the charge transport have been investigated by various methods.[5,6] The results clearly demonstrate the dominant influence of the Fe^{2+} and Fe^{3+} centers, though the Fe concentrations are one to two orders of magnitude smaller than deviations from the stoichiometric composition of the crystals.

The photovoltaic current is proportional to the concentration $c_{Fe^{2+}}$ of Fe^{2+} ions[5] and results from an excitation of electrons with preferred momentum. When the electrons are scattered they lose their directional properties and contribute afterwards to drift and diffusion only.

The photoconductivity is determined by the generation rate and the lifetime of excited carriers. The generation rate is proportional to $c_{Fe^{2+}}$, the concentration of filled traps. The lifetime on the other hand is inversely proportional to $c_{Fe^{3+}}$, the concentration of empty traps. For this reason we expect a relation $\sigma_p \sim c_{Fe^{2+}} \cdot (c_{Fe^{3+}})^{-1}$ to be valid. This simple model is confirmed experimentally, Fig. 2 shows the results for $LiNbO_3$:Fe. In the near UV region additional contributions to the photoconductivity occur due to electronic excitations from the valence band and hole migration. The participation of the holes in the light-induced charge transport has been confirmed by a new holographic method.[6]

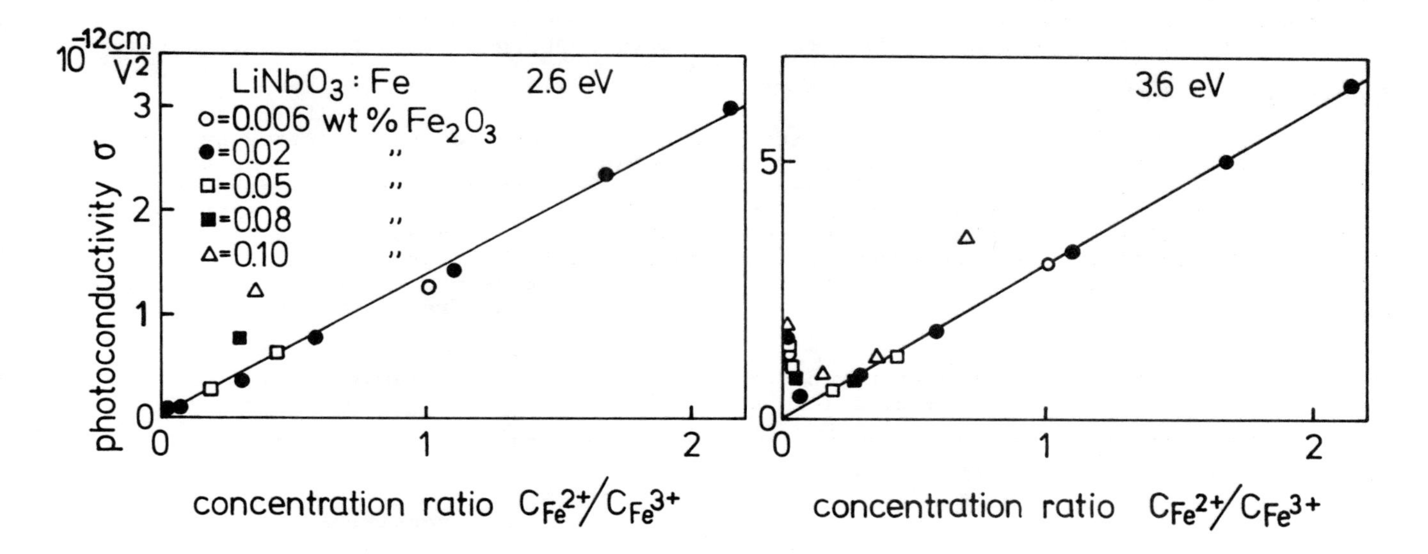

Figure 2: Concentration dependence of the photoconductivity

A schematic illustration of the light-induced charge transport in $LiNbO_3$:Fe is given in Fig. 3. Space charge fields may be generated by electron or hole processes. In both cases similar photorefractive properties are obtained.

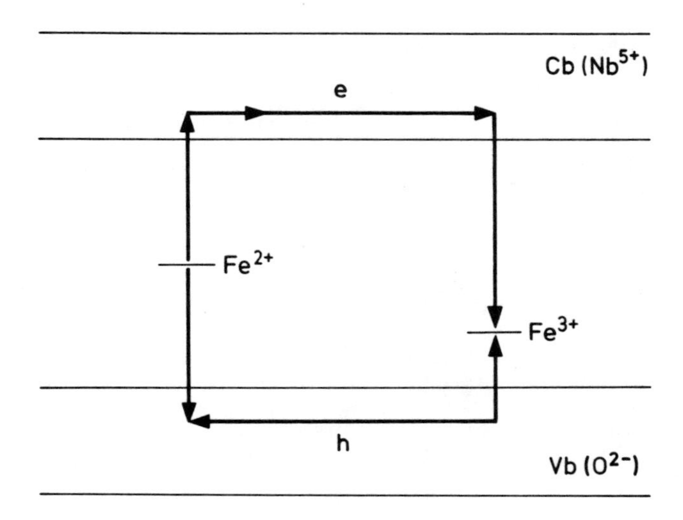

Figure 3: Light-induced charge transport in $LiNbO_3$

Besides the standard examples $LiNbO_3$ and $LiTaO_3$ a large number of photorefractive crystals has been studied. Of particular interest are the ferroelectric perovskites $KNbO_3$, $KTa_{1-x}Nb_xO_3$ (KTN) and $BaTiO_3$, the tungsten-bronze-type crystals $Sr_{1-x}Ba_xNb_2O_6$ (SBN) and $Ba_2NaNb_5O_{15}$ and the non ferroelectric sillénites $Bi_{12}SiO_{20}$ (BSO) and $Bi_{12}GeO_{20}$ (BGO). Several centers have been identified, but the influence on the light-induced charge transport and the involvement in photorefractive processes remain unclear in many cases.

Storage properties

The generation of the space charge fields is described by the Maxwell equations. A system of coupled partial differential equations has to be solved which can be done numerically only. Crude approximations yield an exponential increase and decay of the field amplitude with time[2]. The calculated saturation value for the field amplitude and the time constant $\gamma = \varepsilon\varepsilon_o/(\sigma_p+\sigma_d)$ for recording and erasure are confirmed by the measurements. For this reason the approximations may be utilized for an estimation of the storage properties from measured material constants. - The refractive index change Δn results from the linear electrooptic effect: $\Delta n = -0.5\, n^3 r E$. Here r is the corresponding electrooptic coefficient.

Sensitivity

The holographic sensitivity of electrooptic crystals can be characterized by the energy density W(1 %) for recording an elementary volume phase holograms with 1 % readout efficiency. For practical reasons - small lasers, short recording times - this energy density has to be as small as possible. A theoretical estimate of this quantity yields in most favourable cases attractive values between 10 and 100 $\mu J/cm^2$ depending on resolution. The decisive quantity is the mean free path of the charge carriers after excitation; large mean free paths yield small recording densities. Table 1 summarizes the experimental results for various electrooptic crystals. For $LiNbO_3$:Fe crystals mean free paths of the order of

Crystal	Recording energy density for η = 1 % (measure for sensitivity) mJ/cm^2	Remarks
$LiNbO_3$:Fe [7,8]	300 (λ = 488 nm) 200 (λ = 351 nm, E_{ext} = 15 kV/cm)	
$LiTaO_3$:Fe [9]	11 (λ = 351 nm, E_{ext} = 15 kV/cm)	sensitivity insufficient
$BaTiO_3$ [10]	1 - 10 (λ = 458 nm, E_{ext} = 10 kV/cm)	
$Sr_xBa_{1-x}Nb_2O_6$:Ce [11]	1.5 (λ = 488 nm)	storage time insufficient
$Bi_{12}SiO_{20}$ [12]	0.3 (λ = 514 nm)	
$KTa_{1-x}Nb_xO_3$ [13,14]	0.05 - 0.1 (λ = 350 nm, E_{ext} = 10 kV/cm)	crystal quality insufficient

Table 1: Recording energy density of various electrooptic crystals

1 Å only have been measured leading to rather poor sensitivity values.[7,8] The isomorphous crystal $LiTaO_3$:Fe exhibits mean free paths up to 20 Å and recording energy densities of 11 mJ/cm^2.[9] Relatively good values have been obtained for $BaTiO_3$[10], SBN[11] and BSO[12]. Most sensitive are KTN crystals.[13,14] Here the theoretical limits are nearly reached. - However, in all cases difficulties arise when light of the most attractive near IR region is used.

Storage density

A limit for the storage density is obtained when one bit is contained in a cube of light wavelength. By these means storage densities up to 10^{14} bit/cm^3 have been estimated.

For practical applications, however, the following problems have to be solved: Only a limited number of holograms can be superimposed in the same volume. Three main arguments have to be taken into account:
- The angular selectivity of the holograms depends on geometrical conditions.
- During the recording process holograms previously written are partially erased.
- Refractive index changes resulting from already stored holograms influence object and reference beam of the following holograms.

Difficulties resulting from the last two points may be largely avoided by recording at elevated temperatures thus utilizing the thermal fixing technique which will be described later. By these means 500 holograms have been superimposed in the same volume, each hologram with a readout efficiency larger than 2.5 %.[15]

Besides the number of stored holograms the bit density of a single hologram is of interest. The storage of pictures containing many details requires rather perfect crystals. Already small inhomogeneities produce undesired straylight. By interference with the original beam holographic scattering patterns[16] are generated which diminish the image quality and the resolution considerably.

$LiNbO_3$ and $LiTaO_3$ crystals of good optical quality have been grown. For this reason these crystals have been chosen for investigations of the storage capacity.[7,17] A page of a telephone book containing about $5 \cdot 10^6$ bit has been stored in an area of 3 x 3 mm^2 of a 5 mm thick crystal with a signal to noise ratio of 10:1.[17] More than hundred pages have been superimposed at the same volume. This yields an experimental value of the storage density of about 10^{10} bit/cm^3. - For other crystals numbers for the storage density are not yet available at this point of time.

Storage time

A recorded refractive index pattern relaxes with a time constant $\varepsilon\varepsilon_0/(\sigma_p + \sigma_d)$. Without illumination the darkconductivity determines the relaxation. Large dark storage times of about 10 years have been extrapolated from the decay of volume phase gratings in $LiTaO_3$:Fe.[9] Doped $LiNbO_3$ and KTN crystals still yield values of the order of one year. In the case of $BaTiO_3$, SBN or BSO only dark storage times of about one day or smaller have been measured.

Hologram stabilization versus the readout light

During the readout process a hologram is illuminated homogeneously and the stored information is partially erased. For this reason methods are required to stabilize the holograms versus the readout light and to avoid undesired erasure.

Unfortunately a simple method cannot be applied to volume phase holograms: Recording with e.g. blue light with high photon energy and readout with e.g. red light, whose photon energy is too small to generate mobile charge carriers. In this case the image quality is deteriorated considerably, because the Bragg-condition cannot be fulfilled for a hologram containing various spatial frequencies.

A partial solution for readout without erasure offers the thermal fixing technique.[15] $LiNbO_3$:Fe or $LiTaO_3$:Fe crystals are heated to temperatures between 100 and 200°C during hologram recording or afterwards. Mobile protons migrate in the crystals[18] and neutralize the space charge fields. After cooling down the crystal to room temperature further illumination causes a redistribution of the electrons due to the photovoltaic effect and the generation of an ionic pattern.

Thermally fixed holograms cannot be erased optically but only by heating to still higher temperatures. To preserve the possibility of desired optical erasure the use of two-photon excitations for hologram recording has been proposed.[13] Then readout without erasure is possible with the help of one-photon excitations. The Bragg-condition is fulfilled because the wavelength remains unchanged, but the energy of one photon is not sufficient to

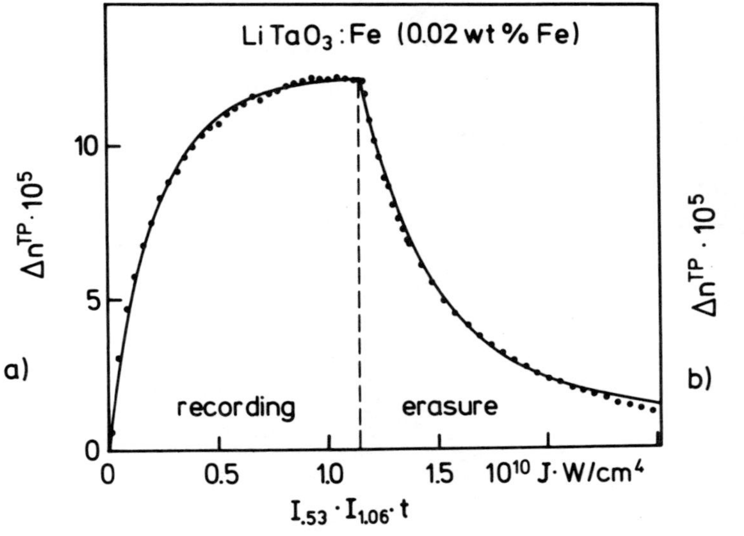

Figure 4a: A two-photon record-erase cycle: The refractive index amplitude Δn^{TP} is plotted versus the product of the intensities of green and IR light and time $I_{.53} \cdot I_{1.06} \cdot t$.

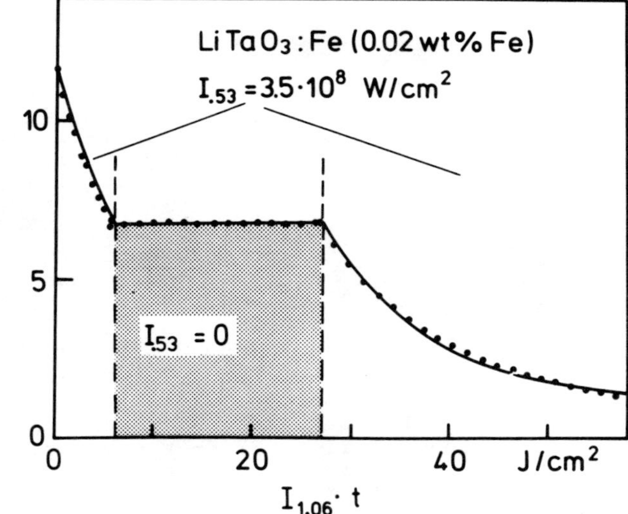

Figure 4b: The same experiment as in a), but now Δn^{TP} is plotted versus the product of IR intensity and time only. So it is possible to show that without a green beam ($I_{.053}=0$; hatched region) no erasure occurs.

generate mobile charge carriers. - Readout without erasure of holograms recorded by two-photon processes has been demonstrated for $LiNbO_3$:Fe[19] and $LiNbO_3$:Cr[20] crystals. Fig. 4 shows two-photon record-erase cycles for $LiTaO_3$:Fe.

Conclusions

In doped $LiNbO_3$ and $LiTaO_3$ crystals photorefractive processes are satisfactorily understood. Furthermore the superposition of several hundred holograms in the same volume under different angles has been successfully demonstrated and storage densities up to 10^{10} bit/cm^3 have been obtained. However, the sensitivity values of these crystals remain inadequate. Other materials offer more promising sensitivities, but in all cases difficulties arise when light of the most attractive IR region is used. In addition further unsolved problems exist with respect to capacity, resolution, storage time and hologram stabilization. For future investigations mixed crystals like KTN are of particular interest. However, an improvement of crystal quality and a better understanding of the physical processes involved is necessary to obtain more reliable informations on the limits of these materials.

Acknowledgement

This work has been performed within the program of the Sonderforschungsbereich 225 'Oxide Crystals for Electro- and Magnetooptic Applications' supported by the Deutsche Forschungsgemeinschaft.

References

1. Ashkin, A., Boyd, G.D., Dziedzic,J.M., Smith, R.G., Ballmann, A.A., Levinstein, H.J., Nassau, K., "Optically-Induced Refractive Index Inhomogeneities in $LiNbO_3$ and $LiTaO_3$", Appl. Phys. Lett., Vol. 9, pp. 72-74. 1966.
2. Kurz, H., Krätzig, E., Keune, W., Engelmann, H., Gonser, U., Dischler, B., Räuber, A., "Photo-refractive Centers in $LiNbO_3$,Studied by Optical- Mössbauer- and EPR-Methods", Appl. Phys., Vol. 12, pp. 355-368. 1977.
3. Glass, A.M., von der Linde, D., Negran, T.J., "High-Voltage Bulk Photovoltaic Effect and the Photorefractive Process in $LiNbO_3$", Appl. Phys. Lett., Vol. 25, pp. 233-235. 1974.
4. Festl, H.G., Hertel, P., Krätzig, E., von Baltz, R., "Investigations of the Photovoltaic Tensor in Doped $LiNbO_3$", phys. stat. sol. (b), Vol. 113, pp. 157-164. 1982.
5. Krätzig, E., "Photorefractive Effects and Photoconductivity in $LiNbO_3$:Fe", Ferroelectrics, Vol. 21, pp. 635-636. 1978.
6. Orlowski, R., Krätzig, E., "Holographic Methods for the Determination of Photo-Induced Electron and Hole Transport in Electro-Optic Crystals", Solid State Commun., Vol. 27, pp. 1351-1354. 1978.
7. Kurz, A., Doormann, V., Kobs, R., Applications of Holography and Optical Data Processing, (ed. by Marom, E. and Friesem, A.A.), Pergamon Press, Oxford and New York. 1977.
8. Orlowski, R., private communication
9. Krätzig, E., Orlowski, R., "$LiTaO_3$ as Holographic Storage Material", Appl. Phys., Vol. 15, pp. 133-139. 1978.
10. Krätzig, E., Welz, F., Orlowski, R., Doormann, V., Rosenkranz, M., "Holographic Storage Properties of $BaTiO_3$", Solid State Commun., Vol. 34, pp. 817-819. 1980.
11. Megumi, K., Kozuka, H., Kobayashi, M., Furwahta, Y., "High-Sensitive Holographic Storage in Ce-doped SBN", Appl. Phys. Lett., Vol. 30, pp. 631-633. 1977.
12. Huignard, J.P., Micheron, F., "High-Sensitivity Read-Write Volume Holographic Storage in $Bi_{12}SiO_{20}$ and $Bi_{12}GeO_{20}$-Crystals", Appl. Phys. Lett., Vol. 29, pp. 591-593. 1976.
13. von der Linde, D., Glass, A.M., Rodgers, K.T., "High-Sensitivity Optical Recording in KTN by Two-Photon Absorption", Appl. Phys. Lett., Vol. 26, pp. 22-24. 1975.
14. Boatner, L.A., Krätzig, E., Orlowski, R., "KTN as a Holographic Storage Material", Ferroelectrics, Vol. 27, pp. 247-259. 1980.
15. Staebler, D.L., Burke, W.J., Phillips, W., Amodei, J.J., "Multiple Storage and Erasure of Fixed Holograms in Fe-doped $LiNbO_3$", Appl. Phys. Lett., Vol. 26, pp. 182-184. 1975.
16. Rupp, R.A., Drees, F.W., "Light-induced Scattering in Photorefractive Crystals", Appl. Phys. B, Vol 39, pp. 223-229. 1986.
17. Krätzig, E., Orlowski, R., Doormann, V., Rosenkranz, M., "Optical Information Storage in $LiTaO_3$:Fe-Crystals", SPIE-Proceedings, Vol. 164, pp. 33-37. 1978.
18. Vormann, H., Weber, G., Kapphan, S., Krätzig, E., "Hydrogen as Origin of Thermal Fixing in $LiNbO_3$:Fe", Solid State Commun., Vol. 40, pp. 543-545. 1981.
19. Vormann, H., Krätzig, E., "Two-Step Excitation in $LiTaO_3$:Fe for Optical Data Storage", Solid State Commun., Vol. 49, pp. 843-847. 1984.
20. Ye Ming, Krätzig, E., Orlowski, R., "Photorefractive Effect in $LiNbO_3$:Cr Induced by Two-Step Excitation", phys. stat. sol. (a), Vol. 92, pp. 221-229. 1985.

Fe-doped $LiNbO_3$ crystal for applications in white-light information processing

Wang Shuying, Zhang Jingjiang, Peng Yulan,
Zhao Yunying and Bao Cailong

Department of Physics, Beijing Normal University, Beijing, China

Abstract

Fe-doped $LiNbO_3$ crystal used as a real-time recording medium for white-light information processing is described. Both of the writing and reading stages used white-light source. Coupled wave theory and some of the basis experiment are given.

Iron-doped $LiNbO_3$ crystal used as a kind of recording medium in coherent light information processings have been reported in many articles, but its application in the white-light information processings are scarcely reported. This paper describes Fe:$LiNbO_3$ crystal used as a real-time recording medium for white-light source, and both writing and reading used white light.

The crystal slice was cut parallel to the crystallographic c axis, thick 0.6-1.2mm, the extent of iron-doping is about 0.0016-0.16 mol%, and reduced properly in argon atmosphere or in powdered lithium carbonate. The spectral transmittance curve of the reduced crystal is shown in Fig.1. The diffraction efficiency of the Fe:$LiNbO_3$ crystal depends on wavelength and intensity of the incident light, but the diffraction efficiency will be affected eventually by existing apparent absorption, particularly around the wavelength 5000Å. So in the crystal growth and reducing precedures it is necessary to judicious control both the concentration and valence state of the impurity ions making trade-off between the transmittance and the diffraction efficiency.

In order to increase the spatial and temporal coherence, we adopted a point source of white-light and a Ronchi or sinusoidal rulling to modulate the input object during the recording stage. So the recording stage can be called encoding stage too. By virtue of the photorefractive effect of the Fe: $LiNbO_3$, the encoded crystal slice is quite similar to a thick phase hologram. Each monochromatic light of the white-light source can be treated as the coherent quasi-monochromatic light which can be explained satisfactorily by the coupled wave theory.

During the white-light illuminates the encoded crystal for read-out, a linear polarizer is placed after the convergent lens with its transmitting axis parallel to the optical axis (i.e. the c axis) of the crystal. The orientation is chosen because the change of the extraordinary index of refraction is much larger than the ordinary index of refraction in Fe:$LiNbO_3$ crystal. Since it is clear that continued readout will eventually erase the phase information recorded in the crystal, the intensity of the reading beam must be less compared with the recording light. In addition, there should be a small angle between the z axis and the incident beam so as to satisfy the Bragg condition.

According to the Kogelnik's coupled wave theory the amplitude of the diffraction wave on the other surface (z=d) of the crystal can be expressed as

$$S_i(d, \lambda) = -j\left(\frac{C_R}{C_s}\right)^{1/2} \exp\left(-\frac{d\alpha_0}{C_R}\right) e^{\xi} \frac{\nu}{(\nu^2 - \xi^2)^{\frac{1}{2}}} \sin(\nu^2 - \xi^2)^{\frac{1}{2}} \qquad (1)$$

where d is the thickness of the crystal slice, $C_R=K_{RZ}/K$, $C_s=K_{sz}/K$ are the obliquity factors of the incident and diffractive waves respectively, K_{RZ}, K_{sz} are the component of the incident wave vector $\bar{K}_R$ and diffractive wave vector $\bar{K}_s$ on the horizontal axis Z (see Fig. 2a), $\nu = \chi d/(C_R C_s)^{\frac{1}{2}}$, $\xi = \frac{1}{2}d\left(\frac{\alpha_0}{C_R} - \frac{\alpha_0}{C_S} - j\frac{K^2 - K_s^2}{2KC_s}\right)$, where $\chi = \frac{\pi n_1}{\lambda} - j\frac{\alpha_1}{2}$ is the coupled constant, n_1 is the amplitude of the refractive index; α_0 is the average absorptivity and α_1 is the amplitude of the absorptivity. Since the reading beam obeys the Bragg condition, $K_R=K_S=K$, $C_R=C_S=\cos\theta$ (see Fig. 2b). Neglecting absorption, $\alpha_0=\alpha_1=0$, and because the angle θ is very small $\cos\theta \approx 1$, thus $\xi=0$, $\nu=\pi n_1 d/\lambda$,

* Projects Supported by the Science Fund of the Chinese Academy of Sciences

then equation (1) is transformed to

$$S_i(d, \lambda) = -j\mathrm{SIN}\left(\frac{\pi n_1 d}{\lambda}\right) \tag{2}$$

the intensity is given by

$$I_i(d,\lambda) = S_i^2(d, \lambda) = \mathrm{SIN}^2\left(\frac{\pi n_1 d}{\lambda}\right) \tag{3}$$

when the illuminating light is a white-light source, in spite of the selectivity of wave length for the time being, the intensity of the diffractive light can be given as follows

$$I(d) = \int_{\lambda r}^{\lambda v} I_i(d, \lambda)\, d\lambda \tag{4}$$

Obviously, using the Fe:$LiNbO_3$ crystal as a recording medium in white-light information processing one of the advantage is that the out-put information can be obtained without any development, and another advantage is that the out-put images can be displayed in color according to their original optical density.

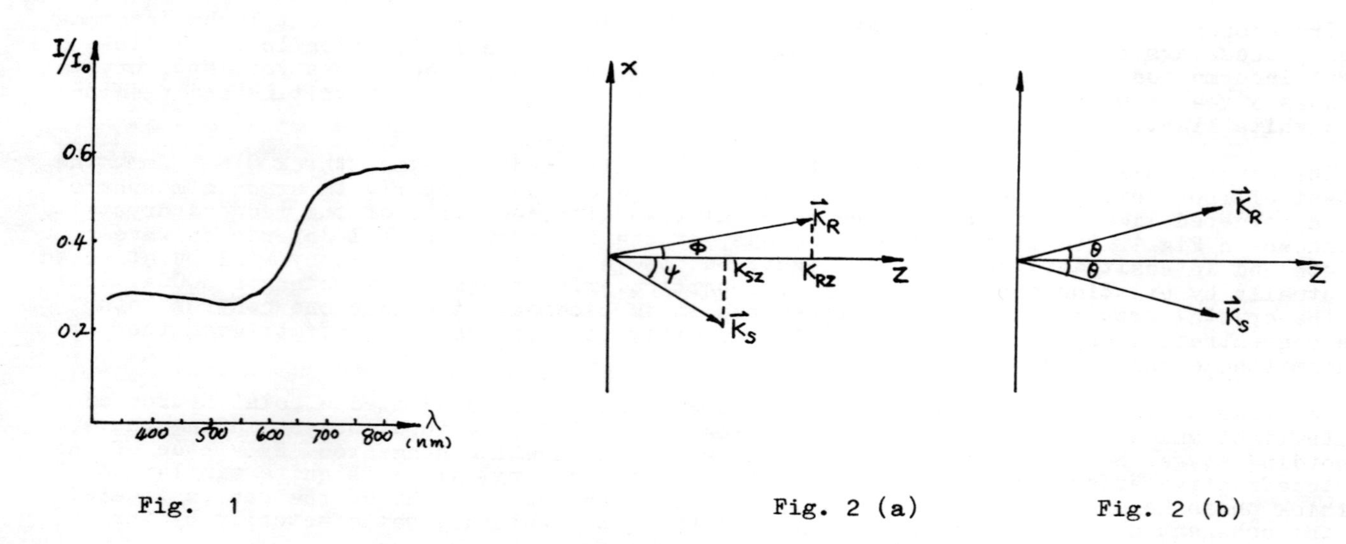

Fig. 1

Fig. 2 (a)

Fig. 2 (b)

Microscopic and macroscopic measurement of holographic emulsions

R. Aliaga, H. Chuaqui

Laboratorio de Optica y Plasma, Facultad de Física, Universidad Católica de Chile
Casilla 6177, Santiago 22, Chile

Abstract

Microscopic properties of holographic emulsions are studied using electron microscopy. Absorption spectra, diffraction efficiency and signal to noise ratio are measured on plane wave holograms. Good qualitative agreement is observed between microscopic and macroscopic measurements. From these measurements the main feature of holographic emulsions for solution physical development are obtained.

Introduction

The current interest on holographic as an artistic medium, as well as its multiple applications in fields such as engineering or medicine has put more stringent demands on holographic materials. High diffraction efficiency, low noise and high sensitivity are the minimum requirements of a practical holographic material. Archival properties of holographic materials are becoming an important consideration. Silver halide materials are processed using a developer followed by bleach or solution physical development. The latter process, followed by a fixing stage, results in a colloidal suspension of nearly spherical pure silver grains in the gelatin.

The absence of silver halides or dye guarantees the stability of holograms processed using a solution physical development when followed by a fixing state. The results obtained with photographic materials would suggest that the stability of holograms obtained with this type of processing ought to be stable for over a century.

The present work is a study at microscopic level of the different stages during processing and its correlation with macroscopic optical properties. In previous work[1,2] emulsions with initial grain size of 10 and 26 nm have been studied. The approach used here to study 8E75HD and 10E75 emulsions (35 and 90 nm respectively[3]) uses measurements based on Transmission Electron Microscopy (TEM), optical absorption, diffraction efficiency and signal to noise ratios.

Microscopic measurements

The microscopic measurements were obtained on a Zeiss EM 109 transmission electron microscope. Good quality micrographs are obtained if sample thickness is less than 80 nm, the best thickness is 50 nm. Most of the results were obtained with his latter thickness. Samples of unexposed emulsion with different processing stages are analyzed.

Emulsions 8E75HD and 10E75 were exposed using two parallel He-Ne beams forming an angle of 30º. The transmission hologram obtained by processing in 4 distinct ways:

a) using a solution physical developer, CPA-1[4]
b) the above developer followed by a fixing bath (Agfa G-321, 4:1 dilution for 4 min at 20 ºC).
c) a cathecol developer CW-C2[5] followed by fixing.
d) a cathecol developer CW-C2 followed by a bleach bath PBQ-2.

Measured grain size on unprocessed samples agrees with manufacturers specification for 8E75HD, whereas 10E75 measured size is smaller by a third. This difference could be attributed to sample thickness smaller than the largest grains. Thicker samples would result in out of focus TEM images. However, grain size measurement after different processing stages are consistent with the original measured size. Fig. 1 shows samples obtained using the CPA-1 on 8E75HD and 10E75 emulsions. The main differences with unprocessed samples is the presence of a growth structure out of the original grains. The diameter of this growth structure is of the order of 20 nm.

A significative change is observed after fixing CPA-1 processed samples, as shown in Fig. 2, 8E75HD grain size increases, whereas 10E75 grains decrease in size. A side by side comparison shows that the only noticeable difference is grain density.

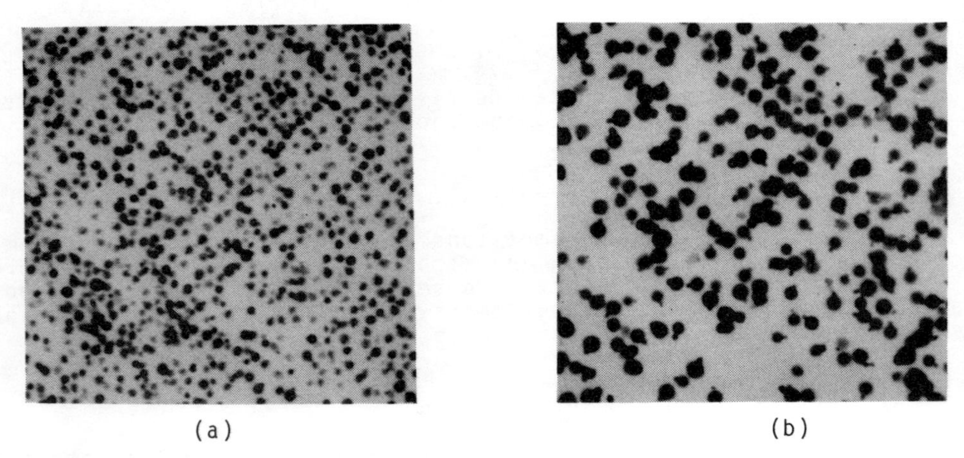

(a) (b)

Fig. 1. Micrographs of a) 8E75HD and b) 10E75 processed on CPA-1, the area show is 2x2μm

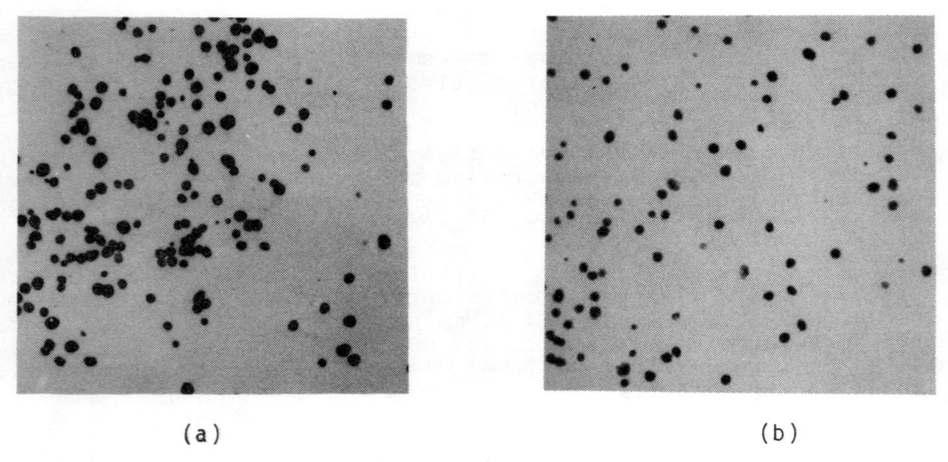

(a) (b)

Fig. 2. Micrographs of a) 8E75HD and b) 10E75 processed on CPA-1 followed by 4 min fixing, the area shown is 2x2 μm.

A conventional developer produces a very different structure, as illustrated on Fig. 3. CW-C2 developing followed by a fixing bath results in grains converted into a filamentary structure. If a bleach is used instead of the fixing bath, the grains are fairly regular in shape that is not very different from the original grains.

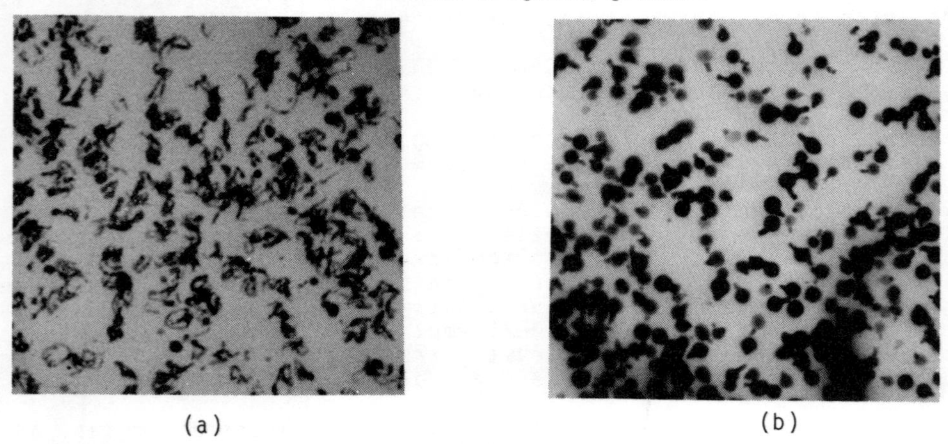

(a) (b)

Fig.3. 10E75 developer on CW-C2 followed by a) fixing b) PBQ-2 bleach. Area shown is 2x2 μm

Modulation of grain distribution is observed depending on the angle of the sample relative to microtome. Fig. 4 shows an example of modulation on 8E75HD.

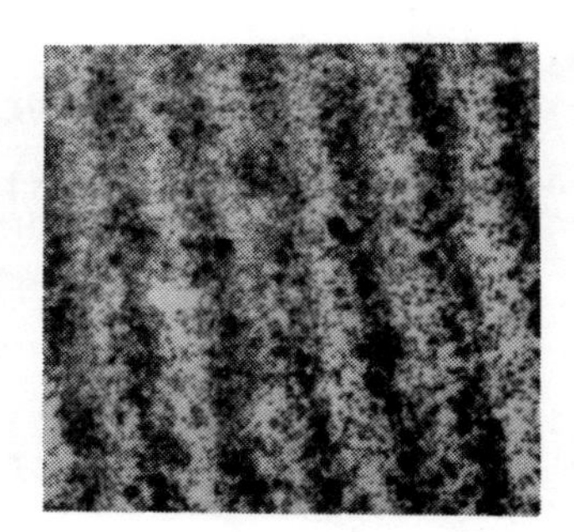

Fig. 4. Modulation of grain distribution on 8E75HD processed on CW-C2 and bleached in PBQ-2. Area shown in 10x10μm

Macroscopic measurements

Diffraction efficiency, signal to noise ratio, measured as in Ref together with spectral transmission were obtained for CPA-1 processed samples. Absorption measurements show that there is a broad absorption band between 440 and 520 nm.

Table 1 shows results for diffraction efficiency, signal to noise ratio and grain size measurements.

Table 1. Macroscopic measurement and grain sizes

	8E75HD			10E75		
	grain size (nm)	Diffraction efficiency	S/N	grain size (nm)	Diffraction efficiency	S/N
CPA-1	35±10	16.4±3	32.7±4.1	11.8±1.4	12.3±3	6±1.1
CPA-1 fixing	52±12	28±2.5	12.9±0.9	51±11	15.2±2	13.1±1

Conclusions

Solution physical development followed by fixing produces the same grain size in both emulsions. The final size, approximately 50 nm, is probably due to the excess in solution whilst fixing. The silver concentration on these emulsions is very nearly equal,[3] which might be the reason for obtaining the same final grain size in both emulsions. An estimate of grain density shows that diffraction efficiency is proportional to density, that is, for equal grain size. No comparison can be carried out for CPA-1 because there is no way to distinguish metallic silver from silver halide. Signal to noise ratio depends only on particle size.

A wide absorption band is indicative of a fairly broad size distribution of silver grains.[6] This is indeed the case, as the dispersion in size obtained from micrographs is roughly 20%. This dispersion in size is similar to the dispersion in the original emulsion.

Results obtained on differents emulsions[1,2] just as the present one show that, in general, the size for silver particles developed using solutions physical development is of the order of 20 nm. It would appear that an optimum holographic emulsion for solution physical development ought to have a monodisperse 20 nm grain suspension on gelatine. This would ensure that the noise level is lowest, as developed particle would not be any smaller.

Clearly, noise level is very dependent on size. On the other hand it seems that there is no real advantage on going to smaller sizes. A reduction in size below 20 nm would only decrease the sensitivity of the emulsions, keeping noise level fairly constant.

Acknowledgements

We are grateful to the Anatomy Pathology Department for allowing the use of the TEM, particularly Drs. B. Chuaqui, R. Rosenberg and Mr. L. Andrade.

The present work has received funding from the National Fund for Development of Science and Technology, Chile, Research Fund (DIUC), Pontificia Universidad Católica de Chile and PNUD.

References

1. Usanov Y. E., Kosobokova N. L. and Tikhomirov G. P. "Investigation of the dependence of the diffraction efficiency of holograms on the sizes of the developed silver particles" Sov. J. Opt. Technol. 44(9) pp. 528-530. 1977.
2. Sainov V. C., "Basic characteristics and applications of reflection holograms" International Symposium on Display Holographic, Lake Forest, pp. 55-68. 1982
3. Agfa-Gevaert, private communication, 1984.
4. Aliaga R., Chuaqui H., Pedraza P., "Solution physical development of Agfa-Gevaert emulsions for holographs". Optica Acta, 30 (12), pp. 1743-1748. 1983.
5. Cooke D. J. and Ward A. A. "Reflection-hologram processing for high efficiency in silver-halide emulsions" Applied Optics, Vol. 23, Nº 6, pp. 934-941, 1984.
6. Skillmand D. C. and Berry C. R., "Effect of particle shape on the spectral absorption of colloidal silver in gelatin", The Journal of Chemical Physics, 48(7), pp. 3297-3304. 1968.

STUDY ON THE PERFORMANCE OF THE LIQUID CRYSTAL LIGHT VALVE

Zhi-yong Chen, Rong-cai Ji, Huan-qing Zhao

Lab of Laser Physics & Optics, Fudan University
Shanghai, China

Abstract

The performance of a liquid crystal light valve (LCLV) is studied experimentally. The effects of applied voltage across the LCLV, its driven frequencies and the input polarization of the reading light are investigated for both the liquid crystal (LC) being in the 45° and in the 90° twisted modes.

Introduction

It is well known that the parallel processing capability of the optical information processing as combined with a real-time device will offer great potential applications. Incoherent-to-coherent conversions in real-time are often required for such a processing. During the past decade, a number of two dimensional spatial optical modulator capable of such type of image conversion have been developed and among them the liquid crystal light valve (LCLV)[1] has emerged as one of the most viable device and has been used frequently.[2,3]

We began on the fabricating and developing of a laboratory-made LCLV few years ago and based on the current technical level of our laboratory, our LCLV could meet the practical needs for optical information processing carried on in our laboratory researches. Those LCLVs fabricated in our laboratory have the following characteristics of performance:

Aperture size	25×35 mm^2
Threshold sensitivity	2 μW/cm^2
Resolution	30 lines/mm
Contrast	90:1
Speed of response -- rise	50 ms
-- fall	100 ms

Figure 1 shows the image displayed on our LCLV where 1(a) corresponds to the coherently converted image and 1(b) its contrast reversal image.

In this paper, we will present the study of the performance of our LCLV both with the liquid crystal (LC) molecules at the 45° and 90° twisted modes against the effects of applied voltages, its driven frequencies and the input polarization of the reading images.

The structure of the LCLV is basically the same as that of those LCLVs made by Hughes Company in the United States, which is composed with a CdS photoconductor layer, a light blocking layer of CdTe and a high reflective mirror with the nematic twisted LC layer all sandwiched between two transparent electrode layers. The LC in the LCLV acts as a modulator to the coherent reading light, its behavior is determined by the applied voltage across it and therefore directly related to the input intensity of the reading light. The performance of the LCLV depends on the characteristics of the liquid crystal predominantly. A detail study of the transmission characteristics of the liquid crystal as well as the LCLV is feasible.

The Transmission Characteristics

It is already known that the LC is operated under the hybrid field effects: that is the off-state (no voltage across on the LC) is the conventional nematic effect and the on-state (voltage across on the LC) is the field induced birefringence effect. However, in the conventional structure of the LCLV, the LC is employed with the 45° twisted mode because in such a configuration, the LC could introduce a maximum birefringence effect. But we found that the 90° twisted mode of the LC, which is commonly used in varieties of LC devices, also possesses its advantages in a LCLV.

Figure 2 shows the transmission characteristics of a simple layer of LC as measured in the reflective mode, that is the light passing through the LC layer, retro-reflected by a high reflective mirror and then passing once again through the LC and with its intensity measured.

In figure 2, it is shown that the 45° twisted mode and the 90° twisted mode of the LC layers possess different characteristics under the applied voltage with the driven frequency of 1 kHz: the threshold for operating the LC of the 90° mode is lower than that of the 45° mode, its transmission peak is narrower and the maximum transmissibility is lower than that of the 45° mode. This means that the 90° mode LC might be better for operating of the LCLV for contrast reversal of images and the 45° mode LC will possess larger dynamic range of operation.

Figure 3 shows the measured transmission characteristics for two LCLV as made in our laboratory with the LC layer in the 45° and 90° modes. The LCLV is illuminated with a constant intensity of an incoherent light source, the reading light is a He-Ne laser of 100 $\mu W/cm^2$ in intensity. The output intensity is measured by a photodiode radiometric detector through a polarizer in the direction perpendicular to that of the incident polarized reading input light. It is seen in figure 3 that as the applied voltage V increases, the output light intensity exhibits the same characteristics as that of the LC as shown in figure 2. But in the case of different ac driven frequency f of the applied voltage, the transmission curves shift towards to higher voltage and the transmission peak will be broader as the frequency f increases. The measured results show that the 90° mode of the LC in the LCLV possesses lower threshold of operation and has the advantage of easy to obtain a contrast reversal image than that of the 45° mode, however the 45° mode LC in the LCLV has the advantage of large dynamic range of operation and higher transmissibility, these results are in consistant with the characteristics of the single LC layer for thecorresponding modes as shown in figure 2.

Sensitometry for the LCLV

A further examination of the performaance for the 45° mode and the 90° mode LC layer in the LCLV is to measure the output intensity of the LCLV versus the input incoherent light intensity, which is known as the sensitometry of the LCLV.

Figure 4 shows the measured relative output intensity of the LCLV against the input intensity for two different modes of 45° and 90° respectively. The measurements are carried out with two optical fiber probes for sampling the intensities and fed to two photomultipliers. All the measurements are fixed at a constant applied voltage but with different driven frequencies of 0.1 kHz, 1kHz and 10kHz. These curves show that higher driven frequency will cause poorer sensitivity but the transmission sensitometric curves become broader as the frequency increases, a larger dynamic range at higher frequency will result and this is true for both cases of the two different modes of the twisted LC.

Figure 5 shows the sensitometric curves of the LCLV of two modes at different operating voltage V but keep the driven frequency f constant. These curves show that the higher the applied voltage the better the sensitivity and its lower threshold. The different operating voltages also give different shapes of the sensitometric curves, thereby it will effect the dynamic range and the contrast ratio of the output images, so it is advisable to adjust the applied voltage over the LCLV for controlling the conversion characteristics as to meet the specific purposes.

As for the differences between the two different modes of the twisted LC layer, it shows that the 90° mode LCLV is featured by a sharp sensitometric curve and we can then easily get contrast reversal images for the sensitometric curve gives a very steep slope. The 45° mode LCLV, however, is characterized by a broad sensitometric curve and we can then use it in the case where a large dynamic range is required. Hence, both of the 90° and the 45° modes of the LC in the LCLV could be selected for different purposes of applications.

Effects of the Input Polarization

In normal operation conditions, the polarization of the input light is always chosen to be either parallel or perpendicular to the long axis of the LC molecules at the entrance interface. However, if one rotates the LCLV along the normal as to vary the input polarization of the reading light, one can find that the behavior of the LCLV will be different.

Figure 6 gives the transmission curves of a 90° mode LCLV at different applied voltages of driven frequency of 1 kHz for the input polarization directions at 20° and 45° against the long axis of the liquid crystal molecules. Figure 7 shows the sensitometric curves of different input polarizations of 45°, 20° and 0° for a 45° mode and a 90° mode LCLV. It can be seen from the transmission curves that the transmission peaks become sharper and increases in numbers as the polarization of the input changes from 20° to 45°; this feature can also be shown in figure 7 as the sensitometric curves are getting sharper as the polarization angle increases. Hence, it shows that in such a case, the LCLV will exhibit good sensitivity and high contrast reversal characteristics. However, a drawback for the sharper curve will cause the degrade of the spatial uniformity of the LCLV, because the spatial variation of the voltage across the LCLV is predominantly influenced by the uniformity of the thickness of the LC layer and which is seldom in uniform. The best spatial uniformity of response of the LCLV is obtained only when the input polarization of the light is parallel or perpendicular to the long axis of the LC molecules at the entrance interface.

By rotating the LCLV along the normal to vary the input polarization of the incident light, the performance of the LCLV could be controlled with its sensitivity, its dynamic range and its contrast reversal of image. This is surely a very useful technique when one applies the LCLV for his research investigations.

Conclusion

We have shown that the performance of our laboratory-made LCLV could be monitored by adjusting the applied ac voltages, its driven frequency, the direction of polarization of the incident input light. Besides, we also find that the 90° mode LC of the LCLV has low voltage threshold and easy to get contrast reversal processing and the 45° mode will give a larger dynamic range and high output and hence both of these divices are acceptable.

Acknowledgment

The authors would like to express their gratituder to Professor Zhi-ming Zhang, for valuable discussions and gratitues to the staffs who help us for fabricating the LCLV devices.

This work is supported by the China Science Foundation and authors would like to express the gratitude.

References

1. Grinberg, J. et.al, IEEE Trans. on Electron Devices, ED-22, 775, 1975
2. Lawis, R.W., Applied Optics, 17, 161, 1978
3. Bleha, W.P., et.al, Optical Engineering 17, 371, 1978
4. Collins, S.A., et al, Applied Optics, 20, 2250, 1981
5. Lee, S.H., Applied Optics, 20, 1424, 1981
6. Collins, S.A., et.al, Applied Optics, 23, 2163, 1984

(a) The coherently converted image (b) The contrast reversal image of (a)

Figure 1. The Images displayed on our LCLV.

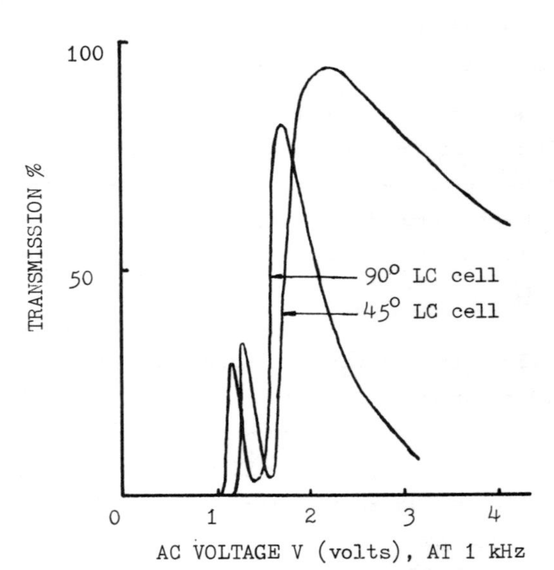

Figure 2. Transmission curves for the LC cell in reflection mode.

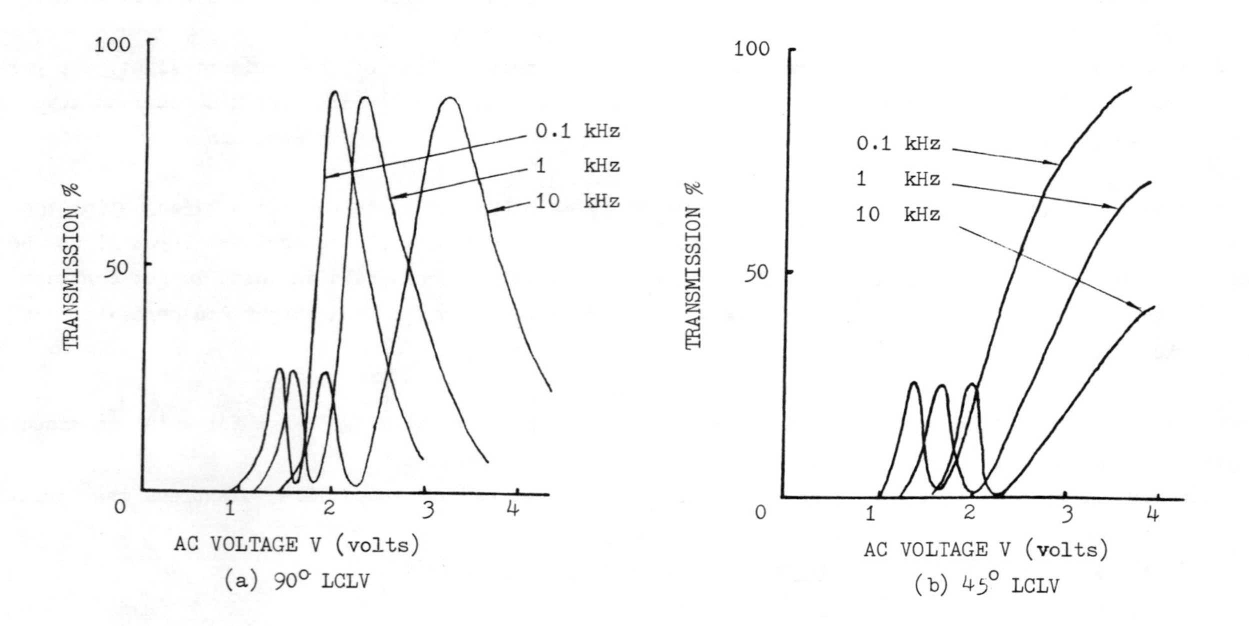

Figure 3. Transmission curves for the LCLV at different ac frequencies.

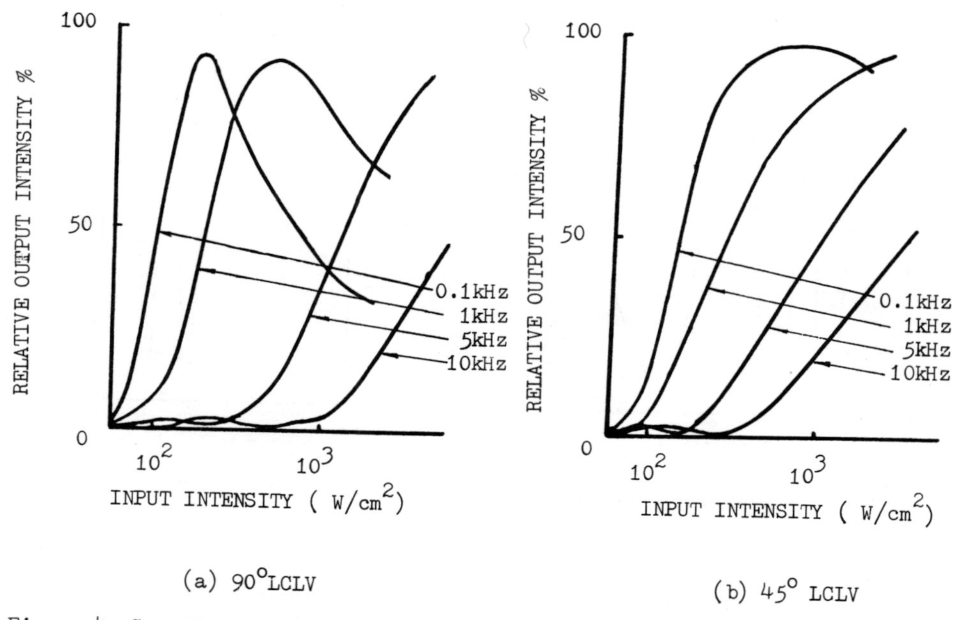

Figure 4 Sensitometry curves for the LCLV at different ac frequencies

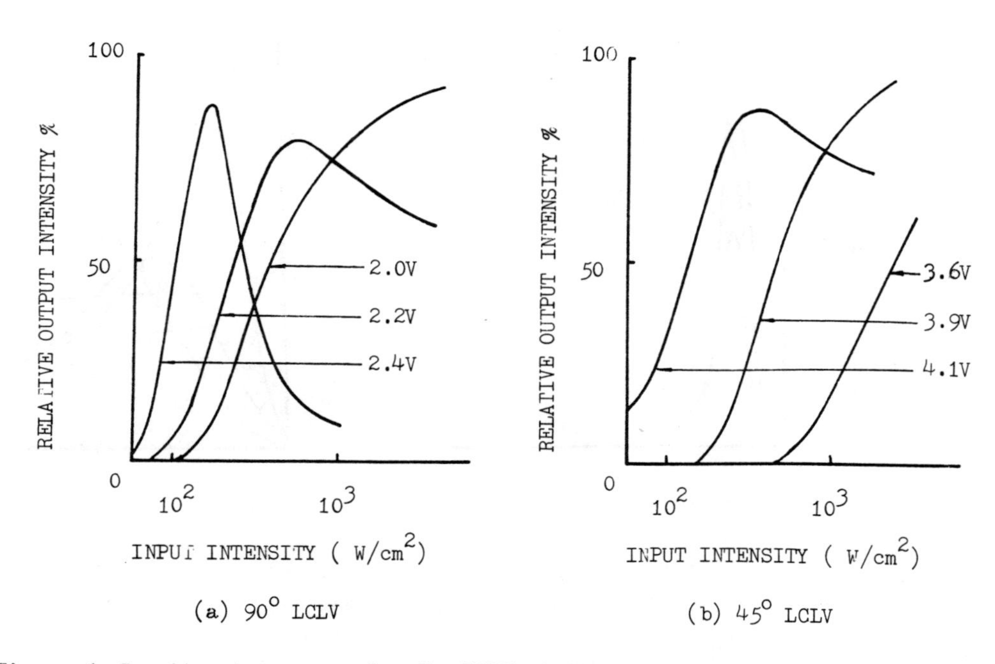

Figure 5. Sensitometry curves for the LCLV at different operating voltages.

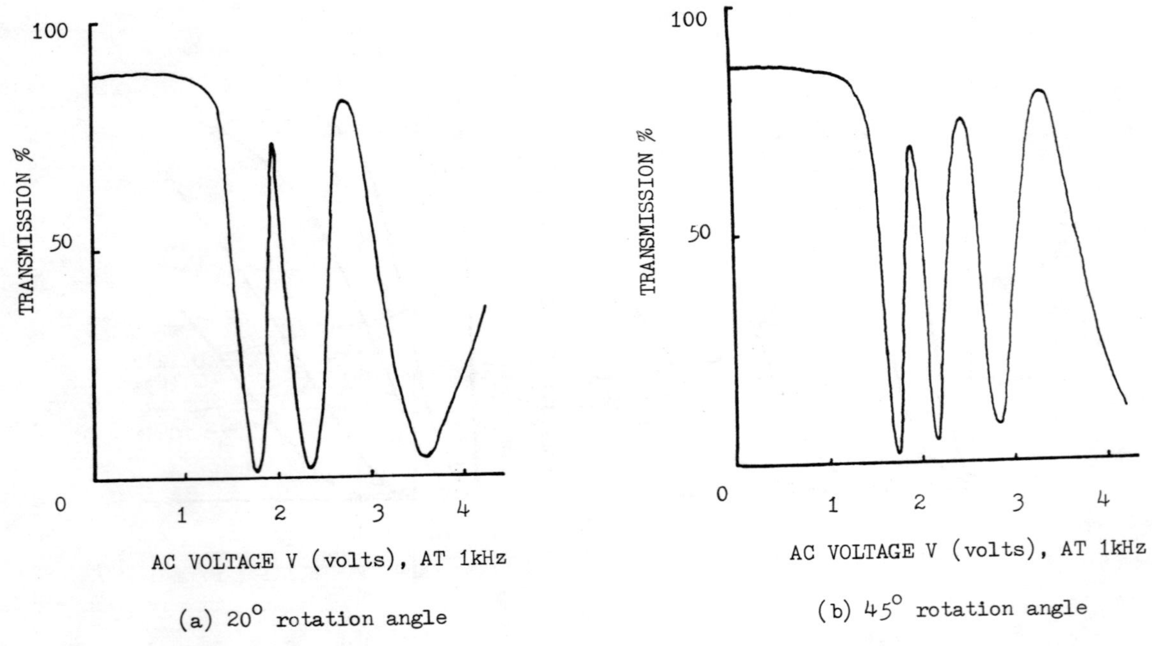

Figure 6. Transmission curves for the LCLV at different LCLV rotation angles.

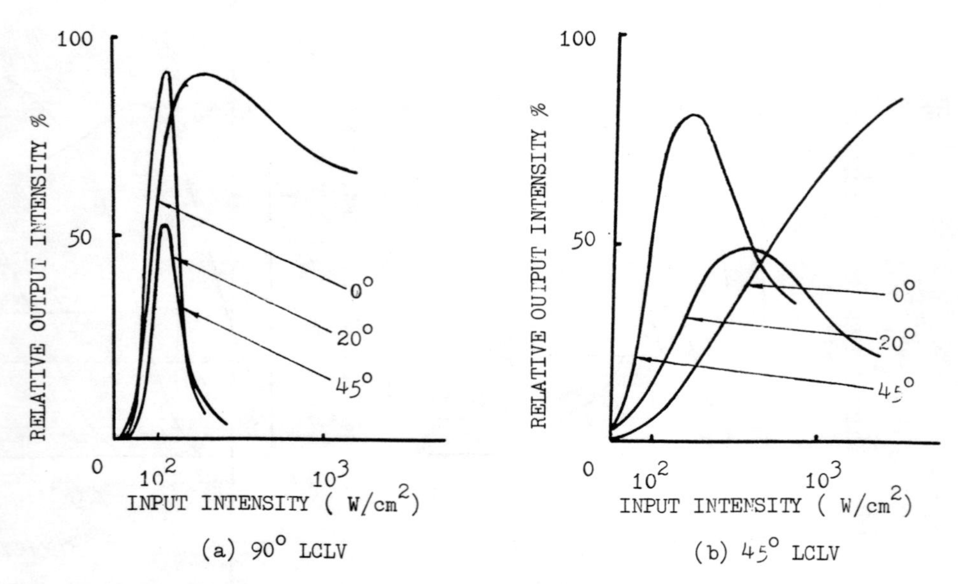

Figure 7. Sensitometry curves for the LCLV at different LCLV rotation angles.

Session 14

Optical Information Processing I

Chairmen
G. von Bally
University of Münster, FRG
Zhi-Ming Zhang
Fudan University, China

Coherent optics in environmental monitoring

K. Hinsch

Fachbereich 8 - Physik, Universität Oldenburg
Pf 2503, D-2900 Oldenburg, Fed. Rep. Germany

Abstract

Major environmental problems all over the world require immediate efforts. Often, it is necessary to obtain quantitative data on pollution-induced effects on natural or man-made systems exposed to a variety of environmental parameters. Useful information about the object may be acquired in the speckle-fields produced by illumination with coherent light. Changes in the object require a corresponding correlation of the speckle fields, a task for which optical methods are employed. Examples of such applications are shown which demonstrate the usefulness of optical metrology in a field challenging scientific engagement.

Introduction

The impact of industrial societies on nature is creating severe problems worldwide. Delicate ecological systems and man-made cultural achievements are threatened likewise. Examples are the effects of polluted soil, water or air on plants, or the deterioration of historical buildings and monuments by aggressive components in the air. In both cases the effects have accelerated greatly in recent years and immediate efforts are requested.

Monitoring such processes and studying according cause-effect relations requires a quantitative measure for changes in the objects under investigation. In addition to a variety of other physical or chemical techniques, a detailed analysis of changes in scattered coherent light fields allows to detect the minute responses of the object to external environmental parameters. The optical techniques employed possess two unique properties:

- they utilize a large amount of information about the object, thus they are quite sensitive to the changes to be measured,
- they process this information economically by a specific reduction of data.

Let us illustrate the question by the following example: Under the influence of environmental parameters like rain, sunshine, frost, SO_2, NO_x etc. a limestone building block of a medieval church experiences all kinds of deteriorations, some of which change its light scattering rough surface. Information about the details of this surface are gained in speckle fields which must be compared quantitatively to determine the similarity of the stone surface at different instants of time. It is the purpose of this paper to give an overview of some solutions to this kind of tasks.

Real time optical correlation

A well-known optical technique to achieve a correlation of two light fields is by matched filtering.[1] Its principle is recalled in Fig. 1a: In the recording step, a holographic Fourier transform filter is produced of the initial signal S_1. In the measuring step illumination of this filter with light from an altered signal S_2 produces a two-dimensional output correlation function; its central peak intensity may be used to measure the similarity of S_1 and S_2. A change of S_1 can thus be monitored in real time by recording this peak intensity. Several provisions, however, must be made for a successful practical application:

- The hologram must be recorded almost instantaneously.
- Accidental misalignments like displacement or tilt of the scattering surface must be controlled. These produce according displacements in the output or filter plane, respectively.

Fig. 1b gives the schematic of a buildup[2] employed in the study of deterioration effects in stone. The stone sample is exposed to a simulated environment in a glass container. The holographic filter is recorded on photo-thermoplastic material permitting almost instantaneous availability of correlation data. Misalignments are controlled by adjusting the position of the detector pinhole and the tilt of a parallel glass plate in front of the filter plane for maximum output intensity.

Two exemplary measurements are shown in the following figures which present the correlation peak intensity vs. time. One hour periods from various days of long-term experiments are shown. Fig. 2 demonstrates the effect of rinsing a rough surface with water: Initially, there is a marked decrease in the intensity due to the water on the surface. During the process of drying, the intensity increases again. In limestone, however, it does not resume its original value indicating that some material has been washed off. In the stainless steel sample, on the other hand, there is no permanent change in the surface. Fig. 3 indicates the combined effect of sulphuric acid and humidity on a limestone sample. When the light intensity in the correlation peak reaches the noise level, the measurement can be continued easily by recording a new hologram.

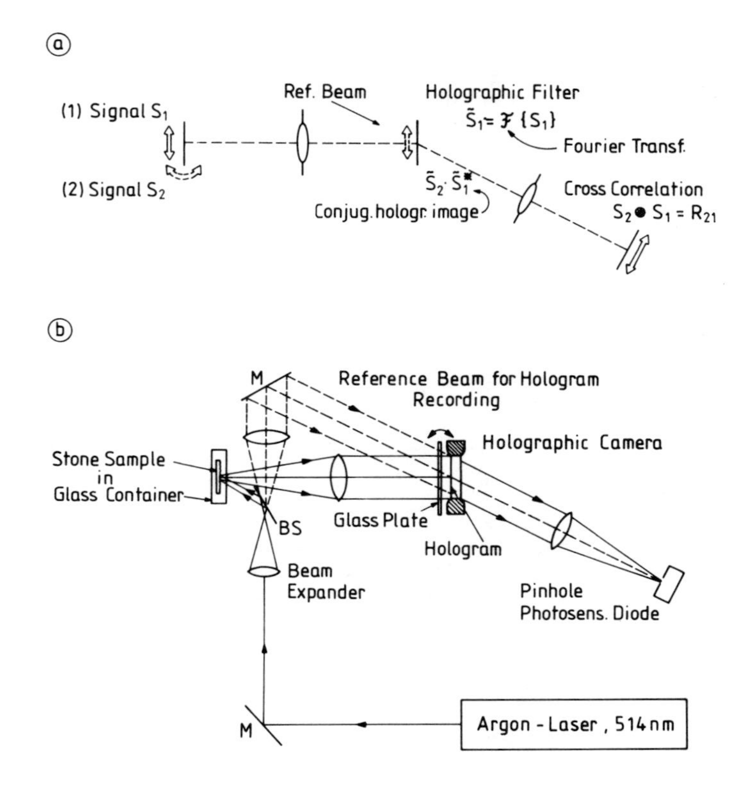

Figure 1. Real time optical correlation of scattered light fields; (a) principle of optical correlation; (b) schematic of optical correlator in study of stone decay.

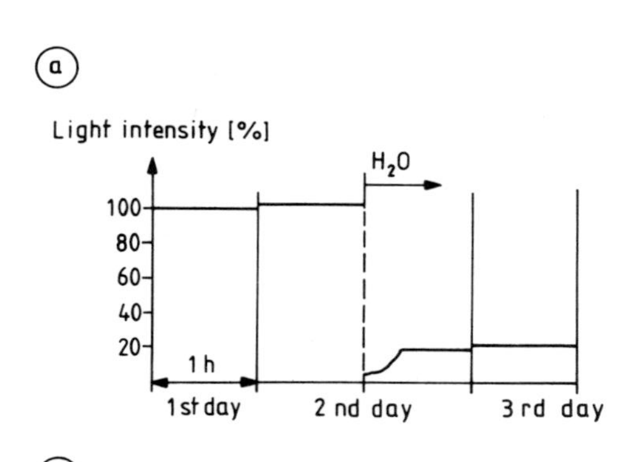

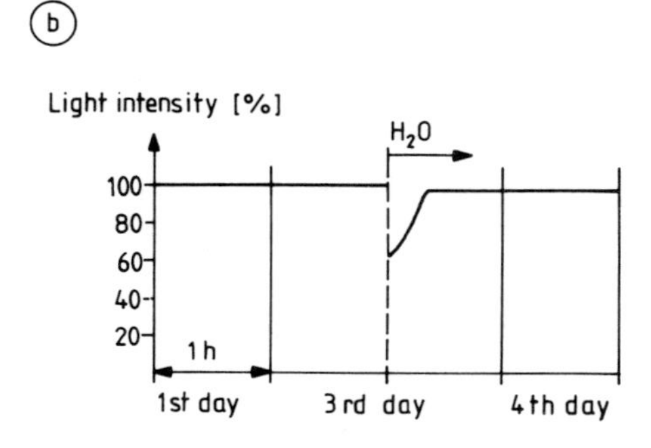

Figure 2. Optical correlator output during rinsing of rough surface; (a) limestone; (b) stainless steel.

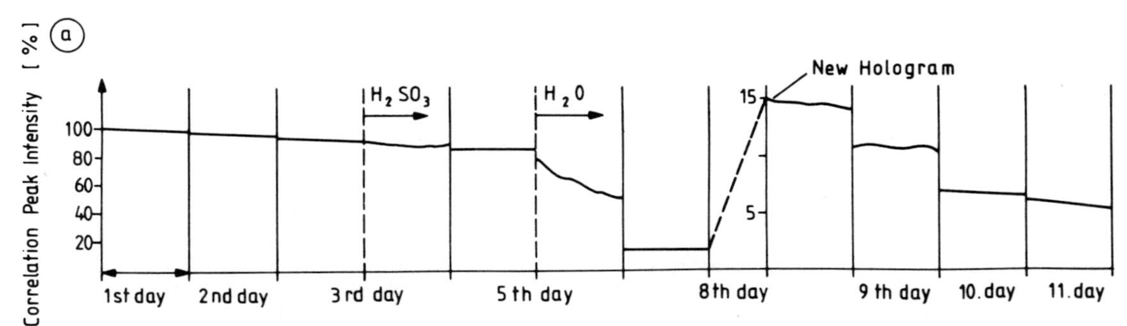

Figure 3. Optical correlator output for a limestone sample under the influence of sulphuric acid and air humidity.

Double exposure speckle correlation

The previously introduced technique has the great advantage of real time capability. However, it is rather vulnerable to external misadjustments and gives just a single correlation value averaged over the illuminated region. In applications requiring a more robust design or when the amount of change depends on the position in the sample (heterogeneous materials, biological processes), a correlation by double exposure speckle photography is more appropriate. More commonly, this method is used for displacement measurements of rough objects.[3] It produces a double exposure speckle image of the object consisting of two slightly displaced speckle patterns. When this transparency (a "specklegram") is illuminated "pointwise" with a narrow laser beam, the familiar Young's diffraction fringes are produced from which the displacement can be calculated. Usually it is assumed that both the superimposed speckle patterns are identical which results in fringes of visibility $V = 1$. When the speckle pattern changes during the interval between exposures, the visibility of the fringes is affected. The analytical treatment establishes relations between the transmittance of the double exposure transparency, its power spectrum (the Young's fringes), and its autocorrelation function which, of course, is the Fourier transform of the power spectrum. It is shown easily that the complete autocorrelation function consists of a central peak (the autocorrelation R_{11} of a speckle pattern) and the two crosscorrelation terms R_{12} shifted accordingly by the original displacement Δx between the exposures

$$R(x',y') = 2R_{11}(x',y') + R_{12}(-x'+\Delta x,y') + R_{12}(x'+\Delta x,y') \quad (1)$$

Thus, a two-dimensional Fourier transformation of the Young's fringe system produces the complete cross correlation of the compared speckle patterns. Such a transformation may be performed either optically or electronically in an image processing system. In practical applications, the necessary displacement Δx of the images can be produced anywhere in the optical system.

It can be shown that for many practical conditions already the average visibility of the fringe system is a good measure for the cross-correlation coefficient $R_{12}(0,0)$. Quite generally, of course, visibility will be a function of position in the fringe system.

When visibility data are used for a quantitative characterization of the similarity of light fields, attention must be paid to possible influences of the photographic recording process. Since the light intensity in speckle fields is governed by an exponential probability density function, the recording is principally nonlinear and the specklegram contains corresponding nonlinear terms which influence the visibility. Thus visibility will depend on the average exposure and on the photographic recording material used. This is shown in Fig. 4, where the effect has been studied for AGFA Holotest 10 E 75. In Fig. 4a the characteristic of the material (literature and own data) is reproduced. A double exposure specklegram was then calculated by a computer simulation in which the light intensity for a computed speckle pattern was recorded on the indicated film. The fringe visibility thus determined is plotted vs. the average exposure in Fig. 4b. Obviously, the expected theoretical value $V = 1$ is obtained only for a limited range of exposures, while under or over exposure result in a lower visibility value. Experimental data obtained with a ground glass plate object yield much lower values but indicate the theoretical dependence. Only when additional chemical treatment of the film for noise reduction is applied, the visibility assumes values closer to the expected data.

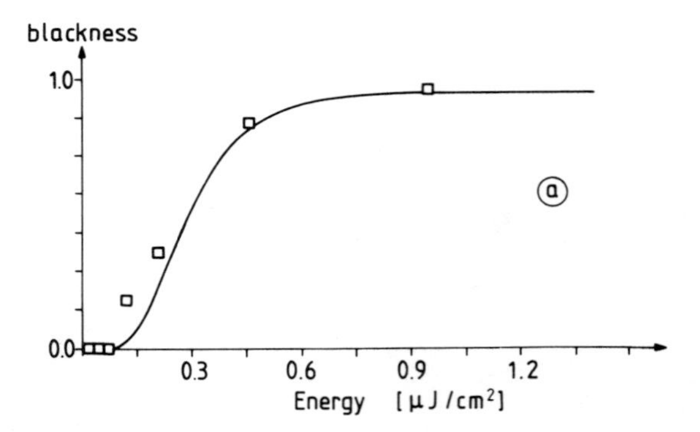

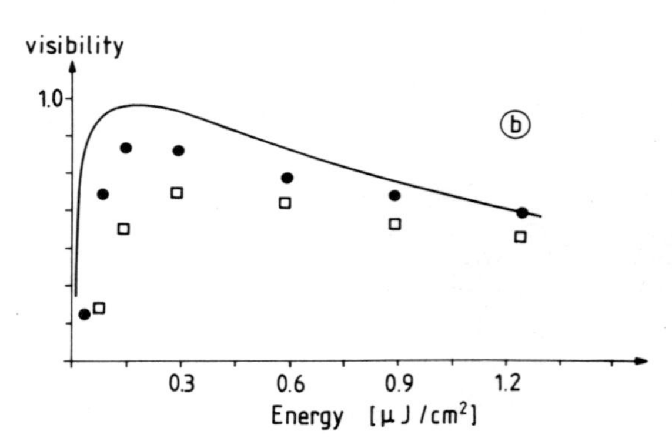

Figure 4. Effect of nonlinearity of photographic material on visibility of Young's fringes; (a) characteristic of holographic material 10 E 75 (curve from literature and own experimental data); (b) visibility vs. exposure (curve calculated from literature data, □ experimental data, ● experimental data after film processing for noise reduction).

Since the exposure may vary within an object (due to changes in brightness, for example) it is important to correct the visibility values accordingly, for which procedure this theoretical analysis forms the basis.

Let us illustrate the power of our method in several applications concerning environmental investigations:

Normal growth of plant leaf[4]

Double exposure speckle images in HeNe-laser light (633 nm) with a time separation of

2 min were made of a green leaf of the plant *Clivia miniata*. Visibility (uncorrected data) was evaluated as a function of distance from the leaf base (Fig. 5a) and reveals clearly three distinct regions of high, medium and low visibility. It is beyond the scope of this paper to describe the related biological growth processes. This, however, has been done successfully as implied by Fig. 5b which shows that a plot of the mean cell area displays just the according regions. Detailed measurements varying the wavelength, polarization, and time separation allow to identify specific biological processes.

Pollution influence on growth in wheat sprout

Wheat sprouts have been subjected to chemical additives in their culture medium. Fig. 6 gives a plot of corrected visibility data for a wheat coleoptile under the influence of a copper solution. Visibility was obtained from double exposure specklegrams of 0.5 s time separation as a function of time after the addition of the copper solution. Obviously, there is an almost instantaneous decrease in visibility, and it takes almost an hour for the system to resume a new stationary state.

Position-resolved deterioration in limestone

The effects of a saturated humid atmosphere on the different mineralogic components constituting a natural limestone material were studied by an according double exposure record obtained with a 30 min time separation. The specklegram was scanned with a narrow laser beam resulting in high spatial resolution. Fig. 7 gives a comparison of visibilities obtained at identical locations in dry and humid atmospheres. Obviously, humidity affects different regions of the stone quite differently which may be set in relation to the local mineralogic composition.

Mobility analysis of microorganisms[5]

To end this chapter, let us review a somewhat different biological application which does not employ an optical double exposure analysis, but, nevertheless, relies on the correlation of speckle patterns.

When a volume containing living microorganisms is illuminated with laser light, a "boiling" scattered-light speckle field is produced. When such speckle fields are recorded as a function of time on 16 mm film, a similarity analysis of adjacent frames as a function of the framing rate yields a time constant characteristic of changes in the positions of the organisms. In the cited paper[5] processing of the optical data has been done completely in an image processing system. A dependence of mobility on environmental factors has been studied.

Double exposure particle imaging

The optical signals treated up to now in this paper were speckle patterns. There is a different class of problems in environmental studies which may be solved with a very similar optical technique. Let us illustrate this in the following example: A population of many identical individuals (bacteria, plankton, e.g.) is observed under a microscope. In studying the influence of environmental parameters on the activity of these organisms, their mobility may be an important indicator. Thus the microscopic picture will change with time and, once more, a quantitative description of similarity is required.

For an optical solution to this task, let us once again record a double exposure transparency of the two images to be compared, introducing an appropriate displacement Δx of the images between exposures. We have simulated this in a computer-drawn image of many identical particles as in Fig. 8, where each pair of particles is characterized by a displacement vector $\vec{d}_i$. The distribution of the random variable d_i can be expressed by its probability density function $f(d_x,d_y)$. Obviously, the crosscorrelation term describing the similarity of the two images to be compared, contains this probability density function[6,7]. We illustrate this for a one-dimensional exponential probability density function as in Fig. 9a. A photographic negative of the corresponding plot (cf. Fig. 8) was illuminated by collimated laser light and the resulting Young's fringe system (Fig. 9b) was Fourier transformed in a digital image processing system. The result in Fig. 9c shows that the probability density function can be regained in the displaced crosscorrelation terms. It is assumed that the autocorrelation of the individual patterns (central peak) is much narrower than the density function; otherwise a deconvolution is necessary.

The analysis shows that quite detailed information about a field of organisms may be obtained in this way. Under the influence of external environmental factors such a study reveals data on the characteristic response of the biological systems concerned. Early experiments[6] relying completely on optical processing were made on 16 mm film recordings of bacteria and allowed to discriminate between different stages of growth of a population. Meanwhile, the method has been employed in studies of technical flows[7,8]. There should, however, be manifold applications in connection with questions which are the reason for this paper.

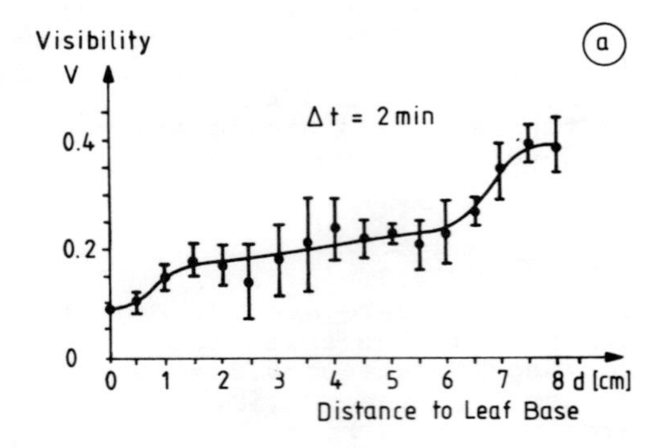

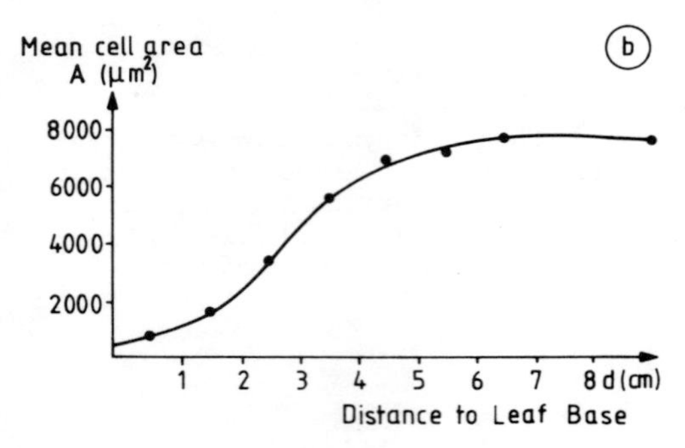

Figure 5. Growth study in plants: leaf of *Clivia miniata*; (a) visibility vs. distance from leaf base; (b) mean cell area for same distance.

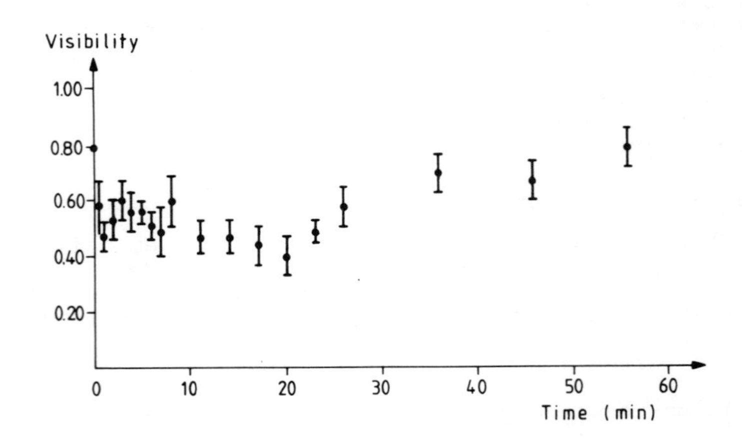

Figure 6. Influence of $CuSO_4$-solution in culture medium on growth of wheat coleoptile. Visibility of double exposure records (Δt=0.5 s) vs. time after the addition of solution.

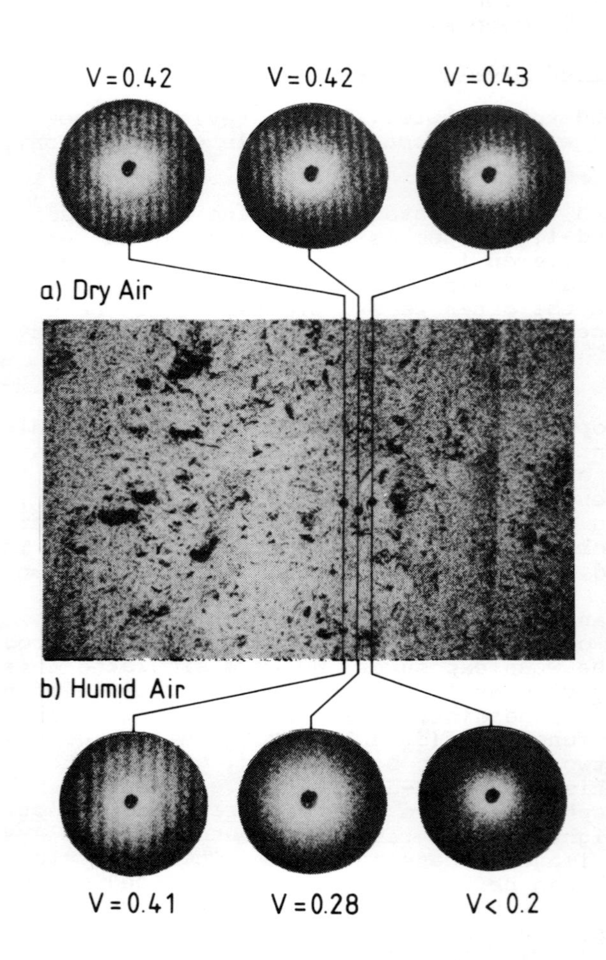

Figure 7. Influence of air humidity on deterioration of limestone. Visibility of Young's fringes (Δt=30 min) at identical locations in dry and humid environment.

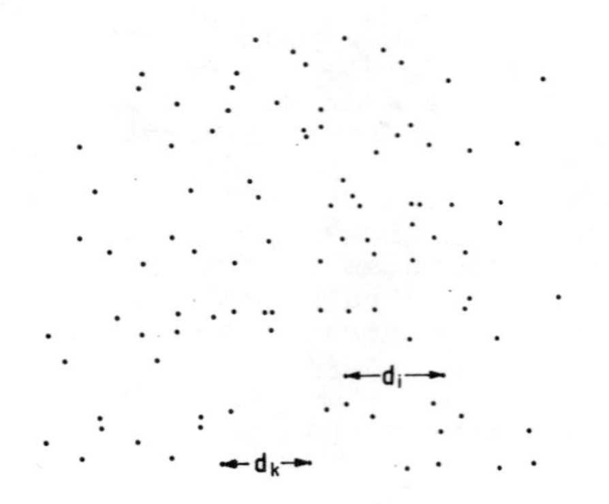

Figure 8. Simulated double exposure particle image ($\vec{d}_i$ displacement of particles).

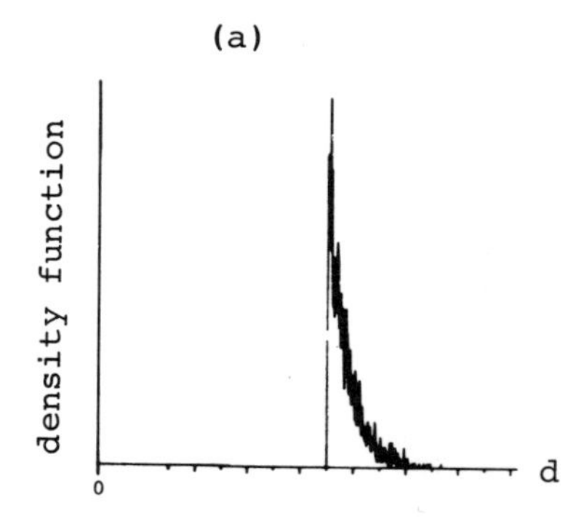

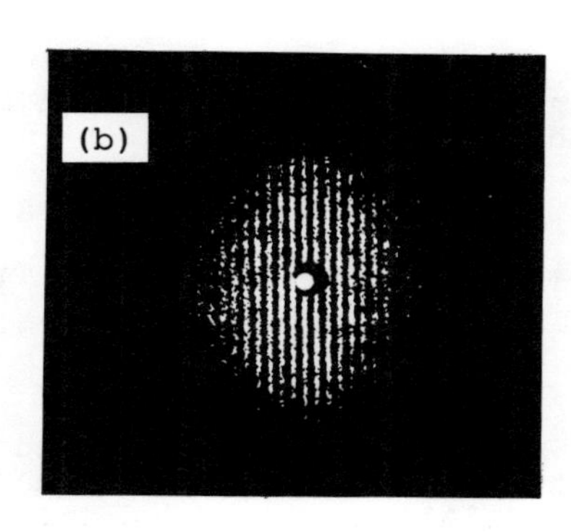

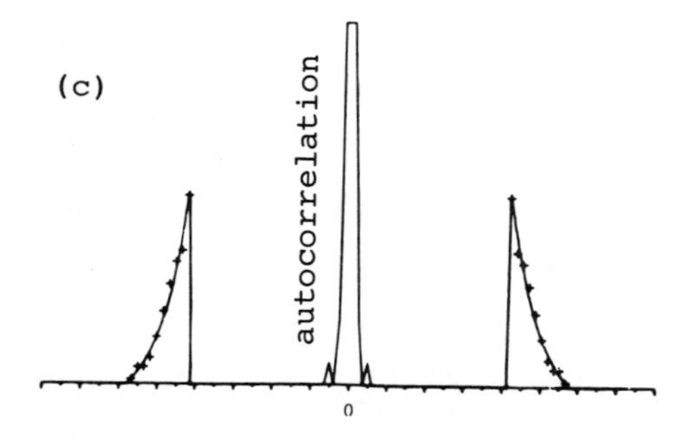

Figure 9. Evaluation of simulated double exposure particle image; (a) probability density function of displacement (exponential); (b) Young's fringe system; (c) autocorrelation reproducing density function (calculated from fringe system).

Conclusions

Coherent optical metrology has found successful applications in many quite different fields. This paper has been occupied with the use of holography, optical processing, and speckle correlation in an area of great common concern. Quite probably, the contribution of the proposed techniques to the handling and solution of the immense problems created by today's load on the environment will be rather small. Yet, knowledge and skill of optical scientists are challenged to participate in solutions. Today, there are many applications of optics which are estimated quite controversial as to their benefit for mankind. The utilization of coherent optics in environmental problems, however, should be a task worth the engagement of responsible scientists.

Acknowledgements

The research communicated in this paper is gratefully supported by grants from Deutsche Forschungsgemeinschaft and Bundesministerium für Forschung und Technologie. The author wishes to thank W. Arnold, D. Dirksen, H. Hinrichs, C. Hölscher, W. Platen, and K. Wolff for their contributions to the results reported. G. Weigelt was so kind to supply information about his work.

References

1. Van der Lugt, A.B., IEEE Trans. Inform. Theory, IT-10, 2. 1964.
2. Hinsch, K., Brokopf, K., Optics Letters 7, 51. 1982.
3. Erf, R.K. (ed.), "Speckle Metrology". Academic Press, New York. 1978.
4. Arnold, W., Hinsch, K., in: "Optics in Modern Science and Technology", ICO-13, Sapporo '84, 668.
5. Ebersberger, J., Weigelt, G., Optics Comm. 58, 89. 1986.
6. Böhm, H., Lohmann, A.W., Weigelt, G., Optics Letters 6, 162. 1981.
7. Hinsch, K., Schipper, W., Mach, D., Appl. Opt. 23, 4460. 1984.
8. Arnold, W., Hinsch, K., Mach, D., Appl. Opt. 25, 330. 1986.

The application of computer-generated oriented speckle screen to image processing

G. G. Mu, X. M. Wang and Z. Q. Wang

Institute of Modern Optics, Nankai University, Tianjin, P. R. C.

Abstract

A new technique of image encoding with a computer-generated oriented speckle screen as the encoding element is presented. The high contrast and high diffraction efficiency of this kind of screen enable the degree of modulation of encoded transparency and the ratio of signal to noise (SNR) of the decoded image to be improved.

Introduction

White-light optical processing is an attractive branch of image processing. And spatial encoding of the input image is commonly necessary. Several papers on white-light image processing with different encoding methods have been reported.[1-6] Image encoding can be done by techniques that utilize coherent speckles,[1] image holography,[2] or utilize Ronchi grating,[3] half-tone screen,[4] multiple-exposure speckle screen,[5] or oriented speckle screen[6] as encoding elements.

White-light processing with an oriented speckle screen as encoding element does not need laser source and shock-proof platform, does not introduce Moire fringes, can process large images with high resolution and give a much better color saturation.

In this report, a new technique of image encoding with a computer-generated oriented speckle screen is presented. There are some extra advantages for this kind of screen, such as high contrast and high diffraction efficiency, due to its binary transmittance. Thus the degree of the encoded transparency and the SNR of the decoded image are improved by this technique.

The making of computer screen

To make the screen, an oriented line segments pattern should be made at first. And reducing the pattern in scale, we get the desired speckle screen. To do so, a computer-controlled plotter is employed.

The initial coordinates of oriented line segments drawn by the plotter are modulated by inner random function of the computer. We can choose, for instance,

$$x = x_0 + a \text{ RAN } (0) \tag{1a}$$

$$y = y + b_0 + b \text{ RAN } (0) \tag{1b}$$

to be the initial coordinates of a line segment with length of l along x direction (generally $l \geq a$). To certain extents of variables, x and y, the computer generates speckle pattern by calling these two functions repeatedly, where RAN is a random function which gives random numbers between 0 and 1, x_0 a reference coordinate whose value will increase by a certain quantity periodically. The modulation amplitudes, a and b, determine the degree of disorder of the whole oriented speckles.

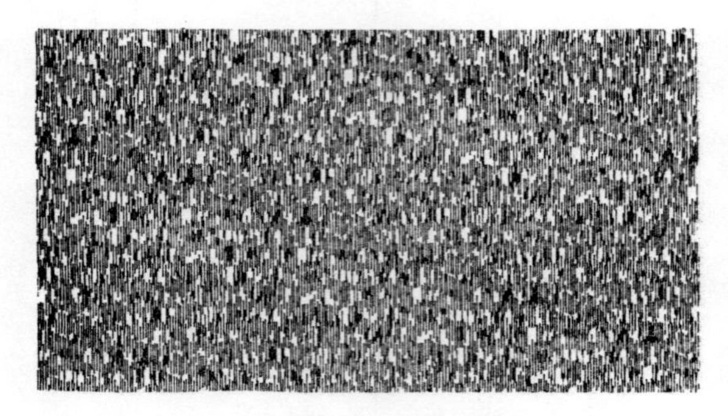

Figure 1. A piece of computer-generated speckle pattern

After reducing in scale, an oriented speckle screen with intensity transmittance

$$H(x,y) = h(x,y) * \text{rect}(x/l_0) \tag{2}$$

is obtained (see Figure 1), where h (x,y) is a binary function expressing the collection of points defined by Equation (1), rect, rectangular function, l_0, the length of any line segment in the screen, and * denotes convolution.

Because of the binary distribution of the computer-generated pattern, an oriented speckle screen with high contrast can be made. This is of advantage for improving the diffraction efficieny of encoded transparency and the SNR of output image.

The application of computer-generated speckle screen

In this paper, we discuss archival storage of color image, and addition and subtraction of images by means of pseudocolor encoding to show the application of computer-generated speckle screen.

Archival storage of color image

Similar to image processing with a laser speckle screen,[6] three-primary-color recordings are sequentially made on a black-and-white photographic plate with respect to three different angular positions of the screen which has superimposed onto the plate (see Figure 2). Then an encoded transparency is created with amplitude transmittance described as

$$t(x,y) = A - B\{T_r(x,y)\cdot[H(x,y) * \mathrm{rect}(x/l_o)] + T_g(x,y)\cdot[H(x',y') * \mathrm{rect}(x'/l_o)] + T_b(x,y)\cdot[H(x'',y'') * \mathrm{rect}(x''/l_o)]\} \quad (3)$$

where T_r, T_g and T_b indicate the intensity transmittances of the original color transparency according to the three primary colors. Coordinates (x,y), (x',y'), (x'',y'') are corresponding to the different positions of the screen, A, B, constants reflecting the property of recording material.

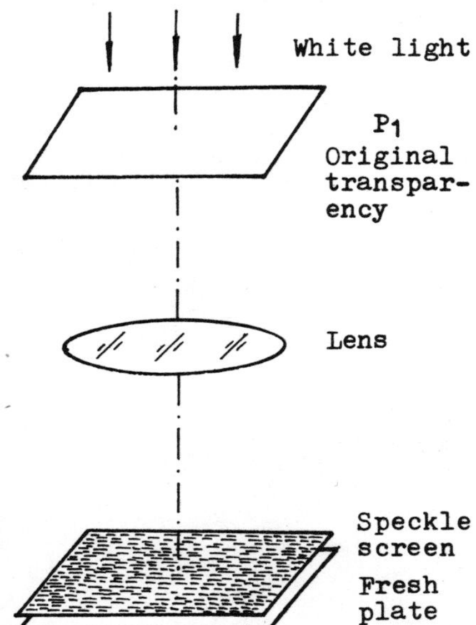

Figure 2. Encoding system

To reconstruct a color image from the corresponding black-and-white encoded transparency, the transparency should be placed at the input plane P_1 of a white-light processor as shown in Figure 3. The diffraction pattern is filtered through three-primary-color filters, as shown in Figure 4, in Fourier plane. Then the complex light distribution for every wavelength λ at the Fourier plane P_2 is

$$E(\alpha,\beta,\lambda) = \tilde{T}_r\left(\frac{\alpha}{\lambda f},\frac{\beta}{\lambda f}\right) * [\tilde{h}\left(\frac{\alpha}{\lambda f},\frac{\beta}{\lambda f}\right) \mathrm{sinc}\left(\frac{l_o\alpha}{\lambda f}\right)]_r + \tilde{T}_g\left(\frac{\alpha}{\lambda f},\frac{\beta}{\lambda f}\right) * [\tilde{h}\left(\frac{\alpha'}{\lambda f},\frac{\beta'}{\lambda f}\right) \mathrm{sinc}\left(\frac{l_o\alpha'}{\lambda f}\right)]_g + \tilde{T}_b\left(\frac{\alpha}{\lambda f},\frac{\beta}{\lambda f}\right) * [\tilde{h}\left(\frac{\alpha''}{\lambda f},\frac{\beta''}{\lambda f}\right) \mathrm{sinc}\left(\frac{l_o\alpha''}{\lambda f}\right)]_b \quad (4)$$

where $\tilde{T}_r$, $\tilde{T}_g$, $\tilde{T}_b$ and $\tilde{h}$ are the Fourier transforms of T_r, T_g, T_b and h respectively.

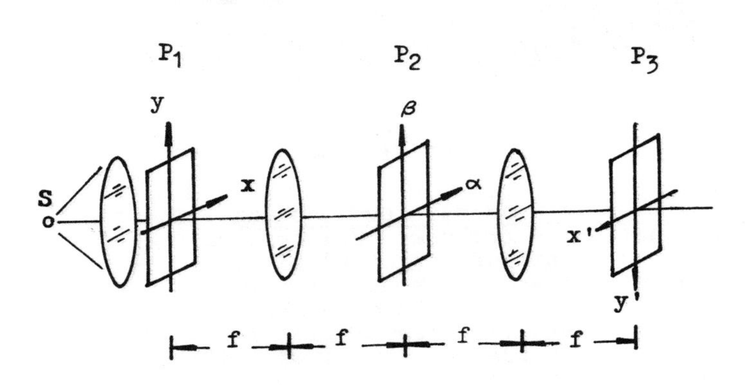

Figure 3. White-light optical processor

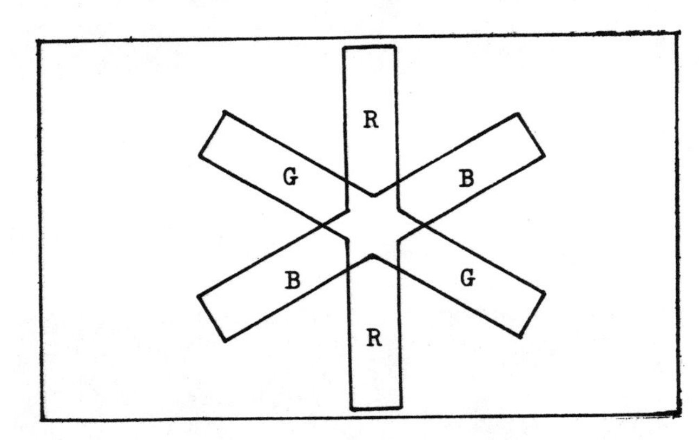

Figure 4. Three-primary-color filters

The complex light distribution in output plane is the inverse Fourier transform of $E(\alpha,\beta,\lambda)$, therefore the intensity distribution of the output image is given by

$$I(x,y) = \{T_r(x,y)[h(x,y) * \mathrm{rect}(x/l_o)]\}_r^2 + \{T_g(x,y)[h(x',y') * \mathrm{rect}(x'/l_o)]\}_g^2 + \{T_b(x,y)[h(x'',y'') * \mathrm{rect}(x''/l_o)]\}_b^2 \quad (5)$$

We can control the size of speckle in order to ensure there is no practical loss in resolution, so Equation (5) can be rewritten as

$$I'(x,y) = [T_r(x,y)]_r^2 + [T_g(x,y)]_g^2 + [T_b(x,y)]_b^2 \tag{6}$$

The color image is approximately reproduced.

Image addition and subtraction. For addition and subtraction of images by means of pseudocolor encoding, two black-and-white transparencies are encoded sequentially through this screen. Between the two exposures, a right angle in-plane rotation of the screen is made. And the decoding process is fulfilled in a white-light processor with two-primary-color filtering in Fourier plane. Then a pseudocolor image is obtained at the output plane. Parts displayed by primary colors denote the difference between the two original transparencies, and the others denote the common parts appear in both of them.

Experimental demonstrations

To experimental demonstrations, an oriented speckle screen with size of 25 X 25 square millimetrs is used. the size of any oriented speckle is made to be 10 micra wide and 170 micra long, and a two-primary-color original transparency is chosen for simplicity.

The result of archival storage of color film

Figure 5 shows the reproduced color image obtained by the technique with computer-generated speckle screen. Figure 6 is what we get with a laser-screen under the same conditions. And the size of the speckle in the screen is about one third of above-mentioned one. There is little difference between the two reproduced images.

Figure 5. Retrieved color image obtained by the technique proposed above

Figure 6. Retrieved color image obtained by technique with laser-screen

The result of image addition and subtraction by means of pseudocolor encoding. The two input images are chosen to be the first three and last three chinese characters of "Nan Kai Da Xue" (Nankai University) respectively. Figure 7 shows the pseudocolor image. The red and green parts show the absolute difference (subtraction) of these two images. And the yellow parts are the result of addition of these transparencies.

Figure 7. Addition and subtraction of two images

Although the two images to be used to do addition and subtraction in our experiment have binary type transmittances, the principle could be extended to continuous tone type transmittances. For continuous tone transparencies, we must measure the normalized color coordinates of pseudocolor image at output plane. Measurement can be performed successfully by colorimeter in terms of colorimetry.[7] Then the ratio of the intensity transmittances of the two transparencies can be determined.

Conclusions

It can be seen that although the computer-generated speckle pattern we made is not perfect, the common camera we used to reduce the speckle pattern is not just right for photoreduction, and the encoding and filtering processes performed in our experiment are not on the best of conditions, the experimental results show clearly that this kind of speckle screen has some advantages over others and it will play an important role in white-light image processing. We believe that super quality speckle screen could be easily made by means of common photoreduction technique, and the quality of processed image could be greatly improved.

References

1. Yu, F. T. S., Ruterbusch, P. H., "Color-image Retrieval from Coherent Speckles by White-light Processing," Appl. Opt., Vol. 21, p. 2300. 1982.
2. Mu, G. G., Wu, F. X., Wang, Z. Q., "White-light Image Processing with Image Holographic Encoding," Applied Laser, Vol. 3, pp. 5-10. 1983.
3. Yu, F. T. S., "White-light Processing Technique for Archival Storage of Color Films," Appl. Opt., Vol. 19, pp. 2457-2460. 1980.
4. Chiang, C. K., Mu, G. G., Liu, H. K., "Contact-screen Pulse-width Encoding of a Polychromatic Image on a Single Piece of Black-and-white Film," Acta Optica Sinica, Vol. 4, pp. 706-713. 1984.
5. Huang Degen, Xu Daxiong, Shen Shuqun, Yu Chongxiu, "Speckle Modulation Screen for White-light Image Processing," Acta Optica Sinica, Vol. 4, p. 139. 1984.
6. Mu, G. G., Wang, Z. Q., Gong, Q., Song, Q. W., "White-light Image Processing using Orientated Speckle Screen Encoding," Acta Optica Sinica, Vol. 5, pp. 113-117. 1985.
7. Mu, G. G., Chiang, C. K., Liu, H. K., "White-light Image Addition and Subtraction by Colorimetric Measurement," Opt. Lett., Vol. 6, pp. 389-391. 1981.

A hybrid real-time correlation system of aerial stereo photographs

R. T. Hong*, J. Tsujiuchi, N. Ohyama and T. Honda

Imaging Science and Engineering Laboratory, Tokyo Institute of Technology
4259 Nagatsuta, Midori-ku, Yokohama, 227, Japan

Abstract

A digital-optical hybrid system composed of a computer and a coherent optical filtering system is successfully applied to perform a real-time correlation of aerial stereo photographs. A LCLV (Liquid Crystal Light Valve) as an incoherent-to-coherent convertor and a thermoplastic photoconductor material for holographic recording of MSF (Matched Spatial Filter) are used, and a sharp correlation peak with a good SNR is obtained.

Introduction

In contour mapping of aerial stereo photogrammetry, a pair of stereo aerial photographs is taken at two view points O_1 and O_2 separated by the base line B as shown in Fig.1. The relative variation of the coordinates of the same object in the stereo pair photographs (the parallax) is used to determine the height of the object. Fig.1 refers to the simple and ideal case of vertical photographs, in which the camera axes are parallel with each other and perpendicular to the reference plane. The height H_c and H_d of two objects C and D with respect to the reference plane are obtained by

$$P_c/B=f/H_c \quad P_d/B=f/H_d, \tag{1}$$

where f is the focal length of the camera lens, P_c and P_d are parallaxes of C and D respectively, and we have finally

$$\Delta H=H_d-H_c=Bf(1/P_d-1/P_c) \tag{2}$$

so the height contour mapping can be done by measuring parallaxes of every point in the stereo pair. In order to determine the parallax, the corresponding position of the same object in the stereo pair should be found. This work has been performed so far mainly by human binocular stereo vision and it is very time-consuming and trublesome. In order to overcome such a difficulty, an automatic operation is promising if the parallax measurement can be performed automatically. One of the effective methods of measuring parallax is to take the cross-correlation of segments of both pictures. The correlation could be well done only by computer but the processing time is too long. A coherent optical matched filtering system can provide a real-time correlation, but direct operation on original pictures seems not effective because of relative distortions in both pictures for slopes and of dominant low frequency components. A computer-preprocessing of the original pictures is very effective and a hybrid operation through an incoherent-to-coherent convertor is employed. In this paper, we shall demonstrate that a real-time correlation peak with high accuracy has been successfully attained by using the hybrid system.

Description of the system

In this system (Fig.2), a LCLV is used as an incoherent-to-coherent convertor. The incoherent input image is from a CRT connected to a computer to provide a preprocessed or original picture. The aerial photograph film is scanned by a drum scanner, the signal thus obtained is sent to a computer through an A/D convertor. The sampling pitch of the drum scanner(one pixel) is 100 μm(corresponding to 4 meters on the ground). The number of pixels in one frame was 512x512 with an 8 bit quantized gray lever. This CRT image is projected onto a LCLV input face and the coherent image is read out from the LCLV by making reflect a He-Ne laser beam, and sending to the coherent optical filtering system(Fig.3), where L_3 is the Fourier transform lens and the MSF is made by recording a Fourier transform hologram of the input image. A CCD detector(CCD_1) is used to check the coherent readout image and another one (CCD_2) to detect the correlation peak.

Spatial resolution and optimization of the incoherent-to-coherent convertor's parameters

a) Aperture 1 (Fig.1) is used to limit the area of image from the CRT into the LCLV down to 120x120 pixels. The resolution of the LCLV is 16 lines/mm (by using a resolution chart) and the usable aperture is 8x8 mm. The SBW(Space Bandwidth) of the LCLV is enough to ensure

* On leave of absence from South-Western Institute of Technological Physics, Chengdu, China

the conversion of 120x120 pixels from incoherent image to coherent one with a good spatial resolution.

b) As incoherent-to-coherent convertor, the linearity of the image conversion of the LCLV from incoherent input intensity I_{in} to coherent output amplitude A_{out} is very important. The optimization of the parameters of the LCLV is performed.

The minimization of $I_{out\ off}$ (coherent output intensity I_{out} of LCLV in off-state and in cross polarizer) is necessary for obtaining the high contrast performance of the LCLV.

The readout light entering the LCLV should be linearly polarized, collimated and perpendicular to the surface of the LCLV. It is most important that the orientation of the polarized light entering the LCLV must be set in the preferred direction of the LCLV for minimizing $I_{out\ off}$.[1,2]

The optimum bias voltage of the LCLV V_{rms} should be the threshold voltage of the LCLV V_s, above which the linearity between I_{in} and A_{out} is kept.

Fig.4 shows the logarithmic relation between incoherent input relative intensity and coherent output relative amplitude of the LCLV ($I_{in}/I_{min\ in}$ & $A_{out}/A_{min\ out}$) under the condition of the optimum linearity and $I_{max\ in}/I_{min\ in}$=const, and Table 1 shows the corresponding operation parameters of these curves. From curve a to g, which have almost the same linearity and dynamic range, the proportional relation between ΔI (the range of incoherent input absolute intensity $I_{max\ in} - I_{min\ in}$) and f_{opt} (the optimum ac frequency of the LCLV) is clear. This is very important for parameters setting of the LCLV. The ac frequency of the LCLV should be determined by the brightness of the CRT screen so as to ensure the optimum linearity.

Table 1 The corresponding parameters of the curves in Fig.4

PARAMETERS	AC FREQUENCY	BIAS VOLTAGE	INCOHERENT INPUT ABSOLUTE INTENSITY ($\mu W/cm^2$)	
CURVE	f_{opt} (KHZ)	v_{rms} (volt)	$I_{max\ in}$ (THE MAXIMUM)	$I_{min\ in}$ (THE MINIMUM)
a	0.25	4.0	2.1	0.02
b	0.50	4.5	4.2	0.04
c	1.00	5.1	8.4	0.08
d	2.00	5.6	16.7	0.17
e	4.00	6.0	33.3	0.33
f	8.00	6.1	66.6	0.66
g	12.00	6.1	100.0	1.00

Real-time correlation of stereo pair

Original image

If the original images to take **correlation with each other** have dominant low frequency components (Fig.5), the correlation gives a broad peak, low SNR and a large correlation aperture area ratio. Fig.6a and 6b show the result of correlation. The most intense peak results from the correlation of aperture of two pictures, the correlation peak of images is weaker. When the aperture area ratio is less than 1/4, mismatching takes place for autocorrelation.

High-pass filtered image by computer-preprocessing

a) Calculation of the high-pass filtered image

The intensity of the high-pass filtered image can be calculated by

$$g'_{ij}=(g_{ij}-\bar{g}_{ij})/\sigma_{ij}, \tag{3}$$

where g_{ij} is intensity transmittance of the photograph at a pixel(i,j), $\bar{g}_{ij}$ is the mean value of g_{ij} over the normalized window of NxN pixels, and σ_{ij} is the standard deviation of g_{kl} from the mean value $\bar{g}_{ij}$,

$$\bar{g}_{ij}=\frac{\sum_{k=i-(N-1)/2}^{i+(N-1)/2}\sum_{l=j-(N-1)/2}^{j+(N-1)/2} g_{kl}}{N^2}, \quad \sigma_{ij}=\sqrt{\frac{\sum_{k=i-(N-1)/2}^{i+(N-1)/2}\sum_{l=j-(N-1)/2}^{j+(N-1)/2}(g_{kl}-\bar{g}_{ij})^2}{N^2}} \tag{4}$$

b) Selection of the normalized window size N

The dominant components frequency f_v of the high-pass filtered image will increase as N decreases and the other components will be attenuated. On the other hand, the MTF of the LCLV will decrease with f_v . Therefore, when the number of pixels of the incoherent input is kept invariant (120x120), there is an optimum window size, with which the high frequency components of coherent output of the LCLV can be enhanced as much as possible. The optimum N can be determined easily by putting the CCD_1 at the position of the MSF and checking the power spectrum of the high-pass filtered image with different N, the optimum value of N is 7 in this case.

c) Experimental results

Fig.7 shows the high-pass filtered image with 7x7 pixels window size. Fig.8 is the cross-correlation with a correlation aperture area ratio of 7/100. The SNR is high, the peak is sharp, a small aperture area ratio is obtained, and mismatching is relieved.

Power spectrum flattened image by computer-preprocessing

a) Calculation of the power spectrum flattened image

Let the original image be g(x,y) (512x512 pixels). Using an FFT algorithm to get the power spectrum of g(x,y), we have

$$\widetilde{F}(g(x,y)) = G(f_x, f_y) = |G(f_x, f_y)| \exp(iG_p(f_x, f_y)), \tag{5}$$

where $\widetilde{F}(\ )$ denotes Fourier transform, f_x and f_y are spatial frequencies along x and y directions and G_p denotes the phase component of G, the Fourier spectrum of g(x,y). The flattening calculation of the power spectrum becomes

$$G'(f_x, f_y) = G(f_x, f_y) / |G(f_x, f_y)| = \exp(iG_p(f_x, f_y)) \tag{6}$$

Then,using inverse FFT to get the power spectrum flattened image

$$\widetilde{F}^{-1}(G'(f_x, f_y)) = \widetilde{F}^{-1}(\exp(iG_p(f_x, f_y))) = g'(x,y) \tag{7}$$

The image $g'(x,y)$ is used as a new incoherent input to the system as shown in Fig.9.

b) Experimental results

Fig.10 demonstrates that the power spectrum flattening preprocessing can give a similar result to that obtained by high-pass filtering preprocessing. A high SNR, a sharp correlation peak and a small aperture ratio is obtained. Mismatching is also relieved.

Correlation aperture area ratio

Although we want to get a small aperture area ratio, the correlation peak intensity will decrease rapidly as the ratio decreases and the SNR also decreases by the edge diffraction of a small aperture. If a hard aperture 2 (Fig.2) is put in the coherent filtering system, a good correlation peak is difficult to obtain when the ratio is less than 1/7 in the correlation of high-pass filtered images and a photographic plate is used for recording the MSF. This phenomenon can be avoided by putting aperture 1 in the incoherent system instead of aperture 2 (Fig.2). An apodized aperture is formed on the LCLV by the blurred image of the edge of the aperture 1, and the edge diffraction effect of the aperture is relieved. As a result, the aperture ratio can be decreased down to 1/25 under the same condition. Furthermore, if using a thermoplastic photoconductor device to record the MSF, the ratio can be decreased down to 1/100 by the high diffraction efficiency of this device($\eta > 30\%$). As a result , a picture of 12x12 pixels can be correlated with 120x120 pixels.

Conclusion and discussion

Parallax measurement in stereo aerial photographs is attained with high operation speed and high accuracy. This is due to the real-time optical correlator connected to a computer with a LCLV as an incoherent-to-coherent convertor. Optimization of operating parameters of the LCLV is performed to get the maximum dynamic range. Computer-preprocessing of images such as high-pass filtering or power spectrum flattening is very effective to increase the SNR and to relieve the mismatching. A thermoplastic photoconductor device provides a MSF with insitu and rapid development and results a high SNR of the output signal with a high diffraction efficiency.

The LCLV used in this experiment was made initially as a device to project a CRT image onto a large size screen with an incoherent projection system. So, the resolution(16 lines /mm) is not very high because of its thick liquid crystal layer (about 12μm), and no special attention was paid for reading out by a laser light. This gives a difficulty of using whole area(40x40 mm) of the LCLV as image plane because of interference fringes produced between two substrates and a residual birefringence, and only 8x8 mm area is possible to use as a coherent read out. If larger size of image plane is available, the area of image to be processed becomes larger and much higher performance is expected.

Compared with full computer operation, the advantage of using hybrid processing is its high speed. On the other hand, if the terrain to be measured has a slant, two images of that place in the stereo pair images will have different shapes and magnifications. Such a local distortion of images causes a difficulty of taking correlation and correction of distortion should be made. The optical method[3] for approaching to this work is inflexible for the variety of distortion, and computer processing[4] is promising as a further preprocessing.

A proposal for contour mapping using this hybrid system is as follows:

1. Registration of a set of stereo images in a computer in consideration of camera position and direction of axis.
2. Preprocessing such as high-pass filtering or power spectrum flattening.
3. Display one of the stereo images, e.g. image L, in CRT with a certain area and convey it into the coherent optical correlator with the aid of LCLV as incoherent-to-coherent convertor.
4. Record the MSF by taking a Fourier transform hologram of image L as a thermoplastic hologram.
5. After correcting distortion with an assumed condition, display a small subimage from image R in CRT, and take cross-correlation with image L recorded in the MSF.
6. If a correlation peak appears, measure the location of the output correlation peak, the parallax, by CCD_2 and store the parallax in a memory of computer together with the coordinates of the center of subimage of R.
7. If no correlation peak appears, change the condition for correcting distortion, and go to the step 5 again.
8. After taking correlation over all image, retrieve all points with a constant parallax and connect them with a smooth curve. This gives a contour line. Continue the same operation with different amount of parallax.

Acknowledgments

The authors would like to extend their thanks to Dr. N. Nishida, NEC Corporation, for giving us the facility of using a LCLV for this experiments.

References

1. Dwight, W. Berreman, J. Opt. Soc. Am., Vol. 63, pp. 1378. 1973.
2. W. P. Bleha, L. T. Lipton and E. Wiener-Avnear, Opt. Eng, Vol. 17, pp. 374. 1978.
3. N. Ohyama, T. Honda and J. Tsujiuchi, Optics Comm, Vol. 37, pp. 339. 1981.
4. T. Honda, M. Hosoi, N. Ohyama and J. Tsujiuchi, The 13 th Congress of the International Commission for Optics, Conference Digest, pp. 186, 1984.

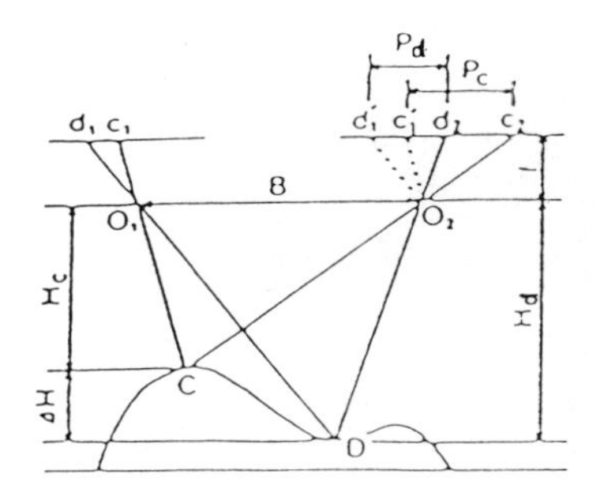

Fig 1. Principle of photogrammetry

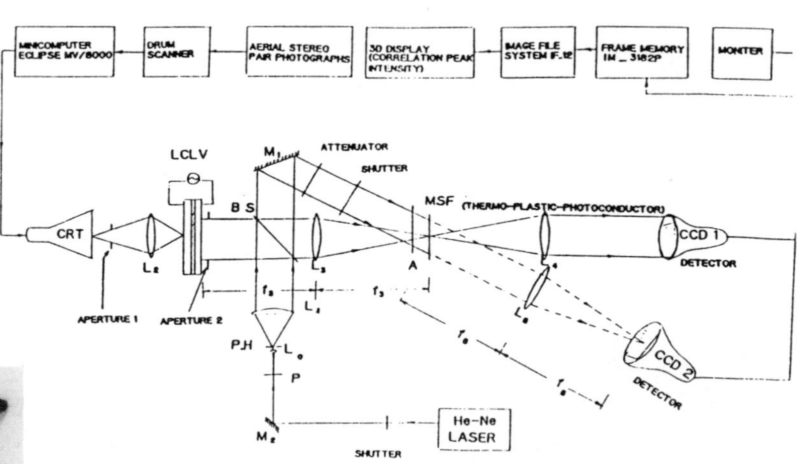

Fig. 2 The schematic diagram of the hybrid real-time correlation system

Fig 3. Coherent optical filtering system

Fig 5. Original image (right aerial photograph, incoherent input)

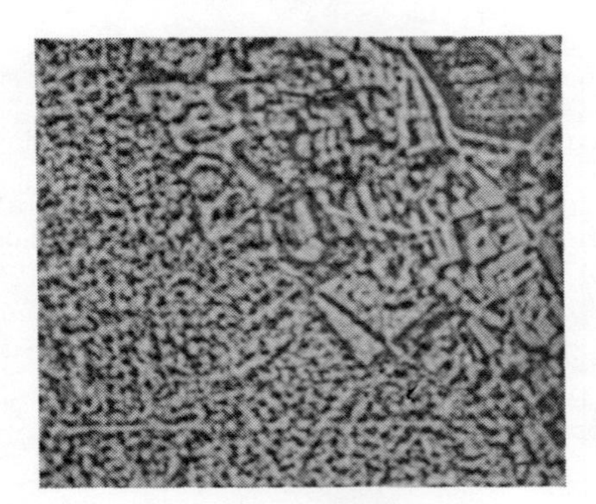

Fig 7. High-pass filtered image(right aerial photograph, incoherent input)
Normalized window size: 7x7 pixels

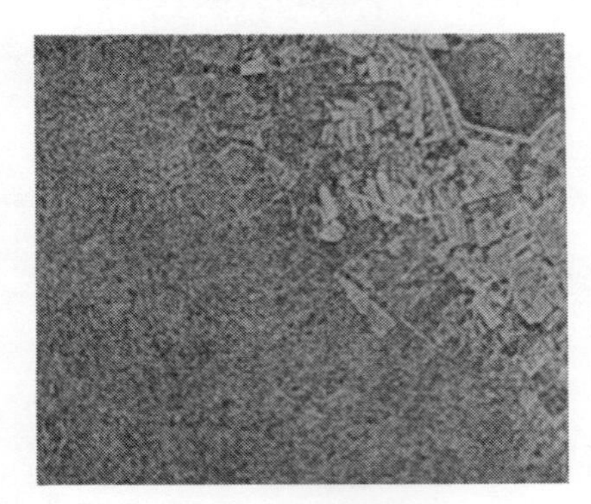

Fig 9. Power spectrum flattened image(right aerial photograph, incoherent input)
The calculation is performed from the original picture with 512x512 pixels.

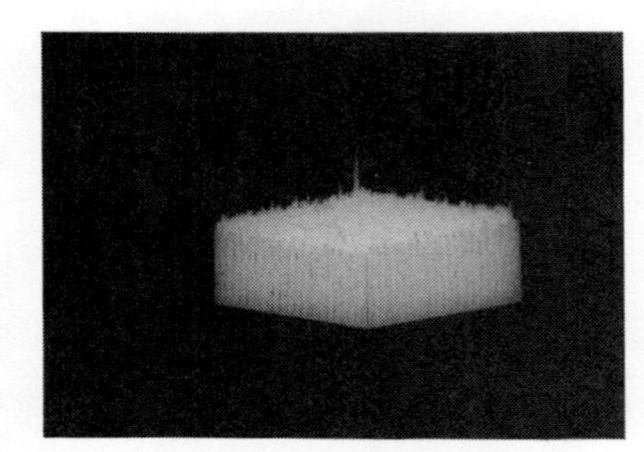

Fig 10. Cross-correlation of the power spectrum flattened image. MSF is recorded using thermoplastic photoconductor. Correlation aperture ratio: 7/100.

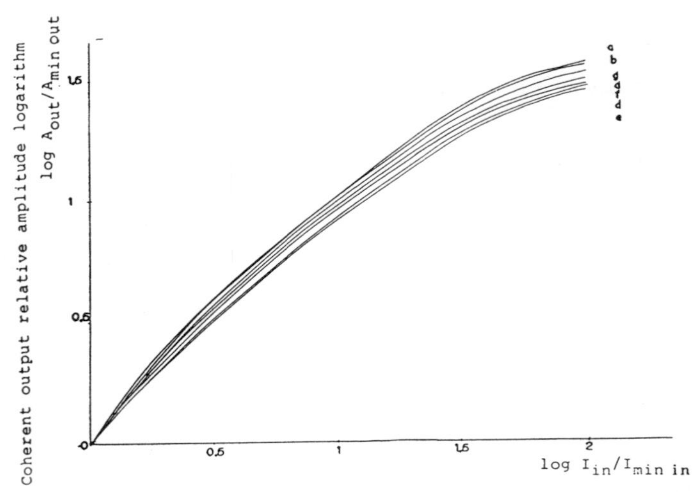

Fig.4 The relation between log $I_{in}/I_{min\ in}$ and log $A_{out}/A_{min\ out}$ of the LCLV under the optimum linearity and dynamic range

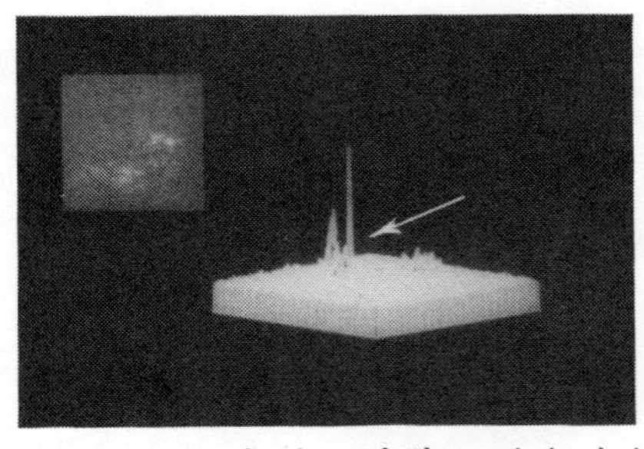

Fig 6a. Autocorrelation of the original image MSF is recorded using thermoplastic photoconductor. Correlation aperture ratio: 1/4. The arrow shows the position of the peak. Mismatching takes place.

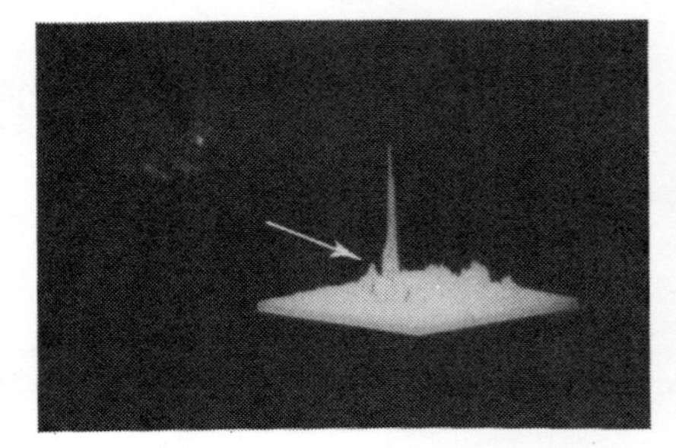

Fig 6b. Cross-correlation of the original image MSF is recorded using thermoplastic photoconductor. Correlation aperture ratio: 1/1. The arrow shows the position of the peak.

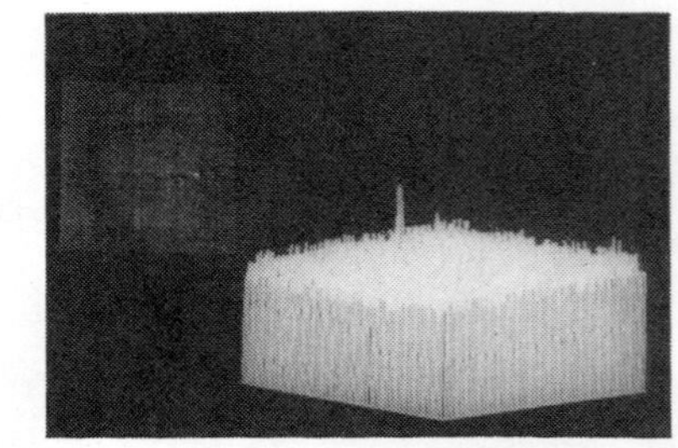

Fig 8. Cross-correlation of the high-pass filtered image. MSF is recorded using thermoplastic photoconductor. Correlation aperture ratio: 7/100.

Real-time Optical Logical Operations Using Grating Filtering

Fu zi-ping, Hsu Da-hsiung and Wang Ben

Department of Applied Physics Beijing Institute of Posts & Telecommunications,
Beijing, P.R. China

Abstract

Without spatial modulator real-time optical logical operations are scarcely reported. However, in this paper a new method to realize 16 different real-time optical logical operations of two variables is proposed and experimentally demomstrated. It employs two cascaded grating filtering of 4f systems as well as auxiliary coherent beams.

Introduction

Optical systems which are able to perform binary logic operations are of importance to optical computing. Optical computing presents advantages over electronic computing in terms of speed, patallelism and interconnection architecture. Experimental approachs to perform such binary logic operations include methods of nonlinear optics, θ modulation and projection. But these last two, in common, are not real-time, in other words, preprocessing for input signals is required. In the following, we propose a new method that can realize 16 possoble logical operations of two varables, employing two cascaded 4F systems. The system is based on spatial grating filtering and can be realized in a simple way of conventional optics.

Optical computing system

Our optical computing system is shown in figure 1. It consists of two typical processing systems F_1 and F_2 with grating filtering. Plane P_1 is the input plane. The black area in it represents signal "0", while white "1". C and D represent auxiliary coherent beam. For certain logical operation, C and D are encoded with a specified binary code of "1" or "0", here "1" denotes the present of the beam whereas "0" the absent.

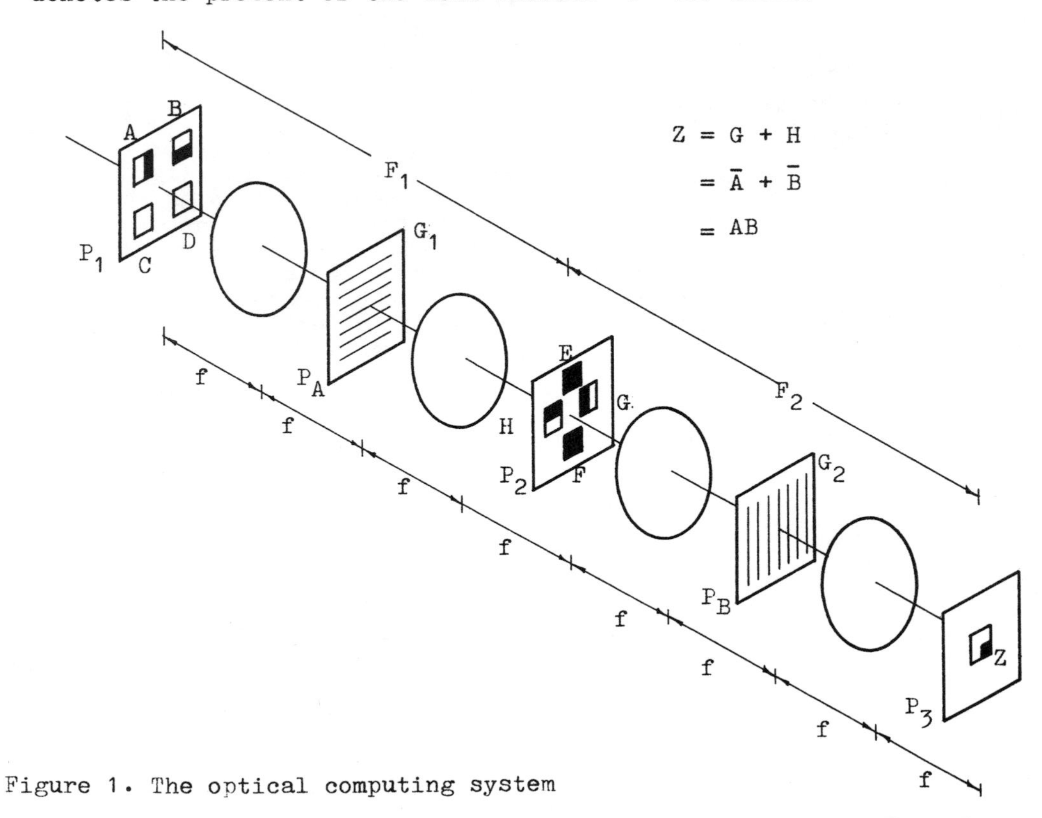

Figure 1. The optical computing system

Plane P_2 is the imaging plane of first grating filtering system F_1. There are four apertures F, E, G, H, in this plane. Here "1" represents an open aperture, for light passing through, and "0" represents a closed aperture for blocking light.

Plane P_3 is the imaging plane of second grating filtering system F_2, meantime, it is the output plane for optical logical operations of signals A and B.

P_A, P_B are spatial filtering plane with two One-dimension gratings G_1, G_2 in each of them respectively. The twogratings are positioned like this: if their frequencies are the same, the directions of them are orthogonal to each other, whereas the directions parallel if the frequencits are different.

The realization of 16 possible binary logic operations of two variables

Table I shows the 16 possible logic functions of the two binary variables. The first two rows on the left-hand side represent the combinations of values of the two binary variables A and B. The rows labeled f_0 - f_{15} show the 16 possible lagic functions resulting from combining two binary variables.

Table I Sixteen Possible Functions of Two Binary Variable

Function		Function Name 1 (True Logic)	0 (TrueLogic
INPUT A	0 0 1 1		
B	0 1 0 1		
OUTPUT			
f_0	0 0 0 0	F	T
f_1	0 0 0 1	AND	OR
f_2	0 0 1 0	$A\bar{B}$	$A + \bar{B}$
f_3	0 0 1 1	A	A
f_4	0 1 0 0	$\bar{A}B$	$\bar{A} + B$
f_5	0 1 0 1	B	B
f_6	0 1 1 0	XOR	$\overline{XOR}$
f_7	0 1 1 1	OR	AND
f_8	1 0 0 0	NOR	NAND
f_9	1 0 0 1	$\overline{XOR}$	XOR
f_{10}	1 0 1 0	$\bar{B}$	$\bar{B}$
f_{11}	1 0 1 1	$A + \bar{B}$	$A\bar{B}$
f_{12}	1 1 0 0	$\bar{A}$	$\bar{A}$
f_{13}	1 1 0 1	$\bar{A} + B$	$\bar{A}B$
f_{14}	1 1 1 0	NAND	NOR
f_{15}	1 1 1 1	T	F

Table II The Rēspodence of 16 Logic Functions to The Operating State of The System

planes / gates	P_1 A	B	C	D	$P_{A(G1)}$ state		P_2 E	F	G	H	$P_{B(G_2)}$ state		P_3 Z
F	◐	◒	0	0	X		0	0	0	0	Y		f_0
AND	◐	○	0	0	X		◒	0	0	0	Y		f_1
$A\bar{B}$	○	◒	0	1	Y	−	0	0	0	◐	X		f_2
A	◐	◒	0	0	Y		0	0	1	0	X		f_3
$\bar{A}B$	◐	○	1	0	Y	−	0	0	◒	0	X		f_4
B	◐	◒	0	0	Y		0	0	0	1	X		f_5
XOR	◐	◒	0	0	X	−	1	0	0	0	Y		f_6
OR	◐	◒	0	0	X	+	1	0	0	0	Y		f_7
NOR	◐	◒	1	1	X	+	1	1	0	0	Y	−	f_8
$\overline{XOR}$	◐	◒	0	1	Y	−	0	0	1	1	X	−	f_9
$\bar{B}$	◐	◒	0	1	Y	−	0	0	0	1	X		f_{10}
$A + \bar{B}$	◐	◒	0	1	Y	−	0	0	1	1	X	+	f_{11}
$\bar{A}$	◐	◒	1	0	Y	−	0	0	1	0	X		f_{12}
$\bar{A}$+B	◐	◒	1	0	Y	−	0	0	1	1	X	+	f_{13}
NAND	◐	◒	1	1	Y	−	0	0	1	1	X	+	f_{14}
T	◐	◒	1	1	X	+	0	1	0	0	Y		f_{15}

All the 16 functions for the two binary variables can be real-time realized by means of the optical computing system above with the orintations of the gratings set differently for each of the logical operations. As we know, when two signals of inpot pass through the system F_1 or F_2, the output will be sum or subtraction of their amplitudes, depending on grating position, namely, it is sinusoid or cosine with respect to axis. So each of the operations could be acquired with a specific combinations of grating position in F_1 and F_2. The relation between 16 logic functions to the operating states of system F_1 and F_2 is showed in table II.

Now, we illustrate the NAND gate consisting of F_1 and F_2. For example, in plane P1 of Fig. 1. A and B are apertures of input signal, while C and D are both encoded by "1". In plane P_2, E and F are blocked when orintation of grating one is vertical. When the system F1 operates on subtraction, H and G act as NOT operation of A and B, respectively.

$$G = \bar{A} ; \qquad H = \bar{B}$$

In plane P_3 , OR operation between G and H is obtained when system F_2 is in the state of addition.

$$\begin{aligned} Z &= G + H \\ &= \bar{A} + \bar{B} \\ &= \overline{AB} \end{aligned}$$

So the NAND gate may be realized by the optical computing system. In addition, we also studied two type of error in the outcome of this computing system, they are false alarm and miss related to speckle noise and lens aberration. The transmission loss of the system can be decided mainly by the diffractive efficiency of the two gratings.

Reference

1. B·K·Jenkins et al; Appl.Opt. , 1984, 23, No19, 3465
2. J·Tanida et al; J·Opt. Soc.Am., 1983, 73, No6, 800
3. H·Sartelt et al; J·Opt.Soc.Am.A, 1984, 1, No9, 944
4. C·C·Guest et al; Appl.Opt, 1984, 23, No19, 3444
5. S·H ·Lee et al; J·Opt.Soc.Am. , 1970, 60, No9, 1037

Determining the phase function in a two dimensional Fourier spectrum

Xu Ke-shu, Xu Xiang-dong*

Department of Physics, Fudan University, Shanghai, China

Abstract

By using a coherent optical information processing system with two orthogonal polarization channels we have measured the Fourier phase experimentally for several two dimensional real objects. One is a shifted square. Its Fourier phase values measured by experiment are in agreement with the calculated phase values quite well. Another object is a binary alphabet, the real part and the imaginary part of its Fourier spectrum has been investigated.

Introduction

The optical phase factor plays an important part in inverse scattering and inverse source problem as well as in adaptive optics. But in the optical field the intensity is the only measurable quantity and during the recording process the phase factor will be lost. Therefore many scientists devoted to the phase-retrieval[1-5]. We had measured the phase function of one dimensional Fourier spectrum and part of results has been reported elsewhere[6]. In this paper we will extend further on the experimental determination of the phase function for a two dimensional Fourier spectrum.

By using an optical information processing system with two orthogonal polarization channels we have separated the real part and the imaginary part and measured the phase of the Fourier spectrum for a real object function. The details of this method and the experimental set up will be described in section 2. In section 3, the experimental results are given for two objects. One is a shifted square, its Fourier phase values measured by the experiment is compared with the phase values calculated theoretically. Another object function is a binary alphabet, the real part and the imaginary part of its Fourier spectrum has been investigated. In section 4, we will discuss the uniqueness of the Fourier phase-retrieval in our experiment, and some features of the Fourier phase will be discussed also.

Experimental method and set up

The principle of the experimental method is as follow: for a real and asymmetrical object function $f(x,y)$ (Fig.1a) its Fourier spectrum is a complex function of the spatial frequency f_x and f_y and the phase function of the spectrum is the gool for us which will be measured experimentally. We construct a new function $f'(x,y)$ related to the original function $f(x,y)$ but which is to be symmetrical with respect to the origin of the coordiate axes (Fig. 1b):

$$
\begin{aligned}
f'(x,y) &= f(x,y), & x>0,\ y>0, \\
&= 0, & x<0,\ y>0, \\
&= f(-x,-y), & x<0,\ y<0, \\
&= 0, & x>0,\ y<0,
\end{aligned}
\qquad (1)
$$

* East China Normal University, Shanghai, China

If we illuminate the object function $f(x,y)$ and its symmetrizer $f(-x,-y)$ with linear polarized light along the x- and y-direction respectively, then the symmetrical function $f'(x,y)$ possessing two orthogonal polarization states can be written in a vectorial form (Fig. 1c) of

$$g(x,y) = \vec{i}\, f(x,y) + \vec{j}\, f(-x,-y), \tag{2}$$

where $\vec{i}$ and $\vec{j}$ represent the unit vectors along the x- and y-axis. Now $g(x,y)$ is regarded as the input function for a coherent optical information processing system with two orthogonal polarized channels. Its Fourier transform will be

$$\begin{aligned} G(f_x,f_y) &= \iint_{-\infty}^{+\infty} g(x,y) \exp\left[-2\pi j\,(f_x x + f_y y)\right] dxdy \\ &= (\vec{i} + \vec{j}) \iint_{-\infty}^{+\infty} f(x,y) \cos 2\pi (f_x x + f_y y)\, dxdy \\ &\quad -j(\vec{i} - \vec{j}) \iint_{-\infty}^{+\infty} f(x,y) \sin 2\pi (f_x x + f_y y)\, dxdy \end{aligned} \tag{3}$$

Clearly, the first term of the right side of (3) is the real part of the Fourier spectrum of the object function $f(x,y)$ but polarized along the direction of 45° to the x-axis, while another term is the imaginary part of the Fourier spectrum of the object function but with the linear polarization state along the direction of 135° to the x-axis. It is easy to separate the real part and the imaginary part of the Fourier spectrum because of their polarization states are orthogonal each other. By using a polarization analyser we can measure the intensity distributions of the imaginary part and the real part:

$$I_i(f_x,f_y) = \left[\iint_{-\infty}^{+\infty} f(x,y) \sin 2\pi (f_x x + f_y y)\, dxdy\right]^2 \tag{4}$$

$$I_r(f_x,f_y) = \left[\iint_{-\infty}^{+\infty} f(x,y) \cos 2\pi (f_x x + f_y y)\, dxdy\right]^2 \tag{5}$$

Hance, the phase function of the Fourier spectrum of the object function can be evaluated by

$$\Theta(f_x,f_y) = \tan^{-1} \left[\frac{\pm\sqrt{I_i(f_x,f_y)}}{\pm\sqrt{I_r(f_x,f_y)}} \right] \tag{6}$$

In section 4, we will discuss how to decide the signs of the numerator and the denominator in (6), which will permit us to determine the Fourier phase uniquely.

Fig.2 shows the optical configuration of the experimental set up which is a conventional coherent optical information processing system. In the front focal plane of the Fourier transform lens L, a liquid gate is located which contains the object with the amplitude transmittance $f'(x,y)$ and two polaroids with their polarization direction corresponding to the x- and y-axis respectively. Before the rear focal plane of the lens L, an analyser is located and rotated at 45° and 135° successively with respect to the x-axis for obtaining the intensity distributions of the real part and the imaginary part of the Fourier spectrum. The intensity distributions is measured by scanning the optical fiber and fed to a photomultiplier. In this experimental work, the key point is to reduce the phase difference introduced by the collimated light beam passes through the two orthogonal polarized channels. So we have to select high quality optical substrates and employ appropriate adjusting techniques for setting up the liquid gate. In our case, this phase difference is less than $\lambda/40$.

Experimental results

A. A shifted square as an object

For the purpose of verifing the reliability of our method for two dimensional Fourier phase-retrieval, a shifted square with its center situated at x=2a, y=2a is examined.

The amplitude transmittance of the shifted square $f(x,y)$ is

$$f(x,y) = \begin{cases} 1, & a \le x \le 3a, \quad a \le y \le 3a \\ 0, & \text{otherwise,} \end{cases} \tag{7}$$

then its Fourier transform is

$$F(f_x,f_y) = 4a^2 \text{ sinc } 2\pi f_x a \cdot \text{sinc } 2\pi f_y a \cdot e^{-2\pi(2af_x + 2af_y)\, j}, \tag{8}$$

Conventionally, we can only get the power spectrum experimentally and its phase factor $e^{-2\pi(2af_x + 2af_y)j}$ will be lost, the intensity distribution of the spectrum is

$$I(f_x,f_y) = (4a^2 \text{ sinc } 2\pi f_x a \cdot \text{sinc } 2\pi f_y\, a)^2 \tag{9}$$

Fig. 4a shows the well-known intensity pattern of the Fourier spectrum of the square for any displacement whatever. In our case, let the input function (Fig. 3b) be

$$\begin{aligned} f'(x,y) &= 1, \text{ but polarized along x-axis for } a \le x \le 3a,\ a \le y \le 3a \\ &= 1, \text{ but polarized along y-axis for } -a \ge x \ge -3a,\ -a \ge y \ge -3a \\ &= 0, \text{ otherwise.} \end{aligned}$$

By rotating the analyser at 45° and 135° respect to the x-axis, we can obtain the intensities of the real part and the imaginary part

$$I_r(f_x,f_y) = \left[4a^2 \text{ sinc } 2\pi f_x a \cdot \text{sinc } 2\pi f_y a \cdot \cos 4\pi a(f_x + f_y)\right]^2 \tag{11}$$

$$I_i(f_x,f_y) = \left[4a^2 \text{ sinc } 2\pi f_x a \cdot \text{sinc } 2\pi f_y a \cdot \sin 4\pi a(f_x + f_y)\right]^2 \tag{12}$$

The intensity pattern of the real part and the imaginary part are shown in Fig. 4b and Fig. 4c respectively. It shows that for the pattern of the real part there is a bright stripe in the center of the frequency plane and the number of the stripes is odd. For the pattern of the imaginary part, there is dark in the center and the number of the stripe is even.

For the shifted square, from the measured intensity distributions of real and imaginary parts, we have obtain the phase factor of the Fourier spectrum. They should be equal to the phase factor contained in equation (8) and can be calculated easily. The intensity distributions of the real and the imaginary parts and the observed phase values as well as the calculated phase values are shown in Fig. 5, Fig. 6 and Fig. 7. They are measured by sampling the spectrum with the scanning fiber along different direction for instance, in Fig. 5, we have sampled it along the horizontal frequency axis f_x; for Fig. 6, the sampling is taken along some fixed value of f_x; and in Fig. 7, we have got the sampling along f' axis which follows the direction of 45° respect to the f_x-axis. In these diagrams the calculated phase values are denoted by the solid lines and the crosses denote the measured phase values. By making a comparison between the experimentally measured and the calculated phase values we have found that their maximum deviation is less than 6°. So we have come to the conclusion that our method is suitable for two dimensional Fourier phase-retrieval.

B. Binary English alphabet as an object

Several English letters have been taken as the objects, but here we only give the experimental results for an English letter c. Fig. 8 shows the pattern of the Fourier power spectrum and the intensity patterns of real part and imaginary part. The intensity distributions of the real part and the imaginary part sampled along the f_x- and f_y-axis is shown in Fig. 9 and Fig. 10 respectively.

Discussion

As we have mentioned before the phase function of (6) should be determined uniquely, now we first examine the phase and the amplitude of the Fourier spectrum at zero spatial frequency. In equation (3), the amplitudes of the real part and the imaginary part of the Fourier sprctrum are

$$A_r(f_x,f_y) = \iint_{-\infty}^{+\infty} f(x,y) \cos 2\pi (f_x x + f_y y) \, dxdy \quad (13)$$

$$A_i(f_x,f_y) = -\iint_{-\infty}^{+\infty} f(x,y) \sin 2\pi (f_x x + f_y y) \, dxdy \quad (14)$$

In equation (14), since $f_x=0$, $f_y=0$, thus for any object function, $Ai(0,0)=0$. It includes the meaning in two folds: one is the phase values could be zero at $f_x=0, f_y=0$ for any object function, namely

$$\Theta(0,0) = \tan^{-1} \frac{A_i(0,0)}{A_r(0,0)} = 0, \quad (15)$$

another is that the intensity of the imaginary part could be zero at the center of the frequency plane and hence a dark range appears there. It is shown in Fig.4c and Fig.8c.
In the equation (13), if $f_x=0$, $f_y=0$, for a real object, the amplitude of the real part $A_r(0,0)$ is maximum, as shown in Fig.4b and Fig.8b, the brightest stripe appears at the zero frequency.

for deciding the signs of the amplitude for the real part and the imaginary part of the Fourier spectrum of point (f_x,f_y), take the example as in our case, that the object function $f(x,y)$ is situated in the first quadrant, so the coordinates of x and y are positive. For a real object function, the amplitude transmittance $f(x,y)$ should be positive. According to (13), the amplitude of the real part of the Fourier spectrum will then be positive at and near the center of the frequency plane. After leaping across the dark range, the sign of the amplitude will be changed. In this way, we can dicide the signs of the amplitude for the real part, it is shown in Fig.8b.

For the same reason, the amplitudes of the imaginary part will be nagative for $0<(f_x x + f_y y)<\frac{1}{2}$. In Fig.8c, the points satisfied this condition is located in the first stripe on the right side of the origin on the frequency plane. The signs of the amplitudes for the imaginary part can be determined by the same way as for the real part.

Once the magnitudes and the signs of the real part and the imaginary part for the Fourier spectrum are decided, the Fourier phase factor can be determined uniquely.

This work is supported by the China Science Foundation and authors would like to express the gratitude. The authors also express their gratitude to Professor Zhi-ming Zhang for the valuable discussions.

References

1. H. A. Ferwerda, in Inverse source problems in optics, H. P. Baltes, ed. (Springer-Verlay, Berlin, 1978), Chap. 2.
2. R. W. Gerchberg and W. O. Saxton, Optik 35, 237 (1972)
3. G. Ross, M A. Fiddy , and M. Neito-Vesperinas, in Inverse scattering problems in optics, H. P. Baltes,

ed. (Springer-Verlay, Berlin, 1980)
4. J. R. Fienup, Appl. opt. 21. 2758 (1982)
5. M. A. Fiddy, B. J. Brames and J. C. Dainty, Optics Letters 8, 96 (1983)
6. Xu Ke-shu, Ji Rong-cai and Zhang Zhi-ming, Optics communications 50, 85 (1984)

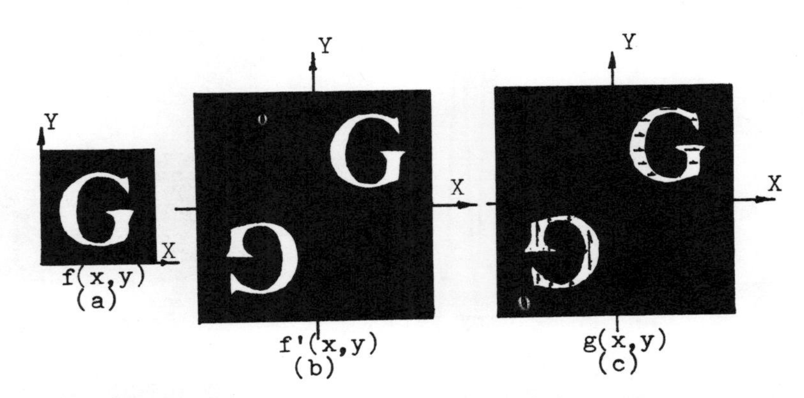

Fig.1 (a) Asymmetrical object function. (b) Symmetrical function formed from (a). (c) Symmetrical function in (b) but under orthogonal linear polarization states.

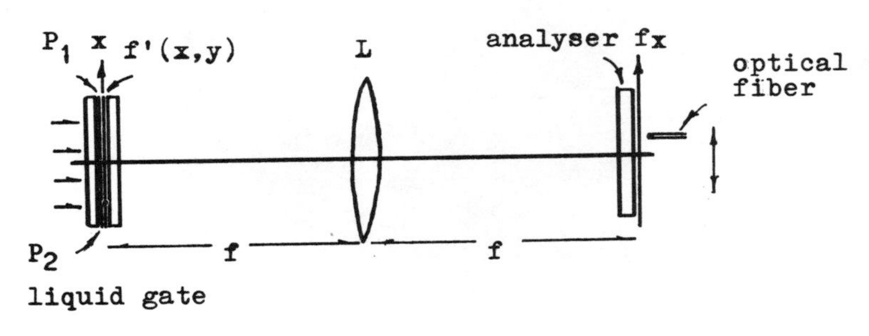

The pass-planes of polaroids P_1 and P_2 corresponding to the x- and y-axis.

Fig.2 Optical system with orthogonal linear polarization channels.

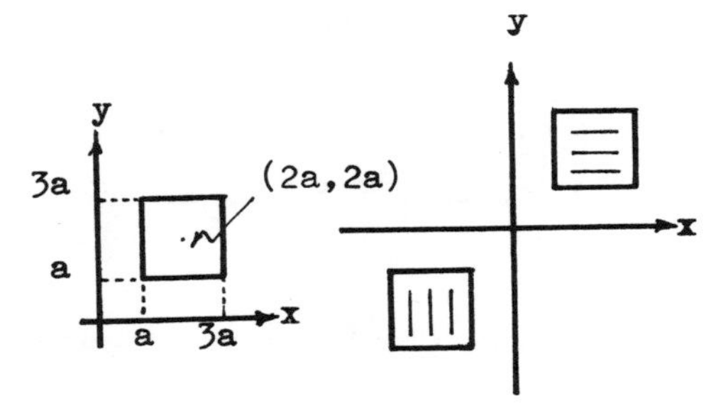

Fig.3 (a) A shifted square as the object. (b) The input function formed from (a).

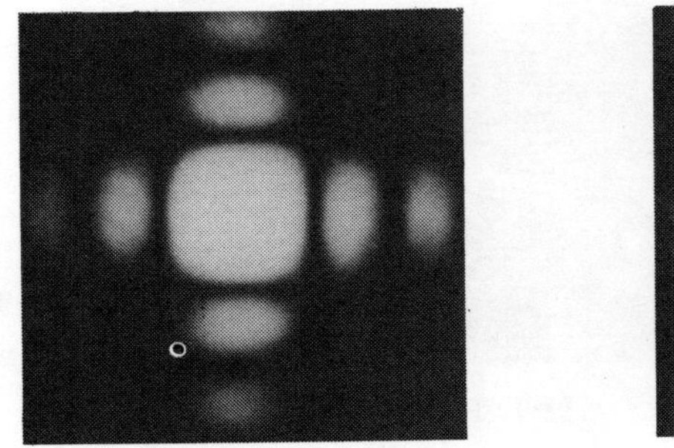

(a) Fourier power spectrum

(b) Real part

(c) Imaginary part

Fig.4 The intensity patterns for the Fourier transform of the shifted square.

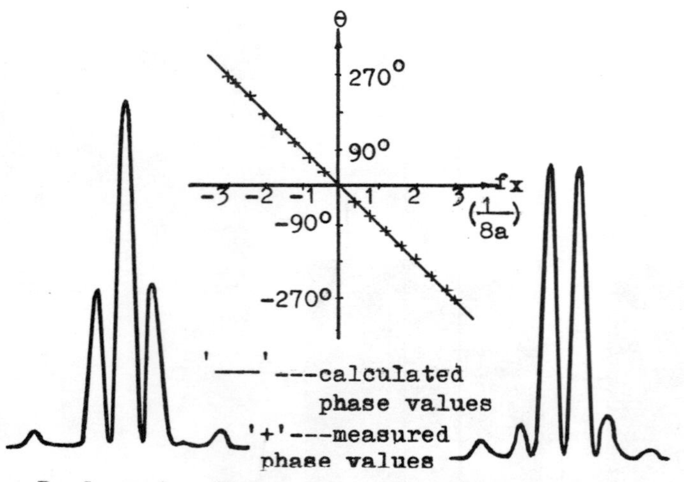

Real part Phase function Imaginary part

Fig.5 The intensity distributions and the phase function of the Fourier spectrum measured along f_x-axis for shifted square.

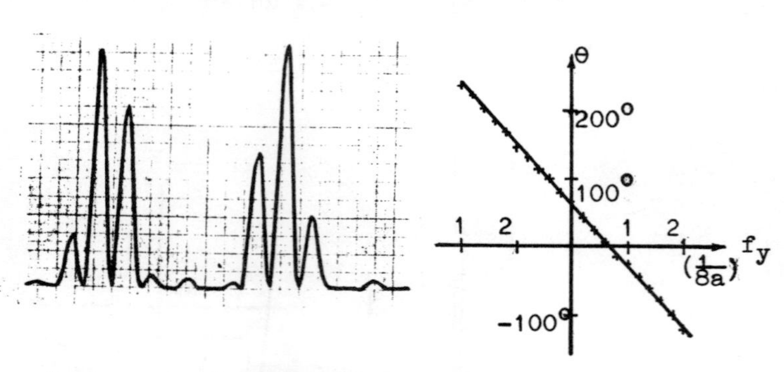

Real part Imaginary part Phase function

Fig.6 The intensity distributions and the phase function of Fourier spectrum sampled as f_x=-0.36mm^{-1} for shifted square.

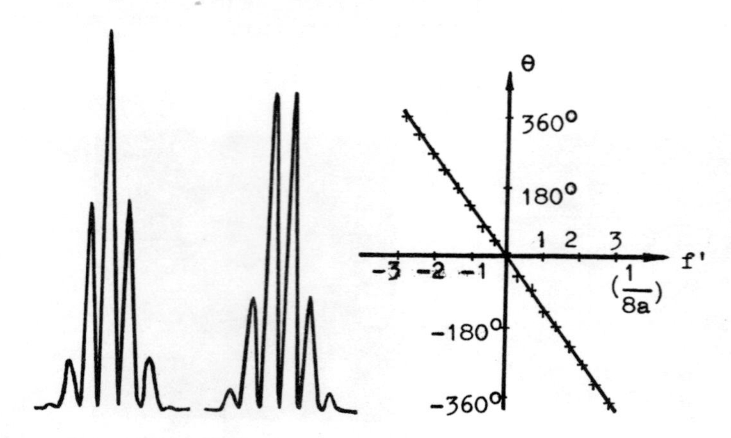

Real part Imaginary part Phase function

Fig.7 The intensity distributions and the phase function of Fourier spectrum sampled along f'-axis (45° to the f_x-axis) for shifted square.

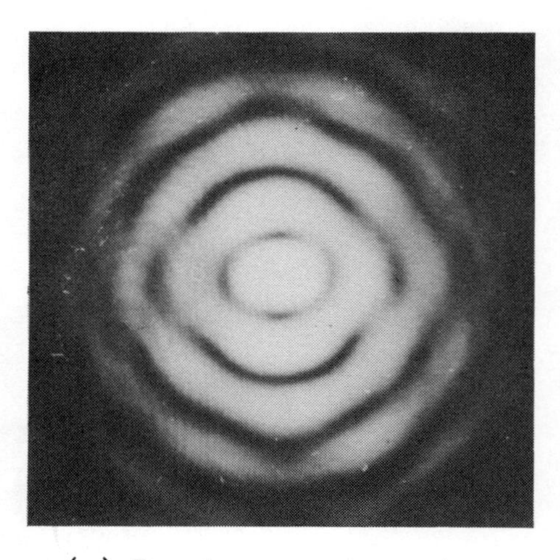

(a) Fourier power spectrum

(b) Real part

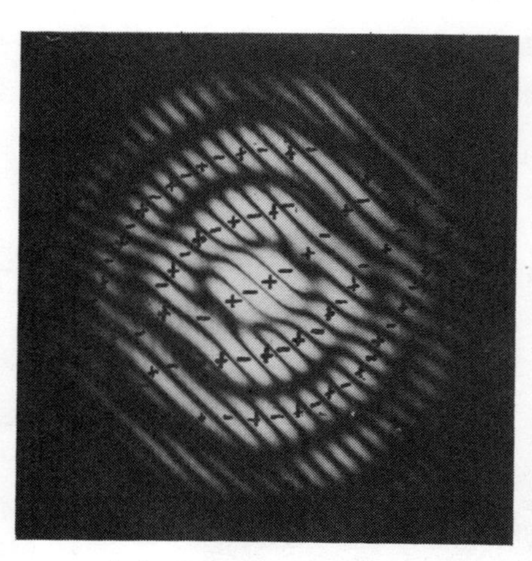

(c) Imaginary part

Fig.8 The intensity patterns and the signs of amplitudes for the Fourier transform of letter 'c'.

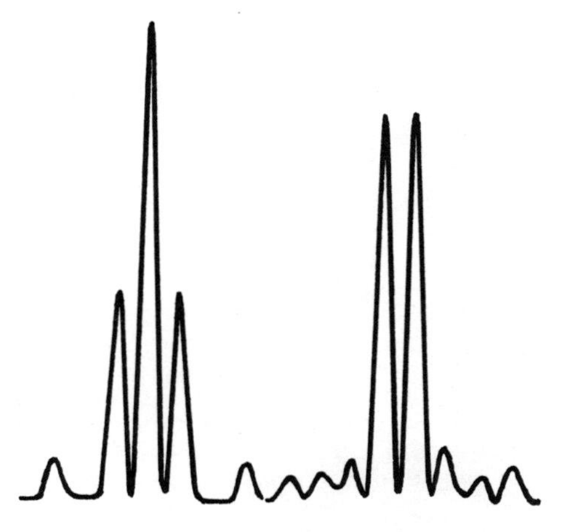

Fig.9 The intensity distributions of the real and imaginary parts of Fourier spectrum measured along f_x-axis for letter 'c'.

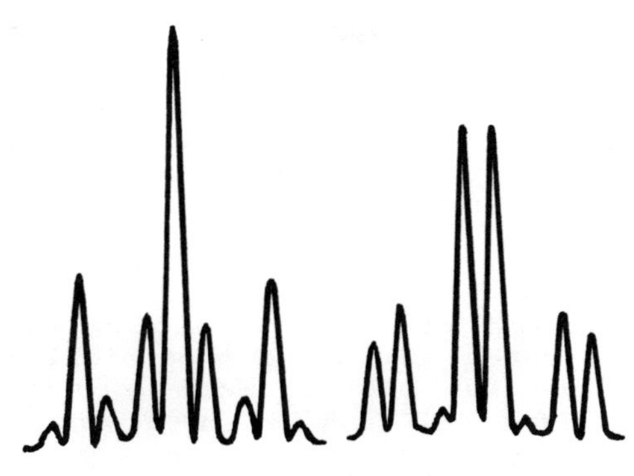

Fig.10 The intensity distributions of the real and imaginary parts of Fourier spectrum sampled along f_y-axis for letter 'c'.

Principle of a photonical computer with biological and medical applications

A.M. Landraud

Laboratoire de Reconnaissance des Formes, Institut de Programmation
Université Pierre et Marie Curie, tour 45-46, 3ème étage,
4, Place Jussieu, 75230 Paris Cedex 05, France

Abstract

A method of high resolution image processing, based upon the theory of projections, is presented. By using photonical devices including holographic filters and an original technique for coding image information from the scanning apparatus, we have shown that the process allows a fast parallel reconstruction with high resolution.

Avantages d'un calculateur photonique

Les techniques de calcul sur ordinateur sont parvenues à un haut degré de flexibilité, de précision et de rapidité grâce à la mise en oeuvre d'algorithmes ce complexités temporelle et spatiale de plus en plus optimales. Elles permettent aussi d'effectuer rapidement des modifications de programme et de procéder à des traitements variés comportant des opérations linéaires ou non linéaires. De ce point de vue, un système analogique classique ne peut rivaliser avec l'ordinateur. Ces techniques sont très performantes lorsqu'elles portent sur un nombre relativement faible d'éléments de résolution dans l'image. Cependant, lorsque le nombre de données devient très important, ce qui est généralement le cas dans le traitement des images, les calculs séquentiels sur ordinateur deviennent rapidement très longs alors qu'un traitement photonique équivalent s'effectuerait de façon pratiquement instantanée. Dans ce cas, l'ordinateur ne peut rivaliser avec ces techniques.

Le premier avantage du traitement photonique est la simultanéité du transfert d'un très grand nombre de données bidimensionnelles, nombre d'autant plus grand que la bande passante du système est plus étendue. Le "produit espace-bande" des lentilles de bonne qualité est d'environ 10^7. La deuxième propriété fondamentale est que les multiplications analogiques, entre autres opérations mathématiques, s'effectuent à la vitesse de la lumière traversant une "transparence" optique (environ 10^{-13} sec.), l'intensité lumineuse transmise par un film photographique uniformément exposé étant le produit analogique de l'intensité incidente par la transmittance du film. En raison de ces deux propriétés combinées, former l'image d'une transparence convenablement codée sur une seconde transparence revient au calcul simultané d'un nombre très supérieur à 10^6 multiplications indépendantes.

Dans le cas des traitements en lumière incohérente, la grandeur physique utilisée est l'intensité lumineuse. Les fonctions mises en jeu doivent donc être partout non négatives, les intensités négatives ou les transmittances négatives n'étant pas réalisables. La méthode consistant à ajouter un terme constant, ou "biais", présente l'inconvénient d'introduire des termes dépendant du signal. Ce problème peut être résolu par d'autres approches.[1] C'est ainsi que la convolution d'une répartition d'intensité avec un masque dont la représentation mathématique est une fonction réelle, mais présentant des lobes négatifs, peut être réalisée en formant deux images: l'une convoluée avec la partie positive du masque, l'autre avec la partie "négative". On procède ensuite à la différence de ces deux images positives. Ce résultat peut être obtenu au moyen d'un dispositif opto-mécanique utilisant deux faisceaux de lumière polarisée à angle droit de façon à effectuer la convolution simultanément sur deux voies indépendantes. En éclairage cohérent laser, les filtres holographiques et les filtres binaires générés par ordinateur permettent d'effectuer des opérations mathématiques analogiques sur des fonctions complexes.

Il est théoriquement possible d'effectuer les principales opérations impliquées par le traitement des images: transformées de Fourier, additions, soustractions, convolutions, etc., par voie photonique au moyen de dispositifs appropriés convenablement éclairés. En particulier, on peut effectuer photoniquement sans problème une transformée de Fourier instantanée à deux dimensions au moyen d'une lentille éclairée par un faisceau laser, alors que les traitements séquentiels correspondants sur ordinateur nécessitent beaucoup de place mémoire et des calculs comparativement beaucoup plus longs. Les opérations de convolution peuvent être effectuées en lumière naturelle. Des réseaux élémentaires permettent d'effectuer un grand nombre d'additions et de soustractions simultanées. Il est également possible d'effectuer photoniquement des opérations de dérivation à différents ordres et d'intégration.[2] On peut aussi générer par ordinateur des filtres de fréquences spatiales, des hologrammes binaires et des micro-circuits à très haute échelle d'intégration utilisables ensuite photoniquement pour des transformations particulières dans un véritable calculateur photonique.[3]

Principe d'un calculateur photonique pour le traitement des projections

La méthode de traitement photonique des images dont les grandes lignes sont rappelées dans cet article, est basée sur la théorie de la reconstruction à partir de projections. Les projections sont définies ici comme des intégrales de la fonction f à reconstruire selon des droites joignant la source S au détecteur P, le balayage s'effectuant dans un plan de coupe perpendiculaire au plus grand axe de l'objet à trois dimensions étudié (géométrie transaxiale). La tomographie transaxiale conventionnelle implique des calculs sur ordinateur. Nous avons montré qu'un processus photonique cohérent par double diffraction donne de bons résultats sans traitement numérique. Notre système de traitement comprend des filtres holographiques et des filtres d'amplitude synthétiques et nous avons développé une technique originale de codage de l'information pour la saisie des données.[4] La mise en évidence d'un modèle unitaire de projection nous a permis d'appliquer la méthode à la reconstruction de divers types d'objets à trois dimensions, en particulier à la reconstruction du profil d'indice des fibres optiques et à la reconstruction d'images biologiques et médicales utilisant des rayonnements X, Gamma, ultra-sons, laser ou lumière naturelle.[5] La méthode peut également s'appliquer à la reconnaissance des trajectoires des particules dans les chambres à bulles, au traitement des images océanographiques et à la reconnaissance des caractères.

Partant de la théorie de la tomographie axiale transverse avec assistance d'ordinateur, nous avons introduit un modèle unitaire de projection. Le principe de la méthode est basé sur une relation mathématique générale entre l'information sur les données prétraitées, ou projections, et l'image à reconstruire. Les fonctions intervenant dans les calculs sont supposées appartenir à l'espace de Hilbert des fonctions de carré sommable possédant une transformée de Fourier. Pour chaque angle θ de projection, les données sont d'abord projetées sur un sous-espace de fonctions à une dimension, $p_\theta(x)$, puis rétroprojetées sur un sous-espace de fonctions à deux dimensions, $p_\theta(x,y)$, de sorte que $p_\theta(x,y) = p_\theta(x)$. Le processus est mis en oeuvre pour tous les angles de projection compris entre 0 et π . En superposant toutes les rétroprojections, on obtient une "image de sommation" à partir de laquelle la répartition du paramètre inconnu caractérisant l'image à reconstruire peut, d'après le théorème de Radon, être déterminée de façon unique. Si l'on désigne par f(x,y) le paramètre de l'objet à reconstruire, les opérations fondamentales du processus de restauration sont:

$$p_\theta(x) = \int_{-\infty}^{+\infty} f(x,y)dy, \quad \text{(projection)} \tag{1}$$

$$p_\theta(x,y) = p_\theta(x), \quad \text{(rétroprojection)} \tag{2}$$

$$h(x,y) = \int^{\pi} p_\theta(x,y)d\theta. \quad \text{(sommation)} \tag{3}$$

Entre l'image de sommation, ou image prétraitée, et la fonction inconnue f(x,y), il existe une relation simple linéaire de convolution:

$$h(x,y) = f(x,y) * (x^2 + y^2)^{-1/2}, \tag{4}$$

(ou le symbole "*" représente l'opérateur de convolution), ou en coordonnées polaires:

$$h(r,\phi) = f(r,\phi) * 1/r \tag{5}$$

et entre les transformées de Fourier associées:

$$H(\rho,\Phi) = F(\rho,\Phi) \;.\; 1/\rho. \tag{6}$$

Les équations (5) et (6) sont fondamentales et utilisées en tomographie avec assistance d'ordinateur. Nous avons montré qu'il est possible d'éviter les traitements numériques en utilisant un dispositif photonique impliquant un codage approprié de l'information. L'ensemble des rétroprojections $p_\theta(x,y)$ est enregistré directement sous la forme de grilles d'amplitude sur une plaque photosensible au moyen d'une lentille cylindrique. La superposition de toutes les grilles ainsi obtenues pour chaque angle de projection fournit une image prétraitée synthétique codée correspondant à l'image de sommation de l'équation (5). Les formulations (5) et (6) montrent qu'il est possible théoriquement de reconstruire $f(r,\phi)$ par un simple filtrage suivi d'une transformée de Fourier. Nous avons réalisé une lentille holographique compacte qui réalise simultanément le filtrage "en ρ" et la transformée de Fourier. On déduit de l'équation (6):

$$f(r,\phi) = h(r,\phi) * T.F^{-1}(\rho). \quad (7)$$

En pratique, pour éviter une telle convolution à deux dimensions, on tire profit de la linéarité du processus. Les différentes grilles, enregistrées sur un film, sont filtrées séparément et subissent une transformée de Fourier en temps réel au moyen d'un filtre holographique disposé dans un dispositif éclairé par laser. L'image résultante peut être visualisée avec une caméra ou un enregistrement photographique. Une très grande précision dans l'image restaurée a été obtenue en apodisant convenablement le filtre.

Rappel de l'algorithme de sommation des projections filtrées et modèle unitaire

La méthode la plus efficace et aussi la plus simple consiste à effectuer la convolution des projections à une dimension par la T.F., $g(x)$, d'une fonction dont les valeurs sont proportionnelles à la coordonnée u du plan de Fourier et à ne procéder à la sommation des projections filtrées qu'en fin de traitement. Si l'on désigne par $g_1(x)$ ce "filtre spatial", le processus de convolution de chaque projection avec cette fonction s'écrit: $h_1(x) = p_\theta(x) * g_1(x)$. Chaque projection "filtrée" est ensuite rétroprojetée selon une fonction à deux dimensions telle que: $h_1(x,y) \equiv h_1(x)$ et la sommation s'effectue en sortie dans le domaine spatial:

$$h_2(r,\phi) = \int_0^{\pi} h_1(x,y)d\theta \equiv \int_0^{\pi} h_1(x)d\theta, \quad (8)$$

au facteur $1/\pi$ près et avec $x = r\cos(\phi - \theta)$. La réponse impulsionnelle correspondante $g(r,\phi)$ s'obtient en considérant comme projection une distribution de Dirac:

$$g_2(r,\phi) = \int_0^{\pi} [g_1(x) * \delta(x)]d\theta = \int_0^{\pi} g_1(x)d\theta. \quad (9)$$

Si la fonction $g_1(x)$ est réelle et paire, il en est de même pour sa transformée de Fourier $G_1(u)$ et la fonction $g_2(r,\phi)$ est une fonction radiale. L'équation (8) peut s'écrire sous la forme d'une transformée de Hankel:

$$g_2(r) = 2\int_0^{\infty} [G_1(u)/u]\, u\, J_o(2\pi ur)du \equiv T.F^{-1}[\,|G_1(u)/u|_{u=\rho}]. \quad (10)$$

La fonction $G_1(u)/u$ doit donc être considérée comme une fonction à deux dimensions radiale. Un filtre à une dimension, $G_1(u) = |u|$, de même forme fonctionnelle que le filtre à deux dimensions, conduit donc au même résultat: $h_2(r,\phi) \equiv f(r,\phi)$. Cette forme de filtre est valable pour la tomographie par transmission où: $F(\rho,\Phi) = \rho H(\rho,\Phi)$.

Pour d'autres types de reconstructions d'images, le modèle unitaire de projection permet d'aboutir à une formulation similaire. En particulier, en tomographie d'émission:

$$P_\theta(u) = F(u,v)_{v=0} = [C(u,v).D(u,v)]_{v=0}, \quad (11)$$

où $D(u,v)$ est la transformée de Fourier d'une fonction $d(x,y)$ représentant à la fois l'auto-absorption et la diffusion subies par les rayons Gamma émis. On montre alors, par un raisonnement analogue, que le filtre spatial est dans ce cas:

$$g_1(x) = T.F^{-1}[\,|u/D(u,v)|_{u= \;,v=0}]. \quad (12)$$

La mise en oeuvre informatique de ces opérations correspond à l'algorithme suivant:

```
{Initialisation}
 θ←0; Δθ←π/N; Δx←2a/M; Δy←2a/M; f(X,Y)←0; {Y=xcosθ-ysinθ, Y=xsinθ+ycosθ}
{Calcul de f(X,Y)}
 Pour K=0 jusqu'à N-1
  θ←θ+Δθ; x←0; h1(x,θ)←0;
  Pour I=0 jusqu'à M-1  {filtrage des projections pour une valeur de θ}
   x←x+Δx; h1(x,θ)←p(x,θ) * g1(x)  {pour -a<x<a};  y←0;
   Pour J=0 jusqu'à M-1  {rétroprojection suivie de sommation des h1}
    y←y+Δy; f(X,Y)←f(X,Y) + h1(X,θ);
 Division du résultat par N.
```
(13)

La forme théorique du filtre est donc une fonction proportionnelle à $|u|$ qui n'est pas réalisable en pratique. Il y a toujours troncature du spectre, ce qui entraîne l'appartition d'oscillations et donc d'imprécisions dans la fonction reconstruite.

Nous avons étudié une méthode d'apodisation des filtres, notamment des "filtres en ρ" que nous avons appliqué à la reconstruction de divers profils. Nous présentons cette étude dans la section suivante.

Recherche d'un procédé d'optimisation au moyen d'un critère de ressemblance objet-image

La troncature des spectres, et du filtre en particulier, s'accompagne non seulement d'oscillations dans le profil synthétisé, ou reconstruit, mais aussi d'une déformation de la pente au niveau des contours et des discontinuités. D'autre part, le filtre étant tronqué, le support de sa T.F ne peut être borné. Comme il n'est pas possible de considérer une T.F. à support infini, il faut donc remplacer la fonction obtenue par un profil qui s'en rapproche le plus possible et dont la T.F soit à support borné: c'est le problème de l'approximation d'une fonction par une fonction qui s'identifie au problème mathématique de l'approximation des fonctions périodiques par des polynômes trigonométriques. Le même problème se pose en spectroscopie interférentielle où la connaissance du spectre du rayonnement est remplacée par celle de sa T.F considérée sur un intervalle fini. La largeur de bande étant l'intervalle sur lequel on connaît la T.F de l'"objet", il est possible de construire une "bande passante" de même largeur et de forme optimale, ou "fonction de filtrage", en utilisant un ensemble de coefficients pondérateurs de façon à former une image la plus proche possible de l'objet étudié. De la même façon, pour construire un profil de forme $f(x)$ donnée à support borné, on peut réaliser l'approximation: $\hat{f}(x) = f(x) * g(x)$, en choisissant une fonction $g(x)$ dont la T.F soit à support borné et telle que l'écart entre $f(x)$ et $\hat{f}(x)$ soit minimum. Cela revient à déterminer la "bande passante" $G(u)$ de largeur $2u_c$. En général, la largeur de bande est imposée par les conditions expérimentales. On détermine la forme de $G(u)$ au moyen d'un critère de ressemblance entre l'objet et l'image qui détermine la qualité de celle-ci.

Le critère de l'écart quadratique moyen est le plus simple: si la fonction f est de carré sommable, si G est bornée pour tout u (alors G et g sont de carré sommable), $\hat{F}$ est sommable, et si g est bornée pour tout x, alors $\hat{f}$ est de carré sommable. Si $[-u_c, +u_c]$ est le support de la bande passante, l'écart quadratique moyen entre l'objet et l'image s'écrit:

$$r(u_c) = \int_{-\infty}^{+\infty} [f(x) - \hat{f}(x)]^2 dx = \int_{-\infty}^{+\infty} [F(u) - \hat{F}(u)]^2 du, \qquad (14)$$

ou:

$$r(u_c) = \int_{-u_c}^{+u_c} |G(u/u_c) - 1|^2 . |F|^2 du + \int_{|u|>u_c} |F|^2 du. \qquad (15)$$

La deuxième intégrale du second membre de l'équation (15) représente la valeur minimum de l'écart quadratique moyen. Ce résultat fondamental montre qu'il est impossible que l'image soit partout identique à l'objet (sauf si le spectre de celui-ci est limité). En minimisant l'écart $r(u_c)$, on obtient une bande passante à valeur centrale constante.

Le critère de l'écart quadratique moyen est insuffisant car il s'agit d'un critère global ne fournissant aucune indication sur l'écart en chaque point entre l'objet et l'image. L'écart peut ainsi être très faible presque partout et très important en un point particulier. Il est donc indispensable d'étudier l'écart local.

Optimisation de l'écart local entre l'image idéale et l'image reconstruite

En normant $G(u)$ à la valeur $G(0) = 1$, l'écart local s'écrit:

$$\hat{f}(x) - f(x) = u_c \int_{-\infty}^{+\infty} g(u_c x')[f(x-x') - f(x)]dx'. \qquad (16)$$

g décroit à l'infini comme $1/x^{\alpha+1}$. On suppose que $g(0)$ est le maximum absolu de g, soit:

$$|g(x)| \leqq g(0) = g_o \qquad \text{pour } |x| \leqq x_o . \qquad (17)$$

La condition:

$$|g(x)| \leqq g_o x_o^{\alpha+1}/|x|^{\alpha+1} \qquad \text{pour } |x| \geqq x_o \qquad (18)$$

permet de définir le paramètre α qui caractérise la vitesse de décroissance de $g(x)$ quand x tend vers l'infini. On suppose que $f(x)$ possède un développement de Taylor limité à l'ordre n, ce qui conduit à une expression de $|\hat{f} - f|$ permettant de dégager, en particulier, la condition des moments et d'en déduire le paramètre $n < \alpha$ tel que:

$$\int_{-\infty}^{+\infty} x^p g(x) dx = 0 \qquad \text{pour } p = 1,2,\ldots,n, \qquad (19)$$

Enfin, les paramètres β et x_1 définissent la régularité de la fonction $\hat{f}(x)$ qui doit vérifier:

$$\int_o^t |f(x - x') - f(x) - P_m(x')|dx' \leqq B\, t^{\beta+1} \qquad \text{pour } |t| \leqq x_1, \qquad (19)$$

où:

$$P_m(x') = -x'f'(x) + \frac{1}{2} x'^2 f''(x) + \ldots + \frac{(-x')^m}{m!} f^{(m)}(x), \qquad m \leq n \tag{20}$$

et où: $B = M_n/(n+1)!$, M_n étant la borne supérieure de $f^{(n)}(x)$ dans l'intervalle $[-x_1, +x_1]$. $\beta = n$ si $f(x)$ possède n dérivées bornées en x. Finalement l'écart entre l'objet et l'image s'écrit:

$$|\hbar - f| \leq 2I_1 + 4M \frac{g_o x_o^{\alpha+1}}{\alpha(u_c x_1)^\alpha} + 2\left(\sum_1^m \frac{|c_p| x_1^p}{\alpha - p}\right) \frac{g_o x_o^{\alpha+1}}{(u_c x_1)^\alpha}, \tag{21}$$

où la valeur de l'intégrale I_1 dépend du signe de $\beta - \alpha$. Pour améliorer l'ordre de l'approximation, il faut rendre α le plus grand possible et réaliser simultanément la condition des moments. On montre qu'une singularité de $f(x)$ n'introduit de défauts dans l'image que sur un petit intervalle autour de la singularité et que si l'on augmente la valeur du paramètre α, on améliore la localisation de l'approximation. En effet, si M est la borne supérieure de $|f(x)|$, et si l'on veut que: $|\hbar(x) - f(x)| \leq \varepsilon M$, où ε est donné à l'avance, l'intervalle nécessaire pour que l'effet d'une singularité disparaisse est tel que:

$$x_1 > \frac{x_o^{\frac{\alpha+1}{\alpha}}}{u_c} \left(\frac{4g_o}{\varepsilon\alpha}\right)^{1/\alpha}. \tag{22}$$

D'autre part, l'étude globale du maximum de l'écart montre qu'il n'y a pas saturation de l'approximation si G(u) est constante autour de l'origine pour $|u| \leq u_o$ (toutes les dérivées sont alors nulles en $u = 0$) et l'on retrouve le résultat correspondant au critère de l'écart quadratique moyen. Nous avons vérifié l'amélioration de l'approximation lorsque la valeur du paramètre α augmente, en comparant les performances de quelques procédés de sommation classiques. Le procédé de Fourier, défini par une bande passante à valeur centrale constante, est tel que $\alpha = 0$, ce qui est lié au fait que G(u) est discontinue en $u = \pm 1$, et g(x) décroît à l'infini comme $1/x$. Ce procédé n'est intéressant que pour les fonctions f(x) dont le spectre est à support borné ou des fonctions très régulières car il ne sature pas. Mais pour des fonctions dont la dérivée peut prendre des valeurs très élevées, voire infinies, il donne lieu au phénomène de Gibbs qui se traduit par d'importantes oscillations au voisinage des singularités et dont la fréquence augmente lorsqu'on augmente la largeur u_c de la bande passante, mais qui ne diminuent pas d'amplitude. Pour éviter ces oscillations, nous avons été conduits à chercher un procédé de sommation optimum qui minimise la borne supérieure de l'erreur tout en réduisant le plus possible l'intervalle x_1 de localisation autour d'une discontinuité de l'objet.

Apodisation du filtre

L'étude de plusieurs procédés couramment utilisés en traitement des images (procédé de Féjer ou fonction "triangle", procédé en $(1-u^2)^2$, etc.) a mis en évidence le fait que l'optimisation de la localisation de l'approximation requiert une fonction G(u) la plus régulière possible, c'est à dire ayant le plus grand nombre possible de dérivées continues et bornées. D'autre part, le fait d'annuler le plus grand nombre possible de dérivées continues à l'origine de G(u) a pour effet de diminuer la valeur de la borne supérieure de l'erreur. Cependant, si la partie centrale de G(u) est constante, l'approximation ne dépend que du nombre de dérivées continues et, en particulier, la localisation est liée au comportement de la fonction G à l'extrémité de son support. Si la dérivée $G^{(n)}(u)$ est continue et bornée, g(x) décroit à l'infini comme $1/x^{n+1}$. Or, la vitesse de décroissance à l'infini est liée à l'"apodisation lointaine", c'est à dire à la réduction des "pieds" de la figure de diffraction loin de l'origine. Nous avons donc cherché une fonction de filtrage qui réalise au mieux toutes ces conditions et en particulier l'effet d'apodisation. Pour cela nous avons formé une fonction G(u) avec des translatées d'une fonction génératrice Φ à support borné $[-m, +m]$:[6]

$$G(u) = \sum_{-n}^{+n} a_k \Phi\left(\frac{u}{k} - k\right), \tag{23}$$

où K est une constante telle que $(n+m)K = 1$ afin que $-1 < u < +1$, et dont la T.F est:

$$g(x) = K\phi(Kx). \sum_{-n}^{+n} a_k \exp(-i2\pi kKx). \tag{24}$$

On démontre que le facteur $\Sigma a_k \exp(-i2\pi kKx)$ est un polynôme trigonométrique de la forme $\cos^2[n\mathrm{Arccos}(z\cos K\pi x)]$, de période $1/K$, qui réalise la meilleure apodisation pour une limite de résolution donnée ou le meilleur pouvoir de résolution pour une apodisation donnée. Le facteur $K\phi(Kx)$ (et si possible plusieurs de ses dérivées) doit s'annuler aux points $x = p/K$; il réalise l'apodisation lointaine permettant une décroissance rapide de g(x) quand x tend vers l'infini. Le polynôme trigonométrique réalise l'apodisation "proche" en réduisant les maximums secondaires qui entourent le maximum central g_o de g(x). Une fonction réalisant au mieux l'apodisation lointaine est:

$$\phi(Kx) = [\sin(K\pi x)/Kx]^q, \qquad (25)$$

qui s'annule en $x = p/K$ et dont la transformée de Fourier a pour support $[-q/2, +q/2]$ de sorte que pour que $(n+m)K = 1$ il faut que $K = 2/(2n+q)$.

Cette étude nous a conduit à choisir une fonction de filtrage qui réalise à la fois le plus faible écart local, la meilleure localisation et l'apodisation. La partie centrale de cette fonction est constante pour permettre l'apodisation lointaine, l'apodisation proche étant obtenue par deux arcs de fonctions en $\sin^2$ situés de chaque côté et tels que:

$$G(u) = \begin{cases} \sin^2[\pi(u_c+u)/2Cu_c] & \text{pour } -u_c \leq u \leq u_o \\ \sin^2[\pi(u_c-u)/2Cu_c] & \text{pour } u_o \leq u \leq u_c \end{cases} \qquad (26)$$

avec $u_o = u_c - Cu_c$ et C un facteur auquel on peut donner différentes valeurs entre 0 et 1. La partie centrale de cette fonction "apodisante" est telle que $G(u) = 1$ pour $-u_o \leq u \leq +u_o$. La dérivée première de $G(u)$ est continue. En effet, la dérivée première du procédé en $\sin^2$ s'annule aux extrémités du support de la partie constante en $u = \pm u_o = \pm u_c(1-C)$ ainsi qu'aux points de raccordement avec l'axe des u en $u = \pm u_c$. La dérivée seconde est discontinue (elle vaut -1 en $u = \pm u_c(1-C)$ pour le $\sin^2$), mais elle est bornée. La valeur du paramètre α est donc égale à 2 et, dans les conditions de régularité maximum de $f(x)$: $\beta = 2$ et $B = M_2/6$. La transformée de Fourier inverse de $G(u)$ est:

$$g(x) = u_c(2-C)\,\frac{\sin[\pi u_c(2-C)]}{\pi x u_c(2-C)} \cdot \frac{\cos(\pi x C u_c)}{1-(2xCu_c)^2}. \qquad (27)$$

A l'origine: $g_o = g(0) = u_c(2-C)$.

Application à la tomographie biologique et médicale

La sommation des projections filtrées, effectuée sur ordinateur en tomographie conventionnelle, peut être réalisée au moyen d'un calculateur photonique en tirant parti du parallélisme intrinsèque permis par l'utilisation de la lumière comme véhicule de l'information. On peut utiliser des dispositifs à ondes acoustiques de surface avec des lentilles intégrées dans une couche mince sur un substrat, tel le Niobate de Lithium, pour réaliser analogiquement et en temps réel les opérations de convolution. Nous avons appliqué le processus d'apodisation décrit ci-dessus à la correction du filtre "en-ρ" qui intervient dans la reconstruction de l'image tomographique. En pratique, le rayon de ce filtre, à symétrie radiale, ne peut être infini. Sa troncature, pour une certaine fréquence de coupure ρ_c, se traduit par l'apparition du phénomène de Gibbs, cause d'imprécisions dans l'image. La multiplication du filtre théorique ρ par la fonction sommatoire décrite mathématiquement par une partie centrale constante se terminant à chaque extrémité par les fonctions (26), a permis d'obtenir une pente et une forme optimales, avec suppression des oscillations, pour la reconstruction d'un profil rectangulaire pris comme référence, en utilisant la valeur 0.2 pour le paramètre C. Les expériences faisabilité avec un filtre holographique convenablement apodisé ont été réalisées au Laboratoire du Groupe de Recherches en Photonique Appliquée (GREPA) de l'Université Louis Pasteur à Strasbourg.

Références

1. Rogers, G.L., "Non coherent optical processing", Optics and Laser Technology, Vol. 7, pp. 153-162. 1975.
2. Lee, S.H., "Mathematical operations by optical processing", Optical Engineering, Vol. 13, pp. 196-207. 1974.
3. Landraud, A.M., Contribution au Traitement Numérique et Photonique de Signaux Spatio-Temporels. Applications Micro-Photoniques, Spectroscopiques et Iconiques, Thèse de Doctorat ès-Sciences, 1982.
4. Landraud, A.M., "Modélisation d'un Tomographe Photonique à Balayage avec Calculateur Analogique, Commandé par Microprocesseur, Basé sur une Méthode Originale de Codage de l'Information par Laser", Innovation et Technologie en Biologie et Médecine., Vol. 5, N° 2, pp.176-188. 1984.
5. Landraud, A.M. et Grosmann, M., "Réalisation de Micro-Circuits Photoniques à Très Haute Echelle d'Intégration et Algorithme de Reconstruction pour le Traitement Ultra-Rapide et en Parallèle d'Images 3-D", High Speed Photography and Photonics, proc. du 16° Congrès International, Vol. S.P.I.E. 1985.
6. Arsac, J. et Simon, J.C., "Représentation d'un Phénomène Physique par des Sommes de Translatées", Annales de Radioélectricité, Vol. 15. 1960.

Data processing on particle field hologram

He Xiangdong, Yao Jie and Zhu Zhulin

Department of Physics, Dalian Institute of Technology,
Dalian, China

Abstract

This paper presents the application of data processing technique to the spray analysis of a diesel nozzle. Faced with a large number of oil droplets, we have used several techniques to improve the image quality and to make the processing more rapid and reliable.

Introduction

Since the middle of 1960's, the application of holographic techniques to micro-particle field analysis has been developed very rapidly. Like other optical methods, holographic techniques are undisturbing method. They are most attractive for their convenience and reliability, for they provide a 3-D display of the particle field and the image of each particle can be viewed. They have been applied to various fields such as agricultural atomizer, aerospace engine and diesel nozzle.

To most nozzle investigators, the spray characteristics, i.e., the inner structure of the spray field, are of great importance. Holographic techniques, however, provide a reliable method to draw such characteristics experimentally.

Meanwhile, data processing has always been the key technique in this field of holographic application, because we are faced with a large number of particles. Manual counting and sizing of the particles is neither convenient nor reliable, and takes too long time. With the help of computer, the speed of processing can be greatly improved. In recent years, semi-automatic and automatic systems have been reported available.

The work of this paper is devoted to the application of holographic technique to the spray field analysis of diesel nozzle, and places stress on data processing. First we have to record the hologram of the spray field jetted by a nozzle which is to be investigated. Then, place the hologram in the reconstructing and data processing system for counting and sizing the particles in the reconstructed image of the spray field. In order to make the processing more rapid and reliable, several techniques have been used in the processing.

Recording system

One of our recording systems is shown in Fig.1. It is an in-line holographic system. It mainly includes a pulsed ruby laser with output energy greater than 200mJ per pulse, a test engine with a nozzle to be investigated, a time sequence device which controls the laser and the fuel pump work in a correct time sequence. As the oil droplets are very small, a magnifying lens system L is always added to the recording system. With the system we have successfully recorded holograms of the spray field.

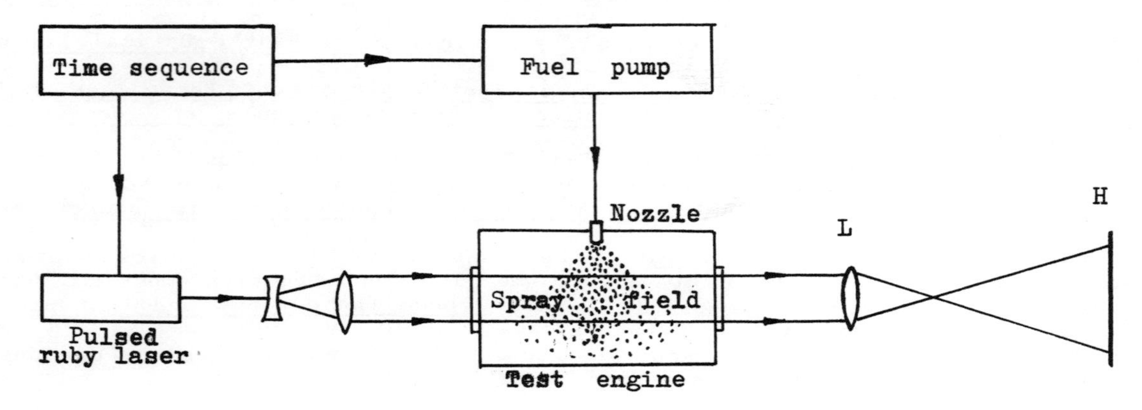

Fig.1 In-line holographic recording system

The reconstructed images are shown in Fig.2. In picture (1), there is a focused image of a calibrated string which was hung inside the engine before recording. As a rule of calibration, it is 40 microns in diameter.

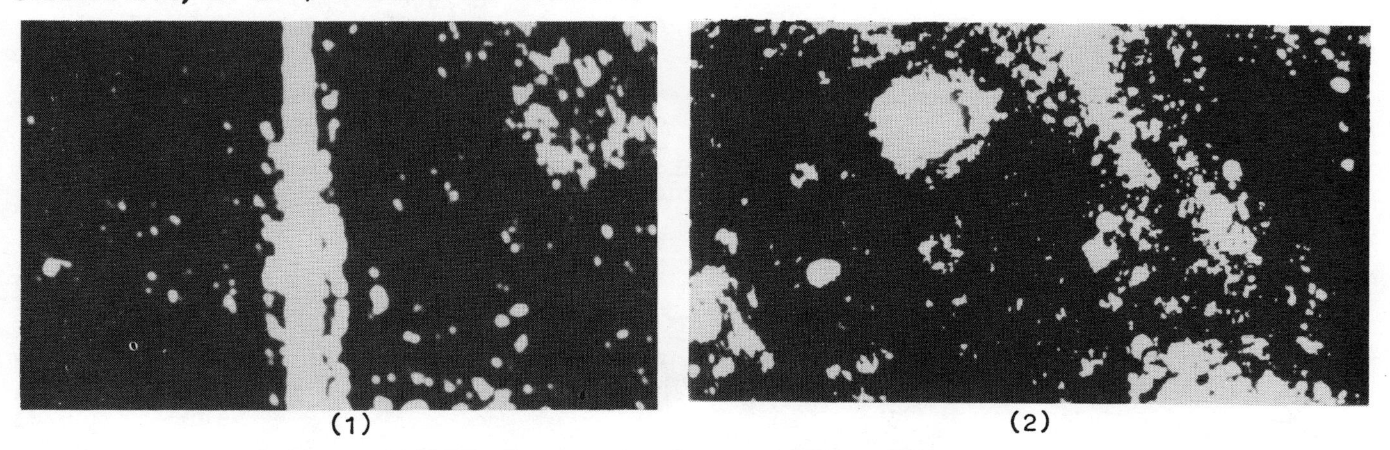

Fig.2 Reconstructed images of the hologram

Reconstructing and data processing system

The reconstructing and data processing system is shown in Fig.3. The system includes the optical path which is adjusted to match the requirements of the processing system, an optoelectric convertor(TV camera), a frame memory device and a PDP-11/24 computer with its peripheral equipment,etc.

The hologram H is held on a 3-axis adjustable mount and is illuminated by a collimated He-Ne laser beam. The reconstructed image of the hologram is formed in front of the lens L and is magnified by the lens. On the cathode of the camera, a planar picture is obtained. The picture consists of a focused image corresponding to a certain layer of the reconstructed 3-D image of the hologram as well as defocused images corresponding to other layers. When moving the hologram along the z direction with the help of the mount, we can obtain different focused images of different layers successively(cf.Fig.4). When the hologram is moved in the x or y direction, the image of other sampling regions is obtained.

The optical high-pass filter is used to improve the picture quality. It is aimed at filtering out the lower spatial frequences present in the image field, i.e., to blacken the bright background, meanwhile leaving the images unaffected.

The cathode of the camera is of the size 10X10mm and has 512X512 pixels altogether. The information about the brightness distribution over the planar image field is turned

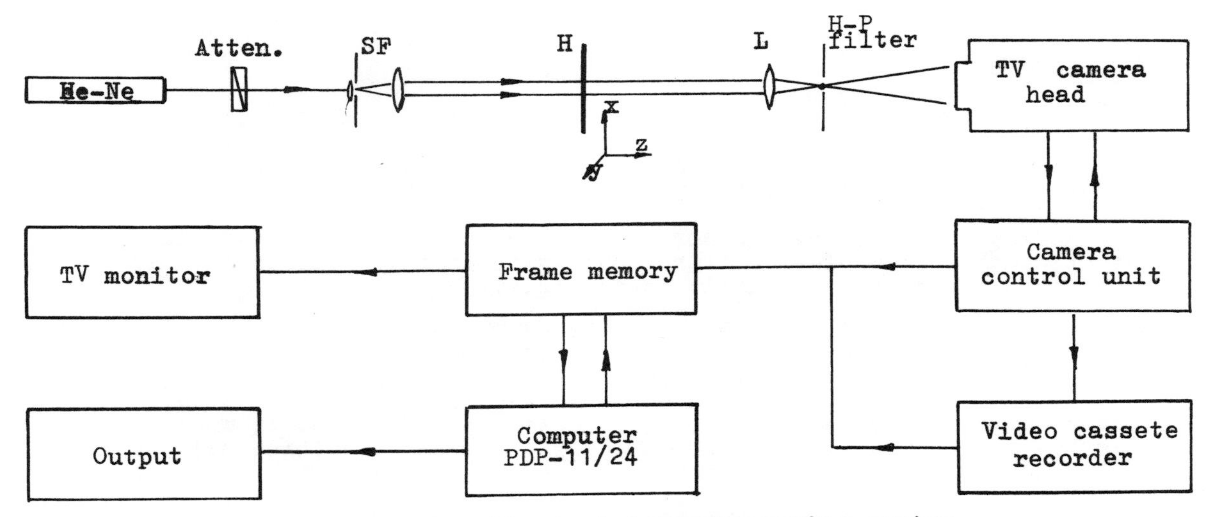

Fig.3 Reconstructing and data processing system

into the intensity of a series of vedio pulses which are sent to the frame memory. The frame memory converts the amplitude signal output from the camera into digital signals and stores them. Therefore, a frame of picture is converted into a numeric array with 512X512 elements and each element is quantitized linearly with 256 brightness levels from 0(darkest) to 255(brightest).

Data stored in the frame memory can be read to the computer for processing, and the processed data, in turn, can be written into the frame memory for picture output on the screen of the monitor.

The vedio tape recorder records real-time pictures scanned and sampled by the TV camera and then plays back to the frame memory for processing, when necessary.

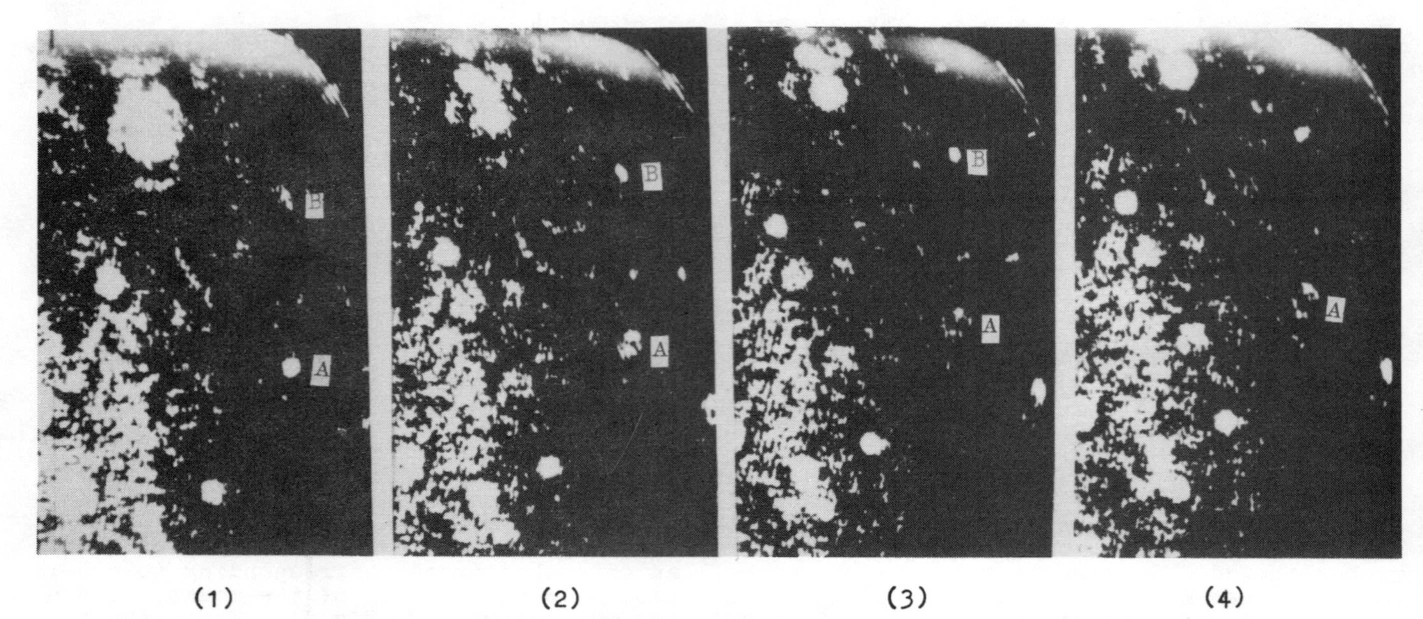

(1) (2) (3) (4)

Fig.4 Pictures from the same region but different layers

For example, the bright spot A in picture (1) is a focused image but is defocused in other pictures. On the other hand, spot B in (3) is focused image and becomes blur in the others, etc.

Several techniques for processing

In order to improve the picture quality and to rule out undesirable factors, i.e., to make the processing more reliable,various threasholding technique and other techniques are needed.

Intensity threashold

First of all, the bright background of the picture has to be blackened. Optical high-pass filtering has to some extent blackened the background, but it is still necessary to be further blackened. As the desirable particle image is brighter than the background, the background can be easily blackened by setting a suitable brightness threashold.

Threashold selection is of great importance in the course of threasholding. An over-threasholded or under-threasholded picture would give us false information about the particle size and even the number of particles(cf.Fig.5).

Threashold selection can be well performed by examining the histogram of the 512X512 picture points with respect to optical density. The sparsely populated area in the histogram, i.e., the valley, should be a resonable threashold for the corresponding picture. Because the number of points whose brightness levels lie between the background and the particle images is relatively small. Difficulty will occur, however, when the histogram has more than one valleys. In this case, one has to select the threashold experimentally. In our experiments, the intensity threashold is usually set at 100 of brightness levels.

Fig.5 shows the effect of different threasholds. Picture (1) is without threasholding. Picture (2) treated with the threashold of 100 brightness level. Picture (3), 150

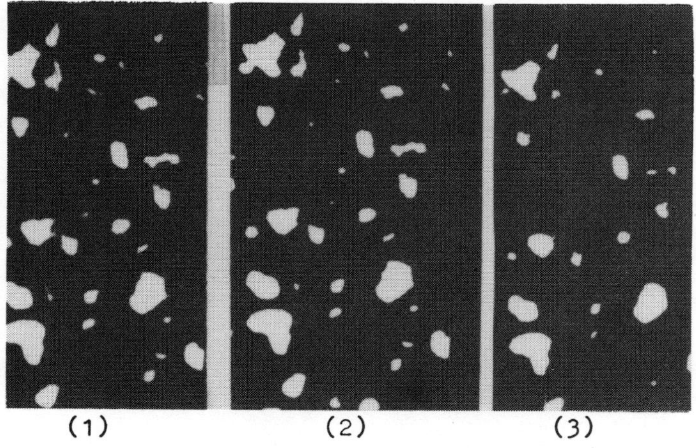

Fig.5 The effect of different threasholds

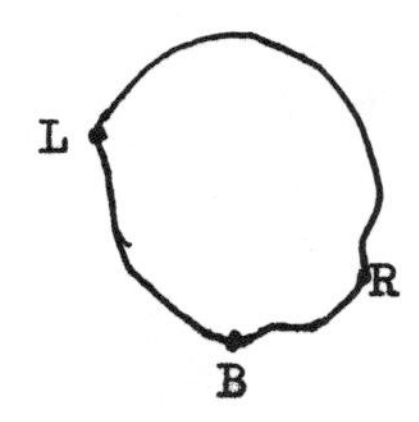

Fig.6 Three-point method for calculating the gradient

brightness level, over-threasholded.

Particle size threashold

After the intensity threasholding, there may be some individual small spots in the picture. Generally they are laser speckle patterns and are much smaller than the images of particles. Therefore, they can be easily threasholded by setting a size threashold. In our experiments, the threashold of size is usually set at 10 pixels.

Gradient threashold

In the picture sampled by the TV camera, as already mentioned above, focused and defocused images are mixed. Focused images, however, are much sharper than those defocused. Therefore, an intensity gradient threashold is always needed so as to rule out those defocused images.

In our processing program, the gradient of brightness is calculated at the three points shown in Fig.6. They are the left-most, the right-most and the bottom points of the image of the particle. The mean value of the three represents the gradient of the particle. The three-point method is more reliable than any one-point method, for it is less vulnerable to accidental disturbation in the picture.

The calculation of gradient is achieved by calculating the differences between the three chosen points and their neighbors.

$$Grad_i Z = \left((Z_i(I,J)-Z_i(I+1,J))^2 + (Z_i(I,J)-Z_i(I,J+1))^2 \right)^{1/2}$$

where Z is the name of the numeric array of the picture, i dentes one of the chosen points, I, J are the line and column respectively where the point i is located.

The gradient threashold is selected experimentally. It is usually set at 10 in our processing program.

Smoothing the particle edge

The actual image of a particle has no smooth edge. Rough edge, as shown in Fig.7, may give us false information about the number of particles.

In our program, rough edges are smoothed by omitting short strings(belonging to a large particle), though they satisfy the intensity threashold. This kind of smoothing, we call it the de-bur technique, does not degrade the gradient of the particle, because the data at the edge are not changed.

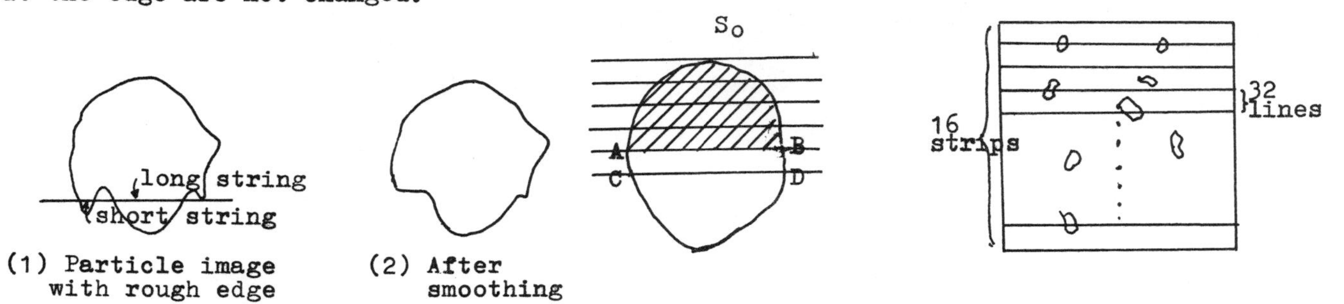

Fig.7 Edge smoothing Fig.8 Scanning mode Fig.9 Dividing of a picture

Though such processing may to some extent under-estimate the area or size of particles, it guarantees reliability for the number of particles, so it is still worth doing.

Calculation and classification of the particles

The calculation of particle size is made by using scanning mode instead of traking mode, for scanning mode is suited to fast processing. The scanning mode checks the points of the picture line by line. The beginning and ending points of a string of consecutive points that satisfy the intensity threashold are stored in two arrays respectively, and such stored data of two adjacent lines are compared to identify which of the strings belong to a same particle. If there are two adjacent strings, say, AB and CD belonging to a same particle(cf.Fig.8), the area of the trapezoid is calculated as (AB+CD)/2. At the top of a particle image, its area is set to zero, then its area is accumulated line by line until the bottom of the particle, by

$$S = S_0 + (AB+CD)/2$$

The diameter of the particle is calculated by

$$d = (S*4/3.1416)^{1/2}$$

for the particle is approximately regarded as a sphere.

As the memory space of the computer is limited, the whole picture cannot be processed simultaneously. In that case, a picture can be divided into several stips, say, 16 strips, with each containing 32 lines(cf.Fig.9). The above processing method is applied to each of the strips. The last line of the upper strip is compared with the first line of the lower strip, and the comparison here is similar to that between two adjacent lines within a same strip. Therefore, the divided picture reserves any property of the original one.

When the processing of a picture is completed, the picture of other layer or other region takes its place. When the processing has finally finished, the computer classifies the particles with respect to their diameters, and prints out the results.

In our experiments, the time it takes for the computer to process one picture is about one minute.

Conclusions

We have applied data processing technique to spray particle analysis. Our work is chiefly intended to analyze the spray field of diesel nozzles. However, the technique applied here can be used to other kinds of particle fields.

In our experiments, the reconstructing system has been calibrated. the relative error is small. The developed software was tested by all means. The results are satisfactory. The system works reliablely and with a high speed of one minute per picture. Various threasholding techniques are applied. Image edges are smoothed, meanwhile their sharpness or gradient are kept.

According to our experiments, scanning mode is a practical method to process large amounts of holographic or other kinds of particle images.

References

1. B.A.Silverman,et al, Journal of Applied Meteorology, 3,792,1964.
2. B.J. Thompson, et al, Applied Optics, 6, 519,1967.
3. B.J. Thompson, et al, SPIE 215, 102, 1980.
4. G.Haussmann, et al, Applied Optics, vol. 19, No.20, 1980.
5. W.K.Witherow, Optical Engineering, vol.18,No.3,1979.

Session 15

Optical Information Processing II

Chairman
Yi-Mo Zhang
Tianjin University, China

Use of synthetic holograms in coherent image processing for high resolution micrographs of a conventional transmission electron microscope (CTEM)

E. Reuber*, S. Boseck**, B. Schmidt**, H. Block**

* Fritz-Haber-Institut der Max-Planck-Gesellschaft, D-1000 Berlin 33
** University of Bremen, Physics Dept., D-2800 Bremen 33

Abstract

A modified Stroke filtering technique is applied to high resolution micrographs of a conventional transmission electron microscope (CTEM). In a coherent optical processor synthetic holograms are used. They are produced by an analog electron-optical device. A microprocessor-controlled electron beam writes the amplitude component on a fluorescent screen. This image is photographed through an alternately shifted grating. The defects of defocus, spherical aberration and axial astigmatism can be reduced.

Introduction

In contrast to optical microscopes, point resolutions of the order of the wavelength cannot be obtained in the electron microscope (EM). A 100 kV EM has an electron wavelength of 0.037 Angström, but the resolution limit is nearly 1 Angström. The reasons are the instrumental limitations, mainly the lens aberration and the blurring. With image processing an improvement is possible.

For high resolution imaging in a conventional transmission electron microscope (CTEM) the specimen must be a very thin slice of app. 50 Angström, and a voltage of 100 kV or more is required. Under these conditions the objects, especially biological objects, behave as phase objects. Pure phase objects would appear as completely transparent when imaged in focus with a perfect microscope. For getting adequate contrast a phase shift is necessary. For the CTEM, the preferred method to image phase objects is to strongly defocus the EM. Then the phase variations directly behind the objective lens are converted into intensity variations which can be recorded in the image.

Because of the defocusing, the information from the object cannot directly be seen in the image. It exists in a coded form and must be reconstructed to be interpreted.

MTF

The blurring of the image is described by the modulation transfer function (MTF). It is a function of the spatial frequency and determines the weight for each spatial frequency component for the transfer. For an ideal transfer system the MFT would be constant for all spatial frequencies.

The Fourier transformation (FT) of the MTF is the point spread function. The image function is the result of a convolution of the object function with the point spread function of the EM. Figure 1.

For retrieving the object function from the image, a deconvolution is necessary. The deconvolution in the image plane can be achieved by a division filter in the Fourier plane (Stroke's division filter).

There exists a number of methods to improve the transfer function. In this paper only the a posteriori deblurring by holographic filtering in an analog optical device will be considered.

A posteriori deblurring by analog optical methods

The first light-optical reconstruction by spatial frequency filtering was made by Maréchal and Croce [1]. In the following years improvements of this method were described in many publications. [2,3,4] Usually the coherent laser light is used in the well known 4-f setup which is sometimes modified, see Figure 2.

Stroke's holographic division filter

The reconstruction with a complex spatial frequency filter, the holographic division filter, was proposed by Stroke and Zech[5]. Stroke's filter is a tranparency consisting basically of two components, for correcting amplitude and phase. The phase component is a diffraction grating. In the following years refinements of this method were done by many authors.

In Figure 2 the schematic set-up is shown. A positive copy of the electron micrograph is made in such a way that after illuminating it with a coherent plane wave the transmitted intensity is the same as the intensity of the original electron micrograph. This copy is put in the object plane of the laser set-up. This plane is the front focal plane of the first transform lens.

In the back focal plane of this lens the spectrum of the object is generated. In this focal plane the division filter will be mounted.

The Fourier plane is also the front focal plane of the second transformation lens. Performing a second FT, this second lens creates in its back focal plane the corrected image. Because the phase component is a diffraction grating the reconstructed image will appear in the two sidebands of +1st and -1st order.

MTF of the electron microscope for weak phase contrast

In order to correct the micrograph the MTF, or more precisely the phase contrast transfer function, must be known for making the correct filter. In the case of bright field axial illumination the phase contrast transfer function, given by Hanßen[6] and Scherzer[7], Figure 3, depends on spherical aberration defocus and diffraction angle.

Figure 4 shows three MTF curves for different defocus values. They are calculated for the 100 kV transmission electron microscope with single-field condensor objective lens (DEEKO 100). The curves are oscillating. With increasing defocus the regions of the same contrast become smaller. The best electron micrographs without reconstruction are made in the optimum or Scherzer focus (first diagram of Figure 4), showing a broad region without contrast reversal.

But with reconstruction it is possible to obtain information for higher spatial frequencies. There is also a gain in the low frequencies with stronger defocused images. And the equalization of the MTF leads to an increase of contrast and imaging faithfulness.

Because the MTF curves pass the zero line several times there always is a lack of information in each single defocused micrograph. Only when two or more micrographs are combined can all the information be retrieved, if the different defocus values are adequately chosen.

For this method two or more micrographs of the same object detail are made with different defocus. They are seperately filtered and then incoherently added.[8,9]

Realization of the holographic filters

We realize the holographic FT division filters mainly by two methods. The first one is very simple but is only applicable for making filters with rotational symmetry to correct micrographs without astigmatism. The second method is more complicated and is used to fabricate filters with elliptical zero lines for reconstructing micrographs with astigmatism. Of course it is also usable for making filters with rotational symmetry.

Considerations for both methods

First some principal remarks for both methods should be made. According to the different operation data of the EM different filters are needed. For each set of parameters a special filter must be fabricated. Therefore an easy production method is important.

Analogous to Stroke's method, all filters are made of two components, with a grating for the phase correction. The MTF is real, therefore the phase shift is only 0 or π .

The basic idea is to make two exposures on the same holographic plate. After the first exposure for the positive areas the grating is shifted by a micrometer drive by half a period before the second exposure for all negative areas is made. The result is a filter with ring-shaped zones of alternately shifted grating lines.

The actual MTF can be either measured or calculated from the operation data of the EM.

A micrograph of a thin amorphous foil, normally a carbon foil, is needed. Its diffractogram, the Thon diffractogram, shows unambiguously the influence of the EM.

In most cases a Thon diffractogram can be made from some regions on the blurred micrograph without specimen, only carbon foil. The density distribution above background along a diameter gives the square of the MTF.

For calculating the MTF the exact defocus value must be known. It can be determined from a Thon diffractogram by the graph, shown in Figure 5, which describes the dependence of phase contrast on spherical aberration and defocusing.[10] From the best fitting with the zero lines of the diffractogram the defocus can be determined to an accuracy of 5 nm.

For micrographs with astigmatism the determination of the focus must be made for the two axes of the zero-line ellipses to obtain the axial astigmatism and the average defocus value.

Noise, Wiener filter

Near the gaps of the MTF the S/N ratio is very poor. Therefore the division filter is modified to a Wiener filter.[11]

Second method, with astigmatism

The first method is described in Herrmann et al.[12]; therefore only the second method will be considered here. The filter is produced by an analog electron-optical device. A fine grain screen of a precise oscillograph (STEM unit of an EM) is imaged through a grating on a holographic plate. The grating is mounted in a micrometer drive very close to the holographic plate. The image on the screen is written by a microprocessor-controlled electron beam.

The exposure is made in four steps:

In the first step the beam writes on the display with constant intensity all positive areas of the MTF. The beam writes elliptical lines with increasing size.

In the second step the grating is shifted half a period and all zones of the negative areas will be written by the electron beam.

In the third step the grating is removed. An additional exposure with variable intensity is given to all those regions where the diffraction efficiency should be attenuated according to the MTF.

In the fourth step an additional exposure only of the central region is made to attenuate the central beam for obtaining a better contrast in the reconstructed image.

Path and intensity of the electron beam are controlled by a PDP 11/05 computer. All points of equal filter function values are situated on ellipses. The generator for these ellipses is a sin-generator with constant amplitude and frequency.

By means of a linear amplifier the size of the ellipses will be determined. Between the x- and y-component is a shift of 90 degrees. The elongation of the y-component is attenuated by a potentiometer to produce the wanted excentricity of the ellipse. The zero lines in the diffractograms are not exact ellipses. But the deviations are less than the diameter of the electron beam.

With the correctly chosen dimensions the zero lines of the filter coincide with the zero lines of the diffractogram . Figure 6 shows on the left side the diffractogram and on the right side the filter for a carbon foil micrograph without astigmatism, defocus - 100 nm, DEEKO 100. Figure 7 shows filter and diffractogram of a carbon foil with astigmatism.

Results

The effect of the filtering is shown with some test objects. A filter with 20 circular zero lines was produced for demonstration purposes, Figure 8 . The grating is shifted half a period on both sides of each zero line. The optical density along all grating lines is constant to get the same amplitude for all spatial frequencies.

With this filter a radial test bar chart with 72 spokes was filtered, Figure 9 . Each spoke changes 20 times from black to white and vice versa depending on the spatial frequency along the radius. Figure 10 shows the filtered image once again filtered with the same filter to recover the original test bar chart. There is no contrast inversion along each spoke, but the image is blurred in the vicinity of the zero lines. The reason is the lack of information, noise artifacts and some nonlinearities of the whole system.

A second test object was a glass ruler with scaling marks of 0.1 mm. In Figure 11 above is shown the original ruler. In the middle of Figure 11 one sees the image filtered with the 20 zones filter. It is almost impossible to recognize the ruler from such a blurred image! In such a manner the micrographs could be blurred. But after filtering this image with the same filter, the last picture of Figure 11 shows the ruler again.

The effect of holographic filtering can also be demonstrated by a superposition diffractogram. If two micrographs are superimposed in a parallel beam of the diffractometer in such a position that identical details nearly coincide, a fringe system appears in the diffractogram.[13] The direction of the fringes is normal to the displacement vector. It has its normal position with the cos-maximum in the origin for all spatial frequencies where the both micrographs have the same contrast. The fringes are shifted by π for all spatial frequencies with opposite contrast.

In the following two figures two correlation diffractograms from micrographs are shown. In Figure 12 the original and filtered micrograph of Murein crystals are correlated. The micrograph was made with the supraconducting lens EM (E. Knapek, Siemens AG., Munich). Operation data: Cs = 1.45 mm, 100 kV, magnification 80 000, defocus -535 nm, without astigmatism.

The correlation diffractogram of Figure 13 was made of the original and the filtered micrograph of carbon foil. The micrograph was taken with the Philips EM 301 H, Cs = 1,6 mm, 100 kV, with astigmatism, defocus 150-200 nm.

Conclusion

It has been shown in this paper that in several cases the very expensive digital methods for the reconstruction of high resolution micrographs can be avoided. By the evaluation of the light-optical holographic filter methods also for micrographs with astigmatism the field of applications is much more extended.

References

1. Maréchal, A., Croce, P., "Un filtre de fréquences spatiales pour l'amélioration du contraste des images optiques", Comptes Rendus Acad. Sci. Paris, Vol. 237, pp. 607-609. 1953.
2. Stroke, G.W., Halioua, M., Thon, F., Willasch, D., "Image Improvement in High-Resolution Electron Microscopy Using Holographic Image Deconvolution", Optik, Vol. 41, pp. 319-343. 1974.
3. Stroke, G.W., Halioua, M., Thon, F., Willasch, D., "Image Improvement and Three-Dimensional Reconstruction Using Holographic Image Processing", Proc. of the IEEE, Vol. 65, pp. 39-62. 1977.
4. Sieber, P., Untersuchungen zur Verbesserung des Auflösungsvermögens von Elektronenmikroskopen und der Interpretierbarkeit elektronenoptischer Aufnahmen, Dissertation, Technische Universität Berlin, D83. 1978.
5. Stroke, G.W. and Zech, R.G., "A posteriori Image Correction Deconvolution by Holographic Fourier Transform Division", Phys. Lett., Vol. 25 A, pp. 89-90. 1967.
6. Hanßen, H.-J. and Morgenstern, B., "Die Phasenkontrast- und Amplitudenkontrast-Übertragung des elektronenmikroskopischen Objektives", Z. angew. Phys., Vol. 19, pp. 215-227. 1965.
7. Scherzer, O., "The Theoretical Resolution Limit of the Electron Microscope", J. Appl. Phys., Vol. 20, pp. 20-29. 1949.
8. Schiske, P., "Zur Frage Bildrekonstruktion durch Fokusreihen", Proc. 4th Europ. Congr. on Electron Microscopy, Rome, Vol. 1, pp. 145-146. 1968.
9. Frank, J., Bußler, P., Langer, R., Hoppe, W., "Einige Erfahrungen mit rechnerischer Analyse und Synthese von elektronenmikroskopischen Bildern hoher Auflösung", Ber. d. Bunsen-Ges. f. Phys. Chemie, Vol. 74, pp. 1105-1115. 1970.
10. Thon, F., Zur Deutung der Bildstrukturen in hochaufgelösten elektronenmikroskopischen Aufnahmen dünner amorpher Objekte, Dissertation, Universität Tübingen. 1968.
11. Wiener, N., Extrapolation, Interpolation and Smooting of Stationary Time Series, Cambridge, Mass., M.I.T. Press. 1949.
12. Herrmann, K.-H., Reuber, E., Schiske, P., "A simple Way for Producing Holographic Filters suitable for Image Improvement", Proc. 9th Intern. Congr. on Electron Microscopy, Toronto, Vol. 1, pp. 226-227. 1978.
13. Frank, J., "Observation of the Relative Phases of Electron Microscopic Phase Contrast Zones with the Aid of the Optical Diffractometer", Optik, Vol. 35, pp. 608-612. 1972.

Figures

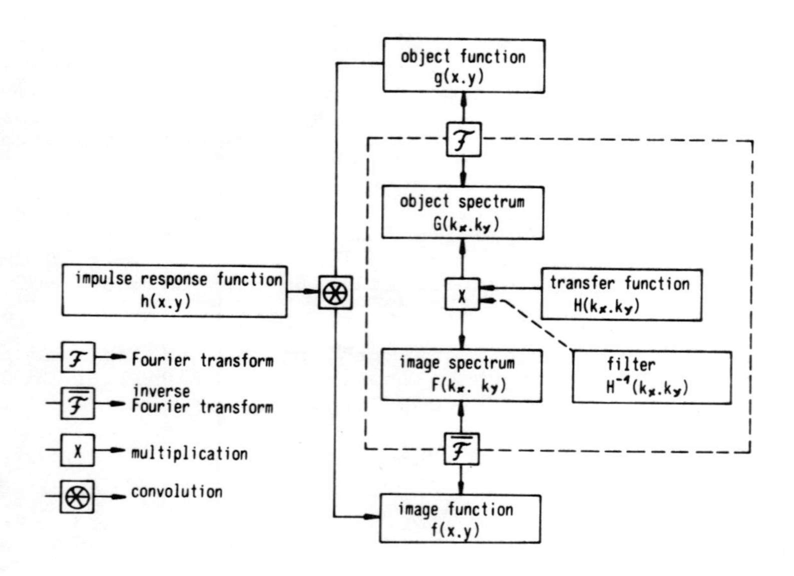

Figure 1. Imaging conditions for a stationary optical system according to the linear transfer theory

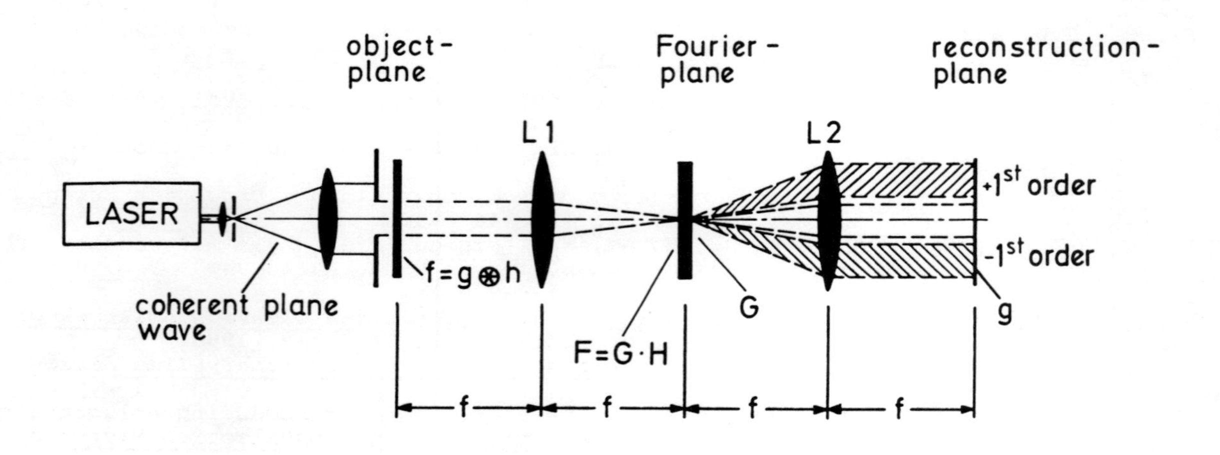

Figure 2. 4-f-set-up

Phase contrast transfer function

$$H\varphi = -2\sin(2\pi W/\lambda)$$

Wave aberration

$$W(\Theta) = \frac{1}{4} C_S \Theta^4 - \frac{1}{2} \Delta z \, \Theta^2$$

C_S = *spherical aberration*

Δz = *defocus*

Θ = $\lambda \cdot R$ = *diffraction angle*

λ = *electron wave length*

R = *spatial frequency*

Figure 3. Phase contrast transfer function

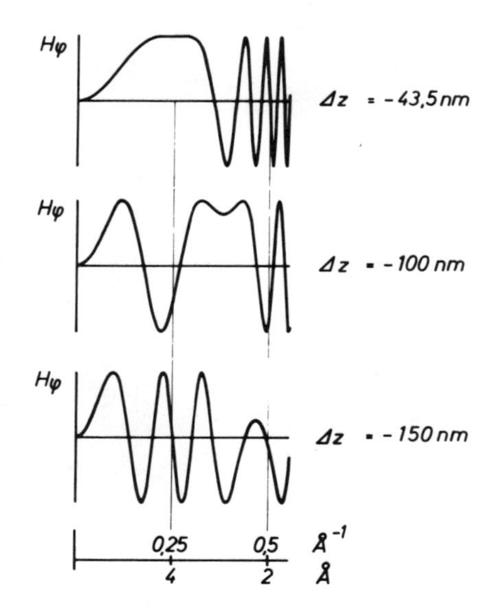

Figure 4. MTF curves for defocusing phase contrast DEEKO 100, c_s=0,5 mm λ=3,7 pm

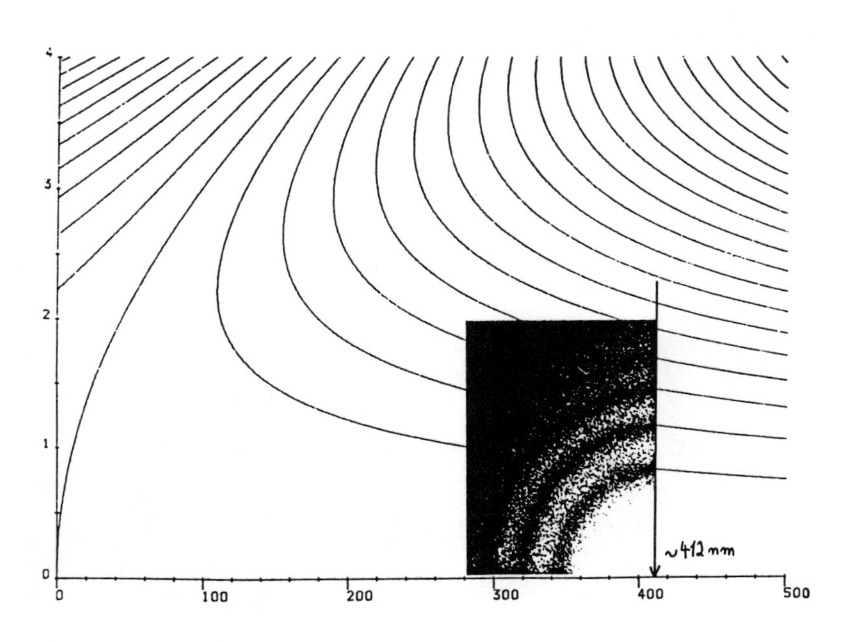

Figure 5. Determination of defocus

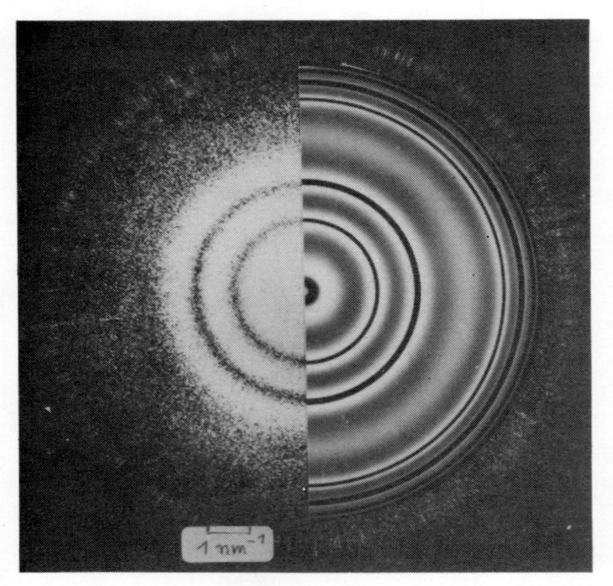

Figure 6. Comparison: diffractogram-filter without astigmatism

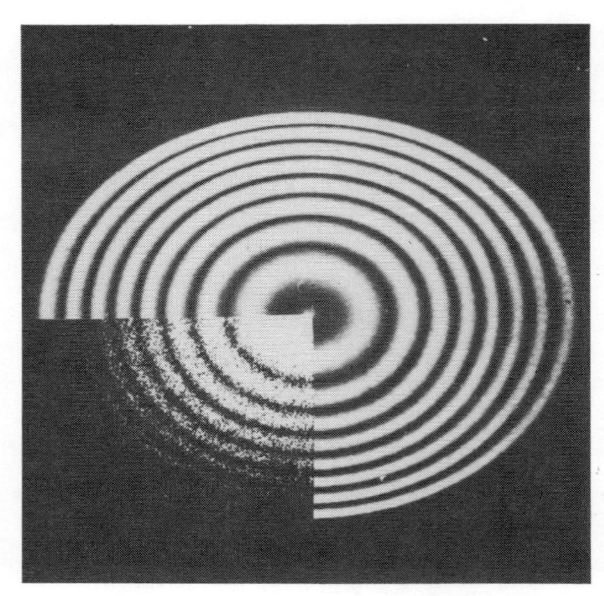

Figure 7. Comparison: diffractogram-filter with astigmatism

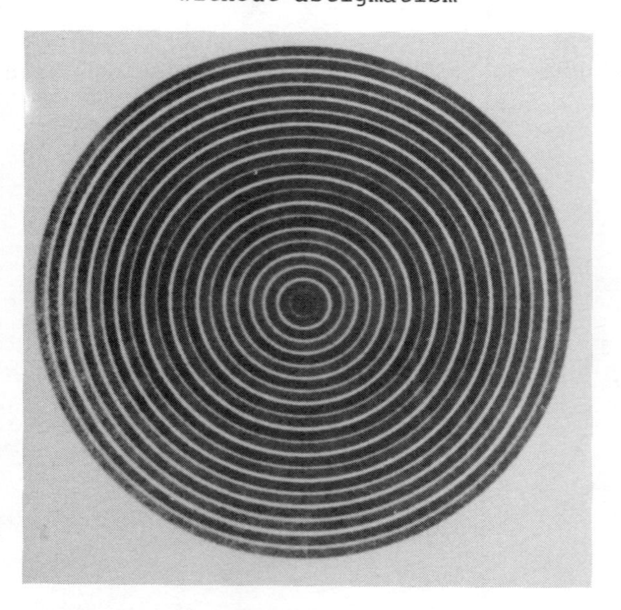

Figure 8. 20-zones-filter

Figure 9. Filtered test bar chart

Figure 10. Filtered test bar chart, once again filtered

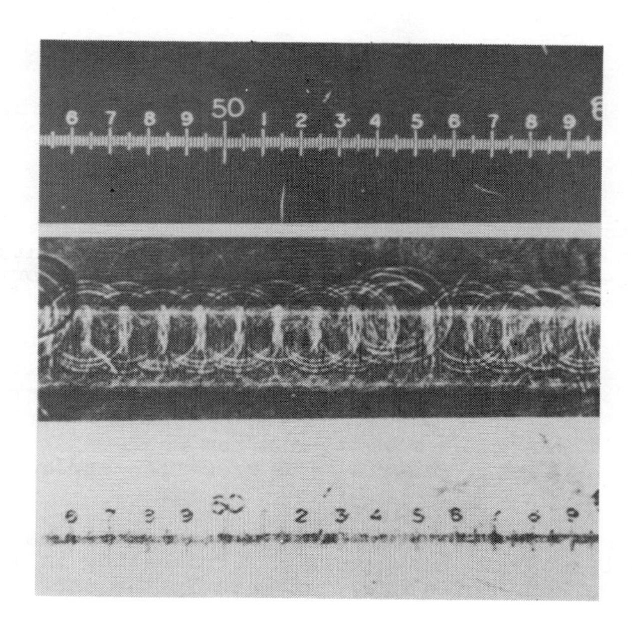

Figure 11. Glass ruler: original, filtered and once again filtered

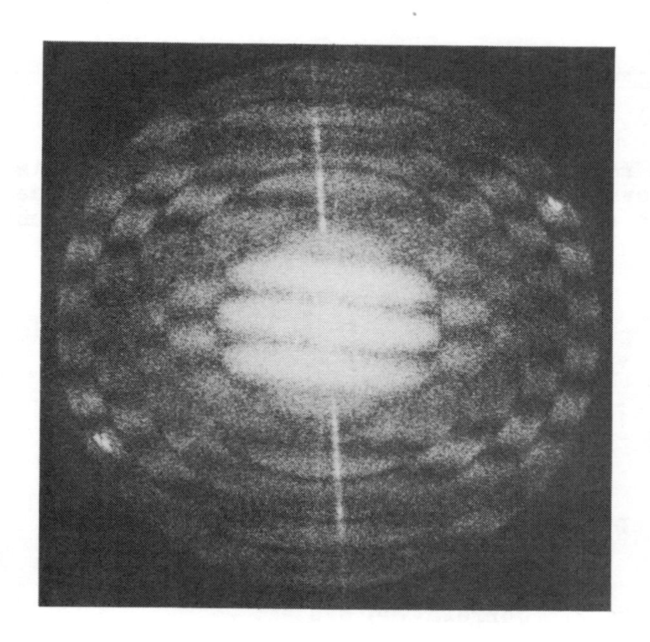

Figure 12. Correlation diffractogram Murein, without astigmatism

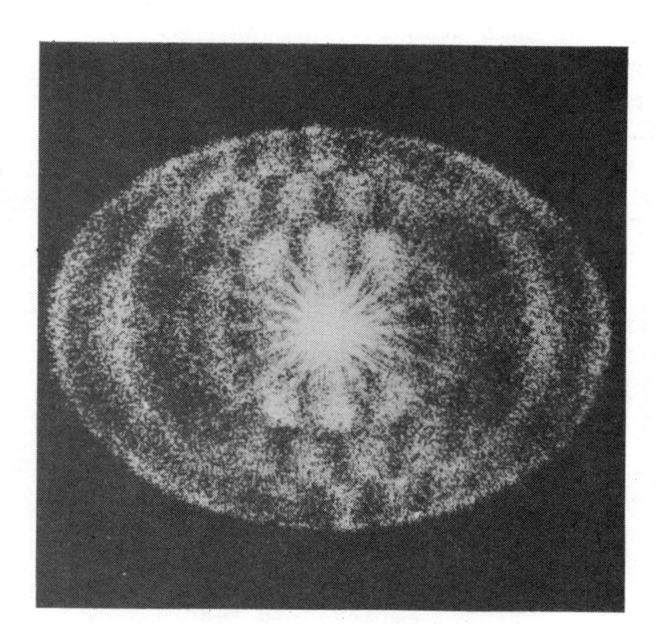

Figure 13. Correlation diffractogram Carbon foil, with astigmatism

A new type of holographic encoding filter for correlation
—— A lensless intensity correlator

G. G. Mu, X. M. Wang and Z. Q. Wang

Institute of Modern Optics, Nankai University, Tianjin, P. R. C.

Abstract

A new type of holographic encoding filter for correlation is presented. This lensless system is especially simple and easy to operate. The low requirements of adjustment of the filter convinience the real-time correlation of opto-electronic hybrid system.

Introduction

Optical information processing by means of spatial frequency filtering have attracted a lot of attentions. But the traditional pattern recognition by Vander Lugt technique requires the matched spatial filter (MSF) to be placed at where it was recorded. Generally the allowable in-plane shift of the filter is no more than a few micrometers, and the out-of-plane one less than a hundred. These limits restrict the practical application of this technique to a certain area.

Since 1968, several types of intensity correlators, based on the principle of geometrical optics, have been reported by Lohmann and others.[1-3] Processing in intensity rather than in amplitude, these systems have some advantages, especially the correlation result is invariant to the shift of the filter.

In this paper, a new type of holographic encoding filter, that is, a lensless intensity correlator, is presented. The in-plane shift of the filter does not influence the correlation result, and the tolerable out-of-plane shift extends to some few millimeters, which is more than one order higher as compared to that of amplitude coherent correlator, besides the technique is characterized by high SNR, intensitivity to phase distortion of the input image, and low requirement of dynamic range of hologram. This permits a wide practical application of this technique.

A lensless intensity correlation system

The recording of holographic filter

The principal scheme of the holographic filter recording system, as shown in Figure 1, is analogous to that of a common hologram recording system, except for the converging reference beam. A characteristic transparency, f(x,y), which will be the pulse response of correlation

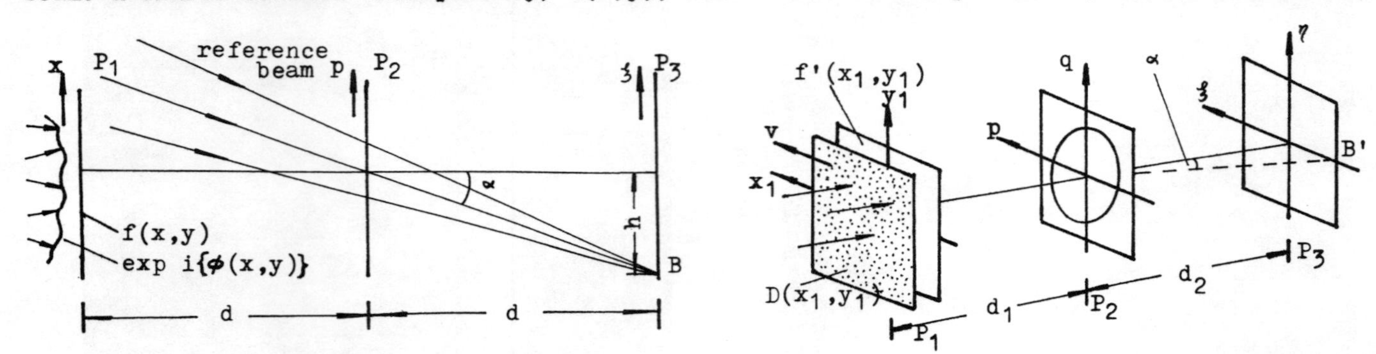

Figure 1. System employed to record holographic filter.

Figure 2. Correlation system.

system, is placed at the input plane P_1 and illuminated by a laser beam uniform in intensity but arbitrary in phase distribution. The converging beam is focused onto plane P_3. A Fresnel hologram of the characteristic image with arbitrary phase, $\phi(x,y)$, due to the illuminating beam, is recorded onto a fresh plate placed at plane P_2, i. e., the desired filter is created. One of the terms of its transmittance function we are interested in is given by

$$R(p,\ q)\ O^*(p,\ q) = e^{-ik\,[(p+h)^2+q^2]/2d} \cdot \int f^*(x,\ y)\ e^{-i\,\phi(x,\ y)}.$$

$$\cdot\, e^{-ik\,[(p-x)^2 + (q-y)^2]/2d}\; dx\, dy \tag{1}$$

where all the parameters are noted in Figure 1. As we show afterwards, this hologram behaves not only as a filter, but also as a lens because of the converging reference.

The correlation process. The correlation system is similar to Figure 1 except for blocking the reference beam, as shown in Figure 2. The whole setup contains a filter besides an input plane and an output plane. An input image, $f'(x_1, y_1)$, is placed at plane P_1, with a contact moving diffuser in front of it. Assume the filter is replaced not exactly at its original position, but with out-of-plane shift of $d_1 - d$, and in-plane one of $(-p_o, -q_o)$. Then the complex field just behind the filter would be

$$C(p, q, t) = R(p + p_o, q + q_o)\cdot O^*(p + p_o, q + q_o)\cdot \int f'(x_1, y_1)\, D(x_1, y_1, t)\cdot$$
$$\cdot\, e^{ik\,[(p-x_1)^2 + (q-y_1)^2]/2d_1}\; dx_1\, dy_1 \tag{2}$$

where

$$D(x_1, y_1, t) = e^{i\,\theta(x_1 - vt,\, y_1)} \tag{3}$$

is the transmittance of the diffuser, v its velocity.

The field distribution in output plane P_3 at an arbitrary time t is described by

$$E(\xi, \eta, t) = \int C(p, q, t)\, e^{ik\,[(\xi - p)^2 + (\eta - q)^2]/2d}\; dp\, dq \tag{4}$$

where d, d_1, d_2 should satisfy $2/d - 1/d_1 - 1/d_2 = 0$. Performing the integrals over input plane and spatial frequency plane, and substituting Equation (3) into (4), we obtain

$$E(\xi, \eta, t) = \int f'(x_1, y_1)\, e^{i\,\theta(x_1 - vt,\, y_1)}\, f^*(x_1\cdot d/d_1 + \xi\cdot d/d_2 + 2p_o + h,$$
$$y_1\cdot d/d_1 + \eta\cdot d/d_2 + 2q_o)\, e^{i\,\varphi(x_1, y_1, \xi, \eta)}\; dx_1\, dy_1 \tag{5}$$

where

$$\varphi(x_1, y_1, \xi, \eta) = \phi(x_1\cdot d/d_1 + \xi\cdot d/d_2 + 2p_o + h,\; y_1\cdot d/d_1 + \eta\cdot d/d_2 + 2q_o)$$
$$+\, k\cdot\{(x_1^2 + y_1^2)/d_1 + (\xi^2 + \eta^2)/d_2$$
$$-\, d\,[(x_1/d_1 + \xi/d_2 + p_o/d + h/d)^2 + (y_1/d_1 + \eta/d_2 + q_o/d)^2]\} \tag{6}$$

By calculating the time average of the mutual intensity function and considering the statistical property of $\theta(x_1 - vt, y_1)$, we get the intensity distribution (time averaging)

$$I(\xi, \eta) = \int |f'(x_1, y_1)|^2 \cdot |f(x_1\cdot d/d_1 + \xi\cdot d/d_2 + 2p_o + h,$$
$$y_1\cdot d/d_1 + \eta\cdot d/d_2 + 2q_o)|^2\; dx_1\, dy_1 \tag{7}$$

Let $T' = |f'|^2$, $T = |f|^2$, the above Equation could be rewritten as

$$I(\xi, \eta) = \int T'(x, y)\cdot T(x\cdot d/d_1 + \xi\cdot d/d_2 + 2p_o + h,\; y\cdot d/d_1 + \eta\cdot d/d_2 + 2q_o)\; dx\, dy \tag{8}$$

If the filter is replaced exactly at its original position, that is, $d_1 = d_2 = d$, $p_o = q_o = 0$, Equation (8) would become as

$$I(\xi, \eta) = \int T'(x, y)\cdot T(x + \xi + h,\; y + \eta)\; dx\, dy \tag{9}$$

which is the expected cross-correlation of the intensity transmittance of the input image and the intensity pulse response of the system.

Analysis and discussion. The described lensless correlator has the same properties as other intensity correlators.

a). Because of the intensity processing, there is no influence of image phase distortion and the phase response, $\phi(x, y)$, of the system.

b). Low requirements of in-plane and out-of-plane adjustments of the holographic filter. If the in-plane shift of the filter is caused only, that is, $d_1 = d_2 = d$, but $p_o \neq 0$ and $q_o \neq 0$, Equation (8) would become as

$$I(\xi, \eta) = \int T'(x, y) \cdot T(x + \xi + 2p_o + h, y + \eta + 2q_o) \, dx \, dy \tag{10}$$

which is also the exact cross-correlation but a whole in-plane shift relative to the coordinate system in output plane. And last, taking out-of-plane shift of the filter only, that is, $p_o = q_o = 0$, but $d_1 \neq d$, we would obtain

$$I(\xi, \eta) = \int T'(x, y) \cdot T(x \cdot d/d_1 + \xi \cdot d/d_2 + h, y \cdot d/d_1 + \eta \cdot d/d_2) \, dx \, dy \tag{11}$$

which is something like a scale-change correlation. For a 3-dB loss of performance, the allowable scale change is about 0.5 %.[4] Equivalently, $1 - d/d_1$ could be 0.5 %. So some few millimeters out-of-plane shift of the filter is permitted to get a harmless-like result.

c). Low requirements of dynamic range of hologram. Because of the Fresnel diffraction, no bright spot appears on the recording plane, so low dynamic range of holographic recording is needed.

d). Increased SNR, etc., due to spatial incoherent light.

Experimental demonstrations

In experiment, a He-Ne laser of 60 mw is used, and d is taken to be about 50 cm. Chinese characters ' Nankai University ' and letters ' A, p, O ' are chosen to be the input images, respectively, and the same ' kai ' and ' p ' to be the corresponding pulse response of the

Figure 3. Input images. ('Kai' and 'p' are to be detected respectively)

Figure 4. Correlation results. (Corresponding to the above two input images respectively)

correlation system, as shown in Figure 3. The correlation results are shown in Figure 4. A computer simulation is also performed which is coincident with the experimental result, as shown in Figure 5.

Conclusion

Figure 5. A computer simulation.

A new type of diffraction intensity correlator is described in this paper. Compared with the diffraction amplitude correlator of Vander Lugt's type, there are a lot of advantages for this kind of correlator along with other known intensity ones that utilize 4-f optical processors, such as, no influences of the input image phase distortion and phase response of the system, low requirements to adjustment accuracy of the filter, low requirements of dynamic range of hologram, and parallelism and multichannelity of high resolution processing. The spatial coherent distortion due to the moving diffuser allows to increase the signal-to-noise ratio.

Meanwhile this lensless correlator we proposed is especially simple, only a holographic filter is used. The experimental demonstrations show that this technique convinience the practical application of correlators.

References

1. Lohmann, A. W., "Matched Filtering with Self-luminous Objects," Appl. Opt., Vol. 7, p. 561. 1968.
2. Potaturkin, O. I., "Incoherent Diffraction Correlator with a Holographic Filter," Appl. Opt., Vol. 18, p. 4203. 1979.
3. Potaturkin, O. I., Nezhevenko, E. S., Khotzkin, V. I., "Coherent Intensity Correlator," J. Optics, Vol. 11, p. 305. 1980.
4. Casasent, D., Furman, A., "Sources of Correlation Degradation," Appl. Opt., Vol. 16, p. 1652. 1977.

Optical tomography-a new holographic inverse filter technique for three-dimensional X-ray imaging

Yaguang Jiang, Linsen Chen

(Suzhou University,Laser Research Section,Suzhou,China)

E.N.Leith

(University of Michigan.Dept.of EECS.Ann Arbor,MI.48104)

Abstract

The application of the advanced holographic inverse filtering technique to the three dimensional X-ray imaging greatly simplifies the decoding process and makes the size of the imaged object twenty times larger than the coded aperture, This might be considered as a stride forward toward the practisal use of flashing tomography.

Introduction

Three-dimensional X-ray imaging technique is a noticeable topic. With computer tomography, it takes several seconds to record the information on the film, but the information of a three-dimensional moving object like the heart beat can only be recorded by means of flashing tomography exposed instantaneously from different angies. Flashing tomography is much cheaper and easier to operate than computer tomography, and it essentially consists of two steps. The first step is the aperture coding, i.e. the convolution of every point on the object with the aperture pattern. The second step is to reconstruct the information of all the layers of the object from the coded image.

In relation to the first step, contributions to the study of the coded aperture pattern such as on-line Fresnal zone plate, off-axis zone plate and others can be found in many papers[1,2]. The nonredundant array[3] is a more desirable coded aperture in which the discerete points are so ordered that the lines between any two points are randomly arranged both in direction and distance, so that the Fourier transform distribution of the array is almost uniform. In the second step, it is necessary to put the coded image in the optical image processing system to make a filtering operation. The key to the solution of this problem is designing and making of a special filter. In the past, the use of a matched filter and autocorrelation operation[4] resulted in a poor signal to noise ratio when there was a strong peak of intensity n and n(n-1) sub-peaks of intensity 1 around the strong peak (n being the number of points in the array). The inverse filter may be an ideal filter for this purpose. In the last twenty years, scientists have been working on the inverse filter, and various methods of making inverse filters such as the sandwich method[5] and holographic method[6] have been developed. Since the function of the coded aperture is complicated, the holographic method is the only way to make a Fourier transform hologram which is to be used as the phase part of the inverse filter. In 1977, H.Weiss replaced the standard matched filter by Stroke's inverse filter in his flashing tomographic experiment and obtained much better results thereof[4]. But the making of Stroke's inverse filter involves much difficulty, since it is still of the sandwich type. As the filter $H^*/|H|^2$ is composed of $1/|H|^2$ and H^*. $1/|H|^2$ being an amplitude part of the filter which has very strong absorption effect, it yields a very low diffraction efficiency and a small dynamic range and is impossible to maintain the inverse relation within a wider dynamic range. Due to the limitation of the filter, no better results of flashing tomography than those of Weiss's obtained in 1977 have yet been discovered in current literature.

Two improvements

In 1982, Jiang and Xu proposed a new holographic inverse filter[7]. In the past four years, we have been working on the improvement of its function and the widening of its applications such as the deblurred images in linear motion, defocus[8], and double vibration[9], Our experimental trsults, all being satisfactory, prove that the new filter has high efficiency, high resolution and can be used in processing very complex blurred images. Unlike what they had been done in the past, our present work has two unique fetures; first, the adoption of a new single holographic inverse filter; second, the introduction of a small angle of view to the coding step.

The making of the new holographic inverse filter involves two main techniques. First, a single inverse filter is made by taking a Fourier transform hologram using a very weak reference beam. Second, a Fourier transform hologram is taken by putting a mask which can attenuate the strong peak intensity of Fourier transform function right in the front of the recoding plate. Thus the average exposure on the plate is nearly uniform and can be controled within the linear range of the plate. The ratio of the object beam and the reference beam remains unchanged, since they are equally attnuated by the mask. The dynamic range of the complex amplitude tranmittance of the filter is about 100:1 or more. The point scale ratio of blurred image to deblurred image reaches 28:1 (for nonredundant array). So, it is highly promising that better results may be achieved by applying this new filter to flashing tomography.

The second improvement on the former work is the using of a small angle of view in the coding step. By angle of view is meant the angle formed between the object and the coded aperture as viewed from the former. It is generally know that the obtainable image depth resolution between the two layers of an object is largely determined by the width of the angle of view. In order to increase the depth resolution, Weiss used ten or more X-ray tubes to compose a coded aperture array (as shown in Fig.1). It is quite evident that should the human body be exposed to the irradiation of the simultaneous flashing of all those X-ray tubes, the radiation dosage would be too strong.

The other alternative is to use just one tube which is to be exposed ten times, with each exposure having a different position and direction arranged in proper order. But this coding method will lengthen the exposure procedure. Moreover, the wide angle coding will cause serious aberration in the processed image. In view of the above disadvantages, we choose the more practical way of using a small coded aperture (about 0.025 rad., or forty times smaller than before) at the cost of somewhat worsening the depth resolution (from the previous 5mm to the present 115mm).

The principal method

The coding setup is shown in Fig.2. The coding source is a lead plate with 43 holes (see Fig.4a) placed in front of an X-ray tube. The size of the aperture is about 50mmX50mm. The X-ray passes through the object and impinges upon the recording film. The distances from layer 1 and 2 to the recording film are 235mm and 120mm respectively. The distance from the coded aperture to the film is 2000mm, so the angie of view is about 0.025arc.. The blurred image recorded on the film is the convolution of the points in the object with coded aperture pattern. Each points in the object will form an array of dots, the size of which is about 6mm for layer 1 and 2.6mm for layer 2 as ditermined by the respective positions of the layers. As the cross section of the layers is 100mmX100mm, the size of the object is about twenty times larger than the coded aperture.

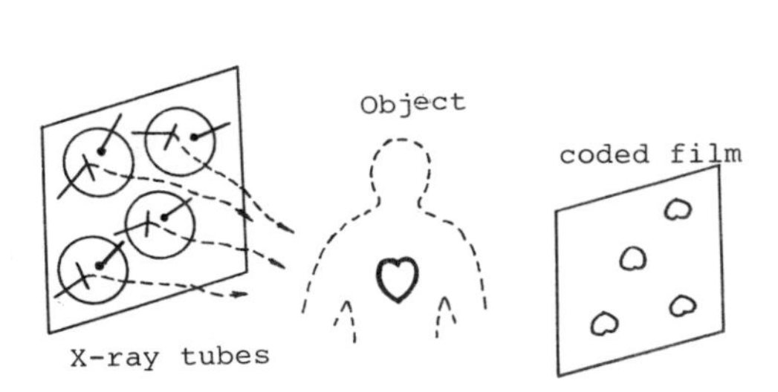

Fig.1. Recording geometry of former work using a large angle of view.

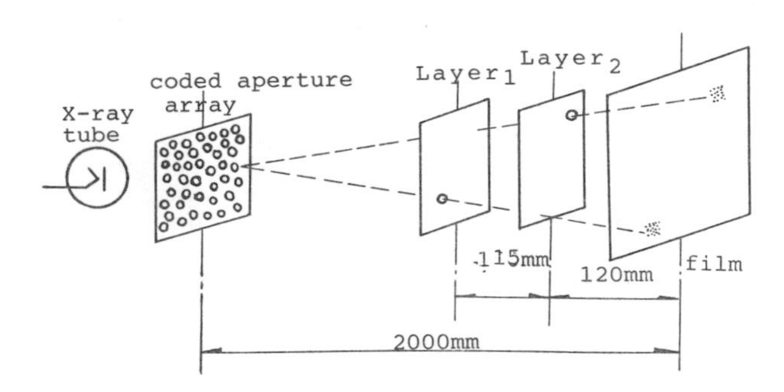

Fig.2. Recording geometry of the present work using a small angle of view.

Let the three-dimensional object be expressed as

$$s = \sum_{i=1}^{n} s_i \qquad (1)$$

The object consists of n layers s_i(i=1 to n).
The coded aperture array is denoted by h_i(i=1 to n).
The coded image s' is given by

$$s' = \sum_{i=1}^{n} s_i * h_i \qquad (2)$$

where "*" denotes the operation of convolution.

The transparent dots array h_1 is used as the sample of the point spread function and put on the input plane of the optical processing system. First, take a photograph of the Fourier trasform of the point spread function to serve as a mask, and then use a very weak reference beam to make a holographic inverse filter. For details of the process refer to (7).
Put the coded image s' on the input plane, (see Fig.3.), Its Fourier transform is

$$S' = \sum_{i=1}^{n} S_i \cdot H_i \tag{3}$$

After filtering the equation (3) is multiplied by $1/H_i$

$$\sum_{i=1}^{n} S_i \cdot \frac{1}{H_1} = S_i + \text{noise} \tag{4}$$

For if $H_i = H_1$, S_i is sharply focused;
if $H_i \neq H_1$, S_i is blurred as the noise.
Since filter $1/H_i$ is made from sample h_1, only the image in layer 1 can be deconvolved into a sparp image, any H_i other than H_1, having scales different from those of H_1 the diffracted light from the filter cannot focus on the output plane but spreads all over the noise.

In order to obtain the other S_i the scaling technique is used. The coded aperture arrays of different layers "h_i"'s are of the same shape, but there scales are different. We put the coded image s' in the convergent illumination beam of lens L_1 and shift it from position E_1 towards E_2, E_3 etc.,the scale of H_i can be reduced until the new H_i' equals to H_1. Let the focal distance of lens L_1 be f, the distance between the coded image and filter be L, then

$$H_i' = H_i \frac{L}{f} = H_1 \tag{5}$$

L is the proper position of the new input plane for deconvoluting the image of layer 1. In this way all the images of the coded layers s_i are deconvolved in the output plane by shifting the coded image s'.

Experimental results

Since the X-ray source is not available in the coding step, we perform an optical simulating experiment. The English and Chinese names of two university "UNIV. OF MI."(University of Michigan) and "   " (Suzhou University) are used to serve as the images of two layers of the three-dimensional object. Chromium photoetched plate on which there is a transparent coded aperture pattern is placed in front of the camara lens. Each point in the object convolves with the coded aperture pattern and forms a blurred image on the imaging plane of the camara. Fig.(4a) shows the coded image and Fig.(4b) is the coded aperture pattern. The blurred image as shown in Fig.(4a) is now deconvolved into Fig.(5a) and Fig.(5b) which show respectively the image of the 1st. and 2nd. layers of the three-dimensional object. The separation between these two layers is 115mm. Althought the depth resolution is low, the results are still considered good since the angle of view of the coded aperture is only 1.5^ (or 0.025 arc.). The adoption of optical simulation in our experiment invariably causes a worsening of the depth resolution by the diffraction effect on account of the comparatively long optical wavelengths. Shold coding be performed with real X-ray which has a much shorter wavelength, the resolution is expected to be further improved.

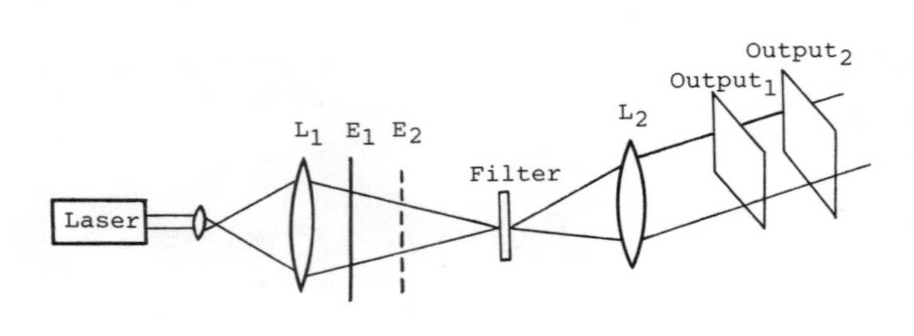

Fig.3. Optical image processing system for deconvolution of the coded image.

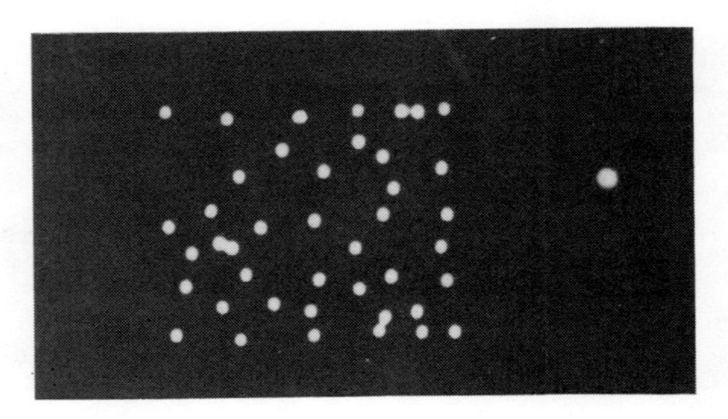

Fig.4b. Coded aperture pattern having 43 holes randomly arranged and its deconvolved dot.

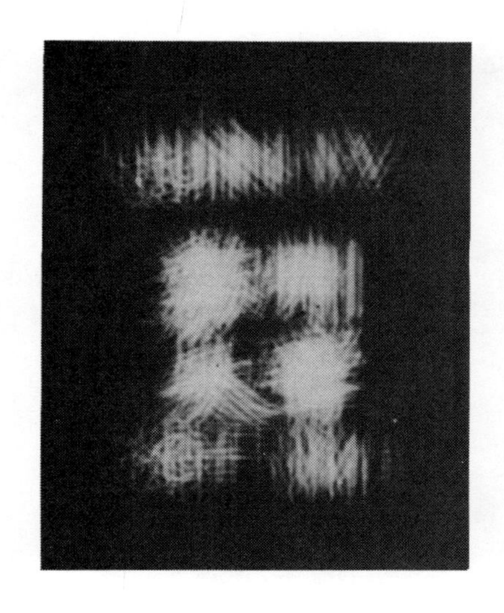

Fig.4a. Coded image composed of two oyerlaping blurred images.

Fig.5a. Deconvolved image of Fig.4a. for layer 1.

Fig.5b. Deconvolved image of Fig.4a. for layer 2.

References

1. Rogers,E.G.W.L.,Han,K.S.,Jones,L.W.,Beierwaltes,W.H.,"Application of a Fresnel Zone Plate to Gamma-Ray Imaging," J. Mucl. Med., Vol. 13, pp. 612-615. 1972.

2. Barrett. H.H.,Gaeewal, K., Wilson, P.T., "A Spatially-Coded X-Ray Source," Radiology, Vol. 104, pp. 429-430. 1972.

3. Golay, M.J.E., "Point Arrays Having Compact Nonredundant Autocorrelation," J. Opt. Soc. Amer. A.,Vol. 61, pp. 272-273. 1971.

4. Weiss, H., Klotz, E., Linde, R., Rabe, G. and Yiemens, U., "Coded Aperture Imaging with X-Rays (Flashing Tomosynthesis)," Optica Acta, Vol. 24, pp. 305-325. 1977.

5. Tsuijuchi, J. Progress in Optics. ed. Wolf, E., North-Holland, Amsterdam. Vol. 2, pp. 133-180. 1963.

6. Stroke, G.W. and Zech, R.G., "A Posteriori Image-Correcting 'Deconvolution' by Holographic Fourier-Transform Division," Physics Letters Vol. 25, pp. 89-90. 1967.

7. Jiang, Y.G., and Xu, Y.R., "A Simple Method for Image Deblurring," Appl. Opt., Vol. 22. pp. 784-788. 1983.

8. Jiang, Y.G. and Chang, C.H., "A New Realization of the Holographic Inverse Filter and its Application," The Special Issue of 83' International Conference on Laser, pp. 426-427. 1983.

9. Jiang, Y.G., Chang, C.H. and Chen, L.S., "Restoration of Blurred Image by Convolution of Arbitrary Point," J. Opt. Soc. Amer. A., pp. 73. 1985.

A holographic match filtering method with high diffraction efficiency

Guoguang Yang, Shaojun Fu, Yonggang Su

Graduate School, University of Science and Technology of China
Beijing, China

Abstract

From physical analysis, we explained that the phase matching is more important than the amplitute in match filtering. Thus the amplitute of Fourier transform of object can be modified to reduce the dynamic range, while the phase information keeps unchanging. The holographic match filter was made by dichromatic gelatin. High efficiency correlation can be obtained by this method.

Introduction

The conventional optical pattern recognition method is the match filtering method. The match filter is made by Vander Lugt [1] Fourier transform hologram. Because amplitude-type hologram is a low diffraction efficiency system, so the match correlation peak is weak. In some applications like air-borne pattern recognition system or robot vision system, this is a serious restriction.

Research in digital imge processing shown [2] that phase information is considerably more inportant than the amplitude information in preserving the visual intelligibility of the picture. Recently, Horner [3] presented a computer simulation results of phase-only match filtering. It shown that phase-only match filter provides high correlation peak and more effective ability of discrimination among different characters. Here we provide an optical method to realize this kind of phase-only match filter and give the experimental results.

Analysis

The concept of match filter has inportant rule in pattern recognition. If the signal to be recognized is $s(x,y)$, the impulse response of match filter system is $h(x,y)= s^*(x,y)$,thus the transfer function of match filter is $H(u,v) = S^*(u,v)$. That is,

$$S(u,v) = |S(u,v)| \exp[j\varphi(u,v)] \tag{1}$$

If the input signal is the pattern to be recognized, the transmitted wavefrant from filter is SS^*. As indicated by Goodman [3], SS^* is a real number, this implies the match filter compansates totally the diviation from plane wave of incident wavefront. Thus the transmitted wavefront is a corrected plane wavefront. This process is shown in Fig.1. Here the match filter is amplitude and phase filter, in which the amplitude filter results in the transmitted field becoming uniform intensity distribution, thus the energy of light is attanuated considerably. In addition, the match filter is made of Fourier transform hologram, the dynamic range of exposure is rather large. This also reduced the correlation efficiency. From above analysis we can see that it is only neccesary to have phase matching, i.e.

$$H(u,v) = \exp[-j\varphi(u,v)] \tag{2}$$

Now

$$S\,H = |S(u,v)|\, e^{j\varphi(u,v)}\, e^{j\varphi(u,v)} = |S(u,v)| \tag{3}$$

is a real number.It is the reason that the phase-only filter can succeed in pattern recognition. Apparently,this is a high efficiency system.

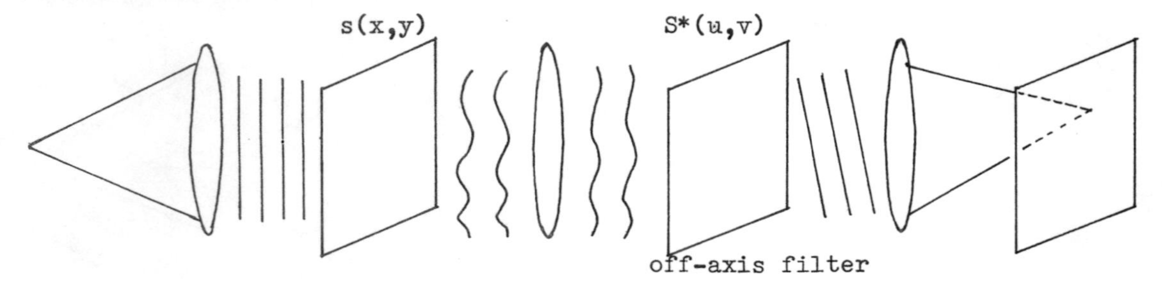

Fig.1. Heuristic analysis system for match filtering

Experiment

In order to realize the phase-only match filter optically, we used analog mathod to make reflection phase holographic match filter. At first, the amplitude spectrum of object is recorded in a holographic plate without reference beam. After developing itbecomes an attanuation mask. As shown in Fig.2, the attanuation mask is put in the back focal plane of lens and the fresh holographic plate is contacted with the mask. The amplitude of Fourier spectrum of object is modified to be an uniform distribution at the recording plate, while the phase of object keeps unchanging. Even the mask has fine structure, because the plate is at Fresnel diffraction deep zone, it ensures the phase after transmitting from the mask is unchanged. In order to record the phase information, the reference beam is introduced from the back of holographic plate. Thus the reflection phase-onlymatch filter was made. In this way the dynamic range is codsiderably reduced.

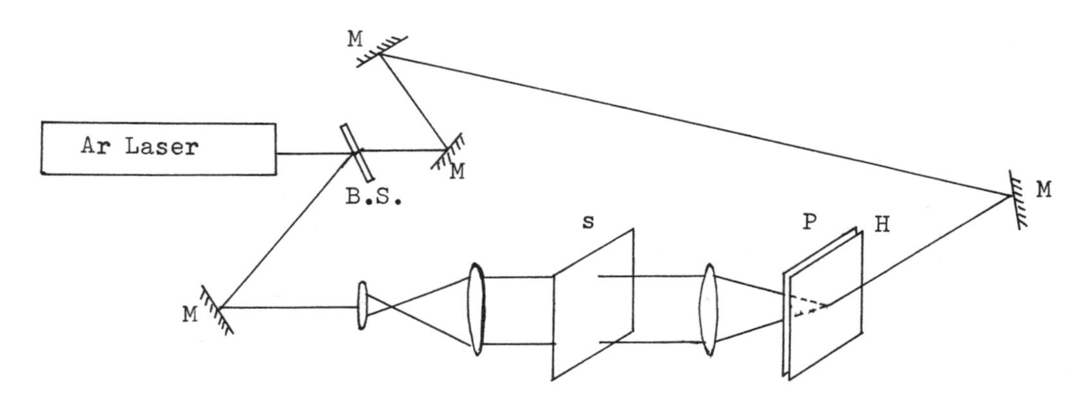

Fig.2. Holographic phase match filter making system;s is the character, P is the attanuation mask,H is recording plate.

To achieve high diffraction efficiency, dichromatic gelatin is used as recording material. and wavelength of 4880 A was exposed. After developing the emulsion of dichromatic gelatin is swollen. It causes the center wavelength of reflection hologram moves towards longer wavelength. In our experiment, the center wavelength moves to 5200-5300 A. Because diffraction efficiency is sensitive to the reconstruction wavelength, thus it requires wavelength matching between recording and reconstruction. Therefore, the emulsion shrinking technique is used. The optical system for recognition is shown in Fig.3.

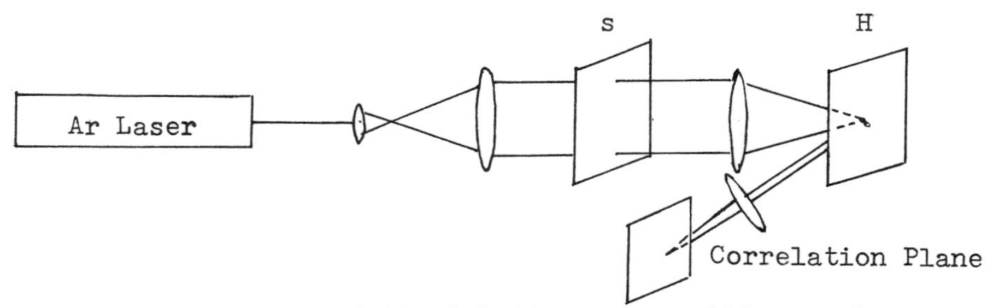

Fig.3. Optical pattern recognition system

In experiment, letter "O" being the character to be recognized was used to make match filter. When letter "O" correlated with this match filter, diffraction efficiency can reach about 45%. For comparing the discrimination ability among similar characters, letter "O", "C" and"G" were used. If input "O" correlated with "O" filter, we normalized the correlation peak as 1. Then letter "C" with filter "O" is 0.8, letter "G" with filter "O" is 0.7. For comparison,if the filter is amplitude and phase match filter, thus correlation of letter "G" with filter "O" is 0.92. Therefore, phase-only match filter provides a good discrimination ability among different characters.

Discussion

In principle, efficiency of reflection hologram of dichromatic grlatin can reach as high as 99%, if developing is carefully processed. Therefore, this is a high efficiency method.

The other advantages are following. The optical system can be more compact by this method. And white light point source can be used for pattern recognition, because the reflection hologram has high power selection of wavelength.

References

1. Lugt, Vander, IEEE Trans, Inf. Theory, Vol. IT-10, pp.139.1964
2. Oppenheim, A.V. and Lim, J.S., Proc. IEEE, Vol.69, pp.520. 1981
3. Horner, J.L. and Gianino, P.D., Appl. Opt. Vol.24, pp.851. 1985
4. Goodman, J.W., Introduction to Fourier Optics John-Willy, 1968

A synthetic phase-only holographic filter for multiobject recognition*

Xian-yu Su, Guan-shen Zhang and Lu-rong Guo

Department of Physics, Sichuan University
Chengdu, P.R.China

Abstract

Combining phase-only filtering with the idea of multiple exposure hologram, we propose a method to make a phase-only synthetic filter for multiobject recognition. When the filter is used in an optical correlation, several correlation responses appear on the output plane. According to the height of the correlation peak, we can easily recognize each of the multiobjects. The results of computer simulation are presented. The characteristic and application prospects of this correlation recognition method are discussed.

Introduction

A.V.Oppenheim has discussed in detail the roles played by phase and amplitude in Fourier domain to characterize a signal and proved that for most images the phase information is more important than the amplitude information. J.L.Horner recently reported a type of correlation-like filter, the phase-only filter (POF) . When this filter applied to optical correlator, a correlation-like response can be obtained in the output plane, which has considerably higher intensity and sharper peak than that of the conventional matched filter. Furthermore, Horner made a phase-only synthetic discriminant function filter and gave out the computer simulation results of using the filter in optical correlation.

Combining phase-only filter with the idea of multiple exposure hologram, we propose, in this paper, a method to make a phase-only synthetic filter using a plane reference wave as the carrier. In the output plane of the correlator, several apparently separable correlation-like responses are obtained simultaneously. Based on the pictures of four different mechanical spares, we make the phase-only synthetic filter, and give out the experimental results of computer simulation of the optical correlation recognition. The experiment shows that the several separate responses simultaneously obtained in the output plane have rather high intensity and obvious difference. At the end of this paper, there is a discussion about the characteristic of this correlation recognition method and its application prospects in multiobject recognition. As a comparison, the correlation recognition results with conventional matched filter are also presented.

Phase-only synthetic filter

The Vanderlugt filter of a function f(x,y), created through conventional holographic techniques, yields the following transmittance function:

$$H(u,v)=A^2+|F(u,v)|^2+A\cdot F(u,v)\exp[j2\pi(au+bv)] + A\cdot F^*(u,v)\exp[-j2\pi(au+bv)] \quad (1)$$

where F(u,v) is the Fourier transform of f(x,y), $A\cdot\exp[-j2\pi(au+bv)]$ is the off-axis reference wave used to create the hologram. It would be possible to synthesize N transmittance functions into a single filter, if N is not too large. This can be done by multiple exposuring and changing the input and orientation of the off-axis reference wave sequentially. When the filter is used in the optical correlator, several separate correlation responses appear in the output plane. However, this method is limited by the dynamic range and diffraction efficiency of the film, and requires a rather sensitive detector and a powerful laser. Whereas a phase-only filter of 100% efficiency has been created by Horner and other researchers. Is it possible to use the idea of multiple exposure hologram to a phase-only synthetic filter, with which several separate correlation responses of rather high intensities and rather sharp peaks can be obtained in the output plane? On this purpose this paper describes the method to make the phase-only synthetic filter. Following is the concrete description: the pictures of four mechanical spares f1 --- f4, as shown in Fig.1,

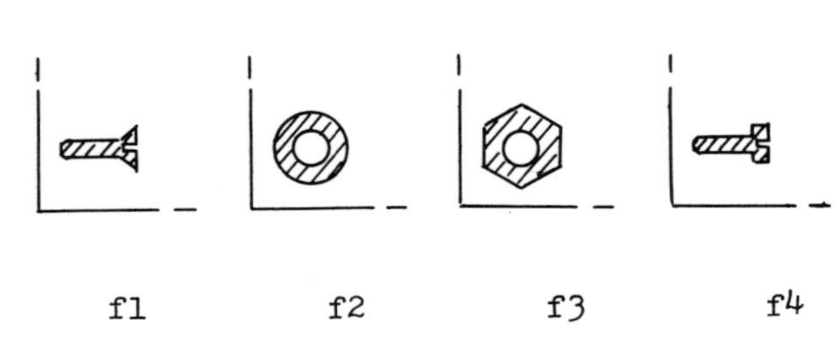

Fig.1 Original images used in correlation experiment

* projects supported by the Science Fund of the Chinese Academy of Sciences

are chosen as the typical original images to make the filter. There are 64×64 pixels in the input image field. The four spares have different shapes and different areas which are 115, 248, 302, 125 respectively; nevertheless, f1 and f4, f2 and f3 are similar in shape to each other. The frequency spectrum of the four original images are symbolized as F1 - F4. Since in the transmittance function (1) the last term is the only one involved in correlation, and because of the flexibility of computer generated hologram techniques, we only consider the last term and then the reference wave appears as frequency spectrum carrier. With different direction plane wave as carrier to each, the frequency spectrum of the said four original images are linearly superposed into a synthetic function. Mathematically, it is represented as

$$H(u,v)=\sum_{i=1}^{4} Ki\cdot Fi(u,v)\cdot \exp[j2\pi(ai\cdot u+bi\cdot v)] \quad (2)$$

where ai and bi are the direction cosine of plane carrier beam i; Ki is proportional factor of the carrier intensity. ai and bi are chosen such that the separation degree of the autocorrelation peaks of the original images would be as large as possible. We chose (16,16), (16,48), (48,16), (48,48) as their anticipated location. And ki is chosen such that autocorrelation of every original image has the same amplitude. The phases of frequency spectrum of image i and the plane carrier of beam i are denoted by ϕi, ϕri respectively. Then Eq.(2) can be rewritten as:

$$\begin{aligned} H(u,v) &= \sum Ki\cdot|Fi(u,v)|\cdot\exp[j\phi_i]\cdot\exp[j\phi_{ri}] \\ &= \sum Ki\cdot|Fi(u,v)|\cdot\exp[j(\phi_i+\phi_{ri})] \\ &= |H(u,v)|\ \exp[j\phi(u,v)] \end{aligned} \quad (3)$$

By setting the modulus of the synthetic function equal to unit, and taking its conjugate, the phase-only synthetic filter function is obtained as follows:

$$H_\phi(u,v)=\exp[-j\phi(u,v)] \quad (4)$$

$$\text{where } \phi(u,v)=\text{arctg}[\sum Ki\cdot|Fi(u,v)|\cdot\sin(\phi_i+\phi_{ri}) / \sum Ki\cdot|Fi(u,v)|\cdot\cos(\phi_i+\phi_{ri})] \quad (5)$$

The filter function expressed in the Eq.(4) contains the information of the modulus and phase of original image frequency spectrum, and of the plane carrier. Because the operation $H_\phi(u,v)=\mathcal{F}[H(u,v)]$ is nonlinear to the original image frequency spectrum, it is difficult to make this filter by optical hologram techniques. However, we can reach our goal with the method of kinoform.

The computer simulation of optical correlation

Using the phase-only filter described above in optical correlator, inputing one of the original images, fi(x,y), the correlation-like response which would appear in the correlator output plane is represented as:

$$g(x,y)=FT^{-1}\{Fi(u,v)\cdot H_\phi(u,v)\} \quad (6)$$

where FT^{-1} denotes inverse Fourier transformation, g(x,y) is the complex amplitude distribution on the output plane. The intensity which can be detected on the output plane is:

$$I(x,y)=|FT^{-1}\{Fi(u,v)\cdot H_\phi(u,v)\}|^2 \quad (7)$$

Inputing the original images into the correlator one by one, the correlation output results of computer simulation are obtained. As comparison, the correlation results of conventional synthetic filter experiment is also given. The eight peak response intensities of the correlation experiments are given in Table 1. The data are normalized, assuming the conventional peak value to be unity. The percentage in the right-hand parentheses refers to the decrease rate of crosscorrelation to autocorrelation, which shows the discrimination power of the correlation system. In Tab.1(a) there are the results when conventional synthetic filter is used, while in

	f1	f2	f3	f4
f1	1.00	0.23(77%)	0.25(75%)	0.89(11%)
f2	0.23(77%)	1.00	0.82(18%)	0.25(75%)
f3	0.25(75%)	0.82(18%)	1.00	0.26(74%)
f4	0.89(11%)	0.25(75%)	0.26(74%)	1.00

(a)

	f1	f2	f3	f4
f1	142.9	5.5(96%)	7.3(95%)	47.6(67%)
f2	3.9(97%)	118.9	13.0(89%)	4.9(96%)
f3	4.9(95%)	8.2(92%)	102.1	4.8(95%)
f4	38.9(67%)	6.2(95%)	6.6(94%)	117.9

(b)

Table 1. Correlation results for
(a) conventional Synthetic filter
(b) phase-only synthetic filter

Tab.1(b) are those when phase-only synthetic filter is used. This paper shows the 3-D graphs of four correlation output out of the eight experiments. Fig.2 depicts those in conventional synthetic filter, with (a) for f1---a screw, (b) for f2--- a washer, respectively. Fig.3 pertains to the phase-only synthetic filter, where (a), (b) are for the f1 and f3. It should be noted that all plots in Fig. 2-3 are normalized to unity by the plotting subroutine.

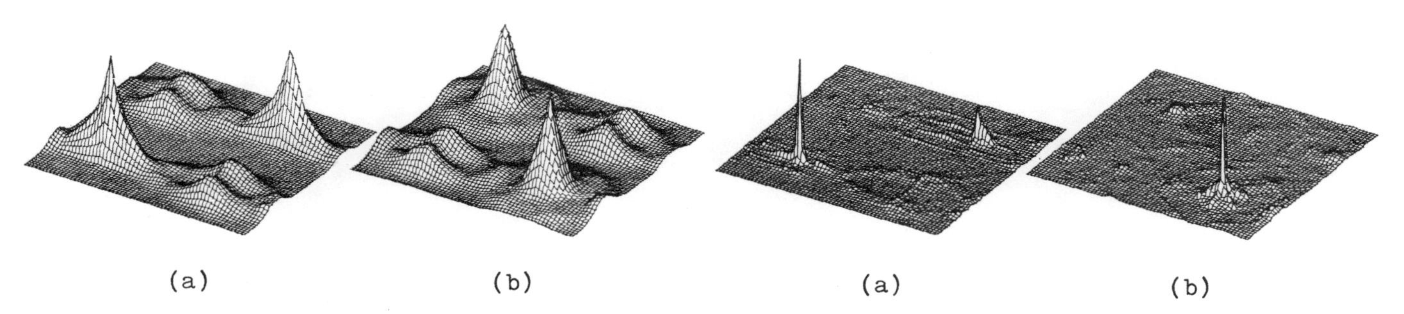

(a) (b) (a) (b)

Fig.2 Correlation response in the output plane for the conventional synthetic filter
(a) f1 as input
(b) f3 as input

Fig.3 Correlation response in the output plane for the phase-only synthetic filter
(a) f1 as input
(b) f3 as input

Discussion

The experimental results show that in the case of multiobject recognition with phase-only synthetic filter, the correlation-like peak intensity is much greater than that of using the conventional filter. From Table 1 we can see that the ratio of autocorrelation peak values of the two kinds of filters are 142.9:1, 118.9:1, 102.1:1 and 117.9:1 when the input images are f1,f2,f3 and f4 respectively. The characteristic of high efficiency output response greatly lowers the requirement of sensitivity of detector, so we can also use the correlation system of low power and compact-structure composed of laser diodes and computer holographic elements.

The value difference between autocorrelation peak and crosscorrelation peaks is an important index of the discrimination power of the correlation recognition system. The results show that there is still obvious difference between the autocorrelation-like and crosscorrelation-like peak intensity when there is multiple output in a common output plane. In Fig.3(a), which refers to the correlation-like output of f1, for example, the autocorrelation-like peak value of f1 is very strong and sharp, while the crosscorrelation-like of f1 with f2 and f3 has no apparent peak values. In comparison with the autocorrelation-like,they decrease 96% and 95% respectively. For the objects of similar shapes, such as f1 and f4, when using a conventional filter, in comparison with the autocorrelation peak value, the crosscorrelation peak value decrease 11%. But when we use phase-only filter, it decreases 67%. It shows the apparent improvement of the discrimination power of the system.

Because of introducing four plane waves of different direction as frequency spectrum carriers to the four original images in making phase-only filter, we get one autocorrelation-like and three crosscorrelation-like responses which are obviously separated. If we set four separate detecting units at the locations, we can easily determine the input image is which of the four original images, and if we use array detector, the location of the input image in the input field can also be determined. It might be important for multiobject recognition.

Conclusions

The phase-only synthetic filter using plane reference wave of different direction as carrier, as presented in this paper, has both the characteristics of the multiple exposure hologram filter and the phase-only filter. The results of computer simulation show that several separate correlation-like responses can be obtained in the output plane of an optical correlator by applying this filter. In comparison with conventional filter, these responses have higher intensity, sharper peaks and more obvious difference. Therefore, it has a bright prospective to be extended in the correlation recognition of more objects and more characteristics.

References

1. A.V.Oppenheim, J.M.Lim, Proc. IEEE 5, 529(1981)
2. J.L.Horner, P.D.Gianino, Appl. Opt. 23, 812(1984)
3. P.D.Gianino, J.L.Horner, Opt. Eng. 23, 695(1984)
4. J.L.Horner, P.D.Gianino, Appl. Opt. 24, 851(1985)

INTERNATIONAL CONFERENCE ON HOLOGRAPHY APPLICATIONS

SPIE Volume 673

Session 16

Optical Information Processing III

Chairmen
H. Podbielska
Technical University of Wroclaw, Poland
Quwu Gu
Changchun Institute of Optics and Fine Mechanics, China

Optical sine transformation

Guoguang Yang, Jianming Gong, Jingjuan Zhang, Junben Chen, Yuping Ho

Graduate School, University of Science and Technology of China
Beijing, China

Abstract

According to the optical general transformation theory, the phase mask distribution of optical sine transformation has been calculated. In order to avoid the diffraction loss of phase mask, optical waveguide method is used. Computation shows that the optical sine transformation can be realized by one phase mask,i.e. one half of cylindrical lens in 1D case and one quarter of spherical lens in 2D case. Experimental results coincide with the theoretical prediction.

Introduction

As well known,optical Fourier transformation has wide applications in optical information processing. But the operations using optical Fourier transformation are only the spatial-invariant operations. It has restriction of information processing cababilities.Recently,Ho[1] proposed a theory of optical general transformation. Realization of the theory will greatly extend the ability of optical information processing. Although there have been some optical linear transformations,such as coordinate transformation,[2] Mellin transformation[3] and cosine transformation[4]. But the optical general transformation theory provides a general method to realize any kind of linear transformation.

Sine transformation has applications in digital image processing,especially in image data compression. For high sampling image, digital technique can not provide real time processing. However,optical method will give the posibility of real time image compression.Therefore,we choose sine transformation as first example of using the theory.

Theory

The optical general transformation system is shown in Fig.1,where H_i is the phase mask (kinoform),l_i is the distance between the masks, f is the input and f' is the output. Thus

Fig.1. Optical general transform system

the optical system can be expressed as the product of the phase masks and free propagation factors between masks, i.e.

$$\tilde{G}= G_o(l_n)\ H_n\ G_o(l_{n-1}) \cdots\cdots G_o(l_1)\ H_1\ G_o(l_o) \tag{1}$$

where $G_o(l_i)$ is the free propagation factor. In near axis approximation, this is

$$G_o(l,x,x')= \frac{1}{j\lambda l} \exp\left[jkl+jk(x-x')^2/2l\right] \tag{2}$$

where l is propagation dictance, $k=2\pi/\lambda$. And H is phase mask matrix. In coodinate represantation, H is diagonal, i.e.

$$H(x,x')= \exp\left[j\varphi(x)\right] \delta(x,x') \tag{3}$$

The theorem of general transformation proven by Ho,Yang and Gu is that: for any unitary transformation G, there are optical system $\tilde{G}$,composed of finite phase masks to realize it approximately, the more accurate we require, the more lenses should be used. In order to design an optical system $\tilde{G}$ to realize an arbitrary given unitary transformation G approximately if the transformation matrix is $N \times N$ matrix, it must calculate N+1 parameters $l_o, l_1, \cdots, l_n$

and N functions $\varphi_1(x,y)$, $\varphi_2(x,y)$, ... $\varphi_n(x,y)$. The total number of parameters are $M \times N^2+M+1$ if M phase masks are used in the optical system. This is a function optimization design problem. We have developed an optimization calculation method to design the optical system.

A serious problem in constrution of an optical system to realize a linear transform is the diffraction loss of phase masks. If the masks are more than 3, because of diffraction loss, the output will be so weak that SNR is too low to detect. Now we proposed an optical waveguide system, as shown in Fig.2. T e system is surrouded by mirrors. It can be proven

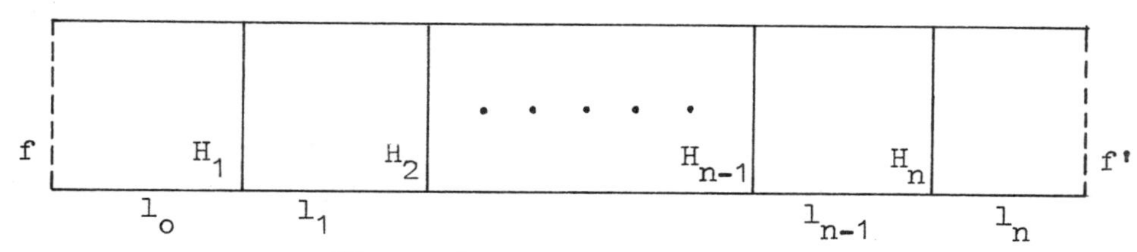

Fig.2. Optical waveguide system

that, the eigenfunction in the waveguide is

$$\phi_{mn} = \frac{1}{\sqrt{L}} \exp\left[-j\pi^2(m^2+n^2)z/2kL^2\right] \sin\left(\frac{m\pi x}{L}\right) \sin\left(\frac{n\pi y}{L}\right) \tag{4}$$

where m,n are integer, L is the size of waveguide. Now the propagation matrix can be deduced as

$$G_o^W(l) = \sum_{n,m} \frac{1}{L} \exp\left[-j\pi^2(m^2+n^2)l/2kL^2\right] \sin\left(\frac{m\pi x'}{L}\right) \sin\left(\frac{m\pi x}{L}\right) \sin\left(\frac{n\pi y'}{L}\right) \sin\left(\frac{n\pi y}{L}\right) \tag{5}$$

By substitution of $G_o^W(l)$ for $G_o(l)$, we have

$$\widetilde{G} = G_o^W(l_n)H_n \cdots\cdots H_1 G_o^W(l_0) \tag{6}$$

According to above waveguide method, computation of sine transformation has been carried out. In result, the optical sine transformation can be realized by only one phase mask. In 1D case, the phase mask is

$$H = \begin{pmatrix} e^{jh_1} & & & \\ & e^{jh_2} & 0 & \\ & 0 & & \\ & & & e^{jh_n} \end{pmatrix} \tag{7}$$

where

$$h_n = -\, n^2\pi/2(N+1) \tag{8}$$

here n is positive integer, N is samling points or order of transformation matrix. And distance is satisfied by the equation

$$l_o = l_1 = l = 2L^2/\lambda(N+1) \tag{9}$$

In principle, mask making should use kinoform technique. But it is easy to recognize, the phase factor is a quadratic factor like a lens. Therefore, we can use a lens to realize a sine transformation. But we should indicate here n is positive integer,thus only one half of cylindrical lens in 1D case or only one quarter of spherical lens in 2D case is used. The optical system is shown in Fig.3. It has been proven that f=l,i.e. the length of waveguide equals to focal length of lens. In other words, the input and output plane of sine transformation are in front and back focal plane of lens respectively, just like Fourier transform.

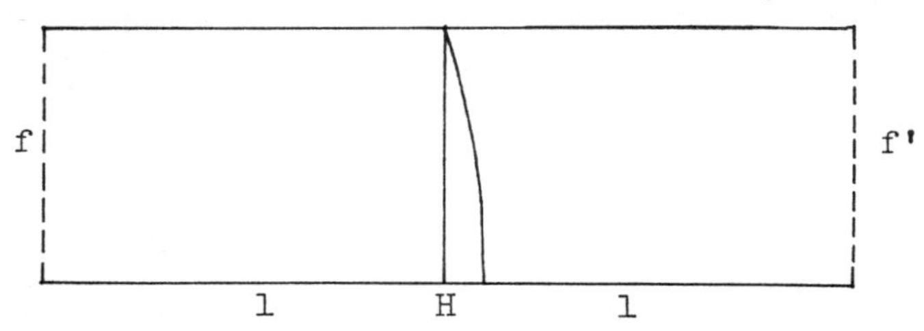

Fig.3. Optical sine transformation system

Analysis

From physical view of point, we can analyse this process as following. If a -function (x-x$_o$) is taken as input, thus the output is $F_s(u) = \sin(2\ ux_o)$. This is the eigenfunction of sine transformation. As shown in Fig.4, the point source at x_o will have a series of images by multiple reflection of mirrors. The input becomes two comb function which are displaced by $2x_o$. i.e.

$$U_o(x') = \sum_{n=-\infty}^{\infty}\delta[x'-(2nL-x_o)] - \sum_{n=-\infty}^{\infty}\delta[x'-(2nL+x_o)] \tag{10}$$

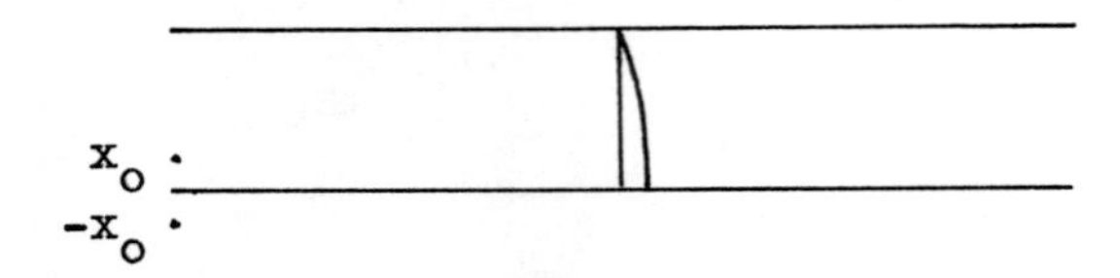

Fig.4. Optical system for analysis

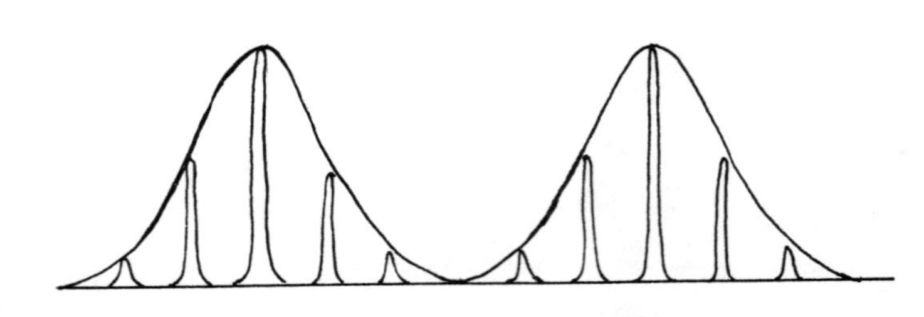

Fig.5. Sampled output of sine transform of -function

The basic interference pattern in the output plane is due to the interference between the sources at x_o and $-x_o$. The minus sign is due to the half wave loss of reflection from the mirror. Thus the output function is

$$U_f(x_f) = \iint_{-\infty}^{\infty} U_1(x)P(x)\ \exp(-j2\pi x_f x/\lambda f)dx = F_1(x_f/\lambda f)*\mathcal{P}(x_f/\lambda f) \tag{11}$$

where P(x) is the aperture, $U_1(x)$ is the wave field at plane P, F_1 is the Fourier transform of $U_1(x)$, $\mathcal{P}$ is the Fourier transform of P(x). From the propagation theorem of angular spectrum[1], we have

$$F_1(x_f/\lambda f) = F_o(x_f/\lambda f)\exp(-j\pi x_f^2/\lambda f) \tag{12}$$

where F_o is the Fourier transform of U_o. And it is easy to obtain

$$F_o(x_f/\lambda f) = -2j\sum_{n=-\infty}^{\infty}\delta(x_f/\lambda f-n/2L)\ \sin(2\pi x_o x_f/\lambda f) \tag{13}$$

and

$$\mathcal{P}(x_f/\lambda f) = 2jL\exp(-j\pi x_f L/\lambda f)\ \mathrm{sinc}(\pi x_f L/\lambda f) \tag{14}$$

Substitute (12),(13),(14) into (11), we obtain

$$U_f(x_f) = \{4L\sum\delta(x_f/\lambda f-n/2L)\ \sin(2\pi x_o x_f/\lambda f)\exp(-j\pi x_f^2/\lambda f)\} * \{\exp(-j\pi x_f L/\lambda f)\ \mathrm{sinc}(\pi x_f L/\lambda f) \tag{15}$$

where * denotes convolution. From equation (15), we notice that $\sin(2\pi x_o x_f/\lambda f)$ is the eigenfunction of sine transform, but this is sampled by a comb function spaced by f/2L. Because the comb function convolves with function $\mathrm{sinc}(\pi x_f L/\lambda f)$, it wides the δ-function to sinc function. The result is shown in Fig.5. Obviously[1], the multiple beam interference due to waveguide reflection results in the automatic sampling effect of output.

Then we can extend this analysis now. If one mirror is taken out from the waveguide, the basic interference pattern of sine transform still exist, but the fine structure will vanish because of no multiple beam interference. The discrete transformation is transited to the continual one.

Experimental results

The 200 mm long waveguide is used. And the focal length of lens also is 200mm. As mentioned above, at first take $f(x)=\delta(x-x_o)$ as input, i.e. a slit is put at some position x_o of input plane in 1D case. Three photos in Fig.6 are the sine transform results corresponding three different x_o respectively. Obviously, the smaller x_o gives low spatial frequency sine grating, and vise versa. From Fig.6 we can see that the sine function is sampled. Fig.7 is the sine transform of -function in continual case. And Fig.8 is the experimental results in 2D case.

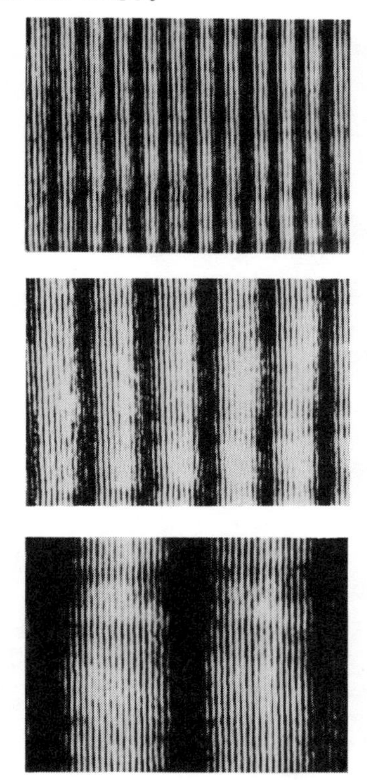

Fig.6. Optical discrete sine transform results of $\delta(x-x_0)$ in 1D case.

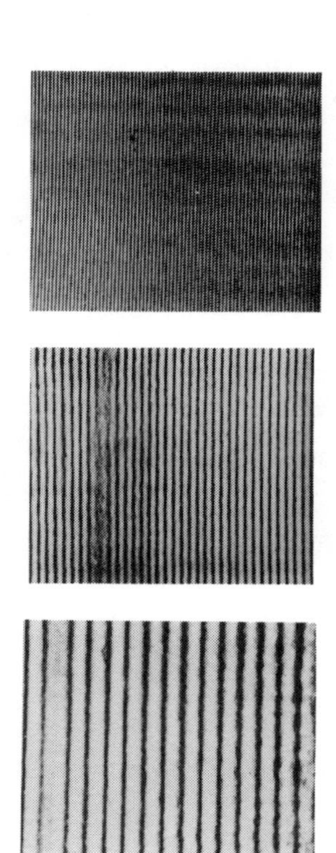

Fig.7. Optical continual sine transform results of $\delta(x-x_0)$ in 1D case.

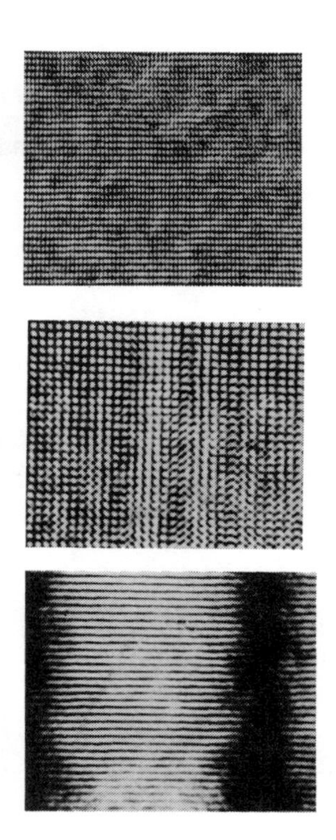

Fig.8. Optical continual sine transform of $\delta(x-x_0,y-y_0)$ in 2D case.

If the input signal is the rectangular function, i.e.

$$f(x) = \begin{cases} 1 & a \leq x \leq b \\ 0 & \text{otherwise} \end{cases}$$

thus its sine transform is

$$F_s(u) = (b-a)\,\text{sinc}[u(b-a)]\,\sin[\pi u(a+b)]$$

It is a high spatial frequency sine function modulated by sinc function as profile. Fig.9 is its experimental results. Fig.10.a is the SMPTE standard picture "a girl" as input, thus the output is shown in Fig.10.b. This result coincides with the digital sine transformation result.

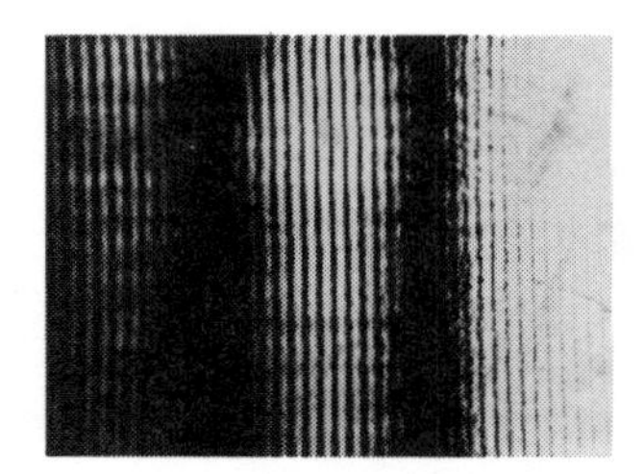

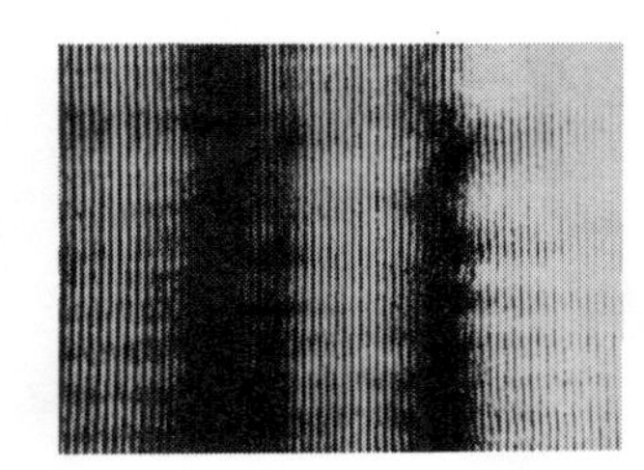

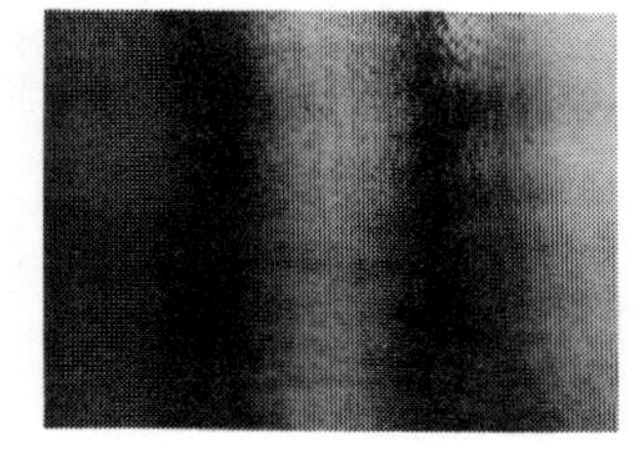

Fig.9. Sine transform of rectangular function,each picture corresponds different position of same slit.

Fig.10.a. SMPTE picture

Fig.10.b. Sine transform of "girl"

Conclusion

According to optical general transformation theory, the optical sine transformation has been realized. This is an example of realization of the theory. It shows the feasibility and effectness of the theory. The optical sine transformation provides the posibility of real time data compression. And the waveguide system avoids the diffraction loss. Therefore, it provides a method to realize other kind of transform optically.

References

1. Ho,Yu-ping, Pattern transformation and holographic optical system, SPIE. Vol.231,pp. 306-313. International Optical Computing Conference,1980.
2. Bryndahl,c,, J.Opt.Soc.Am.,Vol.64, pp.1092. 1974
3. Cassasent,d. and Psaltis,D.,Opt. Eng. Vol.15, pp. 258. 1976
4. Gu,Z.H., Legert,J.R. and Lee,S.H., Opt. Commun. Vol.39, pp.137. 1981

Generalized Laplace`s operators realized by computed - generated spatial filters.

A.Kalestyński

Warsaw Technical University, 00 662 Warsaw, Koszykowa 75, Poland

Abstract

Generalized Laplace operator is defined as $\Delta_{r'r''} = \frac{\partial^{r'}}{\partial x^{r'}} + \frac{\partial^{r''}}{\partial y^{r''}}$,where

r', r'' are numbers.The operation is realized in coherent optical system by means of spatial filters.The transmittance of the filter $T_{r'r''}(u,v)$ has been computed by the Lohmann method.Plotted drawing after suitable deminishing is inserted in the back focal plane in the imaging system. In the image plane of the system we obtain searched derivative of the input function g . Generally the derivatives could be various in different directions with any order. The images of $\Delta_{r'r''}$ for $r',r'' = 1/2,3/4;$ $1,3/2;4/3,7/4$ are realized.

Coherent implementation of derivatives of optical signals

Let $f(x,y)$ be a complex function generated by a transparency placed at the input of the optical system. Then $F(u,v)$ is its Fourier spectrum which appeares in the back focal plane of a lens in this optical system. When a filter with transmittance $T(u)$ $T(u) = \alpha u^r$, where α is a constant and r is a number , is placed in the Fourier plane , then at the output of the optical system the second Fourier transformation appears in the following form :

$$g(x) = \int F(u)\,\alpha\, u^r \exp(2\pi iux)\,du = \frac{\alpha}{(2\pi i)^r}\,\frac{d^r}{dx^r}\int F(u)\exp(2\pi iux)\,dx =$$

$$= \frac{\alpha}{(2\pi i)^r}\,\frac{d^r}{dx^r}\,f(x) \qquad /1/$$

For the sake of simplicity the description is carried out in one dimension, what does not hurt the generality of the description.

Spatial filter with transmittance $T(u) = (2\pi iu)^r$ helps to obtain the light distribution $g(x)$ at the output which mapps the r - th order derivative of the input function $f(x)$.

Spatial filter which should realise two - dimensional derivative $\Delta_r = \frac{\partial^r}{\partial x^r} + \frac{\partial^r}{\partial y^r}$ has following transmittance :

$$T(u,v) = (2\pi iu)^r + (2\pi iv)^r = (2\pi iu)^r + (2\pi i v)^r =$$

$$= (2\pi|u|)^r \exp\left(i\frac{\pi}{2} r \operatorname{sgn} u\right) + (2\pi|v|)^r \exp\left(i\frac{\pi}{2} r \operatorname{sgn} v\right) \qquad /2/$$

Generalized Laplace operator

When availing the definition of the derivative given by eq (2), the order r must not be rational number.Generally $r = p/q$ without any additional conditions ; r could be $r' = p'/q'$ in x direction and an other $r'' = p''/q''$ in y direction. Generalized Laplace operator is defined as $\Delta_{r',r''} = \frac{\partial^{r'}}{\partial x^{r'}} + \frac{\partial^{r''}}{\partial y^{r''}}$ where r',r'' are numbers.

The transmittance of the spatial filter in this case is :

$$T(u,v) = (2\pi iu)^{r'} + (2\pi iv)^{r''} = (2\pi|u|)^{r'}\exp\left(ir'\frac{\pi}{2}\operatorname{sgn} u\right) + (2\pi|v|)^{r''}\exp\left(ir''\frac{\pi}{2}\operatorname{sgn} v\right)$$

For example let $r' = 3/4$ and $r'' = 7/4$ than $T(u,v)$ is :

$$T(u,v) = (2\pi i\,|u|)^{3/4} \exp(i\,3/8\pi \,\mathrm{sgn}\, u) + (2\pi i |v|)^{7/4} \exp(i 7/8\,\pi \,\mathrm{sgn}\, v) \qquad /4/$$

This could be formulated in details :

$$T(u,v) = \begin{cases} +(2\pi|u|)^{3/4} \exp(i\,3/8\pi) + (2\pi|v|)^{7/4} \exp(i\,7/8\pi) & u>0,\ v>0 \\ -(2\pi|u|)^{3/4} \exp(-i\,3/8\pi) + (2\pi|v|)^{7/4} \exp(i\,7/8\pi) & u<0,\ v>0 \\ -(2\pi|u|^{3/4} \exp(-i\,3/8\pi) - (2\pi|v|)^{7/4} \exp(-i\,7/8\pi) & u<0,\ v<0 \\ +(2\pi|u|)^{3/4} \exp(i\,3/8\pi) - (2\pi|v|)^{7/4} \exp(-i\,7/8\pi) & u>0,\ v<0 \end{cases} \qquad /5/$$

The distribution of the changes of the phase in the spatial filter which realises the operation $\Delta_{r',r''} = \Delta_{3/4\ ,\ 7/4}$ is schematically shown in Fig 1.

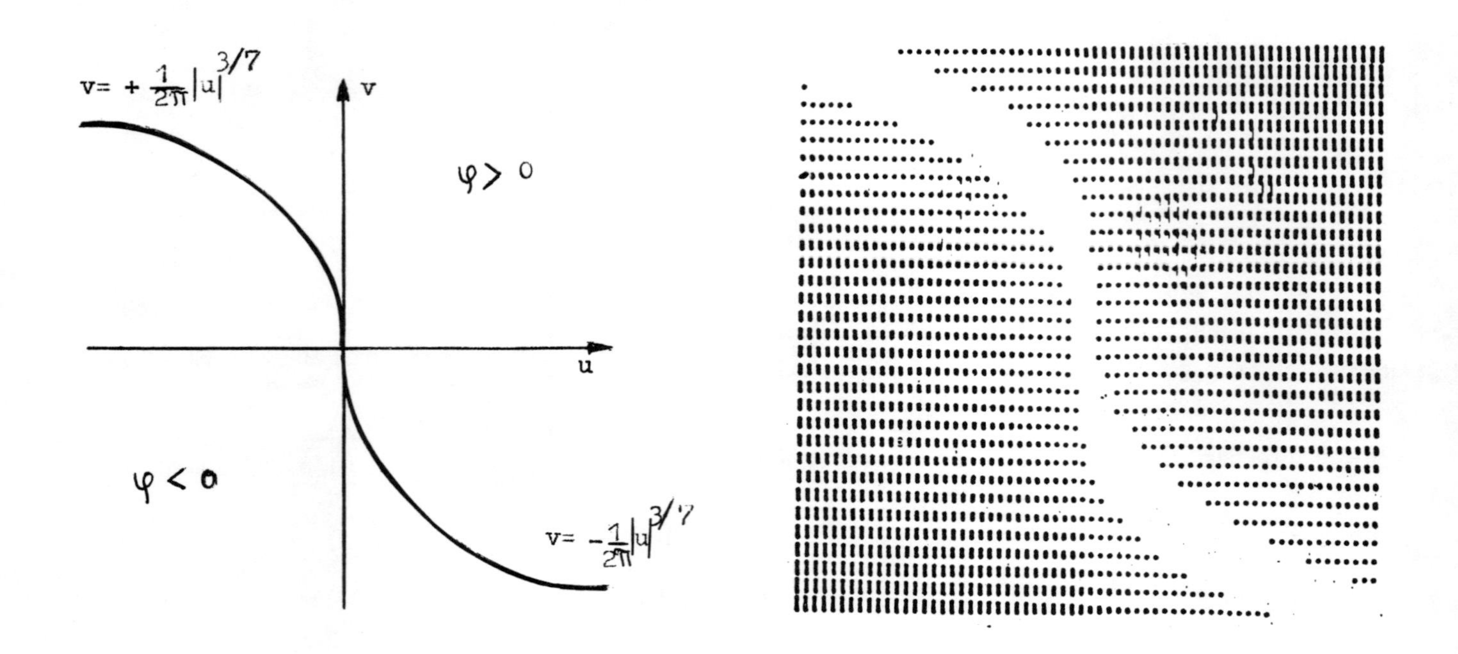

Fig 1. Phase changes in the filter

Fig 2. Computer - generated differentiating filter.

Differentiating spatial filter with transmittance $T(u,v)$, shown in Fig.2. has been implemented by coding the amplitude and the phase using the technique of Lohmann [2]. The filter plane has been divided in $N \times N$ elementary cells. The value of amplitude is coded in the height and the phase in the displacement of transmitting rectangular window in every cell. From simple discussion of the input conditions of differentiating spatial filter results phase $\varphi = 2\pi M P_{n,m}$ from $P_{n,m}\,\delta_u$; δ_u - is the distance between the aperture and the center of the cell and M is diffraction order. From the geometry of elementary cell issues $\left| P_{m,n} \right| + \frac{C}{2} < 1/2$ hence $C \leqslant 1 - 1/M$. The much is is the number of the cells the nearer is C to 1 and the more exact is the coding procedure .

The design of the filter performed in the plotter of the computer is afterwards photographically deminished to the final dimensions 5 mm x 5 mm. Obtained transparency is placed in the Fourier plane of imaging system. Images of generalized derivative

$$\Delta_{r',r''} = \Delta_{4/3,4/3} = \frac{\partial^{4/3}}{\partial x^{4/3}} + \frac{\partial^{4/3}}{\partial y^{4/3}} ,$$

of optical signal is shown in Fig.3

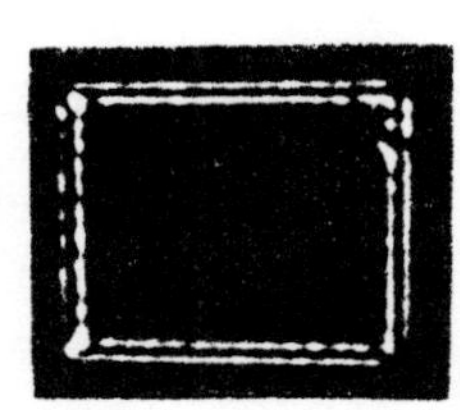

Students of Departament of Technical Physics of Warsaw Technical University have participate in carrying the experiments.

References

1. H.Kasprzak, Appl.Opt. 21, 3287, (1982)
2. A.W.Lohmann, Appl.Opt. ,4 ,1667, (1965)
3. A.W.Lohmann, IEEE Cat.Intern.Opt.Comp.Conf. (1975)
4. L.N.Jarosławski, N.S. Merzljakow ,Appl.Opt., 16 , 2034 ,(1977).

An improved holographic technique for imaging through fog

Yimo Zhang and Zai chun Li

Precision Instruments Department, Tianjin University,
Tianjin, China.

Abstract

An improved holographic technique, the filter-holographic technique for imaging through fog is presented to increase the diffraction efficiency of the hologram and the SNR of the corresponding holographic replica. And also considered is a situation when both the object wave and the reference wave are scattered by fog. The corresponding experimental results are given.

Introduction

Imaging through fog via holographic techniques was proposed and demonstrated successfully in 1967[1,2]. Later real-time holographic techniques were developed by using TV technology for decoding the hologram instantaneously[3,4]. It was also noted that the coherence requirements are not very severe, which can be met in any practical applications[5]. And more recently, A.M.Tai, et al[6] presented an interferometric imaging technique that utilizes a grating interferometer for imaging through scattering media (such as fog). But it is recognized that the effectiveness of all these techniques is limited by the fact that there is a strong background bias produced by the fog-scattered light on the holographic plate, especially when the fog is rather dense.

In this paper, we will present a new technique for imaging through fog—the filter-holographic technique and show our experimental results. And a more practical problem relating to the fog-penetrating techniques is discussed.

The filter-holographic technique

As we know, a very strong background bias produced by the fog-scattered light has to be recorded at the same time when a hologram is made via the previous interferometric techniques (holography or multi-grating interferometric technique). Though the background bias makes no contribution to the hologram, it will severely affect the diffraction efficiency (η) of the hologram and the signal-to-noise ratio (SNR) of the holographic replica.

The complex object light field after penetrating the fog can generally be considered as a sum of two components, i.e., the unscattered object light field and the scattered light field. Assuming that the former is $U_O(x,y)$ and the average intensity of the latter during the exposure time is S, and the plane reference light field without being scattered is U_R. The transmittance of a hologram for the conventional holographic techniques[1-4] can be expressed, providing that the recording is linear, as:

$$T \propto |U_O(x,y)|^2 + |U_R|^2 + S + 2\mathrm{Re}(U_O U_R^*) + 2\mathrm{Re}(U_O^* U_R) \tag{1}$$

where * denotes complex conjugate. From Equation (1), it is easy to find the modulation depth

$$m(x,y) = \frac{2|U_O(x,y)U_R|}{|U_O|^2 + |U_R|^2 + S} \tag{2}$$

It is obvious from Equation (2) that the larger the intensity of the scattered light is, the smaller the modulation depth m(x,y) is. And because the diffraction efficiency η of the hologram and the signal-to-noise ratio (SNR) of the holographic replica are proportional to the square of m(x,y), the fog-scattered light will undoubtedly limit the usefulness of the conventional holographic techniques (and also the grating interferometric technique[6]). On the other hand, if the fog-scattered light can be reduced in one way or another, the diffraction efficiency η and the SNR can be greatly increased. From what follows, we will see the properties of fog particles just provide us the probability to reduce the scattered light.

It is known that fog particles are always moving randomly (corresponding to the Brownian and other motions of the particles). As a result, the light field impinging on the fog will partially undergo random scattering. In terms of Fourier spatial frequency analysis, the fog-scattered light will contain various spatial frequency components, which would be distributed randomly over the whole Fourier transform plane. Meanwhile the spectra of the object light field are distributed concentratedly around the center of the Fourier plane, providing that the object is illuminated by an axial collimated light beam. Thus the fog-scattered light can be greatly reduced by a simple low-pass filter in the Fourier plane. From Equation (2) we see that if the corresponding hologram is made with the filtered fog-penetrating object wave, the modulation depth m(x,y) will be increased everywhere in the hologram. As a result, the diffraction efficiency η of the hologram and the SNR of the holographic replica will be greatly improved over those obtained with the conventional holographic techniques[1-4].

Experiment results

The experimental set-up based on the discussions above is sketched in Figure 1. The fog here was simulated by a mixture of powdered milk and water. The source was a He-Ne laser. Except for a pin-hole filter in the object wave path, the recording geometry is quite similar to Lohmann and shuman's[7].

Our results are shown in Figure 2. The input object is a resolution chart (See Figure 2a). Figure 2b is a common photo of the object through the fog. Figure 2c and 2d are holographic replicas obtained with the conventional holographic and the filter-holographic techniques, respectively. Obviously, the images shown in Figure 2c and 2d are superior to that in Figure 2b, and the image in Figure 2d is better than that in Figure 2c. These results not only demonstrate that holography penetrates fog better than photography, but also show that the filter-holographic technique gives better images than the previous holographic techniques.

Figure 3 shows a group of experimental results for three different input objects through two types of simulated fog. The simulated fog for results in Figure 3a was a mixture of acetic acid and sodium thiosulphate. It should be pointed out that it took as long as 30 minutes for making the hologram. But the result is still quite satisfactory. The inputs in Figure 3b and 3c are a man's face and a TV testing card, respectively. The fog for both Figure 3b and 3c are the mixture of powdered milk and water. These results show that the reproduced images are satisfactory for inputs with both continuous and discrete grey-level object(The black line in the holographic replica in Figure 3b was caused by a crack in the corresponding hologram).

A practical problem

The discussions and experiments above as well as in the references 1-4 were all based on the assumption that the reference wave illuminates the holographic recording plane without being scattered by fog. This is nearly impossible in practical applications because the reference wave have to penetrate fog before it illuminates the recording plane in actual field conditions. To solve this problem, we can perform a low-pass filtering on the fog-penetrating reference wave as we did for the object wave. And it will be more convenient for the filtering because there is little loss in high spatial frequencies for the plane reference wave.

The experimental set-up for this practical problem is shown in Figure 4, where the reference wave also penetrated fog. The fog here was the mixture of powdered milk and water. As we do not have big enough laser power, we used thinner fog for the reference wave.

The experimental results are shown in Figure 5, where the input objects were (a) a resolution chart, (b) Chinese characters, and (c) a man's face, respectively. We see that the holographic replicas are much better than their direct photographic counterparts, though the reference waves were disturbed by fog.

Conclusions

We have shown that the filter-holographic technique gives better reproduced images than those obtained with the conventional holographic techniques. And the proposed mentod makes the holographic fog-penetrating techniques more practical——the assumption that the reference wave illuminates the recording plane without being scattered can be canceled. Similarly, the assumption that the illuminating light illuminates the input object without being scattered can also be canceled.

Meanwhile we recognize that the filter-holographic method is effective essentially for transparent objects. Thus it is more suitable for data or image transmission-type applications. And we should point out that the improvements in the diffraction efficiency η and the SNR are at the sacrifice of the high spatial frequency components of the holographic replica. For different requirements we can select different filters, regarding both the SNR and the high spatial frequency components of the replica.

It is expected that if the new technique is applied to schmalfuss's real-time holographic system[3] and Chang's grating interferometric system[6], the (real-time) replicas will also be improved .

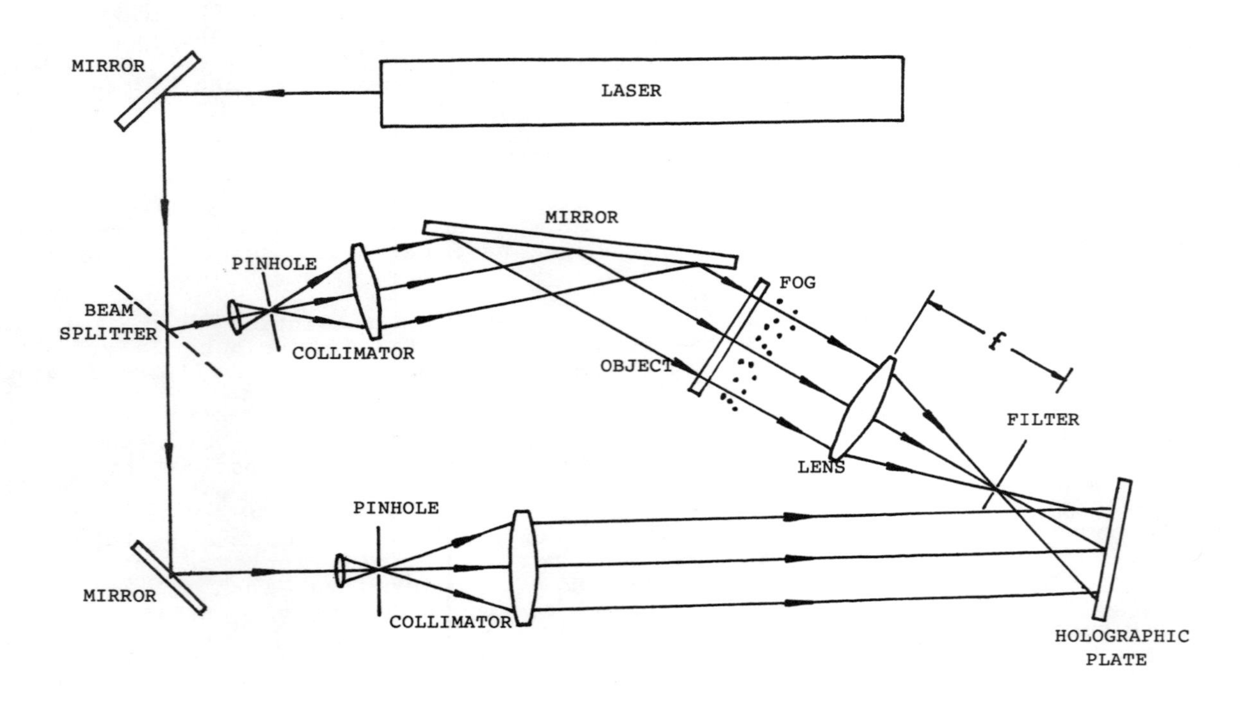

Figure 1. Experimental set-up for imaging through fog via the filter-holographic technique.

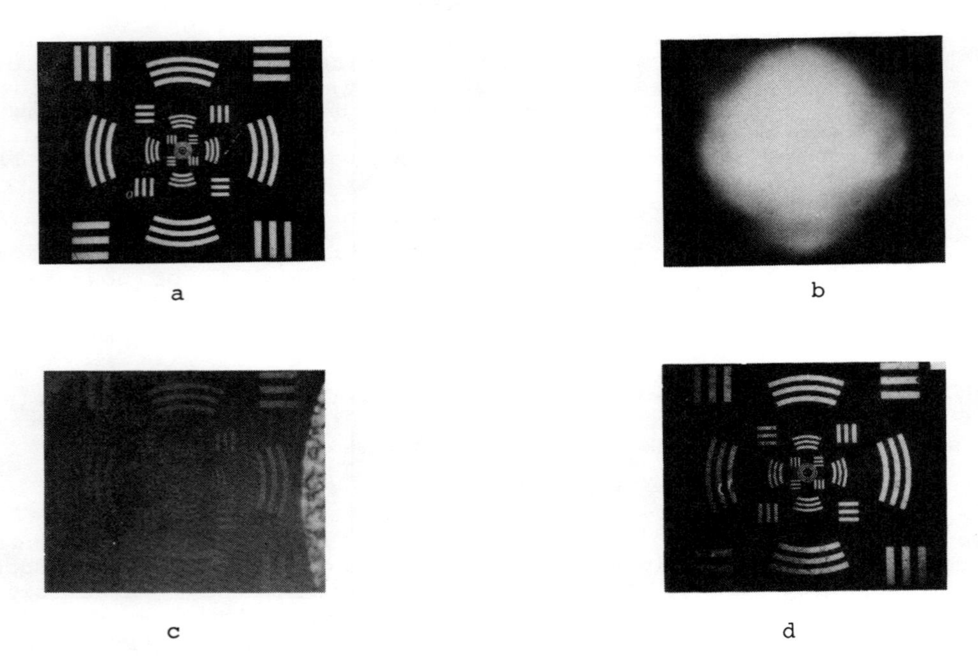

Figure 2. The experimental results.

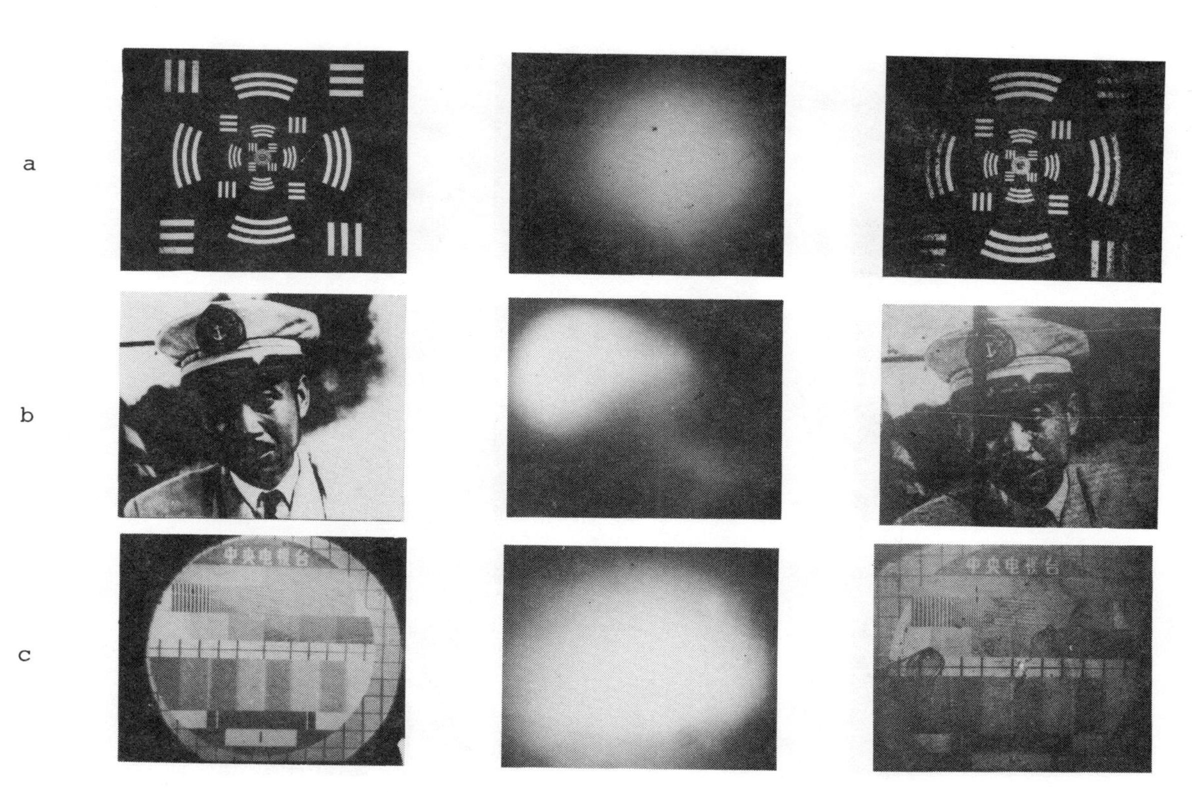

Figure 3. Other experimental results

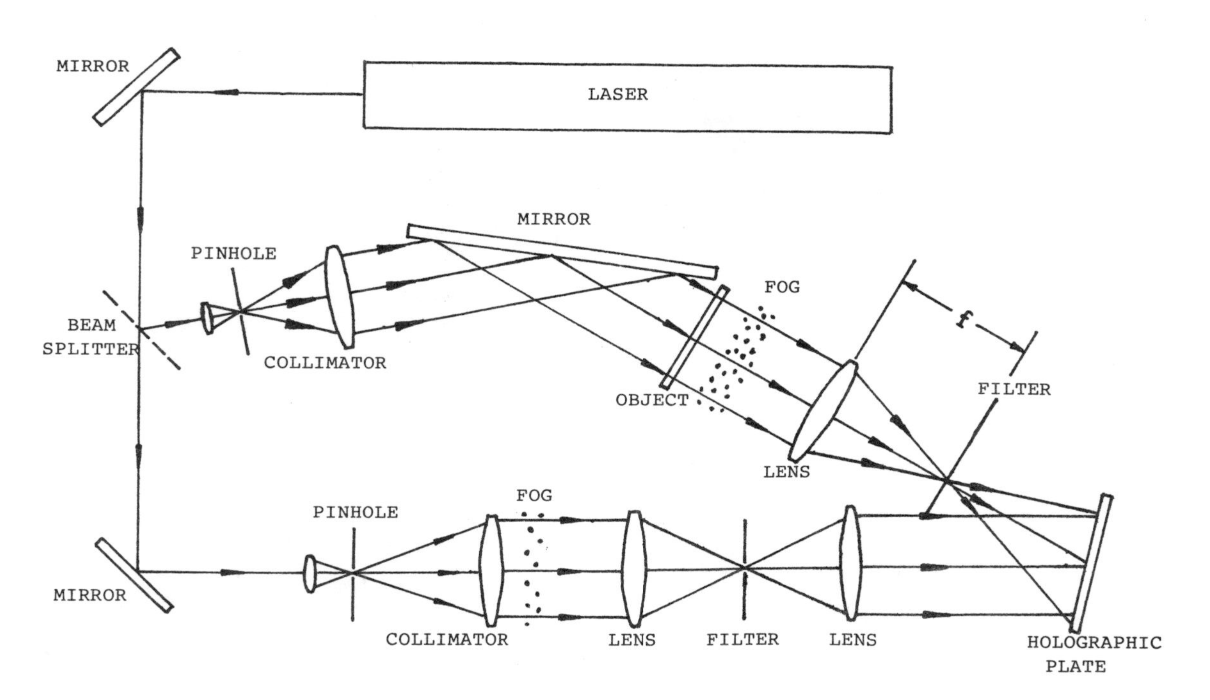

Figure 4. The experimental set-up for imaging through fog when the reference wave was also disturbed.

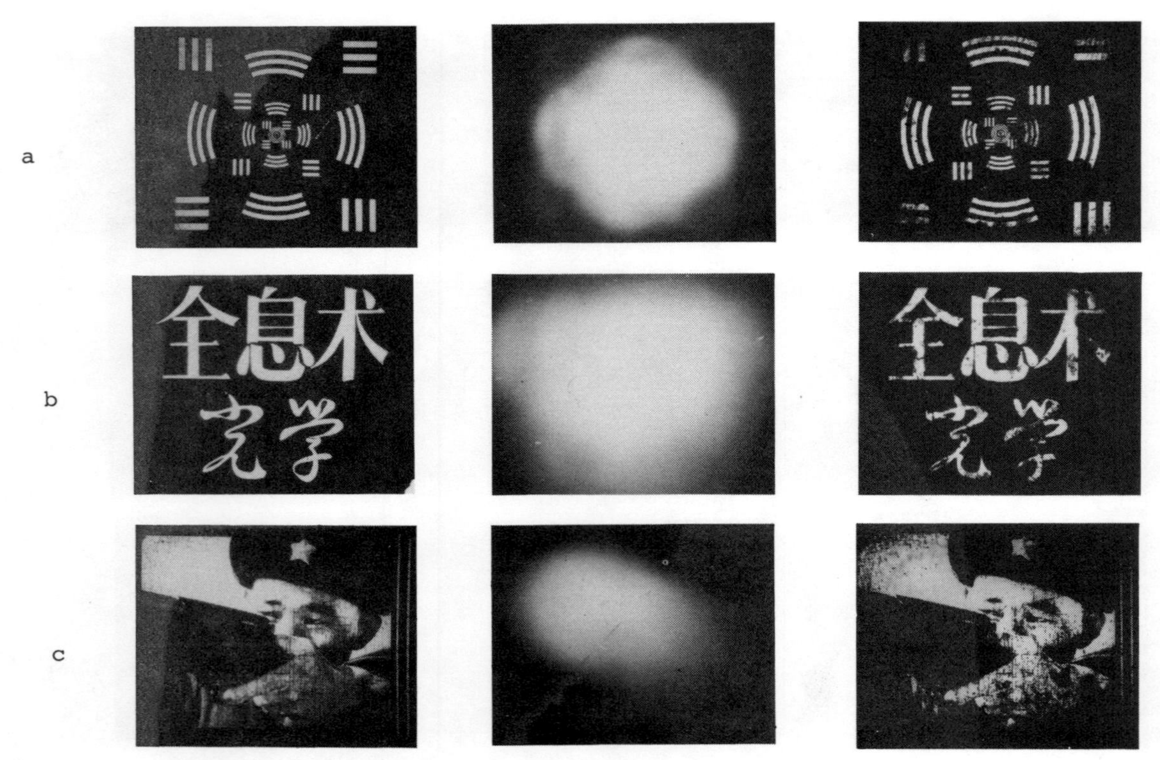

Figure 5. The experimental results for imaging through fog when the reference wave was also disturbed.

References

1. E.Spitz, Compt. Rend. Acad. Sci., 264, 1449 (1967)
2. K.A.Stetson, J.O.S.A., 57 , 1060 (1967)
3. H.Schmalfuss, Opt. Commun., 17, 245 (1976)
4. A.W.Lohmann and H.Schmalfuss, Opt. Commun., 26, 318 (1978)
5. H.Schmulfuss, Optik, 48, 119 (1977)
6. A.M.Tai, C.C.Aleksoff, and B.J.Chang, Appl. Opt., 20, 2484 (1981)
7. A.W.Lohmann and C.A.Shuman, Opt. Commun., 7, 93 (1973)

Density pseudocolor encoder by Bragg effect

Xie, Jinghui; Zhao, Yeling; Zhao, Dazun; Yu, Meiwen

Optical Engineering Department, Beijing Institute of Technology, P.O.Box.327, Beijing, China

Abstract

A new density pseudocolor encoding of black-white photograph by Bragg effect of the reflective grating is presented. The method and process are given. A initial research of the mechanism is dealed with.

Introduction

The method for density pseudocolor encoding of black and white photograph presented in this paper is based on Bargg effect of the reflection grating. In coding procedure of the original draft, any modulation by grating is unneccssary, and in display, the filter operation is unnecessary as well. So it is very simple and flexible. No resolution is lost by the use of this technique. In addition, because of the very high sensitivity to light wavelength, we can get saturated colours on the end plates.

Encoding technique

The recording of the spatial encoded transparency is carried out by the use of contecting exposure shown in Fig.1. The recording material we used is DCG plate. We place the original draft P_1 in contact with the DCG plate P_2 and press them with a special device. Then expose P_2 through P_1 with a collimated Argon laser beam. The wavelength is 4880Å.

The preparation of the DCG suspension and the devecopment of exposed DCG plates are not quite the same as that of the ordinary reflective holograms. Table 1 and Table 2 have listed the procedure for preparing and processing of the DCG plates respectively

Table 1. Procedure for Preparing DCG Plate

Step	Operation
1	Prepare 5% by weight gelatin suspension in water.
2	Thermostat at 50°C and add $(NH_4)_2Cr_2O_7$ to give a ratio of 6.5% to the weight of gelatin. Stir for 5 min with a vibration stirrer.
3	Add formaldehyde to reach a ratio of 1% to the weight of gelatin and stir for another 5 min.
4	Coat the suspension on a piece of carefully cleaned glass substrate uniformly and naturally dry it in horizontal position for 2 hours. The dried DCG layer is about 10 to 12 um in thickness.

Table 2. Procedure for Development of the Exposed DCG Plates

Step	Operation
1	Washing in running water at 22°C for 10 min.
2	Soak in mixture of 70% isopropyl alcohol and 30% distilled water for 1 min.
3	Soak in fresh isopropyl alcohol for 5 min.
4	Dry the plate with a flow of warm air at 30°C to 40°C.

The developed transparency is illuminated by a white light beam. We can see different colours on different isopycnals.

Disucssion on mechanism, colour formation

When we record the transparency P_2 with an arrangement as shown in Fig.1, the object beam and the beam reflected from the back side of the plate interfere each other and construct a reflective volume grating in the DCG layer. In this case, the reflective beam is in the opposite direction to the object beam, so the grating space d_o is only depended on the light wavelength λ_o and the refractive index of the gelatin n_o ,

$$d_o = \frac{\lambda_o}{2n_o} \quad (1)$$

In the washing procedure, the exposed DCG layer will undergo two changes. On the one hand, by taking up large quatities of water, the gelatin swells up uniformly. Its thickness can increase by a factor of 3 to 4 [1]. On the other hand, because of the cross linking between the Cr^{3+} and the gelatin molecules and the removing of the Cr^{6+}, the hardness and the refrative index at different exposed areas are different. At the higher exposed sections, both the hardness and refractive index of the gelatin are higher than those at the lower exposed sections.

In the dehydration procedure, the water molecules are rapidly removed from the gelatin micellae, this results a heavy shrink. But because of the different exposure, the hardness of the gelatin in different section is different, the shrinkage of the gelatin layer is not uniform. At the areas of higher exposure, the hardening effect of the gelatin is stronger, so the shrinkage is smaller; at the areas of lower exposure, the hardening effect is weaker, so the shrinkage is greater. This heavy shrink even results the gelatin a traverse shift from the lower exposed areas to the vicinity area of higher exposure.

It is worth noting that the dehydration of DCG layer is not the case that the alcohol molecules enter the gelatin micellae and replace the water molecules, but it is caused by the hydrogen bonds between the polar parts of water and alcohol molecules. The bonding is strong enough to pull the water molecules out of the gelatin micellae rapidly. In our experiments, we have observed that after developing and dehydrating, the encoded transparency can give a very high efficiency even if it is still soaked in alcohol. As well known that the refractive indexes of water and alcohol are almost the same, if dehydration of DCG layer was based on replacement, the high efficiency would be impossible.

H.M.Smith pointed out that: "In case of hardened layers the gelatin molecules are tied together into a continuous three-dimensional network. During swelling and shrinking, this network is not changed" [2]. This point of view means that during water-alcohol development, although the shrinkage is not uniform, the network structure is not changed but only changes the grating space uniformly.

Thus because of the uniformly swelling and unequally shrinking, a relief pattern on the gelatin layer will be generated (Fig.2). At higher exposed areas, the layer is thicker, the grating space and the mean refractive index is greater.

To increase the disparaty between the swelling and shrinking, the strict control of the prehardening is very important. If hardening is not strong enough, the lower exposed gelatin layer will be dissolved during the washing step. Inversely, if the gelatin is over hardening, the change of the gelatin thickness is too small to show colour difference. For the DCG layer we used the pre-hardening processing was done by adding a small quatity of formaldehyde as a hardening agent as shown in Table 1, step 3, not by baking. The process led to a good result.

Let d(x,y) and n(x,y) be the grating space and the refractive index of the incoded transparency respectively, when the transparency is illuminated along the same direction of the recording beam by a white light beam, the Bragg wavelengths $\lambda_b(x,y)$ diffracted from different isopycnals are given by

$$\lambda_b(x,y) = 2d(x,y) \cdot n(x,y) \quad (2)$$

As we all know, the reflective grating is very sensitive to the wavelength, so the wavelengths in Eq.(2) define the basic hues, of corresponding isopycnals. When the densities at original draft change from low level to high level, the corresponding colours at the encoded transparency will change from red to blue. This leads to density pseudocolor encoder.

Conclusion

The density pseudocolor encoder technique presented in this paper is valuable.

The explaination to the machanism involves complicated photochemistry of DCG. The research on this problem is only a beginning. Further research on this problem would not only improve this technique, but also deepen the understanding of photochemistrical machanism of DCG.

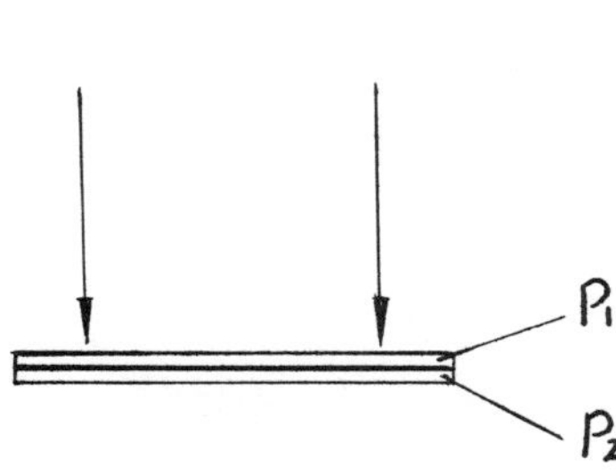

Figure 1. Arrangement for coding

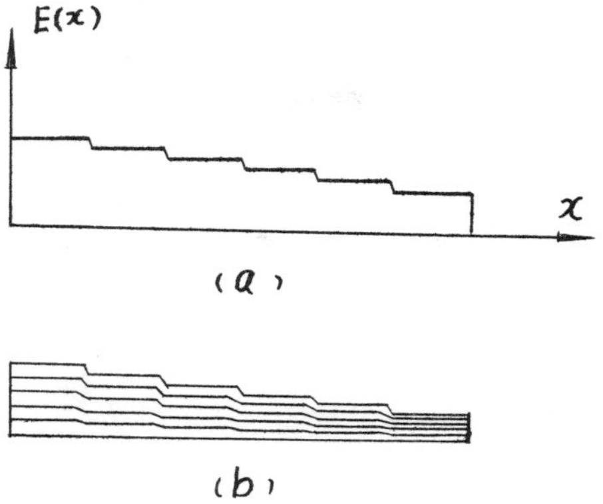

Figure 2. Photochemistry of DCG layer
(a) Exposure distribution.
(b) Changes in grating space.

References

1. Meyerhofer. RCA Rew, Vol.33, p.110, 1972.
2. H.M.Smith, "Holographic Recording Materials". Springer-Verlag Berlin Heidelberg New York p.85, 1977.

Image Quality Improvement on Variable Spatial-Rotation High Resolution Electron Micrograph by Optical Means

X.Shen

Shanghai Institute of Mechanical Engineering, Jun-Gong Street, Shanghai, China

S.H. Zheng, F.H.Li

Institute of Physics, Academia Sinica, Zhong Guan Cun, Beijing, China

Abstract

Using synthetic holograms made by optical and digital method deconvolutions for high resolution electron micrographs of spatial rotational variant system have been developed. This makes possible the restoration of the limited diffraction resolution of the electron microscope.

Introduction

The retrieval and correction of the transfer function of a high resolution electron micrographs are achieved through deconvolution operation for the image, elimination of image aberation and retrival of the ultimate resolution of diffraction by means of an optical filter device. That is to make a composite inverse filter to correct the phase and amplitude of an electron microscope transfer function and make every effort to attain an ideal transfer function independent from frequency. Stroke and Zech(1) first made a suggestion on achieving image reconstruction by using a composite frequency filter, where the filter function is a reciprocal of transfer function. Lohmann et al(2,3) have made use of this method theoretically and practically in partially and fully coherent high resolution micrographs taken from a conventional electron microscope. The image restoration performence mentioned above is to satisfy isoplanatic condition, but when the system itself is provided with rotating image aberrations such as astigmatism and comatic aberration, the impulse response of the system is anisotropic and the image restoration performance becomes a process of linear variable spatial-rotation filtering. Detailed disscussion on the difference between an object and its image as a result of astigmatism are stated in (4) (5). In this paper, Stroke's spatial filtering method is applied for achieving the image quality improvement on the electron micrograph of a variable spatial-rotation system.

The imaging and its deconvolution principle

The construction of the system which we are discussing here is that all the features of information can be transferred linearly onto the electron micrograph, and the electron illuminating beam has very good coherence. Meanwhile, the specimen is a weak phase object and the construct grade of the image is linear function of the phase shift of specimen. The variation of phase is small when the electron wave passes through the phase object (specimen), and the electron wave function can be represented at this time as

$$\psi_0(\vec{r}) = 1 - j\eta(\vec{r}) \tag{1}$$

Generally, an additional phase shift will result from a practical electron imaging system. If it is considered that only the effects of spherical aberration and defocusing are present, then the phase shift is

$$W(\theta) = \frac{2\pi}{\lambda}\left(\frac{1}{4}c_s\theta^4 - \frac{1}{2}\Delta f\theta^2\right) \tag{2}$$

where λ is the electron wavelength, c_s the spherical aberration factor of objective lens, Δf the defocusing magnitude of objective lens, θ the scattering angle of electron wave. If the phase shift caused by the effect of rotating image aberration is considered, the equation of the wave aberration has been given by Thon (6) for a rotating unsymmetrical imaging system as

$$W(\theta) = \frac{2\pi}{\lambda}\left(\frac{1}{4}c_s\theta^4 - \frac{1}{2}\Delta f_m\theta^2 - \frac{1}{2}\theta^2 z_a \sin 2\alpha\right) \tag{3}$$

where α is the rotating azimuth angle, z_a is a coefficient of astigmatism.
For a weak phase object, the effect of phase shift caused by lens on the image can be represented by phase contrast transfer function (PCTF)

$$T = -2\sin W(\theta, \alpha) \tag{4}$$

Owing to the unstabilities of accelerating voltage $\Delta u/u$ of chromatism C_c, temperature T(k) of cathod ray and energy E_b of electron, the amplitude will decrease with the increase of scattering angle of electron wave.

$$\tilde{T} = T(\theta,\alpha)\cdot G1\cdot G2$$

$$G1 = \exp\left[-\left(\frac{\pi}{4} C_c\theta^2 \lambda^{-1}(\ln 2)^{-1/2}\cdot \Delta u/u\right)^2\right]$$

$$G2 = \left[1 + \left(\frac{1}{2.45E_b}\pi\cdot C_c\cdot K\cdot T\cdot\lambda^{-1}\cdot\theta^2\right)^2\right]^{-1} \quad (5)$$

From equation (5) we have the phase shift caused by spherical aberration and the phase contrast transfer function obtained after attenuation when the defocusing magnitude Scherzer is Δf = 88nm with a JEOL-200cx are shown in Figure 1. The PCTF oscillates between $\pm$ 2 with the variation of scattrering

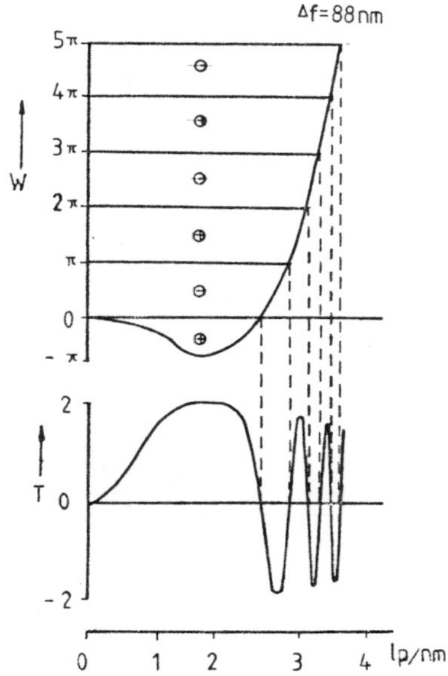

Figure 1.The damped phase contrast transfer function of an electron objective lens from phase shift W.

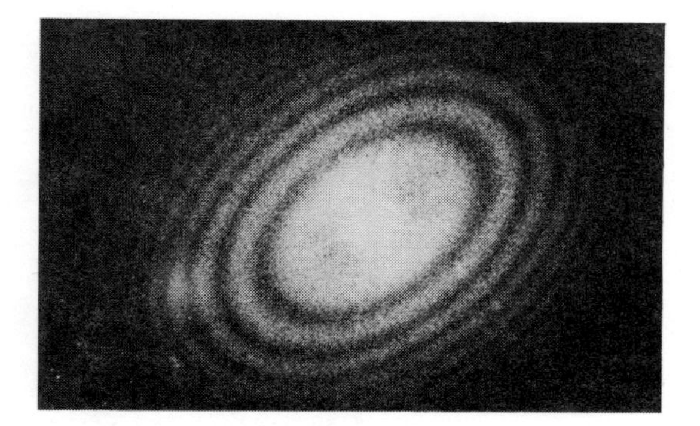

Figure 2. Light optical diffraction pattern of a carbon foil specimen

angle of the specimen. Different frequencies are transferred to the image plane with increasing or decreasing phase contrast grade and some of the frequency bands are transferred with inverse contrast grade so that the corresponding black and white details of the image are inverted. When the azimuth is anisotropic and each determined direction represents a PCTF with different defocusing; especially, those details which are close to the ultimate resolution of diffraction will cause the ratio of signal to electron noise (s/N) to become very small as a result of attenuation of the amplitude of PCTF. The purpose of this paper is to solve the problems of negative phase turnover for PCTF of non-cicular symmetrical system and also the restoration of amplitude.

It is well known that an inverted original image can be attained by placing an inverse filter at the spectral plane, multiplying with the image spectrum and performing Fourier-transform. A filter function composed of amplitude and phase is given by Stroke as

$$H(x, y)^{-1} = \frac{e^{-j\sigma_h}}{H(x, y)} \quad (6)$$

where σ_h describes the phase of the transfer function. for a system with unvaried linear spatial shift and rotation, many different methods are already available for making the filter and determining the impulse response caused by the effects of the instrument and image aberration (7,8,9). However, since the rotation is variable, it is difficult to make the filter and determine the transfer function by means of an ordinary method. The electron micrograph processed by us was taken under the condition of z_a = 50 nm and its diffractogram in a pattern of elliptical ring is shown in Figure.2, where the transfer function is also anisotropic.

Experimental principle

a) Phase-only filter

The phase distribution is determined by solving the scattering angle distribution with the transfer function equal to zero and the positive and negative alternation of the transfer function. Practically, however, there are different influential stochastic factors in the operation of an electron microscope and it is hard to consider them in computation. In the experiment, we have determined the zero point of the phase by means of a film with nonlinear and high contrast characteristics. With the aid of a coherent illumination provided by the optical diffraction device shown in the fore part of Figure.3, a diffractogram on the rear focal plane is recorded linearly on the photographic material after the electron micrograph is Fourier-transformed through the lens. The diffractogram is then placed on the plane P_2 of the imaging system shown in the rear part of Figure.3, and illuminated with a well-distributed incoherent light. A high contrast film is placed on the imaging plane p_3 and the distribution of light intensity passing through the film is shown in Figure.4a. Supposing E_0 is the intensity of an incident

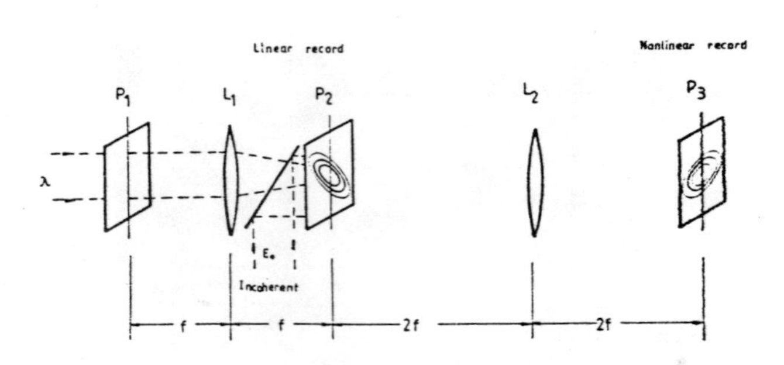

Figure.3 representation of an optical transform and an optical processing system.

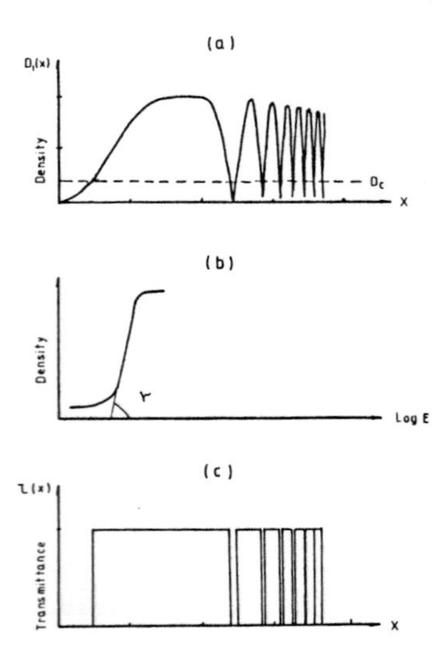

Figure.4 (a) The density of the input image. (b) The response of the high-contrast film. (c) The transmittance $\tau(x)$ of processed high-contrast film.

light, $E = E_0\ 10^{-d_i(x)}$ the exposure on the high-contrast film, $d_i(x)$ the density distribution of input image, the response of the high-contrast film is shown in Figure.4b. The density of amplitude limit D_c provides the maximun density for the duplicating film in Figure.4a and the magnitude of amplitude limit is proportional to the light intensity used. When the threshold exposure of the film used (with a very large γ) is E_0, the exposure of the negative will be darkened only if $E_0 > E_c$, i.e. $D_i(x) < \log E_0/E_c$ is satisfied. The transmittance of the film is

$$\tau(x) = \frac{1}{2}\ [1 + \mathrm{sgn}\ (D_i\ (x) - D_c)\ ] \tag{7}$$

Under ideal condition, the transmittance of a duplicating film is either 1 or 0. (See Figure 4c.)

A phase filter is made by means of a Ronch grating. With the imaging system shown in Figure.5, a photo with unit transmittance is placed on the plane P_1 and a Ronch grating is placed closely to the photo. A band-pass filter attached to the grating is made up of an odd ring and an even ring with respect to an ellipse and then the double exposure method is adopted by letting the odd ring of the ellipse pass in the first exposure and the even ring pass in the second exposure and enabling the grating to shift half a period constant. In this manner, a phase filter is attained by making use of the shift of a grating to achieve π phase variation of the even ring as the odd ring is in zero phase.

Since the phase filter mentioned above is subject to the effect of imaging lens, and the grating spacing cannot be made so dense, we have further processed the filter by applying the method of making an image hologram, that is, a Fourier-transform is performed for the original phase filter of grating construction and only first order of diffraction is allowed to pass the spectral plane and finally a hologram of the original phase filter is obtained through reconstruction by Fourier-transform and record by introducing a plane reference wave. The spatial frequency is determined by the included angle between reference light and optical axis, and a desired phase filter is obtained after development process. Since the increase of spatial frequency keeps the processed image far away from the order zero, the signal to noise ratio is improved.

b) Amplitude filter

The amplitude filter is of granular distribution property in physical construction and functions to extract the transfer function by separating the transfer function from the object spectrum. The high resolution electron micrograph of carbon foil used in our experiment is a convolution of the point spread function in the system and the different spatial distrubution of numerous fine granules, and it can be represented as

$$f(x,y) = \left[\sum_{i=0}^{n} f(d_i,b_i)\cdot\delta(x-d_i\ ,\ y-b_i)\right] \circledast h(x,y) \qquad (8)$$

The spatial positon, magnitude and distribution of the granules are represented by multiplying the function with unit volume and a height corresponding to the volume by different weighting factors $f(d_i, b_i)$. After performing Fourier transform for equation (8), we attain the spectrum of the electron micrograph as

$$F(u,v) = \sum_{i=0}^{n} f(d_i,b_i)\ e^{-j2\pi(ud_i+vb_i)}\cdot H(u,v) \qquad (9)$$

Recording material is used for linearly recording the distribution of its power spectrum

$$F(u,v) = F(u,v)F^*(u,v)$$

$$= \sum_{i=0}^{n} f^2(d_i,b_i)H(u,v)H^*(u,v) + \sum_{\substack{i=0\\k\neq j}}^{n} f^2(d_i,b_i)H(u,v)H^*(u,v)e^{-j2\pi(d_i-d_k)u+(b_i-b_k)v} \qquad (10)$$

To separate the object and the point spread function, we perform Fourier-transform again for the power spectrum.

$$g(x_3,y_3) = \mathcal{F}\left|F(u,v)\right|^2 = \sum_{i=0}^{n} f^2(d_i,b_i)h(x_3,y_3)\circledast h^*(x_3,y_3)\circledast\delta(-x_3,\ -y_3)$$

$$+ \sum_{\substack{i=0\\k\neq j}}^{n} f^2(d_i,\ b_i)h(x_3,y_3)h^*(x_3,y_3)\circledast\delta\left[-(x_3-(d_i-d_k))\ ,\ -(y_3-(b_i-b_k))\right] \qquad (11)$$

A low-pass filter is placed on the transformation plane

$$\mathrm{circ}\left(\sqrt{x_3^2+y_3^2}\right) = \begin{cases} 1 & \sqrt{x_3^2+y_3^2} \leq \dfrac{1}{\pi a r_1} \\ 0 & \text{Otherwise} \end{cases} \qquad (12)$$

here , the radius of the filter is determined by performing Fourier transform for the elliptical ring. Its Fourier transform can be obtained by approximation of the Bessel identity and Bessel series expansion. Now the distribution of the function following the transformation plane filter is

$$g(x_3',y_3') = g(x_3,y_3)\mathrm{circ}\sqrt{x_3^2+y_3^2} \qquad (13)$$

As an ideal case, we only consider that the first term of the equation (11) is allowed to pass while the others are all filtered off. By performing Fourier transform again we have

$$G(u,V) = \left|H(u,v)\right|^2 \sum_{i=0}^{n} f^2(d_i,b_i)$$

where $\sum^{n} f^2(d_i,b_i)$ is a large constant value not varying with (u,v) at the centre point. H(u,v) is the

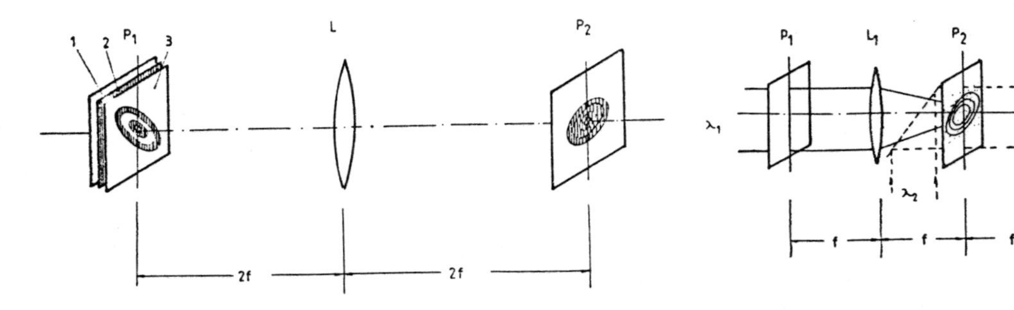

Figure 5. Arrangement for phase filter production using Ronch grating.
1. Original image 2. Ronch grating
3. Band-pass filter

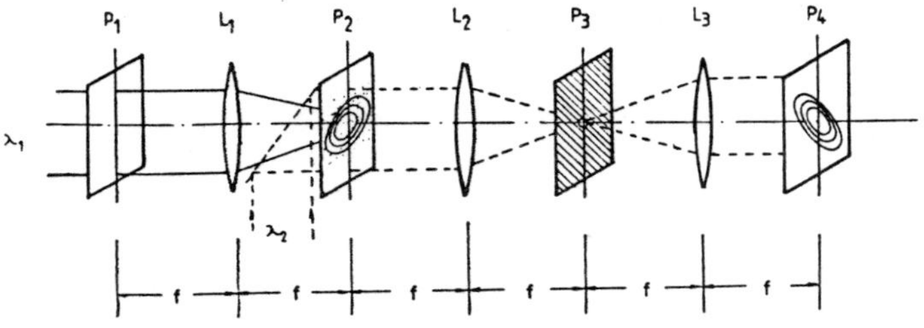

Figure 6. Schematic representation of an optical transform and an optical processing system.

intensity distribution of transfer function. Now, we use Figure 6. to illustrate the process. The electron micrograph is positioned at the front focal plane of the lens and the power spectrogram is obtained and recorded at its back focal plane p_2. The power spectrogram is then positioned again on the plane p_2, the front focal plane of lens L_2. The distribution on plane p_3 is shown in equation (11). Meanwhile, a low-pass filter is placed and the information passing through the filter is inversely transformed by lens L_3, then a substantially smooth curve of the transfer function with the object spectrum filtered off is obtained on the plane p_4. An amplitude filter of $H(u,v)^{-1}$ can be developed by controlling adequately the value γ of the recording material.

Results of experiments

a) Filters

The high resolution electron micrograph used for this experiment was taken with JEOL 200cx from an amorphous carbon foil as shown in Figure 10a, and diffractogram is shown in Figure 2. The theoretical phase filter attained by double exposure through the use of Ronch grating and band-pass filter is illustrated by figure 7a. Holographic imaging method is used for further processing; its spatial frequency is = 900LP/mm as shown in figure 7b. Finer fringe are provided in the fringe spacing of the original phase filter. Furthermore, an amplitude filter is made. The diffractogram recorded linearly is positioned on the input plane of the optical filtersystem. A low-pass filter with a diameter of 0.5mm is placed at the frequency spatial plane. We can obtain a desired filter if the value γ is chosen adequately.

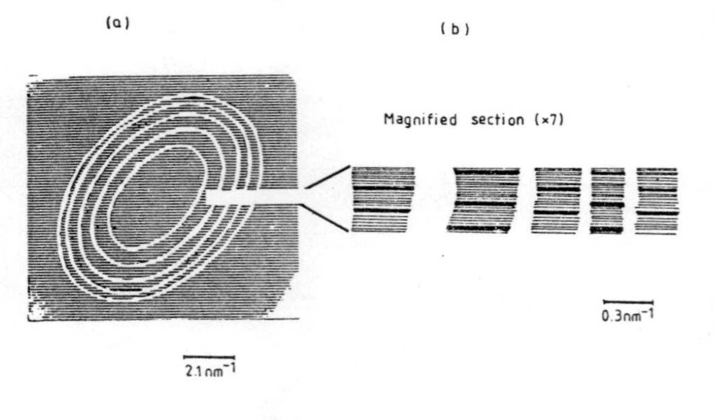

Figure 7. Holographic filter
(a) The theoretical phase filter
(b) The processed phase-only filter

b) Image processing

The deconvolution of an electron micrograph is performed by applying the coherent optical filter device shown in Figure 8. The amplitude filter and the hologram are correctly superpositioned onto the spectral plane P_2. The former acts upon the amplitude of electric field by absorption and the latter acts upon the spatial phase of electric field by the nature of its diffraction grating. The transfer function curves obtained after phase filteration and phase-amplitude composite filteration are given in Figure 9. Figure 10 b and c show the corresponding electron micrograph with rotating image aberration after the phase and phase-amplitude havebeen restored simultaneously.

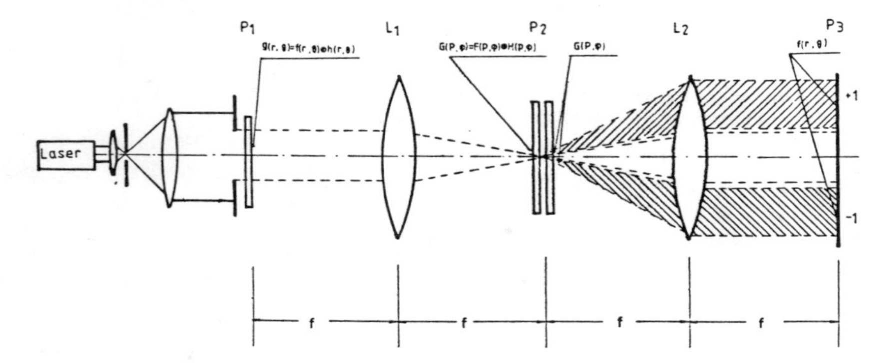

Figure 8. Optical computing arragement used for holographic image deblurring deconvolution.

c) Discussion

Figure 10b is a reconstructed image obtained after the negativ phase of the transfer function has been retrieved. The inverted black and white portions in the original specimen is now inverted again. The enlarged detail at the lower left corner of Figure 10b. will make this clear. From the statistical point of view, the whole distribution remains unchanged after the black and white granules with definite size have retrieved from inversion, i.e. no evident variation in granule size, because the details especially those which approach to the ultimate resolution, still have a low contrast after phase-only filteration, and the S/N of the signal and electron noise is very low due to the amplitude attenuation of the PCTF. Figure 10c is an electron microscopical image obtained after composite filteration. the distinct difference between the reconstructed image obtained after composite filtration and the original image is that the portion of the details is outstanding and the contrast which decreases with the increase of diffraction angle is now increased. By contrasting Figure 10c with Figure 10a, it can be seen that the granular construction becomes finer and the resolution is significantly improved. The resolution of Figure 10c approaches to 2A° . By referring to Figure 9, the PCTF is improved after the original transfer function has been subjected to phase and amplitude filtration, but the zero point of PCTF is still present, that is, the transfer function which is nearby the zero point or has a threshold

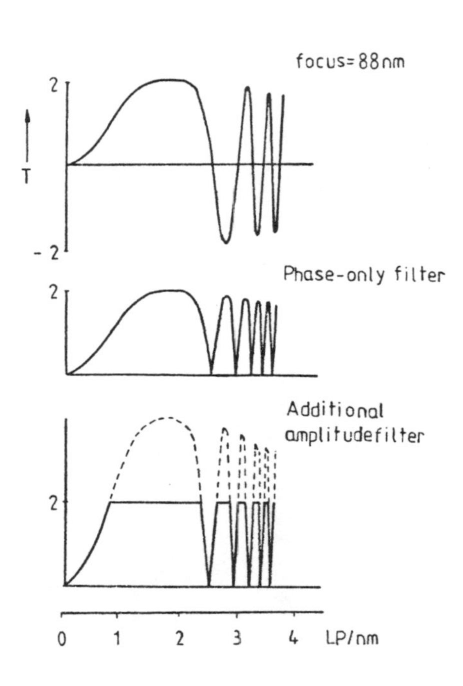

Figure 9. The PCTF after the original transfer function has been subjected to phase and amplitude filtration.

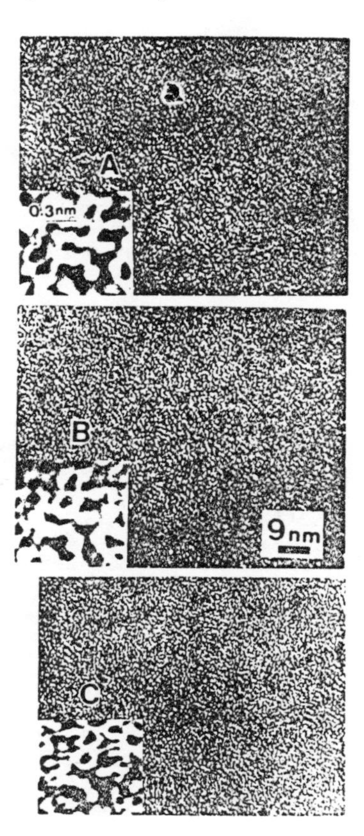

Figure 10. The corresponding electron micrograph with rotating image aberration after the phase and phase-amplitude have been restored.
A. The original micrograph of carbon-foil B. The reconstructed image after phase-Only filter C. The reconstructed image after composite filtration

less than a definite one will not transfer information, different defocusing magnitudes have approximately similar amplitude of transfer function. But, since the position of zero point of a transfer function is different along radial directions, the restored images with different defocusing magnitude will still have some difference between each other. The holographic inverse filter can be widely used in restoring electron micrographs of large molecule specimen and atomic structure which are taken by conventional and super-conductive microscopes.

References

1. Stroke, G.W., Zech, R.G., Phys. Letters 33A(1967)3
2. Lohmann, A.W., Werlich, H.W., Phys. Letters 25A(1967)5700
3. Stroke, G.W., Halioua, M., Phys. Letters 33A(1970)3
4. Shen, X., Boseck, S., Optik 68 No.2(1984)137
5. Shen, X., Boseck, S., Optik. 68 No.1(1984)17
6. Thon, F., <u>Phase Contrast Electron Microscopy, Electron Microscopy in Material Science (U. Valdre, U, Ed)</u> Acad. Press, New York
7. Hanszen, K.J., Optik 32(1970)74
8. Hahn, M.H., Optik 35(1972)326
9. Sroke, G.W., Halioua, M., Thon, F. and Willasch, D.H., Proc IEEE 65(1977)39

A holographic fingerprint lock system

Wang Yaosong,Chang Yingning

The first Institute of Ministry of Public Security
P.O.2808-8,Beijing,China

Abstract

A system of the holographic fingerprint lock to which the optical correlative technique is applied is presented in this paper. The reliability of the system has been improved by adopting unbleached phase hologram.

Introduction

The holographic lock is a safe-guard system with high security.There are two types of digital code and fingerprint code ete. The digital code takes a combination of the five holes which can be freely opened and closed as the encode information, each combination corresponds to one digit from 0 to 9 and a Franhofei's hologram with 3x4 mm in size is prepared for it. Thus , a card of any number can be combined, the door is opened by the reproduction number from the card. The fingerprint code is a fourier's hologram, the fingerprint is taken as the encoded information and the lock is opened with the use of optical correlation technigue. Comparing the former with the later, the security of the later is higher than that of The former, but the reliability of the later is lower. The reason is that, by using an unbleached amplitude hologram,the diffraction efficiency is very low, even if the hologram is changed to a bleached phase hologram, the signal-to-noise ratio of the system is very poor. Here , we have made spatial match filter using unbleached phase hologram, thus the reliability of the system has been improved (Fig1.).

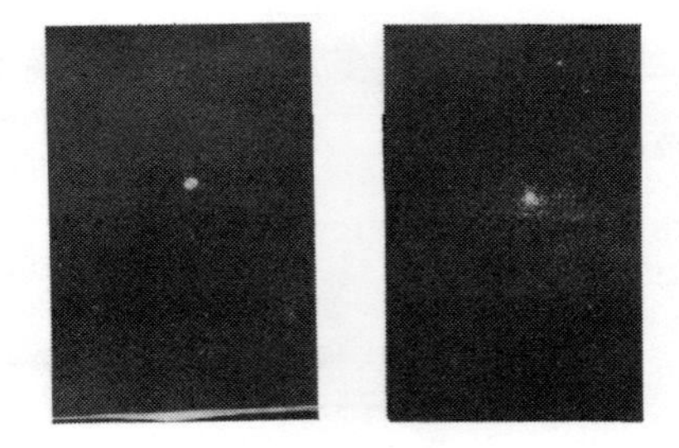

(a) (b)

Fig 1. The autocorrelation point with unbleached phase spatial filter(a) and bleached phase spatial filter(b)

The experiment of the holographic fingerprint lock system ispresented in the paper.

The holographic fingerprint lock is a security apparatus made according to the character of uniqueness and permanency of fingerprint. It consists of an optical correlator, an electron-lock and a photo-electric check circuit. The optical correlator is a unit for comparing the Sumilarity of two fingerprints. When the input fingerprint is identical with specific fingerprint made the spatial match filter. The output becomes delta function and it is changed to an electronic signal by the photo-electroic triode, then it is magnified by an operational amplifier and is used to trigger a silicon controlled rectifier which opens the electron-lock. Its block diagram is shown in fig 2. The size of the whole system is 40x70x20cm, which can be installed on ordinary doors.

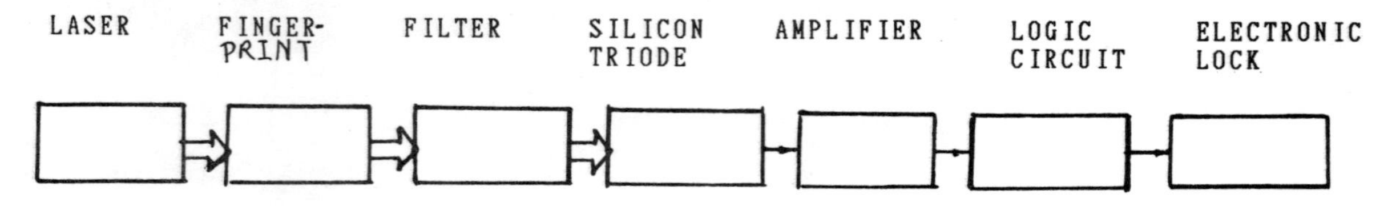

Fig 2.The holographic fingerprint lock block diagram

Spatial match filfer

The key of realization the correlation filter is making a high quality spatial match filter. It can be made with the help of a computer, or by using fourier holography. Here, in our experiment the spatial match filter is made by using the fourier holographic path (Fig 3.).

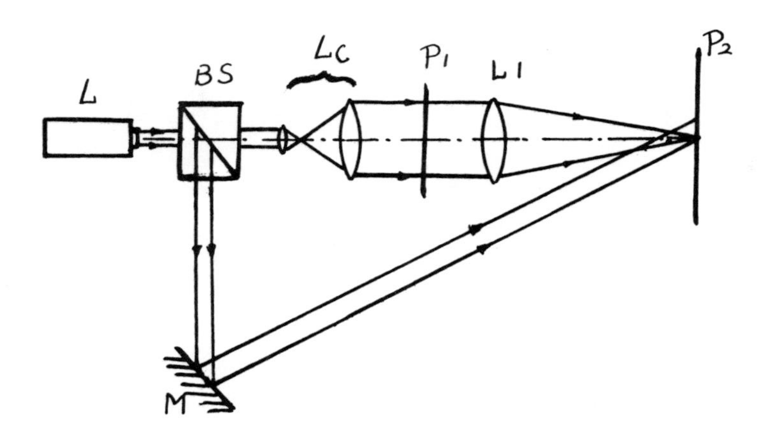

Fig 3. fourier holographic path

The laser beam from the laser L is split into two beams by the beam spliter BS, one beam is spread and collimated by lens Lc, then illuminates the plane P_1 where the transparency of the fingerprint is placed (oily fingerprint),the beam passing though the oily fingerprint is transformed by lens L_1, and the spectrum distribution Uo(u,v) of the fingerprint is obtained in the lock focus plane P_2.The other beam is reflecfed by the mirror M, illuminates the plane P_2 at an angle of the θ ,its distribution is $U_R(u,v)$, which is superposed with the first beam and interference. The holography plate is placed in the plane P_2, and we get the record as follows:

$$I(x,y)=|U_0(u,v)+U_R(u,v)|^2=\gamma_0^2+|H(u,v)|^2+\gamma_0 H^*(u,v)e^{2\pi i\alpha y}+\gamma_0 H(u,v)e^{-2\pi i\alpha y} \quad (1)$$

The third term contains H (u,v),it is the needed filter function:

$$T(u,v)=\frac{H^*(u,v)}{|N(u,v)|^2} \quad (2)$$

Therefore,the so acquired Fourier hologram is the spatial match filter that we want to get. But in order to take the high quality spatial match filter, we must take into consideration the following factors:

(1) Selaction of the aperture of the transform lens.

In an optical system, the lens aperture must be large enough to avoid the loss of the high frequency elements generally, the maximal spatial freguency of a optical system is

$$\nu_{max}=\frac{D_2-D_1}{\lambda_1 D_1} \quad (Fig\ 4)$$

Radius D_1 of the fingerprint is about 10mm. If we select d_1=120mm, the reguired lens aperture is abut 21mm. So we selected the lens of which aperture is 25mm, and the focus length is 120mm as our transform lens.

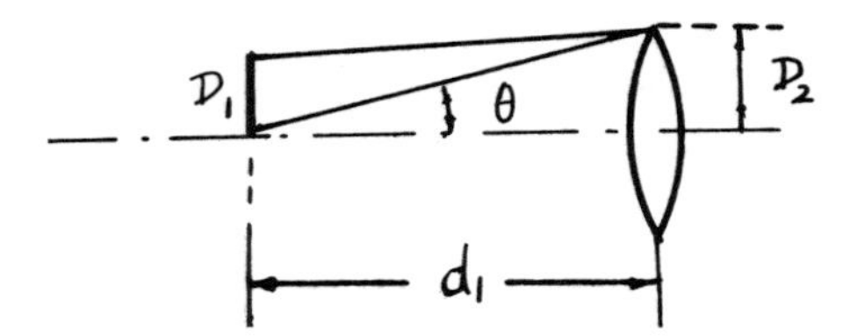

Fig 4. aperture of the transform lens

(2) Reference to-object ratio selection

Appropriate reference-to-object ratio must be selected with the aims of making the spatial match filfer with the highest diffraction efficency and obtaining the maximal autocorrelation peak .Appropriate comparison point must be selected due to the difference intensity of each order of the spectrums distribution of the fingerprint. The optimum spatial filters is produced if we select R/O≃1:1--2:1, base on the comparison point.

(3) Reference beam's area selection.

To record all the spatial frequency of the fingerprint, the whole spectrums must be covered fully by the reference beam. For the fingerprint of which spatial frequency is 15 ℓ/mm, a lens of which focus length is 120mm is used, the area of spectrum in the focus plane should be $\pm x = \pm y = |\lambda f \, \psi_{max}| = 0.63 \times 10^{-3} \times 120 \times 15 = 1.2mm$, thus generally the selected radius of the reference beam is 1.5mm.

(4) Holographic plate selection.

Because the spatial match filter is a effective area is only about 2mm, in order to get the brightest autocorrelation point, the holographic plate with high diffraction efficency and a special developing method must be adopted to obtain the unbleached phase hologram. Here, we used the HP-633P holographic plate and SMD-3 developer. The spatial match filter with high diffraction efficency and high signal-to-noise ratio has been obtainced.

Optical correlator

The nucleus of the holographic fingerprint lock is the optical correlator. It compares two fingerprints and changes the result to physical parameter--intensity of light, which can be measured exactly, its light path is shown in Fig5.

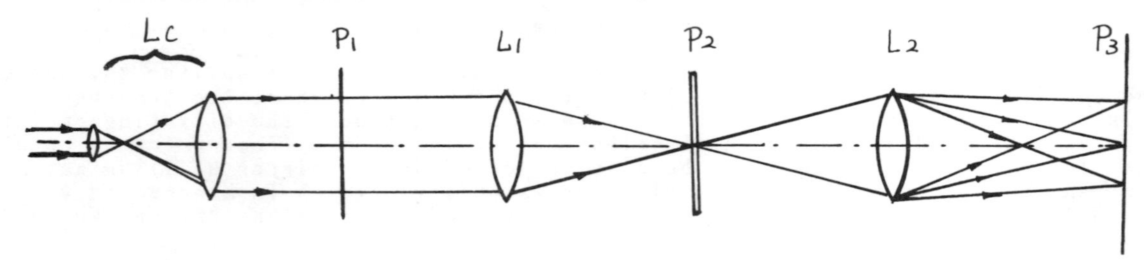

Fig5. Optical correlator'

To place the spatial match filter to the exact position of the record plane P_2, and put a lens behind the plane P_2 to countertransform. The light amplitude distribution at P_3 is obtained as follows:

$$U(-x_3, -y_3) = C g(x_3, y_3) + \iint_{-\infty}^{+\infty} h(x_3-\xi, y_3-\eta+\alpha\lambda f) g(\xi, \eta) d\xi d\eta + \iint_{-\infty}^{+\infty} g(\xi, \eta) h^*(\xi-x_3, \eta-y_3-\alpha\lambda f) d\xi d\eta \quad (3)$$

The three terms on the right in equation (3) correspond to the three points. On plane P_3 of Fig5, the first term is the attenuation image of the $g(x_1, y_1)$ located on the origin point of plane P_3; The second term is convolution of the $g(x,y)$ and $h(x,y)$ which is located on plane P_3 $(0, -\alpha\lambda f)$; The third term is the correlation term which is located on plane P_3 $(0, \alpha\lambda f)$. It is a most useful term for us. When the input fingerprint $g(x,y)$ is identical with the fingerprint made the spatial filter, a sharp bright point (delta function) will appear at the place $(0, \alpha\lambda f)$ (Fig6a), otherwise it is crosscorrelation function, a disperred spot will appear at the place $(0, \alpha\lambda f)$ (Fig6b).

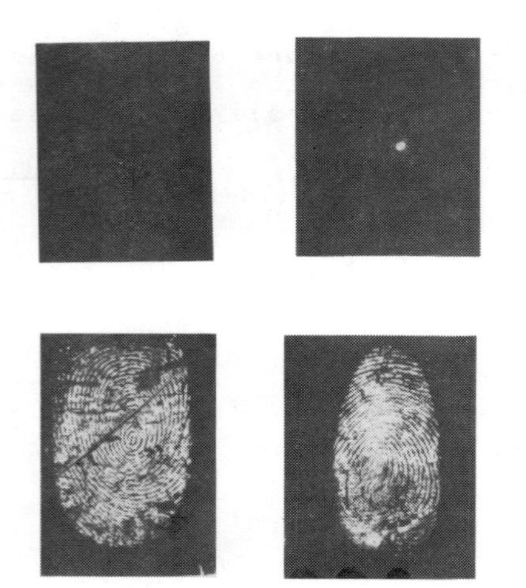

(b) (a)

Fig6 autocorrelation (a) and crosscorrelation (b)

Photoelectric apparatus for opening the lock and its operation proecedure

The photoelectroic apparatus for opening the lock is shown in Fig7.

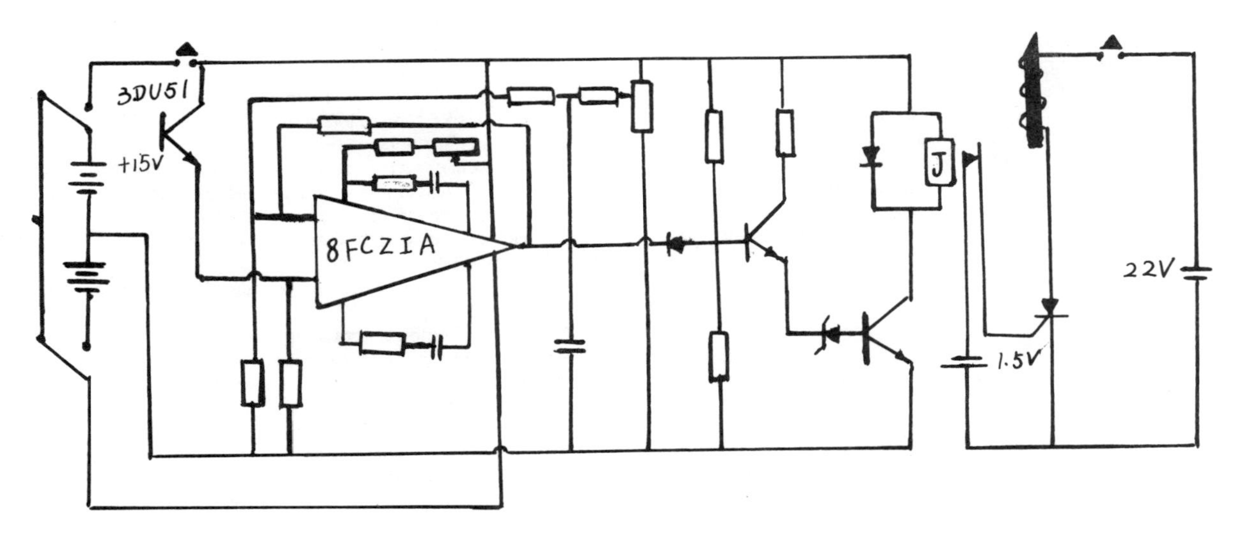

Fig7 Photoelectric appartus
for opening the lock

The operation procedure of the lock is as follows: A storekeeper impressed his certain fingerprint which is selected as "key" upon a glass plate which is then placed to the plane P_2 (shown in Fig3) as the input for making the spatial match filfer. Then the filter is repositioned exactly to the recorded position. After that, the fingerprint on the glass plate is erased and the door is locked by the electron-lock, now the storekeeper can leave. When the storekeeper wants to enter this room, he impressed his specific fingerprint upon the finger locator of the input window and swithed on the power, the door will be opened automatically. If it is not the storekeeper's fingprint, the door can not be opened. This operation procedure is fulfilled by the photoelectric apparatus of the lock. First of all the optical correlator compares the similarity of the impressed fingerprint with that was made match filter stored in system. If the impressed fingerprint is the storekeeper's one,an autocorrelation small bright point will appear and be recieved by the silicon phototriode 3DU51 after passing through the pinhole and transformed into electrical signal which is then amplified by operational amplifier and trigger The SCR through a logic circuit unlocks the electronic lock. Qtherwise, autocorrelation point can not be created, the door will not be opened and an alarm will be given.

The sensitivity of the lock can be changed by regulating the output level of the operational amplifier. Therfore, the threshold of the lock can be selected according to the characterstics of the system and the security requirement.

Conclusions

The parameter of the system was measured with silicon photocell and the reliability of the system was tested. the measured correlation efficency of the spatial match filter is one percent (the illumination of origin fingerprint made the spatial filter). When the power illuminating filter is 400μW, autocorrelation peak power is 4μW. After this fingerprint is erased on the window of the system, if the same fingerpint is reimpressed, the gained autocorrelation peak power is 3.2μW, reposition precision is 80 percent. When different fingerprint $g_1(x,y)$, $g_2(x,y)$... $g_n(x,y)$ are inputed to the light path, all the crosscorrelation peak power is lower than 0.1μW. because the threshold for opening the lock is 1μW, the lock is safe and reliable.

The system is suitable for use at the entrance of the treasure house of a bank, center of the confidentical documents and of a forbidden zone. Also, it can be use for control of the precise instruments and computer terminals etc.

Reference

1. Howard Biesnam, Microwave, Vol.14, No1, P10, 1975.
2. Yuan Weiben, Applied Laser, Vol.2, No3, 1985, Chinese

AUTHOR INDEX

Albe, F., Holographic cinematography and its applications, 22
Aliaga, R., Microscopic and macroscopic measurement of holographic emulsions, 491
Bao, Cailong, Fe-doped $LiNbO_3$ crystal for applications in white-light information processing, 489
Bao, Naikeng, New method of 3D quantitative analysis of holographic interferometry in the applications of solid biomechanics and vibration, 144
Bazargan, Kaveh, New method of color holography, 68
Block, H., Use of synthetic holograms in coherent image processing for high resolution micrographs of a conventional transmission electron microscope (CTEM), 538
Bloisi, F., Heat transfer studies by microholographic interferometry, 167
Borkova, V. N., Holographic registration and spectroscopy of a modulated laser radiation, 100
Boseck, S., Use of synthetic holograms in coherent image processing for high resolution micrographs of a conventional transmission electron microscope (CTEM), 538
Cai, Anni, Profile measurement on IC wafers by holographic interference, 135
Cai, Xueqang, Method for making large rainbow holograms, 93
Cao, Hongsheng, New experimental technique to separate principal stresses—proportional exposure piled-plate holography, 120
Cao, Zhengyuan, Measurement for dynamic deformation by mismatch white speckle method, 354
Cavaliere, P., Heat transfer studies by microholographic interferometry, 167
Celaya, Martin, Analysis of an in-line object beam hologram interferometer, 124
———, Hologram interferometer for film deposits evaluation, 163
Chałasińska-Macukow, K., Phase-only Fourier hologram as an optical matched spatial filter, 188
Chang, Yingning, Holographic fingerprint lock system, 584
Chen, Fang, Measurement for dynamic deformation by mismatch white speckle method, 354
Chen, Guojian, Dynamic laser speckles and refractive measurements of the eye, 317
Chen, Huangming, Application of holographic lens in head-up display, 89
Chen, Junben, Optical sine transformation, 562
Chen, Linsen, Optical tomography—a new holographic inverse filter technique for 3D x-ray imaging, 550
Chen, Pingping, Measurement for dynamic deformation by mismatch white speckle method, 354
Chen, Shiming, New experimental technique to separate principal stresses—proportional exposure piled-plate holography, 120
Chen, Wei-Feng, New proposal for phase-only CGH and its application, 44
Chen, Ying-Li, Formation of amplitude grating in real-time holographic recording medium BSO crystal, 451
Chen, Zhi-Yong, Study on the performance of the liquid crystal light valve, 495
Chiang, F. P., Polychromatic speckle interference for edge enhancement of image, 263
Chin, Kuo-Fan, Making CGH by using electron beam, 107
Chou, Jing, Study of the narrow band holographic reflection filters, 463
Chuaqui, H., Microscopic and macroscopic measurement of holographic emulsions, 491
Dai, Fulong, Shearing holographic moire for strain patterns, 184
Ding, Hanquan, Holographic measurement of an arc plasma field, 130
Ding, Zuquan, New method of 3D quantitative analysis of holographic interferometry in the applications of solid biomechanics and vibration, 144
Dong, Wei, New experimental technique to separate principal stresses—proportional exposure piled-plate holography, 120
Endo, Junji, Use of fringe scanning method in electron holographic interferometry, 19
Fagot, H., Holographic cinematography and its applications, 22
Fang, Jing, Shearing holographic moire for strain patterns, 184
Fang, Ruhua, Measurement for dynamic deformation by mismatch white speckle method, 354
Fei, Zhuzeng, CGH made by dot matrix printer, 105
Fernandez, José F., Moire evaluation holography, 198
Fontaine, J., Thermoplastic holographic camera for industrial nondestructive testing: data acquisition and image analysis, 472
Fu, Shao-Jun, Development of high resolution zone plate by holography and ion-etching techniques, 459
———, Holographic match filtering method with high diffraction efficiency, 554
Fu, Zi-Ping, Real-time optical logical operations using grating filtering, 517
Galiński, J. J., Some aspects of application of a double frequency interferometer for distance measurements, 207
Geng, Wanzhen, Laser holographic inspection of solder joints on printed circuit board (PCB), 176
Gong, Jianming, Optical sine transformation, 562
Grant, I., Speckle velocimetry applied to wake flows, 358
Greguss, Pal, Manifestation of Gabor's holographic principle in various evolutionary stages of the living material, 402
Gu, Quwu, Interferometric method for measuring optical spherical surfaces using holographic phase conjugate compensation, 268
Gu, Yunan, Interferometric method for measuring optical spherical surfaces using holographic phase conjugate compensation, 268
Guo, Jia-Rong, Spatial moire hologram, 66
Guo, Lu-Rong, Synthetic phase-only holographic filter for multiobject recognition, 557
Guoquan, Ren, Dynamic process study by preset light pulse method in holographic system with Faraday Rotator, 239
Han, Lei, Holographic interferometry using digital image processing technique for the measurement of 3D axisymmetric temperature field, 192
Hariharan, P., Applications of holography 1947-86, 2
Hasegawa, Shuji, Use of fringe scanning method in electron holographic interferometry, 19
He, Shi-Ping, Holographic interferometry using digital image processing technique for the measurement of 3D axisymmetric temperature field, 192
He, Wan Xiang, Application of holography in the distribution measurement of fuel spraying field in diesel engines, 212
He, Xiangdong, Data processing on particle field hologram, 532
He, Yuming, Oblique-optical-axis speckle photography used for measuring 3D displacements of practical engineering structures, 366
Himeno, Shun-Ichi, Numerical method for holographic optical fiber diagnostics, 154
Hinsch, K., Coherent optics in environmental monitoring, 502
———, Industrial holography combined with image processing, 303
Hirosawa, A., Ohya large holography studio in a huge rock cavern, 95
Ho, Yuping, Optical sine transformation, 562
Honda, T., Holographic testing of gear tooth surface, 283
———, Hybrid real-time correlation system of aerial stereo photographs, 512
Hong, Jing, Laser holographic inspection of solder joints on printed circuit board (PCB), 176
Hong, R. T., Hybrid real-time correlation system of aerial stereo photographs, 512
Hong, Yi-Lin, Development of high resolution zone plate by holography and ion-etching techniques, 459
Hsu, Da-Hsiung, CGH made by dot matrix printer, 105
———, Holography in China, 8
———, Real-time optical logical operations using grating filtering, 517
Huang, Wanyun, Study of the narrow band holographic reflection filters, 463
Huang, Wei-Shi, Spatial moire hologram, 66

Huang, Xiang-Yang, New proposal for phase-only CGH and its application, 44
Ian, Richard, Holographic aquaculture, 479
Ih, C. S., Advanced holographic scanning and applications, 296
Ishii, Yukihiro, Optimum holographic disk scanners with bow-locus corrections, 426
Iwaasa, Y., Automatic processing of holographic interference fringes to analyze the deflection of a thin plate, 236
Iwasaki, Shigeo, Use of fringe scanning method in electron holographic interferometry, 19
Iwata, Koichi, Deformation measurement of lumbar vertebra by holographic interferometry, 340
Ji, Rong-Cai, Study on the performance of the liquid crystal light valve, 495
Jia, Yu-Run, New proposal for phase-only CGH and its application, 44
Jiang, Lingzhen, Laser holographic inspection of solder joints on printed circuit board (PCB), 176
Jiang, Yaguang, Optical tomography—a new holographic inverse filter technique for 3D x-ray imaging, 550
Kaarli, R., Picosecond time- and space-domain holography by photochemical hole burning, 53
Kalestyński, A., Generalized Laplace's operators realized by computer-generated spatial filters, 567
Kasprzak, Henryk, Mechanical reaction of human skull bones to external load examined by holographic interferometry, 321
Ke, Jingtang, Investigation of strains of objects by means of sandwich holographic interferometry, 71
———, Study of speckle multiple-shearing interferometry, 416
Khyzniak, A. I., Influence of medium nonlinearity on the properties of the steady-state hologram, 77
King, Elisabeth, Holographic aquaculture, 479
Kojima, Arata, Deformation measurement of lumbar vertebra by holographic interferometry, 340
Komarnicki, M., Holographic testing of human vision—optical correlator matched with human eye, 412
Kondo, T., Holographic testing of gear tooth surface, 283
Kowalczyk, M., Forming phase diffusers using speckle, 382
Kowalski, H. Z., Some aspects of application of a double frequency interferometer for distance measurements, 207
Krajskij, A. V., Holographic registration and spectroscopy of a modulated laser radiation, 100
Krätzig, E., Holographic storage properties of electro-optic crystals, 483
Ku, Chungshien, Oblique-optical-axis speckle photography used for measuring 3D displacements of practical engineering structures, 366
Lai, Guanming, Proposals of two reference beam double exposure holographic interferometer with double pulsed laser, 36
Landraud, A. M., Principle of a photonical computer with biological and medical applications, 526
Laude, J. P., Noteworthy qualities of holographic grating in the design of analytical spectrometer, 455
Leith, E. N., Optical tomography—a new holographic inverse filter technique for 3D x-ray imaging, 550
Li, F. H., Image quality improvement on variable spatial-rotation high resolution electron micrograph by optical means, 578
Li, Jianyi, Experimental study of the holographic technique applied to supersonic cascade wind-tunnel, 62
Li, Junyue, Application of monowavelength pulsed laser holometry in the measurement of arc plasma, 243
Li, Li, Fatique detection based on the change of laser-produced diffraction patterns, 258
Li, Meiyue, Method for making large rainbow holograms, 93
Li, Yishan, Application of monowavelength pulsed laser holometry in the measurement of arc plasma, 243
Li, Zai Chun, Improved holographic technique for imaging through fog, 570
Li, Zhi Xiong, Application of holography in the distribution measurement of fuel spraying field in diesel engines, 212
Lian, Jinrui, Application of monowavelength pulsed laser holometry in the measurement of arc plasma, 243
Liao, Jianghong, Interferometric method for measuring optical spherical surfaces using holographic phase conjugate compensation, 268
Liegeois, Christian, Dichromated gelatin holographic scanner, 439
———, Thermoplastic holographic camera for industrial nondestructive testing: data acquisition and image analysis, 472
Liu, Dahe, Study of the narrow band holographic reflection filters, 463
Liu, Fuxiang, Method of determining minute deformation and displacement of skull by speckle photography, 396
Liu, Jingsheng, New methods of making flat holographic stereograms, 28
Liu, Jinhe, Application of monowavelength pulsed laser holometry in the measurement of arc plasma, 243
Liu, Pei-Sen, New method of surface roughness measurement using far field speckle, 377
Liu, Rong-Ping, New method of surface roughness measurement using far field speckle, 377
Løkberg, Ole J., Electronic speckle pattern interferometry, 346
Long, Chang-Ying, Study of optometry apparatus of laser speckles, 338
Lu, B.-X., Holography in China, 8
Lu, Bo, Application of holographic lens in head-up display, 89
Lu, Zhenwu, Optical testing using a point diffraction holographic interferometer, 289
Lui, Jiajun, Dynamic laser speckles and refractive measurements of the eye, 317
Ma, Jing, Laser holographic inspection of solder joints on printed circuit board (PCB), 176
Ma, Yaowu, Study of speckle multiple-shearing interferometry, 416
Martellucci, S., Heat transfer studies by microholographic interferometry, 167
Masamori, Ichiro, Vibration analysis of automotive bodies using continuous wave laser holographic interferometry, 272
Matsumoto, Toshiro, Deformation measurement of lumbar vertebra by holographic interferometry, 340
Mei, Wenhui, Spatial multichannel holography: longitudinal multiplexed Fourier storage, 14
Meucci, R., Heat transfer studies by microholographic interferometry, 167
Meyrueis, Patrick, Dichromated gelatin holographic scanner, 439
———, Thermoplastic holographic camera for industrial nondestructive testing: data acquisition and image analysis, 472
Mochizuki, Hitoshi, Numerical method for holographic optical fiber diagnostics, 154
Morie, T., Ohya large holography studio in a huge rock cavern, 95
Mormile, P., Heat transfer studies by microholographic interferometry, 167
Motonori, Kanaya, Oblique incidence interferometer using holographic optical elements, 446
Mu, G. G., Application of computer-generated oriented speckle screen to image processing, 508
———, New type of holographic encoding filter for correlation—a lensless intensity correlator, 546
Murata, Kazumi, Hologram filters for optical image processing, 434
———, Optimum holographic disk scanners with bow-locus corrections, 426
Nagata, Ryo, Deformation measurement of lumbar vertebra by holographic interferometry, 340
Nakano, Yoshiaki, Automatic measurements of the small angle variation using a holographic moire interferometry and a computer processing, 180
Nishida, H., Automatic processing of holographic interference fringes to analyze the deflection of a thin plate, 236
Nitka, T., Phase-only Fourier hologram as an optical matched spatial filter, 188

Niu, Lin, Method for measuring strain directly by laser objective speckles, 293
Ogawa, Ryoukei, Deformation measurement of lumbar vertebra by holographic interferometry, 340
Ohmura, Katsuyuki, Use of fringe scanning method in electron holographic interferometry, 19
Ohyama, N., Holographic testing of gear tooth surface, 283
———, Hybrid real-time correlation system of aerial stereo photographs, 512
Okada, Katsuyuki, Holographic 3D display of x-ray tomogram, 84
Ose, Teruji, Holographic 3D display of x-ray tomogram, 84
Ozols, A. O., Self-enhancement effect of dynamic amplitude-phase holograms and its applications, 41
Peng, Yulan, Fe-doped $LiNbO_3$ crystal for applications in white-light information processing, 489
Pierattini, G., Heat transfer studies by microholographic interferometry, 167
Podbielska, Halina, Mechanical reaction of human skull bones to external load examined by holographic interferometry, 321
Preater, Richard W. T., Measuring rotating component strains using ESPI, 373
Pryputniewicz, Ryszard J., Computer-aided fringe analysis, 250
Qu, Zhimin, Method for making large rainbow holograms, 93
———, New methods of making flat holographic stereograms, 28
Quartieri, J., Heat transfer studies by microholographic interferometry, 167
Rebane, A., Picosecond time- and space-domain holography by photochemical hole burning, 53
Reuber, E., Use of synthetic holograms in coherent image processing for high resolution micrographs of a conventional transmission electron microscope (CTEM), 538
Roid, W., Industrial holography combined with image processing, 303
Rottenkolber, H., Industrial holography combined with image processing, 303
Rubtsova, I. L., Influence of medium nonlinearity on the properties of the steady-state hologram, 77
Rupp, R. A., Holographic storage properties of electro-optic crystals, 483
Saari, P., Picosecond time- and space-domain holography by photochemical hole burning, 53
Saito, Takayuki, Oblique incidence interferometer using holographic optical elements, 446
Sanecki, J., Some aspects of application of a double frequency interferometer for distance measurements, 207
Sasanishi, Kazuto, Vibration analysis of automotive bodies using continuous wave laser holographic interferometry, 272
Schmidt, B., Use of synthetic holograms in coherent image processing for high resolution micrographs of a conventional transmission electron microscope (CTEM), 538
Schörner, J., Industrial holography combined with image processing, 303
Seki, Masaharu, Numerical method for holographic optical fiber diagnostics, 154
Shen, S. Q., Polychromatic speckle interference for edge enhancement of image, 263
Shen, X., Image quality improvement on variable spatial-rotation high resolution electron micrograph by optical means, 578
Shi, Jia-Qi, Study of optometry apparatus of laser speckles, 338
Shi, Shi-Zhong, Study of optometry apparatus of laser speckles, 338
Shifan, Wang, Method of determining minute deformation and displacement of skull by speckle photography, 396
Smigielski, P., Holographic cinematography and its applications, 22
Smith, G. H., Speckle velocimetry applied to wake flows, 358
Smolińska, B., Holographic testing of human vision—shapes differentiation, 412
———, Holographic testing of human vision—optical correlator matched with human eye, 412
Soares, Olivério D. D., Moire evaluation holography, 198
Song, Yaozu, Experimental study of the holographic technique applied to supersonic cascade wind-tunnel, 62
Su, Xian-Yu, Synthetic phase-only holographic filter for multiobject recognition, 557
Su, Yong-Gang, Development of high resolution zone plate by holography and ion-etching techniques, 459
———, Holographic match filtering method with high diffraction efficiency, 554
Sun, Guanchao, Infrared holography using a CO_2 laser and its application to Si-wafer inspection, 49
Sun, Jing Ao, Profile measurement on IC wafers by holographic interference, 135
Sun, Qi-Yue, Spatial moire hologram, 66
Suzuki, M., Ohya large holography studio in a huge rock cavern, 95
Suzuki, Masane, Oblique incidence interferometer using holographic optical elements, 446
Tan, Yushan, Oblique-optical-axis speckle photography used for measuring 3D displacements of practical engineering structures, 366
Tang, Hong, Dynamic laser speckles and refractive measurements of the eye, 317
Tang, Weiguo, Study of the narrow band holographic reflection filters, 463
Tao, Xiao-Min, Development of high resolution zone plate by holography and ion-etching techniques, 459
Tentori, Diana, Analysis of an in-line object beam hologram interferometer, 124
———, Hologram interferometer for film deposits evaluation, 163
Tonomura, Akira, Use of fringe scanning method in electron holographic interferometry, 19
Toyooka, S., Automatic processing of holographic interference fringes to analyze the deflection of a thin plate, 236
Tsujiuchi, Jupei, Holographic testing of gear tooth surface, 283
———, Hybrid real-time correlation system of aerial stereo photographs, 512
———, Multiplex holograms and their applications in medicine, 312
Vicari, L., Heat transfer studies by microholographic interferometry, 167
Villagomez, Ricardo, Analysis of an in-line object beam hologram interferometer, 124
von Bally, G., Holography in biomedical sciences, 327
———, Holography and the freedom of science—a welcome address to ICHA '86, 337
———, Mechanical reaction of human skull bones to external load examined by holographic interferometry, 321
Wade, Glen, Profile measurement on IC wafers by holographic interference, 135
Wang, B. X., Real time grating shearing interferometry applied to investigating of evaporative convection in liquid drops, 276
Wang, Bao-Cheng, Study of optometry apparatus of laser speckles, 338
Wang, Ben, Real-time optical logical operations using grating filtering, 517
Wang, D.-H., Holography in China, 8
Wang, Linli, Investigation of strains of objects by means of sandwich holographic interferometry, 71
Wang, Shengqing, Dynamic process study by preset light pulse method in holographic system with Faraday Rotator, 239
Wang, Shuying, Fe-doped $LiNbO_3$ crystal for applications in white-light information processing, 489
Wang, Wuyi, Experimental study of the holographic technique applied to supersonic cascade wind-tunnel, 62
Wang, X. M., Application of computer-generated oriented speckle screen to image processing, 508
———, New type of holographic encoding filter for correlation—a lensless intensity correlator, 546
Wang, Yaosong, Holographic fingerprint lock system, 584

Wang, Z. Q., Application of computer-generated oriented speckle screen to image processing, 508
———, New type of holographic encoding filter for correlation—a lensless intensity correlator, 546
Wang, Z.-J., Holography in China, 8
Watanabe, K., Ohya large holography studio in a huge rock cavern, 95
Wu, Wei-Shu, Formation of amplitude grating in real-time holographic recording medium BSO crystal, 451
Wu, Xiao-Ping, Fatique detection based on the change of laser-produced diffraction patterns, 258
———, Holographic interferometry using digital image processing technique for the measurement of 3D axisymmetric temperature field, 192
———, Polychromatic speckle interference for edge enhancement of image, 263
———, Study of speckle pattern and surface roughness measurement with image processing technique, 388
Wu, Xiu-Qing, Study of optometry apparatus of laser speckles, 338
Xiang, L. Q., Advanced holographic scanning and applications, 296
Xie, Jinghui, Density pseudocolor encoder by Bragg effect, 575
Xing, Wu, Dynamic process study by preset light pulse method in holographic system with Faraday Rotator, 239
Xu, Bo-Qin, Fatique detection based on the change of laser-produced diffraction patterns, 258
———, Study of speckle pattern and surface roughness measurement with image processing technique, 388
Xu, Ke-Shu, Determining the phase function in a 2D Fourier spectrum, 519
Xu, Xiang-Dong, Determining the phase function in a 2D Fourier spectrum, 519
Xu, Yingming, Method for making large rainbow holograms, 93
Xue, Wei, Laser holographic inspection of solder joints on printed circuit board (PCB), 176
Yamaguchi, Ichirou, Holography and speckle techniques for deformation measurements, 112
Yamashita, Hajime, Ohya large holography studio in a huge rock cavern, 95
———, Vibration analysis of large-sized structure using cw laser holographic interferometry, 150
———, Vibration analysis of automotive bodies using continuous wave laser holographic interferometry, 272
Yan, Yingbai, Making CGH by using electron beam, 107
Yan, Yuan, Formation of amplitude grating in real-time holographic recording medium BSO crystal, 451
Yang, Guoguang, Holographic match filtering method with high diffraction efficiency, 554
———, Optical sine transformation, 562
Yang, Jingfeng, Laser holographic inspection of solder joints on printed circuit board (PCB), 176
Yang, Jun, Experimental study of the holographic technique applied to supersonic cascade wind-tunnel, 62
Yao, Jie, Data processing on particle field hologram, 532
Yao, Kun, Study of optometry apparatus of laser speckles, 338
Yasuda, Kenji, Oblique incidence interferometer using holographic optical elements, 446
Yatagai, Toyohiko, Proposals of two reference beam double exposure holographic interferometer with double pulsed laser, 36
———, Use of fringe scanning method in electron holographic interferometry, 19
Ye, Zhisheng, Dynamic process study by preset light pulse method in holographic system with Faraday Rotator, 239
Ying, Xuan-Tong, New proposal for phase-only CGH and its application, 44
Youren, Xu, Real time grating shearing interferometry applied to investigating of evaporative convection in liquid drops, 276
Yu, D. Y., Holographic testing of gear tooth surface, 283
Yu, Dongxiao, Making CGH by using electron beam, 107
Yu, F. T. S., White-light color holography and its applications, 222
Yu, Meiwen, Density pseudocolor encoder by Bragg effect, 575
Yuan, Zijuen, Infrared holography using a CO_2 laser and its application to Si-wafer inspection, 49
Zhang, Guan-Shen, Synthetic phase-only holographic filter for multiobject recognition, 557
Zhang, Jingjiang, Fe-doped $LiNbO_3$ crystal for applications in white-light information processing, 489
Zhang, Jingjuan, Optical sine transformation, 562
Zhang, Nengli, Real time grating shearing interferometry applied to investigating of evaporative convection in liquid drops, 276
Zhang, Qing-Chuan, Study of speckle pattern and surface roughness measurement with image processing technique, 388
Zhang, Yimo, Improved holographic technique for imaging through fog, 570
Zhang, Yun-Wu, Development of high resolution zone plate by holography and ion-etching techniques, 459
Zhang, Zunlin, Dynamic process study by preset light pulse method in holographic system with Faraday Rotator, 239
Zhao, Caifu, Investigation of strains of objects by means of sandwich holographic interferometry, 71
Zhao, Dazun, Density pseudocolor encoder by Bragg effect, 575
———, Spatial multichannel holography: longitudinal multiplexed Fourier storage, 14
Zhao, Huan-Qing, Study on the performance of the liquid crystal light valve, 495
Zhao, Mingxi, CGH made by dot matrix printer, 105
Zhao, Shijie, Laser holographic inspection of solder joints on printed circuit board (PCB), 176
Zhao, Yeling, Density pseudocolor encoder by Bragg effect, 575
Zhao, Yunying, Fe-doped $LiNbO_3$ crystal for applications in white-light information processing, 489
Zheng, S. H., Image quality improvement on variable spatial-rotation high resolution electron micrograph by optical means, 578
Zhou, Genyu, Application of holography to research reason of spurting water of the antique "Water Spurting Basin", 160
Zhou, Liang, Spatial moire hologram, 66
Zhou, Wanzhi, Optical testing using a point diffraction holographic interferometer, 289
Zhou, Xingeng, Method for measuring strain directly by laser objective speckles, 293
Zhu, Hongfan, Application of holography to research reason of spurting water of the antique "Water Spurting Basin", 160
Zhu, Zhulin, Data processing on particle field hologram, 532
Zou, Zhenshu, Interferometric method for measuring optical spherical surfaces using holographic phase conjugate compensation, 268
Zubov, V. A., Holographic registration and spectroscopy of a modulated laser radiation, 100